# Modern Particle Physics

Unique in its coverage of all aspects of modern particle physics, this textbook provides a clear connection between the theory and recent experimental results, including the discovery of the Higgs boson at CERN. It provides a comprehensive and self-contained description of the Standard Model of particle physics suitable for upper-level undergraduate students and graduate students studying experimental particle physics.

Physical theory is introduced in a straightforward manner with step-by-step mathematical derivations throughout. Fully worked examples enable students to link the mathematical theory to results from modern particle physics experiments. End-of-chapter exercises, graded by difficulty, provide students with a deeper understanding of the subject

Online resources available at www.cambridge.org/MPP feature password-protected fully worked solutions to problems for instructors, numerical solutions and hints to the problems for students and PowerPoint slides and JPEGs of figures from the book.

**Mark Thomson** is Professor in Experimental Particle Physics at the University of Cambridge. He is an experienced teacher and has lectured particle physics at introductory and advanced levels. His research interests include studies of the electroweak sector of the Standard Model and the properties of neutrinos.

# Modern Particle Physics

MARK THOMSON
University of Cambridge

CAMBRIDGE
UNIVERSITY PRESS

CAMBRIDGE
UNIVERSITY PRESS

Shaftesbury Road, Cambridge CB2 8EA, United Kingdom

One Liberty Plaza, 20th Floor, New York, NY 10006, USA

477 Williamstown Road, Port Melbourne, VIC 3207, Australia

314–321, 3rd Floor, Plot 3, Splendor Forum, Jasola District Centre, New Delhi – 110025, India

103 Penang Road, #05–06/07, Visioncrest Commercial, Singapore 238467

Cambridge University Press is part of Cambridge University Press & Assessment,
a department of the University of Cambridge.

We share the University's mission to contribute to society through the pursuit of
education, learning and research at the highest international levels of excellence.

www.cambridge.org
Information on this title: www.cambridge.org/MPP



First published 2013 (version 10, March 2024)

Printed in the United Kingdom by TJ Books Limited, Padstow Cornwall, March 2024

*A catalogue record for this publication is available from the British Library*

*Library of Congress Cataloguing-in-Publication data*
Thomson, Mark, 1966–
Modern particle physics / Mark Thomson.
pages cm
ISBN 978-1-107-03426-6 (Hardback)
1. Particles (Nuclear physics)–Textbooks. I. Title.
QC793.2.T46 2013
539.7′2–dc23 2013002757

ISBN 978-1-107-03426-6 Hardback

Additional resources for this publication at www.cambridge.org/MPP

**To**

**Sophie, Robert and Isabelle**

***for their love, support and endless patience***

# Contents

# Preface

The Standard Model of particle physics represents one of the triumphs of modern physics. With the discovery of the Higgs boson at the LHC, all of the particles in the Standard Model have now been observed. The main aim of this book is to provide a broad overview of our current understanding of particle physics. It is intended to be suitable for final-year undergraduate physics students and also can serve as an introductory graduate-level text. The emphasis is very much on the modern view of particle physics with the aim of providing a solid grounding in a wide range of topics.

Our current understanding of the sub-atomic Universe is based on a number of profound theoretical ideas that are embodied in the Standard Model of particle physics. However, the development of the Standard Model would not have been possible without a close interplay between theory and experiment, and the structure of this book tries to reflects this. In most chapters, theoretical concepts are developed and then are related to the current experimental results. Because particle physics is mostly concerned with fundamental objects, it is (in some sense) a relatively straightforward subject. Consequently, even at the undergraduate level, it is quite possible to perform calculations that can be related directly to the recent experiments at the forefront of the subject.

## Pedagogical approach

In writing this textbook I have tried to develop the subject matter in a clear and accessible manner and thought long and hard about what material to include. Whilst the historical development of particle physics is an interesting topic in its own right, it does not necessarily provide the best pedagogical introduction to the subject. For this reason, the focus of this book is on the contemporary view of particle physics and earlier experimental results are discussed only to develop specific points. Similarly, no attempt is made to provide a comprehensive review of the many experiments, instead a selection of key measurements is used to illustrate the theoretical concepts; the choice of which experimental measurements to include is primarily motivated by the pedagogical aims of this book.

This textbook is intended to be self-contained, and only a basic knowledge of quantum mechanics and special relativity is assumed. As far as possible, I have tried to derive everything from first principles. Since this is an introductory textbook, the mathematical material is kept as simple as possible, and the derivations show all the main steps. I believe that this approach enables students relatively new to the subject to develop a clear understanding of the underlying physical principles; the more sophisticated mathematical trickery can come later. Calculations are mostly performed using helicity amplitudes based on the explicit Dirac–Pauli representation of the particle spinors. I believe this treatment provides a better connection to the underlying physics, compared to the more abstract trace formalism (which is also described). Some of the more-challenging material is included in optional *starred* sections. When reading these sections, the main aim should be to understand the central concepts, rather than the details.

The general structure of this book is as follows: Chapters 1–5 introduce the underlying concepts of relativistic quantum mechanics and interaction by particle exchange; Chapters 6–12 describe the electromagnetic, strong and weak interactions; and Chapters 13–18 cover major topics in modern particle physics. This textbook includes an extensive set of problems. Each problem is graded according to the *relative* time it is likely to take. This does not always reflect the difficulty of the problem and is meant to provide a guide to students, where for example a shorter graded problem should require relatively little algebra. Hints and outline solutions to many of the problems are available at www.cambridge.org/MPP.

## For instructors

This book covers a wide range of topics and can form the basis of a long course in particle physics. For a shorter course, it may not be possible to fit all of the material into a single semester and certain sections can be omitted. In this case, I would recommend that students read the introductory material in Chapters 1–3 as preparation for a lecture course. Chapters 4–8, covering the calculations of the $e^+e^- \to \mu^+\mu^-$ annihilation and $e^-p$ scattering cross sections, should be considered essential. Some of the material in Chapter 9 on the quark model can be omitted, although not the discussion of symmetries. The material in Chapter 14 stands alone and could be omitted or covered only partially. The material on electroweak unification and the tests of the Standard Model, presented in Chapters 15 and 16, represents one of the highlights of modern particle physics and should be considered as core. The chapter describing the Higgs mechanism is (necessarily) quite involved and it would be possible to focus solely on the properties of the Higgs boson and its discovery, rather than the detailed derivations.

Fully worked solutions to all problems are available to instructors, and these can be found at www.cambridge.org/MPP. In addition, to aid the preparation of new courses, PowerPoint slides covering most of the material in this book are available at the same location, as are all of the images in this book.

# Acknowledgements

I would like to thank colleagues in the High Energy Physics group at the Cavendish Laboratory for their comments on early drafts of this book. This book is based on my final-year undergraduate lecture course in the Physics Department at the University of Cambridge and as such it represents an evolution of earlier courses; for this reason I am indebted to R. Batley and M. A. Parker who taught the previous incarnations. For their specific comments on a number of the more technical chapters, I am particularly grateful to A. Bevan, B. Webber and J. Wells.

For the permissions to reproduce figures and to use experimental data I am indebted to the following authors and experimental collaborations:

R. Felst and the JADE Collaboration for Figure 6.7;
S. Wojcicki and the DELCO Collaboration for Figure 6.12;
M. Breidenbach for Figure 8.3;
S. Schmitt and the H1 Collaboration for Figures 8.13 and 12.14;
D. Plane and the OPAL Collaboration for Figures 10.12, 10.19, 16.8 and 16.9;
S. Bethke for Figure 10.14;
C. Kiesling and the CELLO Collaboration for Figure 10.16;
C. Vellidis, L. Ristori and the CDF Collaboration for Figures 10.29, 16.14 and 16.21;
J. Incandela and the CMS Collaboration for Figures 10.32, 17.18 and 17.19;
F. Gianotti and the ATLAS Collaboration for Figures 10.30, 17.18 and 17.19;
J. Steinberger, F. Dydak and the CDHS Collaboration for Figure 12.10;
R. Bernstein and the NuTeV Collaboration for Figure 12.12;
M. Pinsonneault for Figure 13.3;
Y. Suzuki and the Super-Kamiokande Collaboration for Figures 13.4 and 13.6;
N. Jelley, R.G.H. Robertson and the SNO Collaboration for Figure 13.8;
K-B. Luk, Y. Wang and the Daya Bay Collaboration for Figure 13.19;
K. Inoue and the KamLAND Collaboration for Figure 13.20;
R. Patterson for Figure 13.21;
R. Plunkett, J. Thomas and the MINOS Collaboration for Figure 13.22;
P. Bloch, N. Pavlopoulos and the CPLEAR Collaboration for Figures 14.14–14.16;
L. Piilonen, H. Hayashii, Y. Sakai and the Belle Collaboration for Figure 14.21;
M. Roney and the BaBar Collaboration for Figure 14.24;

The LEP Electroweak Working Group for Figures 16.2, 16.5, 16.6 and 16.10; S. Mele and the L3 Collaboration for Figure 16.13.

I am grateful to the Durham HepData project, which is funded by the UK Science and Technologies Facilities Council, for providing the online resources for access to high energy physics data that greatly simplifying the production of a number of the figures in this book.

Every effort has been made to obtain the necessary permissions to reproduce or adapt copyrighted material and I acknowledge:

*Annual Reviews* for Figure 8.4;

The American Physical Society for Figure 6.12 from Bacino *et al.* (1978), Figure 7.7 from Hughes *et al.* (1965), Figure 7.8 from Sill *et al.* (1993) and Walker *et al.* (1994), Figure 8.3 from Breidenbach *et al.* (1969), Figure 8.4 from Friedman and Kendall (1972) and Bodek *et al.* (1979), Figure 8.14 from Beringer *et al.* (2012), Figure 10.29 from Abe *et al.* (1999), Figure 12.12 from Tzanov *et al.* (2006), Figure 13.3 from Bahcall and Pinsonneault (2004), Figure 13.6 from Fukada *et al.* (2001), Figure 13.8 from Ahmad *et al.* (2002), Figure 13.19 from An *et al.* (2012), Figure 13.20 from Abe *et al.* (2008), Figure 13.22 from Adamson *et al.* (2011), Figure 14.21 from Abe *et al.* (2005), Figure 14.24 from Aubert *et al.* (2007), Figure 16.14 from Aaltonen *et al.* (2012) and Figure 16.21 from Aaltonen *et al.* (2011);

Elsevier for Figure 8.2 from Bartel *et al.* (1968), Figures 8.11, 8.12 and 8.18 from Whitlow *et al.* (1992), Figure 10.16 from Behrend *et al.* (1987), Figures 16.2, 16.5 and 16.6 from LEP and SLD Collaborations (2006);

Springer for Figure 6.7 from Bartel *et al.* (1985), Figure 10.12 from Abbiendi *et al.* (2004), Figure 10.14 from Bethke (2009), Figure 12.10 from de Groot *et al.* (1979), Figure 12.14 from Aaron *et al.* (2012), Figure 14.15 from Angelopoulos *et al.* (2001), Figure 14.16 from Angelopoulos *et al.* (2000), Figure 16.8 from Abbiendi *et al.* (2001) and Figure 16.13 from Achard *et al.* (2006);

CERN Information Services for Figures 10.32, 17.18 and 17.19.

# 1 Introduction

The purpose of this chapter is to provide a brief introduction to the Standard Model of particle physics. In particular, it gives an overview of the fundamental particles and the relationship between these particles and the forces. It also provides an introduction to the interactions of particles in matter and how they are detected and identified in the experiments at modern particle colliders.

## 1.1 The Standard Model of particle physics

Particle physics is at the heart of our understanding of the laws of nature. It is concerned with the fundamental constituents of the Universe, the *elementary particles*, and the interactions between them, the *forces*. Our current understanding is embodied in the Standard Model of particle physics, which provides a unified picture where the forces between particles are themselves described by the exchange of particles. Remarkably, the Standard Model provides a successful description of all current experimental data and represents one of the triumphs of modern physics.

### 1.1.1 The fundamental particles

In general, physics aims to provide an effective mathematical description of a physical system, appropriate to the energy scale being considered. The world around us appears to be formed from just a few different particles. Atoms are the bound states of negatively charged electrons ($e^-$) which orbit around a central nucleus composed of positively charged protons (p) and electrically neutral neutrons (n). The electrons are bound to the nucleus by the electrostatic attraction between opposite charges, which is the low-energy manifestation of the fundamental theory of electromagnetism, namely Quantum Electrodynamics (QED). The rich structure of the properties of the elements of the periodic table emerges from quantum mechanics, which dictates the precise electronic structure of the different atoms. In the atomic nucleus, the protons and neutrons are bound together by the strong nuclear force, which is a manifestation of the fundamental theory of strong interactions,

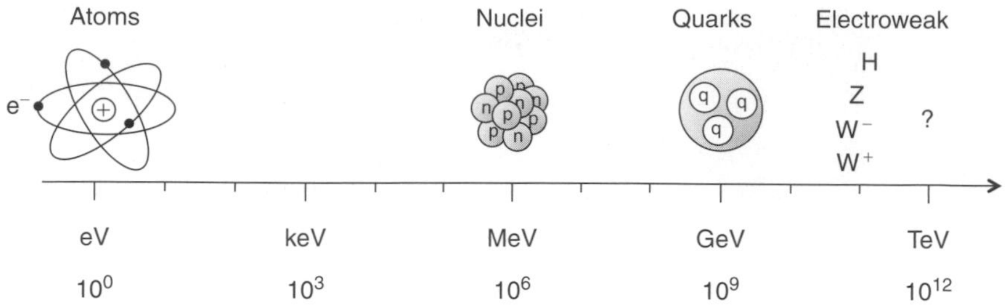


Fig. 1.1 The Universe at different energy scales, from atomic physics to modern particle physics at the TeV scale.

called Quantum Chromodynamics (QCD). The fundamental interactions of particle physics are completed by the weak force, which is responsible for the nuclear $\beta$-decays of certain radioactive isotopes and the nuclear fusion processes that fuel the Sun. In both nuclear $\beta$-decay and nuclear fusion, another particle, the nearly massless electron neutrino ($\nu_e$) is produced. Almost all commonly encountered physical phenomena can be described in terms of the electron, electron neutrino, proton and neutron, interacting by the electromagnetic, strong and weak forces. The picture is completed by gravity, which although extremely weak, is always attractive and is therefore responsible for large-scale structure in the Universe. This is an appealingly simple physical model with just four "fundamental" particles and four fundamental forces. However, at higher energy scales, further structure is observed, as indicated in Figure 1.1. For example, the protons and neutrons are found to be bound states of (what are believed to be) genuinely fundamental particles called *quarks*, with the proton consisting of two up-quarks and a down-quark, p(uud), and the neutron consisting of two down-quarks and an up-quark, n(ddu).

The electron, the electron neutrino, the up-quark and down-quark are known collectively as the *first generation*. As far as we know, they are elementary particles, rather than being composite, and represent the basic building blocks of the low-energy Universe. However, when particle interactions are studied at the energy scales encountered in high-energy particle colliders, further complexity is revealed. For each of the four first-generation particles, there are exactly two copies which differ only in their masses. These additional eight particles are known as the *second* and *third generations*. For example, the muon ($\mu^-$) is essentially a heavier version of the electron with mass $m_\mu \approx 200\, m_e$, and the third generation tau-lepton ($\tau^-$) is an even heavier copy with $m_\tau \approx 3500\, m_e$. Apart from the differences in masses, which have physical consequences, the properties of the electron, muon and tau-lepton are the same in the sense that they possess exactly the same fundamental interactions.

It is natural to ask whether this pattern is repeated and that there are further generations of particles. Perhaps surprisingly, this seems not to be the case; there is

**Table 1.1** The twelve fundamental fermions divided into quarks and leptons. The masses of the quarks are the current masses.

| | Leptons | | | | Quarks | | | |
|---|---|---|---|---|---|---|---|---|
| | Particle | | $Q$ | mass/GeV | Particle | | $Q$ | mass/GeV |
| First generation | electron | ($e^-$) | −1 | 0.0005 | down | (d) | −1/3 | 0.003 |
| | neutrino | ($\nu_e$) | 0 | $< 10^{-9}$ | up | (u) | +2/3 | 0.005 |
| Second generation | muon | ($\mu^-$) | −1 | 0.106 | strange | (s) | −1/3 | 0.1 |
| | neutrino | ($\nu_\mu$) | 0 | $< 10^{-9}$ | charm | (c) | +2/3 | 1.3 |
| Third generation | tau | ($\tau^-$) | −1 | 1.78 | bottom | (b) | −1/3 | 4.5 |
| | neutrino | ($\nu_\tau$) | 0 | $< 10^{-9}$ | top | (t) | +2/3 | 174 |

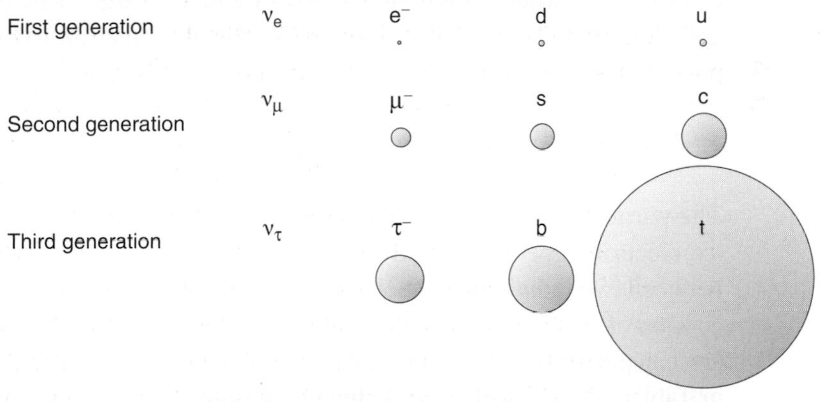


Fig. 1.2 The particles in the three generations of fundamental fermions with the masses indicated by imagined spherical volumes of constant density. In reality, fundamental particles are believed to be point-like.

strong experimental evidence that there are just three generations; hence the matter content of the Universe appears to be in the form of the twelve fundamental spin-half particles listed in Table 1.1. There is a subtlety when it comes to the description of the neutrinos; the $\nu_e$, $\nu_\mu$ and $\nu_\tau$ are in fact quantum-mechanical mixtures of the three fundamental neutrino states with well-defined masses, labelled simply $\nu_1$, $\nu_2$ and $\nu_3$. This distinction is only important in the discussion of the behaviour of neutrinos that propagate over large distances, as described in Chapter 13. Whilst it is known that the neutrinos are not massless, the masses are sufficiently small that they have yet to be determined. From the upper limits on the possible neutrino masses, it is clear that they are at least nine orders of magnitude lighter than the other fermions. Apart from the neutrinos, the masses of the particles within a particular generation are found to be rather similar, as illustrated in Figure 1.2. Whilst it is likely that there is some underlying reason for this pattern of masses, it is not currently understood.

**Table 1.2** The forces experienced by different particles.

| | | | | | strong | electromagnetic | weak |
|---|---|---|---|---|---|---|---|
| Quarks | down-type | d | s | b | ✓ | ✓ | ✓ |
| | up-type | u | c | t | | | |
| Leptons | charged | $e^-$ | $\mu^-$ | $\tau^-$ | | ✓ | ✓ |
| | neutrinos | $\nu_e$ | $\nu_\mu$ | $\nu_\tau$ | | | ✓ |

The dynamics of each of the twelve fundamental fermions are described by the Dirac equation of relativistic quantum mechanics, which is the subject of Chapter 4. One important consequence of the Dirac equation is that for each of the twelve fermions there exists an antiparticle state with exactly the same mass, but opposite charge. Antiparticles are denoted either by their charge or by a bar over the corresponding particle symbol. For example, the anti-electron (which is known as the positron) is denoted by $e^+$, and the anti-up-quark is written $\overline{u}$.

### Quarks and leptons

The particles interact with each other through the four fundamental forces, gravity, electromagnetism, the strong force and the weak force. The gravitational force between two individual particles is extremely small and can be neglected in the discussion of particle interactions. The properties of the twelve fundamental fermions are categorised by the types of interaction that they experience, as summarised in Table 1.2. All twelve fundamental particles "feel" the weak force and undergo weak interactions. With the exception of the neutrinos, which are electrically neutral, the other nine particles are electrically charged and participate in the electromagnetic interaction of QED. Only the quarks carry the QCD equivalent of electric charge, called *colour charge*. Consequently, only the quarks feel the strong force. Because of the nature of the QCD interaction, quarks are never observed as free particles, but are always confined to bound states called *hadrons*, such as the proton and neutron. Because the quarks feel the strong force, their properties are very different from those of the electron, muon, tau-lepton and the neutrinos, which are collectively referred to as the *leptons*.

## 1.1.2 The fundamental forces

In classical electromagnetism, the electrostatic force between charged particles can be described in terms of a scalar potential. This classical description of a force arising from a potential is unsatisfactory on a number of levels. For example, when an electron scatters in the electrostatic potential of a proton, there is a transfer of momentum from one particle to the other without any apparent mediating body.

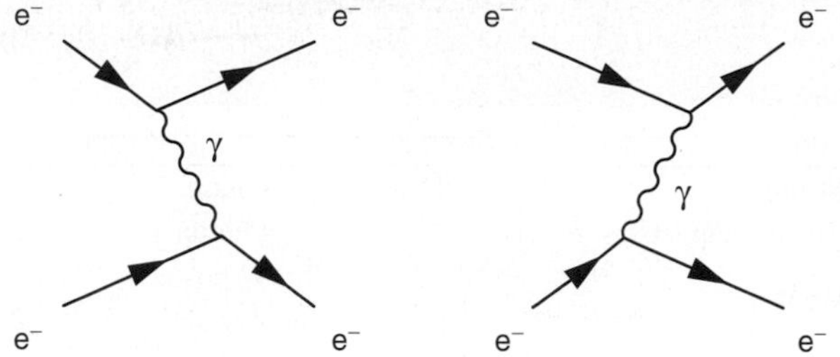


Fig. 1.3 The scattering of two electrons in QED by the exchange of a photon. With time running from left to right, the diagrams indicate the two possible time-orderings.

Regarding this apparent action-at-a-distance, Newton famously wrote "*It is inconceivable that inanimate brute matter should, without the mediation of something else which is not material, operate upon and affect other matter without mutual contact*". Whilst it is convenient to express classical electromagnetism in terms of potentials, it hides the fundamental origin of the electromagnetic interaction.

In modern particle physics, each force is described by a Quantum Field Theory (QFT). In the case of electromagnetism this is the theory of Quantum Electrodynamics (QED), where the interactions between charged particles are mediated by the exchange of *virtual* photons; the meaning of the term virtual is explained in Chapter 5. By describing a force in terms of particle exchange, there is no longer any mysterious action at a distance. As an example, Figure 1.3 shows the interaction between two electrons by the exchange of a photon. In the first diagram, the upper electron emits a photon, which at a later time is absorbed by the lower electron. The effect is to transfer momentum from one electron to the other, and it is this transfer of momentum which manifests itself as a force. The second diagram shows the other possible time-ordering with the lower electron emitting the photon that is subsequently absorbed by the upper electron. Since the exchanged particle is not observed, only the combined effect of these two time-ordered diagrams is physically meaningful.

Each of the three forces of relevance to particle physics is described by a QFT corresponding to the exchange of a spin-1 force-carrying particle, known as a *gauge boson*. The familiar spin-1 photon is the gauge boson of QED. In the case of the strong interaction, the force-carrying particle is called the *gluon* which, like the photon, is massless. The weak charged-current interaction, which is responsible for nuclear $\beta$-decay and nuclear fusion, is mediated by the charged $W^+$ and $W^-$ bosons, which are approximately eighty times more massive than the proton. There is also a weak neutral-current interaction, closely related to the charged current, which is mediated by the electrically neutral Z boson. The relative strengths of the forces associated with the different gauge bosons are indicated in Table 1.3. It should be noted that these numbers are only indicative as the strengths of the forces depend on the distance and energy scale being considered.

**Table 1.3** The four known forces of nature. The relative strengths are approximate indicative values for two fundamental particles at a distance of 1 fm $= 10^{-15}$ m (roughly the radius of a proton).

| Force | Strength | Boson | | Spin | Mass/GeV |
|---|---|---|---|---|---|
| Strong | 1 | Gluon | g | 1 | 0 |
| Electromagnetism | $10^{-3}$ | Photon | $\gamma$ | 1 | 0 |
| Weak | $10^{-8}$ | W boson | $W^{\pm}$ | 1 | 80.4 |
| | | Z boson | Z | 1 | 91.2 |
| Gravity | $10^{-37}$ | Graviton? | G | 2 | 0 |

### 1.1.3 The Higgs boson

The final element of the Standard Model is the Higgs boson, which was discovered by the ATLAS and CMS experiments at the Large Hadron Collider (LHC) in 2012. The Higgs boson, which has a mass

$$m_{\mathrm{H}} \approx 125\,\mathrm{GeV},$$

differs from all other Standard Model particles. Unlike, the fundamental fermions and the gauge bosons, which are respectively spin-half and spin-1 particles, the Higgs boson is spin-0 scalar particle. As conceived in the Standard Model, the Higgs boson is the only fundamental scalar discovered to date.

The Higgs boson plays a special rôle in the Standard Model; it provides the mechanism by which all other particles acquire mass. Without it the Universe would be a very different, all the particles would be massless and would propagate at the speed of light! In QFT, the Higgs boson can be thought of as an excitation of the Higgs field. Unlike the fields associated with the fundamental fermions and bosons, which have zero expectation values in the vacuum, the Higgs field is believed to have a non-zero vacuum expectation value. It is the interaction of the initially massless particles with this non-zero Higgs field that gives them their masses. The discovery of a Higgs-like particle at the LHC represented a remarkable validation of the theoretical ideas which constitute the Standard Model. The mathematical details of the Higgs mechanism, which are subtle, are discussed in detail in Chapter 17. The masses of the $W^{\pm}$, Z and H bosons are all of the order of 100 GeV, which is known as the electroweak scale. This doesn't happen by chance; in the Standard Model, the masses of the weak gauge bosons are intimately connected to the Higgs mechanism.

### 1.1.4 The Standard Model vertices

The nature of the strong, electromagnetic and weak forces are determined by the properties of the bosons of the associated quantum field theory, and the way in

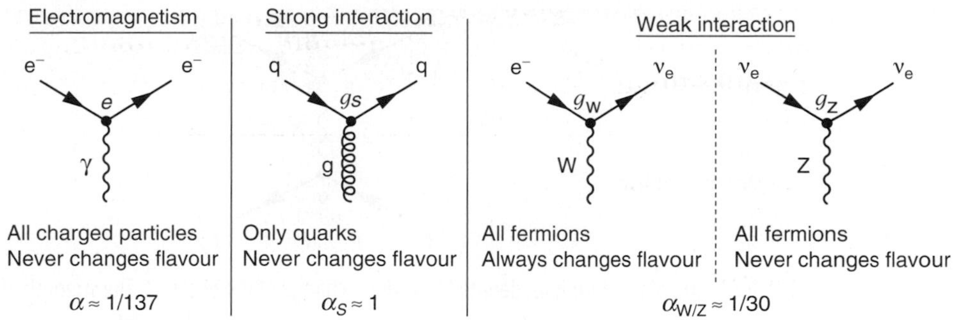


Fig. 1.4 The Standard Model interaction vertices.

which the gauge bosons couple to the spin-half fermions. The coupling of the gauge bosons to the fermions is described by the Standard Model interaction vertices, shown in Figure 1.4. In each case, the interaction is a three-point vertex of the gauge boson and an incoming and outgoing fermion. For each type of interaction there is an associated coupling strength $g$. For QED the coupling strength is simply the electron charge, $g_{\text{QED}} = e \equiv +|e|$.

A particle couples to a force-carrying boson only if it carries the charge of the interaction. For example, only electrically charged particles couple to the photon. Only the quarks carry the colour charge of QCD, and hence only quarks participate in the strong interaction. All twelve fundamental fermions carry the charge of the weak interaction, known as weak isospin, and therefore they all participate in the weak interaction. The weak charged-current interaction does not correspond to the usual concept of a force as it couples together different flavour fermions. Since the $\text{W}^+$ and $\text{W}^-$ bosons have charges of $+e$ and $-e$ respectively, in order to conserve electric charge, the weak charged-current interaction only couples together pairs of fundamental fermions that differ by one unit of electric charge. In the case of the leptons, by definition, the weak interaction couples a charged lepton with its corresponding neutrino,

$$\begin{pmatrix} \nu_e \\ e^- \end{pmatrix}, \begin{pmatrix} \nu_\mu \\ \mu^- \end{pmatrix}, \begin{pmatrix} \nu_\tau \\ \tau^- \end{pmatrix}.$$

For the quarks, the weak interaction couples together all possible combinations differing by one unit of charge,

$$\begin{pmatrix} \text{u} \\ \text{d} \end{pmatrix}, \begin{pmatrix} \text{u} \\ \text{s} \end{pmatrix}, \begin{pmatrix} \text{u} \\ \text{b} \end{pmatrix}, \begin{pmatrix} \text{c} \\ \text{d} \end{pmatrix}, \begin{pmatrix} \text{c} \\ \text{s} \end{pmatrix}, \begin{pmatrix} \text{c} \\ \text{b} \end{pmatrix}, \begin{pmatrix} \text{t} \\ \text{d} \end{pmatrix}, \begin{pmatrix} \text{t} \\ \text{s} \end{pmatrix}, \begin{pmatrix} \text{t} \\ \text{b} \end{pmatrix}.$$

The strength of the weak charged-current coupling between the charge $+\frac{2}{3}$ up-type quarks (u, c, t) and the charge $-\frac{1}{3}$ down-type quarks (d, s, b) is greatest for quarks of the same generation. Since the weak interaction is the only known force

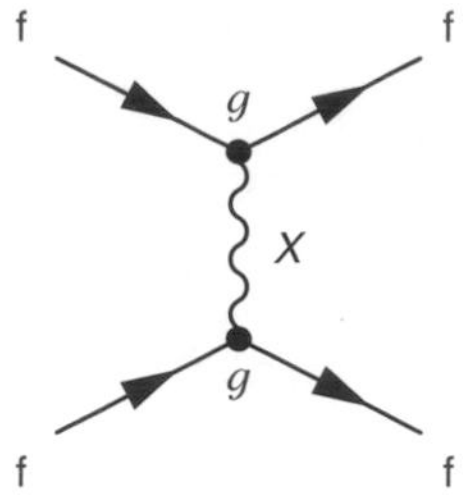


Fig. 1.5 The scattering of two fermions, denoted f, by the exchange of the boson, $X$. The strength of the fundamental interaction at each of the two three-point ff$X$ vertices is denoted by the coupling constant $g$.

for which the incoming and outgoing fermions are different, the weak charged-current interaction is particularly important when considering particle decays as it introduces a change of flavour.

The strength of the fundamental interaction between the gauge boson and a fermion is determined by the coupling constant $g$, which can be thought of as a measure of the probability of a spin-half fermion emitting or absorbing the boson of the interaction. Put more precisely, the quantum-mechanical transition matrix element for an interaction process includes a factor of the coupling constant $g$ for each interaction vertex. For example, the matrix element for the scattering process indicated by Figure 1.5 contains two factors of $g$, one at each vertex, and therefore

$$\mathcal{M} \propto g^2.$$

Hence, the interaction probability, which is proportional to the matrix element squared, $|\mathcal{M}|^2 = \mathcal{M}\mathcal{M}^*$, contains a factor $g^2$ from *each* interaction vertex, thus in this example

$$|\mathcal{M}|^2 \propto g^4.$$

Rather than working with the coupling constant itself, it is often more convenient to use the associated dimensionless constant, $\alpha \propto g^2$. In the case of electromagnetism this is the familiar fine-structure constant

$$\alpha = \frac{e^2}{4\pi\varepsilon_0 \hbar c}.$$

One advantage of writing the coupling strength in terms of a dimensionless constant is that the numerical value is independent of the system of units used for a calculation. In addition, the quantum-mechanical probability of the interaction includes a single factor of $\alpha$ for each interaction vertex. The intrinsic strength of the electromagnetic interaction is given by the size of fine-structure constant $\alpha = 1/137$. The QCD interaction is intrinsically stronger with $\alpha_S \sim 1$. The *intrinsic* strength of the weak interaction, with $\alpha_W \sim 1/30$, is in fact greater than that

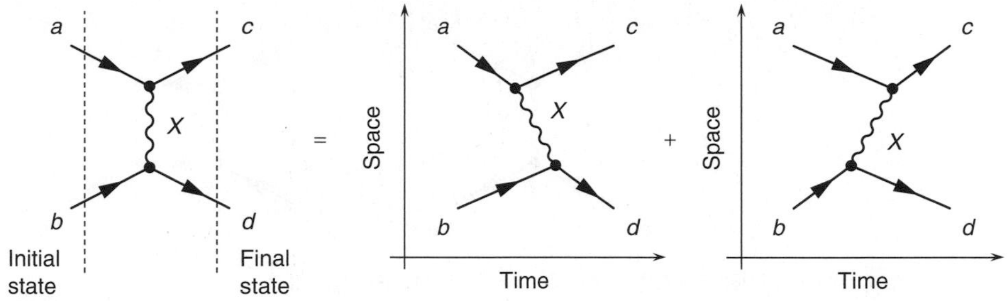


Fig. 1.6 The Feynman diagram for the scattering process $a + b \to c + d$ and the two time-ordered processes that it represents.

of QED. However, the large mass of the associated W boson means that at relatively low-energy scales, such as those encountered in particle decays, the weak interaction is (as its name suggests) very much weaker than QED.

### 1.1.5 Feynman diagrams

Feynman diagrams are an essential part of the language of particle physics. They are a powerful representation of transitions between states in quantum field theory and represent all possible time-orderings in which a process can occur. For example, the generic Feynman diagram for the process $a + b \to c + d$, involving the exchange of boson $X$, shown in Figure 1.6, represents the sum of the quantum mechanical amplitudes for the two possible time-orderings. It should be remembered that in a Feynman diagram time runs from left to right but only in the sense that the left-hand side of a Feynman diagram represents the initial state, in this case particles $a$ and $b$, and the right-hand side represents the final state, here $c$ and $d$. The central part of the Feynman diagram shows the particles exchanged and the Standard Model vertices involved in the interaction, but not the order in which these processes occurred. Feynman diagrams are much more than a pictorial representation of the fundamental physics underlying a particular process. From Quantum Field Theory it is possible to derive simple Feynman rules associated with the vertices and virtual particles in a Feynman diagram. Once the Feynman diagram has been drawn, it is straightforward to write down the quantum-mechanical transition matrix element using the relevant Feynman rules, thus avoiding the need to calculate each process from first principles in Quantum Field Theory.

In general, for each process considered, there will be an infinite number of Feynman diagrams that can be drawn. For example, Figure 1.7 shows Feynman diagrams for the scattering of two electrons by the exchange of either one or two photons. Both diagrams have the same initial and final state, and therefore correspond to the same physical process, $e^-e^- \to e^-e^-$. Each interaction vertex is associated with a factor $e$ in the matrix element, or equivalently a factor of $\alpha$ in the matrix

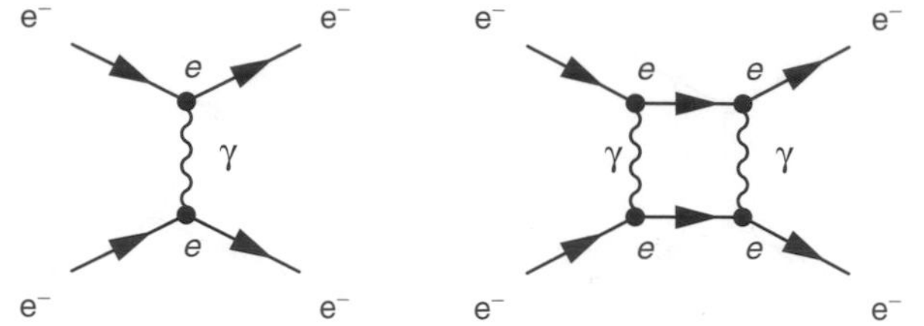


Fig. 1.7 Two Feynman diagrams for $e^-e^- \to e^-e^-$ scattering.

element squared. Thus, the matrix element squared for the diagram involving a single photon exchange and two vertices is proportional to $\alpha^2$, and that involving two photons and four vertices is proportional to $\alpha^4$,

$$|\mathcal{M}_\gamma^2| \propto \alpha^2 \quad \text{and} \quad |\mathcal{M}_{\gamma\gamma}^2| \propto \alpha^4.$$

Because the coupling strength of the electromagnetic interaction is relatively small, $\alpha \sim 1/137$, the diagram with four vertices is suppressed by a factor $O(10^4)$ relative to the diagram with two vertices. In the language of perturbation theory, only the lowest-order term is significant. Consequently, for almost all processes that will be encountered in this book, only the simplest (i.e. lowest-order) Feynman diagram needs to be considered.

For reasons that will become clear in Chapter 4, antiparticles are drawn in Feynman diagrams with arrows pointing in the "backwards in time" direction. In the Standard Model, particles and antiparticles can be created or annihilated only in pairs. This means that the arrows on the incoming and outgoing fermion lines in Standard Model vertices are always in the same sense and flow through the vertex; they never both point towards or away from the vertex.

### 1.1.6 Particle decays

Most particles decay with a very short lifetime. Consequently, only the relatively few stable and long-lived types of particle are detected in particle physics experiments. There are twelve fundamental spin-half particles (and the twelve corresponding antiparticles), but they are not all stable. For a particle to decay there must be a final state with lower total rest mass that can be reached by a process with a Feynman diagram constructed from the Standard Model vertices. Decays of the fundamental particles all involve the weak charged current which has the only interaction vertex that allows for a change in flavour. For example, since $m_\mu > m_e$ and the neutrinos are almost massless, the muon can decay via $\mu^- \to e^-\overline{\nu}_e\nu_\mu$ through the weak charged-current process with the Feynman diagram of Figure 1.8. Similar diagrams can be drawn for the tau-lepton. Since the electron is the lightest charged lepton, there is no corresponding weak decay process which conserves energy and momentum and consequently the electron is stable.

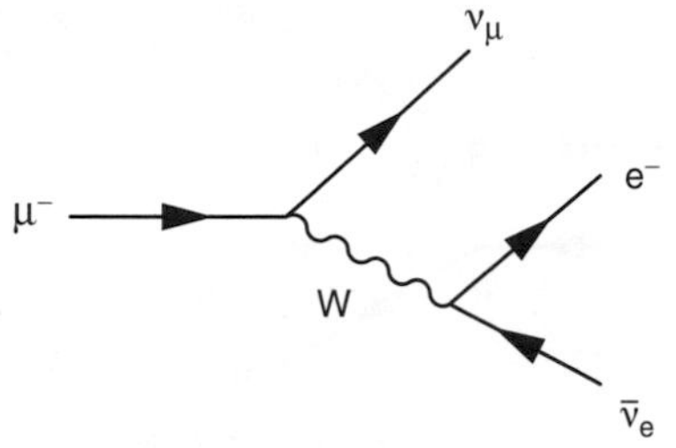


Fig. 1.8 The Feynman diagram for muon decay. The arrow in the "negative time direction" denotes an antiparticle, in this case an electron antineutrino ($\overline{\nu}_e$).

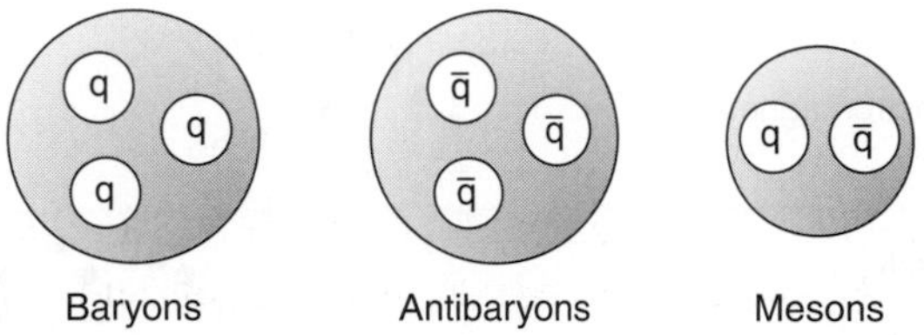


Fig. 1.9 The three types of observed hadronic states.

Because of the nature of the QCD interaction, quarks are never observed as free particles but are always found confined in bound states, known as hadrons. Consequently their decays need to be considered in the context of these bound states. The only hadronic states that have been observed to date, indicated in Figure 1.9, are the *mesons* which consist of a quark and an antiquark ($q\overline{q}$), the *baryons* which consist of three quarks ($qqq$), and the *antibaryons* consisting of three antiquarks ($\overline{q}\,\overline{q}\,\overline{q}$).

Many hadronic states have been observed. These correspond to different combinations of quark flavours and different internal angular momenta states. Each of these distinct states is observed as a particle with a particular mass, which is not just the sum of the masses of the constituent quarks, but includes a large contribution from the QCD binding energy. The total angular momentum of a hadron, which is referred to as its spin, depends on the orbital angular momentum between the constituent quarks and the overall spin state. Hadronic states can be labelled by their flavour content, i.e. the type of quarks they contain, their total angular momentum $J$, and their parity $P$, which is an observable quantum number reflecting the symmetry of the wavefunction under the transformation $\mathbf{r} \to -\mathbf{r}$. For example, the positively charged pion $\pi^+(u\overline{d})$, which is the lightest meson state consisting of an up-quark and an anti-down-quark, has spin-parity $J^P = 0^-$. The masses and lifetimes for a number of commonly encountered hadrons are given in Appendix C.

The only stable hadron is the proton, which is the lightest system of three quarks with $m_p = 938.3\,\text{MeV} \equiv 1.673 \times 10^{-27}\,\text{kg}$. As a free particle, the neutron with mass $m_n = 939.6\,\text{MeV}$, decays with a lifetime of about 15 min via the weak interaction

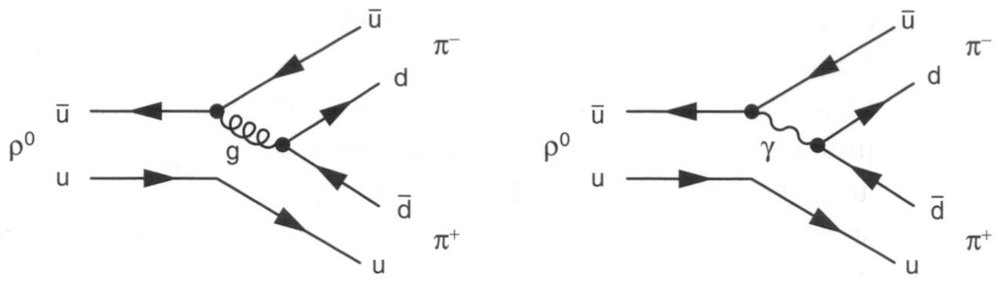


Fig. 1.10 Two possible Feynman diagrams for the decay $\rho^0 \to \pi^+\pi^-$.

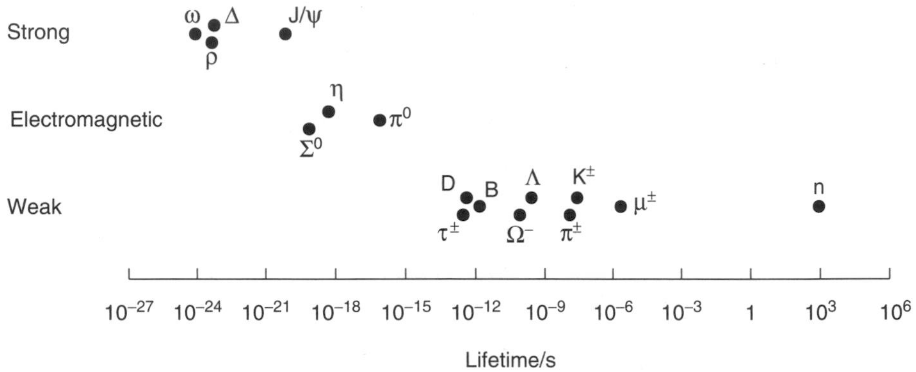


Fig. 1.11 The lifetimes of a number of common hadronic states grouped into the type of decay. Also shown are the lifetimes of the muon and tau-lepton, both of which decay weakly.

process $n \to p e^- \overline{\nu}_e$. Although a free neutron can decay, when bound within a nucleus, the change in nuclear binding energy is usually larger than the proton–neutron mass difference, and under these circumstances the neutron behaves as a stable particle. All other hadronic states decay, usually very rapidly.

Whilst particle decay rates depend on a number of factors, the most important is the type of fundamental interaction involved in the decay. For example, Figure 1.10 shows two possible Feynman diagrams for the decay of the $\rho^0$ meson, $\rho^0 \to \pi^+\pi^-$. The first diagram is a strong decay involving the exchange of a gluon. The second diagram is an electromagnetic process. The respective matrix elements depend on the coupling strengths of the strong and electromagnetic forces,

$$|\mathcal{M}_g|^2 \propto \alpha_S^2 \quad \text{and} \quad |\mathcal{M}_\gamma|^2 \propto \alpha^2.$$

Because $\alpha_S$ is two orders of magnitude greater than $\alpha$, the contribution from the strong decay Feynman diagram dominates.

The above example illustrates an important point; if a particle can decay by the strong interaction this will almost always dominate over any possible electromagnetic or weak decay processes. Similarly, electromagnetic decay modes will dominate over weak interaction processes. To illustrate this point, Figure 1.11 shows the lifetimes of a selection of hadrons divided according to whether the dominant decay mode is a strong, electromagnetic or weak interaction. Particles where only weak decay processes are possible are relatively long-lived (at least in the context of particle physics). Nevertheless, because the charged-current weak interaction produces a change of flavour at the interaction vertex, the weak interaction plays an important role in the decays of many particles for which electromagnetic and strong decay modes are not possible. Because many particles have very short lifetimes, only their decay products are observed in particle physics experiments.

## 1.2 Interactions of particles with matter

Particle physics experiments are designed to detect and identify the particles produced in high-energy collisions. Of the particles that can be produced, only the electron, proton, photon and the effectively undetectable neutrinos are stable. Unstable particles will travel a distance of order $\gamma v \tau$ before decaying, where $\tau$ is the mean lifetime (in the rest frame of the particle) and $\gamma = 1/\sqrt{1 - v^2/c^2}$ is the Lorentz factor accounting for relativistic time dilation. Relativistic particles with lifetimes greater than approximately $10^{-10}$ s will propagate over several metres when produced in high-energy particle collisions and thus can be directly detected. These relatively long-lived particles include the muon $\mu^\pm$, the neutron n(ddu), the charged pions $\pi^+(\mathrm{u\overline{d}})/\pi^-(\mathrm{d\overline{u}})$, and the charged kaons $\mathrm{K}^+(\mathrm{u\overline{s}})/\mathrm{K}^-(\mathrm{s\overline{u}})$. Short-lived particles with lifetimes of less than $10^{-10}$ s will typically decay before they travel a significant distance from the point of production and only their decay products can be detected.

The stable and relatively long-lived particles form the observables of particle physics collider experiments. The techniques employed to detect and identify the different particles depends on the nature of their interactions in matter. Broadly speaking, particle interactions can be divided into three categories: (i) the interactions of charged particles; (ii) the electromagnetic interactions of electrons and photons; and (iii) the strong interactions of charged and neutral hadrons.

### 1.2.1 Interactions and detection of charged particles

When a relativistic charged particle passes through a medium, it interacts electromagnetically with the atomic electrons and loses energy through the ionisation of the atoms. For a singly charged particle with velocity $v = \beta c$ traversing a medium with atomic number $Z$ and number density $n$, the ionisation energy loss per unit length traversed is given by the Bethe–Bloch equation,

$$\frac{\mathrm{d}E}{\mathrm{d}x} \approx -4\pi\hbar^2 c^2 \alpha^2 \frac{nZ}{m_\mathrm{e} v^2}\left\{\ln\left[\frac{2\beta^2\gamma^2 c^2 m_\mathrm{e}}{I_\mathrm{e}}\right] - \beta^2\right\}. \tag{1.1}$$

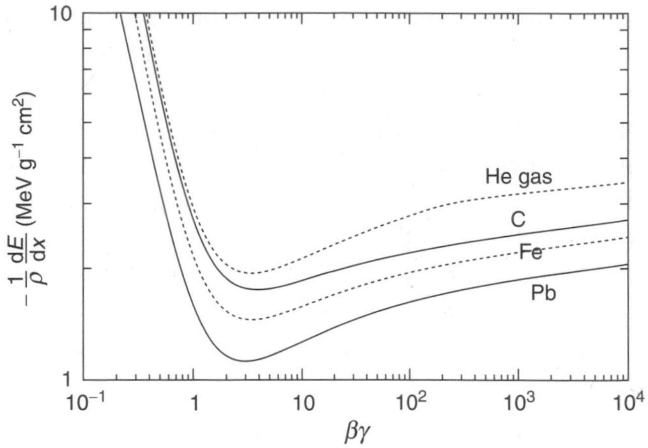


Fig. 1.12 The ionisation energy loss curves for a singly charged particle traversing lead, iron, carbon and gaseous helium. Adapted from Beringer *et al.* (2012).

Here $I_e$ is the effective ionisation potential of the material averaged over all atomic electrons, which is very approximately given by $I_e \sim 10\,Z\,\mathrm{eV}$. For a particular medium, the rate of the ionisation energy loss of a charged particle is a function of its velocity. Owing to the $1/\mathrm{v}^2$ term in the Bethe–Bloch equation, $\mathrm{d}E/\mathrm{d}x$ is greatest for low-velocity particles. Modern particle physics is mostly concerned with highly relativistic particles where $\mathrm{v} \approx c$. In this case, for a given medium, $\mathrm{d}E/\mathrm{d}x$ depends logarithmically on $(\beta\gamma)^2$, where

$$\beta\gamma = \frac{\mathrm{v}/c}{\sqrt{1-(\mathrm{v}/c)^2}} = \frac{p}{mc},$$

resulting in a slow "relativistic rise" of the rate of ionisation energy loss that is evident in Figure 1.12.

The rate of ionisation energy loss does not depend significantly on the material except through its density $\rho$. This can be seen by expressing the number density of atoms as $n = \rho/(Am_u)$, where $A$ is the atomic mass number and $m_u = 1.66\times10^{-27}$ kg is the unified atomic mass unit. Hence (1.1) can be written

$$\frac{1}{\rho}\frac{\mathrm{d}E}{\mathrm{d}x} \approx -\frac{4\pi\hbar^2c^2\alpha^2}{m_e\mathrm{v}^2m_u}\frac{Z}{A}\left\{\ln\left[\frac{2\beta^2\gamma^2m_ec^2}{I_e}\right]-\beta^2\right\}, \tag{1.2}$$

and it can be seen that $\mathrm{d}E/\mathrm{d}x$ is proportional to $Z/A$. Because nuclei consist of approximately equal numbers of protons and neutrons, $Z/A$ is roughly constant and thus the rate of energy loss by ionisation is proportional to density but otherwise does not depend strongly on the material. This can be seen from Figure 1.12, which shows the ionisation energy loss (in units of MeV g$^{-1}$ cm$^2$) as a function of $\beta\gamma$ for a singly charged particle in helium, carbon, iron and lead. Particles with $\beta\gamma \approx 3$, which corresponds to the minimum in the ionisation energy loss curve, are referred to as minimum ionising particles.

All charged particles lose energy through the ionisation of the medium in which they are propagating. Depending on the particle type, other energy-loss mechanisms maybe present. Nevertheless, for muons with energies below about 100 GeV, ionisation is the dominant energy-loss process. As a result, muons travel significant distances even in dense materials such as iron. For example, a 10 GeV muon loses approximately 13 MeV $cm^{-1}$ in iron and therefore has a range of several metres. Consequently, the muons produced at particle accelerators are highly penetrating particles that usually traverse the entire detector, leaving a trail of ionisation. This feature can be exploited to identify muons; all other charged particles have other types of interactions in addition to ionisation energy loss.

## Tracking detectors

The detection and measurement of the momenta of charged particles is an essential aspect of any large particle physics experiment. Regardless of the medium through which a charged particle travels, it leaves a trail of ionised atoms and liberated electrons. By detecting this ionisation it is possible to reconstruct the trajectory of a charged particle. Two main tracking detector technologies are used. Charged particle tracks can detected in a large gaseous tracking volume by drifting the liberating electrons in a strong electric field towards sense wires where a signal can be recorded. However, in recent particle physics experiments, for example the ATLAS and CMS experiments at the LHC, there has been a move to using tracking detectors based on semiconductor technology using silicon pixels or strips.

When a charged particle traverses an appropriately doped silicon wafer, electron–hole pairs are created by the ionisation process, as indicated by Figure 1.13. If a potential difference is applied across the silicon, the holes will drift in the direction of the electric field where they can be collected by p–n junctions. The sensors can be shaped into silicon strips, typically separated by $\mathcal{O}(25\ \mu m)$, or into silicon pixels giving a precise 2D space point. The signals are not small; in crossing a typical silicon wafer, a charged particle will liberate $\mathcal{O}(10\,000)$ electron–hole pairs that,

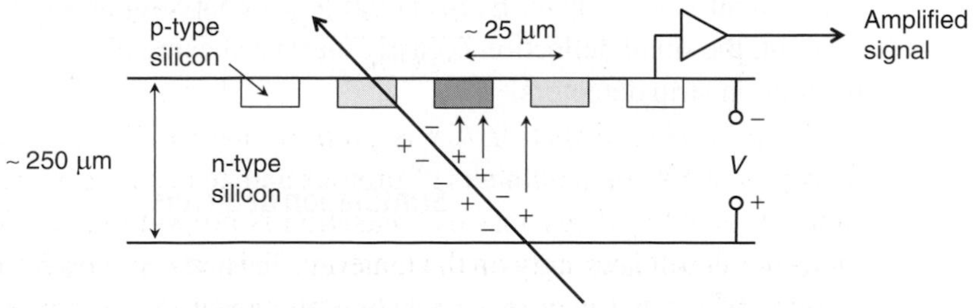


Fig. 1.13 The production and collection of charge in a silicon tracking sensor.

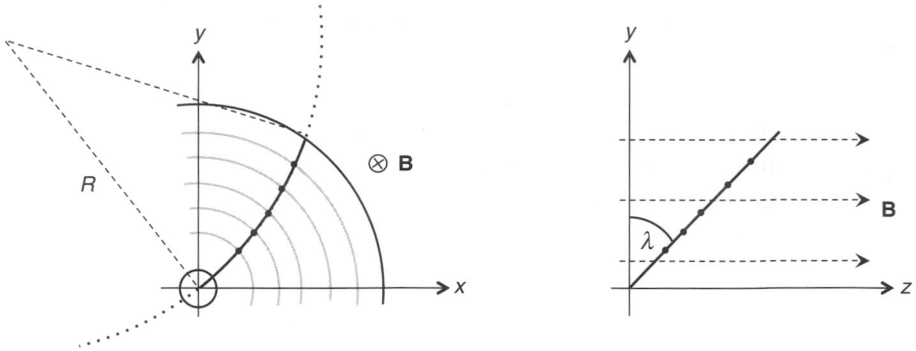


Fig. 1.14 The principle of charged particle track reconstruction from the space points observed in a (five-layer) silicon tracking detector. The curvature in the *xy*-plane determines the transverse momentum.

with appropriate amplification electronics, gives a clear signal associated with the strip/pixel on which the charge was collected.

Silicon tracking detectors typically consist of several cylindrical surfaces of silicon wafers, as indicated in Figure 1.14. A charged particle will leave a "hit" in a silicon sensor in each cylindrical layer from which the trajectory of the charged particle *track* can be reconstructed. The tracking system is usually placed in a large solenoid producing an approximately uniform magnetic field in the direction of axis of the colliding beams, taken to be the $z$-axis. Owing to the $\mathbf{v} \times \mathbf{B}$ Lorentz force, the trajectory of a charged particle in the axial magnetic field is a helix with a radius of curvature $R$ and a pitch angle $\lambda$, which for a singly charged particle ($|q| = e$) are related to its momentum by

$$\mathrm{p} \cos \lambda = 0.3\,\mathrm{B}R,$$

where the momentum p is given in GeV/$c$, B is the magnetic flux density in tesla and $R$ is in metres. Hence by determining the parameters of the helical trajectory from the measured hits in the tracking detectors, $R$ and $\lambda$ can be obtained and thus the momentum of the particle can be reconstructed. For high-momentum particles, the radius of curvature can be large. For example, the radius of curvature of a 100 GeV $\pi^{\pm}$ in the 4 T magnetic field of the super-conductor solenoid of the CMS experiment is $R \sim 100$ m. Even though such charged particle tracks appear almost straight, the small deflection is easily measured using the precise space-points from the silicon strip detectors.

## Scintillation detectors

Organic scintillators are used extensively in modern particle physics experiments as a cost effective way to detect the passage of charged particles where precise spatial information is not required. In particular, detectors based on plastic and liquid scintillators have been used in a number of recent neutrino experiments. In an organic scintillator, the passage of a charged particle leaves some of the molecules in an excited state. In a scintillator, the subsequent decay of the excited state results in the emission of light in the ultraviolet (UV) region. By adding fluorescent dyes to the scintillator, the molecules of the dye absorb the UV light and re-emit it as photons in the blue region. The blue light can be detected by using photomultiplier devices which are capable of detecting single optical photons.

### Čerenkov radiation

Charged particles can also be detected through their emission of Čerenkov radiation. When a charged particle traverses a dielectric medium of refractive index $n$ it polarises the molecules in the medium. After its passage, the molecules return to the unpolarised state through the emission of photons. If the velocity of the particle is greater than the speed of light in that medium, $\mathrm{v} > c/n$, constructive interference occurs and Čerenkov radiation is emitted as a coherent wavefront at a fixed angle $\theta$ to the trajectory of the charged particle, analogous to the sonic boom produced by supersonic aircraft. The angle at which the radiation is emitted is given by the geometrical construction shown in Figure 1.15. In a time $t$, the particle travels a distance $\beta ct$. In this time the wavefront emitted at $t = 0$ has travelled a distance $ct/n$ and therefore the angle $\theta$ at which the radiation is produced is given by

$$\cos\theta = \frac{1}{n\beta}.$$

The photons emitted as Čerenkov radiation can be detected using photo-multiplier tubes (PMTs), capable of detecting a single photon with reasonable efficiency. Čerenkov radiation can be used to detect relativistic particles in large volumes of transparent liquid (for example water) as has been used extensively in the detection

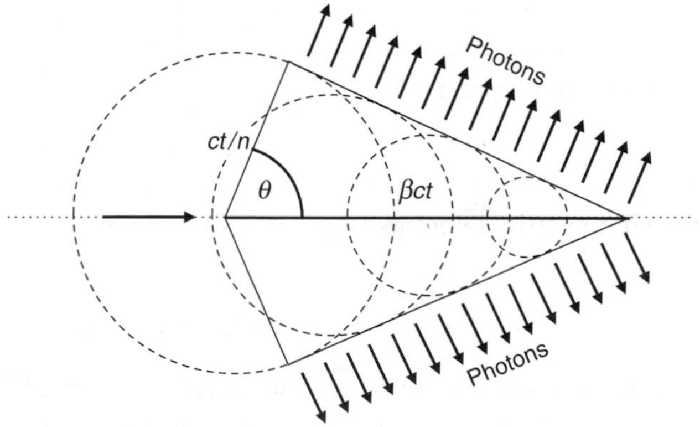


Fig. 1.15 The geometry of the emission of Čerenkov radiation.

of neutrinos. Furthermore, Čerenkov radiation is emitted only when $\beta > 1/n$. This threshold behaviour can be utilised to aid the identification of particles of a given momentum p; for a relativistic particle $\beta = \mathrm{p}c/E = \mathrm{p}/(\mathrm{p}^2 + m^2c^2)^{1/2}$ and therefore only particles with mass

$$mc < (n^2 - 1)^{1/2}\mathrm{p},$$

will produce Čerenkov radiation.

### 1.2.2 Interactions and detection of electrons and photons

At low energies, the energy loss of electrons is dominated by ionisation. However, for energies above a "critical energy" $E_c$, the main energy loss mechanism is bremsstrahlung (German for braking radiation), whereby the electron radiates a photon in the electrostatic field of a nucleus, as shown in Figure 1.16. The critical energy is related to the charge $Z$ of the nucleus and is approximately

$$E_c \sim \frac{800}{Z}\,\mathrm{MeV}.$$

The electrons of interest in most particle physics experiments are in the multi-GeV range, significantly above the critical energy, and therefore interact with matter primarily through bremsstrahlung. The bremsstrahlung process can occur for all charged particles, but the rate is inversely proportional to the square of the mass of the particle. Hence, for muons the rate of energy loss by bremsstrahlung is suppressed by $(m_\mathrm{e}/m_\mu)^2$ relative to that for electrons. It is for this reason that bremsstrahlung is the dominant energy-loss process for electrons, but ionisation energy loss dominates for muons (except at very high energies, $E_\mu > 100\,\mathrm{GeV}$, where bremsstrahlung also contributes).

At low energies, photons interact in matter primarily by the photoelectric effect, whereby the photon is absorbed by an atomic electron that is ejected from the atom. At somewhat higher energies, $E_\gamma \sim 1\,\mathrm{MeV}$, the Compton scattering process $\gamma\mathrm{e}^- \to \gamma\mathrm{e}^-$ becomes significant. At higher energies still, $E_\gamma > 10\,\mathrm{MeV}$, the interactions of photons are dominated by $\mathrm{e}^+\mathrm{e}^-$ pair production in the field of the nucleus, as shown in Figure 1.16.

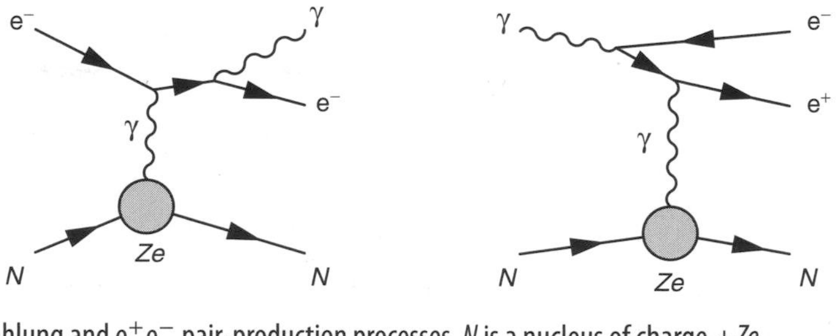


Fig. 1.16 The bremsstrahlung and $\mathrm{e}^+\mathrm{e}^-$ pair-production processes. $N$ is a nucleus of charge $+Ze$.

The electromagnetic interactions of high energy electrons and photons in matter are characterised by the *radiation length* $X_0$. The radiation length is the average distance over which the energy of an electron is reduced by bremsstrahlung by a factor of $1/e$. It is also approximately 7/9 of the mean free path of the $e^+e^-$ pair-production process for a high-energy photon. The radiation length is related to the atomic number $Z$ of the material, and can be approximated by the expression

$$X_0 \approx \frac{1}{4\alpha n Z^2 r_e^2 \ln(287/Z^{1/2})},$$

where $n$ is the number density of nuclei and $r_e$ is the "classical radius of the electron" defined as

$$r_e = \frac{e^2}{4\pi\epsilon_0 m_e c^2} = 2.8 \times 10^{-15}\,\text{m}.$$

For high-$Z$ materials the radiation length is relatively short. For example, iron and lead have radiation lengths of $X_0(\text{Fe}) = 1.76\,\text{cm}$ and $X_0(\text{Pb}) = 0.56\,\text{cm}$.

## Electromagnetic showers

When a high-energy electron interacts in a medium it radiates a bremsstrahlung photon, which in turn produces an $e^+e^-$ pair. The process of bremsstrahlung and pair production continues to produce a cascade of photons, electrons and positrons, referred to as an electromagnetic shower, as indicated in Figure 1.17. Similarly, the primary interaction of a high-energy photon will produce an $e^+e^-$ pair that will then produce an electromagnetic shower.

The number of particles in an electromagnetic shower approximately doubles after every radiation length of material traversed. Hence, in an electromagnetic

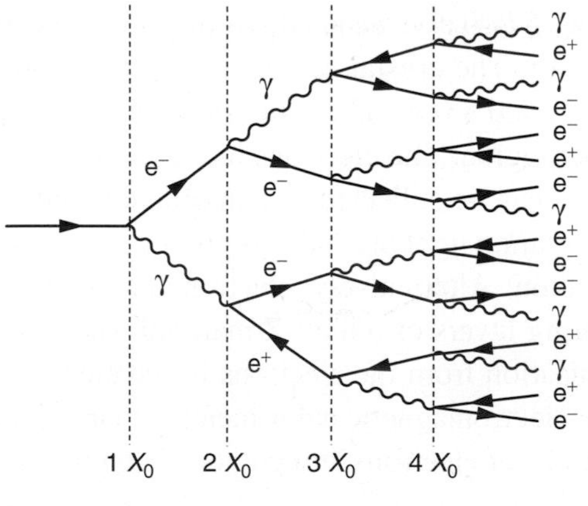


Fig. 1.17 The development of an electromagnetic shower where the number of particles roughly doubles after each radiation length.

shower produced by an electron or photon of energy $E$, the average energy of the particles after $x$ radiation lengths is

$$\langle E \rangle \approx \frac{E}{2^x}. \tag{1.3}$$

The shower continues to develop until the average energy of the particles falls below the critical energy $E_c$, at which point the electrons and positrons in the cascade lose energy primarily by ionisation. The electromagnetic shower therefore has the maximum number of particles after $x_{max}$ radiation lengths, given by the condition $\langle E \rangle \approx E_c$. From (1.3) it can be seen that this point is reached after

$$x_{max} = \frac{\ln (E/E_c)}{\ln 2}$$

radiation lengths. In a high-$Z$ material, such as lead with $E_c \sim 10\,\text{MeV}$, a 100 GeV electromagnetic shower reaches is maximum after $x_{max} \sim 13\,X_0$. This corresponds to less than 10 cm of lead. Consequently, electromagnetic showers deposit most of their energy in a relatively small region of space. The development of a shower is a stochastic process consisting of a number of discrete interactions. However, because of the large numbers of particles involved, which is of order $2^{x_{max}}$, the fluctuations in the development of different electromagnetic showers with the same energy are relatively small and individual electromagnetic showers of the same energy are very much alike.

## Electromagnetic calorimeters

In high-energy particle physics experiments, the energies of electrons and photons are measured using an electromagnetic calorimeter constructed from high-$Z$ materials. A number of different technologies can be used. For example, the electromagnetic calorimeter in the CMS detector at the LHC is constructed from an array of 75 000 crystals made from lead tungstate ($PbWO_4$), which is an inorganic scintillator. The crystals are both optically transparent and have a short radiation length $X_0 = 0.83\,\text{cm}$, allowing the electromagnetic showers to be contained in a compact region. The electrons in the electromagnetic shower produce scintillation light that can be collected and amplified by efficient photon detectors. The amount of scintillation light produced is proportional to the total energy of the original electron/photon. Alternatively, electromagnetic calorimeters can be constructed from alternating layers of a high-$Z$ material, such as lead, and an active layer in which the ionisation from the electrons in the electromagnetic shower can be measured. For the electromagnetic calorimeters in large particle physics detectors, the energy resolution for electrons and photons is typically in the range

$$\frac{\sigma_E}{E} \sim \frac{3\% - 10\%}{\sqrt{E/\text{GeV}}}.$$

### 1.2.3 Interactions and detection of hadrons

Charged hadrons (for example, protons and charged pions) lose energy continuously by the ionisation process as they traverse matter. In addition, both charged and neutral hadrons can undergo a strong interaction with a nucleus of the medium. The particles produced in this primary hadronic interaction will subsequently interact further downstream in the medium, giving rise to a cascade of particles. The development of hadronic showers is parameterised by the nuclear interaction interaction length $\lambda_I$ defined as the mean distance between hadronic interactions of relativistic hadrons. The nuclear interaction length is significantly larger than the radiation length. For example, for iron $\lambda_I \approx 17$ cm, compared to its radiation length of 1.8 cm.

Unlike electromagnetic showers, which develop in a uniform manner, hadronic showers are inherently more variable because many different final states can be produced in high-energy hadronic interactions. Furthermore, any $\pi^0$s produced in the hadronic shower decay essentially instantaneously by $\pi^0 \to \gamma\gamma$, leading to an electromagnetic component of the shower. The fraction of the energy in this electromagnetic component will depend on the number of $\pi^0$s produced and will vary from shower to shower. In addition, not all of the energy in a hadronic shower is detectable; on average 30% of incident energy is effectively lost in the form of nuclear excitation and break-up.

#### Hadron calorimeters

In particle detector systems, the energies of hadronic showers are measured in a hadron calorimeter. Because of the relatively large distance between nuclear interactions, hadronic showers occupy a significant volume in any detector. For example, in a typical hadron calorimeter, the shower from a 100 GeV hadron has longitudinal and lateral extents of order 2 m and 0.5 m respectively. Therefore a hadron calorimeter necessarily occupies a large volume. A number of different technologies have been used to construct hadron calorimeters. A commonly used technique is to use a sandwich structure of thick layers of high-density absorber material (in which the shower develops) and thin layers of active material where the energy depositions from the charged particles in the shower are sampled. For example, the hadron calorimeter in the ATLAS experiment at the LHC consists of alternating layers of steel absorber and plastic scintillator tiles. The signals in the different layers of the scintillator tiles are summed to give a measure of the energy of the hadronic shower. Fluctuations in the electromagnetic fraction of the shower and the amount of energy lost in nuclear break-up limits the precision to which the energy can be measured to

$$\frac{\sigma_E}{E} \gtrsim \frac{50\%}{\sqrt{E/\mathrm{GeV}}},$$

which is roughly an order of magnitude worse than the energy resolution for electromagnetic showers.

## 1.3 Collider experiments

At a particle accelerator, the colliding beams produce individual interactions referred to as *events*. The large particle physics detector systems use a wide range of technologies to detect and measure the properties of the particles produced in these high-energy collisions with the aim of reconstructing the primary particles produced in the interaction. In essence, one tries to go from the signals in the different detector systems back to the Feynman diagram responsible for the interaction.

The basic structure of a modern particle physics detector is indicated in Figure 1.18. In general, a detector consists of a cylindrical (or polygonal) barrel part, with its axis parallel to the incoming colliding beams. The cylindrical structure is closed by two flat end caps, providing almost complete solid angle coverage down to the beam pipe. The inner region of the detector is devoted to the tracking of charged particles. The tracking volume is surrounded by an electromagnetic calorimeter (ECAL) for detecting electrons and photons. The relatively large-volume hadronic calorimeter (HCAL) for detecting and measuring the energies of hadrons is located outside the ECAL. Dedicated detectors are positioned at the outside of the experiment to record the signals from any high-energy muons produced in the collisions, which are the only particles (apart from neutrinos) that can penetrate through the HCAL. In order to be able to measure the momenta of

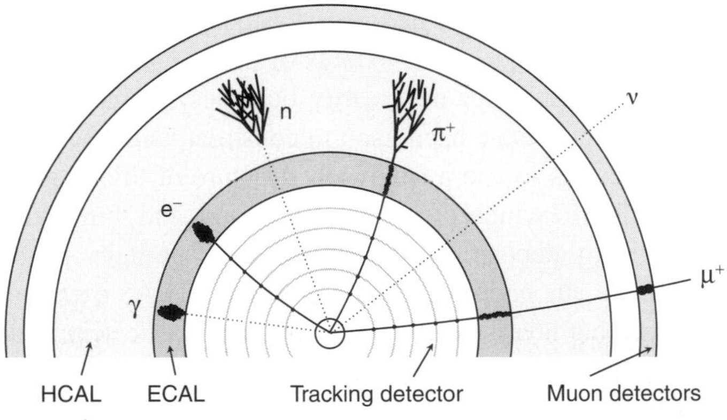


Fig. 1.18 The typical layout of a large particle physics detector consisting of a tracking system (here shown with cylindrical layers of a silicon detector), an electromagnetic calorimeter (ECAL), a hadron calorimeter (HCAL) and muon detectors. The solenoid used to produce the magnetic field is not shown. The typical signatures produced by different particles are shown.

charged particles, a detector usually has a solenoid which produces a strong axial magnetic field in the range $B = 1{-}4\,T$. The solenoid may be located between the tracking volume and the calorimeters.

The design of a collider experiment is optimised for the identification and energy measurement of the particles produced in high-energy collisions. The momenta of charged particles are obtained from the curvature of the reconstructed tracks. The energies of neutral particles are obtained from the calorimeters. Particle identification is achieved by comparing the energy deposits in the different detector systems as indicated in Figure 1.18. Photons appear as isolated energy deposits in the ECAL. Electrons are identified as charged-particle tracks that are associated with an electromagnetic shower in the ECAL. Neutral hadrons will usually interact in the HCAL and charged hadrons are identified as charged-particle tracks associated with a small energy deposit in the ECAL (from ionisation energy loss) and a large energy deposition in the HCAL. Finally, muons can be identified as charged-particle tracks associated with small energy depositions in both the ECAL and HCAL and signals in the muon detectors on the outside of the detector system.

Whilst neutrinos leave no signals in the detector, their presence often can be inferred from the presence of *missing momentum*, which is defined as

$$\mathbf{p}_{mis} = -\sum_i \mathbf{p}_i,$$

where the sum extends over the measured momenta of all the observed particles in an event. If all the particles produced in the collision have been detected, this sum should be zero (assuming the collision occurs in the centre-of-mass frame). Significant missing momentum is therefore indicative of the presence of an undetected neutrino.

The ultimate aim in collider experiments is to reconstruct the fundamental particles produced in the interaction. Electrons, photons and muons give clear signatures and are easily identified. Tau-leptons, which decay in $2.9 \times 10^{-13}$ s, have to be identified from their observed decay products. The main tau-lepton decay modes are $\tau^- \to e^-\overline{\nu}_e\nu_\tau$ (17.8%), $\tau^- \to \mu^-\overline{\nu}_\mu\nu_\tau$ (17.4%), $\tau^- \to \pi^-(n\pi^0)\nu_\tau$ (48%) and $\tau^- \to \pi^-\pi^+\pi^-(n\pi^0)\nu_\tau$ (15%). The hadronic decay modes typically lead to final states with one or three charged pions and zero, one or two $\pi^0$s which decay to photons $\pi^0 \to \gamma\gamma$. Tau-leptons can therefore be identified as narrowly collimated jets of just a few particles and the presence of missing momentum in the event, associated with the neutrino.

### 1.3.1 Detection of quarks

Owing to the nature of QCD, quarks are never observed as free particles, but are always found confined within hadrons. However, in high-energy collisions it is

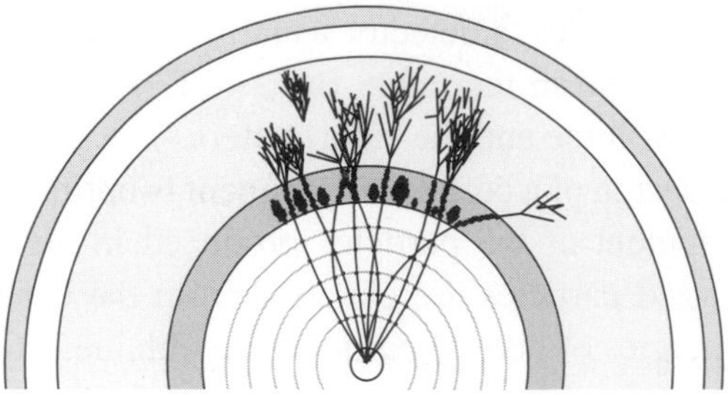

Fig. 1.19 An illustration of the appearance of a jet in a detector. In practice, the individual particles are not resolved.

quarks that are produced, not hadrons. For example, in the process $e^+e^- \to q\overline{q}$ the two quarks will be produced flying apart at relativistic velocities. As a result of the QCD interaction, the energy in the strong interaction field between the two quarks is converted into further pairs of quarks and antiquarks through a process call hadronisation (described in Chapter 10) that occurs over a distance scale of $10^{-15}$ m. As a result of hadronisation, each quark produced in a collision produces a jet of hadrons, as indicated in Figure 1.19. Hence a quark is observed as an energetic jet of particles. On average, approximately 60% of the energy in a jet is in the form of charged particles (mostly $\pi^\pm$), 30% of the energy is in the form of photons from $\pi^0 \to \gamma\gamma$ decays, and 10% is in the form of neutral hadrons (mostly neutrons and $K_L$s). In high-energy jets, the separation between the individual particles is typically smaller than the segmentation of the calorimeters and not all of the particles in the jet can be resolved. Nevertheless, the energy and momentum of the jet can be determined from the total energy deposited in the calorimeters.

## Tagging of b-quarks

In general, it is not possible to tell which flavour of quark was produced, or even whether the jet originated from a quark or a gluon. However, if a b-quark is produced, the hadronisation process will create a jet of hadrons, one of which will contain the b-quark, for example a $\overline{B}^0(b\overline{d})$ meson. It turns out that b-quark hadrons are relatively long-lived with lifetimes of order $1.5 \times 10^{-12}$ s. When produced in high-energy collisions, this relatively long lifetime, combined with the Lorentz time-dilation factor, means that B hadrons travel on average a few millimetres before decaying. The decays of B hadrons often produce more than one charged particle. Because of the relatively large mass of the b-quark, the decay products can be produced at a relatively large angle to the original b-quark direction. Therefore the experimental signature for a b-quark is a jet of particles emerging from the point of the collision (the primary vertex) and a secondary vertex from the b-quark decay,

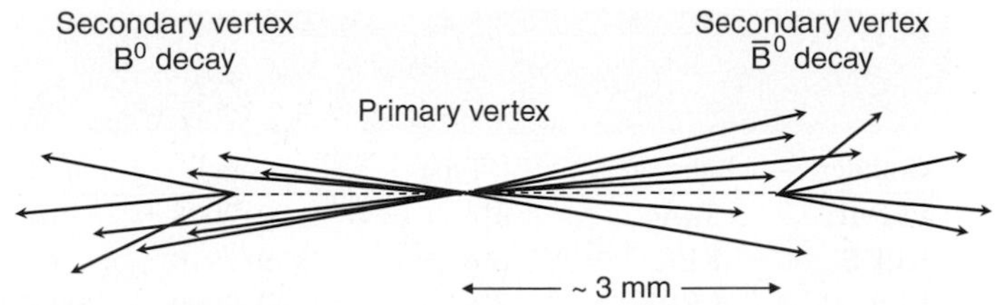


Fig. 1.20 An illustration of the principle of b-quark tagging in a $e^+e^- \rightarrow Z \rightarrow b\bar{b}$ event.

which is displaced from the primary vertex by several millimetres, as indicated in Figure 1.20.

The identification of b-quark jets relies on the ability to resolve the secondary vertices from the primary vertex. In practice, this is achieved by using high-precision silicon microvertex detectors consisting of several concentric layers of silicon at radii of a few centimetres from the axis of the colliding beams. Such detectors can achieve a single hit resolution of $O(10\,\mu\text{m})$, sufficient to be able to identify and reconstruct the secondary vertices, even in a dense jet environment. The ability to tag b-quarks has played an important role in a number of recent experiments.

## 1.4 Measurements at particle accelerators

With the exception of the measurements of the properties of the neutrino, most of the recent breakthroughs in particle physics have come from experiments at high-energy particle accelerators. Particle accelerators can be divided into two types: (i) colliding beam machines where two beams of accelerated particles are brought into collision; and (ii) fixed-target experiments where a single beam is fired at a stationary target. In order to produce massive particles, such as the $W^\pm$, Z and H bosons, high energies are required. More precisely, the energy available in the centre-of-mass frame has to be greater than the sum of the masses of the particles being produced. The centre-of-mass energy $\sqrt{s}$ is given by the square root of the Lorentz invariant quantity $s$ formed from the total energy and momentum of the two initial-state particles, which in natural units with $c = 1$ is

$$s = \left(\sum_{i=1}^{2} E_i\right)^2 - \left(\sum_{i=1}^{2} \mathbf{p}_i\right)^2 .$$

In a fixed-target experiment, momentum conservation implies that the final-state particles are always produced with significant kinetic energy and much of the initial

**Table 1.4** The basic parameters of the recent particle accelerators. At the time of writing the LHC was operating at $\sqrt{s} = 8$ TeV.

| Collider | Laboratory | Type | Date | $\sqrt{s}$/GeV | Luminosity/$cm^{-2}s^{-1}$ |
|---|---|---|---|---|---|
| PEP-II | SLAC | $e^+e^-$ | 1999–2008 | 10.5 | $1.2 \times 10^{34}$ |
| KEKB | KEK | $e^+e^-$ | 1999–2010 | 10.6 | $2.1 \times 10^{34}$ |
| LEP | CERN | $e^+e^-$ | 1989–2000 | 90–209 | $10^{32}$ |
| HERA | DESY | $e^-p/e^+p$ | 1992–2007 | 320 | $8 \times 10^{31}$ |
| Tevatron | Fermilab | $p\overline{p}$ | 1987–2012 | 1960 | $4 \times 10^{32}$ |
| LHC | CERN | pp | 2009– | 14 000 | $10^{34}$ |

energy is effectively wasted. For example, if an $E = 7$ TeV proton collides with a proton at rest,

$$s = (E + m_p)^2 - p^2 = 2m_p^2 + 2m_pE \approx 2m_pE,$$

giving a centre-of-mass energy of just 115 GeV. Colliding beam machines have the advantage that they can achieve much higher centre-of-mass energies since the collision occurs in the centre-of-mass frame. For example, the LHC will ultimately collide two beams of 7 TeV protons giving a centre-of-mass energy of 14 TeV. For this reason, almost all high-energy particle physics experiments are based on large particle colliders.

Only charged stable particles can be accelerated to high energies, and therefore the possible types of accelerator are restricted to $e^+e^-$ colliders, hadron colliders (pp or $p\overline{p}$) and electron–proton colliders ($e^-p$ or $e^+p$). The most recent examples have been the Tevatron $p\overline{p}$ collider, the LHC pp collider, the LEP $e^+e^-$ collider, the PEP-II and KEKB $e^+e^-$ b-factories, and the HERA electron–proton collider. The main parameters of these machines are summarised in Table 1.4. The two most important features of an accelerator are its centre-of-mass energy, which determines the types of particles that can be studied/discovered, and its instantaneous luminosity $\mathcal{L}$, which determines the event rates. For a given process, the number of interactions is the product of the luminosity integrated over the lifetime of the operation of the machine and the *cross section* for the process in question,

$$N = \sigma \int \mathcal{L}(t)\,dt. \tag{1.4}$$

The cross section (defined in Chapter 3) is a measure of quantum mechanical probability for the interaction. It depends on the fundamental physics involved in the Feynman diagram(s) contributing to the process.

In order to convert the observed numbers of events of a particular type to the cross section for the process, the integrated luminosity needs to be known. In principle, this can be calculated from the knowledge of the parameters of the colliding beams. Typically, the particles in an accelerator are grouped into bunches that are brought into collision at one or more interaction points where the detectors are located. In the case of the LHC, the bunches are separated by 25 ns, corresponding to a collision frequency of $f = 40\,\text{MHz}$. The instantaneous luminosity of the machine can be expressed in terms of the numbers of particles in the colliding bunches, $n_1$ and $n_2$, the frequency at which the bunches collide, and the root-mean-square (rms) horizontal and vertical beam sizes $\sigma_x$ and $\sigma_y$. Assuming that the beams have a Gaussian profile and collide head-on, the instantaneous luminosity is given by

$$\mathcal{L} = f\frac{n_1 n_2}{4\pi\,\sigma_x \sigma_y}. \tag{1.5}$$

In practice, the exact properties of the colliding beams, such as the transverse profiles, are not known precisely and it is not possible to accurately calculate the instantaneous luminosity. For this reason, cross section measurements are almost always made with reference to a process where the cross section is already known. Hence, a cross section measurement is performed by counting the number of events of interest $N$, and the number of observed events for the reference process $N_{\text{ref}}$, such that the measured cross section is given by

$$\sigma = \sigma_{\text{ref}}\frac{N}{N_{\text{ref}}}.$$

Corrections may needed to account for the detection efficiency and possible sources of background events. Nevertheless, ultimately many experimental particle physics measurements reduce to counting events, where the event type is identified using the experimental techniques described in Section 1.3. Of course, this is not always quite as easy as it sounds.

## Summary

The intention of this chapter was to introduce some of the basic ideas of particle physics. At this point you should be familiar with the types of particles and forces in the Standard Model and you should have a qualitative understanding of how to use the Standard Model vertices associated with the electromagnetic, strong and weak interactions to construct Feynman diagrams for particle interactions and decays. The second part of the chapter introduced the experimental techniques of particle physics and is intended to provide the context for the experimental measurements used to demonstrate the theoretical ideas developed in the following chapters. At this point you should understand how the different particles appear in the large detector systems employed in collider experiments.

## Problems

**1.1** Feynman diagrams are constructed out of the Standard Model vertices shown in Figure 1.4. Only the weak charged-current ($W^{\pm}$) interaction can change the flavour of the particle at the interaction vertex. Explaining your reasoning, state whether each of the sixteen diagrams below represents a valid Standard Model vertex.

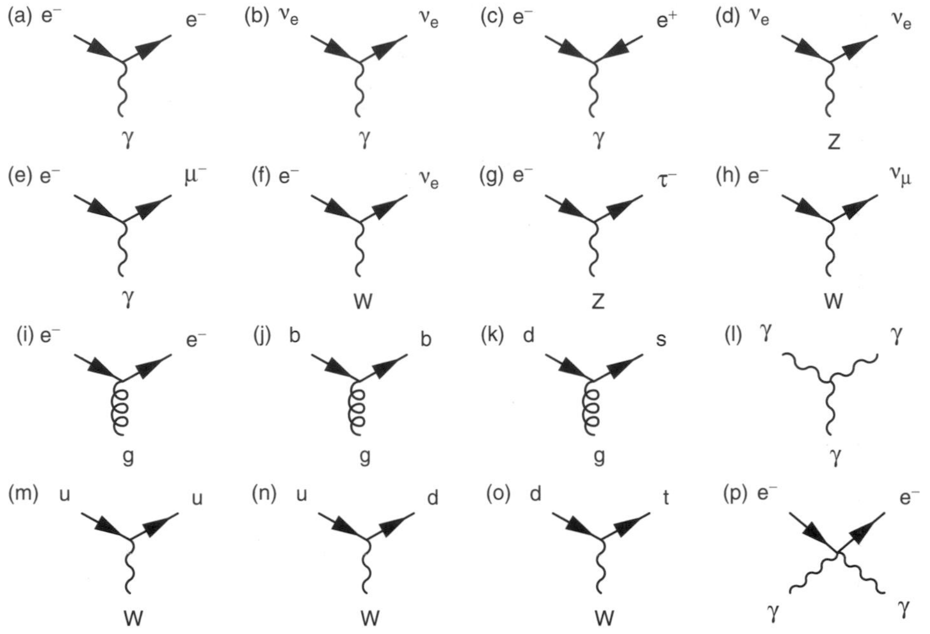


**1.2** Draw the Feynman diagram for $\tau^- \to \pi^- \nu_\tau$ (the $\pi^-$ is the lightest $d\bar{u}$ meson).

**1.3** Explain why it is *not* possible to construct a valid Feynman diagram using the Standard Model vertices for the following processes:

(a) $\mu^- \to e^+ e^- e^+$,
(b) $\nu_\tau + p \to \mu^- + n$,
(c) $\nu_\tau + p \to \tau^+ + n$,
(d) $\pi^+(u\bar{d}) + \pi^-(d\bar{u}) \to n(udd) + \pi^0(u\bar{u})$.

**1.4** Draw the Feynman diagrams for the decays:

(a) $\Delta^+(uud) \to n(udd)\,\pi^+(u\bar{d})$,
(b) $\Sigma^0(uds) \to \Lambda(uds)\,\gamma$,
(c) $\pi^+(u\bar{d}) \to \mu^+ \nu_\mu$,

and place them in order of increasing lifetime.

**1.5** Treating the $\pi^0$ as a $u\bar{u}$ bound state, draw the Feynman diagrams for:

(a) $\pi^0 \to \gamma\gamma$,
(b) $\pi^0 \to \gamma e^+ e^-$,

(c) $\pi^0 \to e^+e^-e^+e^-$,
(d) $\pi^0 \to e^+e^-$.

By considering the number of QED vertices present in each decay, *estimate* the relative decay rates taking $\alpha = 1/137$.

**1.6** Particle interactions fall into two main categories, scattering processes and annihilation processes, as indicated by the Feynman diagrams below.

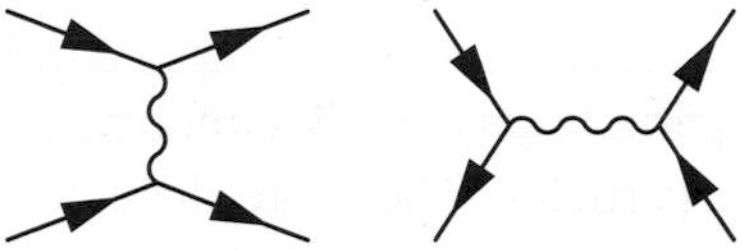

Draw the lowest-order Feynman diagrams for the scattering and/or annihilation processes:

(a) $e^-e^- \to e^-e^-$,
(b) $e^+e^- \to \mu^+\mu^-$,
(c) $e^+e^- \to e^+e^-$,
(d) $e^-\nu_e \to e^-\nu_e$,
(e) $e^-\overline{\nu}_e \to e^-\overline{\nu}_e$.

In some cases there may be more than one lowest-order diagram.

**1.7** High-energy muons traversing matter lose energy according to

$$-\frac{1}{\rho}\frac{dE}{dx} \approx a + bE,$$

where $a$ is due to ionisation energy loss and $b$ is due to the bremsstrahlung and $e^+e^-$ pair-production processes. For standard rock, taken to have $A = 22$, $Z = 11$ and $\rho = 2.65\ \mathrm{g\ cm^{-3}}$, the parameters $a$ and $b$ depend only weakly on the muon energy and have values $a \approx 2.5\ \mathrm{MeV\ g^{-1}\ cm^2}$ and $b \approx 3.5 \times 10^{-6}\ \mathrm{g^{-1}\ cm^2}$.

(a) At what muon energy are the ionisation and bremsstrahlung/pair production processes equally important?
(b) Approximately how far does a 100 GeV cosmic-ray muon propagate in rock?

**1.8** Tungsten has a radiation length of $X_0 = 0.35$ cm and a critical energy of $E_c = 7.97$ MeV. Roughly what thickness of tungsten is required to fully contain a 500 GeV electromagnetic shower from an electron?

**1.9** The CPLEAR detector (see Section 14.5.2) consisted of: tracking detectors in a magnetic field of 0.44 T; an electromagnetic calorimeter; and Čerenkov detectors with a radiator of refractive index $n = 1.25$ used to distinguish $\pi^\pm$ from $K^\pm$.

A charged particle travelling perpendicular to the direction of the magnetic field leaves a track with a measured radius of curvature of $R = 4$ m. If it is observed to give a Čerenkov signal, is it possible to distinguish between the particle being a pion or kaon? Take $m_\pi \approx 140\ \mathrm{MeV}/c^2$ and $m_K = 494\ \mathrm{MeV}/c^2$.

**1.10** In a fixed-target pp experiment, what proton energy would be required to achieve the same centre-of-mass energy as the LHC, which will ultimately operate at 14 TeV.

**1.11** At the LEP $e^+e^-$ collider, which had a circumference of 27 km, the electron and positron beam currents were both 1.0 mA. Each beam consisted of four equally spaced bunches of electrons/positrons. The bunches had an effective area of $1.8 \times 10^4\ \mu\mathrm{m}^2$. Calculate the instantaneous luminosity on the assumption that the beams collided head-on.

# 2 Underlying concepts

Much of particle physics is concerned with the high-energy interactions of relativistic particles. Therefore the calculation of interaction and decay rates requires a relativistic formulation of quantum mechanics. Relativistic quantum mechanics (RQM) is founded on the two pillars of "modern" physics, Einstein's theory of special relativity and the wave mechanics developed in the early part of the twentieth century. It is assumed that you are already familiar with special relativity and non-relativistic quantum mechanics. The purpose of this chapter is to review the specific aspects of special relativity and quantum mechanics used in the subsequent development of relativistic quantum mechanics. Before discussing these important topics, the system of units commonly used in particle physics is introduced.

## 2.1 Units in particle physics

The system of S.I. units [kg, m, s] forms a natural basis for the measurements of mass, length and time for everyday objects and macroscopic phenomena. However, it is not a natural choice for the description of the properties of particles, where we are almost always dealing with very small quantities, such as the mass of the electron, which in S.I. units is $9.1 \times 10^{-31}$ kg. One way to avoid carrying around large exponents is to use S.I. based units. For example, interaction cross sections (which have the dimension of area) are usually quoted in *barns*, where

$$1\,\text{barn} \equiv 10^{-28}\,\text{m}^2.$$

The cross sections for the more interesting physical processes at the highest energies are typically in the picobarn (pb) to femtobarn (fb) range, where 1 pb = $10^{-12}$ barn and 1 fb = $10^{-15}$ barn. The use of derived S.I. units solves the problem of large exponents, nevertheless, it is more convenient to work with a system of units that from the outset reflects the natural length and time scales encountered in particle physics.

### 2.1.1 Natural units

The system of units used in particle physics is known as natural units. It is based on the fundamental constants of quantum mechanics and special relativity. In natural units, [kg, m, s] are replaced by [$\hbar$, $c$, GeV], where $\hbar = 1.055 \times 10^{-34}$ J s is the unit of action in quantum mechanics, $c = 2.998 \times 10^8$ m s$^{-1}$ is the speed of light in vacuum, and 1 GeV = $10^9$ eV = $1.602 \times 10^{-10}$ J, which is very approximately the rest mass energy of the proton. Table 2.1 lists the units used for a number of commonly encountered quantities expressed in terms of both [kg, m, s] and [$\hbar$, $c$, GeV], where the conversion can be obtained from dimensional analysis.

Natural units provide a well-motived basis for expressing quantities in particle physics and can be simplified by *choosing*

$$\hbar = c = 1.$$

In this way, all quantities are expressed in powers of GeV, as shown in the rightmost column of Table 2.1. Setting $\hbar = c = 1$ has the advantage of simplifying algebraic expressions as there is no longer the need to carry around (possibly large) powers of $\hbar$ and $c$. For example, the Einstein energy–momentum relation

$$E^2 = p^2c^2 + m^2c^4 \quad \text{becomes} \quad E^2 = p^2 + m^2.$$

At first sight it might appear that information has been lost in setting $\hbar = c = 1$. However, the factors of $\hbar$ and $c$ have not simply vanished; they are still present in the dimensions of quantities. The conversion back to S.I. units is simply a question of reinserting the necessary missing factors of $\hbar$ and $c$, which can be identified from dimensional analysis. For example, the result of a calculation using natural units might determine the root-mean-square charge radius of the proton to be

$$\langle r^2 \rangle^{1/2} = 4.1\,\text{GeV}^{-1}.$$

**Table 2.1** Relationship between S.I. and natural units.

| Quantity | [kg, m, s] | [$\hbar$, $c$, GeV] | $\hbar = c = 1$ |
|---|---|---|---|
| Energy | kg m$^2$ s$^{-2}$ | GeV | GeV |
| Momentum | kg m s$^{-1}$ | GeV/$c$ | GeV |
| Mass | kg | GeV/$c^2$ | GeV |
| Time | s | (GeV/$\hbar$)$^{-1}$ | GeV$^{-1}$ |
| Length | m | (GeV/$\hbar c$)$^{-1}$ | GeV$^{-1}$ |
| Area | m$^2$ | (GeV/$\hbar c$)$^{-2}$ | GeV$^{-2}$ |

To convert this to back into S.I. units the correct dimensions are obtained by multiplying by $\hbar c$, giving

$$\begin{aligned}\langle r^2\rangle^{1/2} &= 4.1 \times \frac{1.055\times 10^{-34}\times 2.998\times 10^{8}}{1.602\times 10^{-10}}\,\mathrm{m}\\ &= 4.1\times(0.197\times 10^{-15})\,\mathrm{m} = 0.8\times 10^{-15}\,\mathrm{m}.\end{aligned}$$

In converting from natural units to S.I. units, it is useful to remember the conversion factor

$$\hbar c = 0.197\,\mathrm{GeV\,fm},$$

where one femtometre (fm) = $10^{-15}$ m.

## Heaviside–Lorentz units

The equations of classical electromagnetism can be simplified by adopting Heaviside–Lorentz units. The value of the electron charge is defined by the magnitude of the Coulomb force between two electrons separated by a distance r,

$$\mathrm{F} = \frac{e^2}{4\pi\varepsilon_0 \mathrm{r}^2},$$

where $\varepsilon_0$ is the permittivity of free space. In Heaviside–Lorentz units $\varepsilon_0$ is set to unity, and the expression for the Coulomb force becomes

$$\mathrm{F} = \frac{e^2}{4\pi \mathrm{r}^2}.$$

Effectively $\varepsilon_0$ has been absorbed into the definition of the electron charge. Because $1/(\varepsilon_0\mu_0) = c^2$, choosing $\varepsilon_0 = 1$ and $c = 1$ implies that the permeability of free space $\mu_0 = 1$. Hence, in the combined system of natural units and Heaviside–Lorentz units used in particle physics,

$$\hbar = c = \varepsilon_0 = \mu_0 = 1.$$

With $c = \varepsilon = \mu_0 = 1$, Maxwell's equations take the same form as with S.I. units.

The strength of the QED interaction is defined in terms of the dimensionless fine structure constant,

$$\alpha = \frac{e^2}{4\pi\varepsilon_0\hbar c}. \tag{2.1}$$

Since $\alpha$ is dimensionless, it has the same numerical value regardless of the system of units used,

$$\alpha \approx \frac{1}{137}.$$

In natural units, the relationship between $\alpha$ and the electron charge (which is not dimensionless) is simply

$$\alpha = \frac{e^2}{4\pi} \approx \frac{1}{137}.$$

## 2.2 Special relativity

This section gives a brief overview of the basic concepts of special relativity, with the emphasis on the definition and application of four-vectors and the concept of Lorentz invariance and Lorentz invariant quantities.

### 2.2.1 The Lorentz transformation

Special relativity is based on the space-time transformation properties of physical observables as measured in two or more inertial frames moving relative to each other. For example, Figure 2.1 shows a space-time event that occurs at $(t, \mathbf{r})$ in the inertial frame $\Sigma$ and at $(t', \mathbf{r}')$ in the inertial frame $\Sigma'$ that is moving with a velocity v in the $z$-direction relative to the frame $\Sigma$. For the case where $\mathrm{v} \ll c$ and the origins of two inertial frames coincide at $t = t' = 0$, the two sets of coordinates are related by the Galilean transformation

$$t' = t, \quad x' = x, \quad y' = y \quad \text{and} \quad z' = z - \mathrm{v}t.$$

Einstein postulated that the speed of light in the vacuum is the same in all inertial frames. This primary postulate of special relativity implies that a space-time point on the wavefront of a pulse of light emitted at $t = t' = 0$ satisfies both $x^2 + y^2 + z^2 = c^2t^2$ and $x'^2 + y'^2 + z'^2 = c^2t'^2$. Consequently the space-time interval,

$$c^2t^2 - x^2 - y^2 - z^2 = c^2t'^2 - x'^2 - y'^2 - z'^2, \tag{2.2}$$

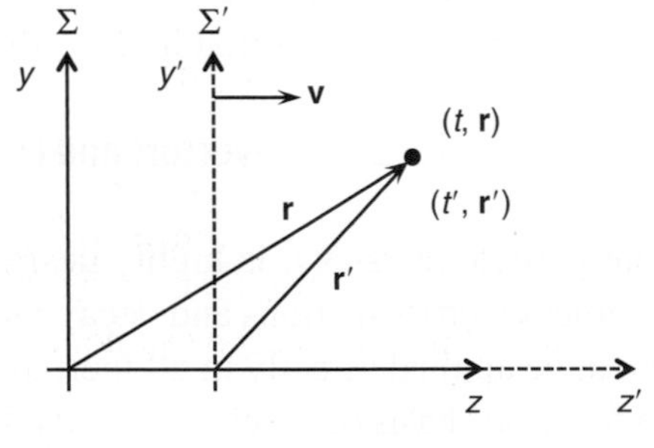


Fig. 2.1 A space-time event as seen into two inertial frames. The frame $\Sigma'$ moves with a velocity $\mathbf{v}$ in the $z$ direction relative to frame $\Sigma$.

is an invariant quantity; it is observed to be the same in all reference frames. Equation (2.2) is satisfied if the coordinates in $\Sigma$ and $\Sigma'$ are related by the Lorentz transformation

$$t' = \gamma\left(t - \frac{\mathrm{v}}{c^2}z\right), \quad x' = x, \quad y' = y \quad \text{and} \quad z' = \gamma(z - \mathrm{v}t),$$

where the Lorentz factor $\gamma$ is given by

$$\gamma = (1 - \beta^2)^{-\frac{1}{2}},$$

and $\beta = \mathrm{v}/c$. In the low velocity limit $\mathrm{v} \ll c$, the Lorentz factor reduces to unity and the Galilean transformations are recovered. In natural units, where $c = 1$, the Lorentz transformation of the space-time coordinates becomes

$$t' = \gamma\,(t - \beta z)\,, \quad x' = x, \quad y' = y \quad \text{and} \quad z' = \gamma(z - \beta t). \tag{2.3}$$

This can be written in matrix form as $\mathbf{X}' = \mathbf{\Lambda}\mathbf{X}$,

$$\begin{pmatrix} t' \\ x' \\ y' \\ z' \end{pmatrix} = \begin{pmatrix} \gamma & 0 & 0 & -\gamma\beta \\ 0 & 1 & 0 & 0 \\ 0 & 0 & 1 & 0 \\ -\gamma\beta & 0 & 0 & \gamma \end{pmatrix} \begin{pmatrix} t \\ x \\ y \\ z \end{pmatrix}, \tag{2.4}$$

where $\mathbf{X}$ is the four-component vector $\{t, \mathbf{x}\}$. The inverse Lorentz transformation, from $\Sigma'$ to $\Sigma$, is obtained by reversing the sign of the velocity in (2.3) such that

$$t = \gamma\,(t' + \beta z')\,, \quad x = x', \quad y = y' \quad \text{and} \quad z = \gamma(z' + \beta t'). \tag{2.5}$$

In matrix form this can be written $\mathbf{X} = \mathbf{\Lambda}^{-1}\mathbf{X}'$,

$$\begin{pmatrix} t \\ x \\ y \\ z \end{pmatrix} = \begin{pmatrix} \gamma & 0 & 0 & +\gamma\beta \\ 0 & 1 & 0 & 0 \\ 0 & 0 & 1 & 0 \\ +\gamma\beta & 0 & 0 & \gamma \end{pmatrix} \begin{pmatrix} t' \\ x' \\ y' \\ z' \end{pmatrix}. \tag{2.6}$$

It is straightforward to confirm that the matrices appearing in (2.4) and (2.6) are the inverse of each other, $\mathbf{\Lambda}\mathbf{\Lambda}^{-1} = \mathbf{I}$. The matrix equations of (2.4) and (2.6) define the Lorentz transformation between the space-time coordinates measured in two inertial frames with relative motion in the $z$-direction.

### 2.2.2 Four-vectors and Lorentz invariance

Throughout particle physics it is highly desirable to express physical predictions, such as interaction cross sections and decay rates, in an explicitly Lorentz-invariant form that can be applied directly in all inertial frames. Although the Lorentz transformation forms the basis of special relativity, Lorentz invariance is the more important concept for much that follows. Lorentz invariance is best expressed in terms of four-vectors. A *contravariant* four-vector is defined to be a set of quantities that when measured in two inertial frames are related by the Lorentz transformation of (2.4). For example, the contravariant four-vector $x^\mu$ is defined as

$$x^\mu = (t, x, y, z),$$

where the indices $\mu = \{0, 1, 2, 3\}$ label the space-time coordinates with the zeroth component representing time. In tensor form, the Lorentz transformation of (2.4) now can be expressed as

$$x'^\mu = \Lambda^\mu{}_\nu x^\nu, \tag{2.7}$$

where $\Lambda^\mu{}_\nu$ can be thought of as the elements of the matrix $\mathbf{\Lambda}$ and Einstein's summation convention for repeated indices is used to express the matrix multiplication.

The magnitude of a normal three-vector, which is given by the three-vector scalar product $\mathbf{x}\cdot\mathbf{x}$, is invariant under rotations. The Lorentz invariance of the space-time interval, $t^2-x^2-y^2-z^2$, can be expressed as a four-vector scalar product by defining the *covariant* space-time four-vector,

$$x_\mu = (t, -x, -y, -z).$$

With this notation, the Lorentz-invariant space-time interval can be written as the four-vector scalar product

$$x^\mu x_\mu = x^0x_0 + x^1x_1 + x^2x_2 + x^3x_3 = t^2 - x^2 - y^2 - z^2.$$

The main reason for introducing covariant four-vectors, which are denoted with a "downstairs" index to distinguish them from the corresponding contravariant four-vectors, is to keep account of the minus signs in Lorentz-invariant products. The Lorentz transformation of the space-time coordinates (2.3) can be written in terms of the components of the covariant four-vector as

$$\begin{pmatrix} t' \\ -x' \\ -y' \\ -z' \end{pmatrix} = \begin{pmatrix} \gamma & 0 & 0 & +\gamma\beta \\ 0 & 1 & 0 & 0 \\ 0 & 0 & 1 & 0 \\ +\gamma\beta & 0 & 0 & \gamma \end{pmatrix} \begin{pmatrix} t \\ -x \\ -y \\ -z \end{pmatrix}. \tag{2.8}$$

The sign changes in this matrix relative to that of (2.4) compensate for the changes of sign in the definition of $x_\mu$ relative to $x^\mu$. Both (2.8) and (2.4) are *equivalent* expressions of the same Lorentz transformation originally defined in (2.3). The transformation matrix appearing in (2.8) is the inverse of that of (2.4). To make this distinction explicit in tensor notation the transformation of a covariant four-vector is written as

$$x'_\mu = \Lambda_\mu{}^\nu x_\nu, \tag{2.9}$$

where the downstairs index appears first in $\Lambda_\mu{}^\nu$ which represents the elements of $\mathbf{\Lambda^{-1}}$.

In tensor notation, the relationship between covariant and contravariant four-vectors in special relativity can be expressed as

$$x_\mu = g_{\mu\nu} x^\nu,$$

where summation over repeated indices is again implicit and the diagonal *metric tensor* $g_{\mu\nu}$ is defined as

$$g_{\mu\nu} \equiv \begin{pmatrix} 1 & 0 & 0 & 0 \\ 0 & -1 & 0 & 0 \\ 0 & 0 & -1 & 0 \\ 0 & 0 & 0 & -1 \end{pmatrix}. \tag{2.10}$$

By definition, only quantities with the Lorentz transformation properties of (2.4) are written as contravariant four-vectors. For such a set of quantities $a^\mu$ the scalar product with the corresponding covariant four-vector $a_\mu$ is guaranteed to be Lorentz invariant. Furthermore, if $a^\mu$ and $b^\mu$ are both (contravariant) four-vectors, then the scalar product

$$a^\mu b_\mu = a_\mu b^\mu = g_{\mu\nu} a^\mu b^\nu,$$

is automatically Lorentz invariant. Again this follows directly from the form of the Lorentz transformation for contravariant and covariant four-vectors. Hence *any* expression that can be written in terms of four-vector scalar products is guaranteed to be Lorentz invariant. From the linearity of the Lorentz transformation, it also follows that the sum of any number of contravariant four-vectors also transforms according to (2.4) and therefore is itself a four-vector.

## Four-momentum

The relativistic expressions for the energy and momentum of a particle of mass $m$ can be identified as $E = \gamma mc^2$ and $\mathbf{p} = \gamma m\mathbf{v}$, which when expressed in natural units are

$$E = \gamma m \quad \text{and} \quad \mathbf{p} = \gamma m\boldsymbol{\beta}. \tag{2.11}$$

By considering the transformation properties of velocity, $\mathrm{d}\mathbf{x}/\mathrm{d}t$, it can be shown that relativistic energy and momentum, defined in this way, transform according to (2.4) and therefore form a contravariant four-vector,

$$p^\mu = (E, p_x, p_y, p_z),$$

referred to as four-momentum. Because momentum and energy are separately conserved, four-momentum is also conserved. Furthermore, since four-momentum is a four-vector, the scalar product

$$p^\mu p_\mu = E^2 - \mathbf{p}^2,$$

is a Lorentz-invariant quantity.

From (2.11) it can be seen that a single particle at rest has four-momentum $p^\mu = (m, 0, 0, 0)$ and therefore $p^\mu p_\mu = m^2$. Since $p^\mu p_\mu$ is Lorentz invariant, the relation

$$E^2 - \mathbf{p}^2 = m^2$$

holds in *all* inertial frames. This is, of course, just the Einstein energy–momentum relationship. For a system of $n$ particles, the total energy and momentum

$$p^\mu = \sum_{i=1}^{n} p_i^\mu$$

is also a four-vector. Therefore for a system of particles the quantity

$$p^\mu p_\mu = \left(\sum_{i=1}^{n} E_i\right)^2 - \left(\sum_{i=1}^{n} \mathbf{p}_i\right)^2$$

is a Lorentz-invariant quantity, which gives the squared *invariant mass* of the system. In a particle decay $a \to 1+2$, the invariant mass of the decay products is equal to the mass of the decaying particle,

$$(p_1 + p_2)^\mu (p_1 + p_2)_\mu = p_a^\mu p_{a\mu} = m_a^2.$$

## Four-derivative

The transformation properties of the space-time derivatives can be found by using the Lorentz transformation of (2.3) to express the coordinates of an event in the frame $\Sigma'$ as functions of the coordinates measured in the frame $\Sigma$, for example $z'(t, x, y, z)$ and $t'(t, x, y, z)$. Hence, for a Lorentz transformation in the $z$-direction, the derivatives in the primed-frame can be expressed as

$$\frac{\partial}{\partial z'} = \left(\frac{\partial z}{\partial z'}\right)\frac{\partial}{\partial z} + \left(\frac{\partial t}{\partial z'}\right)\frac{\partial}{\partial t} \quad \text{and} \quad \frac{\partial}{\partial t'} = \left(\frac{\partial z}{\partial t'}\right)\frac{\partial}{\partial z} + \left(\frac{\partial t}{\partial t'}\right)\frac{\partial}{\partial t}.$$

From (2.5), the relevant partial derivatives are

$$\left(\frac{\partial z}{\partial z'}\right) = \gamma, \quad \left(\frac{\partial t}{\partial z'}\right) = +\gamma\beta, \quad \left(\frac{\partial z}{\partial t'}\right) = +\gamma\beta \quad \text{and} \quad \left(\frac{\partial t}{\partial t'}\right) = \gamma,$$

and therefore,

$$\frac{\partial}{\partial z'} = \gamma\frac{\partial}{\partial z} + \gamma\beta\frac{\partial}{\partial t} \quad \text{and} \quad \frac{\partial}{\partial t'} = \gamma\beta\frac{\partial}{\partial z} + \gamma\frac{\partial}{\partial t}. \tag{2.12}$$

From (2.12) it can be seen that Lorentz transformation properties of the partial derivatives are

$$\begin{pmatrix} \partial/\partial t' \\ \partial/\partial x' \\ \partial/\partial y' \\ \partial/\partial z' \end{pmatrix} = \begin{pmatrix} \gamma & 0 & 0 & +\gamma\beta \\ 0 & 1 & 0 & 0 \\ 0 & 0 & 1 & 0 \\ +\gamma\beta & 0 & 0 & \gamma \end{pmatrix} \begin{pmatrix} \partial/\partial t \\ \partial/\partial x \\ \partial/\partial y \\ \partial/\partial z \end{pmatrix},$$

and comparison with (2.8) shows that

$$\left(\frac{\partial}{\partial t}, \frac{\partial}{\partial x}, \frac{\partial}{\partial y}, \frac{\partial}{\partial z}\right)$$

transforms as a *covariant* four-vector, which is written as

$$\partial_\mu = \frac{\partial}{\partial x^\mu},$$

and has components

$$\partial_0 = \frac{\partial}{\partial t}, \quad \partial_1 = +\frac{\partial}{\partial x}, \quad \partial_2 = +\frac{\partial}{\partial y} \quad \text{and} \quad \partial_3 = +\frac{\partial}{\partial z}.$$

The corresponding contravariant four-derivative is therefore

$$\partial^\mu = \left(\frac{\partial}{\partial t}, -\frac{\partial}{\partial x}, -\frac{\partial}{\partial y}, -\frac{\partial}{\partial z}\right),$$

and it should be noted that here the space-like coordinates enter with minus signs. The equivalent of the Laplacian for the four-derivative, which is known as the d'Alembertian, is therefore

$$\Box = \partial^\mu \partial_\mu = \frac{\partial^2}{\partial t^2} - \frac{\partial^2}{\partial x^2} - \frac{\partial^2}{\partial y^2} - \frac{\partial^2}{\partial z^2}.$$

In this book the symbol $\Box$ is used to represent the d'Alembertian, in some textbooks you may see it written as $\Box^2$.

## Vector and four-vector notation

This is a convenient place to introduce the notation used in this book. Unless otherwise stated, quantities written simply as $x$ and $p$ always should be interpreted as four-vectors. Three-vectors, such as the three-momentum of a particle, are always written in boldface, for example $\mathbf{p}$, with three-vector scalar products written as

$$\mathbf{p}_1 \cdot \mathbf{p}_2.$$

The magnitude of a three-vector is written either as $|\mathbf{p}|$ or simply p. Four-vector scalar products are written either as $a^\mu b_\mu$ or $a{\cdot}b$, with

$$a{\cdot}b \equiv a^\mu b_\mu \equiv g_{\mu\nu} a^\mu b^\nu = a^0 b^0 - a^1 b^1 - a^2 b^2 - a^3 b^3.$$

Just as $\mathbf{p}^2$ is shorthand for $\mathbf{p} \cdot \mathbf{p}$, then for a four-vector $a$, the expression $a^2$ is shorthand for the four-vector scalar product $a \cdot a$. For example, the Einstein energy–momentum relationship for a single particle can be expressed as $p^2 = m^2$, since $p^2 = p{\cdot}p = E^2 - \mathbf{p}^2$. Finally, it will sometimes be convenient to work with quantities measured in the centre-of-mass frame of a system of particles, and such quantities are denoted by a star. For example, p* is the magnitude of the three-momentum of a particle evaluated in the centre-of-mass frame, which for a system of particles is the inertial frame in which there is no net three-momentum.

### 2.2.3 Mandelstam variables

Feynman diagrams, involving the exchange of a single force mediating particle, can be placed in the three categories shown in Figure 2.2. The first two diagrams represent the $s$-channel annihilation process and the $t$-channel scattering process. The third diagram represents $u$-channel scattering and is only relevant when there are identical particles in the final state. In Chapter 5 it will be shown that four-momentum is conserved at each vertex in a Feynman diagram. In a process involving two initial-state and two final-state particles, the Mandelstam variables

$$\begin{aligned} s &= (p_1 + p_2)^2 = (p_3 + p_4)^2, \\ t &= (p_1 - p_3)^2 = (p_2 - p_4)^2, \\ u &= (p_1 - p_4)^2 = (p_2 - p_3)^2, \end{aligned}$$

are equivalent to the four-momentum squared $q^2$ of the exchanged boson in the respective class of diagram. For identical final-state particles the distinction between $u$- and $t$-channel diagrams is necessary because the final-state particle with four-momentum $p_3$ can originate from either interaction vertex, and the four-momentum $q$ of the virtual particle is different for the two cases.

Since the Mandelstam variables are four-vector scalar products, they are manifestly Lorentz invariant and can be evaluated in any frame. For example, in the centre-of-mass frame where there is no net momentum, the four-momenta of two colliding particles are $p_1 = (E_1^*, \mathbf{p}^*)$ and $p_2 = (E_2^*, -\mathbf{p}^*)$, from which

$$s = (p_1 + p_2)^2 = (E_1^* + E_2^*)^2 - (\mathbf{p}^* - \mathbf{p}^*)^2 = (E_1^* + E_2^*)^2. \tag{2.13}$$

Hence, the Lorentz-invariant quantity $\sqrt{s}$ can be identified as the total energy available in the centre-of-mass frame. It is worth noting that for the process

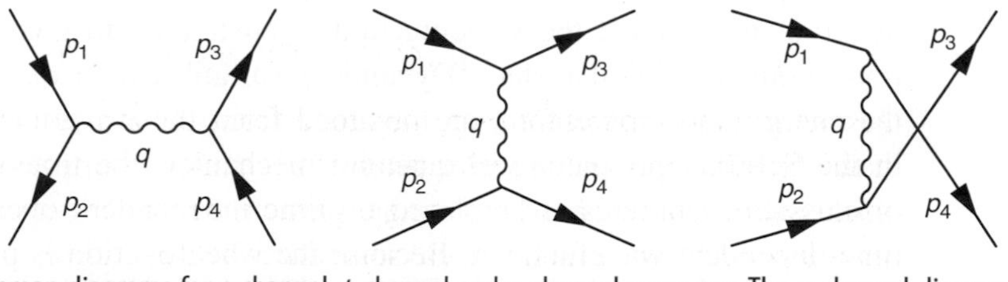


Fig. 2.2 The Feynman diagrams for $s$-channel, $t$-channel and $u$-channel processes. The $u$-channel diagram applies only when there are identical particles in the final state.

$1+2\to 3+4$, the sum of $s+u+t$ can be shown to be equal to the sum of the squares of the masses of the four particles (see Problem 2.12),

$$s+u+t=m_1^2+m_2^2+m_3^2+m_4^2. \tag{2.14}$$

## 2.3 Non-relativistic quantum mechanics

This section gives a brief overview of topics in non-relativistic quantum mechanics which are of direct relevance to the development of the relativistic treatment of spin-half particles in Chapter 4. It also reviews of the algebraic treatment of angular momentum that serves as an introduction to the algebra of the SU(2) symmetry group.

### 2.3.1 Wave mechanics and the Schrödinger equation

In quantum mechanics it is postulated that free particles are described by wave packets which can be decomposed into a Fourier integral of plane waves of the form

$$\psi(\mathbf{x},t)\propto \exp\{i(\mathbf{k}\cdot\mathbf{x}-\omega t)\}. \tag{2.15}$$

Following the de Broglie hypothesis for wave–particle duality, the wavelength of a particle in quantum mechanics can be related to its momentum by $\lambda=h/\mathrm{p}$, or equivalently, the wave vector $\mathbf{k}$ is given by $\mathbf{k}=\mathbf{p}/\hbar$. The angular frequency of the plane wave describing a particle is given by the Planck–Einstein postulate, $E=\hbar\omega$. In natural units with $\hbar=1$, the de Broglie hypothesis and Planck–Einstein postulate imply $\mathbf{k}=\mathbf{p}$ and $\omega=E$, and thus the plane wave of (2.15) becomes

$$\psi(\mathbf{x},t)=N\exp\{i(\mathbf{p}\cdot\mathbf{x}-Et)\}, \tag{2.16}$$

where $N$ is the normalisation constant.

In classical physics, the energy and momentum of a particle are dynamical variables represented by time-dependent real numbers. In the Schrödinger picture of quantum mechanics, the wavefunction is postulated to contain all the information about a particular state. Dynamical variables of a quantum state, such as the energy and momentum, are obtained from the wavefunction. Consequently, in the Schrödinger picture of quantum mechanics, the time-dependent variables of classical dynamics are replaced by time-independent operators acting on the time-dependent wavefunction. Because the wavefunction is postulated to contain all the information about a system, a physical observable quantity $A$ corresponds to the action of a quantum mechanical *operator* $\hat{A}$ on the wavefunction. A further postulate of quantum mechanics is that the result of the measurement of the observable $A$ will be one of the eigenvalues of the operator equation

$$\hat{A}\psi = a\psi.$$

For $A$ to correspond to a physical observable, the eigenvalues of the corresponding operator must be real, which implies that the operator is Hermitian. This is formally defined by the requirement

$$\int \psi_1^* \hat{A}\psi_2 \, \mathrm{d}\tau = \int \left[\hat{A}\psi_1\right]^* \psi_2 \, \mathrm{d}\tau.$$

Because the plane wave of (2.16) is intended to represent a free particle with energy $E$ and momentum $\mathbf{p}$, it is reasonable to identify the momentum and energy operators, $\hat{\mathbf{p}}$ and $\hat{E}$, as

$$\hat{\mathbf{p}} = -i\boldsymbol{\nabla} \quad \text{and} \quad \hat{E} = i\frac{\partial}{\partial t}, \tag{2.17}$$

such that $\hat{\mathbf{p}}$ and $\hat{E}$ acting on the plane wave of (2.16) give the required eigenvalues,

$$\hat{\mathbf{p}}\psi = -i\boldsymbol{\nabla}\psi = \mathbf{p}\psi,$$
$$\hat{E}\psi = i\frac{\partial\psi}{\partial t} = E\psi.$$

In classical dynamics, the total energy of a non-relativistic particle can be expressed as the sum of its kinetic and potential energy terms,

$$E = H = T + V = \frac{\mathbf{p}^2}{2m} + V,$$

where $H = T + V$ is the Hamiltonian. The equivalent quantum mechanical expression is obtained by replacing each of the terms with the corresponding operators defined in (2.17) acting on the wavefunction. This gives rise to the time-dependent Schrödinger equation,

$$i\frac{\partial\psi(\mathbf{x},t)}{\partial t} = \hat{H}\psi(\mathbf{x},t), \tag{2.18}$$

where, for a non-relativistic particle, the Hamiltonian operator is

$$\hat{H}_{NR} = \frac{\hat{\mathbf{p}}^2}{2m} + \hat{V} = -\frac{1}{2m}\nabla^2 + \hat{V}. \tag{2.19}$$

For a one-dimensional system (2.18) and (2.19) reduce to the familiar one-dimensional time-dependent Schrödinger equation,

$$i\frac{\partial\psi(\mathbf{x},t)}{\partial t} = -\frac{1}{2m}\frac{\partial^2\psi(\mathbf{x},t)}{\partial x^2} + \hat{V}\psi(\mathbf{x},t).$$

### 2.3.2 Probability density and probability current

The physical interpretation of the wavefunction $\psi(\mathbf{x}, t)$ is that $\psi^*\psi\, d^3\mathbf{x}$ is the probability of finding the particle represented by the wavefunction in the volume element $d^3\mathbf{x}$. This is equivalent to identifying the probability density $\rho(\mathbf{x}, t)$ as

$$\rho(\mathbf{x}, t) = \psi^*(\mathbf{x}, t)\,\psi(\mathbf{x}, t).$$

Assuming the particle does not decay or interact, its associated total probability will be constant. This conservation of probability can be expressed in terms of a continuity equation by defining the probability current density (sometimes referred to as the probability flux density), denoted $\mathbf{j}(\mathbf{x}, t)$, such that the flux of probability across an elemental surface $d\mathbf{S}$ is given by $\mathbf{j} \cdot d\mathbf{S}$. The rate of change of the total probability contained within a volume $V$, shown in Figure 2.3, is related to the net flux leaving the surface by

$$\frac{\partial}{\partial t}\int_V \rho\, dV = -\int_S \mathbf{j} \cdot d\mathbf{S}.$$

Using the divergence theorem this can be written as

$$\frac{\partial}{\partial t}\int_V \rho\, dV = -\int_V \boldsymbol{\nabla} \cdot \mathbf{j}\, dV.$$

Because this holds for an arbitrary volume, the continuity equation for the conservation of quantum mechanical probability can be written

$$\boldsymbol{\nabla} \cdot \mathbf{j} + \frac{\partial \rho}{\partial t} = 0. \tag{2.20}$$

In non-relativistic quantum mechanics, the expression for the probability current can be obtained from the free particle time-dependent Schrödinger equation,

$$i\frac{\partial \psi}{\partial t} = -\frac{1}{2m}\nabla^2\psi, \tag{2.21}$$

and the corresponding equation for the complex conjugate of $\psi$,

$$-i\frac{\partial \psi^*}{\partial t} = -\frac{1}{2m}\nabla^2\psi^*. \tag{2.22}$$

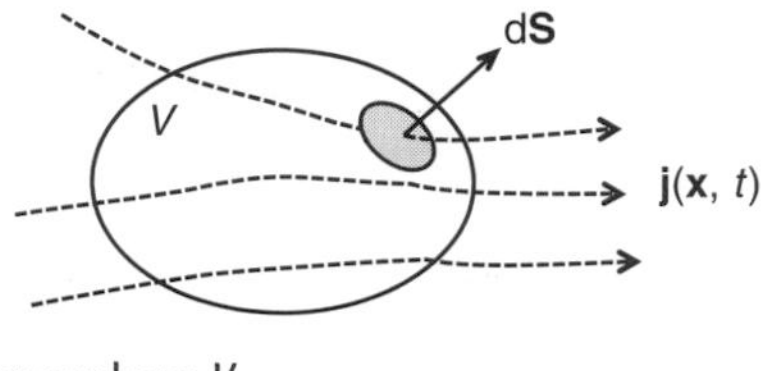


Fig. 2.3 The net flux of probability leaving a volume $V$.

Taking $\psi^* \times (2.21) - \psi \times (2.22)$ then gives

$$-\frac{1}{2m}\left(\psi^*\nabla^2\psi - \psi\nabla^2\psi^*\right) = i\left(\psi^*\frac{\partial\psi}{\partial t} + \psi\frac{\partial\psi^*}{\partial t}\right)$$
$$\Rightarrow \quad -\frac{1}{2m}\boldsymbol{\nabla}\cdot(\psi^*\boldsymbol{\nabla}\psi - \psi\boldsymbol{\nabla}\psi^*) = i\frac{\partial}{\partial t}(\psi^*\psi) = i\frac{\partial\rho}{\partial t}. \tag{2.23}$$

Comparing (2.23) with the general form of the continuity equation of (2.20) leads to the identification of probability current as

$$\mathbf{j} = \frac{1}{2im}\left(\psi^*\boldsymbol{\nabla}\psi - \psi\boldsymbol{\nabla}\psi^*\right). \tag{2.24}$$

The plane wave

$$\psi(\mathbf{x}, t) = Ne^{i(\mathbf{p}\cdot\mathbf{x}-Et)},$$

is therefore associated with a constant probability density of $\psi\psi^* = |N|^2$ and can be interpreted as representing a region of space with a number density of particles $n = |N|^2$. The corresponding expression for the probability current density of (2.24) gives

$$\mathbf{j} = |N|^2\frac{\mathbf{p}}{m} \equiv n\mathbf{v},$$

where $\mathbf{v}$ is the (non-relativistic) velocity. Thus, the plane wave $\psi(\mathbf{x}, t)$ represents a region of space with number density of $n = |N|^2$ particles per unit volume moving with velocity $\mathbf{v}$, such that flux of particles passing through a unit area per unit time is $\mathbf{j} = n\mathbf{v}$.

### 2.3.3 Time dependence and conserved quantities

The time evolution of a quantum mechanical state is given by the time-dependent Schrödinger equation of (2.18). If $\psi_i$ is an eigenstate of the Hamiltonian $\hat{H}$ with energy $E_i$ such that

$$\hat{H}\psi_i(\mathbf{x}, t) = E_i\psi_i(\mathbf{x}, t),$$

then, from (2.18), the time evolution of the wavefunction is given by

$$i\frac{\partial\psi_i(\mathbf{x}, t)}{\partial t} = E_i\psi_i(\mathbf{x}, t).$$

Hence, the time dependence of an eigenstate of the Hamiltonian is given by

$$\psi_i(\mathbf{x}, t) = \phi_i(\mathbf{x})e^{-iE_it}. \tag{2.25}$$

For a system in a quantum mechanical state[1] $|\psi(\mathbf{x},t)\rangle$, the expectation value of an operator $\hat{A}$ is given by

$$\langle \hat{A} \rangle = \langle \psi | \hat{A} | \psi \rangle = \int \psi^\dagger \hat{A} \psi \, \mathrm{d}^3\mathbf{x},$$

where the complex conjugate used up to this point has been replaced by the Hermitian conjugate, $\psi^\dagger = (\psi^*)^T$. The time evolution of the expectation value $\langle \hat{A} \rangle$ can therefore be expressed as

$$\frac{\mathrm{d}\langle \hat{A} \rangle}{\mathrm{d}t} = \int \left[ \frac{\partial \psi^\dagger}{\partial t} \hat{A} \psi + \psi^\dagger \hat{A} \frac{\partial \psi}{\partial t} \right] \mathrm{d}^3\mathbf{x}, \tag{2.26}$$

where it has been assumed that there is no explicit time dependence in the operator itself, i.e. $\partial \hat{A}/\partial t = 0$. The time derivatives in (2.26) can be expressed using (2.18) and its Hermitian conjugate, giving

$$\frac{\mathrm{d}\langle \hat{A} \rangle}{\mathrm{d}t} = \int \left[ \left\{ \tfrac{1}{i} \hat{H} \psi \right\}^\dagger \hat{A} \psi + \psi^\dagger \hat{A} \left\{ \tfrac{1}{i} \hat{H} \psi \right\} \right] \mathrm{d}^3\mathbf{x} \tag{2.27}$$

$$= i \int \left[ \psi^\dagger \hat{H}^\dagger \hat{A} \psi - \psi^\dagger \hat{A} \hat{H} \psi \right] \mathrm{d}^3\mathbf{x}$$

$$= i \int \psi^\dagger (\hat{H}\hat{A} - \hat{A}\hat{H}) \psi \, \mathrm{d}^3\mathbf{x}. \tag{2.28}$$

The last step follows from the fact that the Hamiltonian is Hermitian (which must be the case for it to have real eigenvalues). The relation of (2.28) implies that for *any* state

$$\frac{\mathrm{d}\langle \hat{A} \rangle}{\mathrm{d}t} = i \left\langle [\hat{H}, \hat{A}] \right\rangle, \tag{2.29}$$

where $[\hat{H}, \hat{A}] = \hat{H}\hat{A} - \hat{A}\hat{H}$ is the commutator of the Hamiltonian and the operator $\hat{A}$. Hence, if the operator $\hat{A}$ commutes with the Hamiltonian, the corresponding observable $A$ does not change with time and therefore corresponds to a conserved quantity. Furthermore if $\psi_i$ is an eigenstate of the Hamiltonian, then (2.27) immediately reduces to

$$\frac{\mathrm{d}\langle \hat{A} \rangle}{\mathrm{d}t} = \int \left[ [iE_i \psi_i^\dagger] \hat{A} \psi_i + \psi_i^\dagger \hat{A} [-iE_i \psi_i] \right] \mathrm{d}^3\mathbf{x} = 0.$$

Therefore, for an eigenstate of the Hamiltonian, the expectation value of *any* operator is constant. For this reason, the eigenstates of the Hamiltonian are known as the *stationary states* of the system.

[1] The wavefunction $\psi$ has been replaced by the more general state $|\psi\rangle$ written in Dirac ket notation which may have a number of degrees of freedom, for example spin.

In general, a state $|\varphi\rangle$ can be expressed in terms of the complete set of states formed from the eigenstates of the Hamiltonian $|\psi_i\rangle$,

$$|\varphi\rangle = \sum_i c_i|\psi_i\rangle,$$

and the time dependence of the system is determined by the evolution of the stationary states according to (2.25). If at time $t=0$, a system is in the state $|\varphi(\mathbf{x})\rangle$, then the time evolution of the system is determined by the time evolution of the component stationary states

$$|\varphi(\mathbf{x},t)\rangle = \sum_i c_i\,|\phi_i(\mathbf{x})\rangle e^{-iE_it}. \tag{2.30}$$

This relationship between the time evolution of a state and the time dependence of the stationary states will be used extensively in the discussion of neutrino and strangeness oscillations in Chapters 13 and 14.

### 2.3.4 Commutation relations and compatible observables

The commutation relation between the operators for different observables determines whether they can be known simultaneously. Consider two observables corresponding to operators $\hat{A}$ and $\hat{B}$ which commute,

$$[\hat{A},\hat{B}] \equiv \hat{A}\hat{B} - \hat{B}\hat{A} = 0.$$

If $|\phi\rangle$ is an non-degenerate eigenstate of $\hat{A}$ with eigenvalue $a$, such that

$$\hat{A}|\phi\rangle = a|\phi\rangle,$$

then

$$\hat{A}\hat{B}\,|\phi\rangle = \hat{B}\hat{A}\,|\phi\rangle = a\hat{B}\,|\phi\rangle.$$

Therefore the state $\hat{B}\,|\phi\rangle$ is also an eigenstate of $\hat{A}$ with eigenvalue $a$. For this to be true, $\hat{B}\,|\phi\rangle \propto |\phi\rangle$, which implies that $|\phi\rangle$ satisfies

$$\hat{B}|\phi\rangle = b|\phi\rangle.$$

Hence $|\phi\rangle$ is a simultaneous eigenstate of both $\hat{A}$ and $\hat{B}$ and the state corresponds to well-defined values of the two observables, $a$ and $b$. The same conclusion is obtained even if the states are degenerate. If $\hat{A}$ and $\hat{B}$ commute, the corresponding observables are referred to as compatible. In general, a quantum mechanical state can be labelled by the *quantum numbers* specifying the complete set of compatible observables. In the above example $|\phi\rangle$ can be labelled by $|a,b\rangle$. If there is a further operator $\hat{C}$ that commutes with both $\hat{A}$ and $\hat{B}$, the state is labelled by the quantum numbers $|a,b,c\rangle$. In the quantum mechanical description of angular momentum, described in Section 2.3.5, the states are labelled in terms of the eigenvalues of angular momentum squared and the $z$-component of angular momentum, $|\ell,m\rangle$.

Similar arguments can be applied to show that if $\hat{A}$ and $\hat{B}$ do not commute,

$$[\hat{A}, \hat{B}] \equiv \hat{A}\hat{B} - \hat{B}\hat{A} \neq 0,$$

then it is not in general possible to define a simultaneous eigenstate of the two operators. In this case, it is not possible to know simultaneously the exact values of the physical observables $A$ and $B$ and the limit to which $A$ and $B$ can be known is given by the generalised uncertainty principle

$$\Delta A\, \Delta B \geq \tfrac{1}{2}\left|\langle i[\hat{A}, \hat{B}]\rangle\right|, \tag{2.31}$$

where $(\Delta A)^2 = \langle \hat{A}^2 \rangle - \langle \hat{A} \rangle^2$.

### Position–momentum uncertainty relation

An important example of incompatible variables is that of the position and momentum uncertainty principal. The operators corresponding to the $x$ position of a particle and the $x$ component of its momentum are respectively given by

$$\hat{x}\psi = x\psi \quad \text{and} \quad \hat{p}_x \psi = -i\frac{\partial}{\partial x}\psi.$$

The commutator $[\hat{x}, \hat{p}_x]$ can be evaluated from its action on a wavefunction $\psi$,

$$\begin{aligned}[\hat{x}, \hat{p}_x]\psi &= -ix\frac{\partial}{\partial x}\psi + i\frac{\partial}{\partial x}(x\psi) \\ &= -ix\frac{\partial \psi}{\partial x} + i\psi + ix\frac{\partial \psi}{\partial x} = +i\psi,\end{aligned}$$

giving

$$[\hat{x}, \hat{p}_x] = +i.$$

The usual expression of the Heisenberg uncertainty principle for position and momentum is then obtained by substituting this commutation relation into (2.31) giving (after reinserting the hidden factor of $\hbar$)

$$\Delta x\, \Delta p_x \geq \frac{\hbar}{2}.$$

## 2.3.5 Angular momentum in quantum mechanics

The concept of angular momentum and its quantum mechanical treatment plays an important role in particle physics. In classical dynamics, the angular momentum $\mathbf{L}$ of a body is defined by the moment of its momentum,

$$\mathbf{L} = \mathbf{r} \times \mathbf{p} = (yp_z - zp_y, zp_x - xp_z, xp_y - yp_x).$$

The corresponding quantum mechanical operator $\hat{\mathbf{L}}$ is obtained by replacing the position and momentum coordinates by their operator equivalents. Hence, in quantum mechanics, the components of angular momentum operator are given by

$$\hat{L}_x = \hat{y}\hat{p}_z - \hat{z}\hat{p}_y, \quad \hat{L}_y = \hat{z}\hat{p}_x - \hat{x}\hat{p}_z \quad \text{and} \quad \hat{L}_z = \hat{x}\hat{p}_y - \hat{y}\hat{p}_x.$$

Because the position operator does not commute with the corresponding component of momentum,

$$[\hat{x}, \hat{p}_x] = \left[\hat{y}, \hat{p}_y\right] = [\hat{z}, \hat{p}_z] = +i,$$

the angular momentum operators do not commute with each other and it is straightforward to show that

$$\left[\hat{L}_x, \hat{L}_y\right] = i\hat{L}_z, \quad \left[\hat{L}_y, \hat{L}_z\right] = i\hat{L}_x \quad \text{and} \quad \left[\hat{L}_z, \hat{L}_x\right] = i\hat{L}_y. \tag{2.32}$$

It is important to realise that the commutation relations of (2.32) are sufficient to fully define the algebra of angular momentum in quantum mechanics. This is significant because exactly the same commutation relations arise naturally in the discussion of other symmetries, such as flavour symmetry which is described in Chapter 9. For this reason, the development of the algebra defined by (2.32) and the subsequent identification of the angular momentum states is directly applicable to the more abstract symmetry concepts encountered in context of the quark model and QCD.

Because $\hat{L}_x$, $\hat{L}_y$ and $\hat{L}_z$ do not commute, they correspond to incompatible observables and (unless the state has zero angular momentum) it is not possible to define a simultaneous eigenstate of more than one of the components of angular momentum. However, it is relatively straightforward to show (see Problem 2.15) that the operator for the total squared angular momentum defined by

$$\hat{L}^2 = \hat{L}_x^2 + \hat{L}_y^2 + \hat{L}_z^2,$$

commutes with each of the components of angular momentum,

$$\left[\hat{L}^2, \hat{L}_x\right] = \left[\hat{L}^2, \hat{L}_y\right] = \left[\hat{L}^2, \hat{L}_z\right] = 0.$$

Hence it is possible to express the angular momentum states in terms of the simultaneous eigenstates of $\hat{L}^2$ and any *one* of the components of angular momentum which, by convention, is chosen to be $\hat{L}_z$.

It is also useful to define angular momentum raising and lowering ladder operators,

$$\hat{L}_+ = \hat{L}_x + i\hat{L}_y,$$
$$\hat{L}_- = \hat{L}_x - i\hat{L}_y,$$

for which $\hat{L}_+^\dagger = \hat{L}_-$ and $\hat{L}_-^\dagger = \hat{L}_+$. Because $\hat{L}^2$ commutes with both $\hat{L}_x$ and $\hat{L}_y$, it also commutes with both the ladder operators,

$$\left[\hat{L}^2, \hat{L}_\pm\right] = 0.$$

The commutator of the ladder operators with $\hat{L}_z$ is given by

$$\begin{aligned}\left[\hat{L}_z, \hat{L}_\pm\right] &= \left[\hat{L}_z, \hat{L}_x\right] \pm i\left[\hat{L}_z, \hat{L}_y\right] \\ &= i\hat{L}_y \pm \hat{L}_x,\end{aligned}$$

and therefore

$$\left[\hat{L}_z, \hat{L}_\pm\right] = \pm\hat{L}_\pm. \tag{2.33}$$

Furthermore, using $\left[\hat{L}_x, \hat{L}_y\right] = i\hat{L}_z$, it can be shown that $\hat{L}^2$ can be expressed as (see Problem 2.15)

$$\hat{L}^2 = \hat{L}_-\hat{L}_+ + \hat{L}_z + \hat{L}_z^2. \tag{2.34}$$

The simultaneous eigenstates of $\hat{L}^2$ and $\hat{L}_z$ can be obtained using the relations of (2.33) and (2.34). Suppose the state $|\lambda, m\rangle$ is a simultaneous eigenstate of both $\hat{L}_z$ and $\hat{L}^2$, with eigenvalues given by

$$\hat{L}_z\,|\lambda, m\rangle = m\,|\lambda, m\rangle \quad \text{and} \quad \hat{L}^2\,|\lambda, m\rangle = \lambda\,|\lambda, m\rangle\,. \tag{2.35}$$

Now consider the state $\psi = \hat{L}_+\,|\lambda, m\rangle$, defined by the action of the angular momentum raising operator on the original state. Because $\hat{L}^2$ commutes with $\hat{L}_+$,

$$\hat{L}^2\psi = \hat{L}^2\hat{L}_+\,|\lambda, m\rangle = \hat{L}_+\hat{L}^2\,|\lambda, m\rangle = \lambda\hat{L}_+\,|\lambda, m\rangle = \lambda\psi.$$

Furthermore from (2.33), $\hat{L}_z\hat{L}_+ = \hat{L}_+\hat{L}_z + \hat{L}_+$ and therefore

$$\begin{aligned}\hat{L}_z\psi = \hat{L}_z\left[\hat{L}_+\,|\lambda, m\rangle\right] &= (\hat{L}_+\hat{L}_z + \hat{L}_+)\,|\lambda, m\rangle \\ &= (m+1)\left[\hat{L}_+\,|\lambda, m\rangle\right] = (m+1)\psi.\end{aligned}$$

Hence, the state $\psi = \hat{L}_+\,|\lambda, m\rangle$ is also a simultaneous eigenstate of $\hat{L}^2$ and $\hat{L}_z$, with respective eigenvalues of $\lambda$ and $m + 1$. Therefore, the effect of the angular momentum raising operators is to step along states with the same value of total angular momentum squared but with one unit more of the $z$-component of angular momentum. The angular momentum lowering operator has the opposite effect, lowering the $z$-component of angular momentum by one unit.

The magnitude of the $z$-component of angular momentum can be no greater than the total angular momentum itself,

$$\left\langle\hat{L}_z^2\right\rangle \le \left\langle\hat{L}^2\right\rangle.$$

This implies that, for a particular value of $\lambda$, there must be maximum and minimum values of $m$ and that the action of $\hat{L}_+$ on the state with the largest value of $m$ gives

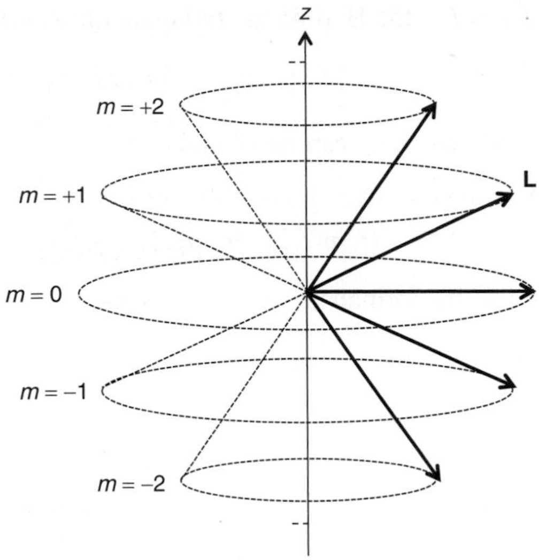


Fig. 2.4 A pictorial representation of the $2\ell + 1$ states for $\ell = 2$.

zero. Suppose the state with the largest $z$-component of angular momentum has $m = \ell$ such that

$$\hat{L}_+ |\lambda, \ell\rangle = 0,$$

then total angular momentum squared of this state is

$$\hat{L}^2 |\lambda, \ell\rangle = \left(\hat{L}_-\hat{L}_+ + \hat{L}_z + \hat{L}_z^2\right) |\lambda, \ell\rangle$$
$$\lambda |\lambda, \ell\rangle = (0 + \ell + \ell^2) |\lambda, \ell\rangle .$$

Hence, for the $m = \ell$ extreme state, the eigenvalue of $\hat{L}^2$ is $\lambda = \ell(\ell + 1)$. The same arguments can be applied to show at the other extreme, $m = -\ell$. Hence, for each value of $\lambda$ (or equivalently for each value of $\ell$), there are $2\ell + 1$ states (see Figure 2.4), differing by one unit of the $z$-component of angular momentum,

$$m = -\ell, -\ell + 1, \ldots, +\ell - 1, +\ell.$$

This implies that $\ell$ is quantised, and can take only integer or half-integer values. Expressing the states in terms of the quantum number $\ell$ rather than $\lambda$, the eigenvalue equations of (2.35) can be written as

$$\hat{L}_z |\ell, m\rangle = m |\ell, m\rangle \quad \text{and} \quad \hat{L}^2 |\ell, m\rangle = \ell(\ell + 1) |\ell, m\rangle .$$

The effect of the angular momentum raising operator on the state $|\ell, m\rangle$ is to generate the state $|\ell, m + 1\rangle$ with a coefficient $\alpha_{\ell,m}$ which still needs to be determined,

$$\hat{L}_+ |\ell, m\rangle = \alpha_{\ell,m} |\ell, m + 1\rangle . \tag{2.36}$$

Since $\hat{L}_+^\dagger = \hat{L}_-$, the Hermitian conjugate of (2.36) is

$$\left[\hat{L}_+ \left|\ell, m\right\rangle\right]^\dagger = \langle \ell, m|\hat{L}_- = \alpha^*_{\ell,m}\langle \ell, m+1|. \tag{2.37}$$

The coefficient $\alpha_{\ell,m}$ can be obtained by taking the product of (2.36) and (2.37) giving

$$\langle \ell, m|\,\hat{L}_-\hat{L}_+\,|\ell, m\rangle = |\alpha_{\ell,m}|^2\,\langle \ell, m+1|\ell, m+1\rangle.$$

Hence, for the normalised states $|\ell, m\rangle$ and $|\ell, m+1\rangle$,

$$\begin{aligned}
|\alpha_{\ell,m}|^2 &= \langle \ell, m|\,\hat{L}_-\hat{L}_+\,|\ell, m\rangle \\
&= \langle \ell, m|\,\hat{L}^2 - \hat{L}_z - \hat{L}_z^2\,|\ell, m\rangle \\
&= (\ell(\ell+1) - m - m^2)\,\langle \ell, m|\ell, m\rangle \\
&= \ell(\ell+1) - m(m+1),
\end{aligned}$$

and therefore,

$$\hat{L}_+\,|\ell, m\rangle = \sqrt{\ell(\ell+1) - m(m+1)}\,|\ell, m+1\rangle. \tag{2.38}$$

The corresponding relation for the angular momentum lowering operator, which can be obtained in the same manner, is

$$\hat{L}_-\,|\ell, m\rangle = \sqrt{\ell(\ell+1) - m(m-1)}\,|\ell, m-1\rangle. \tag{2.39}$$

The relations given in (2.38) and (2.39) will be used to construct the angular momentum (and flavour states) formed from the combination of more than one particle.

### 2.3.6 Fermi's golden rule

Particle physics is mainly concerned with decay rates and scattering cross sections, which in quantum mechanics correspond to transitions between states. In non-relativistic quantum mechanics, calculations of transition rates are obtained from Fermi's golden rule. The derivation of Fermi's golden rule is far from trivial, but is included here for completeness.

Let $\phi_k(\mathbf{x}, t)$ be the normalised solutions to the Schrödinger equation for the unperturbed time-independent Hamiltonian $\hat{H}_0$, where

$$\hat{H}_0\phi_k = E_k\phi_k \quad \text{and} \quad \langle \phi_j|\phi_k\rangle = \delta_{jk}.$$

In the presence of an interaction Hamiltonian $\hat{H}'(\mathbf{x}, t)$, which can induce transitions between states, the time-dependent Schrödinger equation becomes

$$i\frac{\mathrm{d}\psi}{\mathrm{d}t} = \left[\hat{H}_0 + \hat{H}'(\mathbf{x}, t)\right]\psi. \tag{2.40}$$

The wavefunction $\psi(\mathbf{x}, t)$ can be expressed in terms of complete set of states of the unperturbed Hamiltonian as

$$\psi(\mathbf{x}, t) = \sum_k c_k(t)\phi_k(\mathbf{x})e^{-iE_kt}, \tag{2.41}$$

where the time-dependent coefficients $c_k(t)$ allow for transitions between states. Substituting (2.41) into (2.40) gives a set of differential equations for the coefficients $c_k(t)$,

$$i\sum_k \left[\frac{\mathrm{d}c_k}{\mathrm{d}t}\phi_k e^{-iE_kt} - iE_kc_k\phi_k e^{-iE_kt}\right] = \sum_k c_k\hat{H}_0\phi_k e^{-iE_kt} + \sum_k \hat{H}'c_k\phi_k e^{-iE_kt}$$

$$\Rightarrow \qquad i\sum_k \frac{\mathrm{d}c_k}{\mathrm{d}t}\phi_k e^{-iE_kt} = \sum_k \hat{H}'c_k(t)\phi_k e^{-iE_kt}. \tag{2.42}$$

Suppose at time $t = 0$, the initial-state wavefunction is $|i\rangle = \phi_i$ and the coefficients are $c_k(0) = \delta_{ik}$. If the perturbing Hamiltonian, which is constant for $t > 0$, is sufficiently small that at all times $c_i(t) \approx 1$ and $c_{k\neq i}(t) \approx 0$, then to a first approximation (2.42) can be written

$$i\sum_k \frac{\mathrm{d}c_k}{\mathrm{d}t}\phi_k e^{-iE_kt} \approx \hat{H}'\phi_i e^{-iE_it}. \tag{2.43}$$

The differential equation for the coefficient $c_f(t)$, corresponding to transitions to a particular final state $|f\rangle = \phi_f$, is obtained by taking the inner product of both the LHS and RHS of (2.43) with $\phi_f(\mathbf{x})$ and using $\langle\phi_f|\phi_k\rangle = \delta_{fk}$ to give

$$\frac{\mathrm{d}c_f}{\mathrm{d}t} = -i\langle f|\hat{H}'|i\rangle e^{i(E_f-E_i)t}, \tag{2.44}$$

where

$$\langle f|\hat{H}'|i\rangle = \int_V \phi_f^*(\mathbf{x})\hat{H}'\phi_i(\mathbf{x})\,\mathrm{d}^3\mathbf{x}.$$

The *transition matrix element* $T_{fi} = \langle f|\hat{H}'|i\rangle$ has dimensions of energy because both $\phi_i$ and $\phi_f$ are normalised by a volume integral. At time $t = T$, the amplitude for transitions to the state $|f\rangle$ is given by the integral of (2.44)

$$c_f(T) = -i\int_0^T T_{fi}\,e^{i(E_f-E_i)t}\,\mathrm{d}t.$$

If the perturbing Hamiltonian is time-independent, so is the term $\langle f|\hat{H}'|i\rangle$ and thus

$$c_f(T) = -iT_{fi}\int_0^T e^{i(E_f-E_i)t}\,\mathrm{d}t. \tag{2.45}$$

The probability for a transition to the state $|f\rangle$ is given by

$$P_{fi} = c_f(T)c_f^*(T) = |T_{fi}|^2\int_0^T\int_0^T e^{i(E_f-E_i)t}e^{-i(E_f-E_i)t'}\,\mathrm{d}t\,\mathrm{d}t'.$$

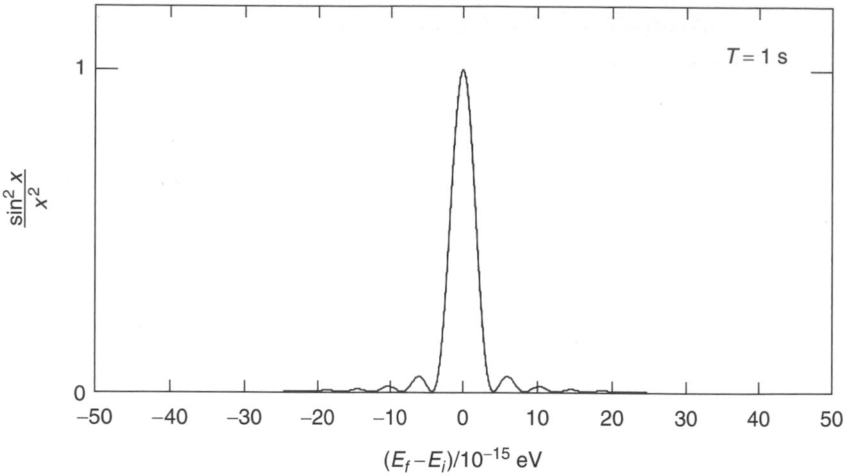


Fig. 2.5 The functional form of the integral of Equation (2.46) for $T = 1$ s.

The transition *rate* $d\Gamma_{fi}$ from the initial state $|i\rangle$ to the single final state $|f\rangle$ is therefore

$$d\Gamma_{fi} = \frac{P_{fi}}{T} = \frac{1}{T}|T_{fi}|^2 \int_{-\frac{T}{2}}^{+\frac{T}{2}} \int_{-\frac{T}{2}}^{+\frac{T}{2}} e^{i(E_f - E_i)t} e^{-i(E_f - E_i)t'} \, dt \, dt', \tag{2.46}$$

where the limits of integration are obtained by the substitutions $t \to t + T/2$ and $t' \to t' + T/2$. The exact solution to the integral in (2.46) has the form

$$\frac{\sin^2 x}{x^2} \quad \text{with} \quad x = \frac{(E_f - E_i)T}{2\hbar},$$

where the factor of $\hbar$ is included for clarity. This solution is shown for $T = 1$ s in Figure 2.5, from which it can be seen that the transition rate is only significant for final states where $E_f \approx E_i$ and that energy is conserved within the limits of the energy–time uncertainty relation

$$\Delta E \Delta t \sim \hbar. \tag{2.47}$$

The narrowness of the functional form of (2.46) means that for all practical purposes, it can be written as

$$d\Gamma_{fi} = |T_{fi}|^2 \lim_{T \to \infty} \left\{ \frac{1}{T} \int_{-\frac{T}{2}}^{+\frac{T}{2}} \int_{-\frac{T}{2}}^{+\frac{T}{2}} e^{i(E_f - E_i)t} e^{-i(E_f - E_i)t'} \, dt \, dt' \right\}.$$

Using the definition of the Dirac delta-function given by (A.4) in Appendix A, the integral over $\mathrm{d}t'$ can be replaced by $2\pi\,\delta(E_f - E_i)$ and thus

$$\mathrm{d}\Gamma_{fi} = 2\pi\,|T_{fi}|^2 \lim_{T\to\infty}\left\{\frac{1}{T}\int_{-\frac{T}{2}}^{+\frac{T}{2}} e^{i(E_f-E_i)t}\delta(E_f - E_i)\,\mathrm{d}t\right\}.$$

If there are $\mathrm{d}n$ accessible final states in the energy range $E_f \to E_f + \mathrm{d}E_f$, then the total transition rate $\Gamma_{fi}$ is given by

$$\Gamma_{fi} = 2\pi\int |T_{fi}|^2 \frac{\mathrm{d}n}{\mathrm{d}E_f} \lim_{T\to\infty}\left\{\frac{1}{T}\int_{-\frac{T}{2}}^{+\frac{T}{2}} e^{-i(E_f-E_i)t}\delta(E_f - E_i)\,\mathrm{d}t\right\}\mathrm{d}E_f. \tag{2.48}$$

The delta-function in the integral implies that $E_f = E_i$ and therefore (2.48) can be written

$$\begin{aligned}\Gamma_{fi} &= 2\pi\int |T_{fi}|^2 \frac{\mathrm{d}n}{\mathrm{d}E_f}\delta(E_f - E_i)\lim_{T\to\infty}\left\{\frac{1}{T}\int_{-\frac{T}{2}}^{+\frac{T}{2}}\mathrm{d}t\right\}\mathrm{d}E_f\\ &= 2\pi\int |T_{fi}|^2 \frac{\mathrm{d}n}{\mathrm{d}E_f}\delta(E_f - E_i)\,\mathrm{d}E_f\\ &= 2\pi\,|T_{fi}|^2\left|\frac{\mathrm{d}n}{\mathrm{d}E_f}\right|_{E_i}.\end{aligned} \tag{2.49}$$

The term $\left|\frac{\mathrm{d}n}{\mathrm{d}E_f}\right|_{E_i}$ is referred to as the *density of states*, and is often written as $\rho(E_i)$ where

$$\rho(E_i) = \left|\frac{\mathrm{d}n}{\mathrm{d}E_f}\right|_{E_i}.$$

Fermi's golden rule for the total transition rate is therefore

$$\Gamma_{fi} = 2\pi\,|T_{fi}|^2\rho(E_i),$$

where, to *first order*, $T_{fi} = \langle f|\hat{H}'|i\rangle$.

In the above derivation, it was assumed that $c_{k\neq i}(t) \approx 0$. An improved approximation can be obtained by again taking $c_i(t) \approx 1$ and substituting the expression for $c_{k\neq i}(t)$ from (2.45) back into (2.42), which after taking the inner product with a particular final state $\phi_f(\mathbf{x})$ gives

$$\frac{\mathrm{d}c_f}{\mathrm{d}t} \approx -i\langle f|\hat{H}|i\rangle e^{i(E_f-E_i)t} + (-i)^2\sum_{k\neq i}\langle f|\hat{H}'|k\rangle e^{i(E_f-E_k)t}\int_0^t \langle k|\hat{H}'|i\rangle e^{i(E_k-E_i)t'}\,\mathrm{d}t'. \tag{2.50}$$

Because the perturbation is not present at $t=0$, and for $t>0$ it is constant, the integral in (2.50) can be written

$$\int_0^t \langle k|\hat{H}'|i\rangle e^{i(E_k-E_i)t'}\,\mathrm{d}t' = \langle k|\hat{H}'|i\rangle \frac{e^{i(E_k-E_i)t}}{i(E_k-E_i)}.$$

Therefore, the improved approximation for the evolution of the coefficients $c_f(t)$ is given by

$$\frac{\mathrm{d}c_f}{\mathrm{d}t} = -i\left(\langle f|\hat{H}|i\rangle + \sum_{k\neq i}\frac{\langle f|\hat{H}'|k\rangle\langle k|\hat{H}'|i\rangle}{E_i-E_k}\right)e^{i(E_f-E_i)t}.$$

Comparison with (2.44) shows that, to second order, the transition matrix element $T_{fi}$ is given by

$$T_{fi} = \langle f|\hat{H}|i\rangle + \sum_{k\neq i}\frac{\langle f|\hat{H}'|k\rangle\langle k|\hat{H}'|i\rangle}{E_i-E_k}.$$

The second-order term corresponds to the transition occurring via some intermediate state $|k\rangle$. The full perturbation expansion can be obtained by successive substitutions. Provided the perturbation is sufficiently small, the successive terms in the perturbation expansion decrease rapidly, and it is possible to obtain accurate predictions using only the lowest-order term that contributes to a particular process.

## Summary

Three main topics have been presented in this chapter. Firstly, the system of natural units with

$$\hbar = c = \varepsilon_0 = \mu_0 = 1$$

was introduced. It is used widely in particle physics and is adopted throughout this book. You should be comfortable with the concept of natural units and should be able to convert between natural units and S.I. units.

Because almost all of particle physics deals with relativistic particles, a sound understanding of special relativity and, in particular, the use of four-vectors is essential for much of what follows. Four-vector notation is used throughout this book with the conventions that the metric tensor is

$$g_{\mu\nu} = g^{\mu\nu} \equiv \begin{pmatrix} 1 & 0 & 0 & 0 \\ 0 & -1 & 0 & 0 \\ 0 & 0 & -1 & 0 \\ 0 & 0 & 0 & -1 \end{pmatrix},$$

such that zeroth component of a four-vector is the time-like quantity, for example

$$x^\mu = (t, x, y, z) \quad \text{and} \quad p^\mu = (E, p_x, p_y, p_z).$$

The scalar product of *any* two four-vectors,

$$a{\cdot}b \equiv a^\mu b_\mu \equiv g_{\mu\nu} a^\mu b^\mu \equiv a^0 b^0 - a^1 b^1 - a^2 b^2 - a^3 b^3 = \text{invariant},$$

forms a Lorentz-invariant quantity that does not depend on the frame of reference. The results of the calculations that follow are usually presented in a frame independent manner using Lorentz invariant quantities.

A number of concepts in quantum mechanics are central to the theoretical ideas developed in the following chapters and it is important that you are familiar with the material reviewed in this chapter. Here the four most important concepts are: (i) the operator formulation of quantum mechanics, where physical observables are described by time-independent operators acting on time-dependent wavefunctions; (ii) the idea of stationary states of the Hamiltonian and the time development of a quantum mechanical system; (iii) the treatment of angular momentum in quantum mechanics and the algebra defined by the commutation relations between the angular momentum operators; and (iv) Fermi's golden rule to describe transition rates.

## Problems

**2.1** When expressed in natural units the lifetime of the W boson is approximately $\tau \approx 0.5\,\text{GeV}^{-1}$. What is the corresponding value in S.I. units?

**2.2** A cross section is measured to be 1 pb; convert this to natural units.

**2.3** Show that the process $\gamma \to e^+e^-$ can not occur in the vacuum.

**2.4** A particle of mass 3 GeV is travelling in the positive $z$-direction with momentum 4 GeV; what are its energy and velocity?

**2.5** In the laboratory frame, denoted $\Sigma$, a particle travelling in the $z$-direction has momentum $\mathbf{p} = p_z\hat{\mathbf{z}}$ and energy $E$.

(a) Use the Lorentz transformation to find expressions for the momentum $p'_z$ and energy $E'$ of the particle in a frame $\Sigma'$, which is moving in a velocity $\mathbf{v} = +v\hat{\mathbf{z}}$ relative to $\Sigma$, and show that $E^2 - p_z^2 = (E')^2 - (p'_z)^2$.

(b) For a system of particles, prove that the total four-momentum squared,

$$p^\mu p_\mu \equiv \left(\sum_i E_i\right)^2 - \left(\sum_i \mathbf{p}_i\right)^2,$$

is invariant under Lorentz transformations.

**2.6** For the decay $a \to 1 + 2$, show that the mass of the particle $a$ can be expressed as

$$m_a^2 = m_1^2 + m_2^2 + 2E_1E_2(1 - \beta_1\beta_2 \cos\theta),$$

where $\beta_1$ and $\beta_2$ are the velocities of the daughter particles ($\beta_i = v_i/c$) and $\theta$ is the angle between them.

**2.7** In a collider experiment, $\Lambda$ baryons can be identified from the decay $\Lambda \to \pi^- p$, which gives rise to a displaced vertex in a tracking detector. In a particular decay, the momenta of the $\pi^+$ and p are measured to be 0.75 GeV and 4.25 GeV respectively, and the opening angle between the tracks is 9°. The masses of the pion and proton are 139.6 MeV and 938.3 MeV.

(a) Calculate the mass of the $\Lambda$ baryon.
(b) On average, $\Lambda$ baryons of this energy are observed to decay at a distance of 0.35 m from the point of production. Calculate the lifetime of the $\Lambda$.

**2.8** In the laboratory frame, a proton with total energy $E$ collides with proton at rest. Find the minimum proton energy such that process

$$p + p \to p + p + \bar{p} + \bar{p}$$

is kinematically allowed.

**2.9** Find the maximum opening angle between the photons produced in the decay $\pi^0 \to \gamma\gamma$ if the energy of the neutral pion is 10 GeV, given that $m_{\pi^0} = 135$ MeV.

**2.10** The maximum of the $\pi^- p$ cross section, which occurs at $p_\pi = 300$ MeV, corresponds to the resonant production of the $\Delta^0$ baryon (i.e. $\sqrt{s} = m_\Delta$). What is the mass of the $\Delta$?

**2.11** Tau-leptons are produced in the process $e^+e^- \to \tau^+\tau^-$ at a centre-of-mass energy of 91.2 GeV. The angular distribution of the $\pi^-$ from the decay $\tau^- \to \pi^-\nu_\tau$ is

$$\frac{dN}{d(\cos\theta^*)} \propto 1 + \cos\theta^*,$$

where $\theta^*$ is the polar angle of the $\pi^-$ in the tau-lepton rest frame, relative to the direction defined by the $\tau$ (tau) spin. Determine the laboratory frame energy distribution of the $\pi^-$ for the cases where the tau-lepton spin is (i) *aligned with* or (ii) *opposite to* its direction of flight.

**2.12** For the process $1 + 2 \to 3 + 4$, the Mandelstam variables $s$, $t$ and $u$ are defined as $s = (p_1 + p_2)^2$, $t = (p_1 - p_3)^2$ and $u = (p_1 - p_4)^2$. Show that

$$s + u + t = m_1^2 + m_2^2 + m_3^2 + m_4^2.$$

**2.13** At the HERA collider, 27.5 GeV electrons were collided head-on with 820 GeV protons. Calculate the centre-of-mass energy.

**2.14** Consider the Compton scattering of a photon of momentum $\mathbf{k}$ and energy $E = |\mathbf{k}| = k$ from an electron *at rest*. Writing the four-momenta of the scattered photon and electron respectively as $k'$ and $p'$, conservation of four-momentum is expressed as $k + p = k' + p'$. Use the relation $p'^2 = m_e^2$ to show that the energy of the scattered photon is given by

$$E' = \frac{E}{1 + (E/m_e)(1 - \cos\theta)},$$

where $\theta$ is the angle through which the photon is scattered.

**2.15** Using the commutation relations for position and momentum, prove that

$$\left[\hat{L}_x, \hat{L}_y\right] = i\hat{L}_z.$$

Using the commutation relations for the components of angular momenta prove

$$\left[\hat{L}^2, \hat{L}_x\right] = 0,$$

and

$$\hat{L}^2 = \hat{L}_-\hat{L}_+ + \hat{L}_z + \hat{L}_z^2.$$

**2.16** Show that the operators $\hat{S}_i = \frac{1}{2}\sigma_i$, where $\sigma_i$ are the three Pauli spin-matrices,

$$\hat{S}_x = \frac{1}{2}\begin{pmatrix} 0 & 1 \\ 1 & 0 \end{pmatrix}, \quad \hat{S}_y = \frac{1}{2}\begin{pmatrix} 0 & -i \\ i & 0 \end{pmatrix} \quad \text{and} \quad \hat{S}_z = \frac{1}{2}\begin{pmatrix} 1 & 0 \\ 0 & -1 \end{pmatrix},$$

satisfy the same algebra as the angular momentum operators, namely

$$\left[\hat{S}_x, \hat{S}_y\right] = i\hat{S}_z, \quad \left[\hat{S}_y, \hat{S}_z\right] = i\hat{S}_x \quad \text{and} \quad \left[\hat{S}_z, \hat{S}_x\right] = i\hat{S}_y.$$

Find the eigenvalue(s) of the operator $\hat{\mathbf{S}}^2 = \frac{1}{4}(\hat{S}_x^2 + \hat{S}_y^2 + \hat{S}_z^2)$, and deduce that the eigenstates of $\hat{S}_z$ are a suitable representation of a spin-half particle.

**2.17** Find the third-order term in the transition matrix element of Fermi's golden rule.

# 3 Decay rates and cross sections

This chapter describes the methodology for the calculations of cross sections and decay rates in relativistic quantum mechanics. In particular, it introduces the ideas of Lorentz-invariant phase space, the Lorentz-invariant matrix element and the treatment of kinematics in particle decays and interactions. The end product is a set of master formulas which, once the quantum mechanical matrix element for a process is known, can be used to obtain expressions for decays rates and cross sections. Provided the main concepts are understood, it is possible to skip the details of the derivations.

## 3.1 Fermi's golden rule

Much of particle physics is based on the experimental measurements of particle decay rates and particle interaction cross sections. These experimentally observable phenomena represent transitions between different quantum mechanical states. In *non-relativistic* quantum mechanics, transition rates are obtained using Fermi's golden rule, which was derived in Section 2.3.6. Fermi's golden rule for the transition rate $\Gamma_{fi}$ from an initial state $|i\rangle$ to a final state $|f\rangle$ is usually expressed as

$$\Gamma_{fi} = 2\pi|T_{fi}|^2\rho(E_i), \tag{3.1}$$

where $T_{fi}$ is the transition matrix element and $\rho(E_i)$ is the *density of states*. The transition matrix element is determined by the Hamiltonian for the interaction which causes the transitions $\hat{H}'$. In the limit where the perturbation is weak, the transition matrix element is given by a perturbation expansion in terms of the interaction Hamiltonian,

$$T_{fi} = \langle f|\hat{H}'|i\rangle + \sum_{j\neq i}\frac{\langle f|\hat{H}'|j\rangle\langle j|\hat{H}'|i\rangle}{E_i - E_j} + \cdots .$$

The transition rate of (3.1) depends on the density of states $\rho(E_i)$,

$$\rho(E_i) = \left|\frac{\mathrm{d}n}{\mathrm{d}E}\right|_{E_i},$$

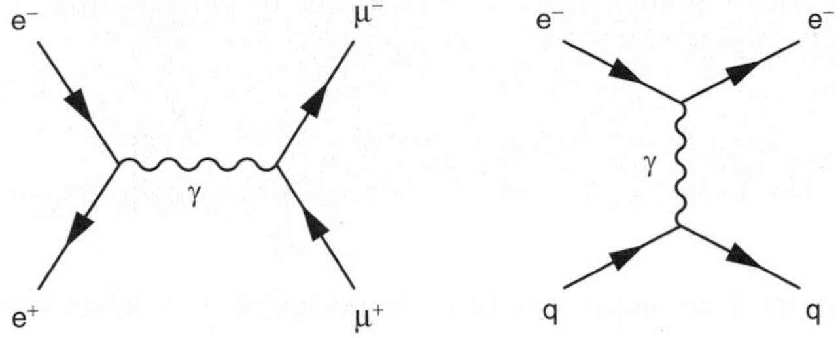


Fig. 3.1 Feynman diagrams for $e^+e^- \to \mu^+\mu^-$ annihilation and $e^-q \to e^-q$ scattering.

where $\mathrm{d}n$ is the number of accessible states in the energy range $E \to E + \mathrm{d}E$. Alternatively, the density of states can be written as an integral over *all* final-state energies using the Dirac delta-function to impose energy conservation,

$$\left|\frac{\mathrm{d}n}{\mathrm{d}E}\right|_{E_i} = \int \frac{\mathrm{d}n}{\mathrm{d}E}\delta(E_i - E)\,\mathrm{d}E,$$

giving the alternative form of Fermi's golden rule

$$\Gamma_{fi} = 2\pi \int |T_{fi}|^2 \delta(E_i - E)\,\mathrm{d}n, \tag{3.2}$$

which appeared as an intermediate step in the derivation of Fermi's golden rule, see (2.49).

The transition rate between two states depends on two components, (i) the *transition matrix element*, which contains the fundamental particle physics, and (ii) the *density of accessible states*, which depends on the kinematics of the process being considered. The aim of the first part of this book is to develop the methodology for the calculation of decay rates and interaction cross sections for particle annihilation and scattering processes such as those represented by the Feynman diagrams of Figure 3.1. In modern particle physics the most complete theoretical approach to such calculations is to use quantum field theory. Nevertheless, the same results can be obtained using perturbation theory in relativistic quantum mechanics (RQM). This requires a relativistic formulation of Fermi's golden rule where the density of states is based on relativistic treatments of phase space and the normalisation of the plane waves used to represent the particles.

## 3.2 Phase space and wavefunction normalisation

Before discussing the relativistic wavefunction normalisation and phase space, it is worth briefly reviewing the non-relativistic treatment. In non-relativistic quantum mechanics, the decay rate for the process $a \to 1 + 2$ can be calculated using

Fermi's golden rule. To first order in perturbation theory, the transition matrix element is

$$T_{fi} = \langle\psi_1\psi_2|\hat{H}'|\psi_a\rangle \tag{3.3}$$

$$= \int_V \psi_1^*\psi_2^*\hat{H}'\psi_a \mathrm{d}^3\mathbf{x}. \tag{3.4}$$

In the Born approximation, the perturbation is taken to be small and the initial- and final-state particles are represented by plane waves of the form

$$\psi(\mathbf{x},t) = Ae^{i(\mathbf{p}\cdot\mathbf{x}-Et)}, \tag{3.5}$$

where $A$ determines the wavefunction normalisation. The integral in (3.4) extends over the volume in which the wavefunctions are normalised. It is usual to adopt a scheme where each plane wave is normalised to one particle in a cubic volume of side $a$. Using the non-relativistic expression for probability density $\rho = \psi^*\psi$, this is equivalent to writing

$$\int_0^a\int_0^a\int_0^a \psi^*\psi\,\mathrm{d}x\,\mathrm{d}y\,\mathrm{d}z = 1,$$

which implies that the normalisation constant in (3.5) is given by

$$A^2 = 1/a^3 = 1/V,$$

where $V$ is the volume of the box.

The normalisation of one particle in a box of volume $a^3$ implies that the wavefunction satisfies the periodic boundary conditions[1]

$$\psi(x+a,y,z) = \psi(x,y,z), \text{ etc.},$$

as illustrated in Figure 3.2. The periodic boundary conditions on the wavefunction, for example $e^{ip_x x} = e^{ip_x(x+a)}$, imply that the components of momentum are quantised to

$$(p_x,p_y,p_z) = (n_x,n_y,n_z)\frac{2\pi}{a},$$

[1] In terms of counting the number of states, the periodic boundary conditions are equivalent to requiring that the wavefunction is zero at the boundaries of the volume. This condition implies that the wavefunction consists of standing waves of the form $\psi(x,y,z) = A\sin(p_x x)\sin(p_y y)\sin(p_z z)$, with $p_x$, $p_y$ and $p_z$ such that there are a half-integer number of wavelengths along each side of the box. Since $\sin(p_x x) = (e^{ip_x x} - e^{-ip_x x})/2i$, the wavefunction expressed in this way has forward-going and backward-going components and the integration over phase space is restricted to positive values of $p_x$, $p_y$ and $p_z$. The same number of states are obtained with periodic boundary conditions, with an integer number of wavelengths in each direction. In this case, the phase space integral extends over both positive and negative values of $p_x$, $p_y$ and $p_z$.

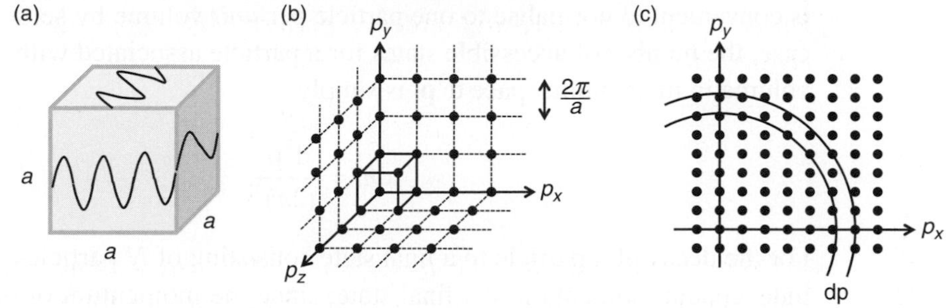


Fig. 3.2 The non-relativistic treatment of phase space: (a) the wavefunction of a particle confined to a box of side *a* satisfies the periodic boundary conditions such that there are an integer number of wavelengths in each direction; (b) the allowed states in momentum space; and (c) the number of states in a range $\mathrm{p} \rightarrow \mathrm{p} + \mathrm{dp}$ in two dimensions.

where $n_x$, $n_y$ and $n_z$ are integers. This restricts the allowed momentum states to the discrete set indicated in Figure 3.2b. Each state in momentum space occupies a cubic volume of

$$\mathrm{d}^3\mathbf{p} = \mathrm{d}p_x\mathrm{d}p_y\mathrm{d}p_z = \left(\frac{2\pi}{a}\right)^3 = \frac{(2\pi)^3}{V}.$$

As indicated in Figure 3.2c, the number of states $\mathrm{d}n$ with magnitude of momentum in the range $\mathrm{p} \rightarrow \mathrm{p} + \mathrm{dp}$, is equal to the momentum space volume of the spherical shell at momentum p with thickness dp divided by the average volume occupied by a single state, $(2\pi)^3/V$, giving

$$\mathrm{d}n = 4\pi\mathrm{p}^2\mathrm{dp} \times \frac{V}{(2\pi)^3},$$

and hence

$$\frac{\mathrm{d}n}{\mathrm{dp}} = \frac{4\pi\mathrm{p}^2}{(2\pi)^3}V.$$

The density of states in Fermi's golden rule then can be obtained from

$$\rho(E) = \frac{\mathrm{d}n}{\mathrm{d}E} = \frac{\mathrm{d}n}{\mathrm{dp}}\left|\frac{\mathrm{dp}}{\mathrm{d}E}\right|.$$

The density of states corresponds to the number of momentum states accessible to a particular decay and increases with the momentum of the final-state particle. Hence, all other things being equal, decays to lighter particles, which will be produced with larger momentum, are favoured over decays to heavier particles.

The calculation of the decay rate will *not* depend on the normalisation volume; the volume dependence in the expression for phase space is cancelled by the factors of $V$ associated with the wavefunction normalisations that appear in the square of transition matrix element. Since the volume will not appear in the final result, it is convenient to normalise to one particle per *unit* volume by setting $V = 1$. In this case, the number of accessible states for a particle associated with an infinitesimal volume in momentum space $d^3\mathbf{p}_i$ is simply

$$dn_i = \frac{d^3\mathbf{p}_i}{(2\pi)^3}.$$

For the decay of a particle to a final state consisting of $N$ particles, there are $N - 1$ independent momenta in the final state, since the momentum of one of the final-state particles can always be obtained from momentum conservation. Thus, the number of independent states for an $N$-particle final state is

$$dn = \prod_{i=1}^{N-1} dn_i = \prod_{i=1}^{N-1} \frac{d^3\mathbf{p}_i}{(2\pi)^3}.$$

This can be expressed in a more democratic form including the momentum space volume element for the $N$th particle $d^3\mathbf{p}_N$ and using a three-dimensional delta-function to impose momentum conservation

$$dn = \prod_{i=1}^{N-1} \frac{d^3\mathbf{p}_i}{(2\pi)^3} \delta^3\left(\mathbf{p}_a - \sum_{i=1}^{N} \mathbf{p}_i\right) d^3\mathbf{p}_N, \tag{3.6}$$

where $\mathbf{p}_a$ is the momentum of the decaying particle. Therefore the general non-relativistic expression for $N$-body phase space is

$$dn = (2\pi)^3 \prod_{i=1}^{N} \frac{d^3\mathbf{p}_i}{(2\pi)^3} \delta^3\left(\mathbf{p}_a - \sum_{i=1}^{N} \mathbf{p}_i\right). \tag{3.7}$$

### 3.2.1 Lorentz-invariant phase space

The wavefunction normalisation of one particle per unit volume is not Lorentz invariant since it only applies to a particular frame of reference. In a different reference frame, the original normalisation volume will be Lorentz contracted by a factor of $1/\gamma$ along its direction of relative motion, as shown in Figure 3.3. Thus, the original normalisation of one particle per unit volume corresponds to a normalisation of $\gamma = E/m$ particles per unit volume in the boosted frame of reference. A Lorentz-invariant choice of wavefunction normalisation must therefore be proportional to $E$ particles per unit volume, such that the increase in energy accounts for the effect of Lorentz contraction. The usual convention is to normalise to $2E$ particles per unit volume. The reason for this particular choice is motivated in Section 3.2.3 and also in Chapter 4.

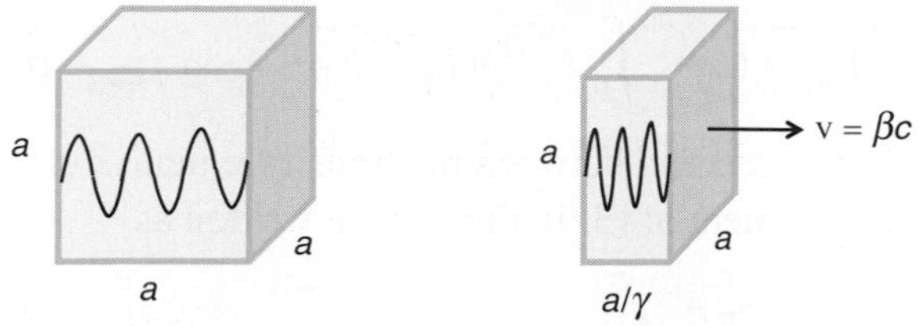


Fig. 3.3 The normalisation volume in a particular frame is length contracted along the direction of motion for a general rest frame.

The wavefunctions $\psi$ appearing in the transition matrix element $T_{fi}$ of Fermi's golden rule are normalised to one particle per unit volume,

$$\int_V \psi^* \psi \, \mathrm{d}^3\mathbf{x} = 1.$$

Wavefunctions with the appropriate Lorentz-invariant normalisation, here written as $\psi'$, are normalised to $2E$ particles per unit volume

$$\int_V \psi'^* \psi' \mathrm{d}^3\mathbf{x} = 2E,$$

and therefore

$$\psi' = (2E)^{1/2}\psi.$$

For a general process, $a + b + \cdots \to 1 + 2 + \cdots$, the *Lorentz-invariant matrix element*, using wavefunctions with a Lorentz-invariant normalisation, is defined as

$$\mathcal{M}_{fi} = \langle \psi_1' \psi_2' \cdots | \hat{H}' | \psi_a' \psi_b' \cdots \rangle. \tag{3.8}$$

The Lorentz-invariant matrix element is therefore related to the transition matrix element of Fermi's golden rule by

$$\mathcal{M}_{fi} = \langle \psi_1' \psi_2' \cdots | \hat{H}' | \psi_a' \psi_b' \cdots \rangle = (2E_1 \cdot 2E_2 \cdots 2E_a \cdot 2E_b \cdots)^{1/2} T_{fi}, \tag{3.9}$$

where the product on the RHS of (3.9) includes all intial- and final-state particles.

### 3.2.2 Fermi's golden rule revisited

For a two-body decay $a \to 1 + 2$, the quantum mechanical transition rate is given by Fermi's golden rule, which in the form of (3.2) can be written

$$\Gamma_{fi} = 2\pi \int |T_{fi}|^2 \delta(E_a - E_1 - E_2) \, \mathrm{d}n,$$

where $\mathrm{d}n$ is given by (3.7), and hence

$$\Gamma_{fi} = (2\pi)^4 \int |T_{fi}|^2 \delta(E_a - E_1 - E_2)\delta^3(\mathbf{p}_a - \mathbf{p}_1 - \mathbf{p}_2)\frac{d^3\mathbf{p}_1}{(2\pi)^3}\frac{d^3\mathbf{p}_2}{(2\pi)^3}. \tag{3.10}$$

Using the relation between the transition matrix element and the Lorentz invariant matrix element of (3.9), this can be written as

$$\Gamma_{fi} = \frac{(2\pi)^4}{2E_a} \int |\mathcal{M}_{fi}|^2 \delta(E_a - E_1 - E_2)\delta^3(\mathbf{p}_a - \mathbf{p}_1 - \mathbf{p}_2)\frac{d^3\mathbf{p}_1}{(2\pi)^3 2E_1}\frac{d^3\mathbf{p}_2}{(2\pi)^3 2E_2}, \tag{3.11}$$

with $|\mathcal{M}_{fi}|^2 = (2E_a 2E_1 2E_2)|T_{fi}|^2$. One consequence of using wavefunctions with a Lorentz invariant normalisation, is that the phase space integral over $d^3\mathbf{p}/(2\pi)^3$ has been replaced by an integral over terms like

$$\frac{d^3\mathbf{p}}{(2\pi)^3 2E},$$

which is known as the Lorentz-invariant phase space factor. To prove this is Lorentz invariant, consider a Lorentz transformation along the $z$-axis, where the element $d^3\mathbf{p}$ transforms to $d^3\mathbf{p}'$ given by

$$d^3\mathbf{p}' \equiv dp_x' dp_y' dp_z' = dp_x dp_y \cdot \frac{dp_z'}{dp_z} dp_z = \frac{dp_z'}{dp_z} d^3\mathbf{p}. \tag{3.12}$$

From the Einstein energy–momentum relation, $E^2 = p_x^2 + p_y^2 + p_z^2 + m^2$, and the Lorentz transformation of the energy–momentum four-vector,

$$p_z' = \gamma(p_z - \beta E) \quad \text{and} \quad E' = \gamma(E - \beta p_z),$$

it follows that

$$\frac{dp_z'}{dp_z} = \gamma\left(1 - \beta\frac{\partial E}{\partial p_z}\right) = \gamma\left(1 - \beta\frac{p_z}{E}\right) = \frac{1}{E}\gamma(E - \beta p_z) = \frac{E'}{E},$$

which when substituted into (3.12) demonstrates that

$$\frac{d^3\mathbf{p}'}{E'} = \frac{d^3\mathbf{p}}{E},$$

and hence $d^3\mathbf{p}/E$ is Lorentz invariant.

The matrix element $\mathcal{M}_{fi}$ in (3.11) is defined in terms of wavefunctions with a Lorentz-invariant normalisation, and the elements of integration over phase space $d^3\mathbf{p}_i/E_i$ are also Lorentz invariant. Consequently, the integral in (3.11) is Lorentz invariant and thus (3.11) expresses Fermi's golden rule in a Lorentz-invariant form. This is an important result, it is exactly the required relativistic treatment of transition rates needed for the calculation of decay rates. The resulting transition rate for the decay $a \to 1 + 2$ given in (3.11) is inversely proportional to the energy of the decaying particle in the frame in which it is observed, $E_a = \gamma m_a$, as expected from relativistic time dilation.

### 3.2.3 *Lorentz-invariant phase space

The expression for the decay rate given (3.11) can be extended to an $N$-body decay, $a \to 1 + 2 + \cdots + N$. In this more general case, the phase space integral involves the three-momenta of all final-state particles

$$\mathrm{d}LIPS = \prod_{i=1}^{N} \frac{\mathrm{d}^3\mathbf{p}_i}{(2\pi)^3 2E_i},$$

where $\mathrm{d}LIPS$ is known as the element of Lorentz-invariant phase space (LIPS). The factors $1/2E_i$ can be rewritten in terms of a delta-function using (A.6) of Appendix A and the constraint from the Einstein energy–momentum relationship, $E_i = \mathbf{p}_i^2 + m_i^2$, which implies that

$$\int \delta(E_i^2 - \mathbf{p}_i^2 - m^2)\,\mathrm{d}E_i = \frac{1}{2E_i}.$$

Hence, the integral over Lorentz-invariant phase space can be written as

$$\int \cdots \,\mathrm{d}LIPS = \int \cdots \prod_{i=1}^{N} (2\pi)^{-3} \delta(E_i^2 - \mathbf{p}_i^2 - m_i^2)\,\mathrm{d}^3\mathbf{p}_i\,\mathrm{d}E_i,$$

which, in terms of the four-momenta of the final-state particles is

$$\int \cdots \,\mathrm{d}LIPS = \int \cdots \prod_{i=1}^{N} (2\pi)^{-3} \delta(p_i^2 - m_i^2)\,\mathrm{d}^4 p_i.$$

Similarly, the transition rate for the two-body decay $a \to 1 + 2$, given in (3.11), can be written as

$$\Gamma_{fi} = \frac{(2\pi)^4}{2E_a} \int (2\pi)^{-6} |\mathcal{M}_{fi}|^2 \delta^4(p_a - p_1 - p_2) \delta(p_1^2 - m_1^2) \delta(p_2^2 - m_2^2)\,\mathrm{d}^4 p_1 \mathrm{d}^4 p_2.$$

The integral now extends over *all* values of the energies and momenta of each of the final-state particles. The delta-functions ensure that the decay rate only has contributions from values of the four-momenta of the final-state particles compatible with overall energy and momentum conservation and the Einstein energy–momentum relation $p_i^2 = m_i^2$. This form of the expression for the decay rate elucidates clearly the point that all the fundamental physics lives in the matrix element. It also provides a deeper insight into the origin of the phase space integral.

## 3.3 Particle decays

In general, a given particle may decay by more than one decay mode. For example, the tau-lepton can decay into a number of final states, $\tau^- \to e^-\overline{\nu}_e\nu_\tau$, $\tau^- \to \mu^-\overline{\nu}_\mu\nu_\tau$ and $\tau^- \to \nu_\tau + \text{hadrons}$. The transition rate for *each* decay mode $j$ can be calculated independently using Fermi's golden rule. The individual transition rates $\Gamma_j$ are referred to as *partial decay rates* or, for reasons that will become apparent later, *partial widths*.

The total decay rate is simply the sum of the decay rates for the individual decay modes. For example, if there are $N$ particles of a particular type, the number that decay in time $\delta t$ is given by the sum of the numbers of decays into each decay channel,

$$\delta N = -N\Gamma_1\delta t - N\Gamma_2\delta t - \cdots = -N\sum_j \Gamma_j\,\delta t = -N\Gamma\delta t, \tag{3.13}$$

where the total decay rate per unit time $\Gamma$ is the sum of the individual decay rates,

$$\Gamma = \sum_j \Gamma_j.$$

The number of particles remaining after a time $t$ is obtained by integrating (3.13) to give the usual exponential form

$$N(t) = N(0)\,e^{-\Gamma t} = N(0)\,\exp\left(-\frac{t}{\tau}\right),$$

where the lifetime of the particle in its rest frame $\tau$ is referred to as its proper lifetime and is determined from the total decay rate

$$\tau = \frac{1}{\Gamma}.$$

The relative frequency of a particular decay mode is referred to as the *branching ratio* (or branching fraction). The branching ratio for a particular decay mode $BR(j)$ is given by the decay rate to the mode $j$ relative to the total decay rate

$$BR(j) = \frac{\Gamma_j}{\Gamma}.$$

For example, the branching ratio for the tau-lepton decay $\tau^- \to e^-\overline{\nu}_e\nu_\tau$ is 0.17, which means that on average 17% of the time a $\tau^-$ will decay to $e^-\overline{\nu}_e\nu_\tau$. By definition, the branching ratios for all decay modes of a particular particle sum to unity.

### 3.3.1 Two-body decays

The transition rate for each decay mode of a particle can be calculated by using the relativistic formulation of Fermi's golden rule given in (3.11). The rate depends on

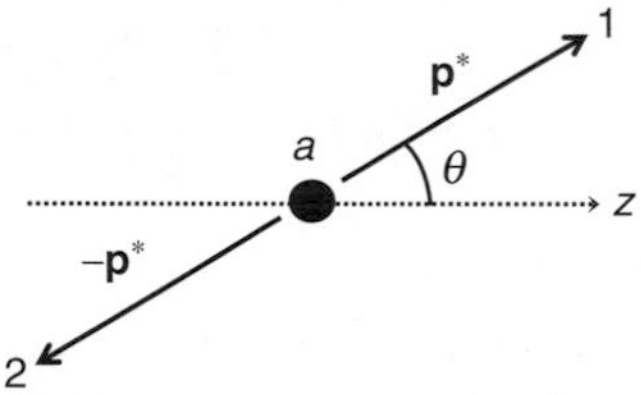


Fig. 3.4 The two-body decay $a \longrightarrow 1 + 2$ in the rest frame of particle $a$.

the matrix element for the process and the phase space integral. The matrix element depends on the nature of the decay and needs to be evaluated for each process. In contrast, the form of the phase space integral depends only on the number of particles in the final state. Furthermore, since the integral of (3.11) is Lorentz invariant, it can be evaluated in any frame.

Consider the two-body decay $a \to 1 + 2$, shown in Figure 3.4. In the centre-of-mass frame, the decaying particle is at rest, $E_a = m_a$ and $\mathbf{p}_a = \mathbf{0}$, and the two daughter particles are produced back to back with three-momenta $\mathbf{p}^*$ and $-\mathbf{p}^*$. In this frame, the decay rate is given by (3.11),

$$\Gamma_{fi} = \frac{1}{8\pi^2 m_a} \int |\mathcal{M}_{fi}|^2 \delta(m_a - E_1 - E_2)\delta^3(\mathbf{p}_1 + \mathbf{p}_2)\, \frac{\mathrm{d}^3\mathbf{p}_1}{2E_1}\frac{\mathrm{d}^3\mathbf{p}_2}{2E_2}. \tag{3.14}$$

It is not straightforward to evaluate the phase space integral in this expression, but fortunately the calculation applies to all two-body decays and has to be performed only once. The $\delta^3(\mathbf{p}_1 + \mathbf{p}_2)$ term in (3.14) means that the integral over $\mathrm{d}^3\mathbf{p}_2$ has the effect of relating the three-momenta of the final-state particles giving $\mathbf{p}_2 = -\mathbf{p}_1$ and hence

$$\Gamma_{fi} = \frac{1}{8\pi^2 m_a} \int |\mathcal{M}_{fi}|^2 \frac{1}{4E_1E_2} \delta\left(m_a - E_1 - E_2\right) \mathrm{d}^3\mathbf{p}_1, \tag{3.15}$$

where $E_2$ is now given by $E_2^2 = (m_2^2 + \mathrm{p}_1^2)$. In spherical polar coordinates,

$$\mathrm{d}^3\mathbf{p}_1 = \mathrm{p}_1^2 \mathrm{d}\mathrm{p}_1 \sin\theta \,\mathrm{d}\theta \,\mathrm{d}\phi = \mathrm{p}_1^2 \,\mathrm{d}\mathrm{p}_1 \mathrm{d}\Omega,$$

and (3.15) can be written

$$\Gamma_{fi} = \frac{1}{8\pi^2 m_a} \int |\mathcal{M}_{fi}|^2 \delta\left(m_a - \sqrt{m_1^2 + \mathrm{p}_1^2} - \sqrt{m_2^2 + \mathrm{p}_1^2}\right) \frac{\mathrm{p}_1^2}{4E_1E_2}\, \mathrm{d}\mathrm{p}_1 \mathrm{d}\Omega. \tag{3.16}$$

At first sight this integral looks quite tricky. Fortunately the Dirac delta-function does most of the work. Equation (3.16) has the functional form

$$\Gamma_{fi} = \frac{1}{8\pi^2 m_a} \int |\mathcal{M}_{fi}|^2 g(\mathrm{p}_1)\, \delta(f(\mathrm{p}_1))\, \mathrm{d}\mathrm{p}_1 \mathrm{d}\Omega, \tag{3.17}$$

with

$$g(\mathrm{p}_1) = \frac{\mathrm{p}_1^2}{4E_1E_2}, \tag{3.18}$$

and

$$f(\mathrm{p}_1) = m_a - E_1 - E_2 = m_a - \sqrt{m_1^2 + \mathrm{p}_1^2} - \sqrt{m_2^2 + \mathrm{p}_1^2}. \tag{3.19}$$

The Dirac delta-function $\delta(f(\mathrm{p}_1))$ imposes energy conservation and is only non-zero for $\mathrm{p}_1 = \mathrm{p}^*$, where $\mathrm{p}^*$ is the solution of $f(\mathrm{p}^*) = 0$. The integral over $\mathrm{dp}_1$ in (3.17) can be evaluated using the properties of the Dirac delta-function (see Appendix A), whereby

$$\int |\mathcal{M}_{fi}|^2\, g(\mathrm{p}_1)\, \delta\left(f(\mathrm{p}_1)\right)\, \mathrm{dp}_1 = |\mathcal{M}_{fi}|^2\, g(\mathrm{p}^*) \left|\frac{\mathrm{d}f}{\mathrm{dp}_1}\right|^{-1}_{\mathrm{p}^*}. \tag{3.20}$$

The derivative $\mathrm{d}f/\mathrm{d}p_1$ can be obtained from (3.19),

$$\left|\frac{\mathrm{d}f}{\mathrm{dp}_1}\right| = \frac{\mathrm{p}_1}{(m_1^2 + \mathrm{p}_1^2)^{1/2}} + \frac{\mathrm{p}_1}{(m_2^2 + \mathrm{p}_1^2)^{1/2}} = \mathrm{p}_1 \left(\frac{E_1 + E_2}{E_1E_2}\right),$$

which, when combined with the expression for $g(\mathrm{p}_1)$ given in (3.18), leads to

$$g(\mathrm{p}^*) \left|\frac{\mathrm{d}f}{\mathrm{dp}_1}\right|^{-1}_{\mathrm{p}_1=\mathrm{p}^*} = \frac{\mathrm{p}^{*2}}{4E_1E_2} \cdot \frac{E_1E_2}{\mathrm{p}^*(E_1 + E_2)} = \frac{\mathrm{p}^*}{4m_a}.$$

Thus, the integral of (3.20) is

$$\int |\mathcal{M}_{fi}|^2\, g(\mathrm{p}_1)\, \delta(f(\mathrm{p}_1))\, \mathrm{dp}_1 = \frac{\mathrm{p}^*}{4m_a} |\mathcal{M}_{fi}|^2,$$

and therefore,

$$\int |\mathcal{M}_{fi}|^2 \delta(m_a - E_1 - E_2)\delta^3(\mathbf{p}_1 + \mathbf{p}_2) \frac{\mathrm{d}^3\mathbf{p}_1}{2E_1} \frac{\mathrm{d}^3\mathbf{p}_2}{2E_2} = \frac{\mathrm{p}^*}{4m_a} \int |\mathcal{M}_{fi}|^2 \mathrm{d}\Omega, \tag{3.21}$$

and hence (3.14) becomes

$$\Gamma_{fi} = \frac{\mathrm{p}^*}{32\pi^2 m_a^2} \int |\mathcal{M}_{fi}|^2\, \mathrm{d}\Omega. \tag{3.22}$$

Equation (3.22) is the general expression for *any* two-body decay. The fundamental physics is contained in the matrix element and the additional factors arise from the phase space integral. The matrix element, which may depend on the decay angle, remains inside the integral. The centre-of-mass frame momentum of the final-state particles $\mathrm{p}^*$ can be obtained from energy conservation, or equivalently $f(\mathrm{p}^*) = 0$, and is given by (see Problem 3.2)

$$\mathrm{p}^* = \frac{1}{2m_a} \sqrt{\left[(m_a^2 - (m_1 + m_2)^2\right]\left[m_a^2 - (m_1 - m_2)^2\right]}.$$

## 3.4 Interaction cross sections

The calculation of interaction rates is slightly more complicated than that for particle decays because it is necessary to account for the flux of initial-state particles, where flux is defined as the number of particles crossing a unit area per unit time. In the simplest case, one can imagine a beam of particles of type $a$, with flux $\phi_a$, crossing a region of space in which there are $n_b$ particles per unit volume of type $b$. The interaction rate per target particle $r_b$ will be proportional to the incident particle flux and can be written

$$r_b = \sigma\phi_a. \tag{3.23}$$

The fundamental physics is contained in $\sigma$, which has dimensions of area, and is termed the interaction *cross section*. Sometimes it is helpful to think of $\sigma$ as the *effective* cross sectional area associated with each target particle. Indeed, there are cases where the cross section is closely related to the physical cross sectional area of the target, for example, neutron absorption by a nucleus. However, in general, the cross section is simply an expression of the underlying quantum mechanical probability that an interaction will occur.

The definition of the cross section is illustrated by the situation shown in Figure 3.5a, where a single incident particle of type $a$ is travelling with a velocity $\mathbf{v}_a$ in a region defined by the area $A$, which contains $n_b$ particles of type $b$ per unit volume moving with a velocity $\mathbf{v}_b$ in the opposite direction to $\mathbf{v}_a$. In time $\delta t$, the particle $a$ crosses a region containing $\delta N = n_b(\mathrm{v}_a + \mathrm{v}_b)A\delta t$ particles of type $b$. The interaction probability can be obtained from the *effective* total cross sectional area of the $\delta N$ particles divided by the area $A$, which can be thought of as the probability that the incident particle passes through one of the regions of area $\sigma$ drawn around each of the $\delta N$ target particles, as shown in Figure 3.5b. The interaction probability $\delta P$ is therefore

$$\delta P = \frac{\delta N\,\sigma}{A} = \frac{n_b(\mathrm{v}_a + \mathrm{v}_b)A\,\sigma\delta t}{A} = n_b \mathrm{v}\sigma\delta t,$$

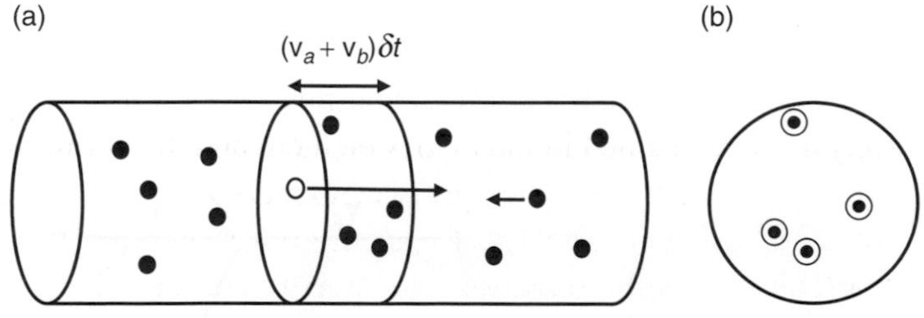


Fig. 3.5 The left-hand plot (a) shows a single incident particle of type $a$ traversing a region containing particles of type $b$. The right-hand plot (b) shows the projected view of the region traversed by the incident particle in time $\delta t$.

where $v = v_a + v_b$. Hence the interaction rate for each particle of type $a$ is

$$r_a = \frac{dP}{dt} = n_b v\sigma.$$

For a beam of particles of type $a$, with number density $n_a$ confined to a volume $V$, the total interaction rate is

$$\text{rate} = r_a\, n_a V = (n_b v\sigma)\, n_a V. \tag{3.24}$$

The expression of (3.24) can be rearranged into

$$\text{rate} = (n_a v)(n_b V)\sigma = \phi\, N_b\, \sigma.$$

Thus the total rate is equal to

$$\text{rate} \;=\; \text{flux} \times \text{number of target particles} \times \text{cross section},$$

which is consistent with the definition of (3.23). More formally, the cross section for a process is defined as

$$\sigma = \frac{\text{number of interactions per unit time per target particle}}{\text{incident flux}}.$$

It should be noted that the flux $\phi$ accounts for the *relative* motion of the particles.

### 3.4.1 Lorentz-invariant flux

The cross section for a particular process can be calculated using the relativistic formulation of Fermi's golden rule and the appropriate Lorentz-invariant expression for the particle flux. Consider the scattering process $a + b \rightarrow 1 + 2$, as observed in the rest frame where the particles of type $a$ have velocity $\mathbf{v}_a$ and those of type $b$ have velocity $\mathbf{v}_b$, as shown in Figure 3.6. If the number densities of the particles are $n_a$ and $n_b$, the interaction rate in the volume $V$ is given by

$$\text{rate} = \phi_a\, n_b V\, \sigma = (v_a + v_b)\, n_a\, n_b\, \sigma\, V, \tag{3.25}$$

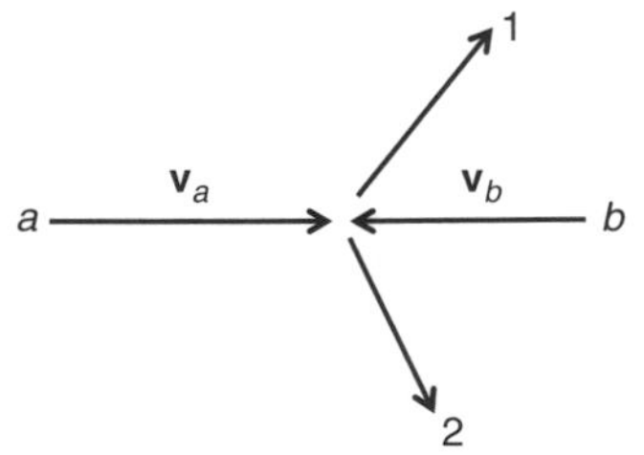


Fig. 3.6 The two-body scattering process $a + b \rightarrow 1 + 2$.

where $\phi_a$ is the flux of particles of type $a$ through a plane moving at velocity $\mathbf{v}_b$,

$$\phi_a = n_a(\mathrm{v}_a + \mathrm{v}_b).$$

Normalising the wavefunctions to one particle in a volume $V$, gives $n_a = n_b = 1/V$, for which the interaction rate in the volume $V$ is

$$\Gamma_{fi} = \frac{(\mathrm{v}_a + \mathrm{v}_b)}{V}\sigma. \tag{3.26}$$

Because the factors of $V$ in the expression for the flux will ultimately be cancelled by the corresponding factors from the wavefunction normalisation and phase space, the volume $V$ will not appear in the final result and it is again convenient to adopt a normalisation of one particle per unit volume. With this choice, the cross section is related to the transition rate by

$$\sigma = \frac{\Gamma_{fi}}{(\mathrm{v}_a + \mathrm{v}_b)}.$$

The transition rate $\Gamma_{fi}$ is given by Fermi's golden rule, which in the form of (3.10) gives

$$\sigma = \frac{(2\pi)^4}{(\mathrm{v}_a + \mathrm{v}_b)} \int |T_{fi}|^2 \delta(E_a + E_b - E_1 - E_2)\delta^3(\mathbf{p}_a + \mathbf{p}_b - \mathbf{p}_1 - \mathbf{p}_2)\frac{\mathrm{d}^3\mathbf{p}_1}{(2\pi)^3}\frac{\mathrm{d}^3\mathbf{p}_2}{(2\pi)^3}.$$

This can be expressed in a Lorentz-invariant form by writing $T_{fi}$ in terms of the Lorentz-invariant matrix element $\mathcal{M}_{fi} = (2E_1\, 2E_2\, 2E_3\, 2E_4)^{1/2}T_{fi}$,

$$\sigma = \frac{(2\pi)^{-2}}{4\,E_aE_b(\mathrm{v}_a + \mathrm{v}_b)} \int |\mathcal{M}_{fi}|^2 \delta(E_a + E_b - E_1 - E_2)\delta^3(\mathbf{p}_a + \mathbf{p}_b - \mathbf{p}_1 - \mathbf{p}_2)\frac{\mathrm{d}^3\mathbf{p}_1}{2E_1}\frac{\mathrm{d}^3\mathbf{p}_2}{2E_2}. \tag{3.27}$$

The integral in (3.27) is now written in a Lorentz-invariant form. The quantity $F = 4E_aE_b(\mathrm{v}_a + \mathrm{v}_b)$ is known as the Lorentz-invariant flux factor. To demonstrate the Lorentz invariance of $F$, first write

$$F = 4E_aE_b(\mathrm{v}_a + \mathrm{v}_b) = 4E_aE_b\left(\frac{\mathrm{p}_a}{E_a} + \frac{\mathrm{p}_b}{E_b}\right) = 4(E_a\mathrm{p}_b + E_b\mathrm{p}_a),$$

$$\Rightarrow \qquad F^2 = 16(E_a^2\mathrm{p}_b^2 + E_b^2\mathrm{p}_a^2 + 2E_aE_b\mathrm{p}_a\mathrm{p}_b), \tag{3.28}$$

and then note that, for the case where the incident particle velocities are collinear,

$$(p_a{\cdot}p_b)^2 = (E_aE_b + \mathrm{p}_a\mathrm{p}_b)^2 = E_a^2E_b^2 + \mathrm{p}_a^2\mathrm{p}_b^2 + 2E_aE_b\mathrm{p}_a\mathrm{p}_b. \tag{3.29}$$

Substituting the expression for $2E_aE_b\mathrm{p}_a\mathrm{p}_b$ from (3.29) into (3.28) then gives

$$F^2 = 16\left[(p_a{\cdot}p_b)^2 - (E_a^2 - \mathrm{p}_a^2)(E_b^2 - \mathrm{p}_b^2)\right].$$

Thus, $F$ can be written in the manifestly Lorentz-invariant form

$$F = 4\left[(p_a \cdot p_b)^2 - m_a^2 m_b^2\right]^{\frac{1}{2}}.$$

Since both $F$ and the integral in (3.27) are Lorentz invariant, it can be concluded that the cross section for an interaction is itself Lorentz invariant.

### 3.4.2 Scattering in the centre-of-mass frame

Because the interaction cross section is a Lorentz-invariant quantity, the cross section for the process $a + b \to 1 + 2$ can be calculated in any frame. The most convenient choice is the centre-of-mass frame where $\mathbf{p}_a = -\mathbf{p}_b = \mathbf{p}_i^*$ and $\mathbf{p}_1 = -\mathbf{p}_2 = \mathbf{p}_f^*$, and the centre-of-mass energy is given by $\sqrt{s} = (E_a^* + E_b^*)$. In the centre-of-mass frame, the Lorentz-invariant flux factor is

$$\begin{aligned} F &= 4E_a^*E_b^*(\mathrm{v}_a^* + \mathrm{v}_b^*) = 4E_a^*E_b^*\left(\frac{\mathrm{p}_i^*}{E_a^*} + \frac{\mathrm{p}_i^*}{E_b^*}\right) = 4\mathrm{p}_i^*(E_a^* + E_b^*) \\ &= 4\mathrm{p}_i^*\sqrt{s}. \end{aligned}$$

Using this expression and the constraint that $\mathbf{p}_a + \mathbf{p}_b = \mathbf{0}$, (3.27) becomes

$$\sigma = \frac{1}{(2\pi)^2}\frac{1}{4\mathrm{p}_i^*\sqrt{s}}\int |\mathcal{M}_{fi}|^2\delta\left(\sqrt{s} - E_1 - E_2\right)\delta^3(\mathbf{p}_1 + \mathbf{p}_2)\frac{\mathrm{d}^3\mathbf{p}_1}{2E_1}\frac{\mathrm{d}^3\mathbf{p}_2}{2E_2}. \tag{3.30}$$

The integral in (3.30) is the same as that of (3.21) with $m_a$ replaced by $\sqrt{s}$. Therefore, applying the results from Section 3.3.1 immediately leads to

$$\sigma = \frac{1}{16\pi^2\mathrm{p}_i^*\sqrt{s}} \times \frac{\mathrm{p}_f^*}{4\sqrt{s}}\int |\mathcal{M}_{fi}|^2\mathrm{d}\Omega^*,$$

where the solid angle element has been written as $\mathrm{d}\Omega^*$ to emphasise that it refers to the centre-of-mass frame. Hence the cross section for *any* two-body $\to$ two-body process is given by

$$\sigma = \frac{1}{64\pi^2 s}\frac{\mathrm{p}_f^*}{\mathrm{p}_i^*}\int |\mathcal{M}_{fi}|^2\mathrm{d}\Omega^*. \tag{3.31}$$

## 3.5 Differential cross sections

In many cases it is not only the total cross section that is of interest, but also the distribution of some kinematic variable. For example, Figure 3.7 shows the inelastic

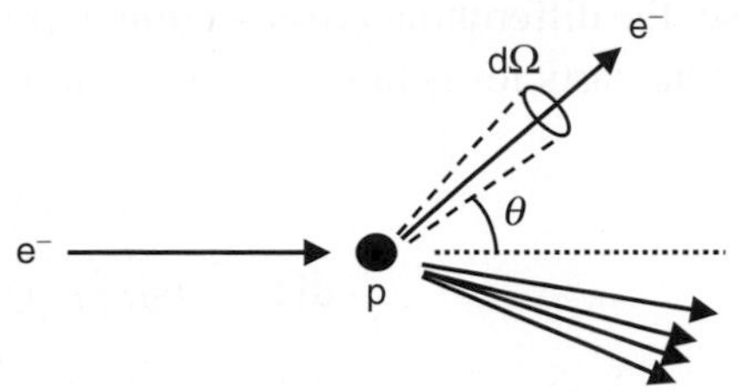


Fig. 3.7 An example of $e^-p \rightarrow e^-p$ scattering where the electron is scattered into a solid angle $d\Omega$.

scattering process $e^-p \rightarrow eX$ where the proton breaks up. Here, the angular distribution of the scattered electron provides essential information about the fundamental physics of the interaction. In this case, the relevant experimental measurement is the *differential* cross section for the scattering rate into an element of solid angle $d\Omega = d(\cos\theta)d\phi$,

$$\frac{d\sigma}{d\Omega} = \frac{\text{number of particles scattered into } d\Omega \text{ per unit time per target particle}}{\text{incident flux}}.$$

The integral of the differential cross section gives the total cross section,

$$\sigma = \int \frac{d\sigma}{d\Omega} d\Omega.$$

Differential cross sections are not restricted to angular distributions. In some situations, it is the energy distribution of the scattered particle that is sensitive to the underlying fundamental physics. In other situations one might be interested in the joint angular and energy distribution of the scattered particles. In each case, it is possible to define the corresponding differential cross section, for example

$$\frac{d\sigma}{dE} \quad \text{or} \quad \frac{d^2\sigma}{dE d\Omega}.$$

### 3.5.1 Differential cross section calculations

Differential cross sections can be calculated from the differential form of (3.31),

$$d\sigma = \frac{1}{64\pi^2 s} \frac{p_f^*}{p_i^*} |\mathcal{M}_{fi}|^2 d\Omega^*. \tag{3.32}$$

The simplest situation is where the laboratory frame corresponds to the centre-of-mass frame, for example $e^+e^-$ annihilation at LEP or pp collisions at the LHC.

In this case, the differential cross section expressed in terms of the angles of one of the final-state particles is immediately obtained from (3.32)

$$\frac{d\sigma}{d\Omega^*} = \frac{1}{64\pi^2 s}\frac{p_f^*}{p_i^*}|\mathcal{M}_{fi}|^2. \tag{3.33}$$

In fixed-target experiments, such as $e^-p \to e^-p$ elastic scattering, where the target proton is at rest, the laboratory frame is not the centre-of-mass frame and the calculation is more involved. Here, the differential cross section is most useful when expressed in terms of the observable laboratory frame quantities, such as the angle through which the electron is scattered, $\theta$. The differential cross section with respect to the laboratory frame electron scattering angle can be obtained by applying the appropriate coordinate transformation to (3.32).

The transformation from the differential cross section in the centre-of-mass frame to that in the laboratory frame is most easily obtained by first writing (3.32) in a Lorentz-invariant form, which is applies in all frames. This is achieved by expressing the element of solid angle $d\Omega^*$ in terms of the Mandelstam variable $t = p_1 - p_3$. For $e^-p \to e^-p$ scattering, $t$ is a function of the initial- and final-state electron four-momenta. Using the definitions of the particle four-momenta shown in Figure 3.8,

$$\begin{aligned} t = (p_1^* - p_3^*)^2 &= p_1^{*2} + p_3^{*2} - 2p_1^*\cdot p_3^* \\ &= m_1^2 + m_3^2 - 2(E_1^*E_3^* - \mathbf{p}_1^*\cdot\mathbf{p}_3^*) \\ &= m_1^2 + m_3^2 - 2E_1^*E_3^* + 2p_1^*p_3^*\cos\theta^*. \end{aligned} \tag{3.34}$$

In the centre-of-mass frame, the magnitude of the momenta and the energies of the final-state particles are fixed by energy and momentum conservation and the only free parameter in (3.34) is the electron scattering angle $\theta^*$, thus

$$dt = 2p_1^*p_3^*\,d(\cos\theta^*),$$

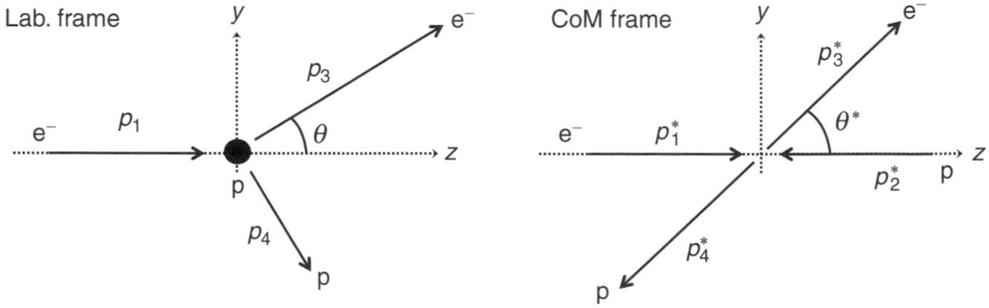


Fig. 3.8 The process of $e^-p \to e^-p$ elastic scattering shown in the laboratory (left) and centre-of-mass (right) frames.

and therefore

$$d\Omega^* \equiv d(\cos\theta^*)\, d\phi^* = \frac{dt\, d\phi^*}{2\mathrm{p}_1^*\mathrm{p}_3^*}. \tag{3.35}$$

Writing $\mathrm{p}_1^*$ and $\mathrm{p}_3^*$ respectively as $\mathrm{p}_i^*$ and $\mathrm{p}_f^*$, and substituting (3.35) into (3.32) leads to

$$d\sigma = \frac{1}{128\pi^2 s\, \mathrm{p}_i^{*2}} |\mathcal{M}_{fi}|^2 d\phi^* dt. \tag{3.36}$$

Assuming that matrix element is independent of the azimuthal angle, the integration over $d\phi^*$ just introduces a factor of $2\pi$ and therefore

$$\frac{d\sigma}{dt} = \frac{1}{64\pi s\, \mathrm{p}_i^{*2}} |\mathcal{M}_{fi}|^2. \tag{3.37}$$

The magnitude of the momentum of the initial-state particles in the centre-of-mass frame can be shown to be

$$\mathrm{p}_i^{*2} = \frac{1}{4s}[s - (m_1 + m_2)^2][s - (m_1 - m_2)^2]. \tag{3.38}$$

Since $\sigma$, $s$, $t$ and $|\mathcal{M}_{fi}|^2$ are all Lorentz-invariant quantities, Equation (3.37) gives a general Lorentz-invariant expression for the differential cross section for the two-body $\to$ two-body scattering process.

### 3.5.2 Laboratory frame differential cross section

Because (3.37) is valid in all rest frames, it can be applied directly to the example of $e^-p \to e^-p$ elastic scattering in the laboratory frame, shown in Figure 3.8. In the limit where the incident and scattered electron energies are much greater than the electron rest mass, the laboratory frame four-momenta of the particles are

$$\begin{aligned} p_1 &\approx (E_1, 0, 0, E_1), \\ p_2 &= (m_\mathrm{p}, 0, 0, 0), \\ p_3 &\approx (E_3, 0, E_3\sin\theta, E_3\cos\theta), \\ \text{and}\quad p_4 &= (E_4, \mathbf{p}_4). \end{aligned}$$

The momenta of the initial-state particles in the centre-of-mass frame are given by (3.38) and since $m_\mathrm{e} \ll m_\mathrm{p}$,

$$\mathrm{p}_i^{*2} \approx \frac{(s - m_\mathrm{p}^2)^2}{4s}, \tag{3.39}$$

where $s$ is given by

$$s = (p_1 + p_2)^2 = p_1^2 + p_2^2 + 2p_1{\cdot}p_2 \approx m_p^2 + 2p_1{\cdot}p_2$$
$$= m_p^2 + 2E_1 m_p,$$

and therefore

$$\mathrm{p}_i^{*2} = \frac{E_1^2 m_p^2}{s}. \tag{3.40}$$

The differential cross section in terms of the laboratory frame scattering angle of the electron can be obtained from

$$\frac{d\sigma}{d\Omega} = \frac{d\sigma}{dt}\left|\frac{dt}{d\Omega}\right| = \frac{1}{2\pi}\frac{dt}{d(\cos\theta)}\frac{d\sigma}{dt}, \tag{3.41}$$

where the factor $2\pi$ arises from the integral over $d\phi$ (again assuming azimuthal symmetry). An expression for $dt/d(\cos\theta)$ can be obtained by writing the Mandelstam variable $t$ in terms of the laboratory frame four-momenta, defined above,

$$t = (p_1 - p_3)^2 \approx -2E_1E_3(1 - \cos\theta), \tag{3.42}$$

where $E_3$ is itself a function of $\theta$. Conservation of energy and momentum imply that $p_1 + p_2 = p_3 + p_4$, and $t$ can also be expressed in terms of the four-momenta of the initial and final-state proton,

$$t = (p_2 - p_4)^2 = 2m_p^2 - 2p_2{\cdot}p_4 = 2m_p^2 - 2m_pE_4 = -2m_p(E_1 - E_3), \tag{3.43}$$

where the last step follows from energy conservation, $E_4 = E_1 + m_p - E_3$. Equating (3.42) and (3.43) gives the expression for $E_3$ as a function of $\cos\theta$,

$$E_3 = \frac{E_1 m_p}{m_p + E_1 - E_1\cos\theta}. \tag{3.44}$$

Because $E_1$ is the fixed energy of the initial-state electron, differentiating (3.43) with respect to $\cos\theta$ gives

$$\frac{dt}{d(\cos\theta)} = 2m_p\frac{dE_3}{d(\cos\theta)}. \tag{3.45}$$

Differentiating the expression for $E_3$ of (3.44), gives

$$\frac{dE_3}{d(\cos\theta)} = \frac{E_1^2 m_p}{(m_p + E_1 - E_1\cos\theta)^2} = \frac{E_3^2}{m_p},$$

which when substituted into (3.45) leads to

$$\frac{dt}{d(\cos\theta)} = 2E_3^2. \tag{3.46}$$

Substituting (3.46) into (3.41), and using the Lorentz-invariant expression for the differential cross section of (3.37) gives

$$\frac{d\sigma}{d\Omega} = \frac{1}{2\pi} 2E_3^2 \frac{d\sigma}{dt} = \frac{E_3^2}{64\pi^2 s\, p_i^{*2}} |\mathcal{M}_{fi}|^2.$$

The momentum of the intial-state particles in the centre-of-mass frame can be eliminated using (3.40) and thus

$$\frac{d\sigma}{d\Omega} = \frac{1}{64\pi^2} \left( \frac{E_3}{m_p E_1} \right)^2 |\mathcal{M}_{fi}|^2. \tag{3.47}$$

Finally, the energy of the scattered electron $E_3$ can be expressed in terms of $\cos\theta$ alone using (3.44). Therefore the differential cross section can be written as an explicit function of $\cos\theta$ and the energy of the incident electron

$$\frac{d\sigma}{d\Omega} = \frac{1}{64\pi^2} \left( \frac{1}{m_p + E_1 - E_1 \cos\theta} \right)^2 |\mathcal{M}_{fi}|^2. \tag{3.48}$$

The same calculation including the mass of the electron is algebraically more involved, although the steps are essentially the same.

## Summary

The general expression for the decay rate $a \to 1 + 2$ is

$$\Gamma = \frac{p^*}{32\pi^2 m_a^2} \int |\mathcal{M}_{fi}|^2 d\Omega, \tag{3.49}$$

where $p^*$ is the magnitude of the momentum of the final-state particles in the rest frame of the decaying particle, which is given by

$$p^* = \frac{1}{2m_i} \sqrt{\left[(m_i^2 - (m_1 + m_2)^2\right]\left[m_i^2 - (m_1 - m_2)^2\right]}.$$

The expression for the differential cross section for the process $a + b \to c + d$ in the centre-of-mass frame is

$$\frac{d\sigma}{d\Omega^*} = \frac{1}{64\pi^2 s} \frac{p_f^*}{p_i^*} |\mathcal{M}_{fi}|^2, \tag{3.50}$$

where $p_i^*$ and $p_f^*$ are respectively the magnitudes of the initial- and final-state momenta in the centre-of-mass frame. In the limit where the electron mass can be neglected, the differential cross section for $e^-p \to e^-p$ elastic scattering, in the proton rest frame is

$$\frac{d\sigma}{d\Omega} = \frac{1}{64\pi^2}\left(\frac{E_3}{m_p E_1}\right)^2 |\mathcal{M}_{fi}|^2, \tag{3.51}$$

where $E_3$ is a function of the electron scattering angle.

## Problems

**3.1** Calculate the energy of the $\mu^-$ produced in the decay at rest $\pi^- \to \mu^-\overline{\nu}_\mu$. Assume $m_\pi = 140$ GeV, $m_\mu = 106$ MeV and take $m_\nu \approx 0$.

**3.2** For the decay $a \to 1 + 2$, show that the momenta of both daughter particles in the centre-of-mass frame $p^*$ are

$$p^* = \frac{1}{2m_a}\sqrt{\left[(m_a^2 - (m_1 + m_2)^2\right]\left[m_a^2 - (m_1 - m_2)^2\right]}.$$

**3.3** Calculate the branching ratio for the decay $K^+ \to \pi^+\pi^0$, given the partial decay width $\Gamma(K^+ \to \pi^+\pi^0) = 1.2 \times 10^{-8}$ eV and the mean kaon lifetime $\tau(K^+) = 1.2 \times 10^{-8}$ s.

**3.4** At a future $e^+e^-$ linear collider operating as a Higgs factory at a centre-of-mass energy of $\sqrt{s} = 250$ GeV, the cross section for the process $e^+e^- \to HZ$ is 250 fb. If the collider has an instantaneous luminosity of $2 \times 10^{34}$ cm$^{-2}$ s$^{-1}$ and is operational for 50% of the time, how many Higgs bosons will be produced in five years of running?

Note: 1 femtobarn $\equiv 10^{-15}$ b.

**3.5** The total $e^+e^- \to \gamma \to \mu^+\mu^-$ annihilation cross section is $\sigma = 4\pi\alpha^2/3s$, where $\alpha \approx 1/137$. Calculate the cross section at $\sqrt{s} = 50$ GeV, expressing your answer in both natural units and in barns (1 barn $= 10^{-28}$ m$^2$). Compare this to the total pp cross section at $\sqrt{s} = 50$ GeV which is approximately 40 mb and comment on the result.

**3.6** A 1 GeV muon neutrino is fired at a 1 m thick block of iron ($^{56}_{26}Fe$) with density $\rho = 7.874 \times 10^3$ kg m$^{-3}$. If the average neutrino–nucleon interaction cross section is $\sigma = 8 \times 10^{-39}$ cm$^2$, calculate the (small) probability that the neutrino interacts in the block.

**3.7** For the process $a + b \to 1 + 2$ the Lorentz-invariant flux term is

$$F = 4\left[(p_a \cdot p_b)^2 - m_a^2 m_b^2\right]^{\frac{1}{2}}.$$

In the non-relativistic limit, $\beta_a \ll 1$ and $\beta_b \ll 1$, show that

$$F \approx 4m_a m_b\,|\mathbf{v}_a - \mathbf{v}_b|,$$

where $\mathbf{v}_a$ and $\mathbf{v}_b$ are the (non-relativistic) velocities of the two particles.

**3.8** The Lorentz-invariant flux term for the process $a + b \to 1 + 2$ in the centre-of-mass frame was shown to be $F = 4p_i^*\sqrt{s}$, where $p_i^*$ is the momentum of the intial-state particles. Show that the corresponding expression in the frame where $b$ is at rest is

$$F = 4m_b p_a.$$

**3.9** Show that the momentum in the centre-of-mass frame of the initial-state particles in a two-body scattering process can be expressed as

$$p_i^{*2} = \frac{1}{4s}[s - (m_1 + m_2)^2][s - (m_1 - m_2)^2].$$

**3.10** Repeat the calculation of Section 3.5.2 for the process $e^- p \to e^- p$ where the mass of the electron is no longer neglected.

(a) First show that

$$\frac{dE_3}{d(\cos\theta)} = \frac{p_1 p_3^2}{p_3(E_1 + m_p) - E_3 p_1 \cos\theta}.$$

(b) Then show that

$$\frac{d\sigma}{d\Omega} = \frac{1}{64\pi^2} \cdot \frac{p_3^2}{p_1 m_p} \cdot \frac{1}{p_3(E_1 + m_p) - E_3 p_1 \cos\theta} \cdot |\mathcal{M}_{fi}|^2,$$

where $(E_1, p_1)$ and $(E_3, p_3)$ are the respective energies and momenta of the initial-state and scattered electrons as measured in the laboratory frame.

# 4 The Dirac equation

This chapter provides an introduction to the Dirac equation, which is the relativistic formulation of quantum mechanics used to describe the fundamental fermions of the Standard Model. Particular emphasis is placed on the free-particle solutions to the Dirac equation that will be used to describe fermions in the calculations of cross sections and decay rates in the following chapters.

## 4.1 The Klein–Gordon equation

One of the requirements for a relativistic formulation of quantum mechanics is that the associated wave equation is Lorentz invariant. The Schrödinger equation, introduced in Section 2.3.1, is first order in the time derivative and second order in the spatial derivatives. Because of the different dependence on the time and space coordinates, the Schrödinger equation is clearly not Lorentz invariant, and therefore cannot provide a description of relativistic particles. The non-invariance of the Schrödinger equation under Lorentz transformations is a consequence its construction from the non-relativistic relationship between the energy of a free particle and its momentum

$$E = \frac{\mathbf{p}^2}{2m}.$$

The first attempt at constructing a relativistic theory of quantum mechanics was based on the Klein–Gordon equation. The Klein–Gordon wave equation is obtained by writing the Einstein energy–momentum relationship,

$$E^2 = \mathbf{p}^2 + m^2,$$

in the form of operators acting on a wavefunction,

$$\hat{E}^2\psi(\mathbf{x},t) = \hat{\mathbf{p}}^2\psi(\mathbf{x},t) + m^2\psi(\mathbf{x},t).$$

Using the energy and momentum operators identified in Section 2.3.1,

$$\hat{\mathbf{p}} = -i\boldsymbol{\nabla} \quad \text{and} \quad \hat{E} = i\frac{\partial}{\partial t},$$

this leads to the Klein–Gordon wave equation,

$$\frac{\partial^2 \psi}{\partial t^2} = \boldsymbol{\nabla}^2 \psi - m^2 \psi. \tag{4.1}$$

The Klein–Gordon equation, which is second order in both space and time derivatives, can be expressed in the manifestly Lorentz-invariant form

$$(\partial^\mu \partial_\mu + m^2)\psi = 0, \tag{4.2}$$

where

$$\partial^\mu \partial_\mu \equiv \frac{\partial^2}{\partial t^2} - \frac{\partial^2}{\partial x^2} - \frac{\partial^2}{\partial y^2} - \frac{\partial^2}{\partial z^2},$$

is the Lorentz-invariant scalar product of two four-vectors.

The Klein–Gordon equation has plane wave solutions,

$$\psi(\mathbf{x}, t) = N e^{i(\mathbf{p}\cdot\mathbf{x} - Et)}, \tag{4.3}$$

which when substituted into (4.2) imply that

$$E^2 \psi = \mathrm{p}^2 \psi + m^2 \psi,$$

and thus (by construction) the plane wave solutions to the Klein–Gordon equation satisfy the Einstein energy–momentum relationship, where the energy of the particle is related to its momentum by

$$E = \pm\sqrt{\mathrm{p}^2 + m^2}.$$

In classical mechanics, the negative energy solutions can be dismissed as being unphysical. However, in quantum mechanics all solutions are required to form a complete set of states, and the negative energy solutions simply cannot be discarded. Whilst it is not clear how the negative energy solutions should be interpreted, there is a more serious problem with the associated probability densities. The expressions for the probability density and probability current for the Klein–Gordon equation can be identified following the procedure used in Section 2.3.2. Taking the difference $\psi^* \times (4.1) - \psi \times (4.1)^*$ gives

$$\psi^* \frac{\partial^2 \psi}{\partial t^2} - \psi \frac{\partial^2 \psi^*}{\partial t^2} = \psi^*(\nabla^2 \psi - m^2 \psi) - \psi(\nabla^2 \psi^* - m^2 \psi^*)$$

$$\Rightarrow \quad \frac{\partial}{\partial t}\left(\psi^* \frac{\partial \psi}{\partial t} - \psi \frac{\partial \psi^*}{\partial t}\right) = \boldsymbol{\nabla} \cdot (\psi^* \boldsymbol{\nabla} \psi - \psi \boldsymbol{\nabla} \psi^*). \tag{4.4}$$

Comparison with the continuity equation of (2.20) leads to the identification of the probability density and probability current for solutions to the Klein–Gordon equation as

$$\rho = i\left(\psi^* \frac{\partial \psi}{\partial t} - \psi \frac{\partial \psi^*}{\partial t}\right) \quad \text{and} \quad \mathbf{j} = -i(\psi^* \boldsymbol{\nabla} \psi - \psi \boldsymbol{\nabla} \psi^*), \tag{4.5}$$

where the factor of $i$ is included to ensure that the probability density is real. For a plane wave solution, the probability density and current are

$$\rho = 2|N|^2 E \quad \text{and} \quad \mathbf{j} = 2|N|^2 \mathbf{p},$$

which can be written as a four-vector $j^{\mu}_{KG} = 2|N|^2 p^{\mu}$. The probability density is proportional to the energy of the particle, which is consistent with the discussion of relativistic length contraction of Section 3.2.1. However, this implies that the negative energy solutions have unphysical negative probability densities. From the presence of negative probability density solutions, it can be concluded that the Klein–Gordon equation does not provide a consistent description of single particle states for a relativistic system. It should be noted that this problem does not exist in quantum field theory, where the Klein–Gordon equation is used to describe *multi-particle* excitations of a spin-0 quantum field.

## 4.2 The Dirac equation

The apparent problems with the Klein–Gordon equation led Dirac (1928) to search for an alternative formulation of relativistic quantum mechanics. The resulting wave equation not only solved the problem of negative probability densities, but also provided a natural description of the intrinsic spin and magnetic moments of spin-half fermions. Its development represents one of the great theoretical breakthroughs of the twentieth century.

The requirement that relativistic particles satisfy $E^2 = \mathbf{p}^2 + m^2$ results in the Klein–Gordon wave equation being second order in the derivatives. Dirac looked for a wave equation that was first order in both space and time derivatives,

$$\hat{E}\psi = (\boldsymbol{\alpha} \cdot \hat{\mathbf{p}} + \beta m)\psi, \tag{4.6}$$

which in terms of the energy and momentum operators can be written

$$i\frac{\partial}{\partial t}\psi = \left(-i\alpha_x\frac{\partial}{\partial x} - i\alpha_y\frac{\partial}{\partial y} - i\alpha_z\frac{\partial}{\partial z} + \beta m\right)\psi. \tag{4.7}$$

If the solutions of (4.7) are to represent relativistic particles, they must *also* satisfy the Einstein energy–momentum relationship, which implies they satisfy the Klein–Gordon equation. This requirement places strong constraints on the possible nature of the constants $\boldsymbol{\alpha}$ and $\beta$ in (4.6). The conditions satisfied by $\boldsymbol{\alpha}$ and $\beta$ can be obtained by "squaring" (4.7) to give

$$-\frac{\partial^2\psi}{\partial t^2} = \left(i\alpha_x\frac{\partial}{\partial x} + i\alpha_y\frac{\partial}{\partial y} + i\alpha_z\frac{\partial}{\partial z} - \beta m\right)\left(i\alpha_x\frac{\partial}{\partial x} + i\alpha_y\frac{\partial}{\partial y} + i\alpha_z\frac{\partial}{\partial z} - \beta m\right)\psi,$$

which, when written out in gory detail, is

$$\begin{aligned}\frac{\partial^2\psi}{\partial t^2} &= \alpha_x^2\frac{\partial^2\psi}{\partial x^2} + \alpha_y^2\frac{\partial^2\psi}{\partial y^2} + \alpha_z^2\frac{\partial^2\psi}{\partial z^2} - \beta^2 m^2\psi \\ &+ (\alpha_x\alpha_y + \alpha_y\alpha_x)\frac{\partial^2\psi}{\partial x\partial y} + (\alpha_y\alpha_z + \alpha_z\alpha_y)\frac{\partial^2\psi}{\partial y\partial z} + (\alpha_z\alpha_x + \alpha_x\alpha_z)\frac{\partial^2\psi}{\partial z\partial x} \\ &+ i(\alpha_x\beta + \beta\alpha_x)m\frac{\partial\psi}{\partial x} + i(\alpha_y\beta + \beta\alpha_y)m\frac{\partial\psi}{\partial y} + i(\alpha_z\beta + \beta\alpha_z)m\frac{\partial\psi}{\partial z}. \end{aligned} \tag{4.8}$$

In order for (4.8) to reduce to the Klein–Gordon equation,

$$\frac{\partial^2\psi}{\partial t^2} = \frac{\partial^2\psi}{\partial x^2} + \frac{\partial^2\psi}{\partial y^2} + \frac{\partial^2\psi}{\partial z^2} - m^2\psi,$$

the coefficients $\boldsymbol{\alpha}$ and $\beta$ must satisfy

$$\alpha_x^2 = \alpha_y^2 = \alpha_z^2 = \beta^2 = I, \tag{4.9}$$

$$\alpha_j\beta + \beta\alpha_j = 0, \tag{4.10}$$

$$\alpha_j\alpha_k + \alpha_k\alpha_j = 0 \quad (j \neq k), \tag{4.11}$$

where $I$ represents unity. The anticommutation relations of (4.10) and (4.11) cannot be satisfied if the $\alpha_i$ and $\beta$ are normal numbers. The simplest mathematical objects that can satisfy these anticommutation relations are matrices. From the cyclic property of traces, $\text{Tr}(ABC) = \text{Tr}(BCA)$, and the requirements that $\beta^2 = I$ and $\alpha_i\beta = -\beta\alpha_i$, it is straightforward to show that the $\alpha_i$ and $\beta$ matrices must have trace zero:

$$\text{Tr}(\alpha_i) = \text{Tr}(\alpha_i\beta\beta) = \text{Tr}(\beta\alpha_i\beta) = -\text{Tr}(\alpha_i\beta\beta) = -\text{Tr}(\alpha_i).$$

Furthermore, it can be shown that the eigenvalues of the $\alpha_i$ and $\beta$ matrices are $\pm 1$. This follows from multiplying the eigenvalue equation,

$$\alpha_i X = \lambda X,$$

by $\alpha_i$ and using $\alpha_i^2 = I$, which implies

$$\alpha_i^2 X = \lambda\alpha_i X \quad \Rightarrow \quad X = \lambda^2 X,$$

and therefore $\lambda = \pm 1$. Because the sum of the eigenvalues of a matrix is equal to its trace, and here the matrices have eigenvalues of either $+1$ or $-1$, the only way the trace can be zero is if the $\alpha_i$ and $\beta$ matrices are of even dimension. Finally, because the Dirac Hamiltonian operator of (4.6),

$$\hat{H}_D = (\boldsymbol{\alpha}\cdot\hat{\mathbf{p}} + \beta m),$$

must be Hermitian in order to have real eigenvalues, the $\alpha$ and $\beta$ matrices also must be Hermitian,

$$\alpha_x = \alpha_x^\dagger, \quad \alpha_y = \alpha_y^\dagger, \quad \alpha_z = \alpha_z^\dagger \quad \text{and} \quad \beta = \beta^\dagger. \tag{4.12}$$

Hence $\alpha_x$, $\alpha_y$, $\alpha_z$ and $\beta$ are four mutually anticommuting Hermitian matrices of even dimension and trace zero. Because there are only three mutually anticommuting $2 \times 2$ traceless matrices, for example the Pauli spin-matrices, the lowest dimension object that can represent $\alpha_x$, $\alpha_y$, $\alpha_z$ and $\beta$ are $4 \times 4$ matrices. Therefore, the Dirac Hamiltonian of (4.6) is a $4 \times 4$ matrix of operators that must act on a four-component wavefunction, known as a *Dirac spinor*,

$$\psi = \begin{pmatrix} \psi_1 \\ \psi_2 \\ \psi_3 \\ \psi_4 \end{pmatrix}.$$

The consequence of requiring the quantum-mechanical wavefunctions for a relativistic particle satisfy the Dirac equation and be consistent with the Klein–Gordon equation, is that the wavefunctions are forced to have four degrees of freedom. Before leaving this point, it is worth noting that, if all particles were massless, there would be no need for the $\beta$ term in (4.7) and the $\alpha$ matrices could be represented by the three Pauli spin-matrices. In this Universe without mass, it would be possible to describe a particle by a two-component object, known as a Weyl spinor.

The algebra of the Dirac equation is fully defined by the relations of (4.9)–(4.11) and (4.12). Nevertheless, it is convenient to introduce an explicit form for $\alpha_x$, $\alpha_y$, $\alpha_z$ and $\beta$. The conventional choice is the *Dirac–Pauli representation*, based on the familiar Pauli spin-matrices,

$$\beta = \begin{pmatrix} I & 0 \\ 0 & -I \end{pmatrix} \quad \text{and} \quad \alpha_i = \begin{pmatrix} 0 & \sigma_i \\ \sigma_i & 0 \end{pmatrix}, \tag{4.13}$$

with

$$I = \begin{pmatrix} 1 & 0 \\ 0 & 1 \end{pmatrix}, \quad \sigma_x = \begin{pmatrix} 0 & 1 \\ 1 & 0 \end{pmatrix}, \quad \sigma_y = \begin{pmatrix} 0 & -i \\ i & 0 \end{pmatrix} \quad \text{and} \quad \sigma_z = \begin{pmatrix} 1 & 0 \\ 0 & -1 \end{pmatrix}.$$

This is only one possible representation of the $\alpha$ and $\beta$ matrices. The matrices $\alpha_i' = U\alpha_i U^{-1}$ and $\beta = U\beta U^{-1}$, generated by any $4 \times 4$ unitary matrix $U$, are Hermitian and also satisfy the necessary anticommutation relations. The physical predictions obtained from the Dirac equation will not depend on the specific representation used; the physics of the Dirac equation is defined by the algebra satisfied by $\alpha_x$, $\alpha_y$, $\alpha_z$ and $\beta$, not by the specific representation.

## 4.3 Probability density and probability current

The expressions for the probability density and probability current for solutions of the Dirac equation can be obtained following a similar procedure to that used for the Schrödinger and Klein–Gordon equations. Since the wavefunctions are now four-component spinors, the complex conjugates of wavefunctions have to be replaced by Hermitian conjugates, $\psi^* \to \psi^\dagger = (\psi^*)^T$. The Hermitian conjugate of the Dirac equation,

$$-i\alpha_x\frac{\partial\psi}{\partial x} - i\alpha_y\frac{\partial\psi}{\partial y} - i\alpha_z\frac{\partial\psi}{\partial z} + m\beta\psi = +i\frac{\partial\psi}{\partial t}, \tag{4.14}$$

is simply

$$+i\frac{\partial\psi^\dagger}{\partial x}\alpha_x^\dagger + i\frac{\partial\psi^\dagger}{\partial y}\alpha_y^\dagger + i\frac{\partial\psi^\dagger}{\partial z}\alpha_z^\dagger + m\psi^\dagger\beta^\dagger = -i\frac{\partial\psi^\dagger}{\partial t}. \tag{4.15}$$

Using the fact that the $\alpha$ and $\beta$ matrices are Hermitian, the combination of $\psi^\dagger \times$ (4.14) – (4.15) $\times\, \psi$ gives

$$\begin{aligned} &\psi^\dagger\left(-i\alpha_x\frac{\partial\psi}{\partial x} - i\alpha_y\frac{\partial\psi}{\partial y} - i\alpha_z\frac{\partial\psi}{\partial z} + \beta m\psi\right) \\ &-\left(i\frac{\partial\psi^\dagger}{\partial x}\alpha_x + i\frac{\partial\psi^\dagger}{\partial y}\alpha_y + i\frac{\partial\psi^\dagger}{\partial z}\alpha_z + m\psi^\dagger\beta\right)\psi = i\psi^\dagger\frac{\partial\psi}{\partial t} + i\frac{\partial\psi^\dagger}{\partial t}\psi. \end{aligned} \tag{4.16}$$

Equation (4.16) can be simplified by writing

$$\psi^\dagger\alpha_x\frac{\partial\psi}{\partial x} + \frac{\partial\psi^\dagger}{\partial x}\alpha_x\psi \equiv \frac{\partial(\psi^\dagger\alpha_x\psi)}{\partial x} \quad \text{and} \quad \psi^\dagger\frac{\partial\psi}{\partial t} + \frac{\partial\psi^\dagger}{\partial t}\psi \equiv \frac{\partial(\psi^\dagger\psi)}{\partial t},$$

giving

$$\boldsymbol{\nabla}\cdot(\psi^\dagger\boldsymbol{\alpha}\psi) + \frac{\partial(\psi^\dagger\psi)}{\partial t} = 0,$$

where $\psi^\dagger = (\psi_1^*, \psi_2^*, \psi_3^*, \psi_4^*)$. By comparison with the continuity equation of (2.20), the probability density and probability current for solutions of the Dirac equation can be identified as

$$\rho = \psi^\dagger\psi \quad \text{and} \quad \mathbf{j} = \psi^\dagger\boldsymbol{\alpha}\psi. \tag{4.17}$$

In terms of the four components of the Dirac spinors, the probability density is

$$\rho = \psi^\dagger\psi = |\psi_1|^2 + |\psi_2|^2 + |\psi_3|^2 + |\psi_4|^2,$$

and thus, all solutions of the Dirac equation have positive probability density. By requiring that the wavefunctions satisfy a wave equation linear in both space and time derivatives, in addition to being solutions of the Klein–Gordon equation, Dirac solved the perceived problem with negative probability densities. The price is that particles now have to be described by four-component wavefunctions. The Dirac equation could have turned out to be a purely mathematical construction without physical relevance. However, remarkably, it can be shown that the additional degrees of freedom of the four-component wavefunctions naturally describe the intrinsic angular momentum of spin-half particles and antiparticles. The proof that the Dirac equation provides a natural description of spin-half particles is given in the following starred section. It is fairly involved and the details are not essential to understand the material that follows.

## 4.4 *Spin and the Dirac equation

In quantum mechanics, the time dependence of an observable corresponding to an operator $\hat{O}$ is given by (2.29),

$$\frac{\mathrm{d}O}{\mathrm{d}t} = \frac{\mathrm{d}}{\mathrm{d}t}\langle\hat{O}\rangle = i\langle\psi|[\hat{H},\hat{O}]|\psi\rangle.$$

Therefore, if the operator for an observable commutes with the Hamiltonian of the system, it is a constant of the motion. The Hamiltonian of the free-particle Schrödinger equation,

$$\hat{H}_{SE} = \frac{\hat{\mathbf{p}}^2}{2m},$$

commutes with the angular momentum operator $\hat{\mathbf{L}} = \hat{\mathbf{r}} \times \hat{\mathbf{p}}$, and thus angular momentum is a conserved quantity in non-relativistic quantum mechanics. For the free-particle Hamiltonian of the Dirac equation,

$$\hat{H}_D = \boldsymbol{\alpha} \cdot \hat{\mathbf{p}} + \beta m, \tag{4.18}$$

the corresponding commutation relation is

$$[\hat{H}_D, \hat{\mathbf{L}}] = [\alpha \cdot \hat{\mathbf{p}} + \beta m, \hat{\mathbf{r}} \times \hat{\mathbf{p}}] = [\alpha \cdot \hat{\mathbf{p}}, \hat{\mathbf{r}} \times \hat{\mathbf{p}}]. \tag{4.19}$$

This can be evaluated by considering the commutation relation for a particular component of $\hat{\mathbf{L}}$, for example

$$[\hat{H}_D, \hat{L}_x] = [\boldsymbol{\alpha} \cdot \hat{\mathbf{p}}, (\hat{\mathbf{r}} \times \hat{\mathbf{p}})_x] = [\alpha_x \hat{p}_x + \alpha_y \hat{p}_y + \alpha_z \hat{p}_z,\ \hat{y}\hat{p}_z - \hat{z}\hat{p}_y]. \tag{4.20}$$

The only terms in (4.20) that are non-zero arise from the non-zero position–momentum commutation relations

$$[\hat{x}, \hat{p}_x] = [\hat{y}, \hat{p}_y] = [\hat{z}, \hat{p}_z] = i,$$

giving

$$\begin{aligned}[\hat{H}_D, \hat{L}_x] &= \alpha_y[\hat{p}_y, \hat{y}]\hat{p}_z - \alpha_z[\hat{p}_z, \hat{z}]\hat{p}_y \\ &= -i(\alpha_y\hat{p}_z - \alpha_z\hat{p}_y) \\ &= -i(\boldsymbol{\alpha} \times \hat{\mathbf{p}})_x,\end{aligned}$$

where $(\boldsymbol{\alpha} \times \hat{\mathbf{p}})_x$ is the $x$-component of $\boldsymbol{\alpha} \times \hat{\mathbf{p}}$. Generalising this result to the other components of $\hat{\mathbf{L}}$ gives

$$[\hat{H}_D, \hat{\mathbf{L}}] = -i\boldsymbol{\alpha} \times \hat{\mathbf{p}}. \tag{4.21}$$

Hence, for a particle satisfying the Dirac equation, the "orbital" angular momentum operator $\hat{\mathbf{L}}$ does not commute with the Dirac Hamiltonian, and therefore does not correspond to a conserved quantity.

Now consider the $4 \times 4$ matrix operator $\hat{\mathbf{S}}$ formed from the Pauli spin-matrices

$$\hat{\mathbf{S}} \equiv \tfrac{1}{2}\hat{\boldsymbol{\Sigma}} \equiv \frac{1}{2}\begin{pmatrix} \boldsymbol{\sigma} & 0 \\ 0 & \boldsymbol{\sigma} \end{pmatrix}, \tag{4.22}$$

with

$$\hat{\Sigma}_x = \begin{pmatrix} 0 & 1 & 0 & 0 \\ 1 & 0 & 0 & 0 \\ 0 & 0 & 0 & 1 \\ 0 & 0 & 1 & 0 \end{pmatrix}, \quad \hat{\Sigma}_y = \begin{pmatrix} 0 & -i & 0 & 0 \\ i & 0 & 0 & 0 \\ 0 & 0 & 0 & -i \\ 0 & 0 & i & 0 \end{pmatrix} \quad \text{and} \quad \hat{\Sigma}_z = \begin{pmatrix} 1 & 0 & 0 & 0 \\ 0 & -1 & 0 & 0 \\ 0 & 0 & 1 & 0 \\ 0 & 0 & 0 & -1 \end{pmatrix}.$$

Because the $\boldsymbol{\alpha}$-matrices in the Dirac–Pauli representation and the $\Sigma$-matrices are both derived from the Pauli spin-matrices, they have well-defined commutation relations. Consequently, the commutator $[\alpha_i, \hat{\Sigma}_x]$ can be expressed in terms of the commutators of the Pauli spin-matrices. Writing the $4 \times 4$ matrices in $2 \times 2$ block form,

$$\begin{aligned}[\alpha_i, \hat{\Sigma}_x] &= \begin{pmatrix} 0 & \sigma_i \\ \sigma_i & 0 \end{pmatrix}\begin{pmatrix} \sigma_x & 0 \\ 0 & \sigma_x \end{pmatrix} - \begin{pmatrix} \sigma_x & 0 \\ 0 & \sigma_x \end{pmatrix}\begin{pmatrix} 0 & \sigma_i \\ \sigma_i & 0 \end{pmatrix} \\ &= \begin{pmatrix} 0 & [\sigma_i, \sigma_x] \\ [\sigma_i, \sigma_x] & 0 \end{pmatrix}.\end{aligned} \tag{4.23}$$

The commutation relations,

$$[\sigma_x, \sigma_x] = 0, \quad [\sigma_y, \sigma_x] = -2i\sigma_z \quad \text{and} \quad [\sigma_z, \sigma_x] = 2i\sigma_y,$$

imply that (4.23) is equivalent to

$$[\alpha_x, \Sigma_x] = 0, \tag{4.24}$$

$$[\alpha_y, \Sigma_x] = \begin{pmatrix} 0 & -2i\sigma_z \\ -2i\sigma_z & 0 \end{pmatrix} = -2i\alpha_z, \tag{4.25}$$

$$[\alpha_z, \Sigma_x] = \begin{pmatrix} 0 & 2i\sigma_y \\ 2i\sigma_y & 0 \end{pmatrix} = 2i\alpha_y. \tag{4.26}$$

Now consider the commutator of $\hat{\Sigma}_x$ with the Dirac Hamiltonian

$$[\hat{H}_D, \Sigma_x] = [\boldsymbol{\alpha} \cdot \hat{\mathbf{p}} + \beta m, \Sigma_x].$$

It is straightforward to show that $[\beta, \hat{\Sigma}_x] = 0$ and hence

$$\begin{aligned}[\hat{H}_D, \hat{\Sigma}_x] &= [\boldsymbol{\alpha} \cdot \hat{\mathbf{p}}, \hat{\Sigma}_x] = [\alpha_x \hat{p}_x + \alpha_y \hat{p}_y + \alpha_z \hat{p}_z, \hat{\Sigma}_x] \\ &= \hat{p}_x[\alpha_x, \hat{\Sigma}_x] + \hat{p}_y[\alpha_y, \hat{\Sigma}_x] + \hat{p}_z[\alpha_z, \hat{\Sigma}_x]. \end{aligned} \tag{4.27}$$

Using the commutation relations of (4.24)–(4.26) implies that

$$\begin{aligned}[\hat{H}_D, \hat{\Sigma}_x] &= -2i\hat{p}_y \alpha_z + 2i\hat{p}_z \alpha_y \\ &= 2i(\boldsymbol{\alpha} \times \hat{\mathbf{p}})_x. \end{aligned}$$

Generalising this derivation to the $y$ and $z$ components of $[\hat{H}_D, \hat{\boldsymbol{\Sigma}}]$ and using $\hat{\mathbf{S}} = \frac{1}{2}\hat{\boldsymbol{\Sigma}}$ gives the result

$$[\hat{H}_D, \hat{\mathbf{S}}] = i\,\boldsymbol{\alpha} \times \hat{\mathbf{p}}. \tag{4.28}$$

Because $\hat{\mathbf{S}}$ does not commute with the Dirac Hamiltonian, the corresponding observable is not a conserved quantity. However, from (4.21) and (4.28) it can be seen that the sum $\hat{\mathbf{J}} = \hat{\mathbf{L}} + \hat{\mathbf{S}}$ commutes with the Hamiltonian of the Dirac equation,

$$\left[\hat{H}_D, \hat{\mathbf{J}}\right] \equiv \left[\hat{H}_D, \hat{\mathbf{L}} + \hat{\mathbf{S}}\right] = 0.$$

Hence $\hat{\mathbf{S}}$ can be identified as the operator for the intrinsic angular momentum (the spin) of a particle. The total angular momentum of the particle, associated with the operator $\hat{\mathbf{J}} = \hat{\mathbf{L}} + \hat{\mathbf{S}}$, is a conserved quantity.

Because the $4 \times 4$ matrix operator $\hat{\mathbf{S}}$ is defined in terms of the Pauli spin-matrices,

$$\hat{\mathbf{S}} = \tfrac{1}{2}\hat{\boldsymbol{\Sigma}} = \frac{1}{2}\begin{pmatrix} \boldsymbol{\sigma} & 0 \\ 0 & \boldsymbol{\sigma} \end{pmatrix}, \tag{4.29}$$

its components have the same commutation relations as the Pauli spin-matrices, for example $[\hat{S}_x, \hat{S}_y] = i\hat{S}_z$. These are the same commutation relations satisfied by the operators for orbital angular momentum, $[\hat{L}_x, \hat{L}_y] = i\hat{L}_z$, etc. Therefore, from the arguments of Section 2.3.5, it follows that spin is quantised in exactly the same way as orbital angular momentum. Consequently, the total spin $s$ can be identified from the eigenvalue of the operator,

$$\hat{\mathbf{S}}^2 = \frac{1}{4}(\hat{\Sigma}_x^2 + \hat{\Sigma}_y^2 + \hat{\Sigma}_z^2) = \frac{3}{4}\begin{pmatrix} 1 & 0 & 0 & 0 \\ 0 & 1 & 0 & 0 \\ 0 & 0 & 1 & 0 \\ 0 & 0 & 0 & 1 \end{pmatrix},$$

for which $\hat{\mathbf{S}}^2|s, m_s\rangle = s(s+1)|s, m_s\rangle$. Hence, for any Dirac spinor $\psi$,

$$\hat{\mathbf{S}}^2\psi = s(s+1)\psi = \tfrac{3}{4}\psi,$$

and thus a particle satisfying the Dirac equation has intrinsic angular momentum $s = \frac{1}{2}$. Furthermore, it can be shown (see Appendix B.1) that the operator $\hat{\mu}$ giving the intrinsic magnetic moment of a particle satisfying the Dirac equation is given by

$$\hat{\mu} = \frac{q}{m}\hat{\mathbf{S}}, \tag{4.30}$$

where $q$ and $m$ are respectively the charge and mass of the particle. Hence $\hat{\mathbf{S}}$ has all the properties of the quantum-mechanical spin operator for a Dirac spinor. The Dirac equation therefore provides a natural description of spin-half particles. This is a profound result, spin emerges as a direct consequence of requiring the wavefunction to satisfy the Dirac equation.

## 4.5 Covariant form of the Dirac equation

Up to this point the Dirac equation has been expressed in terms of the $\alpha$- and $\beta$-matrices. This naturally brings out the connection with spin. However, the Dirac equation is usually expressed in the form which emphasises its covariance. This is achieved by first pre-multiplying the Dirac equation of (4.7) by $\beta$ to give

$$i\beta\alpha_x\frac{\partial\psi}{\partial x} + i\beta\alpha_y\frac{\partial\psi}{\partial y} + i\beta\alpha_z\frac{\partial\psi}{\partial z} + i\beta\frac{\partial\psi}{\partial t} - \beta^2 m\psi = 0. \tag{4.31}$$

By defining the four Dirac $\gamma$-matrices as

$$\gamma^0 \equiv \beta, \quad \gamma^1 \equiv \beta\alpha_x, \quad \gamma^2 \equiv \beta\alpha_y \quad \text{and} \quad \gamma^3 \equiv \beta\alpha_z,$$

and using $\beta^2 = I$, equation (4.31) becomes

$$i\gamma^0\frac{\partial\psi}{\partial t} + i\gamma^1\frac{\partial\psi}{\partial x} + i\gamma^2\frac{\partial\psi}{\partial y} + i\gamma^3\frac{\partial\psi}{\partial z} - m\psi = 0.$$

By labelling the four $\gamma$-matrices by the index $\mu$, such that $\gamma^\mu = (\gamma^0, \gamma^1, \gamma^2, \gamma^3)$, and using the definition of the covariant four-derivative

$$\partial_\mu \equiv (\partial_0, \partial_1, \partial_2, \partial_3) \equiv \left(\frac{\partial}{\partial t}, \frac{\partial}{\partial x}, \frac{\partial}{\partial y}, \frac{\partial}{\partial z}\right),$$

the Dirac equation can be expressed in the covariant form

$$(i\gamma^\mu\partial_\mu - m)\psi = 0, \tag{4.32}$$

with the index $\mu$ being *treated* as the Lorentz index of a four-vector and, as usual, summation over repeated indices is implied. Despite the suggestive way in which (4.32) is written, it is important to realise that the Dirac $\gamma$-matrices are not four-vectors; they are constant matrices which are invariant under Lorentz transformations. Hence, the Lorentz covariance of the Dirac equation, which means that it applies in all rest frames, is not immediately obvious from Equation (4.32). The proof of the covariance of the Dirac equation and the derivation of the Lorentz transformation properties of Dirac spinors is quite involved and is deferred to Appendix B.2.

The properties of the $\gamma$-matrices can be obtained from the properties of the $\alpha$- and $\beta$-matrices given in (4.9), (4.11) and (4.12). For example, using $\beta^2 = I$, $\alpha_x^2 = I$ and $\beta\alpha_x = -\alpha_x\beta$, it follows that

$$(\gamma^1)^2 = \beta\alpha_x\beta\alpha_x = -\alpha_x\beta\beta\alpha_x = -\alpha_x^2 = -I.$$

Similarly, it is straightforward to show that the products of two $\gamma$-matrices satisfy

$$\begin{aligned} (\gamma^0)^2 &= I, \\ (\gamma^k)^2 &= -I, \\ \text{and} \quad \gamma^\mu\gamma^\nu &= -\gamma^\nu\gamma^\mu \quad \text{for} \quad \mu \neq \nu, \end{aligned}$$

where the convention used here is that the index $k = 1$, 2 or 3. The above expressions can be written succinctly as the anticommutation relation

$$\{\gamma^\mu, \gamma^\nu\} \equiv \gamma^\mu\gamma^\nu + \gamma^\nu\gamma^\mu = 2g^{\mu\nu}. \tag{4.33}$$

The $\gamma^0$ matrix, which is equivalent to $\beta$, is Hermitian and it is straightforward to show that the other three gamma matrices are anti-Hermitian, for example,

$$\gamma^{1\dagger} = (\beta\alpha_x)^\dagger = \alpha_x^\dagger\beta^\dagger = \alpha_x\beta = -\beta\alpha_x = -\gamma^1,$$

and hence

$$\gamma^{0\dagger} = \gamma^0 \quad \text{and} \quad \gamma^{k\dagger} = -\gamma^k. \tag{4.34}$$

Equations (4.33) and (4.34) fully define the algebra of the $\gamma$-matrices, which in itself is sufficient to define the properties of the solutions of the Dirac equation. Nevertheless, from a practical and pedagogical perspective, it is convenient to consider a particular representation of the $\gamma$-matrices. In the Dirac–Pauli representation, the $\gamma$-matrices are

$$\gamma^0 = \beta = \begin{pmatrix} I & 0 \\ 0 & -I \end{pmatrix} \quad \text{and} \quad \gamma^k = \beta\alpha_k = \begin{pmatrix} 0 & \sigma_k \\ -\sigma_k & 0 \end{pmatrix},$$

where the $\alpha$- and $\beta$-matrices are those defined previously. Hence in the Dirac–Pauli representation,

$$\gamma^0 = \begin{pmatrix} 1 & 0 & 0 & 0 \\ 0 & 1 & 0 & 0 \\ 0 & 0 & -1 & 0 \\ 0 & 0 & 0 & -1 \end{pmatrix}, \quad \gamma^1 = \begin{pmatrix} 0 & 0 & 0 & 1 \\ 0 & 0 & 1 & 0 \\ 0 & -1 & 0 & 0 \\ -1 & 0 & 0 & 0 \end{pmatrix},$$
$$\gamma^2 = \begin{pmatrix} 0 & 0 & 0 & -i \\ 0 & 0 & i & 0 \\ 0 & i & 0 & 0 \\ -i & 0 & 0 & 0 \end{pmatrix}, \quad \gamma^3 = \begin{pmatrix} 0 & 0 & 1 & 0 \\ 0 & 0 & 0 & -1 \\ -1 & 0 & 0 & 0 \\ 0 & 1 & 0 & 0 \end{pmatrix}. \tag{4.35}$$

### 4.5.1 The adjoint spinor and the covariant current

In Section 4.3, it was shown that the probability density and the probability current for a wavefunction satisfying the Dirac equation are respectively given by $\rho = \psi^\dagger\psi$ and $\mathbf{j} = \psi^\dagger\boldsymbol{\alpha}\psi$. These two expressions can be written compactly as

$$j^\mu = (\rho, \mathbf{j}) = \psi^\dagger\gamma^0\gamma^\mu\psi, \tag{4.36}$$

which follows from $(\gamma^0)^2 = 1$ and $\gamma^0\gamma^k = \beta\beta\alpha_k = \alpha_k$. By considering the Lorentz transformation properties of the four components of $j^\mu$, as defined in (4.36), it can be shown (see Appendix B.3) that $j^\mu$ is a four-vector. Therefore, the continuity equation (2.20), which expresses the conservation of particle probability,

$$\frac{\partial\rho}{\partial t} + \boldsymbol{\nabla}\cdot\mathbf{j} = 0,$$

can be written in the manifestly Lorentz-invariant form of a four-vector scalar product

$$\partial_\mu j^\mu = 0.$$

The expression for the four-vector current, $j^\mu = \psi^\dagger\gamma^0\gamma^\mu\psi$, can be simplified by introducing the *adjoint spinor* $\overline{\psi}$, defined as

$$\overline{\psi} \equiv \psi^\dagger\gamma^0.$$

The definition of the adjoint spinor allows the four-vector current $j^\mu$ to be written compactly as

$$j^\mu = \overline{\psi}\gamma^\mu\psi. \tag{4.37}$$

For completeness, it is noted that in the Dirac–Pauli representation of the $\gamma$-matrices, the adjoint spinor is simply

$$\overline{\psi} = \psi^\dagger\gamma^0 = (\psi^*)^T\gamma^0 = (\psi_1^*, \psi_2^*, \psi_3^*, \psi_4^*)\begin{pmatrix} 1 & 0 & 0 & 0 \\ 0 & 1 & 0 & 0 \\ 0 & 0 & -1 & 0 \\ 0 & 0 & 0 & -1 \end{pmatrix} = (\psi_1^*, \psi_2^*, -\psi_3^*, -\psi_4^*).$$

## 4.6 Solutions to the Dirac equation

The ultimate aim of this chapter is to identify explicit forms for the wavefunctions of spin-half particles that will be used in the matrix element calculations that follow. It is natural to commence this discussion by looking for free-particle plane wave solutions of the form

$$\psi(\mathbf{x}, t) = u(E, \mathbf{p})e^{i(\mathbf{p}\cdot\mathbf{x}-Et)}, \tag{4.38}$$

where $u(E, \mathbf{p})$ is a four-component Dirac spinor and the overall wavefunction $\psi(\mathbf{x}, t)$ satisfies the Dirac equation

$$(i\gamma^\mu\partial_\mu - m)\psi = 0. \tag{4.39}$$

The position and time dependencies of the plane waves described by (4.38) occur solely in exponent; the four-component spinor $u(E, \mathbf{p})$ is a function of the energy and momentum of the particle. Hence the derivatives $\partial_\mu\psi$ act only on the exponent and therefore,

$$\partial_0\psi \equiv \frac{\partial\psi}{\partial t} = -iE\psi, \quad \partial_1\psi \equiv \frac{\partial\psi}{\partial x} = ip_x\psi, \quad \partial_2\psi = ip_y\psi \quad \text{and} \quad \partial_3\psi = ip_z\psi. \tag{4.40}$$

Substituting thc relations of (4.40) back into (4.39) gives

$$(\gamma^0E - \gamma^1p_x - \gamma^2p_y - \gamma^3p_z - m)u(E, \mathbf{p})e^{i(\mathbf{p}\cdot\mathbf{x}-Et)} = 0,$$

and therefore the four-component Dirac spinor $u(E, \mathbf{p})$ satisfies

$$(\gamma^\mu p_\mu - m)\, u = 0, \tag{4.41}$$

where, because of the covariance of the Dirac equation, the index $\mu$ on the $\gamma$-matrices can be treated as a four-vector index. Equation (4.41), which contains no derivatives, is the free-particle Dirac equation for the *spinor u* written in terms of the four-momentum of the particle.

### 4.6.1 Particles at rest

For a particle at rest with $\mathbf{p} = \mathbf{0}$, the free-particle wavefunction is simply

$$\psi = u(E, 0)e^{-iEt},$$

and thus (4.41) reduces to

$$E\gamma^0 u = mu.$$

This can be expressed as an eigenvalue equation for the components of the spinor

$$E\begin{pmatrix} 1 & 0 & 0 & 0 \\ 0 & 1 & 0 & 0 \\ 0 & 0 & -1 & 0 \\ 0 & 0 & 0 & -1 \end{pmatrix}\begin{pmatrix} \phi_1 \\ \phi_2 \\ \phi_3 \\ \phi_4 \end{pmatrix} = m\begin{pmatrix} \phi_1 \\ \phi_2 \\ \phi_3 \\ \phi_4 \end{pmatrix}.$$

Because $\gamma^0$ is diagonal, this yields four orthogonal solutions. The first two,

$$u_1(E, 0) = N\begin{pmatrix} 1 \\ 0 \\ 0 \\ 0 \end{pmatrix} \quad \text{and} \quad u_2(E, 0) = N\begin{pmatrix} 0 \\ 1 \\ 0 \\ 0 \end{pmatrix}, \tag{4.42}$$

have positive energy eigenvalues, $E = +m$. The other two solutions,

$$u_3(E, 0) = N\begin{pmatrix} 0 \\ 0 \\ 1 \\ 0 \end{pmatrix} \quad \text{and} \quad u_4(E, 0) = N\begin{pmatrix} 0 \\ 0 \\ 0 \\ 1 \end{pmatrix}, \tag{4.43}$$

have negative energy eigenvalues, $E = -m$. In all cases $N$ determines the normalisation of the wavefunction. These four states are also eigenstates of the $\hat{S}_z$ operator, as defined in Section 4.4. Hence $u_1(E, 0)$ and $u_2(E, 0)$ represent spin-up and spin-down positive energy solutions to the Dirac equation, and $u_3(E, 0)$ and $u_4(E, 0)$ represent spin-up and spin-down negative energy solutions. The four solutions to the Dirac equation for a particle at rest, including the time dependence, are therefore

$$\psi_1 = N\begin{pmatrix} 1 \\ 0 \\ 0 \\ 0 \end{pmatrix} e^{-imt}, \; \psi_2 = N\begin{pmatrix} 0 \\ 1 \\ 0 \\ 0 \end{pmatrix} e^{-imt}, \; \psi_3 = N\begin{pmatrix} 0 \\ 0 \\ 1 \\ 0 \end{pmatrix} e^{+imt} \; \text{and} \; \psi_4 = N\begin{pmatrix} 0 \\ 0 \\ 0 \\ 1 \end{pmatrix} e^{+imt}.$$

### 4.6.2 General free-particle solutions

The general solutions of the free-particle Dirac equation for a particle with momentum $\mathbf{p}$ can be obtained from the solutions for a particle at rest, using the Lorentz transformation properties of Dirac spinors derived in Appendix B.2. However, it is more straightforward to solve directly the Dirac equation for the general plane wave solution of (4.38). The Dirac equation for the spinor $u(E, \mathbf{p})$ given in (4.41) when written in full is

$$(E\gamma^0 - p_x\gamma^1 - p_y\gamma^2 - p_z\gamma^3 - m)u = 0.$$

This can be expressed in matrix form using the Dirac–Pauli representation of the $\gamma$-matrices, giving

$$\left[\begin{pmatrix} I & 0 \\ 0 & -I \end{pmatrix} E - \begin{pmatrix} 0 & \boldsymbol{\sigma}\cdot\mathbf{p} \\ -\boldsymbol{\sigma}\cdot\mathbf{p} & 0 \end{pmatrix} - m\begin{pmatrix} I & 0 \\ 0 & I \end{pmatrix}\right] u = 0, \tag{4.44}$$

where the $4 \times 4$ matrix multiplying the four-component spinor $u$ has been expressed in $2 \times 2$ block matrix form with

$$\boldsymbol{\sigma}\cdot\mathbf{p} \equiv \sigma_x p_x + \sigma_x p_y + \sigma_x p_z = \begin{pmatrix} p_z & p_x - ip_y \\ p_x + ip_y & -p_z \end{pmatrix}.$$

Writing the spinor $u$ in terms of two two-component column vectors, $u_A$ and $u_B$,

$$u = \begin{pmatrix} u_A \\ u_B \end{pmatrix},$$

allows (4.44) to be expressed as

$$\begin{pmatrix} (E-m)I & -\boldsymbol{\sigma}\cdot\mathbf{p} \\ \boldsymbol{\sigma}\cdot\mathbf{p} & -(E+m)I \end{pmatrix}\begin{pmatrix} u_A \\ u_B \end{pmatrix} = 0,$$

giving coupled equations for $u_A$ in terms of $u_B$,

$$u_A = \frac{\boldsymbol{\sigma}\cdot\mathbf{p}}{E-m} u_B, \tag{4.45}$$

$$u_B = \frac{\boldsymbol{\sigma}\cdot\mathbf{p}}{E+m} u_A. \tag{4.46}$$

Two solutions to the free-particle Dirac equation, $u_1$ and $u_2$, can be found by taking the two simplest orthogonal choices for $u_A$,

$$u_A = \begin{pmatrix} 1 \\ 0 \end{pmatrix} \quad \text{and} \quad u_A = \begin{pmatrix} 0 \\ 1 \end{pmatrix}. \tag{4.47}$$

The corresponding components of $u_B$, given by (4.46), are

$$u_B = \frac{\boldsymbol{\sigma}\cdot\mathbf{p}}{E+m} = \frac{1}{E+m}\begin{pmatrix} p_z & p_x - ip_y \\ p_x + ip_y & -p_z \end{pmatrix} u_A,$$

and thus the first two solutions of the free-particle Dirac equation are

$$u_1(E,\mathbf{p}) = N_1\begin{pmatrix}1\\0\\\frac{p_z}{E+m}\\\frac{p_x+ip_y}{E+m}\end{pmatrix} \quad\text{and}\quad u_2(E,\mathbf{p}) = N_2\begin{pmatrix}0\\1\\\frac{p_x-ip_y}{E+m}\\\frac{-p_z}{E+m}\end{pmatrix},$$

where $N_1$ and $N_2$ determine the wavefunction normalisation. It should be noted that whilst the choice of the two orthogonal forms for $u_A$ is arbitrary, any other orthogonal choice would have been equally valid, since a general $(E > 0)$ spinor can be expressed as a linear combination of $u_1$ and $u_2$. Choosing the forms of $u_A$ of (4.47) is analogous to choosing a particular basis for spin where conventionally the $z$-axis is chosen to label the states. The two other solutions of the Dirac equation can be found by writing

$$u_B = \begin{pmatrix}1\\0\end{pmatrix} \quad\text{and}\quad u_B = \begin{pmatrix}0\\1\end{pmatrix},$$

and using (4.45) to give the corresponding components for $u_A$. The four orthogonal plane wave solutions to the free-particle Dirac equation of the form

$$\psi_i = u_i(E,\mathbf{p})e^{i(\mathbf{p}\cdot\mathbf{x}-Et)}$$

are therefore

$$u_1 = N_1\begin{pmatrix}1\\0\\\frac{p_z}{E+m}\\\frac{p_x+ip_y}{E+m}\end{pmatrix},\quad u_2 = N_2\begin{pmatrix}0\\1\\\frac{p_x-ip_y}{E+m}\\\frac{-p_z}{E+m}\end{pmatrix},\quad u_3 = N_3\begin{pmatrix}\frac{p_z}{E-m}\\\frac{p_x+ip_y}{E-m}\\1\\0\end{pmatrix}\quad\text{and}$$

$$u_4 = N_4\begin{pmatrix}\frac{p_x-ip_y}{E-m}\\\frac{-p_z}{E-m}\\0\\1\end{pmatrix}. \tag{4.48}$$

If any one of these four spinors is substituted back into the Dirac equation, the Einstein energy–momentum relation $E^2 = p^2 + m^2$ is recovered. In the limit $\mathbf{p} = \mathbf{0}$, the spinors $u_1$ and $u_2$ reduce to the $E > 0$ spinors for a particle at rest given in (4.42). Hence $u_1$ and $u_2$ can be identified as the positive energy spinors with

$$E = +\left|\sqrt{\mathrm{p}^2 + m^2}\right|,$$

and $u_3$ and $u_4$ are the negative energy particle spinors with

$$E = -\left|\sqrt{\mathrm{p}^2 + m^2}\right|.$$

The same identification of $u_1$ and $u_2$ as being the positive energy spinors, and $u_3$ and $u_4$ as the negative energy spinors, can be reached by transforming the solutions for a particle at rest into the frame where the particle has momentum $\mathbf{p}$ (see Appendix B.2).

At this point it is reasonable to ask whether it is possible to interpret all four solutions of (4.48) as having $E > 0$. The answer is no, as if this were the case, the exponent of the wavefunction,

$$\psi(\mathbf{x}, t) = u(E, \mathbf{p})e^{i(\mathbf{p}\cdot\mathbf{x}-Et)},$$

would be the same for all four solutions. In this case the four solutions no longer would be independent since, for example, it would be possible to express $u_1$ as the linear combination

$$u_1 = \frac{p_z}{E+m}u_3 + \frac{p_x + ip_y}{E+m}u_4.$$

Hence, there are only four *independent* solutions to the Dirac equation when two are taken to have $E < 0$; it is not possible to avoid the need for the negative energy solutions. The same conclusion can be reached from the fact that the Dirac Hamiltonian is a $4 \times 4$ matrix with trace zero, and therefore the sum of its eigenvalues is zero, implying equal numbers of positive and negative energy solutions.

## 4.7 Antiparticles

The Dirac equation provides a beautiful mathematical framework for the relativistic quantum mechanics of spin-half fermions in which the properties of spin and magnetic moments emerge naturally. However, the presence of negative energy solutions is unavoidable. In quantum mechanics, a complete set of basis states is required to span the vector space, and therefore the negative energy solutions cannot simply be discarded as being unphysical. It is therefore necessary to provide a physical interpretation for the negative energy solutions.

### 4.7.1 The Dirac sea interpretation

If negative energy solutions represented accessible negative energy particle states, one would expect that all positive energy electrons would fall spontaneously into these lower energy states. Clearly this does not occur. To avoid this apparent contradiction, Dirac proposed that the vacuum corresponds to the state where all negative energy states are occupied, as indicated in Figure 4.1. In this "Dirac sea" picture, the Pauli exclusion principle prevents positive energy electrons from falling into the fully occupied negative energy states. Furthermore, a photon with energy $E > 2m_\mathrm{e}$ could excite an electron from a negative energy state, leaving a hole in

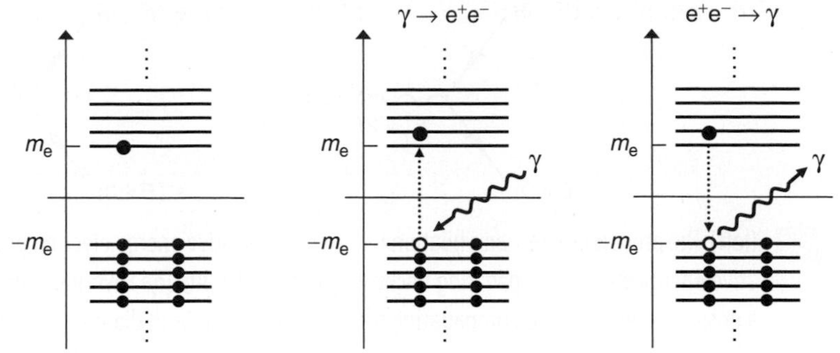


Fig. 4.1 The Dirac interpretation of negative energy solutions as holes in the vacuum that correspond to antiparticle states.

the vacuum. A hole in the vacuum would correspond to a state with more energy (less negative energy) and a positive charge relative to the fully occupied vacuum. In this way, holes in the Dirac sea correspond to positive energy antiparticles with the opposite charge to the particle states. The Dirac sea interpretation thus provides a picture for $e^+e^-$ pair production and also particle–antiparticle annihilation (shown in Figure 4.1). The discovery of positively charged electrons in cosmic-ray tracks in a cloud chamber, Anderson (1933), provided the experimental confirmation that the antiparticles predicted by Dirac corresponded to physical observable states.

Nowadays, the Dirac sea picture of the vacuum is best viewed in terms of historical interest. It has a number of conceptual problems. For example, antiparticle states for bosons are also observed and in this case the Pauli exclusion principle does not apply. Furthermore, the fully occupied Dirac sea implies that the vacuum has infinite negative energy and it is not clear how this can be interpreted physically. The negative energy solutions are now understood in terms of the Feynman–Stückelberg interpretation.

### 4.7.2 The Feynman–Stückelberg interpretation

It is an experimentally established fact that for each fundamental spin-half particle there is a corresponding antiparticle. The antiparticles produced in accelerator experiments have the opposite charges compared to the corresponding particle. Apart from possessing different charges, antiparticles behave very much like particles; they propagate forwards in time from the point of production, ionise the gas in tracking detectors, produce the same electromagnetic showers in the calorimeters of large collider particle detectors, and undergo many of the same interactions as particles. It is not straightforward to reconcile these physical observations with the negative energy solutions that emerge from the abstract mathematics of the Dirac equation.

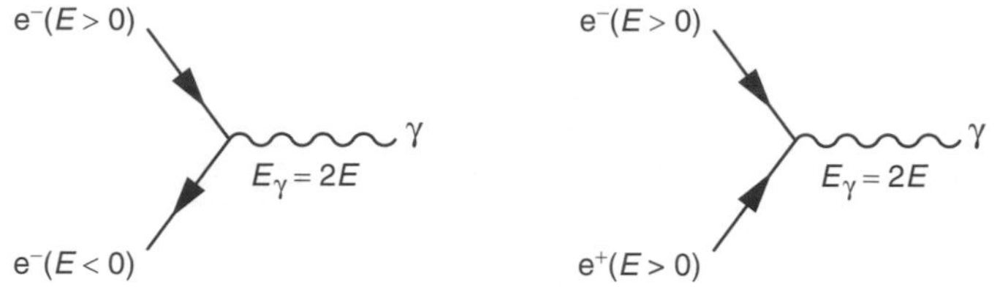


Fig. 4.2 (left) The process of $e^+e^-$ annihilation in terms of a positive energy electron producing a photon and a negative energy electron propagating backwards in time. (right) The Feynman–Stückelberg interpretation with a positive energy positron propagating forwards in time. In both diagrams, time runs from the left to right.

The modern interpretation of the negative energy solutions, due to Stückelberg and Feynman, was developed in the context of quantum field theory. The $E < 0$ solutions are interpreted as *negative* energy *particles* which propagate *backwards* in time. These negative energy particle solutions correspond to physical *positive* energy *antiparticle* states with opposite charge, which propagate *forwards* in time. Since the time dependence of the wavefunction, $\exp(-iEt)$, is unchanged under the simultaneous transformation $E \to -E$ and $t \to -t$ these two pictures are mathematically equivalent,

$$\exp\{-iEt\} \equiv \exp\{-i(-E)(-t)\}.$$

To illustrate this idea, Figure 4.2 shows the process of electron–positron annihilation in terms of negative energy particle solutions and in the Feynman–Stückelberg interpretation of these solutions as positive energy antiparticles. In the left plot, an electron of energy $E$ emits a photon with energy $2E$ and, to conserve energy, produces a electron with energy $-E$, which being a negative energy solution of the Dirac equation propagates backwards in time. In the Feynman–Stückelberg interpretation, shown on the right, a positive energy positron of energy $E$ annihilates with the electron with energy $E$ to produce a photon of energy $2E$. In this case, both the particle and antiparticle propagate forwards in time. It should be noted that although antiparticles propagate forwards in time, in a Feynman diagram they are still drawn with an arrow in the "backwards in time" sense, as shown in the left plot of Figure 4.2.

### 4.7.3 Antiparticle spinors

In principle, it is possible to perform calculations with the negative energy particle spinors $u_3$ and $u_4$. However, this necessitates remembering that the energy which appears in the definition of the spinor is the negative of the physical energy. Furthermore, because $u_3$ and $u_4$ are interpreted as propagating backwards in time, the momentum appearing in the spinor is the negative of the physical momentum. To avoid this possible confusion, it is more convenient to work with antiparticle spinors written in terms of the physical momentum and physical energy, $E = +|\sqrt{\mathbf{p}^2 + m^2}\,|$. Following the Feynman–Stückelberg interpretation, the negative energy particle spinors, $u_3$ and $u_4$, can be rewritten in terms of the physical positive energy *antiparticle spinors*, $v_1$ and $v_2$, simply by reversing the signs of $E$ and $\mathbf{p}$ to give

$$v_1(E, \mathbf{p})e^{-i(\mathbf{p}\cdot\mathbf{x}-Et)} = u_4(-E, -\mathbf{p})e^{i[-\mathbf{p}\cdot\mathbf{x}-(-E)t]}$$

$$v_2(E, \mathbf{p})e^{-i(\mathbf{p}\cdot\mathbf{x}-Et)} = u_3(-E, -\mathbf{p})e^{i[-\mathbf{p}\cdot\mathbf{x}-(-E)t]}.$$

A more formal approach to identifying the antiparticle spinors is to look for solutions of the Dirac equation of the form

$$\psi(\mathbf{x}, t) = v(E, \mathbf{p})e^{-i(\mathbf{p}\cdot\mathbf{x}-Et)}, \tag{4.49}$$

where the signs in the exponent are reversed with respect to those of (4.38). For $E > 0$, the wavefunctions of (4.49) still represent negative energy solutions in the sense that

$$i\frac{\partial}{\partial t}\psi = -E\psi.$$

Substituting the wavefunction of (4.49) into the Dirac equation, $(i\gamma^\mu\partial_\mu - m)\psi = 0$, gives

$$(-\gamma^0 E + \gamma^1 p_x + \gamma^2 p_y + \gamma^3 p_z - m)v = 0,$$

which can be written as

$$(\gamma^\mu p_\mu + m)v = 0.$$

This is the Dirac equation in terms of momentum for the $v$ spinors. Proceeding as before and writing

$$v = \begin{pmatrix} v_A \\ v_B \end{pmatrix},$$

leads to

$$v_A = \frac{\boldsymbol{\sigma}\cdot\mathbf{p}}{E+m}v_B \quad \text{and} \quad v_B = \frac{\boldsymbol{\sigma}\cdot\mathbf{p}}{E-m}v_A,$$

giving the solutions

$$v_1 = N\begin{pmatrix} \frac{p_x - ip_y}{E+m} \\ \frac{-p_z}{E+m} \\ 0 \\ 1 \end{pmatrix}, \quad v_2 = N\begin{pmatrix} \frac{p_z}{E+m} \\ \frac{p_x + ip_y}{E+m} \\ 1 \\ 0 \end{pmatrix}, \quad v_3 = N\begin{pmatrix} 1 \\ 0 \\ \frac{p_z}{E-m} \\ \frac{p_x + ip_y}{E-m} \end{pmatrix} \quad \text{and}$$

$$v_4 = N\begin{pmatrix} 0 \\ 1 \\ \frac{p_x - ip_y}{E-m} \\ \frac{-p_z}{E-m} \end{pmatrix}, \tag{4.50}$$

where

$$E = +\left|\sqrt{\mathbf{p}^2 + m^2}\right|$$

for $v_1$ and $v_2$, and

$$E = -\left|\sqrt{\mathbf{p}^2 + m^2}\right|$$

for $v_3$ and $v_4$. Hence we have now identified eight solutions to the free particle Dirac equation, given in (4.48) and (4.50). Of these eight solutions, only four are independent. In principle it would be possible to perform calculations using only the $u$-spinors, or alternatively using only the $v$-spinors. Nevertheless, it is more natural to work with the four solutions for which the energy that appears in the spinor is the positive physical energy of the particle/antiparticle, namely $\{u_1, u_2, v_1, v_2\}$.

To summarise, in terms of the physical energy, the two *particle* solutions to the Dirac equation are

$$\psi_i = u_i e^{+i(\mathbf{p}\cdot\mathbf{x} - Et)}$$

with

$$u_1(p) = \sqrt{E+m}\begin{pmatrix} 1 \\ 0 \\ \frac{p_z}{E+m} \\ \frac{p_x + ip_y}{E+m} \end{pmatrix} \quad \text{and} \quad u_2(p) = \sqrt{E+m}\begin{pmatrix} 0 \\ 1 \\ \frac{p_x - ip_y}{E+m} \\ \frac{-p_z}{E+m} \end{pmatrix}, \tag{4.51}$$

and the two *antiparticle* solutions are

$$\psi_i = v_i e^{-i(\mathbf{p}\cdot\mathbf{x} - Et)}$$

with

$$v_1(p) = \sqrt{E+m}\begin{pmatrix} \frac{p_x - ip_y}{E+m} \\ \frac{-p_z}{E+m} \\ 0 \\ 1 \end{pmatrix} \quad \text{and} \quad v_2(p) = \sqrt{E+m}\begin{pmatrix} \frac{p_z}{E+m} \\ \frac{p_x + ip_y}{E+m} \\ 1 \\ 0 \end{pmatrix}. \tag{4.52}$$

### Wavefunction normalisation

The spinors in (4.51) and (4.52) have been normalised to the conventional $2E$ particles per unit volume. This required normalisation factor can be found from the definition of probability density, $\rho = \psi^\dagger\psi$, which for $\psi = u_1(p)\exp i(\mathbf{p}\cdot\mathbf{x} - Et)$ is

$$\rho = \psi^\dagger\psi = (\psi^*)^T\psi = u_1^\dagger u_1.$$

Using the explicit form for $u_1$ of (4.48) gives

$$u_1^\dagger u_1 = |N|^2\left(1 + \frac{p_z^2}{(E+m)^2} + \frac{p_x^2 + p_y^2}{(E+m)^2}\right) = |N|^2\frac{2E}{E+m}.$$

Hence, to normalise the wavefunctions to $2E$ particles per unit volume implies

$$N = \sqrt{E+m}.$$

The same normalisation factor is obtained for the $u$ and $v$ spinors.

## 4.7.4 Operators and the antiparticle spinors

There is a subtle, but nevertheless important, point related to using the antiparticle spinors written in terms of the physical energy and momenta,

$$\psi = v(E, \mathbf{p})e^{-i(\mathbf{p}\cdot\mathbf{x} - Et)}.$$

The action of the normal quantum mechanical operators for energy and momentum do not give the physical quantities,

$$\hat{H}\psi = i\frac{\partial\psi}{\partial t} = -E\psi \quad \text{and} \quad \hat{\mathbf{p}}\psi = -i\nabla\psi = -\mathbf{p}\psi.$$

The minus signs should come as no surprise; the antiparticle spinors are still the negative energy particle solutions of the Dirac equation, albeit expressed in terms of the physical (positive) energy $E$ and physical momentum $\mathbf{p}$ of the antiparticle. The operators which give the *physical* energy and momenta of the antiparticle spinors are therefore

$$\hat{H}^{(v)} = -i\frac{\partial}{\partial t} \quad \text{and} \quad \hat{\mathbf{p}}^{(v)} = +i\nabla,$$

where the change of sign reflects the Feynman–Stückelberg interpretation of the negative energy solutions. Furthermore, with the replacement $(E, \mathbf{p}) \rightarrow (-E, -\mathbf{p})$, the orbital angular momentum of a particle

$$\mathbf{L} = \mathbf{r} \times \mathbf{p} \rightarrow -\mathbf{L}.$$

In order for the commutator $[\hat{H}_D, \hat{\mathbf{L}} + \hat{\mathbf{S}}]$ to remain zero for the antiparticle spinors, the operator giving the physical spin states of the $v$ spinors must be

$$\hat{\mathbf{S}}^{(v)} = -\hat{\mathbf{S}},$$

where $\hat{\mathbf{S}}$ is defined in (4.29). Reverting (very briefly) to the the Dirac sea picture, a spin-up hole in the negative energy particle sea, leaves the vacuum in a net spin-down state.

### 4.7.5 *Charge conjugation

Charge conjugation is an important example of a discrete symmetry transformation that will be discussed in depth in Chapter 14. The effect of charge conjugation is to replace particles with the corresponding antiparticles and vice versa. In classical dynamics, the motion of a charged particle in an electromagnetic field $A^\mu = (\phi, \mathbf{A})$ can be obtained by making the minimal substitution

$$E \rightarrow E - q\phi \quad \text{and} \quad \mathbf{p} \rightarrow \mathbf{p} - q\mathbf{A}, \tag{4.53}$$

where $\phi$ and $\mathbf{A}$ are the scalar and vector potentials of electromagnetism and $q$ is the charge of the particle. In four-vector notation, (4.53) can be written

$$p_\mu \rightarrow p_\mu - qA_\mu. \tag{4.54}$$

Following the canonical procedure for moving between classical physics and quantum mechanics and replacing energy and momentum by the operators $\hat{\mathbf{p}} = -i\boldsymbol{\nabla}$ and $\hat{E} = i\partial/\partial t$, Equation (4.54) can be written in operator form as

$$i\partial_\mu \rightarrow i\partial_\mu - qA_\mu. \tag{4.55}$$

The Dirac equation for an electron with charge $q = -e$ (where $e \equiv +|e|$ is the magnitude of the electron charge) in the presence of an electromagnetic field can be obtained by making the minimal substitution of (4.55) in the free-particle Dirac equation, giving

$$\gamma^\mu(\partial_\mu - ieA_\mu)\psi + im\psi = 0. \tag{4.56}$$

The equivalent equation for the positron can be obtained by first taking the complex conjugate of (4.56) and then pre-multiplying by $-i\gamma^2$ to give

$$-i\gamma^2(\gamma^\mu)^*(\partial_\mu + ieA_\mu)\psi^* - m\gamma^2\psi^* = 0. \tag{4.57}$$

In the Dirac–Pauli representation of the $\gamma$-matrices, $(\gamma^0)^* = \gamma^0$, $(\gamma^1)^* = \gamma^1$, $(\gamma^2)^* = -\gamma^2$ and $(\gamma^3)^* = \gamma^3$. Using these relations and $\gamma^2\gamma^\mu = -\gamma^\mu\gamma^2$ for $\mu \neq 2$, Equation (4.57) becomes

$$\gamma^\mu(\partial_\mu + ieA_\mu)i\gamma^2\psi^* + im\, i\gamma^2\psi^* = 0. \tag{4.58}$$

If $\psi'$ is defined as

$$\psi' = i\gamma^2\psi^*,$$

then (4.58) can be written

$$\gamma^\mu(\partial_\mu + ieA_\mu)\psi' + im\psi' = 0. \tag{4.59}$$

The equation satisfied by $\psi'$ is the same as that for $\psi$ (4.56), except that the $ieA_\mu$ term now appears with the opposite sign. Hence, $\psi'$ is a wavefunction describing a particle which has the same mass as the original particle but with opposite charge; $\psi'$ can be interpreted as the antiparticle wavefunction. In the Dirac–Pauli representation, the charge conjugation operator $\hat{C}$, which transforms a particle wavefunction into the corresponding antiparticle wavefunction, therefore can be identified as

$$\psi' = \hat{C}\psi = i\gamma^2\psi^*.$$

The identification of $\hat{C}$ as the charge conjugation operator can be confirmed by considering its effect on the particle spinor

$$\psi = u_1 e^{i(\mathbf{p}\cdot\mathbf{x}-Et)}.$$

The corresponding charge-conjugated wavefunction $\psi'$ is

$$\psi' = \hat{C}\psi = i\gamma^2\psi^* = i\gamma^2 u_1^* e^{-i(\mathbf{p}\cdot\mathbf{x}-Et)}.$$

The spinor part of $\psi'$ is

$$i\gamma^2 u_1^* = i\begin{pmatrix} 0 & 0 & 0 & -i \\ 0 & 0 & i & 0 \\ 0 & i & 0 & 0 \\ -i & 0 & 0 & 0 \end{pmatrix} \sqrt{E+m}\begin{pmatrix} 1 \\ 0 \\ \frac{p_z}{E+m} \\ \frac{p_x+ip_y}{E+m} \end{pmatrix}^* = \sqrt{E+m}\begin{pmatrix} \frac{p_x-ip_y}{E+m} \\ \frac{-p_z}{E+m} \\ 0 \\ 1 \end{pmatrix},$$

which is the antiparticle spinor $v_1$ identified in Section 4.7.3. The effect of the charge-conjugation operator on the $u_1$ particle spinor is

$$\psi = u_1 e^{i(\mathbf{p}\cdot\mathbf{x}-Et)} \xrightarrow{\hat{C}} \psi' = v_1 e^{-i(\mathbf{p}\cdot\mathbf{x}-Et)},$$

and likewise (up to a unobservable overall complex phase) the effect on $u_2$ is

$$\psi = u_2 e^{i(\mathbf{p}\cdot\mathbf{x}-Et)} \xrightarrow{\hat{C}} \psi' = v_2 e^{-i(\mathbf{p}\cdot\mathbf{x}-Et)}.$$

Therefore, the effect of the charge-conjugation operator on the particle spinors $u_1$ and $u_2$ is to transform them respectively to the antiparticle spinors $v_1$ and $v_2$.

## 4.8 Spin and helicity states

For particles at rest, the spinors $u_1(E,0)$ and $u_2(E,0)$ of (4.42) are clearly eigenstates of

$$\hat{S}_z = \tfrac{1}{2}\Sigma_z = \frac{1}{2}\begin{pmatrix} \sigma_z & 0 \\ 0 & \sigma_z \end{pmatrix} = \frac{1}{2}\begin{pmatrix} 1 & 0 & 0 & 0 \\ 0 & -1 & 0 & 0 \\ 0 & 0 & 1 & 0 \\ 0 & 0 & 0 & -1 \end{pmatrix},$$

and therefore represent "spin-up" and "spin-down" eigenstates of the $z$-component of the spin operator. However, from the forms of the $u$ and $v$ spinors, given in (4.51) and (4.52), it is immediately apparent that the $u_1$, $u_2$, $v_1$ and $v_2$ spinors are not in general eigenstates of $\hat{S}_z$. Nevertheless, for particles/antiparticles travelling in the $\pm z$-direction ($\mathbf{p} = \pm \mathrm{p}\hat{\mathbf{z}}$), the $u$ and $v$ spinors are

$$u_1 = N\begin{pmatrix} 1 \\ 0 \\ \frac{\pm\mathrm{p}}{E+m} \\ 0 \end{pmatrix}, \quad u_2 = N\begin{pmatrix} 0 \\ 1 \\ 0 \\ \frac{\mp\mathrm{p}}{E+m} \end{pmatrix}, \quad v_1 = N\begin{pmatrix} 0 \\ \frac{\mp\mathrm{p}}{E+m} \\ 0 \\ 1 \end{pmatrix} \quad \text{and } v_2 = N\begin{pmatrix} \frac{\pm\mathrm{p}}{E+m} \\ 0 \\ 1 \\ 0 \end{pmatrix},$$

and therefore

$$\hat{S}_z\, u_1(E,0,0,\pm\mathrm{p}) = +\tfrac{1}{2}u_1(E,0,0,\pm\mathrm{p}),$$
$$\hat{S}_z\, u_2(E,0,0,\pm\mathrm{p}) = -\tfrac{1}{2}u_2(E,0,0,\pm\mathrm{p}).$$

For antiparticle spinors, the *physical* spin is given by the operator $\hat{S}_z^{(v)} = -\hat{S}_z$ and therefore

$$\hat{S}_z^{(v)} v_1(E,0,0,\pm\mathrm{p}) \equiv -\hat{S}_z v_1(E,0,0,\pm\mathrm{p}) = +\tfrac{1}{2}v_1(E,0,0,\pm\mathrm{p}),$$
$$\hat{S}_z^{(v)} v_2(E,0,0,\pm\mathrm{p}) \equiv -\hat{S}_z v_2(E,0,0,\pm\mathrm{p}) = -\tfrac{1}{2}v_2(E,0,0,\pm\mathrm{p}).$$

Hence for a particle/antiparticle with momentum $\mathbf{p} = (0,0,\pm\mathrm{p})$, the $u_1$ and $v_1$ spinors represent spin-up states and the $u_2$ and $v_2$ spinors represent spin-down states, as indicated in Figure 4.3.

Fig. 4.3 The $u_1$, $u_2$, $v_1$ and $v_2$ spinors for particles/antiparticles travelling in the $\pm z$-direction.

### 4.8.1 Helicity

In the chapters that follow, interaction cross sections will be analysed in terms of the spin states of the particles involved. Since the $u_1$, $u_2$, $v_1$ and $v_2$ spinors only map onto easily identified spin states for particles travelling in the $z$-direction, their use for this purpose is limited. Furthermore, since $\hat{S}_z$ does not commute with the Dirac Hamiltonian, $[\hat{H}_D, \hat{S}_z] \neq 0$, it is not possible to define a basis of simultaneous eigenstates of $\hat{S}_z$ and $\hat{H}_D$. Rather than defining basis states in terms of an external axis, it is more natural to introduce to concept of helicity. As illustrated in Figure 4.4, the helicity $h$ of a particle is defined as the normalised component of its spin along its direction of flight,

$$h \equiv \frac{\mathbf{S}\cdot\mathbf{p}}{\mathrm{p}}. \tag{4.60}$$

For a four-component Dirac spinor, the helicity operator is

$$\hat{h} = \frac{\hat{\boldsymbol{\Sigma}}\cdot\hat{\mathbf{p}}}{2\mathrm{p}} = \frac{1}{2\mathrm{p}}\begin{pmatrix} \boldsymbol{\sigma}\cdot\hat{\mathbf{p}} & 0 \\ 0 & \boldsymbol{\sigma}\cdot\hat{\mathbf{p}} \end{pmatrix}, \tag{4.61}$$

where $\hat{\mathbf{p}}$ is the momentum operator. From the form of the Dirac Hamiltonian (4.18), it follows that $[\hat{H}_D, \hat{\boldsymbol{\Sigma}}\cdot\hat{\mathbf{p}}] = 0$ and therefore $\hat{h}$ commutes with the free-particle Hamiltonian. Consequently, it is possible to identify spinor states which are simultaneous eigenstates of the free particle Dirac Hamiltonian and the helicity operator. For a spin-half particle, the component of spin measured along any axis is quantised to be either $\pm 1/2$. Consequently, the eigenvalues of the helicity operator acting on a Dirac spinor are $\pm 1/2$. The two possible helicity states for a spin-half fermion are termed *right-handed* and *left-handed* helicity states, as shown in Figure 4.5. Whilst helicity is an important concept in particle physics, it is important to remember that helicity is not Lorentz invariant; for particles with mass, it is always possible to transform into a frame in which the direction of the particle is reversed. The related Lorentz-invariant concept of chirality is introduced in Chapter 6.

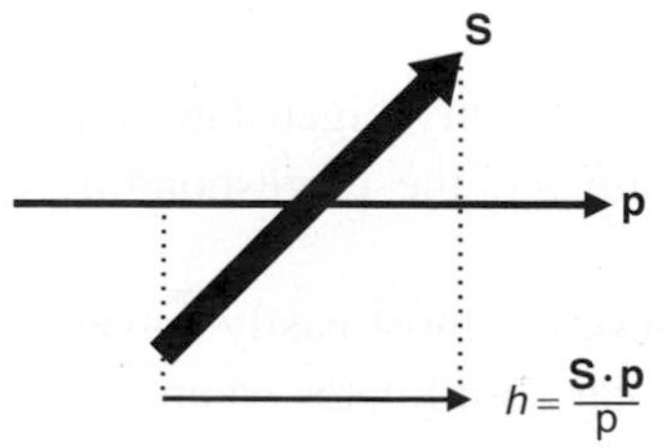


Fig. 4.4 The definition of helicity as the projection of the spin of a particle along its direction of motion.

Fig. 4.5 The two helicity eigenstates for a spin-half fermion. The $h = +1/2$ and $h = -1/2$ states are respectively referred to as right-handed (RH) and left-handed (LH) helicity states.

The simultaneous eigenstates of the free particle Dirac Hamiltonian and the helicity operator are solutions to the Dirac equation which also satisfy the eigenvalue equation,

$$\hat{h}u = \lambda u.$$

Writing the spinor in terms of two two-component column vectors $u_A$ and $u_B$, and using the helicity operator defined above, this eigenvalue equation can be written

$$\frac{1}{2\mathrm{p}}\begin{pmatrix} \boldsymbol{\sigma}\cdot\mathbf{p} & 0 \\ 0 & \boldsymbol{\sigma}\cdot\mathbf{p} \end{pmatrix}\begin{pmatrix} u_A \\ u_B \end{pmatrix} = \lambda\begin{pmatrix} u_A \\ u_B \end{pmatrix},$$

implying that

$$(\boldsymbol{\sigma}\cdot\mathbf{p})u_A = 2\mathrm{p}\,\lambda u_A, \tag{4.62}$$

$$(\boldsymbol{\sigma}\cdot\mathbf{p})u_B = 2\mathrm{p}\,\lambda u_B. \tag{4.63}$$

The eigenvalues of the helicity operator can be obtained by multiplying (4.62) by $\boldsymbol{\sigma}\cdot\mathbf{p}$ and noting (see Problem 4.10) that $(\boldsymbol{\sigma}\cdot\mathbf{p})^2 = \mathrm{p}^2$, from which it follows that

$$\mathrm{p}^2 u_A = 2\mathrm{p}\lambda(\boldsymbol{\sigma}\cdot\mathbf{p})u_A = 4\mathrm{p}^2\lambda^2 u_A,$$

and therefore, as anticipated, $\lambda = \pm 1/2$. Because the spinors corresponding to the two helicity states are also eigenstates of the Dirac equation, $u_A$ and $u_B$ are related by (4.46),

$$(\boldsymbol{\sigma}\cdot\mathbf{p})u_A = (E+m)u_B,$$

which when combined with (4.62) gives

$$u_B = 2\lambda\left(\frac{\mathrm{p}}{E+m}\right)u_A. \tag{4.64}$$

Therefore for a helicity eigenstate, $u_B$ is proportional to $u_A$ and once (4.62) is solved to obtain $u_A$, the corresponding equation for $u_B$ (4.63) is automatically satisfied.

Equation (4.62) is most easily solved by expressing the helicity states in terms of spherical polar coordinates where

$$\mathbf{p} = (\mathrm{p}\sin\theta\cos\phi,\ \mathrm{p}\sin\theta\sin\phi,\ \mathrm{p}\cos\theta),$$

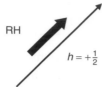

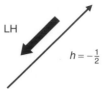

and the helicity operator can be written as

$$\frac{1}{2\mathrm{p}}(\boldsymbol{\sigma}\cdot\mathbf{p}) = \frac{1}{2\mathrm{p}}\begin{pmatrix} p_z & p_x - ip_y \\ p_x + ip_y & -p_z \end{pmatrix}$$
$$= \frac{1}{2}\begin{pmatrix} \cos\theta & \sin\theta e^{-i\phi} \\ \sin\theta e^{i\phi} & -\cos\theta \end{pmatrix}.$$

Writing the components of $u_A$ as

$$u_A = \begin{pmatrix} a \\ b \end{pmatrix},$$

the eigenvalue equation of (4.62) becomes

$$\begin{pmatrix} \cos\theta & \sin\theta e^{-i\phi} \\ \sin\theta e^{i\phi} & -\cos\theta \end{pmatrix}\begin{pmatrix} a \\ b \end{pmatrix} = 2\lambda\begin{pmatrix} a \\ b \end{pmatrix},$$

and therefore the ratio of $b/a$ is equal to

$$\frac{b}{a} = \frac{2\lambda - \cos\theta}{\sin\theta}e^{i\phi}.$$

For the right-handed helicity state with $\lambda = +1/2$,

$$\frac{b}{a} = \frac{1-\cos\theta}{\sin\theta}e^{i\phi} = \frac{2\sin^2\left(\frac{\theta}{2}\right)}{2\sin\left(\frac{\theta}{2}\right)\cos\left(\frac{\theta}{2}\right)}e^{i\phi} = e^{i\phi}\frac{\sin\left(\frac{\theta}{2}\right)}{\cos\left(\frac{\theta}{2}\right)}.$$

Using the relation between $u_A$ and $u_B$ from (4.64), the right-handed helicity particle spinor, denoted $u_\uparrow$, then can be identified as

$$u_\uparrow = N\begin{pmatrix} \cos\left(\frac{\theta}{2}\right) \\ e^{i\phi}\sin\left(\frac{\theta}{2}\right) \\ \frac{\mathrm{p}}{E+m}\cos\left(\frac{\theta}{2}\right) \\ \frac{\mathrm{p}}{E+m}e^{i\phi}\sin\left(\frac{\theta}{2}\right) \end{pmatrix},$$

where $N = \sqrt{E+m}$ is the overall normalisation factor. The left-handed helicity spinor with $h = -1/2$, denoted $u_\downarrow$, can be found in the same manner and thus the right-handed and left-handed helicity particle spinors, normalised to $2E$ particles per unit volume, are

$$u_\uparrow = \sqrt{E+m}\begin{pmatrix} c \\ se^{i\phi} \\ \frac{\mathrm{p}}{E+m}c \\ \frac{\mathrm{p}}{E+m}se^{i\phi} \end{pmatrix} \quad u_\downarrow = \sqrt{E+m}\begin{pmatrix} -s \\ ce^{i\phi} \\ \frac{\mathrm{p}}{E+m}s \\ -\frac{\mathrm{p}}{E+m}ce^{i\phi} \end{pmatrix}, \tag{4.65}$$

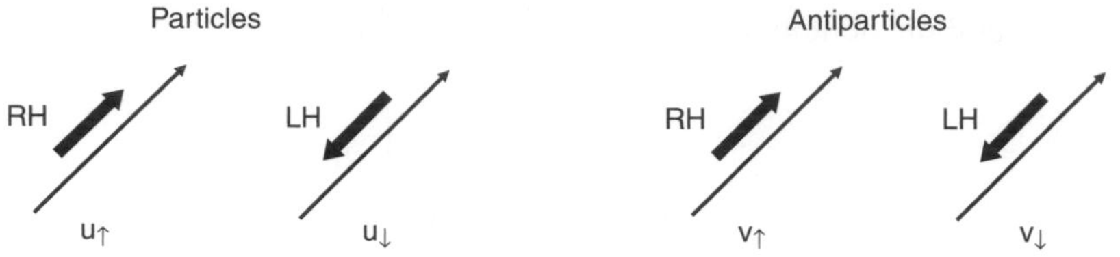


Fig. 4.6 The helicity eigenstates for spin-half particles and antiparticles.

where $s = \sin\left(\frac{\theta}{2}\right)$ and $c = \cos\left(\frac{\theta}{2}\right)$. The corresponding antiparticle states, $v_\uparrow$ and $v_\downarrow$, are obtained in the same way remembering that the physical spin of an antiparticle spinor is given by $\hat{\mathbf{S}}^{(v)} = -\hat{\mathbf{S}}$, and hence for the $h = +1/2$ antiparticle state

$$\left(\frac{\boldsymbol{\Sigma}\cdot\mathbf{p}}{2\mathrm{p}}\right) v_\uparrow = -\tfrac{1}{2} v_\uparrow .$$

The resulting normalised antiparticle helicity spinors are

$$v_\uparrow = \sqrt{E+m}\begin{pmatrix} \frac{\mathrm{p}}{E+m}s \\ -\frac{\mathrm{p}}{E+m}ce^{i\phi} \\ -s \\ ce^{i\phi} \end{pmatrix} \quad v_\downarrow = \sqrt{E+m}\begin{pmatrix} \frac{\mathrm{p}}{E+m}c \\ \frac{\mathrm{p}}{E+m}se^{i\phi} \\ c \\ se^{i\phi} \end{pmatrix}. \tag{4.66}$$

The four helicity states of (4.65) and (4.66), which correspond to the states shown in Figure 4.6, form the helicity basis that is used to describe particles and antiparticles in the calculations that follow. In many of these calculations, the energies of the particles being considered are much greater than their masses. In this ultra-relativistic limit ($E \gg m$) the helicity eigenstates can be approximated by

$$u_\uparrow \approx \sqrt{E}\begin{pmatrix} c \\ se^{i\phi} \\ c \\ se^{i\phi} \end{pmatrix},\; u_\downarrow \approx \sqrt{E}\begin{pmatrix} -s \\ ce^{i\phi} \\ s \\ -ce^{i\phi} \end{pmatrix},\; v_\uparrow \approx \sqrt{E}\begin{pmatrix} s \\ -ce^{i\phi} \\ -s \\ ce^{i\phi} \end{pmatrix} \text{ and } v_\downarrow \approx \sqrt{E}\begin{pmatrix} c \\ se^{i\phi} \\ c \\ se^{i\phi} \end{pmatrix}. \tag{4.67}$$

It should be remembered that the above spinors all can be multiplied by an overall complex phase with no change in any physical predictions.

## 4.9 Intrinsic parity of Dirac fermions

Charge conjugation, discussed in Section 4.7.5, is one example of a discrete symmetry transformation, particle ↔ antiparticle. Another example is the parity transformation, which corresponds to spatial inversion through the origin,

$$x' = -x, \quad y' = -y, \quad z' = -z \quad \text{and} \quad t' = t.$$

Parity is an important concept in particle physics because both the QED and QCD interactions always conserve parity. To understand why this is the case (which is explained in Chapter 11), we will need to use the parity transformation properties of Dirac spinors and will need to identify the corresponding parity operator which acts on solutions of the Dirac equation.

Suppose $\psi$ is a solution of the Dirac equation and $\psi'$ is the corresponding solution in the "parity mirror" obtained from the action of the parity operator $\hat{P}$ such that

$$\psi \to \psi' = \hat{P}\psi.$$

From the definition of the parity operation, the effect of two successive parity transformations is to recover the original wavefunction. Consequently $\hat{P}^2 = I$ and thus

$$\psi' = \hat{P}\psi \quad \Rightarrow \quad \hat{P}\psi' = \psi.$$

The form of the parity operator can be deduced by considering a wavefunction $\psi(x, y, z, t)$ which satisfies the free-particle Dirac equation,

$$i\gamma^1\frac{\partial\psi}{\partial x} + i\gamma^2\frac{\partial\psi}{\partial y} + i\gamma^3\frac{\partial\psi}{\partial z} - m\psi = -i\gamma^0\frac{\partial\psi}{\partial t}. \tag{4.68}$$

The parity transformed wavefunction $\psi'(x', y', z', t') = \hat{P}\psi(x, y, z, t)$ must satisfy the Dirac equation in the new coordinate system

$$i\gamma^1\frac{\partial\psi'}{\partial x'} + i\gamma^2\frac{\partial\psi'}{\partial y'} + i\gamma^3\frac{\partial\psi'}{\partial z'} - m\psi' = -i\gamma^0\frac{\partial\psi'}{\partial t'}. \tag{4.69}$$

Writing $\psi = \hat{P}\psi'$, equation (4.68) becomes

$$i\gamma^1\hat{P}\frac{\partial\psi'}{\partial x} + i\gamma^2\hat{P}\frac{\partial\psi'}{\partial y} + i\gamma^3\hat{P}\frac{\partial\psi'}{\partial z} - m\hat{P}\psi' = -i\gamma^0\hat{P}\frac{\partial\psi'}{\partial t}.$$

Premultiplying by $\gamma^0$ and expressing the derivatives in terms of the primed system (which introduces minus signs for all the space-like coordinates) gives

$$-i\gamma^0\gamma^1\hat{P}\frac{\partial\psi'}{\partial x'} - i\gamma^0\gamma^2\hat{P}\frac{\partial\psi'}{\partial y'} - i\gamma^0\gamma^3\hat{P}\frac{\partial\psi'}{\partial z'} - m\gamma^0\hat{P}\psi' = -i\gamma^0\gamma^0\hat{P}\frac{\partial\psi'}{\partial t'},$$

which using $\gamma^0\gamma^k = -\gamma^k\gamma^0$ can be written

$$i\gamma^1\gamma^0\hat{P}\frac{\partial\psi'}{\partial x'} + i\gamma^2\gamma^0\hat{P}\frac{\partial\psi'}{\partial y'} + i\gamma^3\gamma^0\hat{P}\frac{\partial\psi'}{\partial z'} - m\gamma^0\hat{P}\psi' = -i\gamma^0\gamma^0\hat{P}\frac{\partial\psi'}{\partial t'}. \tag{4.70}$$

In order for (4.70) to reduce to the desired form of (4.69), $\gamma^0\hat{P}$ must be proportional to the $4\times4$ identity matrix,

$$\gamma^0\hat{P} \propto I.$$

In addition, $\hat{P}^2 = I$ and therefore the parity operator for Dirac spinors can be identified as either

$$\hat{P} = +\gamma^0 \quad \text{or} \quad \hat{P} = -\gamma^0.$$

It is conventional to choose $\hat{P} = +\gamma^0$ such that under the parity transformation, the form of the Dirac equation is unchanged provided the Dirac spinors transform as

$$\psi \rightarrow \hat{P}\psi = \gamma^0\psi. \tag{4.71}$$

The intrinsic parity of a fundamental particle is defined by the action of the parity operator $\hat{P} = \gamma^0$ on a spinor for a particle at rest. For example, the $u_1$ spinor for a particle *at rest* given by (4.42), is an eigenstate of the parity operator with

$$\hat{P}u_1 = \gamma^0 u_1 = \begin{pmatrix} 1 & 0 & 0 & 0 \\ 0 & 1 & 0 & 0 \\ 0 & 0 & -1 & 0 \\ 0 & 0 & 0 & -1 \end{pmatrix} \sqrt{2m} \begin{pmatrix} 1 \\ 0 \\ 0 \\ 0 \end{pmatrix} = +u_1.$$

Similarly, $\hat{P}u_2 = +u_2$, $\hat{P}v_1 = -v_1$ and $\hat{P}v_2 = -v_2$. Hence the *intrinsic* parity of a fundamental spin-half particle is opposite to that of a fundamental spin-half antiparticle.

The conventional choice of $\hat{P} = +\gamma^0$ rather than $\hat{P} = -\gamma^0$, corresponds to defining the intrinsic parity of particles to be positive and the intrinsic parity of antiparticles to be negative,

$$\hat{P}u(m,0) = +u(m,0) \quad \text{and} \quad \hat{P}v(m,0) = -v(m,0).$$

Since particles and antiparticles are always created and destroyed in pairs, this choice of sign has no physical consequence. Finally, it is straightforward to verify that the action of the parity operator on Dirac spinors corresponding to a particle with momentum $\mathbf{p}$ reverses the momentum but does not change the spin state, for example

$$\hat{P}u_1(E,\mathbf{p}) = +u_1(E,-\mathbf{p}).$$

# Summary

This chapter described the foundations of relativistic quantum mechanics and it is worth reiterating the main points. The formulation of relativistic quantum mechanics in terms of the Dirac equation, which is linear in both time and space derivatives,

$$\hat{H}_D\psi = (\alpha \cdot \hat{\mathbf{p}} + \beta m)\psi = i\frac{\partial \psi}{\partial t},$$

implies new degrees of freedom of the wavefunction. Solutions to the Dirac equation are represented by four-component Dirac spinors. These solutions provide a natural description of the spin of the fundamental fermions and antifermions. The $E < 0$ solutions to the Dirac equation are interpreted as negative energy particles propagating backwards in time, or equivalently, the physical positive energy antiparticles propagating forwards in time.

The Dirac equation is usually expressed in terms of four $\gamma$-matrices,

$$(i\gamma^\mu \partial_\mu - m)\psi = 0.$$

The properties of the solutions to the Dirac equation are fully defined by the algebra of the $\gamma$-matrices. Nevertheless, explicit free-particle solutions were derived using the Dirac–Pauli representation. The four-vector probability current can be written in terms of the $\gamma$-matrices

$$j^\mu = \psi^\dagger \gamma^0 \gamma^\mu \psi = \overline{\psi}\gamma^\mu \psi,$$

where $\overline{\psi}$ is the adjoint spinor defined as $\overline{\psi} = \psi^\dagger \gamma^0$. The four-vector current will play a central role in the description of particle interactions through the exchange of force-carrying particles.

The solutions to the Dirac equation provide the relativistic quantum mechanical description of spin-half particles and antiparticles. In particular the states $u_\uparrow$, $u_\downarrow$, $v_\uparrow$ and $v_\downarrow$, which are simultaneous eigenstates of the Dirac Hamiltonian and the helicity operator, form a suitable basis for the calculations of cross sections and decay rates that follow.

Finally, two discrete symmetry transformations were introduced, charge conjugation and parity, with corresponding operators

$$\psi \to \hat{C}\psi = i\gamma^2\psi^* \quad \text{and} \quad \psi \to \hat{P}\psi = \gamma^0\psi.$$

The transformation properties of the fundamental interactions under parity and charge-conjugation operations will be discussed in detail in the context of the weak interaction.

## Problems

**4.1** Show that

$$[\hat{\mathbf{p}}^2, \hat{\mathbf{r}} \times \hat{\mathbf{p}}] = 0,$$

and hence the Hamiltonian of the free-particle Schrödinger equation commutes with the angular momentum operator.

**4.2** Show that $u_1$ and $u_2$ are orthogonal, i.e. $u_1^\dagger u_2 = 0$.

**4.3** Verify the statement that the Einstein energy–momentum relationship is recovered if any of the four Dirac spinors of (4.48) are substituted into the Dirac equation written in terms of momentum, $(\gamma^\mu p_\mu - m)u = 0$.

**4.4** For a particle with four-momentum $p^\mu = (E, \mathbf{p})$, the general solution to the free-particle Dirac Equation can be written

$$\psi(p) = [au_1(p) + bu_2(p)]e^{i(\mathbf{p}\cdot\mathbf{x}-Et)}.$$

Using the explicit forms for $u_1$ and $u_2$, show that the four-vector current $j^\mu = (\rho, \mathbf{j})$ is given by

$$j^\mu = 2p^\mu.$$

Furthermore, show that the resulting probability density and probability current are consistent with a particle moving with velocity $\beta = \mathrm{p}/E$.

**4.5** Writing the four-component spinor $u_1$ in terms of two two-component vectors

$$u = \begin{pmatrix} u_A \\ u_B \end{pmatrix},$$

show that in the non-relativistic limit, where $\beta \equiv \mathrm{v}/c \ll 1$, the components of $u_B$ are smaller than those of $u_A$ by a factor $\mathrm{v}/c$.

**4.6** By considering the three cases $\mu = \nu = 0$, $\mu = \nu \neq 0$ and $\mu \neq \nu$ show that

$$\gamma^\mu\gamma^\nu + \gamma^\nu\gamma^\mu = 2g^{\mu\nu}.$$

**4.7** By operating on the Dirac equation,

$$(i\gamma^\mu\partial_\mu - m)\psi = 0,$$

with $\gamma^\nu\partial_\nu$, prove that the components of $\psi$ satisfy the Klein–Gordon equation,

$$(\partial^\mu\partial_\mu + m^2)\psi = 0.$$

**4.8** Show that

$$(\gamma^\mu)^\dagger = \gamma^0\gamma^\mu\gamma^0.$$

**4.9** Starting from

$$(\gamma^\mu p_\mu - m)u = 0,$$

show that the corresponding equation for the adjoint spinor is

$$\bar{u}(\gamma^\mu p_\mu - m) = 0.$$

Hence, without using the explicit form for the $u$ spinors, show that the normalisation condition $u^\dagger u = 2E$ leads to

$$\bar{u}u = 2m,$$

and that

$$\bar{u}\gamma^\mu u = 2p^\mu.$$

**4.10** Demonstrate that the two relations of Equation (4.45) are consistent by showing that

$$(\sigma \cdot \mathbf{p})^2 = \mathbf{p}^2.$$

**4.11** Consider the $e^+e^- \to \gamma \to e^+e^-$ annihilation process in the centre-of-mass frame where the energy of the photon is $2E$. Discuss energy and charge conservation for the two cases where:

(a) the negative energy solutions of the Dirac equation are interpreted as negative energy particles propagating backwards in time;
(b) the negative energy solutions of the Dirac equation are interpreted as positive energy antiparticles propagating forwards in time.

**4.12** Verify that the helicity operator

$$\hat{h} = \frac{\hat{\boldsymbol{\Sigma}} \cdot \hat{\mathbf{p}}}{2p} = \frac{1}{2p}\begin{pmatrix} \sigma \cdot \hat{\mathbf{p}} & 0 \\ 0 & \sigma \cdot \hat{\mathbf{p}} \end{pmatrix},$$

commutes with the Dirac Hamiltonian,

$$\hat{H}_D = \alpha \cdot \hat{\mathbf{p}} + \beta m.$$

**4.13** Show that

$$\hat{P}u_\uparrow(\theta, \phi) = u_\downarrow(\pi - \theta, \pi + \phi),$$

and comment on the result.

**4.14** Under the combined operation of parity and charge conjugation ($\hat{C}\hat{P}$) spinors transform as

$$\psi \to \psi^c = \hat{C}\hat{P}\psi = i\gamma^2\gamma^0\psi^*.$$

Show that up to an overall complex phase factor

$$\hat{C}\hat{P}u_\uparrow(\theta, \phi) = v_\downarrow(\pi - \theta, \pi + \phi).$$

**4.15** Starting from the Dirac equation, derive the identity

$$\bar{u}(p')\gamma^\mu u(p) = \frac{1}{2m}\bar{u}(p')(p + p')^\mu u(p) + \frac{i}{m}\bar{u}(p')\Sigma^{\mu\nu} q_\nu u(p),$$

where $q = p' - p$ and $\Sigma^{\mu\nu} = \frac{i}{4}[\gamma^\mu, \gamma^\nu]$.

# 5 Interaction by particle exchange

In the modern understanding of particle physics, the interactions between particles are mediated by the exchange of force carrying gauge bosons. The rigorous theoretical formalism for describing these interactions is Quantum Field Theory, which is beyond the scope of this book. Here the concepts are developed in the context of relativistic quantum mechanics. The main purpose of this short chapter is to describe how interactions arise from the exchange of virtual particles and to provide an introduction to Quantum Electrodynamics.

## 5.1 First- and second-order perturbation theory

In quantum mechanics, the transition rate $\Gamma_{fi}$ between an initial state $i$ and a final state $f$ is given by Fermi's golden rule $\Gamma_{fi} = 2\pi|T_{fi}|^2\rho(E_f)$, where $T_{fi}$ is the transition matrix element, given by the perturbation expansion

$$T_{fi} = \langle f|V|i\rangle + \sum_{j\neq i} \frac{\langle f|V|j\rangle\langle j|V|i\rangle}{E_i - E_j} + \cdots .$$

The first two terms in the perturbation series can be viewed as "scattering in a potential" and "scattering via an intermediate state $j$" as indicated in Figure 5.1. In the classical picture of interactions, particles act as sources of fields that give rise to a potential in which other particles scatter.

In quantum mechanics, the process of scattering in a static potential corresponds to the first-order term in the perturbation expansion, $\langle f|V|i\rangle$. This picture of

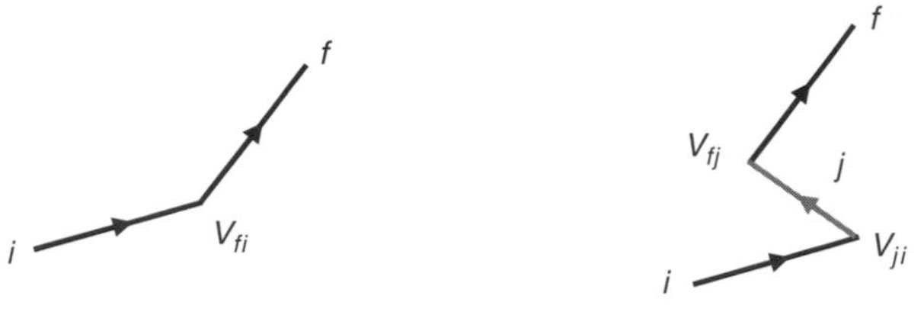


Fig. 5.1 Scattering in an external potential $V_{fi}$ and scattering via an intermediate state, $j$.

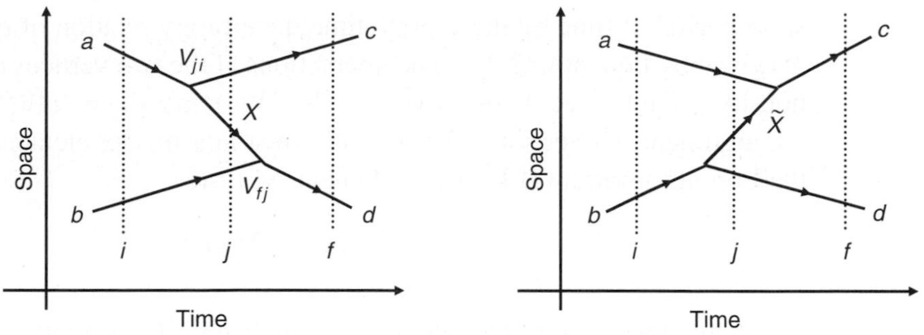


Fig. 5.2 Two possible time-orderings for the process $a + b \to c + d$.

scattering in the potential produced by another particle is unsatisfactory on a number of levels. When a particle scatters in a potential there is a transfer of momentum from one particle to another without any apparent mediating body. Furthermore, the description of forces in terms of potentials seems to imply that if a distant particle were moved suddenly, the potential due to that particle would change instantaneously at all points in space, seemingly in violation the special theory of relativity. In Quantum Field Theory, interactions between particles are mediated by the exchange of other particles and there is no mysterious action at a distance. The forces between particles result from the transfer of the momentum carried by the exchanged particle.

### 5.1.1 Time-ordered perturbation theory

The process of interaction by particle exchange can be formulated by using time-ordered perturbation theory. Consider the particle interaction, $a+b \to c+d$, which can occur via an intermediate state corresponding to the exchange of a particle $X$. There are two possible space-time pictures for this process, shown in Figure 5.2. In the first space-time picture, the initial state $|i\rangle$ corresponds to the particles $a+b$, the intermediate state $|j\rangle$ corresponds to $c + b + X$, and the final state $|f\rangle$ corresponds to $c + d$. In this time-ordered diagram, particle $a$ can be thought of as emitting the exchanged particle $X$, and then at a later time $X$ is absorbed by $b$. In QED this could correspond to an electron emitting a photon that is subsequently absorbed by a second electron. The corresponding term in the quantum-mechanical perturbation expansion is

$$T_{fi}^{ab} = \frac{\langle f|V|j\rangle\langle j|V|i\rangle}{E_i - E_j} = \frac{\langle d|V|X + b\rangle\langle c + X|V|a\rangle}{(E_a + E_b) - (E_c + E_X + E_b)}. \tag{5.1}$$

The notation $T_{fi}^{ab}$ refers to the time ordering where the interaction between $a$ and $X$ occurs before that between $X$ and $b$. It should be noted that the energy of the intermediate state is not equal to that of the initial state, $E_j \neq E_i$, which is allowed for a short period of time by the energy–time uncertainty relation of quantum mechanics given by Equation (2.47). The interactions at the two vertices are defined by the non-invariant matrix elements $V_{ji} = \langle c + X|V|a\rangle$ and $V_{fj} = \langle d|V|X + b\rangle$. Following the arguments of Section 3.2.1, the non-invariant matrix element $V_{ji}$ is related to the Lorentz-invariant (LI) matrix element $\mathcal{M}_{ji}$ by

$$V_{ji} = \mathcal{M}_{ji} \prod_k (2E_k)^{-1/2},$$

where the index $k$ runs over the particles involved. In this case

$$V_{ji} = \langle c + X|V|a\rangle = \frac{\mathcal{M}_{a\to c+X}}{(2E_a 2E_c 2E_X)^{1/2}},$$

where $\mathcal{M}_{a\to c+X}$ is the LI matrix element for the fundamental interaction $a \to c + X$. The requirement that the matrix element $\mathcal{M}_{a\to c+X}$ is Lorentz invariant places strong constraints on its possible mathematical structure. To illustrate the concept of interaction by particle exchange, the simplest possible Lorentz-invariant coupling is assumed here, namely a scalar. In this case, the LI matrix element is simply $\mathcal{M}_{a\to c+X} = g_a$, and thus

$$V_{ji} = \langle c + X|V|a\rangle = \frac{g_a}{(2E_a 2E_c 2E_X)^{1/2}},$$

and the magnitude of the coupling constant $g_a$ is a measure of the strength of the scalar interaction. Similarly

$$V_{fj} = \langle d|V|X + b\rangle = \frac{g_b}{(2E_b 2E_d 2E_X)^{1/2}},$$

where $g_b$ is the coupling strength at the $b + X \to d$ interaction vertex. Therefore, with the assumed scalar form for the interaction, the second-order term in the perturbation series of (5.1) is

$$\begin{aligned} T_{fi}^{ab} &= \frac{\langle d|V|X + b\rangle\langle c + X|V|a\rangle}{(E_a + E_b) - (E_c + E_X + E_b)} \\ &= \frac{1}{2E_X} \cdot \frac{1}{(2E_a 2E_b 2E_c 2E_d)^{1/2}} \cdot \frac{g_a g_b}{(E_a - E_c - E_X)}. \end{aligned} \tag{5.2}$$

The LI matrix element for the process $a + b \to c + d$ is related to the corresponding transition matrix element by (3.9),

$$\mathcal{M}_{fi}^{ab} = (2E_a 2E_b 2E_c 2E_d)^{1/2} T_{fi}^{ab},$$

and thus from (5.2),

$$\mathcal{M}_{fi}^{ab} = \frac{1}{2E_X} \cdot \frac{g_a g_b}{(E_a - E_c - E_X)}. \tag{5.3}$$

The matrix element of (5.3) is Lorentz invariant in the sense that it is defined in terms of wavefunctions with an appropriate LI normalisation and has an LI scalar

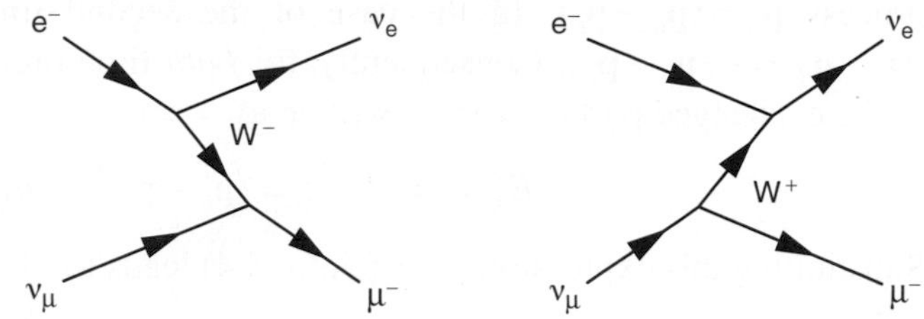


Fig. 5.3 The two lowest-order time-ordered diagrams contributing to $e^-\nu_\mu \to \nu_e\mu^-$ scattering.

form for the interaction. It should be noted that for this second-order process in perturbation theory, momentum is conserved at the interaction vertices but energy is not, $E_j \neq E_i$. Furthermore, the exchanged particle $X$ satisfies the usual energy–momentum relationship, $E_X^2 = \mathbf{p}_X^2 + m_X^2$, and is termed "on-mass shell".

The second possible time-ordering for the process $a + b \to c + d$ is shown in the right-hand plot of Figure 5.2 and corresponds to $b$ emitting $\tilde{X}$ which is subsequently absorbed by $a$. The exchanged particle $\tilde{X}$ in this time-ordering is assumed to have the same mass as $X$ but has opposite charge(s). This must be the case if charge is to be conserved at each vertex. For example, in the process of $e^-\nu_\mu \to \nu_e\mu^-$ scattering, shown Figure 5.3, one of the time-ordered diagrams involves the exchange of a $W^-$ and the other time-ordered diagram involves the exchange of a $W^+$. In the case of a QED process, there is no need to make this distinction for the neutral photon.

By repeating the steps that led to (5.3), it is straightforward to show that the LI matrix element for the second time-ordered diagram of Figure 5.2 is

$$\mathcal{M}_{fi}^{ba} = \frac{1}{2E_X} \cdot \frac{g_a g_b}{(E_b - E_d - E_X)}.$$

In quantum mechanics the different amplitudes for a process need to be summed to obtain the total amplitude. Here the total amplitude (at lowest order) is given by the sum of the two time-ordered amplitudes

$$\begin{aligned}\mathcal{M}_{fi} &= \mathcal{M}_{fi}^{ab} + \mathcal{M}_{fi}^{ba}\\ &= \frac{g_a g_b}{2E_X} \cdot \left(\frac{1}{E_a - E_c - E_X} + \frac{1}{E_b - E_d - E_X}\right),\end{aligned}$$

which, using energy conservation $E_b - E_d = E_c - E_a$, can be written

$$\begin{aligned}\mathcal{M}_{fi} &= \frac{g_a g_b}{2E_X} \cdot \left(\frac{1}{E_a - E_c - E_X} - \frac{1}{E_a - E_c + E_X}\right)\\ &= \frac{g_a g_b}{(E_a - E_c)^2 - E_X^2}.\end{aligned} \tag{5.4}$$

For both time-ordered diagrams, the energy of the exchanged particle $E_X$ is related to its momentum by the usual Einstein energy–momentum relation, $E_X^2 = \mathbf{p}_X^2 + m_X^2$. Since momentum is conserved at each interaction vertex, for the first time-ordered process $\mathbf{p}_X = (\mathbf{p}_a - \mathbf{p}_c)$. In the case of the second time-ordered process $\mathbf{p}_{\tilde{X}} = (\mathbf{p}_b - \mathbf{p}_d) = -(\mathbf{p}_a - \mathbf{p}_c)$. Consequently, for *both* time-ordered diagrams the energy of the exchanged particle can be written as

$$E_X^2 = \mathbf{p}_X^2 + m_X^2 = (\mathbf{p}_a - \mathbf{p}_c)^2 + m_X^2.$$

Substituting this expression for $E_X^2$ into (5.4) leads to

$$\mathcal{M}_{fi} = \frac{g_a g_b}{(E_a - E_c)^2 - (\mathbf{p}_a - \mathbf{p}_c)^2 - m_X^2} \qquad (5.5)$$
$$= \frac{g_a g_b}{(p_a - p_c)^2 - m_X^2},$$

where $p_a$ and $p_c$ are the respective four-momenta of particles $a$ and $c$. Finally writing the four-momentum of the exchanged *virtual* particle $X$ as

$$q = p_a - p_c,$$

gives

$$\mathcal{M}_{fi} = \frac{g_a g_b}{q^2 - m_X^2}. \qquad (5.6)$$

This is a remarkable result. The sum over the two possible time-ordered diagrams in second-order perturbation theory has produced an expression for the interaction matrix element that depends on the four-vector scalar product $q^2$ and is therefore manifestly Lorentz invariant. In (5.6) the terms $g_a$ and $g_b$ are associated with the interaction vertices and the term

$$\frac{1}{q^2 - m_X^2}, \qquad (5.7)$$

is referred to as the propagator, is associated with the exchanged particle.

## 5.2 Feynman diagrams and virtual particles

In Quantum Field Theory, the sum over all possible time-orderings is represented by a *Feynman diagram*. The left-hand side of the diagram represents the initial state, and the right-hand side represents the final state. Everything in between represents the manner in which the interaction happened, regardless of the ordering in time. The Feynman diagram for the scattering process $a + b \to c + d$, shown in Figure 5.4, therefore represents the sum over the two possible time-orderings. The exchanged particles which appear in the intermediate state of a Feynman diagram, are referred to as *virtual particles*. A virtual particle is a mathematical construct representing the effect of summing over all possible time-ordered diagrams and,

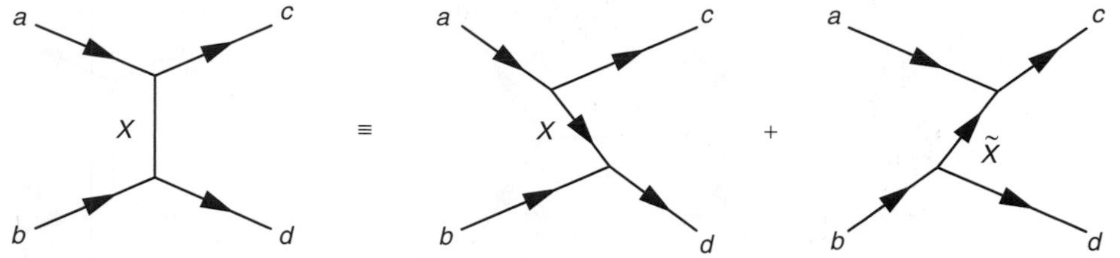


Fig. 5.4 The relation between the Feynman diagram for $a+b \to c+d$ scattering and the two possible time-ordered diagrams.

where appropriate, summing over the possible polarisation states of the exchanged particle.

From (5.5) it can be seen that the four-momentum $q$ which appears in the propagator is given by the difference between the four-momenta of the particles entering and leaving the interaction vertex, $q = p_a - p_c = p_d - p_b$. Hence $q$ can be thought of as the four-momentum of the exchanged virtual particle. By expressing the interaction in terms of the exchange of a virtual particle with four-momentum $q$, both momentum *and* energy are conserved at the interaction vertices of a Feynman diagram. This is not the case for the individual time-ordered diagrams, where energy is not conserved at a vertex. Because the $q^2$-dependence of the propagator is determined by the four-momenta of the incoming and outgoing particles, the virtual particle (which really represents the effect of the sum of all time-ordered diagrams) does not satisfy the Einstein energy–momentum relationship and it is termed off mass-shell, $q^2 \neq m_X^2$. Whilst the effects of the exchanged particles are observable through the forces they mediate, they are not directly detectable. To observe the exchanged particle would require its interaction with another particle and this would be a different Feynman diagram with additional (and possibly different) virtual particles.

The four-momentum $q$ which appears in the propagator can be determined from the conservation of four-momentum at the interaction vertices. For example, Figure 5.5 shows the Feynman diagrams for the $s$-channel annihilation and the $t$-channel scattering processes introduced in Section 2.2.3. For the annihilation process, the four-momentum of the exchanged virtual particle is

$$q = p_1 + p_2 = p_3 + p_4,$$

and therefore $q^2 = (p_1 + p_2)^2$ which is the Mandelstam $s$ variable. Previously (2.13) it was shown that $s = (E_1^* + E_2^*)^2$, where $E_1^*$ and $E_2^*$ are the energies of the initial-state particles in the centre-of-mass frame. Consequently, for an $s$-channel process $q^2 > 0$ and the exchanged virtual particle is termed "time-like" (the square of the time-like component of $q$ is larger than the sum of the squares of the three space-like components). For the $t$-channel scattering diagram of Figure 5.5, the four momentum of the exchanged particle is given by $q = p_1 - p_3 = p_4 - p_2$. In this

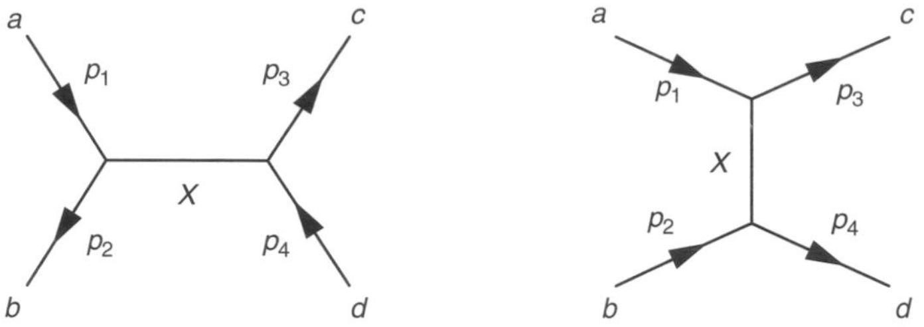


Fig. 5.5 Feynman diagrams for illustrative $s$-channel annihilation and $t$-channel scattering processes.

case $q^2$ is equal to the Mandelstam $t$ variable. In Chapter 8 it will be shown that, for a $t$-channel process, $q^2$ is always less than zero and the exchanged virtual particle is termed "space-like".

### 5.2.1 Scattering in a potential

The covariant formulation of a scalar interaction in terms of the exchange of (virtual) particles leads to a Lorentz-invariant matrix element of the form

$$\mathcal{M}_{fi} = \frac{g_a g_b}{q^2 - m_X^2}. \tag{5.8}$$

This was derived by considering the second-order term in the perturbation expansion for $T_{fi}$. It is reasonable to ask how this picture of interaction by particle exchange relates to the familiar concept of scattering in a potential. For example, the differential cross section for the scattering of non-relativistic electrons (v $\ll c$) in the electrostatic field of a stationary proton can be calculated using first perturbation theory with

$$M = \langle\psi_f|V(\mathrm{r})|\psi_i\rangle = \int \psi_f^* V(\mathrm{r})\,\psi_i\,\mathrm{d}^3\mathbf{r}, \tag{5.9}$$

where $V(\mathrm{r})$ is the effective *static* electrostatic potential due to the proton and $\psi_i$ and $\psi_f$ are the wavefunctions of the initial and final-state electron. In the non-relativistic limit, this approach successfully reproduces the experimental data. However, the concept of scattering from a static potential is intrinsically not Lorentz invariant; the integral in matrix element of (5.9) only involves spatial coordinates.

The covariant picture of scattering via particle exchange applies equally in the non-relativistic and highly relativistic limits. In the non-relativistic limit, the form of the static potential used in first-order perturbation theory is that which reproduces the results of the more general treatment of the scattering process in terms of particle exchange. For example, the form of the potential $V(\mathrm{r})$ that reproduces the low-energy limit of scattering with the matrix element of (5.8) is the

Yukawa potential

$$V(\mathrm{r}) = g_a g_b \frac{e^{-m\mathrm{r}}}{\mathrm{r}}.$$

In this way, it is possible relate the formalism of interaction by particle exchange to the more familiar (non-relativistic) concept of scattering in a static potential. For an interaction involving the exchange of a massless particle, such as the photon, the Yukawa potential reduces to the usual $1/\mathrm{r}$ form of the Coulomb potential.

## 5.3 Introduction to QED

Quantum Electrodynamics (QED) is the Quantum Field Theory of the electromagnetic interaction. A first-principles derivation of the QED interaction from QFT goes beyond the scope of this book. Nevertheless, the basic interaction and corresponding Feynman rules can be obtained following the arguments presented in Section 5.1.1. The LI matrix element for a scalar interaction, given in (5.6), is composed of three parts: the strength of interaction at each of the two vertices, $\langle\psi_c|V|\psi_a\rangle$ and $\langle\psi_d|V|\psi_b\rangle$, and the propagator for the exchanged virtual particle of mass $m_X$, which can be written as

$$\mathcal{M} = \langle\psi_c|V|\psi_a\rangle \frac{1}{q^2 - m_X^2} \langle\psi_d|V|\psi_b\rangle. \tag{5.10}$$

In the previous example, the simplest Lorentz-invariant choice for the interaction vertex was used, namely a scalar interaction of the form $\langle\psi|V|\phi\rangle \propto g$. To obtain the QED matrix element for a scattering process, such as that shown in Figure 5.6, the corresponding expression for the QED interaction vertex is required. Furthermore, for the exchange of the photon, which is a spin-1 particle, it is necessary to sum over the quantum-mechanical amplitudes for the possible polarisation states.

The free photon field $A_\mu$ can be written in terms of a plane wave and a four-vector $\varepsilon^{(\lambda)}$ for the polarisation state $\lambda$,

$$A_\mu = \varepsilon_\mu^{(\lambda)} e^{i(\mathbf{p}\cdot\mathbf{x} - Et)}.$$

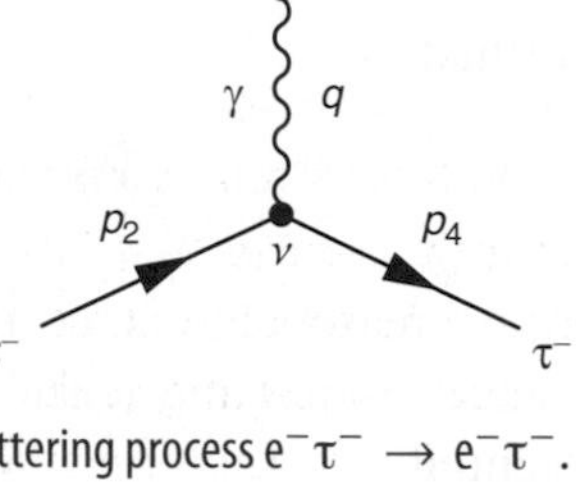

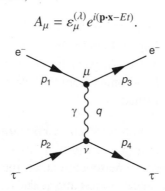

Fig. 5.6 The Feynman diagram for the QED scattering process $e^-\tau^- \to e^-\tau^-$.

The properties of the free photon field in classical electromagnetism are discussed in detail in Appendix D. For a real (as opposed to virtual) photon, the polarisation vector is always transverse to the direction of motion. Thus, a photon propagating in the $z$-direction can be described by two orthogonal polarisation states

$$\varepsilon^{(1)} = (0,1,0,0) \quad \text{and} \quad \varepsilon^{(2)} = (0,0,1,0).$$

The fundamental interaction between a fermion with charge $q$ and an electromagnetic field described by a four-potential $A_\mu = (\phi, \mathbf{A})$ can be obtained by making the minimal substitution (see Section 4.7.5)

$$\partial_\mu \to \partial_\mu + iqA_\mu\,,$$

where $A_\mu = (\phi, -\mathbf{A})$ and $\partial_\mu = (\partial/\partial t, +\boldsymbol{\nabla})$. With this substitution, the free-particle Dirac equation becomes

$$\gamma^\mu \partial_\mu \psi + iq\gamma^\mu A_\mu \psi + im\psi = 0. \tag{5.11}$$

This is the wave equation for a spin-half particle in the presence of the electromagnetic field $A_\mu$. The interaction Hamiltonian can be obtained by pre-multiplying (5.11) by $i\gamma^0$ to give

$$i\frac{\partial \psi}{\partial t} + i\gamma^0 \boldsymbol{\gamma} \cdot \boldsymbol{\nabla}\psi - q\gamma^0\gamma^\mu A_\mu \psi - m\gamma^0\psi = 0,$$

where $\boldsymbol{\gamma} \cdot \boldsymbol{\nabla}$ is shorthand for $\gamma^1 \frac{\partial}{\partial x} + \gamma^2 \frac{\partial}{\partial y} + \gamma^3 \frac{\partial}{\partial z}$. Since

$$\hat{H}\psi = i\frac{\partial \psi}{\partial t},$$

the Hamiltonian for a spin-half particle in an electromagnetic field can be identified as

$$\hat{H} = (m\gamma^0 - i\gamma^0 \boldsymbol{\gamma} \cdot \boldsymbol{\nabla}) + q\gamma^0\gamma^\mu A_\mu. \tag{5.12}$$

The first term on the RHS of (5.12) is just the free-particle Hamiltonian $\hat{H}_D$ already discussed in Chapter 4, and therefore can be identified as the combined rest mass and kinetic energy of the particle. The final term on the RHS of (5.12) is the contribution to the Hamiltonian from the interaction and thus the potential energy operator can be identified as

$$\hat{V}_D = q\gamma^0\gamma^\mu A_\mu. \tag{5.13}$$

This result appears reasonable since the time-like ($\mu = 0$) contribution to $\hat{V}_D$ is $q\gamma^0\gamma^0 A_0 = q\phi$, which is just the normal expression for the energy of a charge $q$ in the scalar potential $\phi$.

The Lorentz-invariant matrix element for the QED process of $e^-\tau^- \to e^-\tau^-$ scattering, shown in Figure 5.6, can be obtained by using the potential of (5.13) for the interaction at the $e^-\gamma$ vertex (labelled by the index $\mu$)

$$\langle\psi(p_3)|\hat{V}_D|\psi(p_1)\rangle \to u_e^\dagger(p_3)\, Q_e e\gamma^0\gamma^\mu\varepsilon_\mu^{(\lambda)}\, u_e(p_1),$$

where the charge $q = Qe$ is expressed in terms of the magnitude of charge of the electron (such that $Q_e = -1$). Since the wavefunctions are four-component spinors, the final-state particle necessarily appears as the Hermitian conjugate $u^\dagger(p_3) \equiv u^{*T}(p_3)$ rather than $u^*(p_3)$. Similarly, the interaction at the $\tau^-\gamma$ vertex (labelled $\nu$) can be written as

$$u_\tau^\dagger(p_4)\, Q_\tau e\gamma^0\gamma^\nu\varepsilon_\nu^{(\lambda)*}\, u_\tau(p_2).$$

The QED matrix element is obtained by summing over both the two possible time orderings and the possible polarisation states of the virtual photon. The sum over the two time-ordered diagrams follows directly from the previous result of (5.10). Hence the Lorentz-invariant matrix element for this QED process, which now includes the additional sum over the photon polarisation, is

$$\mathcal{M} = \sum_\lambda \left[u_e^\dagger(p_3)Q_e e\gamma^0\gamma^\mu u_e(p_1)\right]\varepsilon_\mu^{(\lambda)}\frac{1}{q^2}\varepsilon_\nu^{(\lambda)*}\left[u_\tau^\dagger(p_4)Q_\tau e\gamma^0\gamma^\nu u_\tau(p_2)\right]. \tag{5.14}$$

In Appendix D.4.3, it is shown that the sum over the polarisation states of the virtual photon can be taken to be

$$\sum_\lambda \varepsilon_\mu^{(\lambda)}\varepsilon_\nu^{(\lambda)*} = -g_{\mu\nu},$$

and therefore (5.14) becomes

$$\mathcal{M} = \left[Q_e e\, u_e^\dagger(p_3)\gamma^0\gamma^\mu u_e(p_1)\right]\frac{-g_{\mu\nu}}{q^2}\left[Q_\tau e\, u_\tau^\dagger(p_4)\gamma^0\gamma^\nu u_\tau(p_2)\right]. \tag{5.15}$$

This can be written in a more compact form using the adjoint spinors defined by $\overline{\psi} = \psi^\dagger\gamma^0$,

$$\mathcal{M} = -[Q_e e\,\overline{u}_e(p_3)\gamma^\mu u_e(p_1)]\frac{g_{\mu\nu}}{q^2}[Q_\tau e\,\overline{u}_\tau(p_4)\gamma^\nu u_\tau(p_2)]. \tag{5.16}$$

In Appendix B.3 it is shown that the combination of spinors and $\gamma$-matrices $j^\mu = \overline{u}(p)\gamma^\mu u(p')$ forms as contravariant four-vector under Lorentz boosts. By writing the four-vector currents

$$j_e^\mu = \overline{u}_e(p_3)\gamma^\mu u_e(p_1) \quad \text{and} \quad j_\tau^\nu = \overline{u}_\tau(p_4)\gamma^\nu u_\tau(p_2). \tag{5.17}$$

Equation (5.16) can be written in the manifestly Lorentz-invariant form of a four-vector scalar product

$$\mathcal{M} = -Q_e Q_\tau e^2\,\frac{j_e\cdot j_\tau}{q^2}. \tag{5.18}$$

This demonstrates that the interaction potential of (5.13) gives rise to a Lorentz-invariant description of the electromagnetic interaction.

## 5.4 Feynman rules for QED

A rigorous derivation of the matrix element of (5.16) can be obtained in the framework of quantum field theory. Nevertheless, the treatment described here shares some of the features of the full QED derivation, namely the summation over all possible time-orderings and polarisation states of the massless photon which gives rise to the photon propagator term $g_{\mu\nu}/q^2$, and the $Qe\overline{u}\gamma^\mu u$ form of the QED interaction between a fermion and photon. The expression for the matrix element of (5.16) hides a lot of complexity. If every time we were presented with a new Feynman diagram, it was necessary to derive the matrix element from first principles, this would be extremely time consuming. Fortunately this is not the case; the matrix element for *any* Feynman diagram can be written down immediately by following a simple set of rules that are derived formally from QFT.

There are three basic elements to the matrix element corresponding to the Feynman diagram of Figure 5.6: (i) the Dirac spinors for the external fermions (the initial- and final-state particles); (ii) a propagator term for the virtual photon; and (iii) a vertex factor at each interaction vertex. For each of these elements of the Feynman diagram, there is a *Feynman rule* for the corresponding term in the matrix element. The product of all of these terms is equivalent to $-i\mathcal{M}$. In their simplest form, the Feynman rules for QED, which can be used to calculate lowest-order cross sections, are as follows.

| | | |
|---|---|---|
| initial-state particle: | $u(p)$ |  |
| final-state particle: | $\overline{u}(p)$ |  |
| initial-state antiparticle: | $\overline{v}(p)$ |  |
| final-state antiparticle: | $v(p)$ |  |
| initial-state photon: | $\varepsilon_\mu(p)$ |  |
| final-state photon: | $\varepsilon^*_\mu(p)$ |  |
| photon propagator: | $-\dfrac{ig_{\mu\nu}}{q^2}$ |  |
| fermion propagator: | $-\dfrac{i(\gamma^\mu q_\mu + m)}{q^2 - m^2}$ |  |
| QED vertex: | $-iQe\gamma^\mu$ |  |

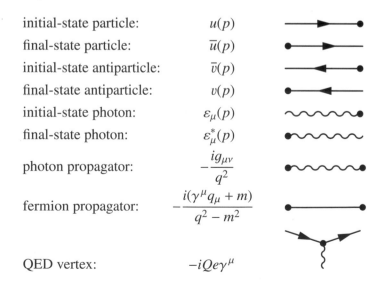

$$\overline{u}(p_3)[ie\gamma^\mu]u(p_1)$$

$$\frac{-ig_{\mu\nu}}{q^2}$$

$$\overline{u}(p_4)[ie\gamma^\nu]u(p_2)$$

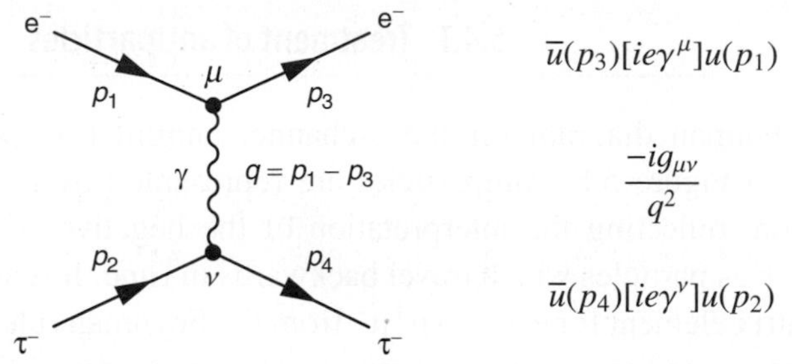

**Fig. 5.7**   The Feynman diagram for the QED scattering process $e^-\tau^- \rightarrow e^-\tau^-$ and the associated elements of the matrix element constructed from the Feynman rules. The matrix element is comprised of a term for the electron current, a term for the tau-lepton current and a term for the photon propagator.

It should be noted that in QED, the fundamental interaction is between a single photon and two spin-half fermions; there is no QED vertex connecting more than three particles. For this reason, all valid QED processes are described by Feynman diagrams formed from the basic three-particle QED vertex.

The use of the Feynman rules is best illustrated by example. Consider again the Feynman diagram for the process $e^-\tau^- \rightarrow e^-\tau^-$, shown in Figure 5.7. The indices $\mu$ and $\nu$ label the two interaction vertices. Applying the Feynman rules to the electron current, gives an adjoint spinor for the final-state electron, a factor $ie\gamma^\mu$ for the interaction vertex labelled by $\mu$, and a spinor for the initial-state electron. The adjoint spinor is always written first and thus the contribution to the matrix element from the electron current is

$$\overline{u}(p_3)[ie\gamma^\mu]u(p_1).$$

The same procedure applied to the tau-lepton current gives

$$\overline{u}(p_4)[ie\gamma^\nu]u(p_2).$$

Finally, the photon propagator contributes a factor

$$\frac{-ig_{\mu\nu}}{q^2}.$$

The product of these three terms gives $-i\mathcal{M}$ and therefore

$$-i\mathcal{M} = [\overline{u}(p_3)\{ie\gamma^\mu\}u(p_1)]\frac{-ig_{\mu\nu}}{q^2}[\overline{u}(p_4)\{ie\gamma^\nu\}u(p_2)], \qquad (5.19)$$

which is equivalent to the expression of (5.16), which was obtained from first principle arguments.

### 5.4.1 Treatment of antiparticles

The Feynman diagram for the $s$-channel annihilation process $e^+e^- \to \mu^+\mu^-$ is shown in Figure 5.8. Antiparticles are represented by lines in the negative time direction, reflecting the interpretation of the negative energy solutions to Dirac equation as particles which travel backwards in time. It is straightforward to obtain the matrix element for $e^+e^- \to \mu^+\mu^-$ from the Feynman rules. The part of the matrix element due to the electron and muon currents are, respectively,

$$\overline{v}(p_2)[ie\gamma^\mu]u(p_1) \quad \text{and} \quad \overline{u}(p_3)[ie\gamma^\nu]v(p_4),$$

where $v$-spinors are used to describe the antiparticles. As before, the photon propagator is

$$\frac{-ig_{\mu\nu}}{q^2}.$$

Hence the matrix element for $e^+e^- \to \mu^+\mu^-$ annihilation is given by

$$-i\mathcal{M} = [\overline{v}(p_2)\{ie\gamma^\mu\}u(p_1)]\frac{-ig_{\mu\nu}}{q^2}[\overline{u}(p_3)\{ie\gamma^\nu\}v(p_4)]. \tag{5.20}$$

The QED matrix element for the $s$-channel annihilation process $e^+e^- \to \mu^+\mu^-$ given by (5.20) is very similar to that for the $t$-channel scattering process $e^-\tau^- \to e^-\tau^-$ given by (5.19). Apart from the presence of the $v$-spinors for antiparticles, the only difference is the order in which the particles appear in the expressions for the currents. Fortunately, it is not necessary to remember the Feynman rules that specify whether a particle/antiparticle appears in the matrix element as a spinor or as an adjoint spinor, there is an easy mnemonic; the first particle encountered when following the line representing a fermion current from the end to the start in the direction *against* the sense of the arrows, always appears as the adjoint spinor. For example, in Figure 5.8, the incoming $e^+$ and outgoing $\mu^-$ are written as adjoint spinors.

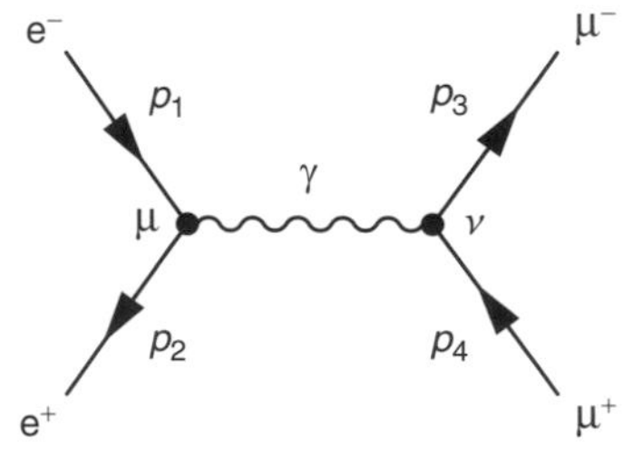


Fig. 5.8 The lowest-order Feynman diagram for the QED annihilation process $e^+e^- \to \mu^+\mu^-$.

## Summary

This chapter described the basic ideas behind the description of particle interactions in terms of particle exchange and provided an introduction to the Feynman rules of QED. A number of important concepts were introduced. The sum of all possible time-ordered diagrams results in a Lorentz-invariant (LI) matrix element including propagator terms for the exchanged virtual particles of the form

$$\frac{1}{q^2 - m_X^2}.$$

The four-momentum appearing in the propagator term was shown to be determined by energy and momentum conservation at the interaction vertices.

The matrix element for a particular process is then constructed from propagator terms for the virtual particles and vertex factors. In QED, the interaction between a photon and a charged fermion has the form

$$ieQ\,\overline{u}_f\gamma^\mu u_i,$$

where $u_i$ is the spinor for the initial-state particle and $\overline{u}_f$ is the adjoint spinor for the final-state particle. Finally, for each element of a Feynman diagram there is a corresponding Feynman rule which can be used to construct the matrix element for the diagram.

## Problems

**5.1** Draw the two time-ordered diagrams for the $s$-channel process shown in Figure 5.5. By repeating the steps of Section 5.1.1, show that the propagator has the same form as obtained for the $t$-channel process.

Hint: one of the time-ordered diagrams is non-intuitive, remember that in second-order perturbation theory the intermediate state does not conserve energy.

**5.2** Draw the *two* lowest-order Feynman diagrams for the Compton scattering process $\gamma e^- \to \gamma e^-$.

**5.3** Draw the lowest-order $t$-channel and $u$-channel Feynman diagrams for $e^+e^- \to \gamma\gamma$ and use the Feynman rules for QED to write down the corresponding matrix elements.

# 6 Electron–positron annihilation

Experimental results from electron–positron colliders have been central to the development and understanding of the Standard Model. In this chapter, the derivation of the cross section for $e^+e^- \to \mu^+\mu^-$ annihilation is used as an example of a calculation in QED. The cross section is first calculated using helicity amplitudes to evaluate the matrix elements, highlighting the underlying spin structure of the interaction. In the final starred section, the more abstract trace formalism is introduced.

## 6.1 Calculations in perturbation theory

In QED, the dominant contribution to a cross section or decay rate is usually the Feynman diagram with the fewest number of interaction vertices, known as the lowest-order (LO) diagram. For the annihilation process $e^+e^- \to \mu^+\mu^-$, there is just a single lowest-order QED diagram, shown in Figure 6.1. In this diagram there are two QED interaction vertices, each of which contributes a factor $ie\gamma^\mu$ to the matrix element. Therefore, regardless of any other considerations, the matrix element squared $|\mathcal{M}|^2$ will be proportional to $e^4$ or equivalently $|\mathcal{M}|^2 \propto \alpha^2$, where $\alpha$ is the dimensionless fine-structure constant $\alpha = e^2/4\pi$. In general, each QED vertex contributes a factor of $\alpha$ to the expressions for cross sections and decay rates.

In addition to the lowest-order diagram of Figure 6.1, there are an infinite number of higher-order-diagrams resulting in the same final state. For example, three of the next-to-leading-order (NLO) diagrams for $e^+e^- \to \mu^+\mu^-$, each with four interaction vertices, are shown in Figure 6.2. Taken in isolation, the matrix element squared for each of these diagrams has a factor $\alpha$ for each of the four QED vertices, and hence $|\mathcal{M}|^2 \propto \alpha^4$. However, in quantum mechanics the individual Feynman diagrams for a particular process can not be taken in isolation; the total amplitude $\mathcal{M}_{fi}$ for a particular process is the sum of all individual amplitudes giving the *same final state*. In the case of $e^+e^- \to \mu^+\mu^-$, this sum can be written as

$$\mathcal{M}_{fi} = \mathcal{M}_{\mathrm{LO}} + \sum_j \mathcal{M}_{1,j} + \cdots, \tag{6.1}$$

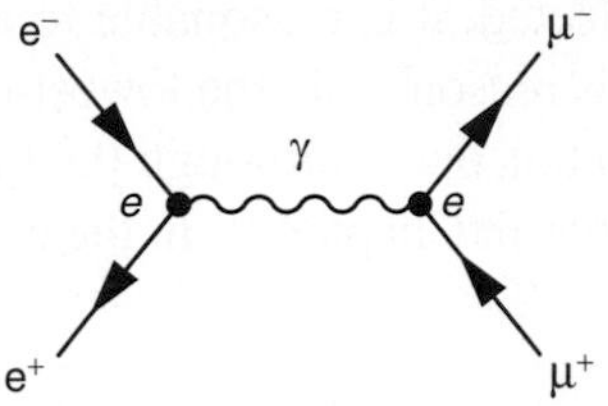


Fig. 6.1 The lowest-order Feynman diagram for the QED annihilation process $e^+e^- \to \mu^+\mu^-$.

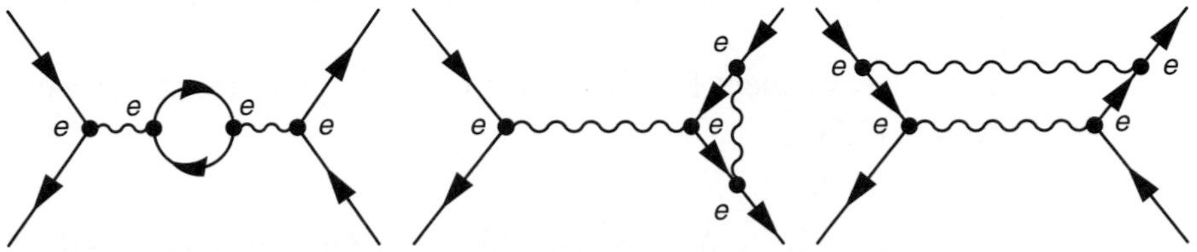


Fig. 6.2 Three of the $\mathcal{O}(\alpha^4)$ Feynman diagrams contributing the QED annihilation process $e^+e^- \to \mu^+\mu^-$.

where $\mathcal{M}_{\rm LO}$ is the matrix element for the single lowest-order (LO) diagram of Figure 6.1, $\mathcal{M}_{1,j}$ are the matrix elements for the NLO diagrams with four interaction vertices, including those of Figure 6.2, and the dots indicate the higher-order diagrams with more than four vertices. The dependence of the each of the terms in (6.1) on $\alpha$ can be shown explicitly by writing it as

$$\mathcal{M}_{fi} = \alpha M_{\rm LO} + \alpha^2 \sum_j M_{1,j} + \cdots ,$$

where the various powers of the coupling constant $\alpha$ have been factored out of the matrix element, such that $\mathcal{M}_{\rm LO}$ is written as $\alpha M_{\rm LO}$, etc.

Physical observables, such as decay rates and cross sections, depend on the matrix element squared given by

$$\begin{aligned} |\mathcal{M}_{fi}|^2 &= \left(\alpha M_{\rm LO} + \alpha^2 \sum_j M_{\rm 1,j} + \cdots\right)\left(\alpha M_{\rm LO}^* + \alpha^2 \sum_k M_{\rm 1,k}^* + \cdots\right) \\ &= \alpha^2 |M_{\rm LO}|^2 + \alpha^3 \sum_j \left(M_{\rm LO}M_{1,j}^* + M_{\rm LO}^* M_{1,j}\right) + \alpha^4 \sum_{jk} M_{1,j}M_{\rm 1,k}^* + \cdots . \end{aligned} \tag{6.2}$$

In general, the individual amplitudes are complex and the contributions from different diagrams can interfere either positively or negatively. Equation (6.2) gives the QED perturbation expansion in terms of powers of $\alpha$. For QED, the dimensionless coupling constant $\alpha \approx 1/137$ is sufficiently small that this series converges rapidly and is dominated by the LO term. The interference between the lowest-order diagram and the NLO diagrams, terms such as $(M_{\rm LO}M_{1,j}^* + M_{\rm LO}^* M_{1,j})$, are suppressed by a factor of $\alpha \approx 1/137$ relative to the lowest-order term. Hence, if all higher-order terms are neglected, it is reasonable to expect QED calculations to be accurate to $\mathcal{O}(1\%)$. For this reason, only the lowest-order diagram(s) will be considered for the calculations in this book, although the impact of the higher-order diagrams will be discussed further in Chapter 10 in the context of renormalisation.

## 6.2 Electron–positron annihilation

The matrix element for the lowest-order diagram for the process $e^+e^- \to \mu^+\mu^-$ is given in (5.20),

$$\mathcal{M} = -\frac{e^2}{q^2} g_{\mu\nu} \left[\overline{v}(p_2)\gamma^\mu u(p_1)\right]\left[\overline{u}(p_3)\gamma^\nu v(p_4)\right] \tag{6.3}$$

$$= -\frac{e^2}{q^2} g_{\mu\nu} j_e^\mu j_\mu^\nu, \tag{6.4}$$

where the electron and muon four-vector currents are defined as

$$j_e^\mu = \overline{v}(p_2)\gamma^\mu u(p_1) \quad \text{and} \quad j_\mu^\nu = \overline{u}(p_3)\gamma^\nu v(p_4). \tag{6.5}$$

The four-momentum of the virtual photon is determined by conservation of energy and momentum at the interaction vertex, $q = p_1 + p_2 = p_3 + p_4$, and therefore $q^2 = (p_1 + p_2)^2 = s$, where $s$ is the centre-of-mass energy squared. Hence the matrix element of (6.4) can be written as

$$\mathcal{M} = -\frac{e^2}{s} j_e \cdot j_\mu. \tag{6.6}$$

Assuming that the electron and positron beams have equal energies, which has been the case for the majority of high-energy $e^+e^-$ colliders, the centre-of-mass energy is simply twice the beam energy, $\sqrt{s} = 2E_{\text{beam}}$.

### 6.2.1 Spin sums

To calculate the $e^+e^- \to \mu^+\mu^-$ cross section, the matrix element of (6.6) needs to be evaluated taking into account the possible spin states of the particles involved. Because each of the $e^+$, $e^-$, $\mu^+$ and $\mu^-$ can be in one of two possible helicity states, there are four possible helicity configurations in the initial state, shown Figure 6.3, and four possible helicity configurations in the $\mu^+\mu^-$ final state. Hence, the process $e^+e^- \to \mu^+\mu^-$ consists of sixteen possible *orthogonal* helicity combinations, each of

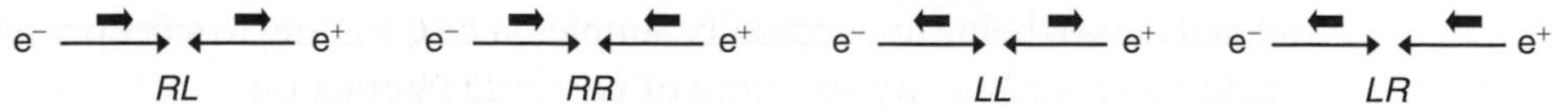


Fig. 6.3 The four possible helicity combinations in the $\mathrm{e}^+\mathrm{e}^-$ initial state.

which constitutes a separate physical process, for example $\mathrm{e}^+_\uparrow\mathrm{e}^-_\uparrow \to \mu^+_\uparrow\mu^-_\uparrow$ (denoted $RR \to RR$) and $\mathrm{e}^+_\uparrow\mathrm{e}^-_\uparrow \to \mu^+_\uparrow\mu^-_\downarrow$. Because the helicity states involved are orthogonal, the processes for the different helicity configurations do not interfere and the matrix element squared for each of the sixteen possible helicity configurations can be considered independently.

For a particular initial-state spin configuration, the total $\mathrm{e}^+\mathrm{e}^- \to \mu^+\mu^-$ annihilation rate is given by the sum of the *rates* for the four possible $\mu^+\mu^-$ helicity states (each of which is a separate process). Therefore, for a given *initial-state* helicity configuration, the cross section is obtained by taking the sum of the four corresponding $|\mathcal{M}|^2$ terms. For example, for the case where the colliding electron and positron are both in right-handed helicity states,

$$\sum |\mathcal{M}_{RR}|^2 = |\mathcal{M}_{RR\to RR}|^2 + |\mathcal{M}_{RR\to RL}|^2 + |\mathcal{M}_{RR\to LR}|^2 + |\mathcal{M}_{RR\to LL}|^2.$$

In most $\mathrm{e}^+\mathrm{e}^-$ colliders, the colliding electron and positron beams are unpolarised, which means that there are equal numbers of positive and negative helicity electrons/positrons present in the initial state. In this case, the helicity configuration for a particular collision is equally likely to occur in any one of the four possible helicity states of the $\mathrm{e}^+\mathrm{e}^-$ initial state. This is accounted for by defining the spin-averaged summed matrix element squared,

$$\begin{aligned}\langle|\mathcal{M}_{fi}|^2\rangle &= \frac{1}{4}\left(|\mathcal{M}_{RR}|^2 + |\mathcal{M}_{RL}|^2 + |\mathcal{M}_{LR}|^2 + |\mathcal{M}_{LL}|^2\right)\\ &= \frac{1}{4}\left(|\mathcal{M}_{RR\to RR}|^2 + |\mathcal{M}_{RR\to RL}|^2 + \cdots + |\mathcal{M}_{RL\to RR}|^2 + \cdots\right),\end{aligned}$$

where the factor $\frac{1}{4}$ accounts for the average over the four possible initial-state helicity configurations. In general, the spin-averaged matrix element is given by

$$\langle|\mathcal{M}_{fi}|^2\rangle = \frac{1}{4}\sum_{\text{spins}} |\mathcal{M}|^2,$$

where the sum corresponds to all possible helicity configurations. Consequently, to evaluate the $\mathrm{e}^+\mathrm{e}^- \to \mu^+\mu^-$ cross section, it is necessary to calculate the matrix element of (6.6) for sixteen helicity combinations. This sum can be performed in two ways. One possibility is to use the trace techniques described in Section 6.5, where the sum is calculated directly using the properties of the Dirac spinors. The second possibility is to calculate each of the sixteen individual helicity amplitudes. This direct calculation of the helicity amplitudes involves more steps, but has the advantages of being conceptually simpler and of leading to a deeper physical understanding of the helicity structure of the QED interaction.

### 6.2.2 Helicity amplitudes

In the limit where the masses of the particles can be neglected, $\sqrt{s} \gg m_\mu$, the four-momenta in the process $e^+e^- \to \mu^+\mu^-$, as shown Figure 6.4, can be written

$$p_1 = (E, 0, 0, E), \tag{6.7}$$
$$p_2 = (E, 0, 0, -E), \tag{6.8}$$
$$p_3 = (E, E\sin\theta, 0, E\cos\theta), \tag{6.9}$$
$$p_4 = (E, -E\sin\theta, 0, -E\cos\theta), \tag{6.10}$$

where, with no loss of generality, the final state $\mu^-$ and $\mu^+$ are taken to be produced with azimuthal angles of $\phi = 0$ and $\phi = \pi$ respectively.

The spinors appearing in the four-vector currents of (6.5) are the ultra-relativistic ($E \gg m$) limit of the helicity eigenstates of (4.67):

$$u_\uparrow = \sqrt{E}\begin{pmatrix} c \\ se^{i\phi} \\ c \\ se^{i\phi} \end{pmatrix}, \quad u_\downarrow = \sqrt{E}\begin{pmatrix} -s \\ ce^{i\phi} \\ s \\ -ce^{i\phi} \end{pmatrix}, \quad v_\uparrow = \sqrt{E}\begin{pmatrix} s \\ -ce^{i\phi} \\ -s \\ ce^{i\phi} \end{pmatrix}, \quad v_\downarrow = \sqrt{E}\begin{pmatrix} c \\ se^{i\phi} \\ c \\ se^{i\phi} \end{pmatrix}, \tag{6.11}$$

where $s = \sin\frac{\theta}{2}$ and $c = \cos\frac{\theta}{2}$. The two possible spinors for initial-state electron with $(\theta = 0, \phi = 0)$ and for the initial-state positron with $(\theta = \pi, \phi = \pi)$ are

$$u_\uparrow(p_1) = \sqrt{E}\begin{pmatrix} 1 \\ 0 \\ 1 \\ 0 \end{pmatrix}, \quad u_\downarrow(p_1) = \sqrt{E}\begin{pmatrix} 0 \\ 1 \\ 0 \\ -1 \end{pmatrix}, \quad v_\uparrow(p_2) = \sqrt{E}\begin{pmatrix} 1 \\ 0 \\ -1 \\ 0 \end{pmatrix}, \quad v_\downarrow(p_2) = \sqrt{E}\begin{pmatrix} 0 \\ -1 \\ 0 \\ -1 \end{pmatrix}.$$

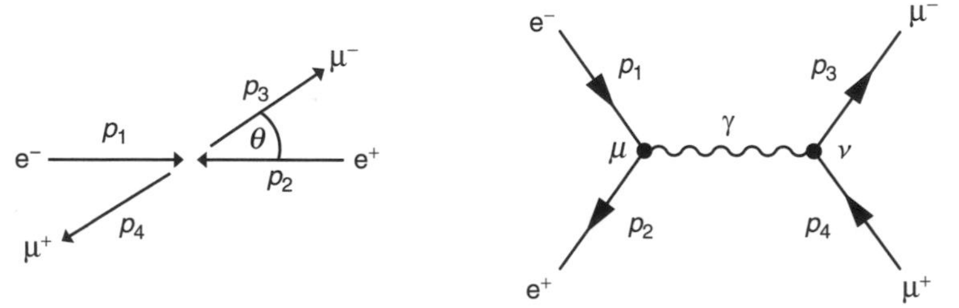


Fig. 6.4 The QED annihilation process $e^+e^- \to \mu^+\mu^-$ viewed in the centre-of-mass frame and the corresponding lowest-order Feynman diagram.

The spinors for the final-state particles are obtained by using the spherical polar angles $(\theta, 0)$ for the $\mu^-$ and $(\pi - \theta, \pi)$ for the $\mu^+$. Using the trigonometric relations

$$\sin\left(\frac{\pi-\theta}{2}\right) = \cos\left(\frac{\theta}{2}\right), \quad \cos\left(\frac{\pi-\theta}{2}\right) = \sin\left(\frac{\theta}{2}\right) \quad \text{and} \quad e^{i\pi} = -1,$$

the spinors for the two possible helicity states of the final-state $\mu^+$ and $\mu^-$ are

$$u_\uparrow(p_3) = \sqrt{E}\begin{pmatrix} c \\ s \\ c \\ s \end{pmatrix}, \; u_\downarrow(p_3) = \sqrt{E}\begin{pmatrix} -s \\ c \\ s \\ -c \end{pmatrix}, \; v_\uparrow(p_4) = \sqrt{E}\begin{pmatrix} c \\ s \\ -c \\ -s \end{pmatrix}, \; v_\downarrow(p_4) = \sqrt{E}\begin{pmatrix} s \\ -c \\ s \\ -c \end{pmatrix}.$$

### 6.2.3 The muon and electron currents

The matrix element for a particular helicity combination is obtained from (6.6),

$$\mathcal{M} = -\frac{e^2}{s} j_e \cdot j_\mu,$$

where the corresponding four-vector currents of (6.5) are defined in terms of the above spinors for the helicity eigenstates. The muon current, $j_\mu^\nu = \overline{u}(p_3)\gamma^\nu v(p_4)$, needs to be evaluated for the four possible final-state helicity combinations shown in Figure 6.5. Using the Dirac–Pauli representation of the $\gamma$-matrices (4.35), it is straightforward to show that, for any two spinors $\psi$ and $\phi$, the components of $\overline{\psi}\gamma^\mu\phi \equiv \psi^\dagger\gamma^0\gamma^\mu\phi$ are

$$\overline{\psi}\gamma^0\phi = \psi^\dagger\gamma^0\gamma^0\phi = \psi_1^*\phi_1 + \psi_2^*\phi_2 + \psi_3^*\phi_3 + \psi_4^*\phi_4, \tag{6.12}$$

$$\overline{\psi}\gamma^1\phi = \psi^\dagger\gamma^0\gamma^1\phi = \psi_1^*\phi_4 + \psi_2^*\phi_3 + \psi_3^*\phi_2 + \psi_4^*\phi_1, \tag{6.13}$$

$$\overline{\psi}\gamma^2\phi = \psi^\dagger\gamma^0\gamma^2\phi = -i(\psi_1^*\phi_4 - \psi_2^*\phi_3 + \psi_3^*\phi_2 - \psi_4^*\phi_1), \tag{6.14}$$

$$\overline{\psi}\gamma^3\phi = \psi^\dagger\gamma^0\gamma^3\phi = \psi_1^*\phi_3 - \psi_2^*\phi_4 + \psi_3^*\phi_1 - \psi_4^*\phi_2. \tag{6.15}$$

Using these relations, the four components of the four-vector current $j_\mu$ can be determined by using the spinors for a particular helicity combination. For example, for the $RL$ combination where the $\mu^-$ is produced in a right-handed helicity state and the $\mu^+$ is produced in a left-handed helicity state, the appropriate spinors are $u_\uparrow(p_3)$ and $v_\downarrow(p_4)$. In this case, from Equations (6.12)–(6.15), the components of

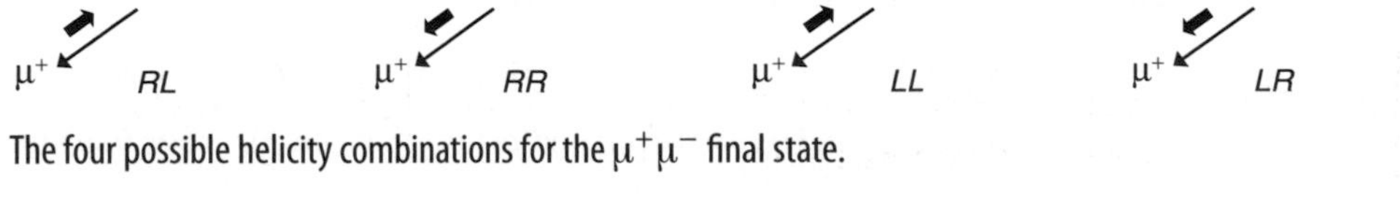

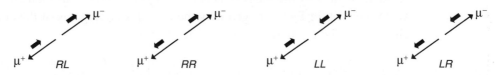

Fig. 6.5 The four possible helicity combinations for the $\mu^+\mu^-$ final state.

the muon current are

$$\begin{aligned}
j_\mu^0 &= \overline{u}_\uparrow(p_3)\gamma^0 v_\downarrow(p_4) = E(cs - sc + cs - sc) = 0,\\
j_\mu^1 &= \overline{u}_\uparrow(p_3)\gamma^1 v_\downarrow(p_4) = E(-c^2 + s^2 - c^2 + s^2) = 2E(s^2 - c^2) = -2E\cos\theta,\\
j_\mu^2 &= \overline{u}_\uparrow(p_3)\gamma^2 v_\downarrow(p_4) = -iE(-c^2 - s^2 - c^2 - s^2) = 2iE,\\
j_\mu^3 &= \overline{u}_\uparrow(p_3)\gamma^3 v_\downarrow(p_4) = E(cs + sc + cs + sc) = 4Esc = 2E\sin\theta.
\end{aligned}$$

Hence, the four-vector current for the helicity combination $\mu^-_\uparrow\mu^+_\downarrow$ is

$$j_{\mu,RL} = \overline{u}_\uparrow(p_3)\gamma^\nu v_\downarrow(p_4) = 2E(0, -\cos\theta, i, \sin\theta).$$

Repeating the calculation for the other three $\mu^+\mu^-$ helicity combinations gives

$$j_{\mu,RL} = \overline{u}_\uparrow(p_3)\gamma^\nu v_\downarrow(p_4) = 2E(0, -\cos\theta, i, \sin\theta), \tag{6.16}$$

$$\begin{aligned}
j_{\mu,RR} &= \overline{u}_\uparrow(p_3)\gamma^\nu v_\uparrow(p_4) = (0,0,0,0),\\
j_{\mu,LL} &= \overline{u}_\downarrow(p_3)\gamma^\nu v_\downarrow(p_4) = (0,0,0,0),
\end{aligned}$$

$$j_{\mu,LR} = \overline{u}_\downarrow(p_3)\gamma^\nu v_\uparrow(p_4) = 2E(0, -\cos\theta, -i, \sin\theta). \tag{6.17}$$

Hence, in the limit where $E \gg m_\mu$, only two of the four $\mu^+\mu^-$ helicity combinations lead to a non-zero four-vector current. This important feature of QED is related to the chiral nature of the interaction, as discussed in Section 6.4.

The electron currents for the four possible initial-state helicity configurations can be evaluated directly using (6.12)–(6.15). Alternatively, the electron currents can be obtained by noting that they differ from the form of the muon currents only in the order in which the particle and antiparticle spinors appear, $j_e^\mu = \overline{v}(p_2)\gamma^\mu u(p_1)$ compared to $j_\mu^\nu = \overline{u}(p_3)\gamma^\nu v(p_4)$. The relationship between $\overline{v}\gamma^\mu u$ and $\overline{u}\gamma^\mu v$ can be found by taking the Hermitian conjugate of the muon current to give

$$\begin{aligned}
[\overline{u}(p_3)\gamma^\mu v(p_4)]^\dagger &= \left[u(p_3)^\dagger\gamma^0\gamma^\mu v(p_4)\right]^\dagger\\
&= v(p_4)^\dagger\gamma^{\mu\dagger}\gamma^{0\dagger}u(p_3) && \text{using } (AB)^\dagger = B^\dagger A^\dagger\\
&= v(p_4)^\dagger\gamma^{\mu\dagger}\gamma^0 u(p_3) && \text{since } \gamma^{0\dagger} = \gamma^0\\
&= v(p_4)^\dagger\gamma^0\gamma^\mu u(p_3) && \text{since } \gamma^{\mu\dagger}\gamma^0 = \gamma^0\gamma^\mu\\
&= \overline{v}(p_4)\gamma^\mu u(p_3).
\end{aligned}$$

The effect of taking the Hermitian conjugate of the QED current is to swap the order in which the spinors appear in the current. Because each element of the four-vector current, labelled by the index $\mu$, is just a complex number, the elements of the four-vector current for $\overline{v}\gamma^\mu u$ are given by the complex conjugates of the corresponding elements of $\overline{u}\gamma^\mu v$. Therefore from (6.16) and (6.17),

$$\begin{aligned}
\overline{v}_\downarrow(p_4)\gamma^\mu u_\uparrow(p_3) &= [\overline{u}_\uparrow(p_3)\gamma^\mu v_\downarrow(p_4)]^* = 2E(0, -\cos\theta, -i, \sin\theta)\\
\overline{v}_\uparrow(p_4)\gamma^\mu u_\downarrow(p_3) &= [\overline{u}_\downarrow(p_3)\gamma^\mu v_\uparrow(p_4)]^* = 2E(0, -\cos\theta, i, \sin\theta).
\end{aligned}$$

By setting $\theta = 0$, it follows that the two non-zero electron currents are

$$j_{e,RL} = \overline{v}_{\downarrow}(p_2)\gamma^{\mu}u_{\uparrow}(p_1) = 2E(0,-1,-i,0), \tag{6.18}$$

$$j_{e,LR} = \overline{v}_{\uparrow}(p_2)\gamma^{\mu}u_{\downarrow}(p_1) = 2E(0,-1,i,0). \tag{6.19}$$

Furthermore, from $j_{\mu,LL} = j_{\mu,RR} = 0$, it follows that $j_{e,LL}$ and $j_{e,RR}$ are also zero.

### 6.2.4 The $e^+e^- \to \mu^+\mu^-$ cross section

In the limit $E \gg m$, only two of the four helicity combinations for both the initial and final state lead to non-zero four-vector currents. Therefore, in the process $e^+e^- \to \mu^+\mu^-$ only the four helicity combinations shown in Figure 6.6 give non-zero matrix elements. For each of these four helicity combinations, the matrix element is obtained from

$$\mathcal{M} = -\frac{e^2}{s} j_e \cdot j_\mu.$$

For example, the matrix element $\mathcal{M}_{RL\to RL}$ for the process $e^-_{\uparrow}e^+_{\downarrow} \to \mu^-_{\uparrow}\mu^+_{\downarrow}$ is determined by the scalar product of the currents

$$j^{\mu}_{e,RL} = \overline{v}_{\downarrow}(p_2)\gamma^{\mu}u_{\uparrow}(p_1) = 2E(0,-1,-i,0),$$

$$\text{and} \quad j^{\nu}_{\mu,RL} = \overline{u}_{\uparrow}(p_3)\gamma^{\nu}v_{\downarrow}(p_4) = 2E(0,-\cos\theta, i, \sin\theta).$$

Taking the four-vector scalar product $j_{e,RL}\cdot j_{\mu,RL}$ and writing $s = 4E^2$ gives

$$\begin{aligned}\mathcal{M}_{RL\to RL} &= -\frac{e^2}{s}\,[2E(0,-1,-i,0)]\cdot[2E(0,-\cos\theta,i,\sin\theta)]\\ &= e^2(1+\cos\theta)\\ &= 4\pi\alpha(1+\cos\theta).\end{aligned}$$

Using the muon and electron currents of (6.16)–(6.19), it follows that the matrix elements corresponding to the four helicity combinations of Figure 6.6 are

$$|\mathcal{M}_{RL\to RL}|^2 = |\mathcal{M}_{LR\to LR}|^2 = (4\pi\alpha)^2(1+\cos\theta)^2, \tag{6.20}$$

$$|\mathcal{M}_{RL\to LR}|^2 = |\mathcal{M}_{LR\to RL}|^2 = (4\pi\alpha)^2(1-\cos\theta)^2, \tag{6.21}$$

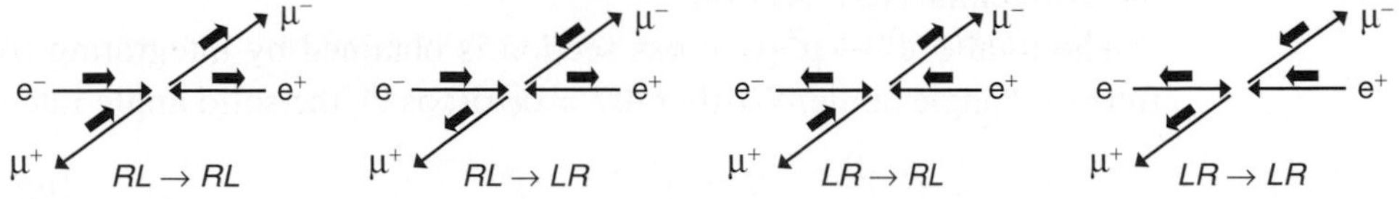


Fig. 6.6 The four helicity combinations for $e^+e^- \to \mu^+\mu^-$ that in the limit $E \gg m$ give non-zero matrix elements.

where $\theta$ is the angle of the outgoing $\mu^-$ with respect to the incoming $e^-$ direction. The spin-averaged matrix element for the process $e^+e^- \to \mu^+\mu^-$ is given by

$$\begin{aligned}\langle|\mathcal{M}_{fi}|^2\rangle &= \frac{1}{4}\times\left(|\mathcal{M}_{RL\to RL}|^2 + |\mathcal{M}_{RL\to LR}|^2 + |\mathcal{M}_{LR\to RL}|^2 + |\mathcal{M}^2_{LR\to LR}|\right)\\ &= \frac{1}{4}e^4\left[2(1+\cos\theta)^2 + 2(1-\cos\theta)^2\right]\\ &= e^4(1+\cos^2\theta).\end{aligned} \tag{6.22}$$

The corresponding differential cross section is obtained by substituting the spin-averaged matrix element squared of (6.22) into the general cross section formula of (3.50) with $p_i^* = p_f^* = E$, giving

$$\frac{d\sigma}{d\Omega} = \frac{1}{64\pi^2 s}e^4(1+\cos^2\theta),$$

where the solid angle is defined in terms of the spherical polar angles of the $\mu^-$ as measured in the centre-of-mass frame. Finally, when written in terms of the dimensionless coupling constant $\alpha = e^2/(4\pi)$, the $e^+e^- \to \mu^+\mu^-$ differential cross section becomes

$$\frac{d\sigma}{d\Omega} = \frac{\alpha^2}{4s}(1+\cos^2\theta). \tag{6.23}$$

Figure 6.7 shows the predicted $(1+\cos^2\theta)$ angular distribution of the $e^+e^- \to \mu^+\mu^-$ differential cross section broken down into the contributions from the different helicity combinations. The distribution is forward–backward symmetric, meaning that equal numbers of $\mu^-$ are produced in the forward hemisphere ($\cos\theta > 0$) as in the backwards hemisphere ($\cos\theta < 0$). This symmetry is a direct consequence of the parity conserving nature of the QED interaction, as explained in Chapter 11.

The right-hand plot of Figure 6.7 shows the measured $e^+e^- \to \mu^+\mu^-$ differential cross section at $\sqrt{s} = 34.4$ GeV from the JADE experiment, which operated between 1979 and 1986 at the PETRA $e^+e^-$ collider at the DESY laboratory in Hamburg. The $(1+\cos^2\theta)$ nature of the dominant QED contribution is apparent. However, the interpretation of these data is complicated the presence of electroweak corrections arising from the interference between the QED amplitude and that from the Feynman diagram involving the exchange of a Z boson (see Chapter 15). This results in a relatively small forward–backward asymmetry in the differential cross section.

The total $e^+e^- \to \mu^+\mu^-$ cross section is obtained by integrating (6.23) over the full solid angle range. Writing $d\Omega = d\phi\, d(\cos\theta)$, the solid angle integral is simply

$$\int (1+\cos^2\theta)\, d\Omega = 2\pi\int_{-1}^{+1}(1+\cos^2\theta)\, d(\cos\theta) = \frac{16\pi}{3}.$$

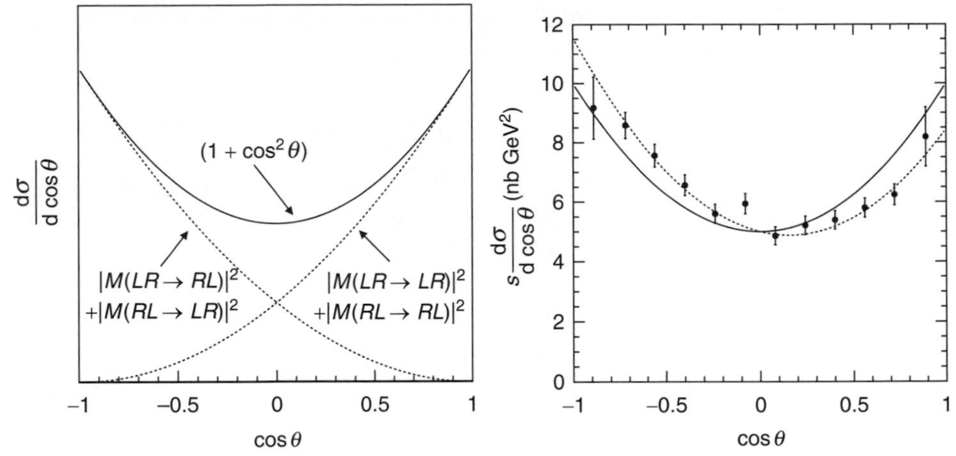


Fig. 6.7 (left) The QED prediction for the distribution of $\cos\theta$ in $e^+e^- \to \mu^+\mu^-$ annihilation, where $\theta$ is the angle of the outgoing $\mu^-$ with respect to the incoming $e^-$ direction. (right) The measured $e^+e^- \to \mu^+\mu^-$ differential cross section at $\sqrt{s} = 34.4$ GeV from the JADE experiment, adapted from Bartel *et al.* (1985). The solid curve is the lowest-order QED prediction. The dotted curve includes electroweak corrections.

Therefore, the lowest-order prediction for the total $e^+e^- \to \mu^+\mu^-$ cross section is

$$\sigma = \frac{4\pi\alpha^2}{3s}. \tag{6.24}$$

Figure 6.8 shows the experimental measurements of the $e^+e^- \to \mu^+\mu^-$ cross section at centre-of-mass energies of $\sqrt{s} < 40$ GeV. In this case, the electroweak corrections are negligible (the effects of interference with the Z boson exchange diagram average to zero in the solid angle integral) and the lowest-order QED prediction provides an excellent description of the data. This is an impressive result, starting from first principles, it has been possible to calculate an expression for the cross section for electron–positron annihilation which is accurate at the $\mathcal{O}(1\%)$ level.

### 6.2.5 Lorentz-invariant form

The spin-averaged matrix element of (6.22) is expressed in terms of the angle $\theta$ as measured in the centre-of-mass frame. However, because the matrix element is Lorentz invariant, it also can be expressed in an explicitly Lorentz-invariant form using four-vector scalar products formed from the four-momenta of the initial- and final-state particles. From the four-momenta defined in (6.7)–(6.10),

$$p_1 \cdot p_2 = 2E^2, \quad p_1 \cdot p_3 = E^2(1-\cos\theta) \quad \text{and} \quad p_1 \cdot p_4 = E^2(1+\cos\theta).$$

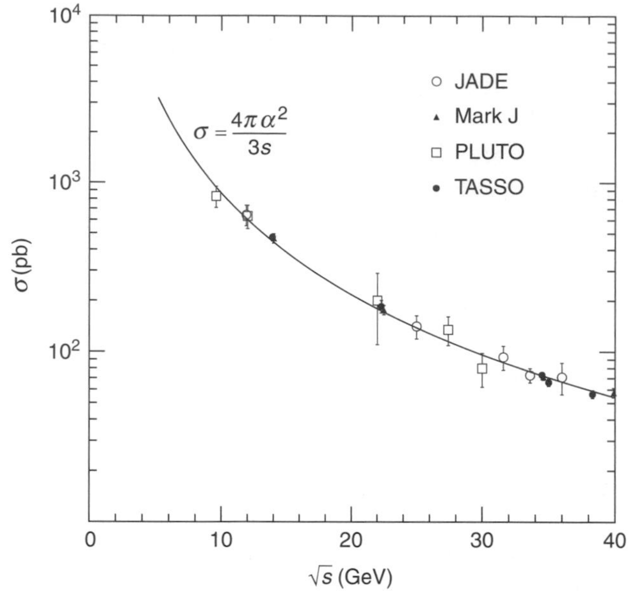


Fig. 6.8 The measured $e^+e^- \to \mu^+\mu^-$ cross section for $\sqrt{s} < 40$ GeV. The curve shows the lowest-order QED prediction of (6.24).

Hence the spin-averaged matrix element of (6.22) can be written as

$$\langle |\mathcal{M}_{fi}|^2 \rangle = 2e^4 \frac{(p_1 \cdot p_3)^2 + (p_1 \cdot p_4)^2}{(p_1 \cdot p_2)^2}. \tag{6.25}$$

The scalar products appearing in (6.25) can be expressed in terms of the Mandelstam variables, where for example

$$s = (p_1 + p_2)^2 = p_1^2 + p_2^2 + 2p_1 \cdot p_2 = m_1^2 + m_2^2 + 2p_1 \cdot p_2.$$

In the limit, where the masses of the particles can be neglected,

$$s = +2p_1 \cdot p_2, \quad t = -2p_1 \cdot p_3 \quad \text{and} \quad u = -2p_1 \cdot p_4,$$

and therefore (6.25) can be written as

$$\langle |\mathcal{M}_{fi}|^2 \rangle = 2e^4 \left( \frac{t^2 + u^2}{s^2} \right). \tag{6.26}$$

This expression, which depends only on Lorentz-invariant quantities, is valid in all frames of reference.

## 6.3 Spin in electron–positron annihilation

The four helicity combinations for the process $e^+e^- \to \mu^+\mu^-$ that give non-zero matrix elements are shown in Figure 6.6. In each case, the spins of the two initial-state particles are aligned, as are the spins of the two final-state particles. Defining the $z$-axis to be in the direction of the incoming electron beam, the $z$-component the combined spin of the $e^+$ and $e^-$ is therefore either $+1$ or $-1$, implying that the non-zero matrix elements correspond to the cases where the electron and positron collide in a state of total spin-1. Therefore, the spin state for the $RL$ helicity combination can be identified as $|S, S_z\rangle = |1, +1\rangle$ and that for the $LR$ combinations as $|1, -1\rangle$. Similarly, the helicity combinations of the $\mu^+\mu^-$ system correspond to spin states of $|1, \pm 1\rangle_\theta$ measured with respect to the axis in the direction of $\mu^-$, as indicated in Figure 6.9.

The angular dependence of the matrix elements for each helicity combination can be understood in terms of these spin states. The operator corresponding to the component of spin along an axis defined by the unit vector $\mathbf{n}$ at an angle $\theta$ to the $z$-axis is $\hat{S}_n = \frac{1}{2}\mathbf{n}\cdot\boldsymbol{\sigma}$. Using this operator, it is possible to express the spin states of the $\mu^+\mu^-$ system in terms of the eigenstates of $\hat{S}_z$ (see Problem 6.6). For example, the spin wavefunction of the $RL$ helicity combination of the $\mu^+\mu^-$ final state, $|1, +1\rangle_\theta$, can be expressed as

$$|1, +1\rangle_\theta = \tfrac{1}{2}(1 - \cos\theta)\,|1, -1\rangle + \tfrac{1}{\sqrt{2}}\sin\theta\,|1, 0\rangle + \tfrac{1}{2}(1 + \cos\theta)\,|1, +1\rangle.$$

The angular distributions of matrix elements of (6.20) and (6.21) can be understood in terms of the inner products of the spin states of initial-state $e^+e^-$ system and the

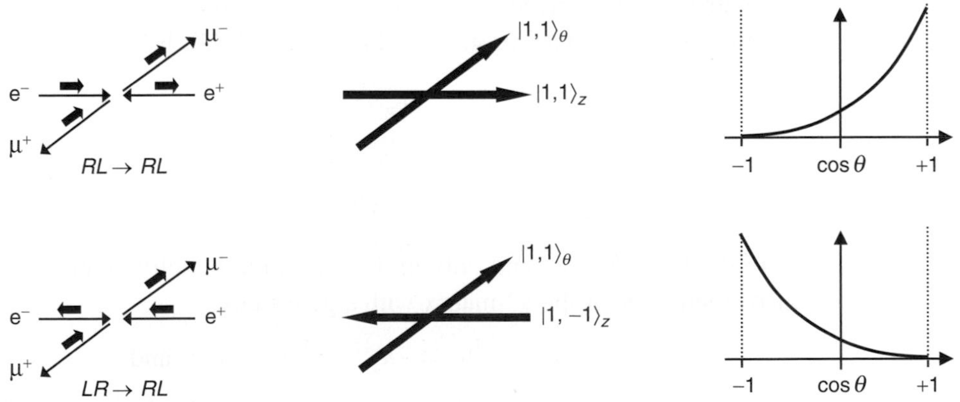


Fig. 6.9 The orientations of the spin-1 system in the $RL \to RL$ and $LR \to RL$ helicity combinations and the angular dependence of the corresponding matrix element in the limit where $E \gg m$.

final-state $\mu^+\mu^-$ system. For example,

$$\mathcal{M}_{RL\to RL} \propto \langle 1,+1|1,+1\rangle_\theta = \tfrac{1}{2}(1+\cos\theta)$$
$$\mathcal{M}_{LR\to RL} \propto \langle 1,-1|1,+1\rangle_\theta = \tfrac{1}{2}(1-\cos\theta).$$

Hence, in the limit $E \gg m$, the spin combinations that give non-zero matrix elements correspond to states of total spin-1 with the spin vector pointing along the direction of the particles motion and the resulting angular distributions can be understood in terms of the quantum mechanics of spin-1. This is consistent with the notion that an interaction of the form $\overline{\phi}\gamma^\mu\psi$ corresponds to the exchange of a spin-1 particle, in this case the photon.

## 6.4 Chirality

In the limit $E \gg m$ only four out of the sixteen possible helicity combinations for the process $e^+e^- \to e^+e^-$ give non-zero matrix elements. This does not happen by chance, but reflects the underlying *chiral* structure of QED. The property of chirality is an important concept in the Standard Model. Chirality is introduced by defining the $\gamma^5$-matrix as

$$\gamma^5 \equiv i\gamma^0\gamma^1\gamma^2\gamma^3 = \begin{pmatrix} 0 & 0 & 1 & 0 \\ 0 & 0 & 0 & 1 \\ 1 & 0 & 0 & 0 \\ 0 & 1 & 0 & 0 \end{pmatrix} = \begin{pmatrix} 0 & I \\ I & 0 \end{pmatrix}. \tag{6.27}$$

The significance of the $\gamma^5$-matrix, whilst not immediately obvious, follows from its mathematical properties and the nature of its eigenstates. The properties of the $\gamma^5$-matrix can be derived (see Problem 6.1) from the commutation and Hermiticity relations of the $\gamma$-matrices given in (4.33) and (4.34), leading to

$$(\gamma^5)^2 = 1, \tag{6.28}$$
$$\gamma^{5\dagger} = \gamma^5, \tag{6.29}$$
$$\gamma^5\gamma^\mu = -\gamma^\mu\gamma^5. \tag{6.30}$$

In the limit $E \gg m$, and only in this limit, the helicity eigenstates of (6.11) are also eigenstates of the $\gamma^5$-matrix with eigenvalues

$$\gamma^5 u_\uparrow = +u_\uparrow, \quad \gamma^5 u_\downarrow = -u_\downarrow, \quad \gamma^5 v_\uparrow = -v_\uparrow \quad \text{and} \quad \gamma^5 v_\downarrow = +v_\downarrow.$$

In general, the eigenstates of the $\gamma^5$-matrix are defined as left- and right-handed *chiral* states (denoted with subscripts $R, L$ to distinguish them from the general helicity eigenstates $\uparrow, \downarrow$) such that

$$\begin{aligned}\gamma^5 u_R &= +u_R \quad \text{and} \quad \gamma^5 u_L = -u_L,\\ \gamma^5 v_R &= -v_R \quad \text{and} \quad \gamma^5 v_L = +v_L.\end{aligned} \tag{6.31}$$

With this convention, when $E \gg m$ the chiral eigenstates are the same as the helicity eigenstates for both particle and antiparticle spinors, for example $u_\uparrow \to u_R$ and $v_\downarrow \to v_L$. Hence, *in general*, the solutions to the Dirac equation which are also eigenstates of $\gamma^5$ are identical to the massless helicity eigenstates of (6.11),

$$u_R \equiv N\begin{pmatrix} c \\ se^{i\phi} \\ c \\ se^{i\phi}\end{pmatrix},\; u_L \equiv N\begin{pmatrix} -s \\ ce^{i\phi} \\ s \\ -ce^{i\phi}\end{pmatrix},\; v_R \equiv N\begin{pmatrix} s \\ -ce^{i\phi} \\ -s \\ ce^{i\phi}\end{pmatrix} \text{ and } v_L \equiv N\begin{pmatrix} c \\ se^{i\phi} \\ c \\ se^{i\phi}\end{pmatrix}, \tag{6.32}$$

where the normalisation is given by $N = \sqrt{E+m}$. Unlike helicity, there is no simple physical interpretation of the property of chirality, it is nevertheless an integral part of the structure of the Standard Model.

## Chiral projection operators

Any Dirac spinor can be decomposed into left- and right-handed chiral components using the chiral projection operators, $P_L$ and $P_R$, defined by

$$\begin{aligned} P_R &= \tfrac{1}{2}(1+\gamma^5),\\ P_L &= \tfrac{1}{2}(1-\gamma^5). \end{aligned} \tag{6.33}$$

Using the properties of the $\gamma^5$-matrix, it is straightforward to show that $P_R$ and $P_L$ satisfy the required algebra of quantum mechanical projection operators, namely,

$$P_R + P_L = 1, \quad P_R P_R = P_R, \quad P_L P_L = P_L \quad \text{and} \quad P_L P_R = 0.$$

In the Dirac–Pauli representation

$$P_R = \frac{1}{2}\begin{pmatrix} 1 & 0 & 1 & 0 \\ 0 & 1 & 0 & 1 \\ 1 & 0 & 1 & 0 \\ 0 & 1 & 0 & 1 \end{pmatrix} \quad \text{and} \quad P_L = \frac{1}{2}\begin{pmatrix} 1 & 0 & -1 & 0 \\ 0 & 1 & 0 & -1 \\ -1 & 0 & 1 & 0 \\ 0 & -1 & 0 & 1 \end{pmatrix}. \tag{6.34}$$

From the definitions of (6.31), it immediately follows that the right-handed chiral projection operator has the properties

$$P_R u_R = u_R, \quad P_R u_L = 0, \quad P_R v_R = 0 \quad \text{and} \quad P_R v_L = v_L.$$

Hence $P_R$ projects out *right-handed* chiral *particle* states and *left-handed* chiral *antiparticle* states. Similarly, for the left-handed chiral projection operator,

$$P_L u_R = 0, \quad P_L u_L = u_L, \quad P_L v_R = v_R \quad \text{and} \quad P_L v_L = 0.$$

Since $P_R$ and $P_L$ project out chiral states, any spinor $u$ can be decomposed into left- and right-handed chiral components with

$$u = a_R\, u_R + a_L\, u_L = \tfrac{1}{2}(1+\gamma^5)u + \tfrac{1}{2}(1-\gamma^5)u,$$

where $a_R$ and $a_L$ are complex coefficients and $u_R$ and $u_L$ are right- and left-handed chiral eigenstates.

### 6.4.1 Chirality in QED

In QED, the fundamental interaction between a fermion and a photon is expressed as a four-vector current $iQ_f e\overline{\psi}\gamma^\mu\phi$, formed from the Dirac spinors $\psi$ and $\phi$. Any four-vector current can be decomposed into contributions from left- and right-handed chiral states using the chiral projection operators defined in (6.33). For example, in the case of QED,

$$\begin{aligned}\overline{\psi}\gamma^\mu\phi &= (a_R^*\overline{\psi}_R + a_L^*\overline{\psi}_L)\gamma^\mu(b_R\phi_R + b_L\phi_L)\\ &= a_R^* b_R\overline{\psi}_R\gamma^\mu\phi_R + a_R^* b_L\overline{\psi}_R\gamma^\mu\phi_L + a_L^* b_R\overline{\psi}_L\gamma^\mu\phi_R + a_L^* b_L\overline{\psi}_L\gamma^\mu\phi_L,\end{aligned}\tag{6.35}$$

where the coefficients, $a$ and $b$, will depend on the spinors being considered. The form of the QED interaction means that two of the chiral currents in (6.35) are always zero. For example, consider the term $\overline{u}_L(p)\gamma^\mu u_R(p')$. The action of $P_R$ on a right-handed chiral spinor leaves the spinor unchanged,

$$u_R(p') = P_R u_R(p').\tag{6.36}$$

Therefore $P_R$ can always be inserted in front of right-handed chiral particle state without changing the expression in which it appears. The equivalent relation for the left-handed adjoint spinor is

$$\begin{aligned}\overline{u}_L(p) &\equiv [u_L(p)]^\dagger\gamma^0 = [P_L u_L(p)]^\dagger\gamma^0 = [\tfrac{1}{2}(1-\gamma^5)u_L(p)]^\dagger\gamma^0\\ &= [u_L(p)]^\dagger\tfrac{1}{2}(1-\gamma^5)\gamma^0 \qquad (\text{using } \gamma^5 = \gamma^{5\dagger})\\ &= [u_L(p)]^\dagger\gamma^0\tfrac{1}{2}(1+\gamma^5) \qquad (\text{using } \gamma^0\gamma^5 = -\gamma^5\gamma^0)\\ &= \overline{u}_L(p)P_R.\end{aligned}$$

From this it follows that

$$\overline{u}_L(p)\gamma^\mu u_R(p') = \overline{u}_L(p)P_R\,\gamma^\mu P_R\, u_R(p').\tag{6.37}$$

But since $\gamma^5\gamma^\mu = -\gamma^\mu\gamma^5$,

$$P_R\gamma^\mu = \tfrac{1}{2}(1+\gamma^5)\gamma^\mu = \gamma^\mu\tfrac{1}{2}(1-\gamma^5) = \gamma^\mu P_L,$$

and thus (6.37) can be written

$$\overline{u}_L(p)\gamma^\mu u_R(p') = \overline{u}_L(p)\gamma^\mu P_L P_R\, u_R(p') = 0,$$

because $P_L P_R = 0$. Therefore, the $\overline{\psi}\gamma^\mu\phi$ form of the QED interaction, implies that only certain combinations of chiral eigenstates give non-zero matrix elements, and the currents of the form

$$\overline{u}_L\gamma^\mu u_R = \overline{u}_R\gamma^\mu u_L = \overline{v}_L\gamma^\mu v_R = \overline{v}_R\gamma^\mu v_L = \overline{v}_L\gamma^\mu u_L = \overline{v}_R\gamma^\mu u_R \equiv 0$$

are always identically zero.

### 6.4.2 Helicity and chirality

It is important not to confuse the concepts of helicity and chirality. Helicity eigenstates are defined by the projection of the spin of a particle onto its direction of motion, whereas the chiral states are the eigenstates of the $\gamma^5$-matrix. The relationship between the helicity eigenstates and the chiral eigenstates can be found by decomposing the general form of the helicity spinors into their chiral components. For example, the right-handed helicity particle spinor of (4.65) can be written as

$$u_\uparrow(\mathrm{p},\theta,\phi) = N\begin{pmatrix} c \\ se^{i\phi} \\ \kappa c \\ \kappa se^{i\phi} \end{pmatrix} \quad \text{with} \quad \kappa = \frac{\mathrm{p}}{E+m} \quad \text{and} \quad N = \sqrt{E+m}.$$

The spinor can be decomposed into its left- and right-handed chiral components by considering the effect of the chiral projection operators,

$$P_R u_\uparrow = \tfrac{1}{2}(1+\kappa)N\begin{pmatrix} c \\ se^{i\phi} \\ c \\ se^{i\phi} \end{pmatrix} \quad \text{and} \quad P_L u_\uparrow = \tfrac{1}{2}(1-\kappa)N\begin{pmatrix} c \\ se^{i\phi} \\ -c \\ -se^{i\phi} \end{pmatrix}.$$

Therefore, the right-handed helicity spinor, expressed in terms of its chiral components, is

$$\begin{aligned} u_\uparrow(\mathrm{p},\theta,\phi) &= \tfrac{1}{2}(1+\kappa)N\begin{pmatrix} c \\ se^{i\phi} \\ c \\ se^{i\phi} \end{pmatrix} + \tfrac{1}{2}(1-\kappa)N\begin{pmatrix} c \\ se^{i\phi} \\ -c \\ -se^{i\phi} \end{pmatrix} \\ &\propto \tfrac{1}{2}(1+\kappa)\,u_R + \tfrac{1}{2}(1-\kappa)\,u_L, \end{aligned} \tag{6.38}$$

where $u_R$ and $u_L$ are chiral eigenstates with $\gamma^5 u_R = +u_R$ and $\gamma^5 u_L = -u_L$. From (6.38) it is clear that it is only in the limit where $E \gg m$ (when $\kappa \to 1$) that the helicity eigenstates are equivalent to the chiral eigenstates. Because only certain combinations of chiral states give non-zero contributions to the QED matrix element, in the ultra-relativistic limit only the corresponding helicity combinations

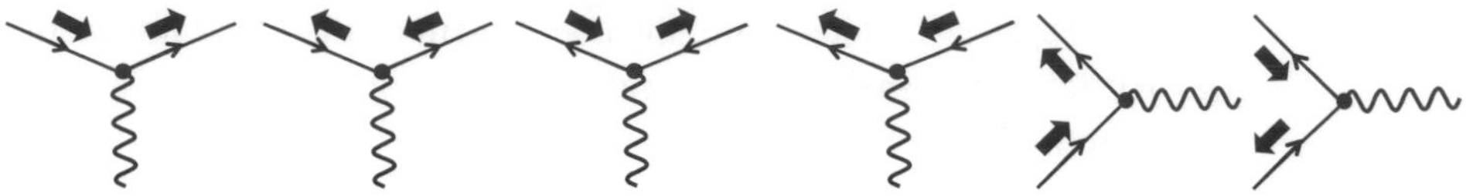

Fig. 6.10 The helicity combinations at the QED vertex which give non-zero four-vector currents in the limit $E \gg m$.

contribute to the QED interaction. The chiral nature of the QED interaction therefore explains the previous observation that only four of the sixteen possible helicity combinations contribute to the $e^+e^- \to \mu^+\mu^-$ annihilation process at high energies.

The correspondence between the helicity and chiral eigenstates in the ultra-relativistic limit implies that for $E \gg m$, the four-vector currents written in terms of the helicity states,

$$\overline{u}_\downarrow \gamma^\mu u_\uparrow = \overline{u}_\uparrow \gamma^\mu u_\downarrow = \overline{v}_\downarrow \gamma^\mu v_\uparrow = \overline{v}_\uparrow \gamma^\mu v_\downarrow = \overline{v}_\downarrow \gamma^\mu u_\downarrow = \overline{v}_\uparrow \gamma^\mu u_\uparrow = 0,$$

are all zero. Therefore in the high-energy QED processes, only the helicity combinations shown in Figure 6.10 give non-zero currents. Consequently, the helicity of the particle leaving the QED vertex is that same as that entering it and helicity is effectively "conserved" in *high-energy* interactions.

## 6.5 *Trace techniques

In the calculation of the $e^+e^- \to \mu^+\mu^-$ cross section described above, the individual matrix elements were calculated for each helicity combination using the explicit representations of the spinors and the $\gamma$-matrices. The resulting squares of the matrix elements were then summed and averaged. This approach is relatively simple and exposes the underlying physics of the interaction. For these reasons, the majority of the calculations that follow will adopt the helicity amplitude approach. In the limit where the masses of the particles can be neglected, these calculations are relatively straightforward as they involve only a limited number of helicity combinations. However, when the particle masses can not be neglected, it is necessary to consider all possible spin combinations. In this case, calculating the individual helicity amplitudes is not particularly efficient (although it is well suited to computational calculations). For more complicated processes, analytic solutions are usually most easily obtained using a powerful technique based on the traces of matrices and the completeness relations for Dirac spinors.

### 6.5.1 Completeness relations

Sums over the spin states of the initial- and final-state particles can be calculated using the completeness relations satisfied by Dirac spinors. The completeness relations are defined by the sum over the two possible spin states of the tensor formed from the product of a spinor with its adjoint spinor,

$$\sum_{s=1}^{2} u_s(p)\overline{u}_s(p),$$

where the sum is for two orthogonal spin states. The sum can be performed using the helicity basis or the spinors $u_1$ and $u_2$, both of which form a complete set of states. Here it is most convenient to work with the spinors $u_1(p)$ and $u_2(p)$, in which case the completeness relation is

$$\sum_{s=1}^{2} u_s(p)\overline{u}_s(p) \equiv u_1(p)\overline{u}_1(p) + u_2(p)\overline{u}_2(p).$$

In the Dirac–Pauli representation, the spinors $u_1$ and $u_2$ can be written as

$$u_s(p) = \sqrt{E+m}\begin{pmatrix} \phi_s \\ \frac{\boldsymbol{\sigma}\cdot\mathbf{p}}{E+m}\phi_s \end{pmatrix} \quad \text{with} \quad \phi_1 = \begin{pmatrix} 1 \\ 0 \end{pmatrix} \quad \text{and} \quad \phi_2 = \begin{pmatrix} 0 \\ 1 \end{pmatrix}.$$

Using $(\boldsymbol{\sigma}\cdot\mathbf{p})^\dagger = \boldsymbol{\sigma}\cdot\mathbf{p}$, the adjoint spinor can be written

$$\overline{u}_s = u_s^\dagger\gamma^0 = \sqrt{E+m}\begin{pmatrix} \phi_s^T & \phi_s^T\frac{(\boldsymbol{\sigma}\cdot\mathbf{p})^\dagger}{E+m} \end{pmatrix}\begin{pmatrix} I & 0 \\ 0 & -I \end{pmatrix} = \sqrt{E+m}\begin{pmatrix} \phi_s^T & -\phi_s^T\frac{(\boldsymbol{\sigma}\cdot\mathbf{p})}{E+m} \end{pmatrix},$$

where $I$ is the $2\times2$ identity matrix. Hence the completeness relation can be written

$$\sum_{s=1}^{2} u_s(p)\overline{u}_s(p) = (E+m)\sum_{s=1}^{2}\begin{pmatrix} \phi_s\phi_s^T & -\frac{\boldsymbol{\sigma}\cdot\mathbf{p}}{E+m}\phi_s\phi_s^T \\ \frac{\boldsymbol{\sigma}\cdot\mathbf{p}}{E+m}\phi_s\phi_s^T & -\frac{(\boldsymbol{\sigma}\cdot\mathbf{p})^2}{(E+m)^2}\phi_s\phi_s^T \end{pmatrix},$$

which using

$$\sum_{s=1}^{2}\phi_s\phi_s^T = \begin{pmatrix} 1 & 0 \\ 0 & 1 \end{pmatrix} \quad \text{and} \quad (\boldsymbol{\sigma}\cdot\mathbf{p})^2 = \mathbf{p}^2 = (E+m)(E-m),$$

gives

$$\sum_{s=1}^{2} u_s(p)\overline{u}_s(p) = \begin{pmatrix} (E+m)I & -\boldsymbol{\sigma}\cdot\mathbf{p} \\ \boldsymbol{\sigma}\cdot\mathbf{p} & (-E+m)I \end{pmatrix}. \tag{6.39}$$

Equation (6.39) can be written in terms of the $\gamma$-matrices as

$$\sum_{s=1}^{2} u_s\overline{u}_s = (\gamma^\mu p_\mu + mI) = \not{p} + m, \tag{6.40}$$

where the "slash" notation is shorthand for $\not{p} \equiv \gamma^\mu p_\mu = E\gamma^0 - p_x\gamma^1 - p_y\gamma^2 - p_z\gamma^3$. Repeating the above derivation, it is straightforward to show that the antiparticle spinors satisfy the completeness relation,

$$\sum_{r=1}^{2} v_r\overline{v}_r = (\gamma^\mu p_\mu - mI) = \not{p} - m, \tag{6.41}$$

where the mass term enters with a different sign compared to the equivalent expression for particle spinors.

### 6.5.2 Spin sums and the trace formalism

The QED, QCD and weak interaction vertex factors all can be written in the form $\overline{u}(p)\,\Gamma\,u(p')$, where $\Gamma$ is a $4\times 4$ matrix constructed out of one or more Dirac $\gamma$-matrices. In index notation, this product of spinors and $\gamma$-matrices can be written

$$\overline{u}(p)\,\Gamma\,u(p') = \overline{u}(p)_j\,\Gamma_{ji}\,u(p')_i,$$

where the indices label the components and summation over repeated indices is implied. It should be noted that $\overline{u}(p)\,\Gamma\,u(p')$ is simply a (complex) number.[1] For the QED vertex $\Gamma = \gamma^\mu$ and the matrix element for the process $e^+e^- \to \mu^+\mu^-$ is given by (6.3),

$$\begin{aligned}\mathcal{M}_{fi} &= -\frac{e^2}{q^2}\,[\overline{v}(p_2)\gamma^\mu u(p_1)]\,g_{\mu\nu}\,[\overline{u}(p_3)\gamma^\nu v(p_4)]\\ &= -\frac{e^2}{q^2}\left[\overline{v}(p_2)\gamma^\mu u(p_1)\right]\left[\overline{u}(p_3)\gamma_\mu v(p_4)\right],\end{aligned} \tag{6.42}$$

where summation over the index $\mu$ is implied. The matrix element squared $|\mathcal{M}_{fi}|^2$ is the product of $\mathcal{M}_{fi}$ and $\mathcal{M}_{fi}^\dagger$, with

$$\mathcal{M}_{fi}^\dagger = \frac{e^2}{q^2}\,[\overline{v}(p_2)\gamma^\nu u(p_1)]^\dagger\,[\overline{u}(p_3)\gamma_\nu v(p_4)]^\dagger\,,$$

where the index $\nu$ has been used for this summation to avoid confusion with the index $\mu$ in the expression for $\mathcal{M}_{fi}$ given in (6.42). Because the components of the

[1] If this is not immediately obvious, consider the $2\times 2$ case of $\mathbf{c}^{\mathrm{T}}\mathbf{B}\mathbf{a}$, where the equivalent product can be written as

$$\begin{aligned}(c_1, c_2)\begin{pmatrix} B_{11} & B_{12} \\ B_{22} & B_{22}\end{pmatrix}\begin{pmatrix} a_1 \\ a_2\end{pmatrix} &= c_1B_{11}a_1 + c_1B_{12}a_2 + c_2B_{21}a_1 + c_2B_{22}\\ &= c_jB_{ji}a_i,\end{aligned}$$

which is just the sum over the product of the components of **a**, **c** and **B**.

currents are simply numbers, the order in which they are written does not matter. Hence $|\mathcal{M}_{fi}|^2 = \mathcal{M}_{fi}\mathcal{M}_{fi}^\dagger$ can be written

$$|\mathcal{M}_{fi}|^2 = \frac{e^4}{q^4}\left[\overline{v}(p_2)\gamma^\mu u(p_1)\right]\left[\overline{v}(p_2)\gamma^\nu u(p_1)\right]^\dagger \times \left[\overline{u}(p_3)\gamma_\mu v(p_4)\right]\left[\overline{u}(p_3)\gamma_\nu v(p_4)\right]^\dagger .$$

The spin-averaged matrix element squared can therefore be written

$$\begin{aligned}\langle|\mathcal{M}_{fi}|^2\rangle &= \frac{1}{4}\sum_{\text{spins}}|\mathcal{M}_{fi}|^2 \\ &= \frac{e^4}{4q^4}\sum_{s,r}\left[\overline{v}^r(p_2)\gamma^\mu u^s(p_1)\right]\left[\overline{v}^r(p_2)\gamma^\nu u^s(p_1)\right]^\dagger \\ &\quad\times\sum_{s',r'}\left[\overline{u}^{s'}(p_3)\gamma_\mu v^{r'}(p_4)\right],\left[\overline{u}^{s'}(p_3)\gamma_\nu v^{r'}(p_4)\right]^\dagger ,\end{aligned} \tag{6.43}$$

where $s$, $s'$, $r$ and $r'$ are the labels for the two possible spin states (or equivalently helicity states) of the four spinors. In this way, the calculation of the spin-averaged matrix element squared has been reduced to the product of two terms of the form

$$\sum_{\text{spins}}\left[\overline{\psi}\Gamma_1\phi\right]\left[\overline{\psi}\Gamma_2\phi\right]^\dagger , \tag{6.44}$$

where $\Gamma_1$ and $\Gamma_2$ are two $4\times4$ matrices, which for this QED process are $\Gamma_1 = \gamma^\mu$ and $\Gamma_2 = \gamma^\nu$. Equation (6.44) can be simplified by writing

$$\left[\overline{\psi}\Gamma\phi\right]^\dagger = \left[\psi^\dagger\gamma^0\Gamma\phi\right]^\dagger = \phi^\dagger\Gamma^\dagger\gamma^{0\dagger}\psi = \phi^\dagger\gamma^0\gamma^0\Gamma^\dagger\gamma^0\psi = \overline{\phi}\gamma^0\Gamma^\dagger\gamma^0\psi,$$

and hence

$$\left[\overline{\psi}\Gamma\phi\right]^\dagger \equiv \overline{\phi}\,\overline{\Gamma}\psi \quad \text{with} \quad \overline{\Gamma} = \gamma^0\Gamma^\dagger\gamma^0.$$

From the properties of the $\gamma$-matrices given (4.33) and (4.34), it can be seen that $\gamma^0\gamma^{\mu\dagger}\gamma^0 = \gamma^\mu$ for all $\mu$. Hence for the QED vertex, with $\Gamma = \gamma^\mu$,

$$\overline{\Gamma} = \gamma^0\gamma^{\mu\dagger}\gamma^0 = \gamma^\mu = \Gamma.$$

Although not shown explicitly, it should be noted that $\overline{\Gamma} = \Gamma$ also holds for the QCD and weak interaction vertices. Hence for all of the Standard Model interactions,

$$\left[\overline{\psi}\Gamma\phi\right]^\dagger \equiv \overline{\phi}\Gamma\psi. \tag{6.45}$$

Using (6.45), the spin-averaged matrix element squared for the process $e^+e^- \to \mu^+\mu^-$ of (6.43) can be written

$$\sum_{\text{spins}} |\mathcal{M}_{fi}|^2 = \frac{e^4}{q^4} \sum_{s,r} [\overline{v}^r(p_2)\gamma^\mu u^s(p_1)] [\overline{u}^s(p_1)\gamma^\nu v^r(p_2)]$$
$$\times \sum_{s',r'} \left[\overline{u}^{s'}(p_3)\gamma_\mu v^{r'}(p_4)\right] \left[\overline{v}^{r'}(p_4)\gamma_\nu u^{s'}(p_3)\right]. \tag{6.46}$$

Denoting the part of (6.46) involving the initial-state $e^+$ and $e^-$ spinors by the tensor $\mathcal{L}_{(e)}^{\mu\nu}$ and writing the matrix multiplication in index form gives

$$\mathcal{L}_{(e)}^{\mu\nu} = \sum_{s,r=1}^{2} \overline{v}_j^r(p_2)\gamma_{ji}^\mu u_i^s(p_1)\ \overline{u}_n^s(p_1)\gamma_{nm}^\nu v_m^r(p_2). \tag{6.47}$$

Since all the quantities in (6.47) are just numbers, with the indices keeping track of the matrix multiplication, this can be written as

$$\mathcal{L}_{(e)}^{\mu\nu} = \left[\sum_{r=1}^{2} v_m^r(p_2)\overline{v}_j^r(p_2)\right]\left[\sum_{s=1}^{2} u_i^s(p_1)\overline{u}_n^s(p_1)\right]\gamma_{ji}^\mu\gamma_{nm}^\nu. \tag{6.48}$$

Using the completeness relations of (6.40), the electron tensor of (6.48) becomes

$$\mathcal{L}_{(e)}^{\mu\nu} = (\not{p}_2 - m)_{mj}(\not{p}_1 + m)_{in}\gamma_{ji}^\mu\gamma_{nm}^\nu, \tag{6.49}$$

where $(\not{p}_2 - m)_{mj}$ is the $(mj)$th element of the $4\times4$ matrix $(\not{p}_2 - mI)$. Equation (6.49) can be put back into normal matrix multiplication order to give

$$\begin{aligned}\mathcal{L}_{(e)}^{\mu\nu} &= (\not{p}_2 - m)_{mj}\gamma_{ji}^\mu(\not{p}_1 + m)_{in}\gamma_{nm}^\nu \\ &= \left[(\not{p}_2 - m)\gamma^\mu(\not{p}_1 + m)\gamma^\nu\right]_{mm} \\ &= \mathrm{Tr}\left([\not{p}_2 - m]\gamma^\mu[\not{p}_1 + m]\gamma^\nu\right).\end{aligned} \tag{6.50}$$

Consequently, the sum over spins of the initial-state particles has been replaced by the calculation of the traces of $4\times4$ matrices, one for each of the sixteen possible combinations of the indices $\mu$ and $\nu$. The order in which the two $\not{p}$ terms appear in trace calculation of (6.50) follows the order in which the spinors appear in the original four-vector currents; the $\not{p}$ term associated with the adjoint spinor appears first (although traces are unchanged by cycling the elements). In constructing the traces associated with a Feynman diagram it is helpful to remember that the order in which different terms appear can be obtained by following the arrows in the fermion currents in the backwards direction. Writing the sum over the spins of final-state particles of (6.46) as the muon tensor,

$$\mathcal{L}_{\mu\nu}^{(\mu)} = \sum_{s',r'} \left[\overline{u}^{s'}(p_3)\gamma_\mu v^{r'}(p_4)\right]\left[\overline{v}^{r'}(p_4)\gamma_\nu u^{s'}(p_3)\right],$$

and expressing this in terms of a trace, leads to

$$\sum_{\text{spins}} |\mathcal{M}_{fi}|^2 = \frac{e^4}{q^4}\mathcal{L}_{(e)}^{\mu\nu}\mathcal{L}_{\mu\nu}^{(\mu)}$$
$$= \frac{e^4}{q^4}\,\mathrm{Tr}\left([\not{p}_2 - m]\gamma^\mu[\not{p}_1 + m]\gamma^\nu\right) \times \mathrm{Tr}\left([\not{p}_3 + M]\gamma_\mu[\not{p}_4 - M]\gamma_\nu\right), \tag{6.51}$$

where the masses of the initial- and final-state particles are respectively written as $m$ and $M$.

### 6.5.3 Trace theorems

The calculation of the spin-summed matrix element has been reduced to a problem of calculating traces involving combinations of $\gamma$-matrices. At first sight this appears a daunting task, but fortunately there are a number of algebraic "tricks" which greatly simplify the calculations. Firstly, traces have the properties

$$\mathrm{Tr}\,(A + B) \equiv \mathrm{Tr}\,(A) + \mathrm{Tr}\,(B), \tag{6.52}$$

and are unchanged by cycling the order of the elements

$$\mathrm{Tr}\,(AB \ldots YZ) \equiv \mathrm{Tr}\,(ZAB \ldots Y). \tag{6.53}$$

Secondly, the algebra of the $\gamma$-matrices is defined by the anticommutation relation of (4.33), namely

$$\gamma^\mu\gamma^\nu + \gamma^\nu\gamma^\mu \equiv 2g^{\mu\nu}I, \tag{6.54}$$

where the presence of the $4 \times 4$ identity matrix has been made explicit. Taking the trace of (6.54) gives

$$\mathrm{Tr}\,(\gamma^\mu\gamma^\nu) + \mathrm{Tr}\,(\gamma^\nu\gamma^\mu) = 2g^{\mu\nu}\,\mathrm{Tr}\,(I),$$

which using $\mathrm{Tr}\,(AB) = \mathrm{Tr}\,(BA)$ becomes $\mathrm{Tr}\,(\gamma^\mu\gamma^\nu) = g^{\mu\nu}\,\mathrm{Tr}\,(I)$, and hence

$$\mathrm{Tr}\,(\gamma^\mu\gamma^\nu) = 4g^{\mu\nu}. \tag{6.55}$$

The trace of any odd number of $\gamma$-matrices can be shown to be zero by inserting $\gamma^5\gamma^5 = I$ into the trace. For example, consider the trace of any three $\gamma$-matrices

$$\begin{aligned}
\mathrm{Tr}\,(\gamma^\mu\gamma^\nu\gamma^\rho) &= \mathrm{Tr}\left(\gamma^5\gamma^5\gamma^\mu\gamma^\nu\gamma^\rho\right) \\
&= \mathrm{Tr}\left(\gamma^5\gamma^\mu\gamma^\nu\gamma^\rho\gamma^5\right) && \text{(traces are cyclical)} \\
&= -\mathrm{Tr}\left(\gamma^5\gamma^5\gamma^\mu\gamma^\nu\gamma^\rho\right) && (\text{since } \gamma^5\gamma^\mu = -\gamma^\mu\gamma^5)
\end{aligned}$$

where the last line follows from commuting $\gamma^5$ through the three $\gamma$-matrices, each time introducing a factor of $-1$. Hence $\mathrm{Tr}\,(\gamma^\mu\gamma^\nu\gamma^\rho) = -\,\mathrm{Tr}\,(\gamma^\mu\gamma^\nu\gamma^\rho)$, which can only be true if

$$\mathrm{Tr}\,(\gamma^\mu\gamma^\nu\gamma^\rho) = 0. \tag{6.56}$$

The same argument can be applied to show that the trace of *any* odd number of $\gamma$-matrices is zero.

Finally, the trace of four $\gamma$-matrices can be obtained from (6.54) which allows $\gamma^a\gamma^b$ to be written as $2g^{ab} - \gamma^b\gamma^a$ and repeated application of this identity gives

$$\begin{aligned}
\gamma^\mu\gamma^\nu\gamma^\rho\gamma^\sigma &= 2g^{\mu\nu}\gamma^\rho\gamma^\sigma - \gamma^\nu\gamma^\mu\gamma^\rho\gamma^\sigma \\
&= 2g^{\mu\nu}\gamma^\rho\gamma^\sigma - 2g^{\mu\rho}\gamma^\nu\gamma^\sigma + \gamma^\nu\gamma^\rho\gamma^\mu\gamma^\sigma \\
&= 2g^{\mu\nu}\gamma^\rho\gamma^\sigma - 2g^{\mu\rho}\gamma^\nu\gamma^\sigma + 2g^{\mu\sigma}\gamma^\nu\gamma^\rho - \gamma^\nu\gamma^\rho\gamma^\sigma\gamma^\mu \\
\Rightarrow \quad \gamma^\mu\gamma^\nu\gamma^\rho\gamma^\sigma + \gamma^\nu\gamma^\rho\gamma^\sigma\gamma^\mu &= 2g^{\mu\nu}\gamma^\rho\gamma^\sigma - 2g^{\mu\rho}\gamma^\nu\gamma^\sigma + 2g^{\mu\sigma}\gamma^\nu\gamma^\rho.
\end{aligned} \tag{6.57}$$

Taking the trace of both sides of (6.57) and using the cyclic property of traces

$$2\mathrm{Tr}\,(\gamma^\mu\gamma^\nu\gamma^\rho\gamma^\sigma) = 2g^{\mu\nu}\,\mathrm{Tr}\,(\gamma^\rho\gamma^\sigma) - 2g^{\mu\rho}\,\mathrm{Tr}\,(\gamma^\nu\gamma^\sigma) + 2g^{\mu\sigma}\,\mathrm{Tr}\,(\gamma^\nu\gamma^\rho),$$

and using (6.55) for the trace of two $\gamma$-matrices gives the identity

$$\mathrm{Tr}\,(\gamma^\mu\gamma^\nu\gamma^\rho\gamma^\sigma) = 4g^{\mu\nu}g^{\rho\sigma} - 4g^{\mu\rho}g^{\nu\sigma} + 4g^{\mu\sigma}g^{\nu\rho}. \tag{6.58}$$

The full set of trace theorems, including those involving $\gamma^5 = i\gamma^0\gamma^1\gamma^2\gamma^3$, are:

(a) $\mathrm{Tr}\,(I) = 4$;
(b) the trace of any odd number of $\gamma$-matrices is zero;
(c) $\mathrm{Tr}\,(\gamma^\mu\gamma^\nu) = 4g^{\mu\nu}$;
(d) $\mathrm{Tr}\,(\gamma^\mu\gamma^\nu\gamma^\rho\gamma^\sigma) = 4g^{\mu\nu}g^{\rho\sigma} - 4g^{\mu\rho}g^{\nu\sigma} + 4g^{\mu\sigma}g^{\nu\rho}$;
(e) the trace of $\gamma^5$ multiplied by an odd number of $\gamma$-matrices is zero;
(f) $\mathrm{Tr}\left(\gamma^5\right) = 0$;
(g) $\mathrm{Tr}\left(\gamma^5\gamma^\mu\gamma^\nu\right) = 0$; and
(h) $\mathrm{Tr}\left(\gamma^5\gamma^\mu\gamma^\nu\gamma^\rho\gamma^\sigma\right) = 4i\varepsilon^{\mu\nu\rho\sigma}$, where $\varepsilon^{\mu\nu\rho\sigma}$ is antisymmetric under the interchange of any two indices.

Armed with these trace theorems, expressions such as that of (6.51) can be evaluated relatively easily; it is worth going through one example of a matrix element calculation using the trace methodology in gory detail.

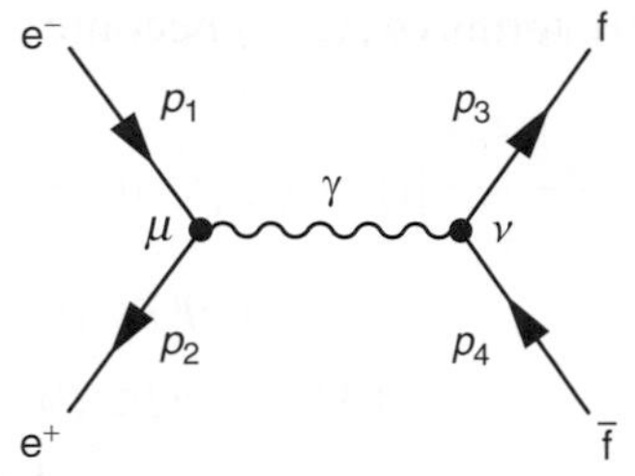


Fig. 6.11 The lowest-order QED Feynman diagram for $e^+e^- \to f\bar{f}$.

### 6.5.4 Electron–positron annihilation revisited

Consider the process $e^+e^- \to f\bar{f}$, shown in Figure 6.11, where f represents any of the fundamental spin-half charged fermions. In the limit where the electron mass can be neglected, but the masses of the final-state fermions cannot, the spin-averaged matrix element squared is given by (6.51) with $m = 0$ and $M = m_f$,

$$\langle|\mathcal{M}_{fi}|^2\rangle = \frac{1}{4}\sum_{\text{spins}}|\mathcal{M}_{fi}|^2 = \frac{Q_f^2 e^4}{4q^4}\,\mathrm{Tr}\left(\not{p}_2\gamma^\mu\not{p}_1\gamma^\nu\right)\mathrm{Tr}\left([\not{p}_3 + m_f]\gamma_\mu[\not{p}_4 - m_f]\gamma_\nu\right). \tag{6.59}$$

This can be evaluated by writing $\not{p}_1 = \gamma^\sigma p_{1\sigma}$ and $\not{p}_2 = \gamma^\rho p_{2\rho}$, in which case the first trace in (6.59) can be written as

$$\begin{aligned}\mathrm{Tr}\left(\not{p}_2\gamma^\mu\not{p}_1\gamma^\nu\right) &= p_{2\rho}p_{1\sigma}\mathrm{Tr}\left(\gamma^\rho\gamma^\mu\gamma^\sigma\gamma^\nu\right)\\ &= 4p_{2\rho}p_{1\sigma}(g^{\rho\mu}g^{\sigma\nu} - g^{\rho\sigma}g^{\mu\nu} + g^{\rho\nu}g^{\mu\sigma})\\ &= 4p_2^\mu p_1^\nu - 4g^{\mu\nu}(p_1\cdot p_2) + 4p_2^\nu p_1^\mu.\end{aligned}$$

Since the trace of an odd number of $\gamma$-matrices is zero and $\mathrm{Tr}(A+B) = \mathrm{Tr}(A) + \mathrm{Tr}(B)$, the second trace in (6.59) can be written

$$\mathrm{Tr}\left([\not{p}_3 + m_f]\gamma_\mu[\not{p}_4 - m_f]\gamma_\nu\right) = \mathrm{Tr}\left(\not{p}_3\gamma_\mu\not{p}_4\gamma_\nu\right) - m_f^2\mathrm{Tr}\left(\gamma_\mu\gamma_\nu\right) \tag{6.60}$$

$$= 4p_{3\mu}p_{4\nu} - 4g_{\mu\nu}(p_3\cdot p_4) + 4p_{3\nu}p_{4\mu} - 4m_f^2 g_{\mu\nu}. \tag{6.61}$$

Hence, the spin-averaged matrix element squared is given by

$$\begin{aligned}\langle|\mathcal{M}_{fi}|^2\rangle = 16\frac{Q_f^2 e^4}{4q^4}&\left[p_2^\mu p_1^\nu - g^{\mu\nu}(p_1\cdot p_2) + p_2^\nu p_1^\mu\right]\\ &\times\left[p_{3\mu}p_{4\nu} - g_{\mu\nu}(p_3\cdot p_4) + p_{3\nu}p_{4\mu} - m_f^2 g_{\mu\nu}\right].\end{aligned} \tag{6.62}$$

This expression can be simplified by contracting the indices, where for example

$$g^{\mu\nu}g_{\mu\nu} = 4, \quad p_2^\mu p_1^\nu g_{\mu\nu} = (p_1\cdot p_2) \quad \text{and} \quad p_2^\mu p_1^\nu p_{3\mu}p_{4\nu} = (p_2\cdot p_3)(p_1\cdot p_4).$$

Thus the twelve terms of (6.62) become

$$\langle|\mathcal{M}_{fi}|^2\rangle = 4\frac{Q_\mathrm{f}^2 e^4}{q^4}\Big[(p_1\cdot p_4)(p_2\cdot p_3) - (p_1\cdot p_2)(p_3\cdot p_4) + (p_1\cdot p_3)(p_2\cdot p_4)$$
$$- (p_1\cdot p_2)(p_3\cdot p_4) + 4(p_1\cdot p_2)(p_3\cdot p_4) - (p_1\cdot p_2)(p_3\cdot p_4)$$
$$+ (p_1\cdot p_3)(p_2\cdot p_4) - (p_1\cdot p_2)(p_3\cdot p_4) + (p_1\cdot p_4)(p_2\cdot p_3)$$
$$- m_\mathrm{f}^2(p_1\cdot p_2) + 4m_\mathrm{f}^2(p_1\cdot p_2) - m_\mathrm{f}^2(p_1\cdot p_2)\Big],$$

which simplifies to

$$\langle|\mathcal{M}_{fi}|^2\rangle = 4\frac{Q_\mathrm{f}^2 e^4}{q^4}\left[2(p_1\cdot p_3)(p_2\cdot p_4) + 2(p_1\cdot p_4)(p_2\cdot p_3) + 2m_\mathrm{f}^2(p_1\cdot p_2)\right].$$

In the limit where the electron mass is neglected, the four-momentum squared of the virtual photon is

$$q^2 = (p_1 + p_2)^2 = p_1^2 + p_2^2 + 2(p_1\cdot p_2) \approx 2(p_1\cdot p_2),$$

and therefore

$$\langle|\mathcal{M}_{fi}|^2\rangle = 2\frac{Q_\mathrm{f}^2 e^4}{(p_1\cdot p_2)^2}\left[(p_1\cdot p_3)(p_2\cdot p_4) + (p_1\cdot p_4)(p_2\cdot p_3) + m_\mathrm{f}^2(p_1\cdot p_2)\right]. \quad (6.63)$$

If the final-state fermion mass is also neglected, (6.63) reduces to the expression for the spin-averaged matrix element squared of (6.25), which was obtained from the helicity amplitudes.

In the above calculation, neither the explicit form of the spinors nor the specific representation of the $\gamma$-matrices is used. The spin-averaged matrix element squared is determined from the completeness relations for the spinors and the commutation and Hermiticity properties of the $\gamma$-matrices alone.

## $e^+e^- \to f\bar{f}$ annihilation close to threshold

The spin-averaged matrix element squared of (6.63) can be used to calculate the cross section for $e^+e^- \to f\bar{f}$ close to threshold. Working in the centre-of-mass frame and writing the momenta of the final-state particles as $\mathrm{p} = \beta E$, where $\beta = \mathrm{v}/c$, the four-momenta of the particles involved can be written

$$p_1 = (E, 0, 0, +E),$$
$$p_2 = (E, 0, 0, -E),$$
$$p_3 = (E, +\beta E\sin\theta, 0, +\beta E\cos\theta),$$
$$p_4 = (E, -\beta E\sin\theta, 0, -\beta E\cos\theta),$$

and the relevant four-vector scalar products are

$$\begin{aligned} p_1 \cdot p_3 &= p_2 \cdot p_4 = E^2(1 - \beta\cos\theta), \\ p_1 \cdot p_4 &= p_2 \cdot p_3 = E^2(1 + \beta\cos\theta), \\ p_1 \cdot p_2 &= 2E^2. \end{aligned}$$

Substituting these expressions into (6.63) gives

$$\begin{aligned} \langle|\mathcal{M}_{fi}|^2\rangle &= 2\frac{Q_f^2 e^4}{4E^4}\left[E^4(1-\beta\cos\theta)^2 + E^4(1+\beta\cos\theta)^2 + 2E^2 m_f^2\right] \\ &= Q_f^2 e^4\left(1 + \beta^2\cos^2\theta + \frac{E^2 - \mathrm{p}^2}{E^2}\right) \\ &= Q_f^2 e^4\left(2 + \beta^2\cos^2\theta - \beta^2\right). \end{aligned} \qquad (6.64)$$

The differential cross section is then obtained by substituting the spin-averaged matrix element squared of (6.64) into the cross section formula of (3.50) to give

$$\begin{aligned} \frac{\mathrm{d}\sigma}{\mathrm{d}\Omega} &= \frac{1}{64\pi^2 s}\frac{\mathrm{p}}{E}\langle|M_{fi}|^2\rangle \\ &= \frac{1}{4s}\beta Q_f^2\alpha^2\left(2 + \beta^2\cos^2\theta - \beta^2\right), \end{aligned}$$

where $e^2 = 4\pi\alpha$. The total cross section is obtained by integrating over $\mathrm{d}\Omega$, giving

$$\sigma(\mathrm{e}^+\mathrm{e}^- \to \mathrm{f}\bar{\mathrm{f}}) = \frac{4\pi\alpha^2 Q_f^2}{3s}\beta\left(\frac{3-\beta^2}{2}\right) \quad \text{with} \quad \beta^2 = \left(1 - \frac{4m_f^2}{s}\right). \qquad (6.65)$$

Close to threshold, the cross section is approximately proportional to the velocity of the final state particles. Figure 6.12 shows the measurements of the total $\mathrm{e}^+\mathrm{e}^- \to \tau^+\tau^-$ cross section at centre-of-mass energies just above threshold. The data are in good agreement with the prediction of (6.65). In the relativistic limit where $\beta \to 1$, the total cross section of (6.65) reduces to the expression of (6.24).

### 6.5.5 Electron–quark scattering

The main topic of the next two chapters is electron–proton scattering. In the case of *inelastic* scattering where the proton breaks up, the underlying QED process is $t$-channel scattering of electrons from the quarks inside the proton. In the limit

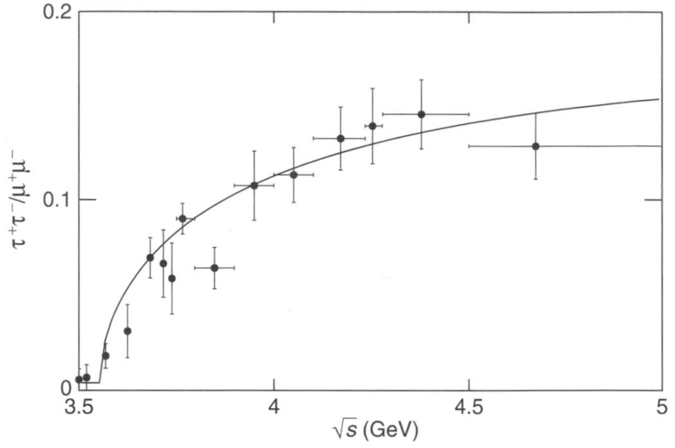


Fig. 6.12 The measured ratio of number of identified $e^+e^- \to \tau^+\tau^-$ events to the number of $e^+e^- \to \mu^+\mu^-$ at centre-of-mass energies close to the $e^+e^- \to \tau^+\tau^-$ threshold. The curve shows the $\beta(3-\beta^2)$ behaviour of (6.65). The normalisation depends on the efficiency for identifying $\tau^+\tau^-$ events and a small background component is included. Adapted from Bacino *et al.* (1978).

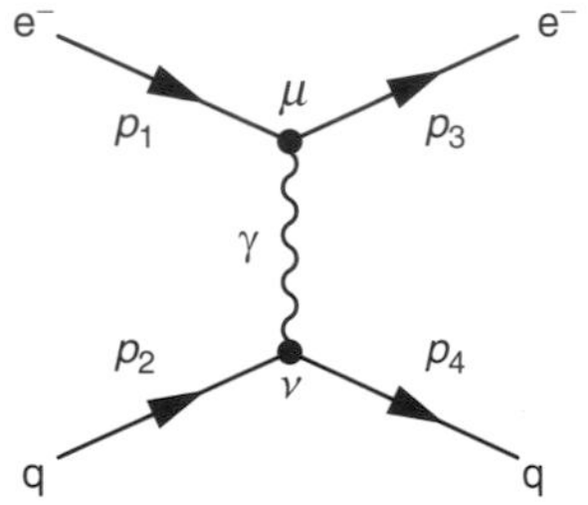


Fig. 6.13 The lowest-order Feynman diagram for QED $t$-channel electron–quark scattering process.

where the masses of the electron and the quark can be neglected, it is relatively straightforward to obtain the expressions for the four non-zero matrix elements using the helicity amplitude approach (see Problem 6.7). However, if the particle masses cannot be neglected, which is the case for low-energy electron–proton scattering, the spin-averaged matrix element is most easily calculated using the trace formalism introduced above.

The QED matrix element for the Feynman diagram of Figure 6.13 is

$$\mathcal{M}_{fi} = \frac{Q_q e^2}{q^2}\left[\overline{u}(p_3)\gamma^\mu u(p_1)\right] g_{\mu\nu} \left[\overline{u}(p_4)\gamma^\nu u(p_2)\right].$$

Noting the order in which the spinors appear in the matrix element (working backwards along the arrows on the fermion lines), the spin-summed matrix element squared is given by

$$\sum_{\text{spins}} |\mathcal{M}_{fi}|^2 = \frac{Q_q^2 e^4}{q^4} \operatorname{Tr}\left([\not{p}_3 + m_e]\gamma^\mu[\not{p}_1 + m_e]\gamma^\nu\right) \operatorname{Tr}\left([\not{p}_4 + m_q]\gamma_\mu[\not{p}_2 + m_q]\gamma_\nu\right). \tag{6.66}$$

Apart from the signs of the mass terms, which are all positive since only particles are involved, the expressions in the traces of (6.66) have the same form as those of (6.60) and can therefore be evaluated using the result of (6.61) with the signs of the $m^2$ terms reversed, giving

$$\sum_{\text{spins}} |\mathcal{M}_{fi}|^2 = \frac{16\, Q_q^2 e^4}{q^4} \left(p_3^\mu p_1^\nu - g^{\mu\nu}(p_1 \cdot p_3) + p_1^\mu p_3^\nu + m_e^2 g^{\mu\nu}\right)$$
$$\times \left(p_{4\mu} p_{2\nu} - g_{\mu\nu}(p_2 \cdot p_4) + p_{2\mu} p_{4\nu} + m_q^2 g_{\mu\nu}\right).$$

From this expression, it follows that

$$\begin{aligned}\langle|\mathcal{M}_{fi}|^2\rangle &= \frac{1}{4}\sum_{\text{spins}} |\mathcal{M}_{fi}|^2 \\ &= \frac{8 Q_q^2 e^4}{(p_1 - p_3)^4} \times \left[(p_1\cdot p_2)(p_3\cdot p_4) + (p_1\cdot p_4)(p_2\cdot p_3)\right. \\ &\qquad \left. - m_e^2 (p_2\cdot p_4) - m_q^2 (p_1\cdot p_3) + 2 m_e^2 m_q^2\right].\end{aligned} \tag{6.67}$$

In the limit where the masses can be neglected, (6.67) reduces to

$$\langle|\mathcal{M}_{fi}|^2\rangle = 2 Q_q^2 e^4 \left(\frac{s^2 + u^2}{t^2}\right). \tag{6.68}$$

Apart from the factor $Q_q^2$ from the quark charge, this spin-averaged matrix element squared for the $t$-channel scattering process of $eq \to eq$ is identical to the corresponding expression for $e^+e^- \to \mu^+\mu^-$ annihilation of (6.26) with $s$ and $t$ interchanged. The similarity between these two expressions is to be expected from the closeness of the forms of the fermion currents for the two processes. This property, known as crossing symmetry, can be utilised to obtain directly the expression for the spin-averaged matrix element squared for a $t$-channel process from that of the corresponding $s$-channel process.

### 6.5.6 Crossing symmetry

The calculations of the spin-averaged squared matrix elements for the $s$-channel $e^+e^- \to f\bar{f}$ annihilation process and the $t$-channel $e^-f \to e^-f$ scattering processes, shown in Figure 6.14, proceed in similar way. In the annihilation process, the two currents are

$$j_e^\mu = \bar{v}(p_2)\gamma^\mu u(p_1) \quad \text{and} \quad j_f^\nu = \bar{u}(p_3)\gamma^\nu v(p_4), \tag{6.69}$$

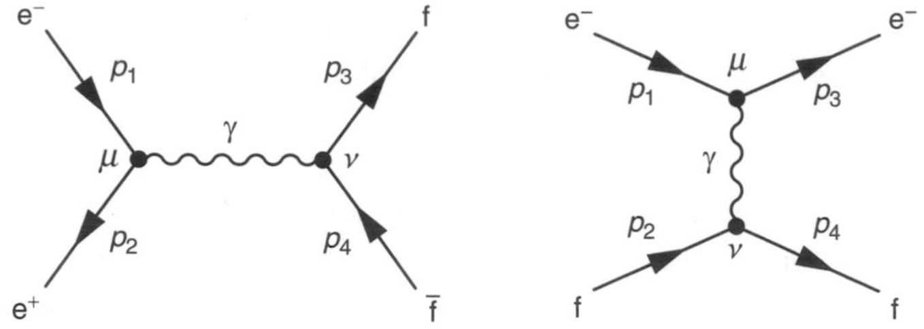


Fig. 6.14 The Feynman diagrams for QED $s$-channel annihilation process $e^+e^- \to f\bar{f}$ and the QED $t$-channel scattering process $e^-f \to e^-f$.

and for the scattering process the corresponding two currents are

$$j_e^\mu = \overline{u}(p_3)\gamma^\mu u(p_1) \quad \text{and} \quad j_f^\nu = \overline{u}(p_4)\gamma^\nu u(p_2). \tag{6.70}$$

By making the replacement $u(p_1) \to u(p_1)$, $\overline{v}(p_2) \to \overline{u}(p_3)$, $\overline{u}(p_3) \to \overline{u}(p_4)$ and $v(p_4) \to u(p_2)$ the currents in the annihilation process (6.69) correspond to those for the scattering process (6.70). In the calculation of traces, this implies making the replacement, $p_1 \to p_1$, $p_2 \to p_3$, $p_3 \to p_4$ and $p_4 \to p_2$. This accounts for the order in which the spinors appear in the four-vector currents, but does not account for the replacement of an antiparticle spinor with a particle spinor. From the completeness relationships of (6.40) and (6.41), the spin sums lead to a term in the trace of $[\not{p} + m]$ for particles and $[\not{p} - m]$ for antiparticles. So when a particle is replaced by an antiparticle, the sign of the mass term in the trace is reversed. Alternatively, the effect of changing the *relative* sign between $\not{p}$ and $m$, can be achieved by changing the sign of the four-momentum when a particle in one diagram is replaced by an antiparticle in the other diagram. Hence, crossing symmetry implies that the matrix element for $e^-f \to e^-f$ can be obtained the matrix element for $e^+e^- \to f\bar{f}$ by making the substitutions,

$$p_1 \to p_1, \quad p_2 \to -p_3, \quad p_3 \to p_4 \quad \text{and} \quad p_4 \to -p_2.$$

The effect on the Mandelstam variables is $s^2 \to t^2$, $t^2 \to u^2$ and $u^2 \to s^2$, and with these replacements the matrix element for the $s$-channel annihilation process $e^+e^- \to f\bar{f}$ of (6.26) transforms to the matrix element for the $t$-channel scattering process $e^-f \to e^-f$ of (6.68)

$$\langle|\mathcal{M}_{fi}|^2\rangle_s = 2Q_f^2 e^4 \left(\frac{t^2 + u^2}{s^2}\right) \quad \longleftrightarrow \quad \langle|\mathcal{M}_{fi}|^2\rangle_t = 2Q_f^2 e^4 \left(\frac{u^2 + s^2}{t^2}\right).$$

## Summary

In this chapter, the $e^+e^- \to \mu^+\mu^-$ annihilation process has been used to introduce the techniques used to perform lowest-order QED calculations. A number of important concepts were introduced. The treatment of the different spin states of the initial- and final-state particles leads to the introduction of the spin-averaged matrix element squared given by

$$\langle|\mathcal{M}_{fi}|^2\rangle = \frac{1}{4}\sum_{\text{spins}}|\mathcal{M}|^2,$$

where the sum extends over the sixteen orthogonal spins states. In the limit where the masses of the particles were neglected, only four of the possible helicity combinations give non-zero matrix elements. This property is due to the chiral nature of the QED interaction, where the left- and right-handed chiral states are eigenstates of the $\gamma^5$-matrix defined as

$$\gamma^5 \equiv i\gamma^0\gamma^1\gamma^2\gamma^3.$$

Because of the $\overline{\phi}\gamma^\mu\psi$ form of the QED interaction vertex, certain combinations of chiral currents are *always* zero, for example $\overline{u}_R\gamma^\mu u_L = 0$. In the limit $E \gg m$, the helicity eigenstates correspond to the chiral eigenstates and twelve of the sixteen possible helicity combinations in the process $e^+e^- \to \mu^+\mu^-$ do not contribute to the cross section and helicity is *effectively* conserved in the interaction. The resulting spin-averaged matrix element squared for $e^+e^- \to \mu^+\mu^-$ is

$$e^+e^- \to \mu^+\mu^- : \quad \langle|\mathcal{M}_{fi}|^2\rangle = 2e^4\left(\frac{t^2+u^2}{s^2}\right). \tag{6.71}$$

In the starred section of this chapter, the method of using traces to perform spin sums was introduced and was then used to calculate the matrix elements for $e^+e^- \to f\overline{f}$ annihilation and $e^-q \to e^-q$ scattering. In the massless limit, the spin-averaged matrix element squared for electron–quark scattering was shown to be

$$e^-q \to e^-q : \quad \langle|\mathcal{M}_{fi}|^2\rangle = 2Q_q^2e^4\left(\frac{s^2+u^2}{t^2}\right). \tag{6.72}$$

## Problems

**6.1** Using the properties of the $\gamma$-matrices of (4.33) and (4.34), and the definition of $\gamma^5 \equiv i\gamma^0\gamma^1\gamma^2\gamma^3$, show that

$$(\gamma^5)^2 = 1, \quad \gamma^{5\dagger} = \gamma^5 \quad \text{and} \quad \gamma^5\gamma^\mu = -\gamma^\mu\gamma^5.$$

**6.2** Show that the chiral projection operators

$$P_R = \tfrac{1}{2}(1+\gamma^5) \quad \text{and} \quad P_L = \tfrac{1}{2}(1-\gamma^5),$$

satisfy

$$P_R + P_L = 1, \quad P_R P_R = P_R, \quad P_L P_L = P_L \quad \text{and} \quad P_L P_R = 0.$$

**6.3** Show that

$$\Lambda^+ = \frac{m + \gamma^\mu p_\mu}{2m} \quad \text{and} \quad \Lambda^- = \frac{m - \gamma^\mu p_\mu}{2m},$$

are also projection operators, and show that they respectively project out particle and antiparticle states, i.e.

$$\Lambda^+ u = u, \quad \Lambda^- v = v \quad \text{and} \quad \Lambda^+ v = \Lambda^- u = 0.$$

**6.4** Show that the helicity operator can be expressed as

$$\hat{h} = -\tfrac{1}{2}\frac{\gamma^0\gamma^5\boldsymbol{\gamma}\cdot\mathbf{p}}{\mathrm{p}}.$$

**6.5** In general terms, explain why *high-energy* electron–positron colliders must also have high instantaneous luminosities.

**6.6** For a spin-1 system, the eigenstate of the operator $\hat{S}_n = \mathbf{n}\cdot\hat{\mathbf{S}}$ with eigenvalue +1 corresponds to the spin being in the direction $\hat{\mathbf{n}}$. Writing this state in terms of the eigenstates of $\hat{S}_z$, i.e.

$$|1,+1\rangle_\theta = \alpha|1,-1\rangle + \beta|1,0\rangle + \gamma|1,+1\rangle,$$

and taking $\mathbf{n} = (\sin\theta, 0, \cos\theta)$ show that

$$|1,+1\rangle_\theta = \tfrac{1}{2}(1-\cos\theta)\,|1,-1\rangle + \tfrac{1}{\sqrt{2}}\sin\theta\,|1,0\rangle + \tfrac{1}{2}(1+\cos\theta)\,|1,+1\rangle.$$

Hint: write $\hat{S}_x$ in terms of the spin ladder operators.

**6.7** Using helicity amplitudes, calculate the differential cross section for $e^-\mu^- \to e^-\mu^-$ scattering in the following steps:

(a) From the Feynman rules for QED, show that the lowest-order QED matrix element for $e^-\mu^- \to e^-\mu^-$ is

$$\mathcal{M}_{fi} = -\frac{e^2}{(p_1-p_3)^2}g_{\mu\nu}\left[\bar{u}(p_3)\gamma^\mu u(p_1)\right]\left[\bar{u}(p_4)\gamma^\nu u(p_2)\right],$$

where $p_1$ and $p_3$ are the four-momenta of the initial- and final-state $e^-$, and $p_2$ and $p_4$ are the four-momenta of the initial- and final-state $\mu^-$.

(b) Working in the centre-of-mass frame, and writing the four-momenta of the initial- and final-state $e^-$ as $p_1^\mu = (E_1, 0, 0, \mathrm{p})$ and $p_3^\mu = (E_1, p\sin\theta, 0, \mathrm{p}\cos\theta)$ respectively, show that the electron currents for the four possible helicity combinations are

$$\begin{aligned}
\bar{u}_\downarrow(p_3)\gamma^\mu u_\downarrow(p_1) &= 2(E_1 c, \mathrm{p}s, -i\mathrm{p}s, \mathrm{p}c),\\
\bar{u}_\uparrow(p_3)\gamma^\mu u_\downarrow(p_1) &= 2(ms, 0, 0, 0),\\
\bar{u}_\uparrow(p_3)\gamma^\mu u_\uparrow(p_1) &= 2(E_1 c, \mathrm{p}s, i\mathrm{p}s, \mathrm{p}c),\\
\bar{u}_\downarrow(p_3)\gamma^\mu u_\uparrow(p_1) &= -2(ms, 0, 0, 0),
\end{aligned}$$

where $m$ is the electron mass, $s = \sin(\theta/2)$ and $c = \cos(\theta/2)$.

(c) Explain why the effect of the parity operator $\hat{P} = \gamma^0$ is

$$\hat{P}u_\uparrow(\mathrm{p}, \theta, \phi) = \hat{P}u_\downarrow(\mathrm{p}, \pi - \theta, \pi + \phi).$$

Hence, or otherwise, show that the muon currents for the four helicity combinations are

$$\begin{aligned}
\bar{u}_\downarrow(p_4)\gamma^\mu u_\downarrow(p_2) &= 2(E_2 c, -\mathrm{p}s, -i\mathrm{p}s, -\mathrm{p}c),\\
\bar{u}_\uparrow(p_4)\gamma^\mu u_\downarrow(p_2) &= 2(Ms, 0, 0, 0),\\
\bar{u}_\uparrow(p_4)\gamma^\mu u_\uparrow(p_2) &= 2(E_2 c, -\mathrm{p}s, i\mathrm{p}s, -\mathrm{p}c),\\
\bar{u}_\downarrow(p_4)\gamma^\mu u_\uparrow(p_2) &= -2(Ms, 0, 0, 0),
\end{aligned}$$

where $M$ is the muon mass.

(d) For the relativistic limit where $E \gg M$, show that the matrix element squared for the case where the incoming $e^-$ and incoming $\mu^-$ are both left-handed is given by

$$|\mathcal{M}_{LL}|^2 = \frac{4e^4 s^2}{(p_1 - p_3)^4},$$

where $s = (p_1 + p_2)^2$. Find the corresponding expressions for $|\mathcal{M}_{RL}|^2$, $|\mathcal{M}_{RR}|^2$ and $|\mathcal{M}_{LR}|^2$.

(e) In this relativistic limit, show that the differential cross section for unpolarised $e^-\mu^- \to e^-\mu^-$ scattering in the centre-of-mass frame is

$$\frac{\mathrm{d}\sigma}{\mathrm{d}\Omega} = \frac{2\alpha^2}{s} \cdot \frac{1 + \frac{1}{4}(1 + \cos\theta)^2}{(1 - \cos\theta)^2}.$$

**6.8*** Using $\gamma^\mu\gamma^\nu + \gamma^\nu\gamma^\mu = 2g^{\mu\nu}$, prove that

$$\gamma^\mu\gamma_\mu = 4, \quad \gamma^\mu \not{a}\gamma_\mu = -2\not{a} \quad \text{and} \quad \gamma^\mu \not{a}\not{b}\gamma_\mu = 4a\cdot b.$$

**6.9*** Prove the relation $\left[\bar{\psi}\gamma^\mu\gamma^5\phi\right]^\dagger = \bar{\phi}\gamma^\mu\gamma^5\psi$.

**6.10*** Use the trace formalism to calculate the QED spin-averaged matrix element squared for $e^+e^- \to f\bar{f}$ including the electron mass term.

**6.11*** Neglecting the electron mass term, verify that the matrix element for $e^-f \to e^-f$ given in (6.67) can be obtained from the matrix element for $e^+e^- \to f\bar{f}$ given in (6.63) using crossing symmetry with the substitutions

$$p_1 \to p_1, \quad p_2 \to -p_3, \quad p_3 \to p_4 \quad \text{and} \quad p_4 \to -p_2.$$

**6.12*** Write down the matrix elements, $\mathcal{M}_1$ and $\mathcal{M}_2$, for the two Feynman diagrams for the Compton scattering process $e^-\gamma \to e^-\gamma$. From first principles, express the spin-averaged matrix element $\langle|\mathcal{M}_1 + \mathcal{M}_2|^2\rangle$ as a trace. You will need the completeness relation for the photon polarisation states (see Appendix D).

# 7 Electron–proton elastic scattering

In $e^+e^-$ collisions, the initial-state particles are fundamental fermions. Consequently, the cross sections for processes such as $e^+e^-$ annihilation are determined by the QED matrix element and the event kinematics (phase space) alone. Calculations of cross sections for collisions involving protons, for example at an electron–proton collider or a hadron collider, also need to account for the composite nature of the proton. This chapter describes low-energy electron–proton *elastic* scattering. The main purpose is to provide an introduction to a number of concepts which form the starting point for the description of the high-energy interactions of protons that is the main topic of the following chapter.

## 7.1 Probing the structure of the proton

Electron–proton scattering provides a powerful tool for probing the structure of the proton. At low energies, the dominant process is elastic scattering where the proton remains intact. Elastic scattering is described by the coherent interaction of a virtual photon with the proton as a whole, and thus provides a probe of the global properties of the proton, such as its charge radius. At high energies, the dominant process is deep inelastic scattering, where the proton breaks up. Here the underlying process is the *elastic* scattering of the electron from one of the quarks within the proton. Consequently, deep inelastic scattering provides a probe of the momentum distribution of the quarks.

The precise nature of the $e^-p \to e^-p$ scattering process depends on the wavelength of the virtual photon in comparison to the radius of the proton. Electron–proton scattering can be broadly categorised into the four classes of process shown schematically in Figure 7.1:

(a) at very low energies, where the electrons are non-relativistic and the wavelength of the virtual photon is large compared to the radius of the proton, $\lambda \gg r_p$, the $e^-p \to e^-p$ process can be described in terms of the elastic scattering of the electron in the static potential of an effectively point-like proton;

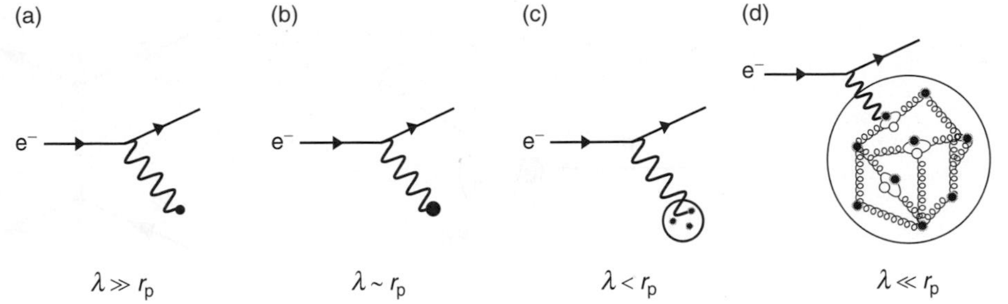


Fig. 7.1 The nature of $e^-p$ scattering depending on the wavelength of the virtual photon.

(b) at higher electron energies, where $\lambda \sim r_{\mathrm{p}}$, the scattering process is no longer purely electrostatic in nature and the cross section calculation also needs to account for the extended charge and magnetic moment distributions of the proton;

(c) when the wavelength of the virtual photon becomes relatively small, $\lambda < r_{\mathrm{p}}$, the elastic scattering cross section also becomes small. In this case, the dominant process is inelastic scattering where the virtual photon interacts with a constituent quark inside the proton and the proton subsequently breaks up;

(d) at very high electron energies, where the wavelength of the virtual photon ($\lambda \ll r_{\mathrm{p}}$) is sufficiently short to resolve the detailed dynamic structure of the proton, the proton appears to be a sea of strongly interacting quarks and gluons.

Whilst we will be interested primarily in the high-energy deep inelastic $e^-p$ scattering, the low-energy $e^-p$ elastic scattering process provides a valuable introduction to a number of important concepts.

## 7.2 Rutherford and Mott scattering

Rutherford and Mott scattering are the low-energy limits of $e^-p$ elastic scattering. In both cases, the electron energy is sufficiently low that the kinetic energy of the recoiling proton is negligible compared to its rest mass. In this case, the proton can be taken to be a *fixed* source of a $1/r$ electrostatic potential. The cross sections for Rutherford and Mott scattering are usually derived from non-relativistic scattering theory using the first-order $\langle\psi_f|V(\mathrm{r})|\psi_i\rangle$ term in the perturbation expansion. Here the cross sections are derived using the helicity amplitude approach of the previous chapter, treating the proton as if it were a point-like Dirac particle. Provided the wavelength of the virtual photon is much larger than the radius of the proton, this is a reasonable approximation.

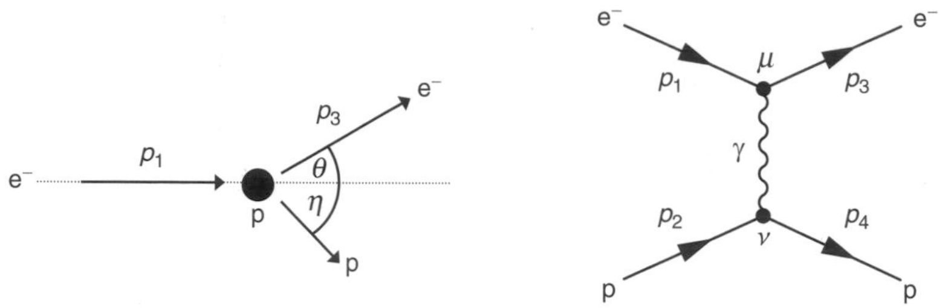


Fig. 7.2 Rutherford scattering of an electron from a proton at rest in the laboratory frame and the corresponding Feynman diagram.

In the limit where the proton is taken to be a point-like Dirac fermion, the matrix element for the Feynman diagram for low-energy $e^-p$ elastic scattering, shown in Figure 7.2, is given by

$$\mathcal{M}_{fi} = \frac{Q_q e^2}{q^2}\left[\overline{u}(p_3)\gamma^\mu u(p_1)\right] g_{\mu\nu}\left[\overline{u}(p_4)\gamma^\nu u(p_2)\right]. \tag{7.1}$$

From (4.65), the Dirac spinors describing the two possible helicity states of the electron can be written in the form

$$u_\uparrow = N_e\begin{pmatrix} c \\ se^{i\phi} \\ \kappa c \\ \kappa s e^{i\phi}\end{pmatrix} \quad\text{and}\quad u_\downarrow = N_e\begin{pmatrix} -s \\ ce^{i\phi} \\ \kappa s \\ -\kappa c e^{i\phi}\end{pmatrix},$$

where $N_e = \sqrt{E + m_e}$, $s = \sin(\theta/2)$ and $c = \cos(\theta/2)$. The parameter $\kappa$ is given by

$$\kappa = \frac{\mathrm{p}}{E + m_e} \equiv \frac{\beta_e\gamma_e}{\gamma_e + 1},$$

where $\beta_e$ and $\gamma_e$ are respectively the speed and Lorentz factor of the electron. Writing the electron spinors in terms of the parameter $\kappa$ clearly differentiates between the non-relativistic ($\kappa \ll 1$) and highly relativistic ($\kappa \approx 1$) limits. If the velocity of the scattered proton is small, its kinetic energy can be neglected, and to a good approximation the energy of the electron does not change in the scattering process. Hence the same value of $\kappa$ applies to both the initial- and final-state electron. For an electron scattering angle $\theta$ (see Figure 7.2) and taking the azimuthal angle for the electrons to be $\phi = 0$, the possible initial- and final-state electron spinors are

$$u_\uparrow(p_1) = N_e\begin{pmatrix} 1 \\ 0 \\ \kappa \\ 0\end{pmatrix},\ u_\downarrow(p_1) = N_e\begin{pmatrix} 0 \\ 1 \\ 0 \\ -\kappa\end{pmatrix} \text{ and } u_\uparrow(p_3) = N_e\begin{pmatrix} c \\ s \\ \kappa c \\ \kappa s\end{pmatrix},\ u_\downarrow(p_3) = N_e\begin{pmatrix} -s \\ c \\ \kappa s \\ -\kappa c\end{pmatrix}.$$

The electron currents for the four possible helicity combinations, calculated from (6.12)–(6.15), are

$$j_{e\uparrow\uparrow} = \overline{u}_\uparrow(p_3)\gamma^\mu u_\uparrow(p_1) = (E+m_e)\left[(\kappa^2+1)c, 2\kappa s, +2i\kappa s, 2\kappa c\right], \tag{7.2}$$

$$j_{e\downarrow\downarrow} = \overline{u}_\downarrow(p_3)\gamma^\mu u_\downarrow(p_1) = (E+m_e)\left[(\kappa^2+1)c, 2\kappa s, -2i\kappa s, 2\kappa c\right], \tag{7.3}$$

$$j_{e\downarrow\uparrow} = \overline{u}_\uparrow(p_3)\gamma^\mu u_\downarrow(p_1) = (E+m_e)\left[(1-\kappa^2)s, 0, 0, 0\right], \tag{7.4}$$

$$j_{e\uparrow\downarrow} = \overline{u}_\downarrow(p_3)\gamma^\mu u_\uparrow(p_1) = (E+m_e)\left[(\kappa^2-1)s, 0, 0, 0\right]. \tag{7.5}$$

Thus, in the relativistic limit where $\kappa \approx 1$, only two of the four helicity combinations give non-zero electron currents, reflecting the chiral nature of the QED interaction vertex. At lower energies, where $\kappa < 1$, all four helicity combinations give non-zero matrix elements; in this limit the helicity eigenstates no longer correspond to the chiral eigenstates and helicity is not conserved in the interaction.

In the limit where the velocity of the recoiling proton is small ($\beta_p \ll 1$), the lower two components of the corresponding particle spinors are approximately zero (since $\kappa \approx 0$). Taking the spherical polar angles defining the direction of the (relatively small) recoil momentum of the proton as ($\theta_p = \eta$, $\phi_p = \pi$), the initial-state and final-state protons can be described respectively by the helicity states

$$u_\uparrow(p_2) = \sqrt{2m_p}\begin{pmatrix}1\\0\\0\\0\end{pmatrix} \equiv u_1(p_2) \quad \text{and} \quad u_\downarrow(p_2) = \sqrt{2m_p}\begin{pmatrix}0\\1\\0\\0\end{pmatrix} \equiv u_2(p_2),$$

and

$$u_\uparrow(p_4) \approx \sqrt{2m_p}\begin{pmatrix}c_\eta\\-s_\eta\\0\\0\end{pmatrix} \quad \text{and} \quad u_\downarrow(p_4) \approx \sqrt{2m_p}\begin{pmatrix}-s_\eta\\-c_\eta\\0\\0\end{pmatrix},$$

where $c_\eta = \cos(\eta/2)$ and $s_\eta = \sin(\eta/2)$. The proton four-vector currents for the four possible combinations of the initial- and final-state helicity states, again calculated using (6.12)–(6.15), are

$$j_{p\uparrow\uparrow} = -j_{p\downarrow\downarrow} = 2m_p\left[c_\eta, 0, 0, 0\right] \quad \text{and} \quad j_{p\uparrow\downarrow} = j_{p\downarrow\uparrow} = -2m_p\left[s_\eta, 0, 0, 0\right]. \tag{7.6}$$

Thus, in the limit where the proton recoil momentum is small, all four spin combinations for the proton current contribute to the scattering process.

From the QED matrix element,

$$\mathcal{M}_{fi} = \frac{e^2}{q^2} j_e \cdot j_p,$$

and the expressions for the electron and proton currents of (7.2)–(7.6), the spin-averaged matrix element squared is

$$\begin{aligned}\langle|\mathcal{M}_{fi}^2|\rangle &= \frac{1}{4}\sum|\mathcal{M}_{fi}^2| \\ &= \frac{1}{4}\frac{e^4}{q^4}\times 4m_p^2(E+m_e)^2\cdot\left[c_\eta^2+s_\eta^2\right]\cdot\left[4(1+\kappa^2)^2c^2+4(1-\kappa^2)^2s^2\right] \\ &= \frac{4m_p^2m_e^2e^4(\gamma_e+1)^2}{q^4}\left[(1-\kappa^2)^2+4\kappa^2c^2\right],\end{aligned} \tag{7.7}$$

where in the last step, the electron energy was written as $E=\gamma_e m_e$. The above expression can be simplified further by writing

$$\kappa=\frac{\beta_e\gamma_e}{\gamma_e+1} \quad \text{and} \quad (1-\beta_e^2)\gamma_e^2=1,$$

in which case, after some algebraic manipulation, (7.7) becomes

$$\langle|\mathcal{M}_{fi}^2|\rangle=\frac{16m_p^2m_e^2e^4}{q^4}\left[1+\beta_e^2\gamma_e^2\cos^2\frac{\theta}{2}\right]. \tag{7.8}$$

In the $t$-channel $e^-p\to e^-p$ scattering process, the square of four-momentum carried by the virtual photon is given by

$$q^2=(p_1-p_3)^2.$$

For the elastic scattering process where the recoil of the proton can be neglected, the energies and momenta of the initial- and final-state electrons are $E_1=E_3=E$ and $\mathrm{p}_1=\mathrm{p}_3=\mathrm{p}$, and hence

$$q^2=(0,\mathbf{p}_1-\mathbf{p}_3)^2=-2\mathrm{p}^2(1-\cos\theta)=-4\mathrm{p}^2\sin^2(\theta/2).$$

Substituting this expression for $q^2$ into (7.8) gives

$$\langle|\mathcal{M}_{fi}^2|\rangle=\frac{m_p^2m_e^2e^4}{\mathrm{p}^4\sin^4(\theta/2)}\left[1+\beta_e^2\gamma_e^2\cos^2\frac{\theta}{2}\right]. \tag{7.9}$$

Provided the proton recoil can be neglected, this matrix element is equally applicable when the electron is either non-relativistic or relativistic.

### 7.2.1 Rutherford scattering

Rutherford scattering is the limit where the proton recoil can be neglected and the electron is non-relativistic, $\beta_e\gamma_e\ll 1$. In this case, the spin-averaged matrix element squared of (7.9) reduces to

$$\langle|\mathcal{M}_{fi}^2|\rangle=\frac{m_p^2m_e^2e^4}{\mathrm{p}^4\sin^4(\theta/2)}. \tag{7.10}$$

The laboratory frame differential cross section is obtained from the cross section formula of (3.48),

$$\frac{\mathrm{d}\sigma}{\mathrm{d}\Omega} = \frac{1}{64\pi^2}\left(\frac{1}{m_\mathrm{p} + E_1 - E_1\cos\theta}\right)^2 \langle|\mathcal{M}_{fi}|^2\rangle. \tag{7.11}$$

In the Rutherford scattering limit, where the electron is non-relativistic, $E_1 \sim m_\mathrm{e} \ll m_\mathrm{p}$, and (7.11) therefore reduces to

$$\frac{\mathrm{d}\sigma}{\mathrm{d}\Omega} = \frac{1}{64\pi^2 m_\mathrm{p}^2}\langle|\mathcal{M}_{fi}|^2\rangle = \frac{m_\mathrm{e}^2 e^4}{64\pi^2 \mathrm{p}^4 \sin^4(\theta/2)}. \tag{7.12}$$

Equation (7.12) can be expressed in the more usual form by writing the kinetic energy of the non-relativistic electron as $E_K = \mathrm{p}^2/2m_\mathrm{e}$ and writing $e^2 = 4\pi\alpha$ to give

$$\left(\frac{\mathrm{d}\sigma}{\mathrm{d}\Omega}\right)_{\mathrm{Rutherford}} = \frac{\alpha^2}{16E_K^2 \sin^4(\theta/2)}. \tag{7.13}$$

The Rutherford scattering cross section of (7.13) is usually derived from first-order perturbation theory by considering the scattering of a non-relativistic electron in the *static* Coulomb potential of the proton, $V(\mathrm{r}) = \alpha/\mathrm{r}$. Therefore, it can be concluded that in the non-relativistic limit, only the interaction between the electric charges of the electron and proton contribute to the scattering process; there is no significant contribution from the magnetic (spin–spin) interaction. It should be noted that the angular dependence of the Rutherford scattering cross section originates solely from the $1/q^2$ propagator term.

### 7.2.2 Mott scattering

Mott scattering is the limit of electron–proton elastic scattering where the electron is relativistic but the proton recoil still can be neglected. These conditions apply when $m_\mathrm{e} \ll E \ll m_\mathrm{p}$. In this case, the parameter $\kappa \approx 1$ and two of the four possible electron currents of (7.2)–(7.5) are zero. Writing $E = \gamma_\mathrm{e} m_\mathrm{e}$ and taking the limit $\beta_\mathrm{e}\gamma_\mathrm{e} \gg 1$ for which $E \approx \mathrm{p}$, the matrix element of (7.9) reduces to

$$\langle|\mathcal{M}_{fi}^2|\rangle \approx \frac{m_\mathrm{p}^2 e^4}{E^2 \sin^4(\theta/2)}\cos^2\frac{\theta}{2},$$

which when substituted into (7.11) gives

$$\left(\frac{\mathrm{d}\sigma}{\mathrm{d}\Omega}\right)_{\mathrm{Mott}} = \frac{\alpha^2}{4E^2 \sin^4(\theta/2)}\cos^2\frac{\theta}{2}. \tag{7.14}$$

The Mott scattering cross section formula of (7.14) could have been derived by considering the scattering of a relativistic electron in the Coulomb potential of a spin-less nucleus. Again it can be concluded that the contribution to the scattering process from a purely magnetic spin–spin interaction is negligible.

## 7.3 Form factors

The Rutherford and Mott scattering formulae of (7.13) and (7.14) can be calculated from first-order perturbation theory for scattering in the Coulomb potential from a *point-like object*. To account for the finite extent of the charge distribution of the proton, this treatment must be modified by introducing a *form factor*. Qualitatively, the form factor accounts for the phase differences between contributions to the scattered wave from different points of the charge distribution, as indicated in Figure 7.3. If the wavelength of the virtual photon is much larger than the radius of the proton, the contributions to the scattered wave from each point in the charge distribution will be in phase and therefore add constructively. When the wavelength is smaller than the radius of the proton, the phases of the scattered waves will have a strong dependence on the position of the part of the charge distribution responsible for the scattering. In this case, when integrated over the entire charge distribution, the negative interference between the different contributions greatly reduces the total amplitude.

The mathematical expression for the form factor (which is not a Lorentz-invariant concept) can be derived in the context of first-order perturbation theory. Consider the scattering of an electron in the static potential from an extended charge distribution, as indicated in Figure 7.4. The charge density can be written as $Q\rho(\mathbf{r}')$,

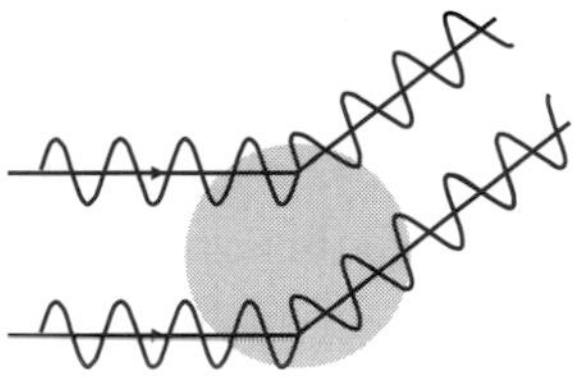

Fig. 7.3 A cartoon indicating the origin of the form factor in elastic scattering.

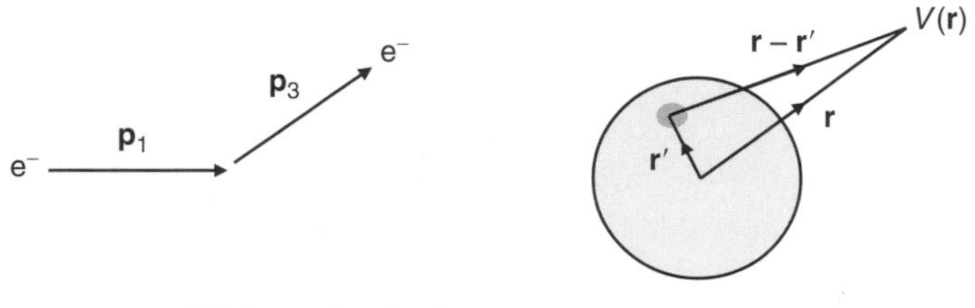


Fig. 7.4 The potential due to an extended charge distribution.

where $Q$ is the total charge and $\rho(\mathbf{r}')$ is the charge distribution normalised to unity

$$\int \rho(\mathbf{r}')\,\mathrm{d}^3\mathbf{r}' = 1\,.$$

The potential at a distance $\mathbf{r}$ from the origin, written in terms of this charge density is simply

$$V(\mathbf{r}) = \int \frac{Q\rho(\mathbf{r}')}{4\pi|\mathbf{r}-\mathbf{r}'|}\mathrm{d}^3\mathbf{r}'\,. \tag{7.15}$$

In the Born approximation, where the wavefunctions of the initial-state and scattered electrons are expressed as the plane waves, $\psi_i = e^{i(\mathbf{p}_1\cdot\mathbf{r}-Et)}$ and $\psi_f = e^{i(\mathbf{p}_3\cdot\mathbf{r}-Et)}$, the lowest-order matrix element for the scattering process is

$$\mathcal{M}_{fi} = \langle\psi_f|V(\mathbf{r})|\psi_i\rangle = \int e^{-i\mathbf{p}_3\cdot\mathbf{r}}V(\mathbf{r})e^{i\mathbf{p}_1\cdot\mathbf{r}}\,\mathrm{d}^3\mathbf{r}.$$

Writing $\mathbf{q} = (\mathbf{p}_1 - \mathbf{p}_3)$ and using the potential of (7.15) leads to

$$\begin{aligned}\mathcal{M}_{fi} &= \iint e^{i\mathbf{q}\cdot\mathbf{r}}\frac{Q\rho(\mathbf{r}')}{4\pi|\mathbf{r}-\mathbf{r}'|}\,\mathrm{d}^3\mathbf{r}'\mathrm{d}^3\mathbf{r}\\ &= \iint e^{i\mathbf{q}\cdot(\mathbf{r}-\mathbf{r}')}e^{i\mathbf{q}\cdot\mathbf{r}'}\frac{Q\rho(\mathbf{r}')}{4\pi|\mathbf{r}-\mathbf{r}'|}\,\mathrm{d}^3\mathbf{r}'\mathrm{d}^3\mathbf{r}\,.\end{aligned} \tag{7.16}$$

By expressing the difference $\mathbf{r}-\mathbf{r}'$ as the vector $\mathbf{R}$, the integral of (7.16) separates into two parts

$$\mathcal{M}_{fi} = \int e^{i\mathbf{q}\cdot\mathbf{R}}\frac{Q}{4\pi|\mathbf{R}|}\mathrm{d}^3\mathbf{R}\int \rho(\mathbf{r}')e^{i\mathbf{q}\cdot\mathbf{r}'}\,\mathrm{d}^3\mathbf{r}'\,.$$

The integral over $\mathrm{d}^3\mathbf{R}$ is simply the equivalent expression for scattering from a potential due to a point charge. Hence the matrix element can be written

$$\mathcal{M}_{fi} = \mathcal{M}_{fi}^{\mathrm{pt}}F(\mathbf{q}^2),$$

where $\mathcal{M}_{fi}^{\mathrm{pt}}$ is the equivalent matrix element for a point-like proton and the form factor $F(\mathbf{q}^2)$ is given by

$$F(\mathbf{q}^2) = \int \rho(\mathbf{r})e^{i\mathbf{q}\cdot\mathbf{r}}\,\mathrm{d}^3\mathbf{r}.$$

Therefore, in order to account for the extended charge distribution of the proton, the Mott scattering cross section of (7.14) has to be modified to

$$\left(\frac{\mathrm{d}\sigma}{\mathrm{d}\Omega}\right)_{\mathrm{Mott}} \rightarrow \frac{\alpha^2}{4E^2\sin^4(\theta/2)}\cos^2\left(\frac{\theta}{2}\right)\left|F(\mathbf{q}^2)\right|^2. \tag{7.17}$$

The form factor $F(\mathbf{q}^2)$ is the three-dimensional Fourier transform of the charge distribution. If the wavelength of the virtual photon is large compared to the size of

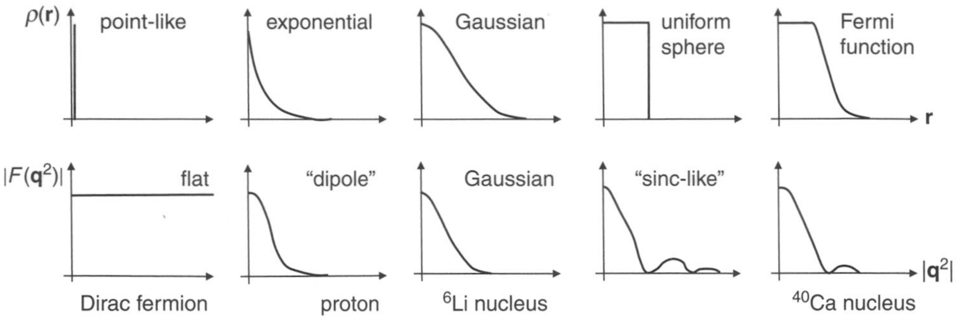


Fig. 7.5 Possible three-dimensional charge distributions and the corresponding form factors plotted as a function of $\mathbf{q}^2$.

the charge distribution then $\mathbf{q}\cdot\mathbf{r} \approx 0$ over the entire volume integral. In this case, the scattering cross section is identical to that for a point-like object and therefore, regardless of the form of the charge distribution, $F(0) = 1$. In the limit where the wavelength is very small compared with the size of the charge distribution, the phases of the contributions from different regions of the charge distribution will vary rapidly and will tend to cancel and $F(\mathbf{q}^2 \to \infty) = 0$. Thus, for *any* finite size charge distribution, the elastic scattering cross section will tend to zero at high values of $\mathbf{q}^2$. The exact form of $F(\mathbf{q}^2)$ depends on the charge distribution; some common examples and the corresponding form factors are shown in Figure 7.5. For a point-like particle, $F(\mathbf{q}^2) = 1$ for all $\mathbf{q}$.

## 7.4 Relativistic electron–proton elastic scattering

In the above calculations of the Rutherford and Mott elastic scattering cross sections, it was assumed that the recoil of the proton could neglected. This a reasonable approximation provided $|\mathbf{q}| \ll m_\mathrm{p}$. In this low-energy limit, it was inferred that the contribution to the scattering process from the pure magnetic spin–spin interaction is negligibly small. For electron–proton elastic scattering at higher energies, the recoil of the proton cannot be neglected and the magnetic spin–spin interaction becomes important.

For the general case, the four-momenta of the initial- and final-state particles, defined in Figure 7.6, can be written as

$$p_1 = (E_1, 0, 0, E_1), \tag{7.18}$$

$$p_2 = (m_\mathrm{p}, 0, 0, 0), \tag{7.19}$$

$$p_3 = (E_3, 0, E_3\sin\theta, E_3\cos\theta), \tag{7.20}$$

$$p_4 = (E_4, \mathbf{p}_4). \tag{7.21}$$

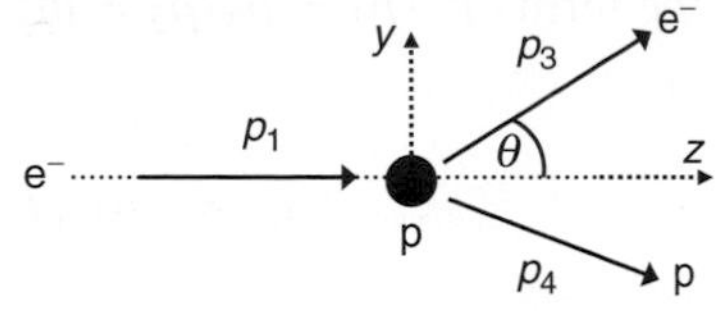


Fig. 7.6 The kinematics of electron–proton scattering in the proton rest frame.

Here the energy of the scattered electron is no longer equal to that of the incident electron. Assuming that the electron energy is sufficiently large that terms of $O(m_e^2)$ can be neglected, and (initially) treating the proton as a point-like Dirac particle, the matrix element for the elastic scattering process $e^-p \to e^-p$ is given by (6.67)

$$\langle|\mathcal{M}_{fi}|^2\rangle = \frac{8e^4}{(p_1-p_3)^4}\left[(p_1.p_2)(p_3.p_4)+(p_1.p_4)(p_2.p_3)-m_p^2(p_1.p_3)\right]. \qquad (7.22)$$

### 7.4.1 Scattering kinematics

In most electron–proton elastic scattering experiments, the final-state proton is not observed. Consequently, the matrix element of (7.22) is most usefully expressed in terms of the experimental observables, which are the energy and scattering angle of the electron. To achieve this, the final-state proton four-momentum $p_4$ can be eliminated using energy and momentum conservation, $p_4 = p_1 + p_2 - p_3$. From the definitions of the four-momenta in (7.18)–(7.20), the four-vector scalar products in (7.22) which do not involve $p_4$ are

$$p_2\cdot p_3 = E_3 m_p, \quad p_1\cdot p_2 = E_1 m_p \quad \text{and} \quad p_1\cdot p_3 = E_1E_3(1-\cos\theta).$$

The two terms involving $p_4$, which can be rewritten using $p_4 = p_1 + p_2 - p_3$, are

$$p_3\cdot p_4 = p_3\cdot p_1 + p_3\cdot p_2 - p_3\cdot p_3 = E_1E_3(1-\cos\theta) + E_3 m_p,$$
$$p_1\cdot p_4 = p_1\cdot p_1 + p_1\cdot p_2 - p_1\cdot p_3 = E_1 m_p - E_1E_3(1-\cos\theta),$$

where the terms $p_1\cdot p_1 = p_3\cdot p_3 = m_e^2$ have been been dropped. Hence, the matrix element of (7.22), expressed in terms of the energy of the final-state electron $E_3$ and the scattering angle $\theta$ is

$$\begin{aligned}\langle|\mathcal{M}_{fi}|^2\rangle &= \frac{8e^4}{(p_1-p_3)^4} m_p E_1 E_3\left[(E_1-E_3)(1-\cos\theta) + m_p[(1+\cos\theta)\right]\\ &= \frac{8e^4}{(p_1-p_3)^4} 2m_p E_1 E_3\left[(E_1-E_3)\sin^2\frac{\theta}{2} + m_p\cos^2\frac{\theta}{2}\right]. \qquad (7.23)\end{aligned}$$

The four-momentum squared of the virtual photon, $q^2 = (p_1 - p_3)^2$, also can be expressed in terms of $E_3$ and $\theta$ using

$$q^2 = (p_1-p_3)^2 = p_1^2 + p_3^2 - 2p_1\cdot p_3 \approx -2E_1E_3(1-\cos\theta), \qquad (7.24)$$

where again the terms $p_1 \cdot p_1 = p_3 \cdot p_3 = m_e^2$ have been neglected. Hence, to a good approximation,

$$q^2 = -4E_1E_3 \sin^2\frac{\theta}{2}. \tag{7.25}$$

Because $q^2$ is always negative, it is more convenient to work in terms of $Q^2$ defined by

$$Q^2 \equiv -q^2 = 4E_1E_3 \sin^2\frac{\theta}{2}, \tag{7.26}$$

which is always positive.

The energy lost by the electron in the scattering process, $E_1 - E_3$, can be expressed in terms of $Q^2$ by first noting that

$$q \cdot p_2 = (p_1 - p_3) \cdot p_2 = m_p(E_1 - E_3). \tag{7.27}$$

A second equation for $q \cdot p_2$ can be obtained by expressing $q$ in terms of the proton four-momenta, $q = p_4 - p_2$, such that

$$p_4^2 = (q + p_2)^2 = q^2 + 2q \cdot p_2 + p_2^2,$$

which, using $p_2^2 = p_4^2 = m_p^2$, gives

$$q \cdot p_2 = -q^2/2. \tag{7.28}$$

Equating (7.27) and (7.28) enables $(E_1 - E_3)$ to be expressed as a function of $Q^2$,

$$E_1 - E_3 = -\frac{q^2}{2m_p} = \frac{Q^2}{2m_p}, \tag{7.29}$$

which (unsurprisingly) demonstrates that the electron always loses energy in the scattering process. Using the relations of (7.25) and (7.29), the spin-averaged matrix element squared of (7.23) can be expressed as

$$\langle|\mathcal{M}_{fi}|^2\rangle = \frac{m_p^2 e^4}{E_1E_3 \sin^4(\theta/2)}\left[\cos^2\frac{\theta}{2} + \frac{Q^2}{2m_p^2}\sin^2\frac{\theta}{2}\right].$$

The differential cross section again can be obtained the cross section formula of (3.47), giving

$$\frac{d\sigma}{d\Omega} \approx \frac{1}{64\pi^2}\left(\frac{E_3}{m_p E_1}\right)^2 \langle|\mathcal{M}_{fi}|^2\rangle.$$

Hence, the differential cross section for the scattering of relativistic electrons from a proton that is initially at rest is

$$\frac{d\sigma}{d\Omega} = \frac{\alpha^2}{4E_1^2 \sin^4(\theta/2)}\frac{E_3}{E_1}\left(\cos^2\frac{\theta}{2} + \frac{Q^2}{2m_p^2}\sin^2\frac{\theta}{2}\right). \tag{7.30}$$

Although (7.30) is expressed in terms of $Q^2$, $E_3$ and $\theta$, it is important to realise that there is only one independent variable; both $Q^2$ and $E_3$ can be expressed in terms of the scattering angle of the electron. This can be seen by firstly equating (7.24) and (7.29) to give

$$-2m_{\mathrm{p}}(E_1 - E_3) = -2E_1E_3(1 - \cos\theta),$$

and hence

$$E_3 = \frac{E_1 m_{\mathrm{p}}}{m_{\mathrm{p}} + E_1(1 - \cos\theta)}. \tag{7.31}$$

Substituting (7.31) back into (7.24) then gives an expression for $Q^2$ in terms of the electron scattering angle

$$Q^2 = \frac{2m_{\mathrm{p}}E_1^2(1 - \cos\theta)}{m_{\mathrm{p}} + E_1(1 - \cos\theta)}. \tag{7.32}$$

Therefore, if the scattering angle of the electron is measured in the elastic scattering process, the entire kinematics of the interaction are determined. In practice, measuring the $e^-p \to e^-p$ differential cross section boils down to counting the number of electrons scattered in a particular direction for a known incident electron flux. Furthermore, because the energy of an elastically scattered electron at a particular angle must be equal to that given by (7.31), by measuring the energy *and* angle of the scattered electron, it is possible to confirm that the interaction was indeed elastic and that the *unobserved* proton remained intact.

In the limit of $Q^2 \ll m_{\mathrm{p}}^2$ and $E_3 \approx E_1$, the expression for the electron–proton differential cross section of (7.30) reduces to that for Mott scattering, demonstrating that the Mott scattering cross section formula applies when $m_{\mathrm{e}} \ll E_1 \ll m_{\mathrm{p}}$. Equation (7.30) differs from the Mott scattering formula by the additional factor $E_3/E_1$, which accounts for the energy lost by electron due the proton recoil, and by the new term proportional to $\sin^2(\theta/2)$, which can be identified as being due to a purely magnetic spin–spin interaction.

## 7.5 The Rosenbluth formula

Equation (7.30) is the differential cross section for elastic $e^-p \to e^-p$ scattering assuming a point-like spin-half proton. The finite size of the proton is accounted for by introducing two form factors, one related to the charge distribution of the proton, $G_E(Q^2)$, and the other related to the magnetic moment distribution within the proton, $G_M(Q^2)$. It can be shown that the most general Lorenz-invariant form for electron–proton scattering via the exchange of a single photon, known as the *Rosenbluth formula*, is

$$\frac{d\sigma}{d\Omega} = \frac{\alpha^2}{4E_1^2 \sin^4(\theta/2)} \frac{E_3}{E_1} \left( \frac{G_E^2 + \tau G_M^2}{(1+\tau)} \cos^2\frac{\theta}{2} + 2\tau G_M^2 \sin^2\frac{\theta}{2} \right), \tag{7.33}$$

where $\tau$ is given by

$$\tau = \frac{Q^2}{4m_p^2}. \tag{7.34}$$

In the Lorentz-invariant Rosenbluth formula, the form factors $G_E(Q^2)$ and $G_M(Q^2)$ are functions of the four-momentum squared of the virtual photon. Unlike the form factor $F(\mathbf{q}^2)$ introduced previously, which was a function of the three-momentum squared, the form factors $G_E(Q^2)$ and $G_M(Q^2)$ cannot be interpreted simply as the Fourier transforms of the charge and magnetic moment distributions of the proton. However, the relation between $G_E(Q^2)$ and $G_M(Q^2)$ and the corresponding Fourier transforms can be obtained by writing

$$Q^2 = -q^2 = \mathbf{q}^2 - (E_1 - E_3)^2,$$

which from (7.29) gives

$$Q^2\left(1 + \frac{Q^2}{4m_p^2}\right) = \mathbf{q}^2.$$

Therefore, in the limit where $Q^2 \ll 4m_p^2$, the time-like component of $Q^2$ is relatively small and $Q^2 \approx \mathbf{q}^2$. Thus, in this low-$Q^2$ limit, the form factors $G_E(Q^2)$ and $G_M(Q^2)$ approximate to functions of $\mathbf{q}^2$ alone and can be interpreted as the Fourier transforms of the charge and magnetic moment distributions of the proton

$$G_E(Q^2) \approx G_E(\mathbf{q}^2) = \int e^{i\mathbf{q}\cdot\mathbf{r}} \rho(\mathbf{r}) d^3\mathbf{r},$$

$$G_M(Q^2) \approx G_M(\mathbf{q}^2) = \int e^{i\mathbf{q}\cdot\mathbf{r}} \mu(\mathbf{r}) d^3\mathbf{r}.$$

There is one further complication. The form of the Rosenbluth equation follows from (7.30), which was obtained from the QED calculation where the proton was treated as a point-like Dirac particle. But the magnetic moment of a point-like Dirac particle (see Appendix B.1) is related to its spin by

$$\boldsymbol{\mu} = \frac{q}{m}\mathbf{S},$$

whereas the experimentally measured value of the *anomalous* magnetic moment of the proton (discussed further in Chapter 9) is

$$\boldsymbol{\mu} = 2.79\frac{e}{m_p}\mathbf{S}.$$

For consistency with this experimental observation, the magnetic moment distribution has to be normalised to +2.79 rather than unity, and therefore

$$G_E(0) = \int \rho(\mathbf{r})\,\mathrm{d}^3\mathbf{r} = 1$$
$$G_M(0) = \int \mu(\mathbf{r})\,\mathrm{d}^3\mathbf{r} = +2.79.$$

It is worth noting that, even taken in isolation, the observation of the anomalous magnetic moment of the proton already provides evidence that the proton is not a point-like particle.

### 7.5.1 Measuring $G_E(Q^2)$ and $G_M(Q^2)$

The $e^-p \to e^-p$ differential cross section is a function of both the charge and magnetic moment distributions of the proton. Whilst it is tempting to assume that magnetic moment distribution follows that of the charge distribution, $G_M(Q^2) = 2.79\,G_E(Q^2)$, there is no *a priori* justification for making this assumption. Fortunately $G_M(Q^2)$ and $G_E(Q^2)$ can be determined separately from experiment. This can be seen by writing the Rosenbluth formula of (7.33) as

$$\frac{\mathrm{d}\sigma}{\mathrm{d}\Omega} = \left(\frac{G_E^2 + \tau G_M^2}{(1+\tau)} + 2\tau G_M^2 \tan^2\frac{\theta}{2}\right)\cdot\left(\frac{\mathrm{d}\sigma}{\mathrm{d}\Omega}\right)_0, \tag{7.35}$$

where

$$\left(\frac{\mathrm{d}\sigma}{\mathrm{d}\Omega}\right)_0 = \frac{\alpha^2}{4E_1^2\sin^4(\theta/2)}\left(\frac{E_3}{E_1}\right)\cos^2\frac{\theta}{2}, \tag{7.36}$$

is the Mott cross section, modified to account for the proton recoil. At low $Q^2$, where $\tau \ll 1$, the electric form factor dominates and (7.35) is approximately

$$\frac{\mathrm{d}\sigma}{\mathrm{d}\Omega}\bigg/\left(\frac{\mathrm{d}\sigma}{\mathrm{d}\Omega}\right)_0 \approx G_E^2.$$

In this limit, $G_E^2$ is equivalent to the form factor $|F(\mathbf{q})|^2$ described previously. At high $Q^2$, where $\tau \gg 1$, the purely magnetic spin–spin term dominates and (7.35) approximates to

$$\frac{\mathrm{d}\sigma}{\mathrm{d}\Omega}\bigg/\left(\frac{\mathrm{d}\sigma}{\mathrm{d}\Omega}\right)_0 \approx \left(1 + 2\tau\tan^2\frac{\theta}{2}\right)G_M^2.$$

In general, the $Q^2$ dependence of $G_M(Q^2)$ and $G_E(Q^2)$ can be inferred from $e^-p \to e^-p$ elastic scattering experiments by varying the electron beam energy. For each beam energy, the differential cross section is measured at the angle corresponding to a particular value of $Q^2$, given by (7.32). For example, Figure 7.7a

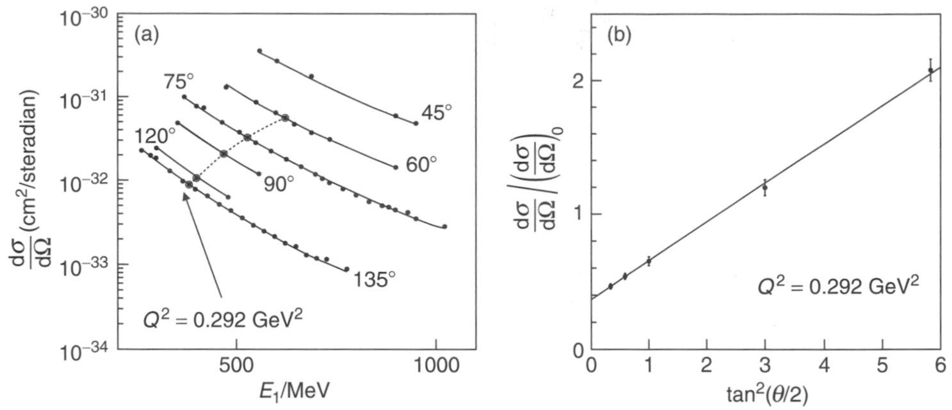


Fig. 7.7 Low energy $e^-p \to e^-p$ elastic scattering data. Data from Hughes *et al.* (1965).

shows the measured $e^-p \to e^-p$ differential cross sections for six different scattering angles and a range of beam energies. The five data points that are highlighted all correspond to $e^-p \to e^-p$ elastic scattering at $Q^2 = 0.292\,\mathrm{GeV}^2$. In this way, the cross section can be measured at fixed $Q^2$ but over a range of scattering angles. Figure 7.7b shows, for the five data points with $Q^2 = 0.292\,\mathrm{GeV}^2$, the measured cross sections normalised to the expected Mott cross section of (7.36), plotted as a function of $\tan^2(\theta/2)$. The observed linear dependence on $\tan^2(\theta/2)$ is expected from (7.35), where it can be seen that the gradient and intercept with the $y$-axis are given respectively by

$$m = 2\tau\left[G_M(Q^2)\right]^2 \quad \text{and} \quad c = \frac{\left[G_E(Q^2)\right]^2 + \tau\left[G_M(Q^2)\right]^2}{(1+\tau)}.$$

Hence, the data shown in Figure 7.7b can be used to extract measurements of both $G_E(Q^2)$ and $G_M(Q^2)$ at $Q^2 = 0.292\,\mathrm{GeV}^2$ (see Problem 7.6). A similar analysis can be applied to cross section measurements corresponding to different values of $Q^2$, providing an experimental determination of the electric and magnetic form factors of the proton over a range of $Q^2$ values, as shown in Figure 7.8a. The fact that the measured form factors decrease with $Q^2$ provides a concrete experimental demonstration that the proton has finite size. The shape of $G_M(Q^2)$ closely follows that of $G_E(Q^2)$, showing that the charge and magnetic moment distributions within the proton are consistent. Furthermore, the measured values extrapolated to $Q^2 = 0$ are in agreement with the expectations of $G_E(0) = 1$ and $G_M(0) = 2.79$. Finally, Figure 7.8b shows measurements of $G_M(Q^2)$ at $Q^2$ values up to $32\,\mathrm{GeV}^2$. For these data recorded at higher values of $Q^2$, the contribution from $G_E(Q^2)$ is strongly suppressed and only $G_M(Q^2)$ can be measured.

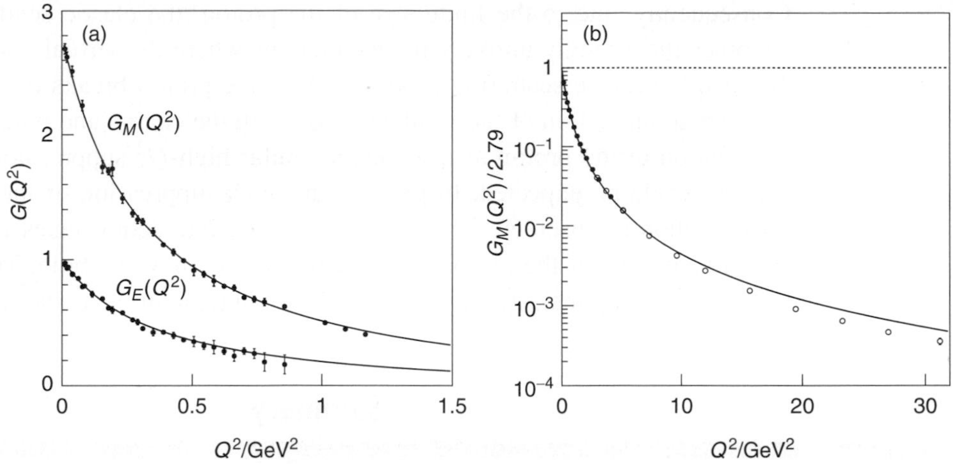


Fig. 7.8 (a) Measurements of $G_E(Q^2)$ and $G_M(Q^2)$ from $e^-p \to e^-p$ elastic scattering data at low $Q^2$, adapted from Hughes *et al*. (1965) and references therein. (b) Measurements of $G_M(Q^2)$ at higher $Q^2$, data from Walker *et al*. (1994) (solid circles) and Sill *et al*. (1993) (open circles). The curves correspond to the dipole function described in the text.

The data shown in Figures 7.8a and 7.8b are reasonably well parameterised by the empirically determined "dipole function"

$$G_M(Q^2) = 2.79 G_E(Q^2) \approx 2.79 \frac{1}{(1 + Q^2/0.71\,\mathrm{GeV}^2)^2}. \tag{7.37}$$

By taking the Fourier transform of the dipole function for $G_E(Q^2)$, which provides a good description of the low $Q^2$ data where $Q^2 \approx \mathbf{q}^2$, the charge distribution of the proton is determined to be

$$\rho(\mathbf{r}) \approx \rho_0 e^{-r/a},$$

with $a \approx 0.24$ fm. This experimentally determined value for $a$ corresponds to a proton root-mean-square charge radius of 0.8 fm.

### 7.5.2 Elastic scattering at high $Q^2$

At high $Q^2$, the electron–proton elastic scattering cross section of (7.35) reduces to

$$\left(\frac{d\sigma}{d\Omega}\right)_{\text{elastic}} \sim \frac{\alpha^2}{4E_1^2 \sin^4(\theta/2)} \frac{E_3}{E_1}\left[\frac{Q^2}{2m_p^2} G_M^2 \sin^2\frac{\theta}{2}\right].$$

From (7.37) it can be seen that in the high-$Q^2$ limit, $G_M(Q^2) \propto Q^{-4}$ and therefore

$$\left(\frac{d\sigma}{d\Omega}\right)_{elastic} \propto \frac{1}{Q^6}\left(\frac{d\sigma}{d\Omega}\right)_{Mott}.$$

Consequently, due to the finite size of the proton, the elastic scattering process becomes increasingly unlikely for interactions where the virtual photon has large $Q^2$. If the inelastic scattering process, where the proton breaks up, also involved a coherent interaction of the virtual photon with the charge and magnetic moment distribution of the proton as a whole, a similar high-$Q^2$ suppression of the cross section would be expected. In practice, no such suppression of the inelastic $e^-p$ cross section is observed. This implies that the interaction takes place with the constituent parts of the proton rather than the proton as a whole. This process of high-energy deep inelastic scattering is the main topic of next chapter.

## Summary

In this chapter, the process of $e^-p \to e^-p$ elastic scattering has been described in some detail. In general, the differential elastic scattering cross section is given by the Rosenbluth formula

$$\frac{d\sigma}{d\Omega} = \frac{\alpha^2}{4E_1^2 \sin^4(\theta/2)} \frac{E_3}{E_1} \left( \frac{G_E^2 + \tau G_M^2}{(1+\tau)} \cos^2\frac{\theta}{2} + 2\tau G_M^2 \sin^2\frac{\theta}{2} \right),$$

where the form factors $G_E(Q^2)$ and $G_M(Q^2)$ describe the charge and magnetic moment distributions of the proton. The techniques used to measure the form factors were described in some detail. It is important that you understand the concepts; they will be used again in the following chapter.

Because of the finite size of the proton, both $G_E(Q^2)$ and $G_M(Q^2)$ become small at high $Q^2$ and the elastic scattering cross section falls rapidly with increasing $Q^2$. Consequently, high-energy electron–proton scattering is dominated by inelastic processes where the virtual photon interacts with the quarks inside the proton, rather than the proton as a coherent whole.

## Problems

**7.1** The derivation of (7.8) used the algebraic relation

$$(\gamma + 1)^2(1 - \kappa^2)^2 = 4,$$

where

$$\kappa = \frac{\beta\gamma}{\gamma + 1} \quad \text{and} \quad (1 - \beta^2)\gamma^2 = 1.$$

Show that this holds.

**7.2** By considering momentum and energy conservation in $e^-p$ elastic scattering from a proton at rest, find an expression for the fractional energy loss of the scattered electron $(E_1 - E_3)/E_1$ in terms of the scattering angle and the parameter

$$\kappa = \frac{p}{E_1 + m_e} \equiv \frac{\beta\gamma}{\gamma + 1}.$$

**7.3** In an $e^-p$ scattering experiment, the incident electron has energy $E_1 = 529.5$ MeV and the scattered electrons are detected at an angle of $\theta = 75°$ relative to the incoming beam.

(a) At this angle, almost all of the scattered electrons are measured to have an energy of $E_3 \approx 373$ MeV. What can be concluded from this observation?
(b) Find the corresponding value of $Q^2$.

**7.4** For a spherically symmetric charge distribution $\rho(r)$, where

$$\int \rho(r)\, d^3\mathbf{r} = 1,$$

show that the form factor can be expressed as

$$F(\mathbf{q}^2) = \frac{4\pi}{q}\int_0^\infty r\sin(qr)\rho(r)\,dr,$$
$$\simeq 1 - \frac{1}{6}q^2\langle R^2\rangle + \cdots,$$

where $\langle R^2\rangle$ Is the mean square charge radius. Hence show that

$$\langle R^2\rangle = -6\left[\frac{dF(\mathbf{q}^2)}{dq^2}\right]_{q^2=0}.$$

**7.5** Using the answer to the previous question and the data in Figure 7.8a, estimate the root-mean-squared charge radius of the proton.

**7.6** From the slope and intercept of the right plot of Figure 7.7, obtain values for $G_M(0.292\,\text{GeV}^2)$ and $G_E(0.292\,\text{GeV}^2)$.

**7.7** Use the data of Figure 7.7 to estimate $G_E(Q^2)$ at $Q^2 = 0.500\,\text{GeV}^2$.

**7.8** The experimental data of Figure 7.8 can be described by the form factor

$$G(Q^2) = \frac{G(0)}{(1 + Q^2/Q_0^2)^2},$$

with $Q_0 = 0.71$ GeV. Taking $Q^2 \approx \mathbf{q}^2$, show that this implies that proton has an exponential charge distribution of the form

$$\rho(\mathbf{r}) = \rho_0 e^{-r/a},$$

and find the value of $a$.

# 8 Deep inelastic scattering

This chapter describes high-energy electron–proton inelastic scattering where the proton breaks up in the interaction. The inelastic scattering process is first discussed in terms of a general Lorentz-invariant extension of the ideas introduced in the previous chapter, with form factors replaced by structure functions. Deep inelastic scattering is then described by the QED interaction of a virtual photon with the constituent quarks inside the proton. The experimental data are then interpreted in the quark–parton model and the measured structure functions are related to parton distribution functions that describe the momentum distributions of the quarks. From the experimental measurements, the proton is found to be a complex dynamical system comprised of quarks, gluons and antiquarks.

## 8.1 Electron–proton inelastic scattering

Because of the finite size of the proton, the cross section for electron–proton elastic scattering decreases rapidly with energy. Consequently, high-energy $e^-p$ interactions are dominated by inelastic scattering processes where the proton breaks up. For $e^-p \to e^-X$ inelastic scattering, shown in Figure 8.1, the hadronic final state resulting from the break-up of the proton usually consists of many particles. The invariant mass of this hadronic system, denoted $W$, depends on the four-momentum of the virtual photon, $W^2 = p_4^2 = (p_2 + q)^2$, and therefore can take a range of values. Compared to the elastic scattering process, where the invariant mass of the final state is always the mass of the proton, this additional degree of freedom in the inelastic scattering process means that the event kinematics must be specified by two quantities. Whereas $e^-p \to e^-p$ elastic scattering was described in terms of the electron scattering angle alone, the two kinematic variables used to describe inelastic scattering are usually chosen from the Lorentz-invariant quantities $W$, $x$, $y$, $\nu$ and $Q^2$, defined below.

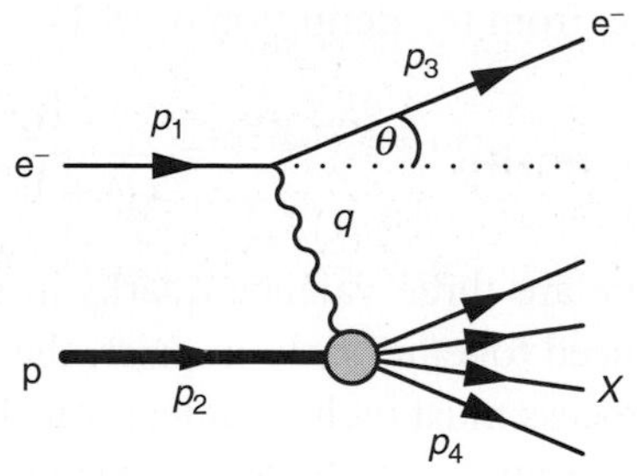


Fig. 8.1 Electron–proton inelastic scattering.

### 8.1.1 Kinematic variables for inelastic scattering

As was the case for elastic scattering, $Q^2$ is defined as the negative four-momentum squared of the virtual photon,

$$Q^2 = -q^2.$$

When written in terms of the four-momenta of the initial- and final-state electrons,

$$Q^2 = -(p_1 - p_3)^2 = -2m_e^2 + 2p_1 \cdot p_3 = -2m_e^2 + 2E_1E_3 - 2\mathbf{p}_1\mathbf{p}_3 \cos\theta.$$

In inelastic scattering, the energies are sufficiently high that the electron mass can be neglected and therefore, to a very good approximation

$$Q^2 \approx 2E_1E_3(1 - \cos\theta) = 4E_1E_3 \sin^2\frac{\theta}{2},$$

implying that $Q^2$ is always positive.

#### Bjorken *x*

The Lorentz-invariant dimensionless quantity

$$x \equiv \frac{Q^2}{2p_2 \cdot q}, \tag{8.1}$$

will turn out to be an important kinematic variable in the discussion of the quark model of deep inelastic scattering. The range of possible values of $x$ can be found by writing the four-momentum of the hadronic system in terms of that of the virtual photon

$$W^2 \equiv p_4^2 = (q + p_2)^2 = q^2 + 2p_2 \cdot q + p_2^2$$
$$\Rightarrow \quad W^2 + Q^2 - m_p^2 = 2p_2 \cdot q,$$

and therefore, from the definition of (8.1),

$$x = \frac{Q^2}{Q^2 + W^2 - m_p^2}. \tag{8.2}$$

Because there are three valence quarks in the proton, and quarks and antiquarks can be produced together only in pairs, the hadronic final state in an $e^-p$ inelastic scattering process must include at least one baryon (qqq). Consequently, the invariant mass of the final-state hadronic system is always greater than the mass of the proton (which is the lightest baryon), thus

$$W^2 \equiv p_4^2 \geq m_p^2.$$

Because $Q^2 \geq 0$ and $W^2 \geq m_p$, the relation of (8.2) implies that $x$ is always in the range

$$0 \leq x \leq 1.$$

The value of $x$ expresses the "elasticity" of the scattering process. The extreme case of $x = 1$ is equivalent to $W^2 = m_p^2$, and therefore corresponds to elastic scattering.

### $y$ and $\nu$

A second dimensionless Lorentz-invariant quantity, the inelasticity $y$, is defined as

$$y \equiv \frac{p_2 \cdot q}{p_2 \cdot p_1}.$$

In the frame where the proton is at rest, $p_2 = (m_p, 0, 0, 0)$, the momenta of the initial-state $e^-$, the final-state $e^-$ and the virtual photon can be written

$$p_1 = (E_1, 0, 0, E_1), \quad p_3 = (E_3, E_3 \sin\theta, 0, E_3 \cos\theta) \text{ and } q = (E_1 - E_3, \mathbf{p}_1 - \mathbf{p}_3),$$

and therefore

$$y = \frac{m_p(E_1 - E_3)}{m_p E_1} = 1 - \frac{E_3}{E_1}. \tag{8.3}$$

Hence $y$ can be identified as the fractional energy lost by the electron in the scattering process in the frame where the proton is initially at rest. In this frame, the energy of the final-state hadronic system is always greater than the energy of the initial-state proton, $E_4 \geq m_p$, which implies the electron must lose energy. Consequently, $y$ is constrained to be in the range

$$0 \leq y \leq 1.$$

Sometimes it is more convenient to work in terms of energies, rather than the fractional energy loss described by $y$. In this case the related quantity

$$\nu \equiv \frac{p_2 \cdot q}{m_\mathrm{p}}, \tag{8.4}$$

is often used. In the frame where the initial-state proton is at rest,

$$\nu = E_1 - E_3,$$

is simply the energy lost by the electron.

### Relationships between kinematic variables

For a given centre-of-mass energy $\sqrt{s}$, the kinematics of inelastic scattering are fully defined by specifying two independent observables which are usually chosen to be two of Lorentz-invariant quantities, $Q^2$, $x$, $y$ and $\nu$. Provided the chosen quantities are independent, the other two quantities then can be determined through the relations that follow from the definitions,

$$Q^2 \equiv -q^2, \quad x \equiv \frac{Q^2}{2p_2 \cdot q}, \quad y \equiv \frac{p_2 \cdot q}{p_2 \cdot p_1} \quad \text{and} \quad \nu \equiv \frac{p_2 \cdot q}{m_\mathrm{p}}. \tag{8.5}$$

For example, it immediately can be seen that $x$ is related to $Q^2$ and $\nu$ by

$$x = \frac{Q^2}{2m_\mathrm{p}\nu}. \tag{8.6}$$

Furthermore, for a fixed centre-of-mass energy,

$$s = (p_1 + p_2)^2 = p_1^2 + p_2^2 + 2p_1 \cdot p_2 = 2p_1 \cdot p_2 + m_\mathrm{p}^2 + m_\mathrm{e}^2.$$

Since $m_\mathrm{e}^2 \ll m_\mathrm{p}^2$, to a good approximation

$$2p_1 \cdot p_2 \simeq s - m_\mathrm{p}^2,$$

and then from the definitions of (8.5), it follows that $y$ is proportional to $\nu$,

$$y = \left(\frac{2m_\mathrm{p}}{s - m_\mathrm{p}^2}\right)\nu. \tag{8.7}$$

Finally from (8.6) and (8.7), it can be seen that $Q^2$ is related to $x$ and $y$ by

$$Q^2 = (s - m_\mathrm{p}^2)xy. \tag{8.8}$$

Hence, for a fixed centre-of-mass energy, the kinematics of inelastic scattering can be described by any two of the Lorentz-invariant quantities $x$, $Q^2$, $y$ and $\nu$, with the exception of $y$ and $\nu$, which are not independent.

### 8.1.2 Inelastic scattering at low $Q^2$

For electron–proton scattering at relatively low electron energies, both elastic and inelastic scattering processes can occur. For example, Figure 8.2 shows the observed energy distribution of electrons scattered through an angle of $\theta = 10^\circ$ at a fixed-target experiment at DESY, where electrons of energy $E_1 = 4.879\,\text{GeV}$ were fired at a liquid hydrogen target (essentially protons at rest). Because two independent variables are required to define the kinematics of inelastic scattering, the corresponding *double-differential* cross section is expressed in terms of two variables, in this case $\mathrm{d}^2\sigma/\mathrm{d}\Omega\,\mathrm{d}E_3$.

Since the kinematics of an individual interaction are fully specified by two independent variables, in this case the angle and energy of the scattered electron, $\theta$ and $E_3$, the invariant mass $W$ of the unobserved final-state hadronic system can be determined on an event-by-event basis using

$$\begin{aligned} W^2 = (p_2 + q)^2 &= p_2^2 + 2p_2 \cdot q + q^2 = m_\mathrm{p}^2 + 2p_2 \cdot (p_1 - p_3) + (p_1 - p_3)^2 \\ &\approx \left[m_\mathrm{p}^2 + 2m_\mathrm{p}E_1\right] - 2\left[m_\mathrm{p} + E_1(1 - \cos\theta)\right]E_3. \end{aligned} \tag{8.9}$$

Hence, for electrons detected at a fixed scattering angle, the invariant mass $W$ of the hadronic system is linearly related to the energy $E_3$ of the scattered electron. Consequently the energy distribution of Figure 8.2 can be interpreted in terms of $W$. The large peak at final-state electron energies of approximately 4.5 GeV corresponds to $W = m_\mathrm{p}$, and these electrons can be identified as coming from elastic scattering. The peak at $E_3 \approx 4.2\,\text{GeV}$ corresponds to resonant production of a single $\Delta^+$ baryon with mass $W = 1.232\,\text{GeV}$ (see Chapter 9). The two smaller peaks at $E_3 \sim 3.85\,\text{GeV}$ and $E_3 \sim 3.55\,\text{GeV}$ correspond to resonant production of other

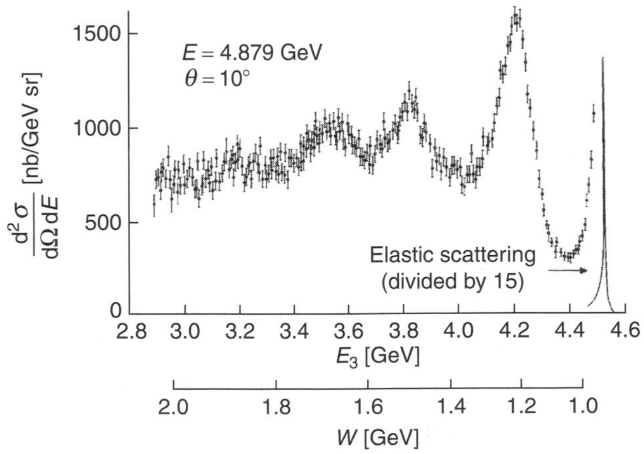


Fig. 8.2 The energy of the scattered electron in low-energy electron–proton scattering and the corresponding invariant mass $W$ of the final state hadronic system. From Bartel *et al*. (1968).

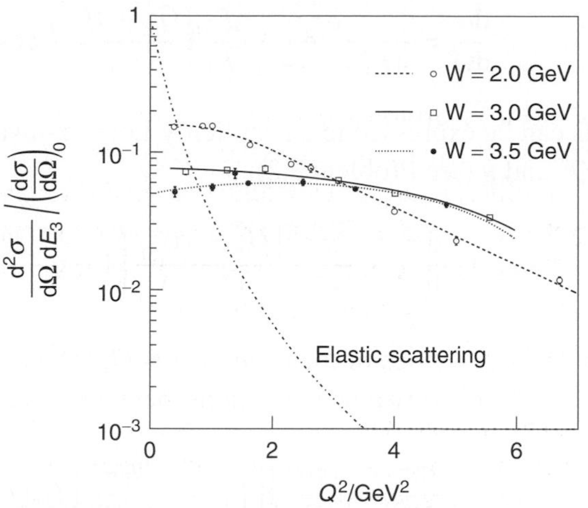


Fig. 8.3 Low-$Q^2$ measurements of the electron–proton inelastic scattering cross section scaled to the Mott cross section. Also shown is the expected dependence for elastic scattering. Adapted from Breidenbach *et al.* (1969).

baryon states. These resonances are essentially excited bound states of the proton (uud), which subsequently decay strongly, for example $\Delta^+ \to p\pi^0$. The full-width-at-half-maximum (FWHM) of a resonance as a function of $W$ is equal to the total decay rate $\Gamma$, which in turn is related to the lifetime of the resonant state by $\Gamma = 1/\tau$. The continuum at higher $W$ is the start of the *deep inelastic* region where the proton is broken up in the collision, resulting in multi-particle final states.

Figure 8.3 shows measurements of the $e^-p \to e^-X$ differential cross section scaled to the Mott scattering cross section of (7.36). The data are plotted as function of $Q^2$ for three different values of $W$. The expected ratio for elastic scattering, assuming the dipole form for $G_E(Q^2)$ and $G_M(Q^2)$ is shown for comparison. The inelastic cross sections are observed to depend only weakly on $Q^2$, in contrast to rapidly falling elastic scattering cross section. In the deep inelastic region (higher values of $W$), the near $Q^2$ independence of the cross section implies a constant form factor, from which it can be concluded that deep inelastic scattering occurs from point-like (or at least very small) entities within the proton.

## 8.2 Deep inelastic scattering

The most general Lorentz-invariant form of the $e^-p \to e^-p$ *elastic* scattering cross section from the exchange of a single photon is given by the Rosenbluth formula of (7.33),

$$\frac{d\sigma}{d\Omega} = \frac{\alpha^2}{4E_1^2 \sin^4(\theta/2)} \frac{E_3}{E_1} \left( \frac{G_E^2 + \tau G_M^2}{(1+\tau)} \cos^2 \frac{\theta}{2} + 2\tau G_M^2 \sin^2 \frac{\theta}{2} \right).$$

This can be expressed in an explicitly Lorentz-invariant form using the definitions of $Q^2$ and $y$ (see Problem 8.2):

$$\frac{d\sigma}{dQ^2} = \frac{4\pi\alpha^2}{Q^4} \left[ \frac{G_E^2 + \tau G_M^2}{(1+\tau)} \left( 1 - y - \frac{m_p^2 y^2}{Q^2} \right) + \frac{1}{2} y^2 G_M^2 \right].$$

The $Q^2$ dependence of the form factors $G_E(Q^2)$ and $G_M(Q^2)$ and $\tau = Q^2/4m_p$ can be absorbed into two new functions, here written as $f_1(Q^2)$ and $f_2(Q^2)$, such that

$$\frac{d\sigma}{dQ^2} = \frac{4\pi\alpha^2}{Q^4} \left[ \left( 1 - y - \frac{m_p^2 y^2}{Q^2} \right) f_2(Q^2) + \frac{1}{2} y^2 f_1(Q^2) \right]. \tag{8.10}$$

Although $y$ appears in this formula, it should be remembered that for elastic scattering $x = 1$ and therefore $y$ is a function of $Q^2$ alone. In this form, $f_1(Q^2)$ is associated with the purely magnetic interaction and $f_2(Q^2)$ has electric and magnetic contributions.

### 8.2.1 Structure functions

Equation (8.10) can be generalised to the inelastic scattering process, where the differential cross section has to be expressed in terms of two independent kinematic quantities. It can be shown that the most general (parity conserving) Lorentz-invariant expression for the cross section for ep $\to$ e$X$ inelastic scattering, mediated by the exchange of a single virtual photon, is

$$\frac{d^2\sigma}{dx\, dQ^2} = \frac{4\pi\alpha^2}{Q^4} \left[ \left( 1 - y - \frac{m_p^2 y^2}{Q^2} \right) \frac{F_2(x, Q^2)}{x} + y^2 F_1(x, Q^2) \right]. \tag{8.11}$$

Here the functions $f_1(Q^2)$ and $f_2(Q^2)$ of (8.10) have been replaced by the two *structure functions*, $F_1(x, Q^2)$ and $F_2(x, Q^2)$, where $F_1(x, Q^2)$ can be identified as being purely magnetic in origin. Because the structure functions depend on both $Q^2$ and $x$, they cannot be interpreted as the Fourier transforms of the proton charge and magnetic moment distributions; as we will see shortly they represent something more fundamental.

For deep inelastic scattering, where $Q^2 \gg m_p^2 y^2$, Equation (8.11) reduces to

$$\frac{d^2\sigma}{dx\, dQ^2} \approx \frac{4\pi\alpha^2}{Q^4} \left[ (1-y) \frac{F_2(x, Q^2)}{x} + y^2 F_1(x, Q^2) \right]. \tag{8.12}$$

In fixed-target electron–proton deep inelastic scattering experiments, the Lorentz-invariant kinematic variables $Q^2$, $x$ and $y$ can be obtained on an event-by-event basis from the observed energy and scattering angle of the electron, $E_3$ and $\theta$,

$$Q^2 = 4E_1E_3\sin^2\frac{\theta}{2}, \quad x = \frac{Q^2}{2m_p(E_1 - E_3)} \quad \text{and} \quad y = 1 - \frac{E_3}{E_1},$$

where $E_1$ is the incident electron energy. The double-differential cross section is measured by counting the numbers of events in the range $x \to x + \Delta x$ and $Q^2 \to Q^2 + \Delta Q^2$. The double-differential cross section at a particular value of $x$ and $Q^2$ can be determined for a range of $y$ values, obtained by varying the incident electron energy (see Problem 8.3). The $y$-dependence of the measured cross sections is then used to disentangle the contributions from $F_1(x, Q^2)$ and $F_2(x, Q^2)$, in much the same way as for the determination of $G_E(Q^2)$ and $G_M(Q^2)$ as described in Section 7.5.1.

### Bjorken scaling and the Callan–Gross relation

The first systematic studies of structure functions in inelastic electron–proton scattering were obtained in a series of experiments at the Stanford Linear Accelerator Center (SLAC) in California. Electrons of energies between 5 GeV and 20 GeV were fired at a liquid hydrogen target. The scattering angle of the electron was measured using a large movable spectrometer, in which the energy of the detected final-state electrons could be selected by using a magnetic field. The differential cross sections, measured over a range of incident electron energies, were used to determine the structure functions. The experimental data revealed two striking features, shown in Figure 8.4. The first observation, known as *Bjorken scaling*, was that both $F_1(x, Q^2)$ and $F_2(x, Q^2)$ are (almost) independent of $Q^2$, allowing the structure functions to be written as

$$F_1(x, Q^2) \to F_1(x) \quad \text{and} \quad F_2(x, Q^2) \to F_2(x).$$

The lack of $Q^2$ dependence of the structure functions is strongly suggestive of scattering from point-like constituents within the proton.

The second observation was that in the deep inelastic scattering regime, $Q^2$ greater than a few $\text{GeV}^2$, the structure functions $F_1(x)$ and $F_2(x)$ are not independent, but satisfy the *Callan–Gross relation*

$$F_2(x) = 2xF_1(x).$$

This observation can be explained by assuming that the underlying process in electron–proton inelastic scattering is the *elastic* scattering of electrons from point-like *spin-half* constituent particles within the proton, namely the quarks. In this case the electric and magnetic contributions to the scattering process are related by the fixed magnetic moment of a Dirac particle.

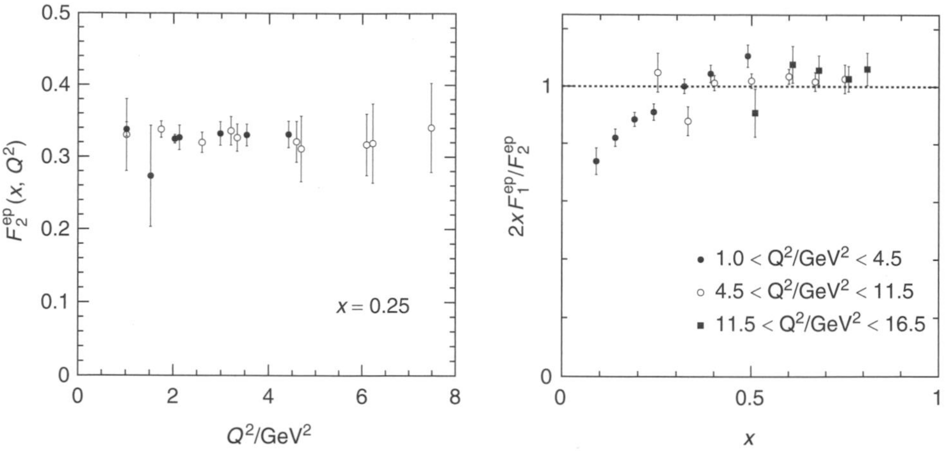


Fig. 8.4 Early structure function measurements from fixed-target electron–proton inelastic scattering at SLAC. Left: measurements of $F_2(x, Q^2)$ showing Bjorken scaling. Right: measurements of $2xF_1/F_2$ showing the Callan–Gross relation. Adapted from Friedman and Kendall (1972) and Bodek *et al.* (1979).

## 8.3 Electron–quark scattering

In the quark model, the underlying interaction in deep inelastic scattering is the QED process of $e^-q \to e^-q$ *elastic* scattering and the deep inelastic scattering cross sections are related to the cross section for this quark-level process. The matrix element for $e^-q \to e^-q$ scattering is obtained from the QED Feynman rules for the Feynman diagram of Figure 8.5. The electron and quark currents are

$$\overline{u}(p_3)[ie\gamma^\mu]u(p_1) \quad \text{and} \quad \overline{u}(p_4)[-iQ_q e\gamma^\nu]u(p_2),$$

and the photon propagator is given by $-ig_{\mu\nu}/q^2$ where $q^2 = p_1 - p_3$. Hence the matrix element can be written

$$\mathcal{M}_{fi} = \frac{Q_q e^2}{q^2}\left[\overline{u}(p_3)\gamma^\mu u(p_1)\right] g_{\mu\nu} \left[\overline{u}(p_4)\gamma^\nu u(p_2)\right]. \tag{8.13}$$

The spin-averaged matrix element squared can be obtained from the helicity amplitudes (see Problem 6.7), or using the trace approach as described in Section 6.5.5. In either case, in the limit where the electron and quark masses can be neglected, the spin-averaged matrix element squared is given by (6.68),

$$\langle|\mathcal{M}_{fi}|^2\rangle = 2Q_q^2 e^4\left(\frac{s^2+u^2}{t^2}\right) = 2Q_q^2 e^4 \frac{(p_1{\cdot}p_2)^2 + (p_1{\cdot}p_4)^2}{(p_1{\cdot}p_3)^2}, \tag{8.14}$$

where as usual, $s = p_1 + p_2$, $t = p_1 - p_3$ and $u = p_1 - p_4$.

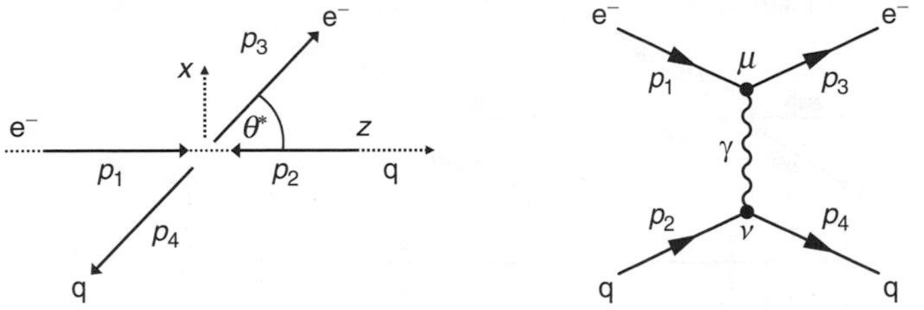


Fig. 8.5 Electron–quark scattering in the centre-of-mass frame and the corresponding lowest-order Feynman diagram.

Here it is convenient to work in the centre-of-mass frame and to express the Lorentz-invariant matrix element of (8.14) in terms of the electron scattering angle, $\theta^*$, as shown in Figure 8.5. Writing the energy of the electron in the centre-of-mass frame as $E = \sqrt{s}/2$, and neglecting the electron and quark masses, the four-momenta of the initial- and final-state particles are given by

$$p_1 = (E, 0, 0, +E), \quad p_3 = (E, +E\sin\theta^*, 0, +E\cos\theta^*),$$
$$p_2 = (E, 0, 0, -E), \quad p_4 = (E, -E\sin\theta^*, 0, -E\cos\theta^*).$$

The four-vector scalar products appearing in (8.14) are

$$p_1{\cdot}p_2 = 2E^2, \quad p_1{\cdot}p_3 = E^2(1 - \cos\theta^*) \quad \text{and} \quad p_1{\cdot}p_4 = E^2(1 + \cos\theta^*).$$

Hence the spin-averaged matrix element squared for the QED process $e^-q \to e^-q$ is

$$\langle|M_{fi}|^2\rangle = 2Q_q^2 e^4 \frac{4E^4 + E^4(1+\cos\theta^*)^2}{E^4(1-\cos\theta^*)^2}.$$

The differential cross section is obtained by substituting this expression for $\langle|\mathcal{M}_{fi}|^2\rangle$ into the cross section formula of (3.50), giving

$$\frac{d\sigma}{d\Omega^*} = \frac{Q_q^2 e^4}{8\pi^2 s} \frac{\left[1 + \frac{1}{4}(1+\cos\theta^*)^2\right]}{(1-\cos\theta^*)^2}. \tag{8.15}$$

The angular dependence in the numerator of (8.15),

$$1 + \tfrac{1}{4}(1+\cos\theta^*)^2, \tag{8.16}$$

reflects the chiral structure of the QED interaction. From the arguments of Section 6.4.2, helicity is conserved in the ultra-relativistic limit of the QED interaction. Therefore, the only non-zero matrix elements originate from spin states where the helicities of the electron and the quark are unchanged in the interaction,

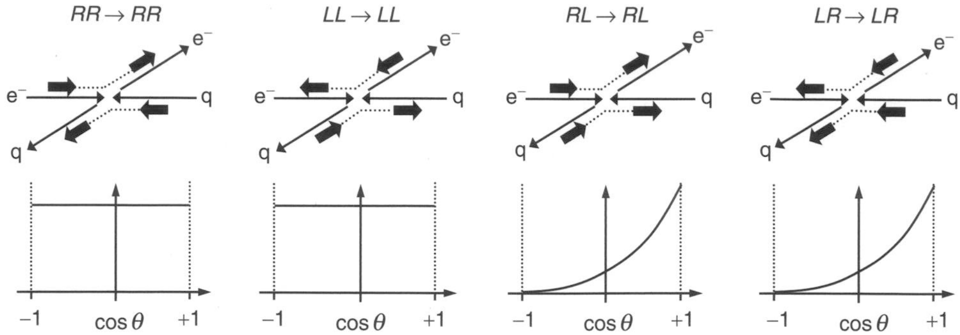


Fig. 8.6 The four helicity combinations contributing to the process $e^-q \to e^-q$ in the limit where $E \gg m$. The first two, $RR \to RR$ and $LL \to LL$, occur in a total spin state with $S_z = 0$. The second two, $RL \to RL$ and $LR \to LR$, take place in $S_z = \pm 1$ states.

as shown in Figure 8.6. The $RR \to RR$ and $LL \to LL$ scattering processes occur in a $S_z = 0$ state, where there is no component of the angular momentum in the $z$-direction. Consequently, there is no preferred polar angle, accounting for the constant term in (8.16). The $RL \to RL$ and $LR \to LR$ scattering processes occur in $S_z = \pm 1$ states and hence (see Section 6.3) result in an angular dependence of

$$\tfrac{1}{4}(1 + \cos\theta^*)^2,$$

explaining the second term in (8.16). The denominator in the expression for the differential cross section of (8.15) arises from the $1/q^2$ propagator term with

$$q^2 = t = (p_1 - p_3)^2 \approx -E^2(1 - \cos\theta^*).$$

When $q^2 \to 0$, in which case the scattering angle $\theta^* \to 0$, the differential cross section tends to infinity. This should not be a surprise. It is analogous to the scattering of a particle in a $1/r$ potential in classical dynamics; regardless of the impact parameter, there is always a finite deflection (however small). The presence of the propagator term implies that in the QED elastic scattering process, the electron is predominantly scattered in the forward direction.

## Lorentz-invariant form

Equation (8.15) gives the $e^-q \to e^-q$ differential cross section in terms of the centre-of-mass scattering angle $\theta^*$. This can be expressed in a Lorentz-invariant form by writing $\cos\theta^*$ in terms of $s$ and $q^2$ and changing variables using

$$\frac{d\sigma}{dq^2} = \frac{d\sigma}{d\Omega^*}\left|\frac{d\Omega^*}{dq^2}\right|.$$

Alternatively, the spin-averaged matrix element squared of (8.14) can be substituted directly into the Lorentz-invariant form for the differential cross section of (3.37) with $t = q^2$, giving

$$\frac{\mathrm{d}\sigma}{\mathrm{d}q^2} = \frac{1}{64\pi s\, \mathrm{p}_i^{*2}}\langle|\mathcal{M}_{fi}|^2\rangle = \frac{Q_\mathrm{q}^2 e^4}{32\pi s\, \mathrm{p}_i^{*2}}\left(\frac{s^2+u^2}{t^2}\right). \tag{8.17}$$

Since $p_i^* = \sqrt{s}/2$ and $t = q^2$, Equation (8.17) can be written as

$$\frac{\mathrm{d}\sigma}{\mathrm{d}q^2} = \frac{Q_\mathrm{q}^2 e^4}{8\pi q^4}\left(\frac{s^2+u^2}{s^2}\right) = \frac{Q_\mathrm{q}^2 e^4}{8\pi q^4}\left[1+\left(\frac{u}{s}\right)^2\right]. \tag{8.18}$$

Finally, this equation can be expressed in terms of $q^2$ and $s$ alone by recalling that the sum of the Mandelstam variables $s + t + u$ is equal to the sum of the masses of the initial- and final-state particles. Therefore, in the high-energy limit where the electron and quark masses can neglected,

$$u \approx -s - t = -s - q^2,$$

and the differential cross section for the $e^-q \to e^-q$ elastic scattering process of (8.18), expressed in terms of $s$ and $q^2$ alone, is

$$\frac{\mathrm{d}\sigma}{\mathrm{d}q^2} = \frac{2\pi\alpha^2 Q_\mathrm{q}^2}{q^4}\left[1+\left(1+\frac{q^2}{s}\right)^2\right]. \tag{8.19}$$

## 8.4 The quark–parton model

Before quarks and gluons were generally accepted, Feynman proposed that the proton was made up of point-like constituents, termed *partons*. In the quark–parton model, the basic interaction in deep inelastic electron–proton scattering is *elastic* scattering from a spin-half quark within the proton, as shown in Figure 8.7. In this process, the quark is treated as a free particle; this assumption will be justified in Chapter 10. The quark–parton model for deep inelastic scattering is formulated in a frame where the proton has very high energy, $E \gg m_\mathrm{p}$, referred to as the *infinite momentum frame*. In the infinite momentum frame the mass of the proton can be neglected, such that its four-momentum can be written $p_2 = (E_2, 0, 0, E_2)$. Furthermore, in this frame any component of the momentum of the struck quark transverse to the direction of motion of the proton also can be neglected. Hence, in the infinite momentum frame, the four-momentum of the struck quark can be written

$$p_\mathrm{q} = \xi p_2 = (\xi E_2, 0, 0, \xi E_2),$$

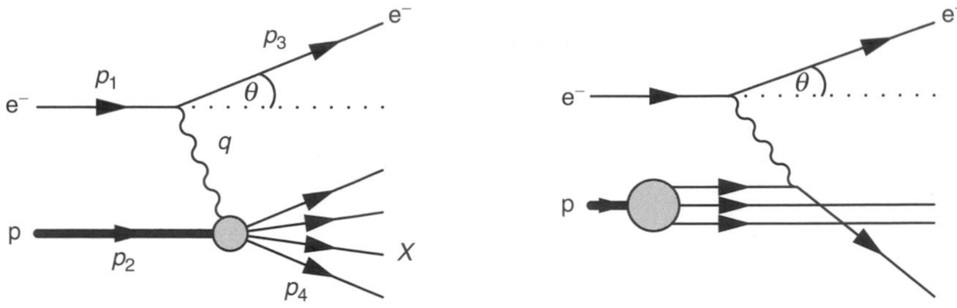


Fig. 8.7 Electron–quark scattering in the centre-of-mass frame and the corresponding lowest-order Feynman diagram.

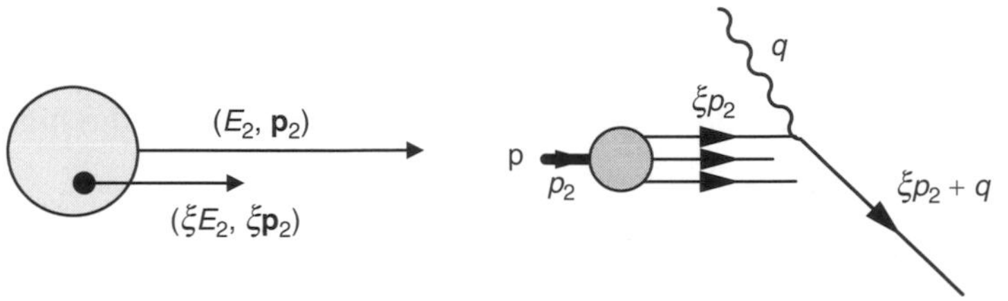


Fig. 8.8 The quark–parton model description of inelastic scattering in terms of the QED interaction between a virtual photon and a quark with the fraction $\xi$ of the momentum of the proton.

where $\xi$ is the fraction of momentum of proton carried by the quark, as indicated by Figure 8.8.

The four-momentum of the quark after the interaction with the virtual photon is simply $\xi p_2 + q$. Since the four-momentum squared of the final-state quark is just the square of its mass,

$$(\xi p_2 + q)^2 = \xi^2 p_2^2 + 2\xi p_2 \cdot q + q^2 = m_\mathrm{q}^2. \tag{8.20}$$

However, $\xi p_2$ is the just four-momentum of the quark before the interaction and therefore $\xi^2 p_2^2 = m_\mathrm{q}^2$. Thus, (8.20) implies that $q^2 + 2\xi p_2 \cdot q = 0$ and the momentum fraction $\xi$ can be identified as

$$\xi = \frac{-q^2}{2p_2 \cdot q} = \frac{Q^2}{2p_2 \cdot q} \equiv x.$$

Hence, in the quark–parton model, Bjorken $x$ can be identified as the fraction of the momentum of the proton carried by the struck quark (in a frame where the proton has energy $E \gg m_\mathrm{p}$). Therefore, the measurements of the $x$-dependence of the structure functions can be related to the momentum distribution of the quarks within the proton.

The kinematic variables for the underlying electron–*quark* scattering process can be related to those for the electron–*proton* collision. Neglecting the electron and proton mass terms, the centre-of-mass energy of the $\mathrm{e}^-\mathrm{p}$ initial state is

$$s = (p_1 + p_2)^2 \approx 2p_1 \cdot p_2.$$

Because the four-momentum of the struck quark is $p_q = xp_2$, the centre-of-mass energy of the initial-state $e^-q$ system is

$$s_q = (p_1 + xp_2)^2 \approx 2xp_1 \cdot p_2 = xs.$$

The kinematic variables $x$ and $y$, defined in terms of the four-momentum of the proton, are

$$y = \frac{p_2 \cdot q}{p_2 \cdot p_1} \quad \text{and} \quad x = \frac{Q^2}{2p_2 \cdot q}.$$

Similarly, for the electron–quark system,

$$y_q = \frac{p_q \cdot q}{p_q \cdot p_1} = \frac{xp_2 \cdot q}{xp_2 \cdot p_1} = y.$$

Finally, because the underlying electron–quark interaction is an elastic scattering process, $x_q = 1$. Hence, the kinematic variables for the $e^-q$ interaction are related to those defined in terms of the $e^-p$ interaction by

$$s_q = xs, \quad y_q = y \quad \text{and} \quad x_q = 1,$$

where $s$, $x$ and $y$ are defined in terms of the electron and proton four-momenta.

The cross section for $e^-q \to e^-q$ scattering, given by (8.19), can now be written

$$\frac{d\sigma}{dq^2} = \frac{2\pi\alpha^2 Q_q^2}{q^4}\left[1 + \left(1 + \frac{q^2}{s_q}\right)^2\right], \tag{8.21}$$

where $s_q$ is the electron–quark centre-of-mass energy squared. From (8.8), the four-momentum squared of the virtual photon $q^2$ can be expressed as

$$q^2 = -Q^2 = -(s_q - m_q^2)x_q y_q,$$

which in the limit where the quark mass is neglected gives

$$\frac{q^2}{s_q} = -x_q y_q = -y.$$

Hence, the differential cross section of (8.21) can be written

$$\frac{d\sigma}{dq^2} = \frac{2\pi\alpha^2 Q_q^2}{q^4}\left[1 + (1 - y)^2\right].$$

Finally, using $q^2 = -Q^2$ and rearranging the terms in the brackets leads to

$$\frac{d\sigma}{dQ^2} = \frac{4\pi\alpha^2 Q_q^2}{Q^4}\left[(1 - y) + \frac{y^2}{2}\right], \tag{8.22}$$

which resembles the form of the deep inelastic scattering cross section expressed in terms of the structure functions, as given by (8.12). Equation (8.22) gives the differential cross section for $e^-q$ elastic scattering where the quark carries a fraction $x$ of the momentum of the proton. Although $x$ does not appear explicitly in this equation, the $x$ dependence is implicit through (8.8) whereby

$$y = \frac{Q^2}{(s - m_p^2)x}.$$

### 8.4.1 Parton distribution functions

The quarks inside the proton will interact with each other through the exchange of gluons. The dynamics of this interacting system will result in a distribution of quark momenta within the proton. These distributions are expressed in terms of *Parton Distribution Functions* (PDFs). For example, the up-quark PDF for the proton $u^p(x)$ is defined such that

$$u^p(x)\,\delta x,$$

represents the number of up-quarks within the proton with momentum fraction between $x$ and $x + \delta x$. Similarly $d^p(x)$ is the corresponding PDF for the down-quarks. In practice, the functional forms of the PDFs depend on the detailed dynamics of the proton; they are not *a priori* known and have to be obtained from experiment. Figure 8.9 shows a few possible forms of the PDFs that correspond to: (i) the proton consists of a single point-like particle which carries all of the momentum of the proton, in this case the PDF is a Dirac delta-function at $x = 1$; (ii) the proton consists of three static quarks each of which carries 1/3 of the momentum of the proton, in this case the PDF has the form of a delta-function at $x = 1/3$ with a normalisation of three; (iii) the three quarks interact with each other and the delta-function at $x = 1/3$ is smeared out as the quarks exchange momentum; and (iv) higher-order processes, such as virtual quark pairs being produced from gluons

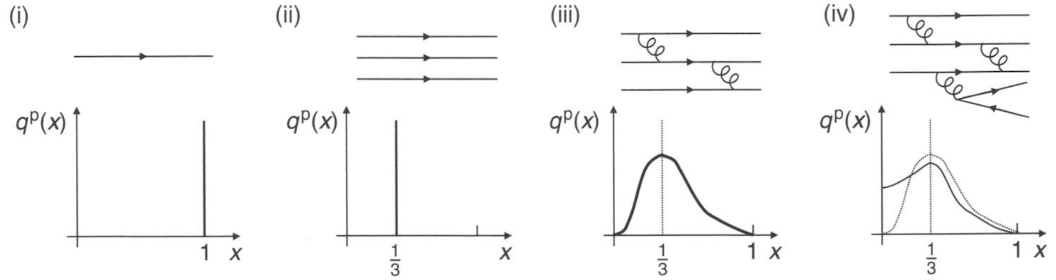


Fig. 8.9 Four possible forms of the quark PDFs within a proton: (i) a single point-like particle; (ii) three static quarks each sharing 1/3 of the momentum of the proton; (iii) three interacting quarks which can exchange momentum; and (iv) interacting quarks including higher-order diagrams. After Halzen and Martin (1984).

inside the proton, tend to result in an enhancement of the PDFs at low $x$, reflecting the $1/q^2$ nature of the gluon propagator.

The electron–proton deep inelastic scattering cross section can be obtained from the definition of the parton distribution functions and the expression for the differential cross section for underlying electron–quark elastic scattering process given in (8.22). The cross section for elastic scattering from a particular flavour of quark $i$ with charge $Q_i$ and momentum fraction in the range $x \to x + \delta x$, is

$$\frac{\mathrm{d}^2\sigma}{\mathrm{d}Q^2} = \frac{4\pi\alpha^2}{Q^4}\left[(1-y) + \frac{y^2}{2}\right] \times Q_i^2 q_i^{\mathrm{p}}(x)\,\delta x\,,$$

where $q_i^{\mathrm{p}}(x)$ is the PDF for that flavour of quark. The double-differential cross section is obtained by dividing by $\delta x$ and summing over all quark flavours

$$\frac{\mathrm{d}^2\sigma^{\mathrm{ep}}}{\mathrm{d}x\,\mathrm{d}Q^2} = \frac{4\pi\alpha^2}{Q^4}\left[(1-y) + \frac{y^2}{2}\right]\sum_i Q_i^2 q_i^{\mathrm{p}}(x). \tag{8.23}$$

This is the parton model prediction for the electron–proton deep inelastic scattering cross section. Comparison with (8.12), which is the general expression for the deep inelastic scattering cross section in terms of the structure functions,

$$\frac{\mathrm{d}^2\sigma}{\mathrm{d}x\mathrm{d}Q^2} = \frac{4\pi\alpha^2}{Q^4}\left[(1-y)\frac{F_2^{\mathrm{ep}}(x,Q^2)}{x} + y^2 F_1^{\mathrm{ep}}(x,Q^2)\right],$$

leads to the parton model predictions for $F_1^{\mathrm{ep}}(x,Q^2)$ and $F_2^{\mathrm{ep}}(x,Q^2)$,

$$F_2^{\mathrm{ep}}(x,Q^2) = 2xF_1^{\mathrm{ep}}(x,Q^2) = x\sum_i Q_i^2 q_i^{\mathrm{p}}(x).$$

The parton model naturally predicts Bjorken scaling; because the underlying process is elastic scattering from *point-like* quarks, no (strong) $Q^2$ dependence is expected. Consequently, both $F_1$ and $F_2$ can be written as functions of $x$ alone, $F_1(x,Q^2) \to F_1(x)$ and $F_2(x,Q^2) \to F_2(x)$. The parton model also predicts the Callan–Gross relation, $F_2(x) = 2xF_1(x)$. This is due to the underlying process being elastic scattering from spin-half Dirac particles; the quark magnetic moment is directly related to its charge and therefore the contributions from the electromagnetic ($F_2$) and the pure magnetic ($F_1$) structure functions are fixed with respect to one another.

### 8.4.2 Determination of the parton distribution functions

The parton distribution functions reflect the underlying structure of the proton. At present they cannot be calculated from first principles. This is because the theory of QCD has a large coupling constant, $\alpha_S \sim O(1)$, and perturbation theory cannot

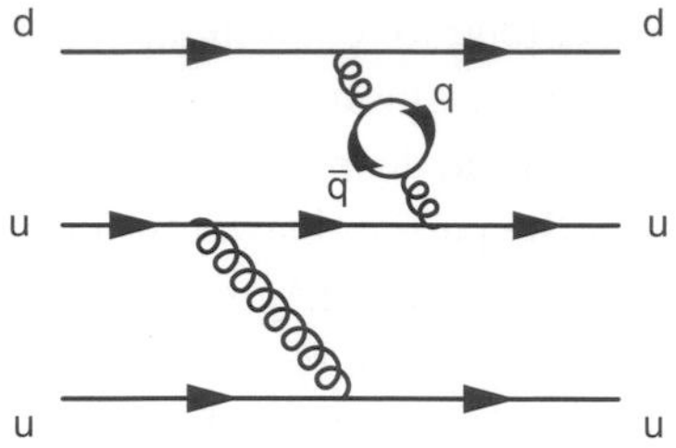


Fig. 8.10 Production of virtual $q\bar{q}$ pairs within the proton.

be applied. The PDFs therefore have to be extracted from measurements of the structure functions in deep inelastic scattering experiments and elsewhere.

For electron–proton deep inelastic scattering, the structure function $F_2^{ep}(x)$ is related to the PDFs by

$$F_2^{ep}(x) = x\sum_i Q_i^2 q_i^p(x)\,. \tag{8.24}$$

In the static model of the proton, it is formed from two up-quarks and a down-quark, and it might be expected that only up- and down-quark PDFs would appear in this sum. However, in reality the proton is a dynamic system where the strongly interacting quarks are constantly exchanging virtual gluons that can fluctuate into virtual $q\bar{q}$ pairs through processes such as that shown in Figure 8.10. Because gluons with large momenta are suppressed by the $1/q^2$ gluon propagator, this *sea* of virtual quarks and antiquarks tend to be produced at low values of $x$. Electron–proton inelastic scattering therefore involves interactions with both quarks and antiquarks. Furthermore, there will be contributions to the scattering process from strange quarks through interactions with virtual $s\bar{s}$ pairs and even very small contributions from off-mass shell heavier quarks. Here, for the sake of clarity, the relatively small contribution from strange quarks is neglected and the sum in (8.24) is restricted to the light flavours, giving the quark–parton model prediction

$$F_2^{ep}(x) = x\sum_i Q_i^2 q_i^p(x) \approx x\left(\frac{4}{9}u^p(x) + \frac{1}{9}d^p(x) + \frac{4}{9}\overline{u}^p(x) + \frac{1}{9}\overline{d}^p(x)\right), \tag{8.25}$$

where $u^p(x)$, $d^p(x)$, $\overline{u}^p(x)$ and $\overline{d}^p(x)$ are respectively the up-, down-, anti-up and anti-down parton distribution functions for the proton. A similar expression can be written down for the structure functions for electron–neutron scattering,

$$F_2^{en}(x) = x\sum_i Q_i^2 q_i^n(x) \approx x\left(\frac{4}{9}u^n(x) + \frac{1}{9}d^n(x) + \frac{4}{9}\overline{u}^n(x) + \frac{1}{9}\overline{d}^n(x)\right), \tag{8.26}$$

where the PDFs now refer to the momentum distributions within the neutron.

With the exception of the relatively small difference in Coulomb interactions between the constituent quarks, the neutron (ddu) would be expected to have the same structure as the proton (uud) with the up- and down-quarks interchanged. This assumed *isospin symmetry* (see Chapter 9) implies that the down-quark PDF in the neutron is the same as the up-quark PDF in the proton and thus

$$d^{\mathrm{n}}(x) = u^{\mathrm{p}}(x) \quad \text{and} \quad u^{\mathrm{n}}(x) = d^{\mathrm{p}}(x).$$

In order to simplify the notation, the PDFs for the *proton* are usually written as $u(x)$, $d(x)$, $\overline{u}(x)$ and $\overline{d}(x)$, in which case the neutron PDFs can be taken to be

$$d^{\mathrm{n}}(x) = u^{\mathrm{p}}(x) \equiv u(x) \quad \text{and} \quad u^{\mathrm{n}}(x) = d^{\mathrm{p}}(x) \equiv d(x).$$

Likewise, the assumed isospin symmetry implies that the neutron antiquark PDFs can be written in terms of the antiquark PDFs of the proton,

$$\overline{d}^{\mathrm{n}}(x) = \overline{u}^{\mathrm{p}}(x) \equiv \overline{u}(x) \quad \text{and} \quad \overline{u}^{\mathrm{n}}(x) = \overline{d}^{\mathrm{p}}(x) \equiv \overline{d}(x).$$

Thus the proton and neutron structure functions of (8.25) and (8.26), can be written in terms of the PDFs of the proton,

$$F_2^{\mathrm{ep}}(x) = 2xF_1^{\mathrm{ep}}(x) = x\left(\frac{4}{9}u(x) + \frac{1}{9}d(x) + \frac{4}{9}\overline{u}(x) + \frac{1}{9}\overline{d}(x)\right), \tag{8.27}$$

$$F_2^{\mathrm{en}}(x) = 2xF_1^{\mathrm{en}}(x) = x\left(\frac{4}{9}d(x) + \frac{1}{9}u(x) + \frac{4}{9}\overline{d}(x) + \frac{1}{9}\overline{u}(x)\right). \tag{8.28}$$

Integrating these expressions for the structure functions over the entire $x$ range gives

$$\int_0^1 F_2^{\mathrm{ep}}(x)\,\mathrm{d}x = \frac{4}{9}f_{\mathrm{u}} + \frac{1}{9}f_{\mathrm{d}} \quad \text{and} \quad \int_0^1 F_2^{\mathrm{en}}(x)\,\mathrm{d}x = \frac{4}{9}f_{\mathrm{d}} + \frac{1}{9}f_{\mathrm{u}}, \tag{8.29}$$

where $f_{\mathrm{u}}$ and $f_{\mathrm{d}}$ are defined by

$$f_{\mathrm{u}} = \int_0^1 [xu(x) + x\overline{u}(x)]\,\mathrm{d}x \quad \text{and} \quad f_{\mathrm{d}} = \int_0^1 [xd(x) + x\overline{d}(x)]\,\mathrm{d}x\,.$$

The quantity $f_{\mathrm{u}}$ is the fraction of the momentum of the proton carried by the up- and anti-up quarks. Similarly $f_{\mathrm{d}}$ is the fraction carried by the down-/anti-down-quarks. The momentum fractions $f_{\mathrm{u}}$ and $f_{\mathrm{d}}$ can be obtained directly from the experimental measurements of the proton and neutron structure functions. For example, Figure 8.11 shows an experimental measurement of $F_2^{\mathrm{ep}}(x, Q^2)$ as a function of $x$ for deep inelastic scattering events with $2\,\mathrm{GeV}^2 < Q^2 < 30\,\mathrm{GeV}^2$ as observed at SLAC. The area defined by the measured data points gives

$$\int F_2^{\mathrm{ep}}(x)\,\mathrm{d}x \approx 0.18\,.$$

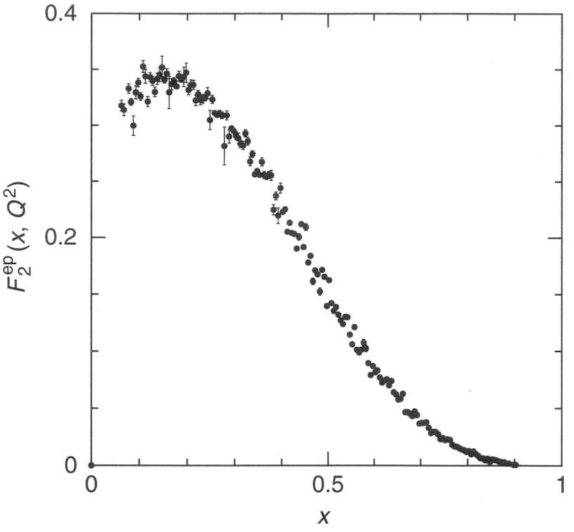


Fig. 8.11 SLAC measurements of $F_2^{ep}(x, Q^2)$ for $2 < Q^2/\,\mathrm{GeV}^2 < 30$. Data from Whitlow *et al.* (1992).

Similarly, $F_2^{en}(x)$ can be extracted from electron–deuterium scattering data (see Problem 8.6), and it is found that

$$\int F_2^{en}(x)\,dx \approx 0.12\,.$$

Using the quark–parton model predictions of (8.29), these experimental results can be interpreted as measurements of the fractions of the momentum of the proton carried by the up-/anti-up- and down-/anti-down-quarks:

$$f_u \approx 0.36 \quad \text{and} \quad f_d \approx 0.18.$$

Given that the proton consists of two up-quarks and one down-quark, it is perhaps not surprising that $f_u = 2f_d$. Nevertheless, the total fraction of the momentum of the proton carried by quarks and antiquarks is just over 50%; the remainder is carried by the gluons that are the force carrying particles of the strong interaction. Because the gluons are electrically neutral, they do not contribute to the QED process of electron–proton deep inelastic scattering.

### 8.4.3 Valence and sea quarks

It is already clear that the proton is a lot more complex than first might have been anticipated. The picture of a proton as a bound state consisting of three "valence" quarks is overly simplistic. The proton not only contains quarks, but also contains of a sea of virtual gluons that give rise to an antiquark component through $g \to q\overline{q}$ pair production. To reflect these two distinct components, the up-quark PDF can be split into the contribution from the two valence quarks, written as $u_V(x)$, and a contribution from the *sea* of up-quarks that are pair-produced from virtual gluons, $u_S(x)$. In this way, the proton light quark PDFs can be decomposed into

$$u(x) = u_V(x) + u_S(x) \quad \text{and} \quad d(x) = d_V(x) + d_S(x).$$

In the case of the antiquark PDFs, there are only sea quark contributions,

$$\overline{u}(x) \equiv \overline{u}_S(x) \quad \text{and} \quad \overline{d}(x) \equiv \overline{d}_S(x).$$

Since the proton consists of two valence up-quarks and one valence down-quark, it is reasonable to expect that the valence quark PDFs are normalised accordingly,

$$\int_0^1 u_V(x)\,\mathrm{d}x = 2 \quad \text{and} \quad \int_0^1 d_V(x)\,\mathrm{d}x = 1\,.$$

Although there is no corresponding *a priori* expectation for the sea quarks, some reasonable assumptions can be made. Firstly, since the sea quarks and the antiquarks of a given flavour are produced in pairs, the sea quark PDF will be the same as the PDF for the corresponding antiquark. Furthermore, since the masses of the up- and down-quarks are similar, it is reasonable to expect that the sea PDFs for the up- and down-quarks will be approximately the same. With these assumptions, the sea PDFs can all be approximated by a single function, written $S(x)$, such that

$$u_S(x) = \overline{u}_S(x) \approx d_S(x) = \overline{d}_S(x) \approx S(x).$$

Writing (8.27) and (8.28) in terms of the valence and sea quark PDFs leads to

$$F_2^{\mathrm{ep}}(x) = x\left(\frac{4}{9}u_V(x) + \frac{1}{9}d_V(x) + \frac{10}{9}S(x)\right),$$

$$F_2^{\mathrm{en}}(x) = x\left(\frac{4}{9}d_V(x) + \frac{1}{9}u_V(x) + \frac{10}{9}S(x)\right).$$

With the above assumptions, the ratio of $F_2^{\mathrm{en}}(x)$ to $F_2^{\mathrm{ep}}(x)$ is predicted to be

$$\frac{F_2^{\mathrm{en}}(x)}{F_2^{\mathrm{ep}}(x)} = \frac{4d_V(x) + u_V(x) + 10S(x)}{4u_V(x) + d_V(x) + 10S(x)}. \tag{8.30}$$

Although the PDFs need to be determined experimentally, some qualitative predictions can be made. For example, since the sea quarks are expected to be produced mainly at low $x$, it is reasonable to hypothesise that the sea quarks will give the dominant contribution to the proton PDFs at low $x$. In this case, the low-$x$ limit of (8.30) would be

$$\frac{F_2^{\mathrm{en}}(x)}{F_2^{\mathrm{ep}}(x)} \to 1 \quad \text{as } x \to 0.$$

This prediction is supported by the data of Figure 8.12, which shows the ratio of the $F_2^{\mathrm{en}}(x)/F_2^{\mathrm{ep}}(x)$ obtained from electron–proton and electron–deuterium deep inelastic scattering measurements.

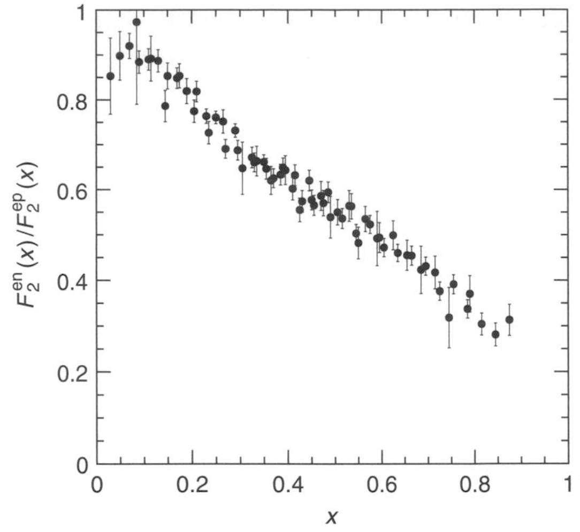


Fig. 8.12 The ratio of $F_2^{en}(x)/F_2^{ep}(x)$ obtained from electron–deuterium and electron–proton deep inelastic scattering measurements at SLAC. Data from Bodek *et al.* (1979).

Owing to the $1/q^2$ gluon propagator, which will suppress the production of sea quarks at high $x$, it might be expected that the high-$x$ PDFs of the proton will be dominated by the valence quarks. In this case,

$$\frac{F_2^{en}(x)}{F_2^{ep}(x)} \rightarrow \frac{4d_V(x) + u_V(x)}{4u_V(x) + d_V(x)} \quad \text{as } x \rightarrow 1\,.$$

If it is also assumed that $u_V(x) = 2d_V(x)$, the ratio of $F_2^{en}(x)/F_2^{ep}(x)$ would be expected to tend to 2/3 as $x \rightarrow 1$. This is in clear disagreement with the data of Figure 8.12, where it can be seen that

$$\frac{F_2^{en}(x)}{F_2^{ep}(x)} \rightarrow \frac{1}{4} \quad \text{as } x \rightarrow 1.$$

This would seem to imply that the ratio $d_V(x)/u_V(x) \rightarrow 0$ as $x \rightarrow 1$. Whilst this behaviour is not fully understood, a qualitative explanation based on the exclusion principle can be made. At high $x$, one of the valence quarks carries most of the momentum of the proton and the other two valence quarks must be in a low momentum state. Since the exclusion principle forbids two like-flavour quarks being in the same state, the configuration where the down-quark in the proton is at high $x$ and both up-quarks have low momentum is disfavoured.

There are a number of conclusions that can be drawn from the above discussion. Firstly, the proton is a complex system consisting of many strongly interacting quarks and gluons. Secondly, whilst qualitative predictions of the properties of the

PDFs can be made, relatively simplistic arguments do not always work. Ultimately, the parton distribution functions have to be inferred directly from experimental data.

## 8.5 Electron–proton scattering at the HERA collider

The studies of deep inelastic scattering at very high $Q^2$ and at very low $x$ were amongst the main goals of the HERA electron–proton collider that operated from 1991 to 2007 at the DESY (Deutsches Elektronen-Synchrotron) laboratory in Hamburg, Germany. It consisted of a 3 km circumference ring where 27.5 GeV electrons (or positrons) were collided with 820 GeV or 920 GeV protons. Two large experiments, H1 and ZEUS, were located at opposite sides of the ring. Each experiment recorded over one million $e^{\pm}p$ deep inelastic collisions at $Q^2 > 200\,\text{GeV}^2$. These large data samples at a centre-of-mass energy of $\sqrt{s} \approx 300\,\text{GeV}$, enabled the structure of the proton to be probed with high precision, both at $Q^2$ values of up to $2 \times 10^4\,\text{GeV}^2$ and at $x$ below $10^{-4}$.

Figure 8.13 shows an example of a very-high-$Q^2$ interaction recorded by the H1 experiment. The final-state hadronic system is observed as a jet of high-energy particles. The energy and direction of this jet of particles is measured less precisely than the corresponding properties of the electron. Consequently, for each observed event, $Q^2$ and $x$ are determined from the energy and scattering angle of the electron. The results from deep inelastic scattering data from the H1 and ZEUS experiments, summarised in Figure 8.14, provide a precise determination of the proton structure functions over a very wide range of $x$ and $Q^2$. The data show a number

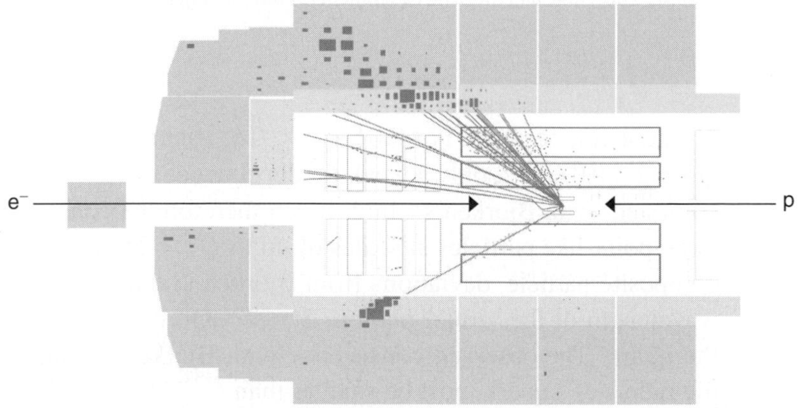


Fig. 8.13 A high-energy electron–proton collision in the H1 detector at HERA. In this event the electron (the particle recorded in the lower part of the detector) is scattered through a large angle and the hadronic system from the break up of the proton forms a jet of particles. Courtesy of the H1 Collaboration.

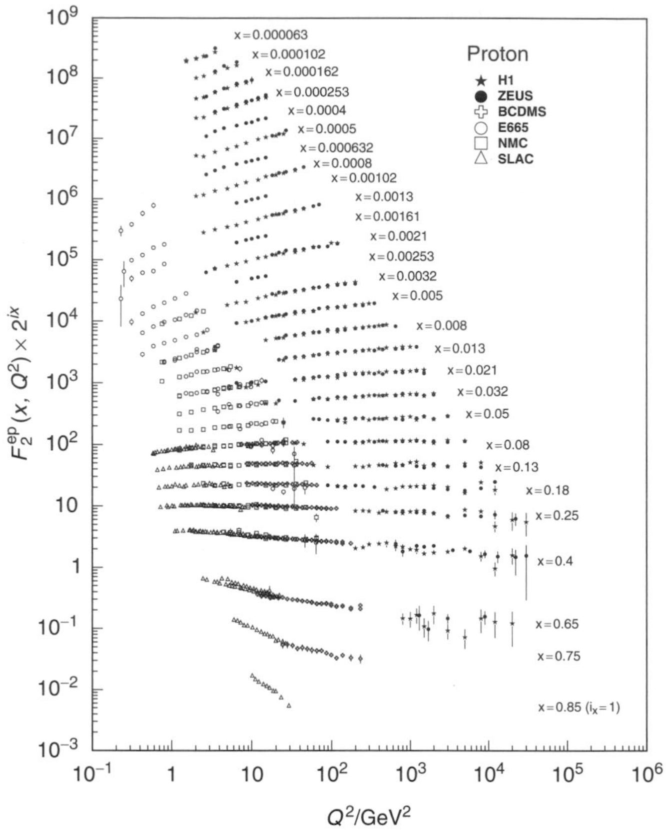


Fig. 8.14 Measurements of $F_2^{ep}(x, Q^2)$ at HERA. Results from both the H1 and ZEUS experiments are shown. The different bands of data points correspond to the $Q^2$ evolution of $F^{ep}$ for different values of $x$. For clarity, the data at different values of $x$ are shifted by $-\log x$. Also shown are lower-$Q^2$ data from earlier fixed-target experiments (BCDMS, E665, NMC and the SLAC experiments). From Beringer *et al.* (2012), ©the American Physical Society.

of interesting features. For $0.01 < x < 0.5$, where the measurements extend out to $Q^2 = 2\times10^4\,\text{GeV}^2$, only a weak $Q^2$ dependence of $F_2^{ep}(x, Q^2)$ is observed, broadly consistent with Bjorken scaling. It can therefore be concluded that quarks appear to be point-like particles at scales of up to $Q^2 = 2 \times 10^4\,\text{GeV}^2$. If the quark was a composite particle, deviations from Bjorken scaling would be expected when the wavelength of the virtual photon, $\lambda \sim hc/|\mathbf{Q}|$, became comparable to the size of the quark. The observed consistency with Bjorken scaling therefore implies that the radius of a quark must be smaller than

$$r_q < 10^{-18}\,\text{m}.$$

### 8.5.1 Scaling violations

Whilst Bjorken scaling holds over a wide range of $x$ values, relatively small deviations are observed at very low and very high values of $x$. For example, at high (low) values of $x$, the proton structure function is observed to decrease (increase) with increasing $Q^2$. Put another way, at high $Q^2$ the measured structure functions are shifted towards lower values of $x$ relative to the structure functions at low $Q^2$, as indicated in Figure 8.15. This behaviour, known as scaling violation, implies that at high $Q^2$, the proton is observed to have a greater fraction of low $x$ quarks. These scaling violations are not only expected, but the observed $Q^2$ dependence is calculable in the theory of the strong interaction, QCD.

The mathematical description of the origin of scaling violations is beyond the scope of this book and only a qualitative description is given here. At low $Q^2$, there is a length scale, determined by the wavelength of the virtual photon, below which it is not possible to resolve any spatial sub-structure, as indicated in Figure 8.16a. At higher values of $Q^2$, corresponding to shorter-wavelengths of the virtual photon, it is possible to resolve finer detail. In this case, the deep inelastic scattering process is sensitive to the effects of quarks radiating virtual gluons, $q \to qg$, over smaller length scales, as indicated in Figure 8.16b. Consequently, more low-$x$ quarks are "seen" in high-$Q^2$ deep inelastic scattering.

Although currently it is not possible to calculate the proton PDFs from first principles within the theory of QCD, the $Q^2$ dependence of the PDFs is calculable

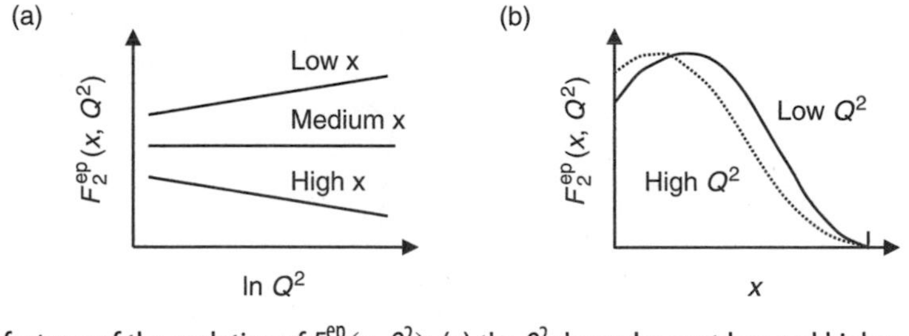


Fig. 8.15 The general features of the evolution of $F_2^{ep}(x, Q^2)$: (a) the $Q^2$ dependence at low and high $x$ and (b) the $x$ dependence at low and high $Q^2$.

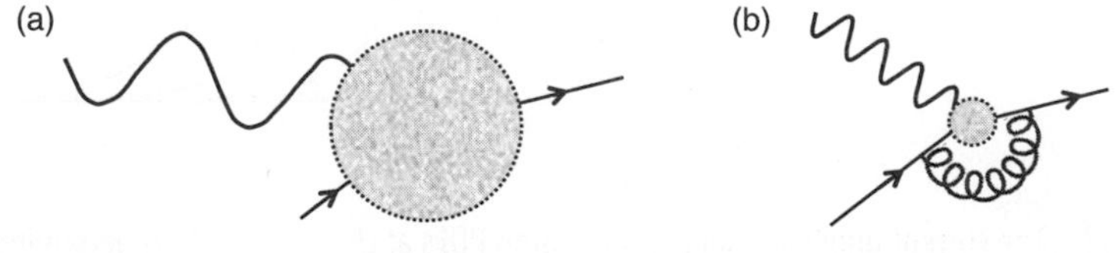


Fig. 8.16 Finer structure within the proton can be resolved by shorter-wavelength virtual photons leading to the observation of lower $x$ partons at higher $Q^2$. The circled regions indicate the length scale below which structure cannot be resolved.

using the parton evolution functions known as the DGLAP (Dokshitzer–Gribov–Lipatov–Altarelli–Parisi) equations. These equations are based on universal parton splitting functions for the QCD processes $q \to qg$ and $g \to q\bar{q}$. The observed scaling violations in deep inelastic scattering therefore provide a powerful validation of the fundamental QCD theory of the strong interaction. A good introduction to the DGLAP evolution equations can be found in Halzen and Martin (1984).

## 8.6 Parton distribution function measurements

Information about the parton distribution functions of the proton can be extracted from high-energy measurements involving protons, such as: fixed-target electron–proton and electron–neutron scattering; high-energy electron–proton collider data; neutrino–nucleon scattering data (discussed in Chapter 12); high-energy $p\bar{p}$ collider data from the Tevatron; and very-high-energy pp collider data from the LHC. The different experimental measurements provide complementary information about the PDFs. For example, neutrino scattering data provide a direct measurement of the $\overline{u}(x)$ and $\overline{d}(x)$ content of the proton and the pp collider data provides information on the gluon PDF, $g(x)$.

The proton PDFs are extracted from a global fit to a wide range of experimental data. Owing to the complementary nature of the different measurements, tight constraints on the PDFs are obtained. In practice, the PDFs are varied, subject to the

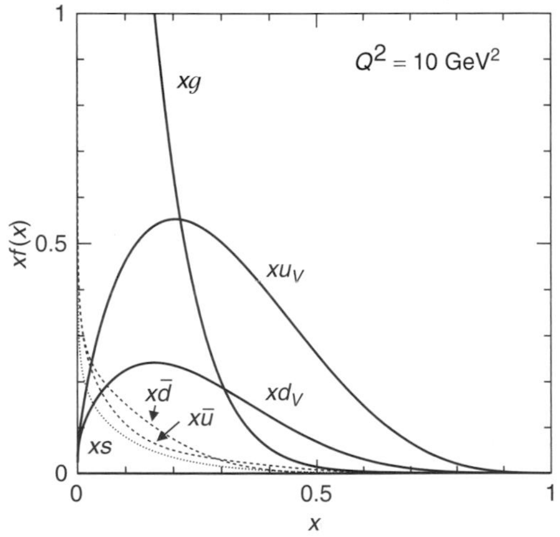


Fig. 8.17 The current understanding of the proton PDFs at $Q^2 = 10\,\text{GeV}^2$ as determined from the MRST fit to a wide range of experimental data. The relatively small strange quark PDF $s(x)$ is shown. PDFs from the Durham HepData project.

constraints imposed by the theoretical framework of QCD such as the DGLAP evolution equations, to obtain the best agreement with experimental data. The output of this procedure is a set of PDFs at a particular $Q^2$ scale. For example, Figure 8.17 shows the extracted PDFs at $Q^2 = 10\,\mathrm{GeV}^2$ obtained from a recent fit to the experimental data, where it is assumed that $u(x) = u_V(x) + \overline{u}(x)$. The contribution from gluons is large and, as expected, is peaked towards low values of $x$. The antiquark PDFs are relatively small and, because the antiquarks originate from $g \to \mathrm{q}\overline{\mathrm{q}}$, also are peaked towards low values of $x$. Apart from at high values of $x$, it is found that $u_V(x) \approx 2d_V(x)$ as expected. Finally, it is worth noting that although the PDFs for $\overline{\mathrm{u}}$ and $\overline{\mathrm{d}}$ are similar, there is a small difference with $\overline{d}(x) > \overline{u}(x)$. This may be explained by a relative suppression of the $g \to \mathrm{u}\overline{\mathrm{u}}$ process due to the exclusion principle and the larger number of up-quark states which are already occupied.

## Summary

In this chapter the process of deep inelastic scattering has been described in terms of the quark–parton model, where the underlying process is the elastic scattering of the electron from the quasi-free spin-half constituent quarks. The kinematics of inelastic scattering were described in terms of two of the kinematic variables defined below

$$Q^2 \equiv -q^2, \quad x \equiv \frac{Q^2}{2p_2 \cdot q}, \quad y \equiv \frac{p_2 \cdot q}{p_2 \cdot p_1} \quad \text{and} \quad \nu \equiv \frac{p_2 \cdot q}{m_\mathrm{p}}.$$

In the quark–parton model, $x$ is identified as the fraction of the momentum of the proton carried by the struck quark in the underlying $\mathrm{e^-q \to e^-q}$ elastic scattering process. The quark–parton model naturally describes the experimentally observed phenomena of Bjorken scaling, $F(x, Q^2) \to F(x)$, and the Callan–Gross relation, $F_2(x) = 2xF_1(x)$.

In the quark–parton model, cross sections can be described in terms of parton distribution functions (PDFs) which represent the momentum distributions of quarks and antiquarks within the proton. The PDFs can not yet be calculated from first principles but are determined from a wide range of experimental data. The resulting PDF measurements reveal the proton to be much more complex than a static bound state of three quarks (uud); it is a dynamic object consisting of three valence quarks and a sea of virtual quarks, antiquarks and gluons, with almost 50% of the momentum of the proton carried by the gluons. The precise knowledge of the PDFs is an essential ingredient to the calculations of cross sections for all high-energy processes involving protons, such as proton–proton collisions at the LHC.

The quark–parton model provides a hugely successful description of the dynamic nature of the proton. However, it does not explain why the only observed hadronic states are baryons and mesons or why the proton is the lowest mass baryon. The static quark model is the subject of the next chapter.

## Problems

**8.1** Use the data in Figure 8.2 to estimate the lifetime of the $\Delta^+$ baryon.

**8.2** In fixed-target electron–proton *elastic* scattering

$$Q^2 = 2m_p(E_1 - E_3) = 2m_pE_1y \quad \text{and} \quad Q^2 = 4E_1E_3\sin^2(\theta/2).$$

**(a)** Use these relations to show that

$$\sin^2\left(\frac{\theta}{2}\right) = \frac{E_1}{E_3}\frac{m_p^2}{Q^2}y^2 \quad \text{and hence} \quad \frac{E_3}{E_1}\cos^2\left(\frac{\theta}{2}\right) = 1 - y - \frac{m_p^2y^2}{Q^2}.$$

**(b)** Assuming azimuthal symmetry and using Equations (7.31) and (7.32), show that

$$\frac{d\sigma}{dQ^2} = \left|\frac{d\Omega}{dQ^2}\right|\frac{d\sigma}{d\Omega} = \frac{\pi}{E_3^2}\frac{d\sigma}{d\Omega}.$$

**(c)** Using the results of (a) and (b) show that the Rosenbluth equation,

$$\frac{d\sigma}{d\Omega} = \frac{\alpha^2}{4E_1^2\sin^4(\theta/2)}\frac{E_3}{E_1}\left(\frac{G_E^2 + \tau G_M^2}{(1+\tau)}\cos^2\frac{\theta}{2} + 2\tau G_M^2\sin^2\frac{\theta}{2}\right),$$

can be written in the Lorentz-invariant form

$$\frac{d\sigma}{dQ^2} = \frac{4\pi\alpha^2}{Q^4}\left[\frac{G_E^2 + \tau G_M^2}{(1+\tau)}\left(1 - y - \frac{m_p^2y^2}{Q^2}\right) + \frac{1}{2}y^2G_M^2\right].$$

**8.3** In fixed-target electron–proton inelastic scattering:

**(a)** show that the laboratory frame differential cross section for deep-inelastic scattering is related to the Lorentz-invariant differential cross section of Equation (8.11) by

$$\frac{d^2\sigma}{dE_3\,d\Omega} = \frac{E_1E_3}{\pi}\frac{d^2\sigma}{dE_3\,dQ^2} = \frac{E_1E_3}{\pi}\frac{2m_px^2}{Q^2}\frac{d^2\sigma}{dx\,dQ^2},$$

where $E_1$ and $E_3$ are the energies of the incoming and outgoing electron.

**(b)** Show that

$$\frac{2m_px^2}{Q^2}\cdot\frac{y^2}{2} = \frac{1}{m_p}\frac{E_3}{E_1}\sin^2\frac{\theta}{2} \quad \text{and} \quad 1 - y - \frac{m_p^2x^2y^2}{Q^2} = \frac{E_3}{E_1}\cos^2\frac{\theta}{2}.$$

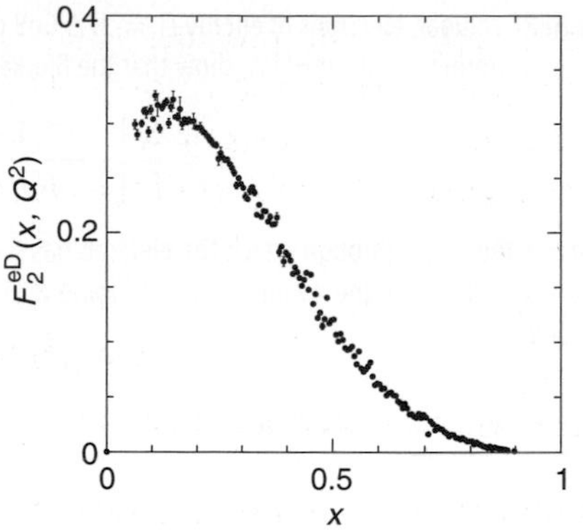


Fig. 8.18 SLAC measurements of $F_2^{eD}(x, Q^2)$ in for $2 < Q^2/\,\mathrm{GeV}^2 < 30$. Data from Whitlow *et al.* (1992).

**(c)** Hence, show that the Lorentz-invariant cross section of Equation (8.11) becomes

$$\frac{d^2\sigma}{dE_3\,d\Omega} = \frac{\alpha^2}{4E_1^2 \sin^4\theta/2}\left[\frac{F_2}{\nu}\cos^2\frac{\theta}{2} + \frac{2F_1}{m_p}\sin^2\frac{\theta}{2}\right].$$

**(d)** A fixed-target ep scattering experiment consists of an electron beam of maximum energy 20 GeV and a variable angle spectrometer that can detect scattered electrons with energies greater than 2 GeV. Find the range of values of $\theta$ over which deep inelastic scattering events can be studied at $x = 0.2$ and $Q^2 = 2\,\mathrm{GeV}^2$.

**8.4** If quarks were spin-0 particles, why would $F_1^{ep}(x)/F_2^{ep}(x)$ be zero?

**8.5** What is the expected value of $\int_0^1 u(x) - \overline{u}(x)\,dx$ for the proton?

**8.6** Figure 8.18 shows the raw measurements of the structure function $F_2(x)$ in low-energy electron–deuterium scattering. When combined with the measurements of Figure 8.11, it is found that

$$\frac{\int_0^1 F_2^{eD}(x)\,dx}{\int_0^1 F_2^{ep}(x)\,dx} \simeq 0.84.$$

Write down the quark–parton model prediction for this ratio and determine the relative fraction of the momentum of proton carried by down-/anti-down-quarks compared to that carried by the up-/anti-up-quarks, $f_d/f_u$.

**8.7** Including the contribution from strange quarks:

**(a)** show that $F_2^{ep}(x)$ can be written

$$F_2^{ep}(x) = \frac{4}{9}\left[u(x) + \overline{u}(x)\right] + \frac{1}{9}\left[d(x) + \overline{d}(x) + s(x) + \overline{s}(x)\right],$$

where $s(x)$ and $\overline{s}(x)$ are the strange quark–parton distribution functions of the proton.

**(b)** Find the corresponding expression for $F_2^{en}(x)$ and show that

$$\int_0^1 \frac{\left[F_2^{ep}(x) - F_2^{en}(x)\right]}{x}\,dx \approx \frac{1}{3} + \frac{2}{3}\int_0^1 \left[\overline{u}(x) - \overline{d}(x)\right] dx,$$

and interpret the measured value of $0.24 \pm 0.03$.

**8.8** At the HERA collider, electrons of energy $E_1 = 27.5$ GeV collided with protons of energy $E_2 = 820$ GeV. In deep inelastic scattering events at HERA, show that the Bjorken $x$ is given by

$$x = \frac{E_3}{E_2}\left[\frac{1-\cos\theta}{2-(E_3/E_1)(1+\cos\theta)}\right],$$

where $\theta$ is the angle through which the electron has scattered and $E_3$ is the energy of the scattered electron. Estimate $x$ and $Q^2$ for the event shown in Figure 8.13 assuming that the energy of the scattered electron is 250 GeV.

# 9 Symmetries and the quark model

Symmetries are a central to our current understanding of particle physics. In this chapter, the concepts of symmetries and conservation laws are first introduced in the general context of quantum mechanics and are then applied to the quark model. The approximate light quark flavour symmetry is used to predict the structure and wavefunctions of the lightest hadronic states. These wavefunctions are used to obtain predictions for the masses and magnetic moments of the observed baryons. The discussion of the quark model provides an introduction to the algebra of the SU(2) and SU(3) symmetry groups that play a central rôle in the Standard Model. No prior knowledge of group theory is assumed; the required properties of the SU(2) and SU(3) symmetry groups are obtained from first principles.

## 9.1 Symmetries in quantum mechanics

In both classical and quantum physics, conservation laws are associated with symmetries of the Hamiltonian. For particle physics it is most natural to introduce these ideas in the context of quantum mechanics. In quantum mechanics, a symmetry of the Universe can be expressed by requiring that all physical predictions are invariant under the wavefunction transformation

$$\psi \rightarrow \psi' = \hat{U}\psi,$$

where, for example, $\hat{U}$ could be the operator corresponding a finite rotation of the coordinate axes. The requirement that all physical predictions are unchanged by a symmetry transformation, constrains the possible form of $\hat{U}$. A necessary requirement is that wavefunction normalisations are unchanged, implying

$$\langle\psi|\psi\rangle = \langle\psi'|\psi'\rangle = \langle\hat{U}\psi|\hat{U}\psi\rangle = \langle\psi|\hat{U}^\dagger\hat{U}|\psi\rangle.$$

From this it can be concluded that the operator corresponding to any acceptable symmetry transformation in quantum mechanics must be unitary

$$\hat{U}^\dagger\hat{U} = I,$$

where $I$ represents unity (which could be 1 or the identity matrix). Furthermore, for physical predications to be unchanged by a symmetry operation, the eigenstates of the system also must be unchanged by the transformation. Hence the Hamiltonian itself must possess the symmetry in question, $\hat{H} \to \hat{H}' = \hat{H}$. The eigenstates of the Hamiltonian satisfy

$$\hat{H}\psi_i = E_i\psi,$$

and because of the invariance of the Hamiltonian, the energies of the transformed eigenstates $\psi_i'$ will be unchanged,

$$\hat{H}'\psi_i' = \hat{H}\psi_i' = E_i\psi_i'.$$

Since $\psi_i' = \hat{U}\psi_i$, this implies

$$\hat{H}\hat{U}\psi_i = E_i\hat{U}\psi_i = \hat{U}E_i\psi_i = \hat{U}\hat{H}\psi_i.$$

Therefore, for all states of the system, $\hat{H}\hat{U}\psi_i = \hat{U}\hat{H}\psi_i$, and it can be concluded that $\hat{U}$ commutes with the Hamiltonian

$$\left[\hat{H}, \hat{U}\right] \equiv \hat{H}\hat{U} - \hat{U}\hat{H} = 0.$$

Hence, for each symmetry of the Hamiltonian there is a corresponding unitary operator which commutes with the Hamiltonian.

A finite continuous symmetry operation can be built up from a series of infinitesimal transformations of the form

$$\hat{U}(\epsilon) = I + i\epsilon\hat{G},$$

where $\epsilon$ is an infinitesimally small parameter and $\hat{G}$ is called the generator of the transformation. Since $\hat{U}$ is unitary,

$$\hat{U}(\epsilon)\hat{U}^\dagger(\epsilon) = (I + i\epsilon\hat{G})(I - i\epsilon\hat{G}^\dagger) = I + i\epsilon(\hat{G} - \hat{G}^\dagger) + O(\epsilon^2).$$

For this infinitesimal transformation terms of $O(\epsilon^2)$ can be neglected, and therefore the requirement that $U^\dagger U = I$ implies that

$$\hat{G} = \hat{G}^\dagger.$$

Thus, for each symmetry of the Hamiltonian there is a corresponding unitary symmetry operation with an associated Hermitian generator $\hat{G}$. The eigenstates of a Hermitian operator are real and therefore the operator $\hat{G}$ is associated with an observable quantity $G$. Furthermore, since $\hat{U}$ commutes with the Hamiltonian, $\left[\hat{H}, I + i\epsilon\hat{G}\right] = 0$, the generator $\hat{G}$ also must commute with the Hamiltonian,

$$\left[\hat{H}, \hat{G}\right] = 0.$$

In quantum mechanics the time evolution of the expectation value of the operator $\hat{G}$ is given by (2.29),

$$\frac{\mathrm{d}}{\mathrm{d}t}\langle\hat{G}\rangle = i\left\langle\left[\hat{H},\hat{G}\right]\right\rangle,$$

and because here $\hat{G}$ commutes with the Hamiltonian,

$$\frac{\mathrm{d}}{\mathrm{d}t}\langle\hat{G}\rangle = 0.$$

Hence, for each symmetry of the Hamiltonian, there is an associated observable *conserved* quantity $G$. Thus in quantum mechanics, symmetries are associated with conservation laws and vice versa. This profound statement is not restricted to quantum mechanics, in classical dynamics symmetries of the Hamiltonian also correspond to conserved quantities. The relationship between symmetries and conservation laws is an expression of Noether's theorem, which associates a symmetry of the Lagrangian with a conserved current (see, for example, Appendix E).

### Translational invariance

As an example of the above arguments, consider the simple case of translational invariance in one dimension. The Hamiltonian for a system of particles depends only on the velocities and the relative distances between particles and therefore does not change if all particles are translated by the same infinitesimal distance $\epsilon$,

$$x \to x + \epsilon.$$

The corresponding wavefunction transformation is

$$\psi(x) \to \psi'(x) = \psi(x+\epsilon).$$

Performing a Taylor expansion of $\psi(x)$ in terms of $\epsilon$ gives

$$\psi'(x) = \psi(x+\epsilon) = \psi(x) + \frac{\partial\psi}{\partial x}\epsilon + O(\epsilon^2).$$

For this infinitesimal transformation, the terms of $O(\epsilon^2)$ can be dropped, giving

$$\psi'(x) = \left(1 + \epsilon\frac{\partial}{\partial x}\right)\psi(x). \tag{9.1}$$

This can be expressed in terms of the quantum-mechanical momentum operator,

$$\hat{p}_x = -i\frac{\partial}{\partial x},$$

giving

$$\psi'(x) = (1 + i\epsilon\hat{p}_x)\psi(x).$$

Comparison with (9.1) shows that the generator of the symmetry transformation, $x \to x + \epsilon$, is the quantum-mechanical momentum operator $\hat{p}_x$. Hence, the translational invariance of Hamiltonian implies momentum conservation.

In general, a symmetry operation may depend on more than one parameter, and the corresponding infinitesimal unitary operator can be written in terms of the set of generators $\hat{\mathbf{G}} = \{\hat{G}_i\}$,

$$\hat{U} = 1 + i\boldsymbol{\epsilon} \cdot \hat{\mathbf{G}},$$

where $\boldsymbol{\epsilon} = \{\epsilon_i\}$. For example, an infinitesimal three-dimensional spatial translation $\mathbf{x} \to \mathbf{x} + \boldsymbol{\epsilon}$ can be associated with the generators $\hat{\mathbf{p}} = (\hat{p}_x, \hat{p}_y, \hat{p}_z)$ with

$$\hat{U}(\boldsymbol{\epsilon}) = 1 + i\boldsymbol{\epsilon} \cdot \hat{\mathbf{p}} \equiv 1 + i\epsilon_x \hat{p}_x + i\epsilon_y \hat{p}_y + \epsilon_z \hat{p}_z.$$

### 9.1.1 Finite transformations

Any finite symmetry transformation can be expressed as a series of infinitesimal transformations using

$$\hat{U}(\boldsymbol{\alpha}) = \lim_{n\to\infty}\left(1 + i\frac{1}{n}\boldsymbol{\alpha} \cdot \hat{\mathbf{G}}\right)^n = \exp\left(i\boldsymbol{\alpha} \cdot \mathbf{G}\right).$$

For example, consider the finite translation $x \to x + x_0$ in one dimension. The corresponding unitary operator, expressed in terms of the generator of the infinitesimal translation $\hat{p}_x$, is

$$\hat{U}(x_0) = \exp\left(ix_0\hat{p}_x\right) = \exp\left(x_0\frac{\partial}{\partial x}\right).$$

Hence for this finite translation, wavefunctions transform according to

$$\begin{aligned}\psi'(x) = \hat{U}\psi(x) &= \exp\left(x_0\frac{\partial}{\partial x}\right)\psi(x)\\ &= \left(1 + x_0\frac{\partial}{\partial x} + \frac{x_0^2}{2!}\frac{\partial^2}{\partial x^2} + \cdots\right)\psi(x)\\ &= \psi(x) + x_0\frac{\partial\psi}{\partial x} + \frac{x_0^2}{2}\frac{\partial^2\psi}{\partial x^2} + \cdots,\end{aligned}$$

which is just the usual Taylor expansion for $\psi(x + x_0)$, and therefore $\hat{U}(x_0)$ results in the transformation

$$\psi(x) \to \psi'(x) = \hat{U}(x_0)\psi = \psi(x + x_0),$$

as required.

## 9.2 Flavour symmetry

In the early days of nuclear physics, it was realised that the proton and neutron have very similar masses and that the nuclear force is approximately charge independent. In other words, the strong force potential is the same for two protons, two neutrons or a neutron and a proton

$$V_{\mathrm{pp}} \approx V_{\mathrm{np}} \approx V_{\mathrm{nn}}.$$

Heisenberg suggested that if you could switch off the electric charge of the proton, there would be no way to distinguish between a proton and a neutron. To reflect this observed symmetry of the nuclear force, it was proposed that the neutron and proton could be considered as two states of a single entity, the *nucleon*, analogous to the spin-up and spin-down states of a spin-half particle,

$$\mathrm{p} = \begin{pmatrix} 1 \\ 0 \end{pmatrix} \quad \text{and} \quad \mathrm{n} = \begin{pmatrix} 0 \\ 1 \end{pmatrix}.$$

This led to the introduction of the idea of *isospin*, where the proton and neutron form an isospin doublet with total isospin $I = 1/2$ and third component of isospin $I_3 = \pm 1/2$. The charge independence of the strong nuclear force is then expressed in terms of invariance under unitary transformations in this isospin space. One such transformation would correspond to replacing all protons with neutrons and vice versa. Physically, isospin has nothing to do with spin. Nevertheless, it will be shown in the following section that isospin satisfies the same SU(2) algebra as spin.

### 9.2.1 Flavour symmetry of the strong interaction

The idea of proton/neutron isospin symmetry can be extended to the quarks. Since the QCD interaction treats all quark flavours equally, the strong interaction possesses a flavour symmetry analogous to isospin symmetry of the nuclear force. For a system of quarks, the Hamiltonian can be broken down into three components

$$\hat{H} = \hat{H}_0 + \hat{H}_{\mathrm{strong}} + \hat{H}_{\mathrm{em}}, \tag{9.2}$$

where $\hat{H}_0$ is the kinetic and rest mass energy of the quarks, and $\hat{H}_{\mathrm{strong}}$ and $\hat{H}_{\mathrm{em}}$ are respectively the strong and electromagnetic interaction terms. If the (effective) masses of the up- and down-quarks are the same, and $\hat{H}_{\mathrm{em}}$ is small compared to $\hat{H}_{\mathrm{strong}}$, then to a good approximation the Hamiltonian possesses an up–down (ud) flavour symmetry; nothing would change if all the up-quarks were replaced by down-quarks and vice versa. One simple consequence of an exact ud flavour symmetry is that the existence of a (uud) bound quark state implies that there will a corresponding state (ddu) with the same mass.

The above idea can be developed mathematically by writing the up- and down-quarks as states in an abstract flavour space

$$\mathrm{u} = \begin{pmatrix} 1 \\ 0 \end{pmatrix} \quad \text{and} \quad \mathrm{d} = \begin{pmatrix} 0 \\ 1 \end{pmatrix}.$$

If the up- and down-quarks were indistinguishable, the flavour independence of the QCD interaction could be expressed as an invariance under a general unitary transformation in this abstract space

$$\begin{pmatrix} \mathrm{u}' \\ \mathrm{d}' \end{pmatrix} = \hat{U} \begin{pmatrix} \mathrm{u} \\ \mathrm{d} \end{pmatrix} = \begin{pmatrix} U_{11} & U_{12} \\ U_{21} & U_{22} \end{pmatrix} \begin{pmatrix} \mathrm{u} \\ \mathrm{d} \end{pmatrix}.$$

Since a general $2 \times 2$ matrix depends on four complex numbers, it can be described by eight real parameters. The condition $\hat{U}\hat{U}^{\dagger} = I$, imposes four constraints; therefore a $2 \times 2$ unitary matrix can be expressed in terms of four real parameters or, equivalently, four linearly independent $2 \times 2$ matrices representing the generators of the transformation

$$\hat{U} = \exp\left(i\alpha_i \hat{G}_i\right).$$

One of the generators can be identified as

$$\hat{U} = \begin{pmatrix} 1 & 0 \\ 0 & 1 \end{pmatrix} e^{i\phi}. \tag{9.3}$$

This $U(1)$ transformation corresponds to multiplication by a complex phase and is therefore not relevant to the discussion of transformations *between* different flavour states. The remaining three unitary matrices form a special unitary SU(2) group with the property[1] $\det U = 1$. The three matrices representing the Hermitian generators of the SU(2) group are linearly independent from the identity and are therefore traceless. A suitable choice[2] for three Hermitian traceless generators of the ud flavour symmetry are the Pauli spin-matrices

$$\sigma_1 = \begin{pmatrix} 0 & 1 \\ 1 & 0 \end{pmatrix}, \quad \sigma_2 = \begin{pmatrix} 0 & -i \\ i & 0 \end{pmatrix} \quad \text{and} \quad \sigma_3 = \begin{pmatrix} 1 & 0 \\ 0 & -1 \end{pmatrix}.$$

The ud flavour symmetry corresponds to invariance under SU(2) transformations leading to three conserved observable quantities defined by the eigenvalues of Pauli

1 The property $\det U = 1$ follows from the properties of determinants, $\det U^{\dagger}U \equiv \det I = \det U^{\dagger} \det U = \det U^{*} \det U = |\det U|^2 = 1$. For the corresponding infinitesimal transformation to be close to the identity, $\det U$ must equal $+1$.

2 The algebra of the SU(2) is determined by the commutation relations of the generators. The use of the Pauli spin-matrices is purely conventional. An equally valid choice of the generators $G_i$ would be $S^{\dagger}\sigma_i S$ where $S$ is an arbitrary unitary matrix. The commutation relations are unchanged by this redefinition, and thus the algebra of SU(2) does not depend on the specific representation of the generators.

spin-matrices. The *algebra* of the ud flavour symmetry is therefore identical to that of spin for a spin-half particle. In analogy with the quantum-mechanical treatment of spin-half particles, isospin $\hat{\mathbf{T}}$ is defined in terms of the Pauli spin-matrices

$$\hat{\mathbf{T}} = \tfrac{1}{2}\boldsymbol{\sigma}.$$

Any finite transformation in the up–down quark flavour space can be written in terms of a unitary transformation

$$\hat{U} = e^{i\boldsymbol{\alpha}\cdot\hat{\mathbf{T}}},$$

such that

$$\begin{pmatrix} \mathrm{u}' \\ \mathrm{d}' \end{pmatrix} = e^{i\boldsymbol{\alpha}\cdot\hat{\mathbf{T}}}\begin{pmatrix} \mathrm{u} \\ \mathrm{d} \end{pmatrix},$$

where $\boldsymbol{\alpha}\cdot\hat{\mathbf{T}} = \alpha_1\hat{T}_1 + \alpha_2\hat{T}_2 + \alpha_3\hat{T}_3$. Hence, the general flavour transformation is a "rotation" in flavour space, not just the simple interchange of up and down quarks. A general unitary transformation in this isospin space would amount to relabelling the up-quark as a linear combination of the up-quark and the down-quark. If the flavour symmetry were exact, and the up- and down-quarks were genuinely indistinguishable, this would be perfectly acceptable. However, because the up- and down-quarks have different charges, it does not make sense to form states which are linear combinations of the two, as this would lead to violations of electric charge conservation. Consequently, the only physical meaningful isospin transformation is that which corresponds to relabelling the states, u $\leftrightarrow$ d.

### 9.2.2 Isospin algebra

Whilst isospin has nothing to do with the physical property of spin, it has exactly the same mathematical structure defined by the generators of the SU(2) symmetry group. In the language of group theory the generators of SU(2) define a non-Abelian (i.e. non-commuting) Lie algebra. The three generators of the group, which correspond to physical observables, satisfy the algebra

$$\left[\hat{T}_1, \hat{T}_2\right] = i\hat{T}_3, \quad \left[\hat{T}_2, \hat{T}_3\right] = i\hat{T}_1 \quad \text{and} \quad \left[\hat{T}_3, \hat{T}_1\right] = i\hat{T}_2.$$

This is exactly the same set of commutators as found for the quantum mechanical treatment of angular momentum, introduced in Section 2.3.5. Consequently, the results obtained for angular momentum can be applied directly to the properties of isospin. The total isospin operator,

$$\hat{T}^2 = \hat{T}_1^2 + \hat{T}_2^2 + \hat{T}_3^2,$$

which commutes with each of the generators, is Hermitian and therefore also corresponds to an observable quantity. Because the three operators $\hat{T}_1$, $\hat{T}_2$ and $\hat{T}_3$ do not

Fig. 9.1 The isospin one-half multiplet consisting of an up-quark and a down-quark.

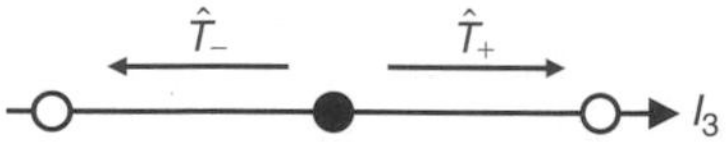


Fig. 9.2 The isospin ladder operators step along the states in $I_3$ within an isospin multiplet.

commute with each other, the corresponding observables cannot be known simultaneously (see Section 2.3.4). Hence, isospin states can be labelled in terms of the total isospin $I$ and the third component of isospin $I_3$. These isospin states $\phi(I, I_3)$ are the mathematical analogues of the angular momentum states $|l, m\rangle$ and have the properties

$$\hat{T}^2\phi(I, I_3) = I(I+1)\phi(I, I_3) \quad \text{and} \quad \hat{T}_3\phi(I, I_3) = I_3\phi(I, I_3).$$

In terms of isospin, the up-quark and down-quark are represented by

$$\mathrm{u} = \begin{pmatrix}1\\0\end{pmatrix} = \phi\left(\tfrac{1}{2}, +\tfrac{1}{2}\right) \quad \text{and} \quad \mathrm{d} = \begin{pmatrix}0\\1\end{pmatrix} = \phi\left(\tfrac{1}{2}, -\tfrac{1}{2}\right).$$

The up- and down-quarks are the two states of an isospin one-half multiplet with respective third components of isospin $+\frac{1}{2}$ and $-\frac{1}{2}$ as indicated in Figure 9.1.

## Isospin ladder operators

The isospin ladder operators, analogous to the quantum mechanical angular momentum ladder operators, defined as

$$\hat{T}_- \equiv \hat{T}_1 - i\hat{T}_2 \quad \text{and} \quad \hat{T}_+ \equiv \hat{T}_1 + i\hat{T}_2,$$

have the effect of moving between the $(2I+1)$ states within an isospin multiplet, as indicated in Figure 9.2. The action the ladder operators on a particular isospin state are

$$\hat{T}_+\phi(I, I_3) = \sqrt{I(I+1) - I_3(I_3+1)}\,\phi(I, I_3+1), \tag{9.4}$$

$$\hat{T}_-\phi(I, I_3) = \sqrt{I(I+1) - I_3(I_3-1)}\,\phi(I, I_3-1), \tag{9.5}$$

where the coefficients were derived in Section 2.3.5. For an isospin multiplet with total isospin $I$, the ladder operators have the effect of raising or lowering the third component of isospin. The action of the ladder operators on the extreme states with $I_3 = \pm I$ yield zero,

$$\hat{T}_-\phi(I, -I) = 0 \quad \text{and} \quad \hat{T}_+\phi(I, +I) = 0.$$

Therefore, the effects of the isospin ladder operators on the u- and d-quarks are

$$\hat{T}_+\mathrm{u} = 0, \quad \hat{T}_+\mathrm{d} = \mathrm{u}, \quad \hat{T}_-\mathrm{u} = \mathrm{d} \quad \text{and} \quad \hat{T}_-\mathrm{d} = 0.$$

## 9.3 Combining quarks into baryons

The strong interaction Hamiltonian does not distinguish between up- and down-quarks, therefore in the limit where the up- and down-quark masses are the same, physical predictions involving the strong interaction alone are symmetric under unitary transformations in this space. The conserved observable quantities, corresponding to the generators of this symmetry are $I_3$ and $I$. Because $I_3$ and $I$ are conserved in strong interactions, the concept of isospin is useful in describing low-energy hadron interactions. For example, isospin arguments can be used to explain the observation that the decay rate for $\Delta^+ \to \mathrm{p}\pi^0$ is twice that for $\Delta^0 \to \mathrm{n}\pi^0$ (see Problem 9.3). Here the concept of isospin will be used to construct the flavour wavefunctions of baryons (qqq) and mesons ($\mathrm{q}\overline{\mathrm{q}}$).

The rules for combining isospin for a system of two quarks are identical to those for the addition of angular momentum. The third component of isospin is added as a scalar and the total isospin is added as the magnitude of a vector. If two isospin states $\phi\left(I^a, I_3^a\right)$ and $\phi\left(I^b, I_3^b\right)$ are combined, the resulting isospin state $\phi\,(I, I_3)$ has

$$I_3 = I_3^a + I_3^b \quad \text{and} \quad |I^a - I^b| \le I \le |I^a + I^b|.$$

These rules can be used to identify the possible isospin states formed from two quarks (each of which can be either an up- or down-quark). The third component of isospin is the scalar sum of $I_3$ for the individual quarks, and hence the $I_3$ assignments of the four possible combinations of two light quarks are those of Figure 9.3. The isospin assignments for the extreme states immediately can be identified as

$$\mathrm{uu} \equiv \phi\left(\tfrac{1}{2}, \tfrac{1}{2}\right)\phi\left(\tfrac{1}{2}, \tfrac{1}{2}\right) = \phi\,(1, +1) \quad \text{and} \quad \mathrm{dd} \equiv \phi\left(\tfrac{1}{2}, -\tfrac{1}{2}\right)\phi\left(\tfrac{1}{2}, -\tfrac{1}{2}\right) = \phi\,(1, -1).$$

This identification is unambiguous, since a state with $I_3 = \pm 1$ must have $I \ge 1$ and the maximum total isospin for a two-quark state is $I = 1$. The quark combinations ud and du, which both have $I_3 = 0$, are not eigenstates of total isospin. The

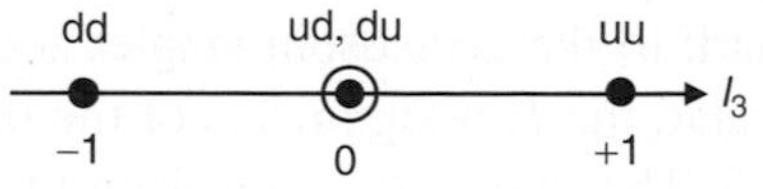


Fig. 9.3 The $I_3$ assignments for the four possible combinations of two up- or down-quarks. There are two states with $I_3 = 0$ (indicated by the point and circle) ud and du.

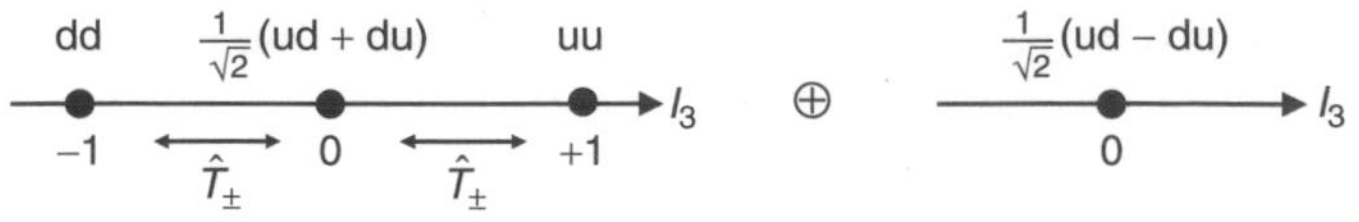


Fig. 9.4 The isospin eigenstates for the combination of two quarks.

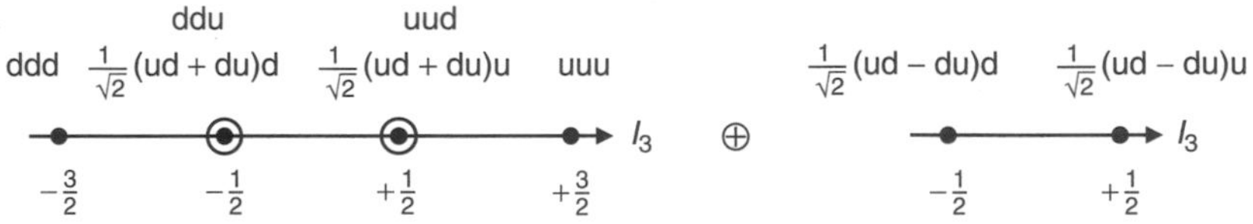


Fig. 9.5 The $I_3$ assignments of three-quark states built from the qq triplet and singlet states.

appropriate linear combination corresponding to the $I = 1$ state can be identified using isospin ladder operators,

$$\hat{T}_-\phi(1,+1) = \sqrt{2}\phi(1,0) = \hat{T}_-(\mathrm{uu}) = \mathrm{ud} + \mathrm{du},$$

and thus

$$\phi(1,0) = \frac{1}{\sqrt{2}}(\mathrm{ud} + \mathrm{du}).$$

The $\phi(0,0)$ state can be identified as the linear combination of ud and du that is orthogonal to $\phi(1,0)$, from which

$$\phi(0,0) = \frac{1}{\sqrt{2}}(\mathrm{ud} - \mathrm{du}). \tag{9.6}$$

Acting on the $I = 0$ singlet state of (9.6) with either $\hat{T}_+$ or $\hat{T}_-$ gives zero, confirming that it is indeed the $\phi(0,0)$ state, for example

$$\begin{aligned}\hat{T}_+\frac{1}{\sqrt{2}}(\mathrm{ud} - \mathrm{du}) &= \frac{1}{\sqrt{2}}\left([\hat{T}_+\mathrm{u}]\mathrm{d} + \mathrm{u}[\hat{T}_+\mathrm{d}] - [\hat{T}_+\mathrm{d}]\mathrm{u} - \mathrm{d}[\hat{T}_+\mathrm{u}]\right)\\ &= \frac{1}{\sqrt{2}}(\mathrm{uu} - \mathrm{uu}) = 0.\end{aligned}$$

The four possible combinations of two isospin *doublets* therefore decomposes into a *triplet* of isospin-1 states and a *singlet* isospin-0 state, as shown in Figure 9.4. This decomposition can be written as $2 \otimes 2 = 3 \oplus 1$. It should be noted that the isospin-0 and isospin-1 states are physically different; the isospin-1 triplet is symmetric under interchange of the two quarks, whereas the isospin singlet is antisymmetric.

The isospin states formed from three quarks can be obtained by adding an up- or down-quark to the qq isospin singlet and triplet states of Figure 9.4. Since $I_3$ adds as a scalar, the $I_3$ assignments of the possible combinations are those shown in Figure 9.5. The two states built from the $I = 0$ singlet will have total isospin $I = 1/2$, whereas those constructed from the $I = 1$ triplet can have either $I = 1/2$ or $I = 3/2$. Of the six combinations formed from the triplet, the extreme ddd and uuu states with $I_3 = -3/2$ and $I_3 = +3/2$ uniquely can be identified as being part of isospin $I = 3/2$ multiplet. The other two $I = 3/2$ states can be identified using the ladder operators. For example, the $\phi\left(\frac{3}{2}, -\frac{1}{2}\right)$ state, which is a linear combination of the ddu and $\frac{1}{\sqrt{2}}(\mathrm{ud} + \mathrm{du})\mathrm{d}$ states, can be obtained from the action of $\hat{T}_+$ on

$$\phi\left(\tfrac{3}{2}, -\tfrac{3}{2}\right) = \mathrm{ddd},$$

from which

$$\begin{aligned}\hat{T}_+\phi\left(\tfrac{3}{2}, -\tfrac{3}{2}\right) = \sqrt{3}\phi\left(\tfrac{3}{2}, -\tfrac{1}{2}\right) &= \hat{T}_+(\mathrm{ddd}) = [\hat{T}_+\mathrm{d}]\mathrm{dd} + \mathrm{d}[\hat{T}_+\mathrm{d}]\mathrm{d} + \mathrm{dd}[\hat{T}_+\mathrm{d}] \\ &= \mathrm{udd} + \mathrm{dud} + \mathrm{ddu},\end{aligned}$$

and therefore

$$\phi\left(\tfrac{3}{2}, -\tfrac{1}{2}\right) = \frac{1}{\sqrt{3}}(\mathrm{udd} + \mathrm{dud} + \mathrm{ddu}). \tag{9.7}$$

From the repeated action of the ladder operators, the four isospin-$\frac{3}{2}$ states, built from the qq triplet, can be shown to be

$$\phi\left(\tfrac{3}{2}, -\tfrac{3}{2}\right) = \mathrm{ddd}, \tag{9.8}$$

$$\phi\left(\tfrac{3}{2}, -\tfrac{1}{2}\right) = \frac{1}{\sqrt{3}}(\mathrm{udd} + \mathrm{dud} + \mathrm{ddu}), \tag{9.9}$$

$$\phi\left(\tfrac{3}{2}, +\tfrac{1}{2}\right) = \frac{1}{\sqrt{3}}(\mathrm{uud} + \mathrm{udu} + \mathrm{duu}), \tag{9.10}$$

$$\phi\left(\tfrac{3}{2}, +\tfrac{3}{2}\right) = \mathrm{uuu}. \tag{9.11}$$

The two states obtained from the qq triplet with total isospin $I = 1/2$ are orthogonal to the $I_3 = \pm 1/2$ states of (9.9) and (9.10). Hence, the $\phi\left(\frac{1}{2}, -\frac{1}{2}\right)$ state can be identified as the linear combination of ddu and $\frac{1}{\sqrt{2}}(\mathrm{ud} + \mathrm{du})\mathrm{d}$ that is orthogonal to the $\phi\left(\frac{3}{2}, -\frac{1}{2}\right)$ state of (9.9), giving

$$\phi_S\left(\tfrac{1}{2}, -\tfrac{1}{2}\right) = -\frac{1}{\sqrt{6}}(2\mathrm{ddu} - \mathrm{udd} - \mathrm{dud}), \tag{9.12}$$

and similarly

$$\phi_S\left(\tfrac{1}{2}, +\tfrac{1}{2}\right) = \frac{1}{\sqrt{6}}(2\mathrm{uud} - \mathrm{udu} - \mathrm{duu}). \tag{9.13}$$

The relative phases of (9.12) and (9.13) ensure that the ladder operators correctly step between the two states. In addition, the two states constructed from the qq isospin singlet of (9.6) are

$$\phi_A\left(\tfrac{1}{2}, -\tfrac{1}{2}\right) = \frac{1}{\sqrt{2}}(\mathrm{udd} - \mathrm{dud}), \tag{9.14}$$

$$\phi_A\left(\tfrac{1}{2}, +\tfrac{1}{2}\right) = \frac{1}{\sqrt{2}}(\mathrm{udu} - \mathrm{duu}). \tag{9.15}$$

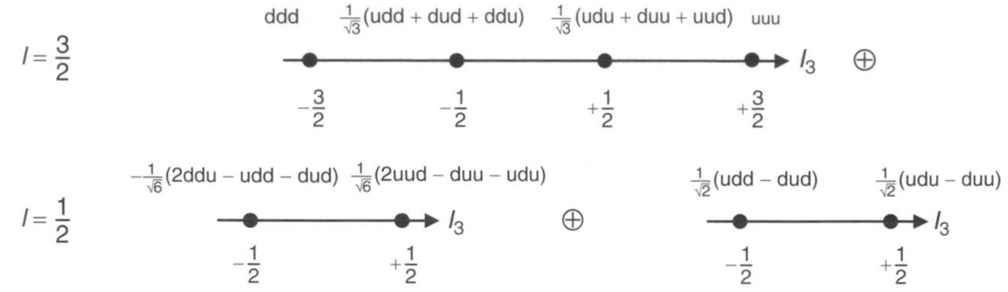


Fig. 9.6 The three-quark $\phi\,(I, I_3)$ states in SU(2) flavour symmetry. The eight combinations decompose into a symmetric quadruplet and two mixed symmetry doublets.

Hence, the eight combinations of three up- and down-quarks, uuu, uud, udu, udd, duu, dud, ddu and ddd, have been grouped into an isospin-$\frac{3}{2}$ *quadruplet* and two isospin-$\frac{1}{2}$ *doublets*, as shown in Figure 9.6. In terms of the SU(2) group structure this can be expressed as

$$2 \otimes 2 \otimes 2 = 2 \otimes (3 \oplus 1) = (2 \otimes 3) \oplus (2 \otimes 1) = 4 \oplus 2 \oplus 2,$$

where $2 \otimes 2 \otimes 2$ represents the combinations of three quarks represented as isospin doublets. The different isospin multiplets have different exchange symmetries. The flavour states in the isospin-$\frac{3}{2}$ quadruplet, (9.8)–(9.11), are symmetric under the interchange of *any* two quarks. The isospin-$\frac{1}{2}$ doublets are referred to as *mixed symmetry* states to reflect the symmetry under the interchange of the first two quarks, but lack of overall exchange symmetry. The doublet states of (9.12) and (9.13), labelled $\phi_S$, are symmetric under the interchange of quarks $1 \leftrightarrow 2$, whereas the doublet states of (9.14) and (9.15), labelled $\phi_A$, are antisymmetric under the interchange of quarks $1 \leftrightarrow 2$. These two isospin doublets have no definite symmetry under the interchange of quarks $1 \leftrightarrow 3$ and $2 \leftrightarrow 3$.

### 9.3.1 Spin states of three quarks

Because the SU(2) algebra for combining spin-half is that same as that for isospin, the possible spin wavefunctions of three quarks, denoted by $\chi$, are constructed in the same manner. Hence the combination of three spin-half particles gives: a spin-$\frac{3}{2}$ quadruplet, with spin states

$$\chi\left(\tfrac{3}{2}, +\tfrac{3}{2}\right) = \uparrow\uparrow\uparrow, \tag{9.16}$$

$$\chi\left(\tfrac{3}{2}, +\tfrac{1}{2}\right) = \tfrac{1}{\sqrt{3}}(\uparrow\uparrow\downarrow + \uparrow\downarrow\uparrow + \downarrow\uparrow\uparrow), \tag{9.17}$$

$$\chi\left(\tfrac{3}{2}, -\tfrac{1}{2}\right) = \tfrac{1}{\sqrt{3}}(\downarrow\downarrow\uparrow + \downarrow\uparrow\downarrow + \uparrow\downarrow\downarrow), \tag{9.18}$$

$$\chi\left(\tfrac{3}{2}, -\tfrac{3}{2}\right) = \downarrow\downarrow\downarrow; \tag{9.19}$$

a mixed symmetry doublet which is symmetric under $1 \leftrightarrow 2$,

$$\chi_S\left(\tfrac{1}{2},-\tfrac{1}{2}\right) = -\tfrac{1}{\sqrt{6}}(2\downarrow\downarrow\uparrow - \uparrow\downarrow\downarrow - \downarrow\uparrow\downarrow), \tag{9.20}$$

$$\chi_S\left(\tfrac{1}{2},+\tfrac{1}{2}\right) = \tfrac{1}{\sqrt{6}}(2\uparrow\uparrow\downarrow - \uparrow\downarrow\uparrow - \downarrow\uparrow\uparrow); \tag{9.21}$$

and a mixed symmetry doublet which is antisymmetric under $1 \leftrightarrow 2$,

$$\chi_A\left(\tfrac{1}{2},-\tfrac{1}{2}\right) = \tfrac{1}{\sqrt{2}}(\uparrow\downarrow\downarrow - \downarrow\uparrow\downarrow), \tag{9.22}$$

$$\chi_A\left(\tfrac{1}{2},+\tfrac{1}{2}\right) = \tfrac{1}{\sqrt{2}}(\uparrow\downarrow\uparrow - \downarrow\uparrow\uparrow). \tag{9.23}$$

## 9.4 Ground state baryon wavefunctions

There are eight possible isospin states for a system of three quarks and eight possible spin states, leading to a total of 64 possible combined flavour and spin states. However, not all combinations satisfy the required fermion exchange symmetry of the total wavefunction. In addition to spin and flavour components, the wavefunction for a qqq state also needs to describe the colour content and the spatial wavefunction. The overall wavefunction for a bound qqq state, accounting for all degrees of freedom, can be written

$$\psi = \phi_{\text{flavour}}\,\chi_{\text{spin}}\,\xi_{\text{colour}}\,\eta_{\text{space}}. \tag{9.24}$$

Because quarks are fermions, the overall wavefunction of (9.24) is required to be antisymmetric under the interchange of *any* two of the quarks. For a system of like fermions, for example uuu, this is simply a statement of the Pauli exclusion principle. However, because of the assumed SU(2) flavour symmetry, when the flavour wavefunction is included, the fermion exchange symmetry applies to the wavefunction as a whole (the argument is given in the starred Addendum in Section 9.7 at the end of this chapter).

The requirement that the wavefunction of (9.24) is totally antisymmetric places restrictions on the individual parts. In Chapter 10, it is shown that the colour wavefunction is necessarily totally antisymmetric. Here the discussion is restricted to the $L = 0$ ground state baryons, in which there is no orbital angular momentum. In this case, the quarks are described by $\ell = 0$ *s*-waves. Since the exchange symmetry of the orbital states is given by $(-1)^{\ell}$, here the orbital wavefunction is symmetric under the interchange of any two quarks. Consequently, for the $L = 0$ baryons the combination $\xi_{\text{colour}}\,\eta_{\text{space}}$ is antisymmetric under the interchange of any two quarks. For the overall wavefunction to be antisymmetric, the combined flavour and spin wavefunctions, $\phi_{\text{flavour}}\,\chi_{\text{spin}}$, must be symmetric.

$$L = 0 \text{ baryons:} \quad \phi_{\text{flavour}}\,\chi_{\text{spin}} = \text{symmetric}.$$

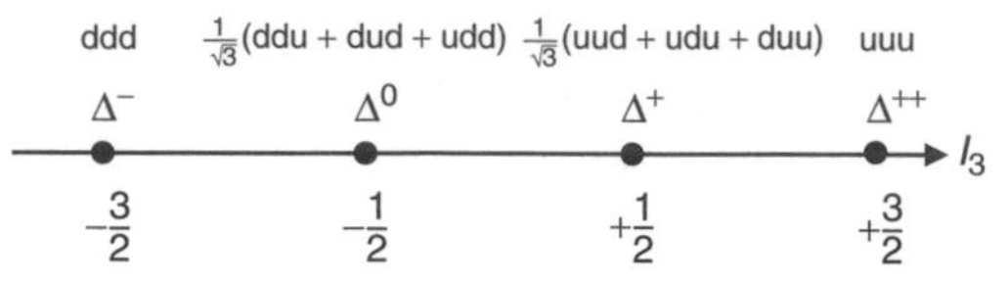


Fig. 9.7 The flavour wavefunctions of the $I = \frac{3}{2}$ light quark $\Delta$-baryons, each of which has total spin $s = \frac{3}{2}$.

The possible forms of the flavour and spin parts of the wavefunction are respectively given by (9.8)–(9.15) and (9.16)–(9.23). There are two ways to construct a totally symmetric combination of $\phi_{\text{flavour}}$ and $\chi_{\text{spin}}$. Firstly, the totally symmetric flavour wavefunctions of (9.8)–(9.11) can be combined with the totally symmetric spin wavefunctions of (9.16)–(9.19) to give four spin-$\frac{3}{2}$, isospin-$\frac{3}{2}$ baryons. These are known as the $\Delta$-baryons with the flavour wavefunctions shown in Figure 9.7.

The second way to construct a totally symmetric $\phi_{\text{flavour}}\chi_{\text{spin}}$ wavefunction is to note that the combinations of mixed symmetry wavefunctions, $\phi_S\chi_S$ and $\phi_A\chi_A$, are both symmetric under interchange of quarks $1 \leftrightarrow 2$. However, neither combination on its own has a definite symmetry under the interchange of quarks $1 \leftrightarrow 3$ and $2 \leftrightarrow 3$. Nevertheless, it is easy to verify that the linear combination

$$\psi = \frac{1}{\sqrt{2}}(\phi_S\chi_S + \phi_A\chi_A) \tag{9.25}$$

is symmetric under the interchange of any two quarks, as required. Here the two possible flavour states correspond to the spin-half proton (uud) and neutron (ddu). Therefore, from (9.25), the wavefunction for a spin-up proton can be identified as

$$\begin{aligned}|\text{p}\uparrow\rangle &= \frac{1}{\sqrt{2}}\left[\phi_S\left(\tfrac{1}{2},+\tfrac{1}{2}\right)\chi_S\left(\tfrac{1}{2},+\tfrac{1}{2}\right) + \phi_A\left(\tfrac{1}{2},+\tfrac{1}{2}\right)\chi_A\left(\tfrac{1}{2},+\tfrac{1}{2}\right)\right]\\ &= \frac{1}{6\sqrt{2}}(2\text{uud} - \text{udu} - \text{duu})(2\uparrow\uparrow\downarrow - \uparrow\downarrow\uparrow - \downarrow\uparrow\uparrow) + \frac{1}{2\sqrt{2}}(\text{udu} - \text{duu})(\uparrow\downarrow\uparrow - \downarrow\uparrow\uparrow),\end{aligned}$$

which when written out in full is

$$\begin{aligned}|\text{p}\uparrow\rangle = \frac{1}{\sqrt{18}}(&2\text{u}\!\uparrow\text{u}\!\uparrow\text{d}\!\downarrow - \text{u}\!\uparrow\text{u}\!\downarrow\text{d}\!\uparrow - \text{u}\!\downarrow\text{u}\!\uparrow\text{d}\!\uparrow\\ &+ 2\text{u}\!\uparrow\text{d}\!\downarrow\text{u}\!\uparrow - \text{u}\!\uparrow\text{d}\!\uparrow\text{u}\!\downarrow - \text{u}\!\downarrow\text{d}\!\uparrow\text{u}\!\uparrow\\ &+ 2\text{d}\!\downarrow\text{u}\!\uparrow\text{u}\!\uparrow - \text{d}\!\uparrow\text{u}\!\uparrow\text{u}\!\downarrow - \text{d}\!\uparrow\text{u}\!\downarrow\text{u}\!\uparrow).\end{aligned} \tag{9.26}$$

The fully antisymmetric version of the proton wavefunction would include the antisymmetric colour wavefunction, which itself has six terms, giving a wavefunction with a total of 54 terms with different combinations of flavour, spin and colour. In practice, the wavefunction of (9.26) is sufficient to calculate the physical properties of the proton.

## 9.5 Isospin representation of antiquarks

In the above description of SU(2) flavour symmetry, the up- and down-quarks were placed in an isospin doublet,

$$\mathrm{q} = \begin{pmatrix} \mathrm{u} \\ \mathrm{d} \end{pmatrix}.$$

A general SU(2) transformation of the quark doublet, $\mathrm{q} \to \mathrm{q}' = U\mathrm{q}$, can be written

$$\begin{pmatrix} \mathrm{u} \\ \mathrm{d} \end{pmatrix} \to \begin{pmatrix} \mathrm{u}' \\ \mathrm{d}' \end{pmatrix} = \begin{pmatrix} a & b \\ -b^* & a^* \end{pmatrix} \begin{pmatrix} \mathrm{u} \\ \mathrm{d} \end{pmatrix}, \tag{9.27}$$

where $a$ and $b$ are complex numbers which satisfy $aa^* + bb^* = 1$. In Section 4.7.5, the charge conjugation operation was identified as $\psi' = \hat{C}\psi = i\gamma^2\psi^*$. Hence taking the complex conjugate of (9.27) gives the transformation properties of the flavour part of the antiquark wavefunctions

$$\begin{pmatrix} \overline{\mathrm{u}}' \\ \overline{\mathrm{d}}' \end{pmatrix} = U^* \begin{pmatrix} \overline{\mathrm{u}} \\ \overline{\mathrm{d}} \end{pmatrix} = \begin{pmatrix} a^* & b^* \\ -b & a \end{pmatrix} \begin{pmatrix} \overline{\mathrm{u}} \\ \overline{\mathrm{d}} \end{pmatrix}. \tag{9.28}$$

In SU(2) it is possible to place the antiquarks in a doublet that transforms in the same way as the quarks, $\overline{\mathrm{q}} \to \overline{\mathrm{q}}' = U\overline{\mathrm{q}}$. If the antiquark doublet is written as

$$\overline{\mathrm{q}} \equiv \begin{pmatrix} -\overline{\mathrm{d}} \\ \overline{\mathrm{u}} \end{pmatrix} = S \begin{pmatrix} \overline{\mathrm{u}} \\ \overline{\mathrm{d}} \end{pmatrix} = \begin{pmatrix} 0 & -1 \\ 1 & 0 \end{pmatrix} \begin{pmatrix} \overline{\mathrm{u}} \\ \overline{\mathrm{d}} \end{pmatrix}, \tag{9.29}$$

then since

$$\begin{pmatrix} \overline{\mathrm{u}} \\ \overline{\mathrm{d}} \end{pmatrix} = S^{-1}\overline{\mathrm{q}} \quad \text{and} \quad \begin{pmatrix} \overline{\mathrm{u}}' \\ \overline{\mathrm{d}}' \end{pmatrix} = S^{-1}\overline{\mathrm{q}}',$$

Equation (9.28) can be written

$$S^{-1}\overline{\mathrm{q}}' = U^* S^{-1}\overline{\mathrm{q}}$$

$$\Rightarrow \qquad \overline{\mathrm{q}}' = S U^* S^{-1}\overline{\mathrm{q}}.$$

Using the definition of the $S$ of (9.29),

$$S U^* S^{-1} = \begin{pmatrix} 0 & -1 \\ 1 & 0 \end{pmatrix} \begin{pmatrix} a^* & b^* \\ -b & a \end{pmatrix} \begin{pmatrix} 0 & 1 \\ -1 & 0 \end{pmatrix} = \begin{pmatrix} a & b \\ -b^* & a^* \end{pmatrix} = U,$$

and therefore, as desired,

$$\overline{\mathrm{q}} \to \overline{\mathrm{q}}' = U\overline{\mathrm{q}}.$$

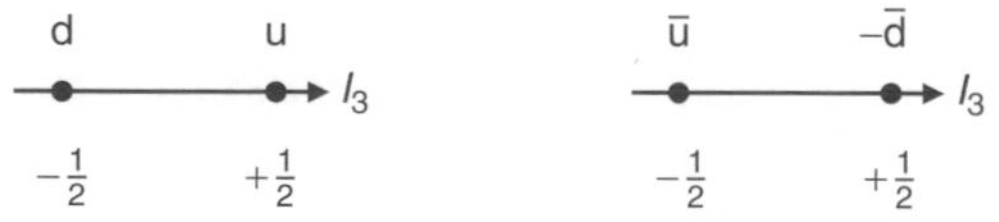


Fig. 9.8 The isospin representation of d and u quarks and $\overline{\text{d}}$ and $\overline{\text{u}}$ antiquarks.

Hence, by placing the antiquarks in an SU(2) doublet defined by

$$\overline{\text{q}} \equiv \begin{pmatrix} -\overline{\text{d}} \\ \overline{\text{u}} \end{pmatrix},$$

the antiquarks transform in exactly the same manner as the quarks. The ordering of the $\overline{\text{d}}$ and $\overline{\text{u}}$ in the doublet and the minus sign in front of the $\overline{\text{d}}$, ensure that quarks and antiquarks behave in the same way under SU(2) flavour transformations and that physical predictions are invariant under the simultaneous transformations of the form $\text{u} \leftrightarrow \text{d}$ and $\overline{\text{u}} \leftrightarrow \overline{\text{d}}$. The $I_3$ assignments of the quark and antiquark doublets are shown in Figure 9.8. The effect of the isospin ladder operators on the antiquark doublet can be seen to be

$$T_+\overline{\text{u}} = -\overline{\text{d}}, \quad T_+\overline{\text{d}} = 0, \quad T_-\overline{\text{u}} = 0 \quad \text{and} \quad T_-\overline{\text{d}} = -\overline{\text{u}}.$$

It is important to note that, in general, it is not possible to place the quarks and antiquarks in the same representation; this is a feature SU(2). It *cannot* be applied to the SU(3) flavour symmetry of Section 9.6.

## Meson states

A meson is a bound state of a quark and an antiquark. In terms of isospin, the four possible states formed from up- and down-quarks/antiquarks can be expressed as the combination of an SU(2) quark doublet and an SU(2) antiquark doublet. Using the isospin assignments of Figure 9.8, the $\text{d}\overline{\text{u}}$ state immediately can be identified as the $\text{q}\overline{\text{q}}$ isospin state, $\phi(1,-1)$. The two other members of the isospin triplet can be identified by application of the isospin ladder operator $\hat{T}_+$ leading to

$$\begin{aligned}
\phi(1,-1) &= \text{d}\overline{\text{u}}, \\
\phi(1,0) &= \tfrac{1}{\sqrt{2}}(\text{u}\overline{\text{u}} - \text{d}\overline{\text{d}}), \\
\phi(1,+1) &= -\text{u}\overline{\text{d}}.
\end{aligned}$$

The isospin singlet, which must be orthogonal to the $\phi(1,0)$ state, is therefore

$$\phi(0,0) = \tfrac{1}{\sqrt{2}}\left(\text{u}\overline{\text{u}} + \text{d}\overline{\text{d}}\right).$$

This decomposition into an isospin triplet and an isospin singlet, shown in Figure 9.9, is expressed as $2 \otimes \overline{2} = 3 \oplus 1$, where the 2 is the isospin representation of the quark doublet and the $\overline{2}$ is the isospin representation of an antiquark

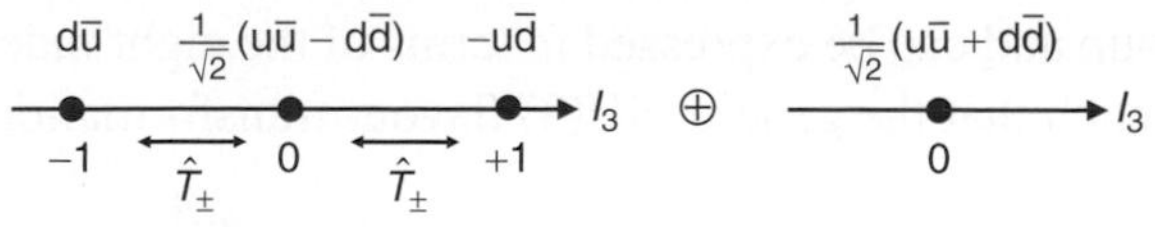


Fig. 9.9 The $q\bar{q}$ isospin triplet and singlet states.

doublet (in the language of group theory the quark doublet is a fundamental representation of SU(2) and the antiquark doublet is the conjugate representation). The action of the isospin raising and lowering operators on the $\phi(0,0)$ state both give zero, confirming that it is indeed a singlet state.

## 9.6 SU(3) flavour symmetry

The SU(2) flavour symmetry described above is almost exact because the difference in the masses of the up- and down-quarks is small and the Coulomb interaction represents a relatively small contribution to the overall Hamiltonian compared to the strong interaction. It is possible to extend the flavour symmetry to include the strange quark. The strong interaction part of the Hamiltonian of (9.2) treats all quarks equally and therefore possesses an exact uds flavour symmetry. However, since the mass of the strange quark is different from the masses of the up- and down-quarks, the overall Hamiltonian is not flavour symmetric. Nevertheless, the difference between $m_s$ and $m_{u/d}$, which is of the order 100 MeV, is relatively small compared to the typical binding energies of baryons, which are of order 1 GeV. It is therefore possible to proceed as if the overall Hamiltonian possessed a uds flavour symmetry. However, the results based on this assumption should be treated with care as, in reality, the symmetry is only approximate.

The assumed uds flavour symmetry can be expressed by a unitary transformation in flavour space

$$\begin{pmatrix} u' \\ d' \\ s' \end{pmatrix} = \hat{U}\begin{pmatrix} u \\ d \\ s \end{pmatrix} = \begin{pmatrix} U_{11} & U_{12} & U_{13} \\ U_{21} & U_{22} & U_{23} \\ U_{31} & U_{32} & U_{33} \end{pmatrix}\begin{pmatrix} u \\ d \\ s \end{pmatrix}.$$

In general, a $3 \times 3$ matrix can be written in terms of nine complex numbers, or equivalently 18 real parameters. There are nine constraints from requirement of unitarity, $\hat{U}^\dagger\hat{U} = I$. Therefore $\hat{U}$ can be expressed in terms of nine linearly independent $3 \times 3$ matrices. As before, one of these matrices is the identity matrix multiplied by a complex phase and is not relevant to the discussion of transformations between different flavour states. The remaining eight matrices form an SU(3) group and can be expressed in terms of the eight independent Hermitian generators $\hat{T}_i$ such that the general SU(3) flavour transformation can be expressed as

$$\hat{U} = e^{i\alpha\cdot\hat{\mathbf{T}}}.$$

The eight generators are written in terms of eight $\lambda$-matrices with

$$\hat{\mathbf{T}} = \frac{1}{2}\lambda,$$

where the matrices act on the SU(3) representations of the u, d and s quarks

$$\mathrm{u} = \begin{pmatrix} 1 \\ 0 \\ 0 \end{pmatrix}, \quad \mathrm{d} = \begin{pmatrix} 0 \\ 1 \\ 0 \end{pmatrix} \quad \text{and} \quad \mathrm{s} = \begin{pmatrix} 0 \\ 0 \\ 1 \end{pmatrix}. \tag{9.30}$$

The SU(3) uds flavour symmetry contains the subgroup of SU(2) u $\leftrightarrow$ d flavour symmetry. Hence, three of the $\lambda$-matrices correspond to the SU(2) ud isospin symmetry and have the Pauli spin-matrices in the top left $2\times2$ block of the $3\times3$ matrix with all other entries zero,

$$\lambda_1 = \begin{pmatrix} 0 & 1 & 0 \\ 1 & 0 & 0 \\ 0 & 0 & 0 \end{pmatrix}, \quad \lambda_2 = \begin{pmatrix} 0 & -i & 0 \\ i & 0 & 0 \\ 0 & 0 & 0 \end{pmatrix} \quad \text{and} \quad \lambda_3 = \begin{pmatrix} 1 & 0 & 0 \\ 0 & -1 & 0 \\ 0 & 0 & 0 \end{pmatrix}.$$

The third component of isospin is now written in terms of the operator

$$\hat{T}_3 = \tfrac{1}{2}\lambda_3,$$

such that

$$\hat{T}_3\mathrm{u} = +\tfrac{1}{2}\mathrm{u}, \quad \hat{T}_3\mathrm{d} = -\tfrac{1}{2}\mathrm{d} \quad \text{and} \quad \hat{T}_3\mathrm{s} = 0.$$

As before, isospin lowering and raising operators are defined as $T_\pm = \frac{1}{2}(\lambda_1 \pm i\lambda_2)$.

The remaining $\lambda$-matrices can be identified by realising that the SU(3) uds flavour symmetry also contains the subgroups of SU(2) u $\leftrightarrow$ s and SU(2) d $\leftrightarrow$ s flavour symmetries, both of which can also be expressed in terms of the Pauli spin-matrices. The corresponding $3\times3$ $\lambda$-matrixes for the u $\leftrightarrow$ s symmetry are

$$\lambda_4 = \begin{pmatrix} 0 & 0 & 1 \\ 0 & 0 & 0 \\ 1 & 0 & 0 \end{pmatrix}, \quad \lambda_5 = \begin{pmatrix} 0 & 0 & -i \\ 0 & 0 & 0 \\ i & 0 & 0 \end{pmatrix} \quad \text{and} \quad \lambda_X = \begin{pmatrix} 1 & 0 & 0 \\ 0 & 0 & 0 \\ 0 & 0 & -1 \end{pmatrix},$$

and for the d $\leftrightarrow$ s symmetry they are

$$\lambda_6 = \begin{pmatrix} 0 & 0 & 0 \\ 0 & 0 & 1 \\ 0 & 1 & 0 \end{pmatrix}, \quad \lambda_7 = \begin{pmatrix} 0 & 0 & 0 \\ 0 & 0 & -i \\ 0 & i & 0 \end{pmatrix} \quad \text{and} \quad \lambda_Y = \begin{pmatrix} 0 & 0 & 0 \\ 0 & 1 & 0 \\ 0 & 0 & -1 \end{pmatrix}.$$

Of the nine $\lambda$-matrices identified above, only eight are independent; one of the three diagonal matrices, $\lambda_3$, $\lambda_X$ and $\lambda_Y$, can be expressed in terms of the other two. Because the u $\leftrightarrow$ d symmetry is nearly exact, it is natural to retain $\lambda_3$ as one of the eight generators of the SU(3) flavour symmetry. The final generator is chosen as the linear combination of $\lambda_X$ and $\lambda_Y$ that treats u and d quarks symmetrically

$$\lambda_8 = \frac{1}{\sqrt{3}}\begin{pmatrix} 0 & 0 & 0 \\ 0 & 1 & 0 \\ 0 & 0 & -1 \end{pmatrix} + \frac{1}{\sqrt{3}}\begin{pmatrix} 1 & 0 & 0 \\ 0 & 0 & 0 \\ 0 & 0 & -1 \end{pmatrix} = \frac{1}{\sqrt{3}}\begin{pmatrix} 1 & 0 & 0 \\ 0 & 1 & 0 \\ 0 & 0 & -2 \end{pmatrix}.$$

The eight matrices used to represent the generators of the SU(3) symmetry, known as the Gell-Mann matrices, are therefore

$$\begin{gathered} \lambda_1 = \begin{pmatrix} 0 & 1 & 0 \\ 1 & 0 & 0 \\ 0 & 0 & 0 \end{pmatrix}, \quad \lambda_4 = \begin{pmatrix} 0 & 0 & 1 \\ 0 & 0 & 0 \\ 1 & 0 & 0 \end{pmatrix}, \quad \lambda_6 = \begin{pmatrix} 0 & 0 & 0 \\ 0 & 0 & 1 \\ 0 & 1 & 0 \end{pmatrix} \\ \lambda_2 = \begin{pmatrix} 0 & -i & 0 \\ i & 0 & 0 \\ 0 & 0 & 0 \end{pmatrix}, \quad \lambda_5 = \begin{pmatrix} 0 & 0 & -i \\ 0 & 0 & 0 \\ i & 0 & 0 \end{pmatrix}, \quad \lambda_7 = \begin{pmatrix} 0 & 0 & 0 \\ 0 & 0 & -i \\ 0 & i & 0 \end{pmatrix} \\ \lambda_3 = \begin{pmatrix} 1 & 0 & 0 \\ 0 & -1 & 0 \\ 0 & 0 & 0 \end{pmatrix}, \quad \lambda_8 = \frac{1}{\sqrt{3}}\begin{pmatrix} 1 & 0 & 0 \\ 0 & 1 & 0 \\ 0 & 0 & -2 \end{pmatrix}. \end{gathered} \tag{9.31}$$

### 9.6.1 SU(3) flavour states

For the case of SU(2) flavour symmetry there are three Hermitian generators, each of which corresponds to an observable quantity. However, since the generators do not commute, they correspond to a set of incompatible variables. Consequently SU(2) states were defined in terms of the eigenstates of the third component of isospin $\hat{T}_3$ and the total isospin $\hat{T}^2 = \hat{T}_1^2 + \hat{T}_2^2 + \hat{T}_3^2$. In SU(3) there is an analogue of total isospin, which for the fundamental representation of the quarks can be written

$$\hat{T}^2 = \sum_{i=1}^{8} \hat{T}_i^2 = \tfrac{1}{4}\sum_{i=1}^{8} \lambda_i^2 = \frac{4}{3}\begin{pmatrix} 1 & 0 & 0 \\ 0 & 1 & 0 \\ 0 & 0 & 1 \end{pmatrix}.$$

Of the eight SU(3) generators, only $T_3 = \frac{1}{2}\lambda_3$ and $T_8 = \frac{1}{2}\lambda_8$ commute and therefore describe compatible observable quantities. Hence, in addition to the analogue of the total isospin, SU(3) states are described in terms of the eigenstates of the $\lambda_3$ and $\lambda_8$ matrices. The corresponding quantum numbers are the third component of isospin and the flavour *hypercharge* defined by the operators

$$\hat{T}_3 = \tfrac{1}{2}\lambda_3 \quad \text{and} \quad \hat{Y} = \frac{1}{\sqrt{3}}\lambda_8.$$

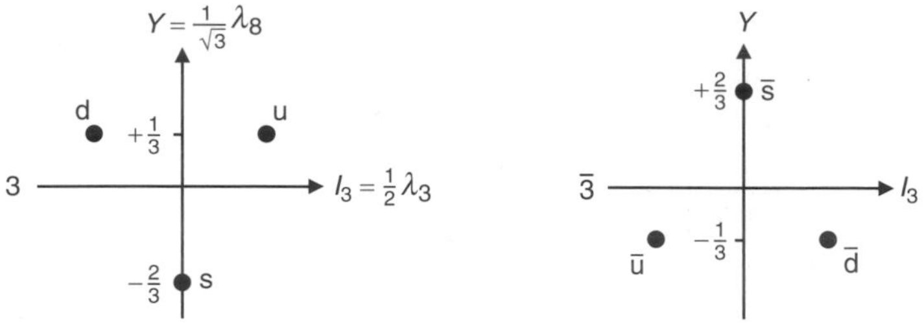


Fig. 9.10 Isospin and hypercharge in SU(3) flavour symmetry for the quarks and antiquarks.

The quarks are the fundamental "3" representation of the SU(3) flavour symmetry. Using the definitions of the quark states of (9.30) it is easy to verify that the isospin and hypercharge assignments of the u, d and s quarks are

$$\begin{aligned}
\hat{T}_3\mathrm{u} &= +\tfrac{1}{2}\mathrm{u} \quad \text{and} \quad \hat{Y}\mathrm{u} = +\tfrac{1}{3}\mathrm{u},\\
\hat{T}_3\mathrm{d} &= -\tfrac{1}{2}\mathrm{d} \quad \text{and} \quad \hat{Y}\mathrm{d} = +\tfrac{1}{3}\mathrm{d},\\
\hat{T}_3\mathrm{s} &= 0 \quad \text{and} \quad \hat{Y}\mathrm{s} = -\tfrac{2}{3}\mathrm{s}.
\end{aligned}$$

The flavour content of a state is uniquely identified by $I_3 = n_\mathrm{u} - n_\mathrm{d}$ and $Y = \frac{1}{3}(n_\mathrm{u} + n_\mathrm{d} - 2n_\mathrm{s})$, where $n_\mathrm{u}$, $n_\mathrm{d}$ and $n_\mathrm{s}$ are the respective numbers of up-, down- and strange quarks. The $I_3$ and $Y$ quantum numbers of the antiquarks have the opposite signs compared to the quarks and they form a $\overline{3}$ multiplet, as shown in Figure 9.10.

Whilst the Gell-Mann $\lambda_3$ and $\lambda_8$ matrices label the SU(3) states, the six remaining $\lambda$-matrices can be used to define ladder operators,

$$\begin{aligned}
\hat{T}_\pm &= \tfrac{1}{2}(\lambda_1 \pm i\lambda_2),\\
\hat{V}_\pm &= \tfrac{1}{2}(\lambda_4 \pm i\lambda_5),\\
\hat{U}_\pm &= \tfrac{1}{2}(\lambda_6 \pm i\lambda_7),
\end{aligned}$$

which respectively step along the d $\leftrightarrow$ u, s $\leftrightarrow$ u and d $\leftrightarrow$ s directions. From the matrix representations of these ladder operators it is straightforward to verify that

$$\hat{V}_+\mathrm{s} = +\mathrm{u},\quad \hat{V}_-\mathrm{u} = +\mathrm{s},\quad \hat{U}_+\mathrm{s} = +\mathrm{d},\quad \hat{U}_-\mathrm{d} = +\mathrm{s},\quad \hat{T}_+\mathrm{d} = +\mathrm{u} \quad \text{and} \quad \hat{T}_-\mathrm{u} = +\mathrm{d},$$

with all other combinations giving zero. In SU(3) flavour symmetry it is not possible to express the antiquarks as a triplet which transforms in the same way as the quark triplet. Nevertheless, following the arguments given in Section 9.5, the effect of a single ladder operator on an antiquark state must reproduce that from the corresponding SU(2) subgroup, such that the states can be obtained from

$$\hat{V}_+\bar{\mathrm{u}} = -\bar{\mathrm{s}},\quad \hat{V}_-\bar{\mathrm{s}} = -\bar{\mathrm{u}},\quad \hat{U}_+\bar{\mathrm{d}} = -\bar{\mathrm{s}},\quad \hat{U}_-\bar{\mathrm{s}} = -\bar{\mathrm{d}},\quad \hat{T}_+\bar{\mathrm{u}} = -\bar{\mathrm{d}} \quad \text{and} \quad \hat{T}_-\bar{\mathrm{d}} = -\bar{\mathrm{u}}.$$

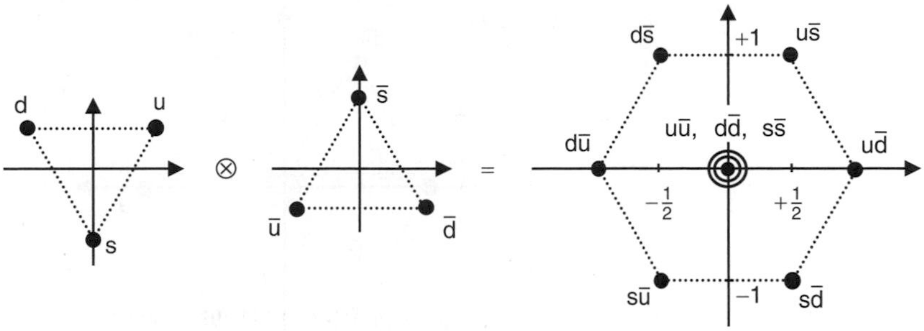


Fig. 9.11 SU(3) isospin and hypercharge assignments of the nine possible $q\overline{q}$ combinations.

## 9.6.2 The light mesons

In the discussion of SU(2) flavour symmetry, the third component of isospin is an additive quantum number, in analogy with angular momentum. In SU(3) flavour symmetry, both $I_3$ and $Y$ are additive quantum numbers, which together specify the flavour content of a state. The light meson ($q\overline{q}$) states, formed from combinations of u, d and s quarks/antiquarks, can be constructed using this additive property to identify the extreme states within an SU(3) multiplet. Having identified the extreme states, the ladder operators can be used to obtain the full multiplet structure. The $I_3$ and $Y$ values for all nine possible combinations of a light quark and a light antiquark are shown in Figure 9.11. The pattern of states can be obtained quickly by drawing triangles corresponding to the antiquark multiplet centred on each of the three positions in the original quark multiplet (this is equivalent to adding the $I_3$ and $Y$ values for all nine combinations).

The states around the edge of the multiplet are uniquely defined in terms of their flavour content. The three physical states with $I_3 = Y = 0$ will be linear combinations of $u\overline{u}$, $d\overline{d}$ and $s\overline{s}$, however, they are not necessarily part of the same multiplet. The $I_3 = Y = 0$ states which are in the same multiplet as the $\{u\overline{s}, u\overline{d}, d\overline{u}, d\overline{s}, s\overline{u}, s\overline{d}\}$ states can be obtained using the ladder operators, as indicated in Figure 9.12,

$$T_+|d\overline{u}\rangle = |u\overline{u}\rangle - |d\overline{d}\rangle \quad \text{and} \quad T_-|u\overline{d}\rangle = |d\overline{d}\rangle - |u\overline{u}\rangle, \tag{9.32}$$

$$V_+|s\overline{u}\rangle = |u\overline{u}\rangle - |s\overline{s}\rangle \quad \text{and} \quad V_-|u\overline{s}\rangle = |s\overline{s}\rangle - |u\overline{u}\rangle, \tag{9.33}$$

$$U_+|s\overline{d}\rangle = |d\overline{d}\rangle - |s\overline{s}\rangle \quad \text{and} \quad U_-|d\overline{s}\rangle = |s\overline{s}\rangle - |d\overline{d}\rangle. \tag{9.34}$$

Of these six states, only two are linearly independent and therefore, of the three physical $I_3 = Y = 0$ states, it can be concluded that one must be in a different SU(3) multiplet. Hence, for the assumed SU(3) flavour symmetry, the $q\overline{q}$ flavour states are decomposed into an *octet* and a *singlet*. The singlet state $\psi_S$ is the linear

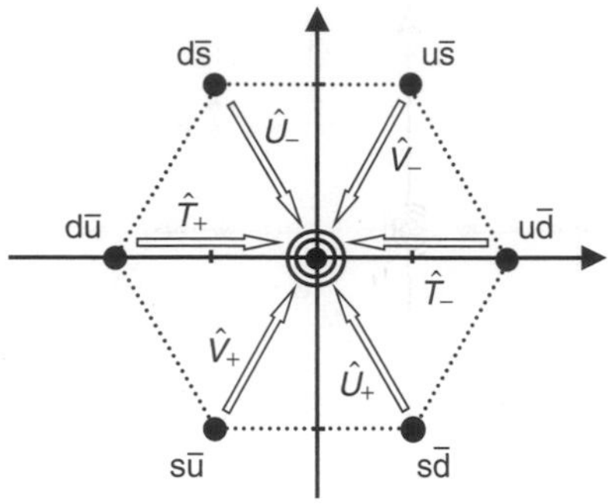


Fig. 9.12 Ladder operators applied to the $q\bar{q}$ states around the edge of the $I_3$, $Y$ diagram.

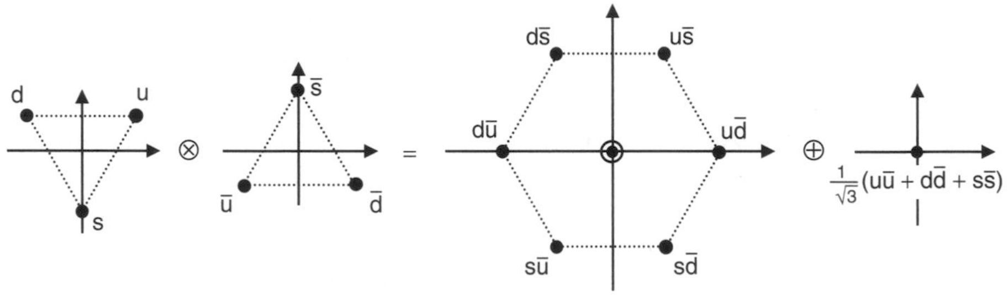


Fig. 9.13 SU(3) flavour $q\bar{q}$ multiplets. The two states at the centre of the octet are linear combinations of $|u\bar{u}\rangle$, $|d\bar{d}\rangle$ and $|s\bar{s}\rangle$ which are orthogonal to the singlet state.

combination of $u\bar{u}$, $d\bar{d}$ and $s\bar{s}$ that is orthogonal to the states of (9.32)–(9.34) and is readily identified as

$$|\psi_S\rangle = \frac{1}{\sqrt{3}}(u\bar{u} + d\bar{d} + s\bar{s}). \tag{9.35}$$

The application of the SU(3) ladder operators on $|\psi_S\rangle$ all give zero, for example

$$\begin{aligned} T_+\psi_S &= \frac{1}{\sqrt{3}}([T_+u]\bar{u} + u[T_+\bar{u}] + [T_+d]\bar{d} + d[T_+\bar{d}] + [T_+s]\bar{s} + s[T_+\bar{s}]) \\ &= \frac{1}{\sqrt{3}}(0 - u\bar{d} + u\bar{d} + 0 + 0 + 0) = 0, \end{aligned}$$

confirming that $|\psi_S\rangle$ is the singlet state.

Figure 9.13 shows the multiplet structure for combining a quark and an antiquark in SU(3) flavour symmetry. In the language of group theory, the combination of a quark 3 representation and an antiquark $\bar{3}$ representation decomposes into an octet and a singlet, $3 \otimes \bar{3} = 8 \oplus 1$. It worth pausing to consider the physical significance of the singlet state. For spin, the corresponding singlet state for the combination of two spin-half states, $|s, m\rangle = |0, 0\rangle$, is a state of zero angular momentum that carries

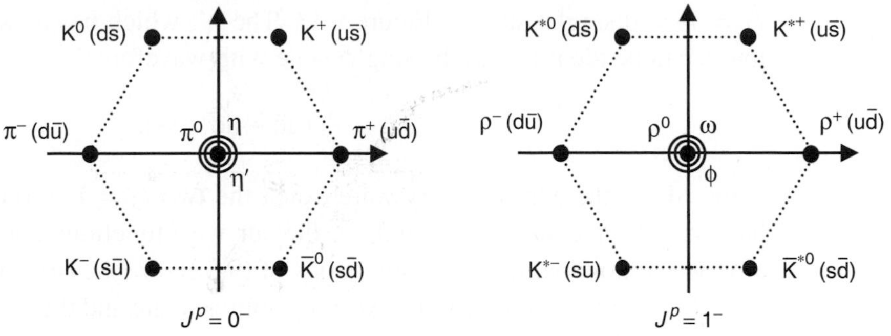


Fig. 9.14 The nine $\ell = 0, s = 0$ pseudoscalar mesons and nine $\ell = 0, s = 1$ vector mesons formed from the light quarks, plotted in terms of $I_3$ and $Y$.

no information about the spins of its constituent particles; it could just have been formed from two scalar particles. Similarly, the SU(3) flavour singlet $|\psi_S\rangle$ can be thought of as a "flavourless" state, carrying no information about the flavours of its constituents.

## The $L = 0$ mesons

In general, the wavefunction for a meson can be written in terms of four components,

$$\psi(\text{meson}) = \phi_{\text{flavour}}\,\chi_{\text{spin}}\,\xi_{\text{colour}}\,\eta_{\text{space}}.$$

Because quarks and antiquarks are distinguishable, there is no restriction on the exchange symmetry of the wavefunction for a $q\overline{q}$ state. For each flavour state, there are two possible spin states, $s = 0$ and $s = 1$. For the lightest mesons, which have zero orbital angular momentum ($\ell = 0$), the total angular momentum $\mathbf{J}$ is determined by the spin state alone. Consequently the lightest mesons divide into the $J = 0$ pseudoscalar mesons and the $J = 1$ (the vector mesons), respectively with $s = 0$ and $s = 1$. Since quarks and antiquarks have opposite *intrinsic* parities, the overall parity is given by

$$P(q\overline{q}) = P(q)P(\overline{q}) \times (-1)^{\ell} = (+1)(-1)(-1)^{\ell},$$

where $(-1)^{\ell}$ is the symmetry of the orbital wavefunction. Hence, the lightest mesons (with $\ell = 0$) have odd intrinsic parities. In Chapter 10, it is shown that there is only one possible colour wavefunction for a bound $q\overline{q}$ system. Therefore, there are nine light $J^P = 0^-$ pseudoscalar mesons and nine $J^P = 1^-$ light vector mesons, corresponding to nine possible flavour states each with two possible spin states.

Figure 9.14 shows the observed $\ell = 0$ meson states plotted in terms of $I_3$ and $Y$. The $\pi^0$, $\eta$ and $\eta'$ can be associated with the two $I_3 = Y = 0$ octet states and the

$I_3 = Y = 0$ singlet state of Figure 9.13. The $\eta'$, which has an anomalously large mass, can be identified as the singlet state with wavefunction

$$|\eta'\rangle \approx \frac{1}{\sqrt{3}}(u\bar{u} + d\bar{d} + s\bar{s}).$$

If the SU(3) flavour symmetry were exact, the two $I_3 = Y = 0$ octet states would have exactly the same mass and the flavour wavefunctions could be taken to be any two orthogonal linear combinations of (9.32)–(9.34). However, because $m_s > m_{u/d}$, the SU(3) flavour symmetry is only approximate and the choice of the flavour wavefunctions for the observed states will lead to different physical predictions. Experimentally, the lightest pseudoscalar mesons, namely the $\pi^+$, $\pi^0$ and $\pi^-$, are observed to have approximately the same mass of about 140 MeV. Since the $\pi^+$ and $\pi^-$ correspond to the $u\bar{d}$ and $d\bar{u}$ states, the $\pi^0$ can be identified as

$$|\pi^0\rangle = \frac{1}{\sqrt{2}}(u\bar{u} - d\bar{d}).$$

The final $I_3 = Y = 0$ pseudoscalar meson, the $\eta$, is the linear combination of $u\bar{u}$, $d\bar{d}$ and $s\bar{s}$ that is orthogonal to both the $|\eta'\rangle$ and the $|\pi^0\rangle$ states,

$$|\eta\rangle = \frac{1}{\sqrt{6}}(u\bar{u} + d\bar{d} - 2s\bar{s}).$$

In the case of the vector mesons, the predictions of the SU(3) flavour symmetry prove to be less useful; the physical $I_3 = Y = 0$ states are mixtures of the octet and singlet states. Experimentally, the observed states are found to correspond to

$$|\rho^0\rangle = \frac{1}{\sqrt{2}}(u\bar{u} - d\bar{d}),$$
$$|\omega\rangle \approx \frac{1}{\sqrt{2}}(u\bar{u} + d\bar{d}),$$
$$|\phi\rangle \approx s\bar{s}.$$

### 9.6.3 Meson masses

The measured masses of the $\ell = 0$ pseudoscalar and vector mesons are listed in Table 9.1. If the SU(3) flavour symmetry were exact, all the states in pseudoscalar meson octet would have the same mass. The observed mass differences can be ascribed to the fact that the strange quark is more massive than the up- and down-quarks. However, this does not explain why the vector mesons are more massive than their pseudoscalar counterparts. For example, the flavour wavefunctions for the $\pi$ and the $\rho$ states are the same, but their masses are very different. The only difference between the pseudoscalar and vector mesons is the spin wavefunction. Therefore, the different masses of the $\pi$ and $\rho$ mesons can be attributed to a spin–spin interaction.

**Table 9.1** The $L = 0$ pseudoscalar and vector meson masses.

| Pseudoscalar mesons | | Vector mesons | |
|---|---|---|---|
| $\pi^0$ | 135 MeV | $\rho^0$ | 775 MeV |
| $\pi^\pm$ | 140 MeV | $\rho^\pm$ | 775 MeV |
| $K^\pm$ | 494 MeV | $K^{*\pm}$ | 892 MeV |
| $K^0, \overline{K}^0$ | 498 MeV | $K^{*0}/\overline{K}^{*0}$ | 896 MeV |
| $\eta$ | 548 MeV | $\omega$ | 783 MeV |
| $\eta'$ | 958 MeV | $\phi$ | 1020 MeV |

In QED, the potential energy between two magnetic dipoles contains a term proportional to scalar product of the two dipole moments, $\boldsymbol{\mu}_i \cdot \boldsymbol{\mu}_j$. For two Dirac particles of masses $m_i$ and $m_j$, this corresponds to a potential energy term of the form

$$U \propto \frac{e}{m_i}\mathbf{S}_i \cdot \frac{e}{m_j}\mathbf{S}_j \propto \frac{\alpha}{m_i m_j}\mathbf{S}_i \cdot \mathbf{S}_j,$$

where $\alpha$ is the fine structure constant. This QED interaction term, which contributes to the hyperfine splitting of the energy levels of the hydrogen atom, is relatively small. In Chapter 10 it is shown that, apart from a numerical constant that accounts for colour, the QCD vertex has the same form as that of QED. Therefore, there will be a corresponding QCD "chromomagnetic" spin–spin interaction giving a term in the $q\overline{q}$ potential of the form

$$U \propto \frac{\alpha_S}{m_i m_j}\mathbf{S}_i \cdot \mathbf{S}_j,$$

where $\alpha_S$ is the coupling constant of QCD. Since $\alpha_S \sim 1$ is much greater than $\alpha \sim 1/137$, the chromomagnetic spin–spin interaction term is relatively large and plays an important role in determining the meson masses. For an $\ell = 0$ meson formed from a quark and an antiquark with masses $m_1$ and $m_2$, the meson mass can be written in terms of the constituent quark masses and the expectation value of the chromomagnetic spin–spin interaction

$$m(q_1 q_2) = m_1 + m_2 + \frac{A}{m_1 m_2}\langle \mathbf{S}_1 \cdot \mathbf{S}_2 \rangle, \tag{9.36}$$

where the parameter $A$ can be determined from experiment.

The scalar product $\mathbf{S}_1 \cdot \mathbf{S}_2$ in (9.36) can be obtained by writing the total spin as the vector sum, $\mathbf{S} = \mathbf{S}_1 + \mathbf{S}_2$, and squaring to give

$$\mathbf{S}^2 = \mathbf{S}_1^2 + 2\mathbf{S}_1 \cdot \mathbf{S}_2 + \mathbf{S}_2^2,$$

which implies that

$$\mathbf{S}_1 \cdot \mathbf{S}_2 = \tfrac{1}{2}\left[\mathbf{S}^2 - \mathbf{S_1}^2 - \mathbf{S_1}^2\right].$$

Therefore, the expectation value of $\mathbf{S}_1 \cdot \mathbf{S}_2$ can be written as

$$\begin{aligned}\langle \mathbf{S}_1 \cdot \mathbf{S}_2 \rangle &= \tfrac{1}{2}\left[\left\langle \mathbf{S}^2 \right\rangle - \left\langle \mathbf{S}_1^2 \right\rangle - \left\langle \mathbf{S}_2^2 \right\rangle\right] \\ &= \tfrac{1}{2}\left[s(s+1) - s_1(s_1+1) - s_2(s_2+1)\right],\end{aligned}$$

where $s_1 = s_2 = \frac{1}{2}$ and $s$ is the total spin of the $\mathrm{q\overline{q}}$ system. For the pseudoscalar mesons $s = 0$ and for the vector mesons $s = 1$ and hence (9.36) can be written

$$\text{Pseudoscalar mesons } (s = 0): \quad m_P = m_1 + m_2 - \frac{3A}{4m_1 m_2}, \tag{9.37}$$

$$\text{Vector mesons } (s = 1): \quad m_V = m_1 + m_2 + \frac{A}{4m_1 m_2}. \tag{9.38}$$

Hence the masses of the spin-0 pseudoscalar mesons are predicted to be lower than the masses of the spin-1 vector mesons. The observed meson masses listed in Table 9.1 are in good agreement with the predictions of the meson mass formulae of (9.37) and (9.38) with the parameters

$$m_\mathrm{d} = m_\mathrm{u} = 0.307\,\mathrm{GeV}, \quad m_\mathrm{s} = 0.490\,\mathrm{GeV} \quad \text{and} \quad A = 0.06\,\mathrm{GeV}^3. \tag{9.39}$$

The one exception is the $\eta'$, where the predicted mass of 355 MeV differs significantly from the anomalously large observed value of 958 MeV. The reason for this discrepancy is attributed to the $\eta'$ being a "flavourless" singlet state that can, in principle, mix with possible purely gluonic flavourless bound states.

### 9.6.4 The $L = 0$ uds baryons

The ground states of the (qqq) baryons are states with no orbital angular momentum in the system. Assuming SU(3) flavour symmetry, the wavefunctions for these $L = 0$ baryons can be obtained by first considering the multiplet structure for the combination of two quarks and then adding the third. This is essentially a repeat of the process used to derive the proton wavefunction in Section 9.4. Here, we will concentrate on the multiplet structure rather than the wavefunctions themselves.

Since $I_3$ and $Y$ are additive quantum numbers, the $(I_3, Y)$ values of the combination of two quarks in SU(3) are just the sums of the individual values. The multiplet structure for the combination of two quarks can be obtained by starting at one of the extreme SU(2) qq states and applying the SU(3) ladder operators. In this way it can be shown that in SU(3) flavour symmetry, the combination of two quarks leads to a symmetric sextet of states and an antisymmetric triplet of states, as shown in Figure 9.15. Since the triplet has the same $(I_3, Y)$ states as the SU(3) representation of a single antiquark, the multiplet structure arising from the combination of two quarks can be written as $3 \otimes 3 = 6 \oplus \overline{3}$.

The multiplet structure for the 27 possible flavour combinations in the qqq system is then obtained by adding a quark triplet to each of the sextet and triplet of

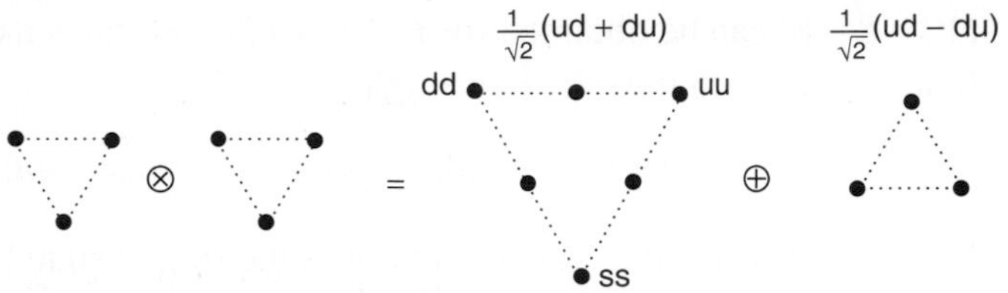


Fig. 9.15 The multiplet structure for the combination of qq in SU(3) flavour symmetry.

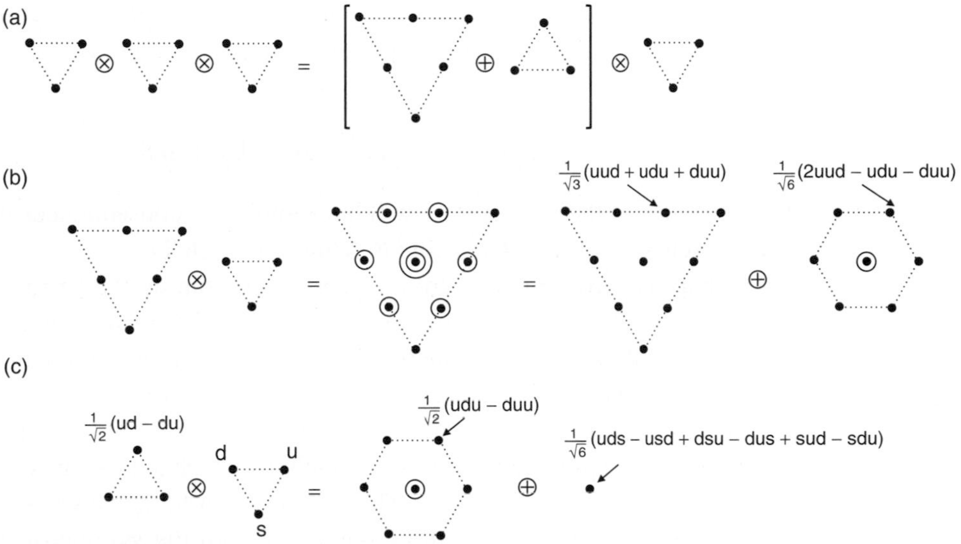


Fig. 9.16 The $I_3$ and $Y$ assignments for the qqq multiplets in SU(3) flavour symmetry broken down into the $6 \otimes 3$ and $\overline{3} \otimes 3$ parts (shown in *b* and *c*).

Figure 9.15. In terms of the group structure, this can be written $3{\otimes}3{\otimes}3 = (6{\oplus}\overline{3}){\otimes}3$ as indicated in Figure 9.16a. Adding an additional quark to the sextet, gives a decuplet of totally symmetric states and a mixed symmetry octet, as shown in Figure 9.16b, where the states without strange quarks are exactly those identified in Section 9.4. This $10 \oplus 8$ multiplet structure can be verified by repeated application of the SU(3) ladder operators to the SU(2) states of (9.8)–(9.11) and to the states of (9.12)–(9.13) to obtain respectively the decuplet and the mixed symmetry octet.

The second set of qqq flavour states are obtained by adding a quark to the qq triplet ($\overline{3}$). In terms of the multiplet structure, this is the same as combining the SU(3) representation of a quark and antiquark ($3 \otimes \overline{3}$), giving a mixed symmetry octet and a *totally antisymmetric* singlet state, as indicated in Figure 9.16c. The wavefunctions for this octet can be obtained from the corresponding SU(2) states of (9.14)–(9.15) using the SU(3) ladder operators. Hence, 26 of the possible states

(10 + 8 + 8) can be obtained from the SU(2) qqq states using ladder operators. The final state, which must be in a singlet, is

$$|\psi_S\rangle = \frac{1}{\sqrt{6}}(\text{uds} - \text{usd} + \text{dsu} - \text{dus} + \text{sud} - \text{sdu}). \tag{9.40}$$

It is straightforward to verify that this is the singlet state by showing that the action of all the SU(3) ladder operators give zero, for example

$$\hat{T}_+|\psi_S\rangle = \frac{1}{\sqrt{6}}(\text{uus} - \text{usu} + \text{usu} - \text{uus} + \text{suu} - \text{suu}) = 0.$$

In summary, the combination of three quarks in SU(3) flavour symmetry gives a symmetric decuplet, two mixed symmetry octets and a totally antisymmetric singlet state,

$$3 \otimes 3 \otimes 3 = 3 \otimes (6 \oplus \overline{3}) = 10 \oplus 8 \oplus 8 \oplus 1.$$

The existence of the singlet state will have important consequences when it comes to the discussion of the SU(3) colour symmetry of QCD.

The baryon wavefunctions are obtained by combining the SU(3) flavour wavefunctions with the spin wavefunctions of Section 9.3.1, respecting the requirement that the overall baryon wavefunction has to be antisymmetric under the exchange of any two of the quarks. Since the colour wavefunction is always antisymmetric and the $\ell = 0$ spatial wavefunction is symmetric, baryon states can be formed from combinations of spin and flavour wavefunctions which are totally symmetric under the interchange of any two quarks. This can be achieved in two ways. Firstly, a symmetric spin-$\frac{3}{2}$ wavefunction can be combined with the symmetric SU(3) flavour decuplet to give ten spin-$\frac{3}{2}$ baryons (including the Δ-particles). Secondly, as in (9.25), the mixed symmetry flavour octet states can be combined with the mixed symmetry spin states to give a spin-$\frac{1}{2}$ octet (including the proton and neutron). It is not possible to construct a totally symmetric *flavour* × *spin* wavefunction from the flavour singlet of (9.40) because there is no corresponding totally antisymmetric spin state formed from the combination three spin-half particles. The experimentally observed $L = 0$ baryons fit neatly into this SU(3) flavour symmetry prediction of an octet of spin-$\frac{1}{2}$ states and a decuplet of spin-$\frac{3}{2}$ states, as shown in Figure 9.17.

## Baryon masses

If the SU(3) flavour symmetry were exact, the masses of all the baryons within the octet would be the same, as would the masses of all the baryons within the decuplet. Because the strange-quark mass is greater than that of the up- and down-quarks, this is not the case. The measured masses of the $L = 0$ baryons are listed in Table 9.2. The patterns of masses within a multiplet largely reflects the number of strange quarks in the state, whereas the difference between the masses of the

**Table 9.2** Measured masses and number of strange quarks for the $L = 0$ light baryons.

| s quarks | Octet | | Decuplet | |
|---|---|---|---|---|
| 0 | p, n | 940 MeV | $\Delta$ | 1230 MeV |
| 1 | $\Sigma$ | 1190 MeV | $\Sigma^*$ | 1385 MeV |
| 1 | $\Lambda$ | 1120 MeV | | |
| 2 | $\Xi$ | 1320 MeV | $\Xi^*$ | 1533 MeV |
| 3 | | | $\Omega$ | 1670 MeV |

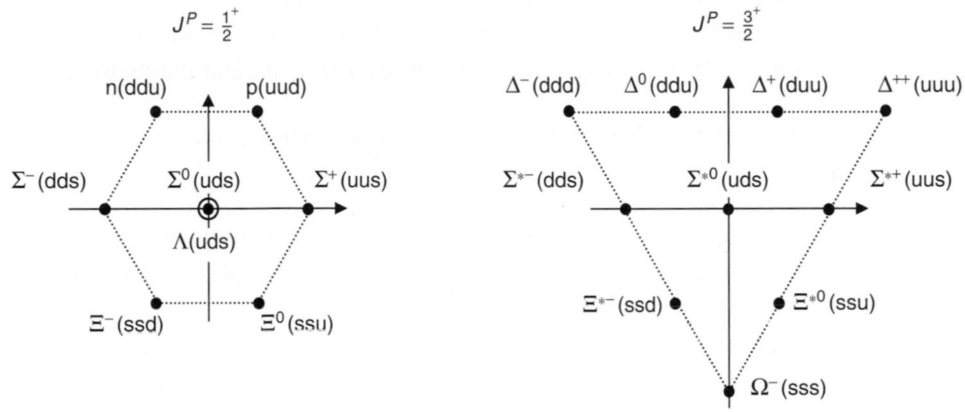


Fig. 9.17 The observed octet and decuplet of light baryon states.

octet and decuplet states is due to the chromomagnetic spin–spin interactions of the individual quarks. Following the argument presented in Section 9.6.3, the $L = 0$ baryon mass formula is

$$m(q_1q_2q_3) = m_1 + m_2 + m_3 + A'\left(\frac{\langle \mathbf{S}_1 \cdot \mathbf{S}_2\rangle}{m_1m_2} + \frac{\langle \mathbf{S}_1 \cdot \mathbf{S}_3\rangle}{m_1m_3} + \frac{\langle \mathbf{S}_2 \cdot \mathbf{S}_3\rangle}{m_2m_3}\right), \qquad (9.41)$$

where $\mathbf{S}_1$, $\mathbf{S}_2$ and $\mathbf{S}_3$ are the spin vectors of the three quarks. This expression is found to give good agreement with the observed baryon masses using

$$m_d = m_u = 0.365\,\text{GeV}, \quad m_s = 0.540\,\text{GeV} \quad \text{and} \quad A' = 0.026\,\text{GeV}^3.$$

It is important to note that the quark masses needed to explain the observed baryon masses are about 50 MeV higher than those used to describe the meson masses, as given in (9.39). Furthermore, they are very different from the fundamental up- and down-quark masses, known as the *current masses*, which are just a few MeV. The quark masses that enter the meson and baryon mass formulae are the *constituent masses*, which can be thought of as the effective masses of the quarks as they move within and interact with the QCD potential inside baryons and mesons. Since the QCD environments within baryons and mesons will be different, it should not be a surprise that the constituent masses are different for baryons and mesons. This distinction between current and constituent quark masses implies that only 1% of the mass of a proton is attributable to the masses of the quarks, the remainder arises from the energy associated with the internal QCD gluon field.

### 9.6.5 Baryon magnetic moments

In Chapter 7 it was seen that the magnetic moment of the proton differs from that expected for a point-like Dirac fermion. The experimentally measured values of the anomalous magnetic moments of the proton and neutron are $2.792\,\mu_N$ and $-1.913\,\mu_N$ respectively, where $\mu_N$ is the nuclear magneton defined as

$$\mu_N = \frac{e\hbar}{2m_\mathrm{p}}.$$

The origin of the proton and neutron anomalous magnetic moments can be explained in terms of the magnetic moments of the individual quarks and the baryon wavefunctions derived above.

Since quarks are fundamental Dirac fermions, the operators for the total magnetic moment and $z$-component of the magnetic moment are

$$\hat{\boldsymbol{\mu}} = Q\frac{e}{m}\hat{\mathbf{S}} \quad \text{and} \quad \hat{\mu}_z = Q\frac{e}{m}\hat{S}_z.$$

For spin-up ($m_s = +\frac{1}{2}$) quarks, the expectation values of the $z$-component of the magnetic moment of the up- and down-quarks are

$$\mu_\mathrm{u} = \langle \mathrm{u}\!\uparrow\!|\hat{\mu}_z|\,\mathrm{u}\!\uparrow\rangle = \left(+\tfrac{2}{3}\right)\frac{e\hbar}{2m_\mathrm{u}} = +\frac{2m_\mathrm{p}}{3m_\mathrm{u}}\mu_N, \tag{9.42}$$

$$\mu_\mathrm{d} = \langle \mathrm{d}\!\uparrow\!|\hat{\mu}_z|\,\mathrm{d}\!\uparrow\rangle = \left(-\tfrac{1}{3}\right)\frac{e\hbar}{2m_\mathrm{d}} = -\frac{m_\mathrm{p}}{3m_\mathrm{d}}\mu_N. \tag{9.43}$$

The corresponding expressions for the spin-down states are

$$\langle \mathrm{u}\!\downarrow\!|\hat{\mu}_z|\,\mathrm{u}\!\downarrow\rangle = -\mu_\mathrm{u} \quad \text{and} \quad \langle \mathrm{d}\!\downarrow\!|\hat{\mu}_z|\,\mathrm{d}\!\downarrow\rangle = -\mu_\mathrm{d}.$$

The total magnetic moment of a baryon is the vector sum of the magnetic moments of the three constituent quarks

$$\hat{\boldsymbol{\mu}} = \hat{\boldsymbol{\mu}}^{(1)} + \hat{\boldsymbol{\mu}}^{(2)} + \hat{\boldsymbol{\mu}}^{(3)},$$

where $\hat{\boldsymbol{\mu}}^{(i)}$ is the magnetic moment operator which acts on the $i$th quark. Therefore, the magnetic moment of the proton can be written

$$\mu_\mathrm{p} = \langle\hat{\mu}_z\rangle = \langle \mathrm{p}\!\uparrow\!|\hat{\mu}_z^{(1)} + \hat{\mu}_z^{(2)} + \hat{\mu}_z^{(3)}\,|\mathrm{p}\!\uparrow\rangle. \tag{9.44}$$

The order that the quarks appear in the proton wavefunction does not affect the calculation of the magnetic moment and it is sufficient to write

$$|\mathrm{p}\uparrow\rangle = \frac{1}{\sqrt{6}}(2\mathrm{u}\uparrow\mathrm{u}\uparrow\mathrm{d}\downarrow - \mathrm{u}\uparrow\mathrm{u}\downarrow\mathrm{d}\uparrow - \mathrm{u}\downarrow\mathrm{u}\uparrow\mathrm{d}\uparrow),$$

and thus (9.44) can be written as

$$\mu_\mathrm{p} = \tfrac{1}{6}\,\langle(2\mathrm{u}\uparrow\mathrm{u}\uparrow\mathrm{d}\downarrow - \mathrm{u}\uparrow\mathrm{u}\downarrow\mathrm{d}\uparrow - \mathrm{u}\downarrow\mathrm{u}\uparrow\mathrm{d}\uparrow)\,|\hat{\mu}_z|\,(2\mathrm{u}\uparrow\mathrm{u}\uparrow\mathrm{d}\downarrow - \mathrm{u}\uparrow\mathrm{u}\downarrow\mathrm{d}\uparrow - \mathrm{u}\downarrow\mathrm{u}\uparrow\mathrm{d}\uparrow)\rangle\,,$$

where $\hat{\mu}_z = \hat{\mu}_z^{(1)} + \hat{\mu}_z^{(2)} + \hat{\mu}_z^{(3)}$. Because of the orthogonality of the quark flavour and spin states, for example $\langle \mathrm{u}\uparrow\mathrm{u}\uparrow\mathrm{d}\downarrow\,|\,\mathrm{u}\downarrow\mathrm{u}\uparrow\mathrm{d}\uparrow\rangle = 0$, the expression for the proton magnetic moment reduces to

$$\begin{aligned}\mu_\mathrm{p} = {} & \tfrac{4}{6}\,\langle\mathrm{u}\uparrow\mathrm{u}\uparrow\mathrm{d}\downarrow|\hat{\mu}_z|\mathrm{u}\uparrow\mathrm{u}\uparrow\mathrm{d}\downarrow\rangle + \tfrac{1}{6}\,\langle\mathrm{u}\uparrow\mathrm{u}\downarrow\mathrm{d}\uparrow|\hat{\mu}_z|\mathrm{u}\uparrow\mathrm{u}\downarrow\mathrm{d}\uparrow\rangle \\ & + \tfrac{1}{6}\,\langle\mathrm{u}\downarrow\mathrm{u}\uparrow\mathrm{d}\uparrow|\hat{\mu}_z|\mathrm{u}\downarrow\mathrm{u}\uparrow\mathrm{d}\uparrow\rangle\,.\end{aligned} \tag{9.45}$$

Equation (9.45) can be evaluated using

$$\begin{aligned}\hat{\mu}_z|\mathrm{u}\uparrow\rangle &= +\mu_\mathrm{u}\,|\mathrm{u}\uparrow\rangle \quad \text{and} \quad \hat{\mu}_z|\mathrm{u}\downarrow\rangle = -\mu_\mathrm{u}\,|\mathrm{u}\downarrow\rangle,\\ \hat{\mu}_z|\mathrm{d}\uparrow\rangle &= +\mu_\mathrm{d}\,|\mathrm{d}\uparrow\rangle \quad \text{and} \quad \hat{\mu}_z|\mathrm{d}\downarrow\rangle = -\mu_\mathrm{d}\,|\mathrm{d}\downarrow\rangle,\end{aligned}$$

giving

$$\mu_\mathrm{p} = \tfrac{4}{6}\,(\mu_\mathrm{u} + \mu_\mathrm{u} - \mu_\mathrm{d}) + \tfrac{1}{6}\,(\mu_\mathrm{u} - \mu_\mathrm{u} + \mu_\mathrm{d}) + \tfrac{1}{6}\,(-\mu_\mathrm{u} + \mu_\mathrm{u} + \mu_\mathrm{d})\,.$$

Therefore, the quark model prediction for the magnetic moment of the proton is

$$\mu_\mathrm{p} = \tfrac{4}{3}\mu_\mathrm{u} - \tfrac{1}{3}\mu_\mathrm{d}.$$

The prediction for the magnetic moment of the neutron can be written down by replacing $\mathrm{u} \to \mathrm{d}$ and vice versa,

$$\mu_\mathrm{n} = \tfrac{4}{3}\mu_\mathrm{d} - \tfrac{1}{3}\mu_\mathrm{u}.$$

Assuming that $m_\mathrm{u} \approx m_\mathrm{d}$, the relations of (9.42) and (9.43) imply that $\mu_\mathrm{u} = -2\mu_\mathrm{d}$. Consequently, the ratio of the proton and neutron magnetic moments is predicted to be

$$\frac{\mu_\mathrm{p}}{\mu_\mathrm{n}} = \frac{4\mu_\mathrm{u} - \mu_\mathrm{d}}{4\mu_\mathrm{d} - \mu_\mathrm{u}} = -\frac{3}{2},$$

which is in reasonable agreement with the experimentally measured value of $-1.46$. The best agreement between the quark model predictions and the measured values of the magnetic moments of the $L = 0$ baryons is obtained with

$$m_\mathrm{u} = 0.338\,\mathrm{GeV}, \quad m_\mathrm{d} = 0.322\,\mathrm{GeV} \quad \text{and} \quad m_\mathrm{s} = 0.510\,\mathrm{GeV}.$$

Using these values in (9.42) and (9.43) gives $\mu_\mathrm{u} = +1.85\mu_N$ and $\mu_\mathrm{d} = -0.97\mu_N$, reproducing the observed values of the proton and neutron magnetic moments.

### 9.6.6 Final words on SU(3) flavour symmetry

Whilst the SU(3) flavour symmetry is only approximate, it is able to account for the observed states of the $L = 0$ mesons and baryons. Furthermore, the hadron wavefunctions derived in the context of SU(3) flavour symmetry can be used to obtain reasonable predictions for baryon and meson masses and the baryon magnetic moments. If anything, it is perhaps surprising that the predictions from SU(3) flavour symmetry give such reasonable results. After all, the SU(3) flavour symmetry can be only approximate because the mass of the strange quark is about 0.1 GeV greater than the masses of the up- and down-quarks, although this mass difference is relatively small compared to the typical QCD binding energy which is of order 1 GeV. A further issue with the static quark model is that the hadronic states have been treated as bound states of valence quarks, whereas from the discussion of deep inelastic scattering it is clear that hadrons are far more complex. To some extent, these additional degrees of freedom are accounted for in the *constituent masses* of the quarks used to obtain the predictions for meson and baryon masses and the baryon magnetic moments. These masses are much larger than the current masses listed in Table 1.1; most of the mass of the hadrons originates from of the energy of the strongly interacting sea of virtual quarks and gluons.

The above discussion was restricted to the approximate SU(3) flavour symmetry of the three light quarks. It is tempting to extend this treatment to an SU(4) flavour symmetry including the charm quark. However, this makes little sense; the difference between the mass of the charm quark and the light quarks is greater than 1 GeV, which is the typical QCD binding energies of hadrons. For this reason, the Hamiltonian for the hadronic states does not possess even an approximate SU(4) flavour symmetry.

## Summary

In this chapter a number of important concepts were introduced. Symmetries of the Hamiltonian were associated with unitary transformations expressed in terms of Hermitian generators

$$\hat{U}(\alpha) = \exp(i\alpha \cdot \hat{G}).$$

In this way, each symmetry of the Hamiltonian is associated with an observable conserved quantity.

The flavour symmetry of the static quark model was used to illustrate these ideas and to introduce the SU(2) and SU(3) groups. Based on symmetry arguments alone, it was possible to derive static wavefunctions for the mesons and baryons formed from u, d and s quarks. The static quark model was shown to provide a good description of the masses and magnetic moments of the light hadrons. In the following chapter, these ideas will be extended to the abstract SU(3) local gauge symmetry that lies at the heart of QCD.

## 9.7 *Addendum: Flavour symmetry revisited

In the derivation of the proton wavefunction, given in Section 9.4, the overall wavefunction,

$$\psi = \phi_{\text{flavour}}\chi_{\text{spin}}\,\xi_{\text{colour}}\,\eta_{\text{space}},$$

was required to be antisymmetric. For cases where the flavour wavefunction describes like particles, for example $\phi_{\text{flavour}} = \text{uuu}$, the requirement of an overall antisymmetric wavefunction is just an expression of Pauli exclusion principle, which arises from the spin-statistics of fermions. It is less obvious why this should also apply to the more general case with different quark flavours; the reasoning is subtle.

In quantum field theory an up-quark state with spin $r$ is expressed by the action of the creation operator $a^{\dagger}_{+r}$ on the vacuum state,

$$|\text{u}\uparrow\rangle = a^{\dagger}_{+r}|0\rangle,$$

where the + sign refers to the creation of the $I_3 = +\frac{1}{2}$ state labelling an up-quark in SU(2) flavour symmetry. The creation operator $a^{\dagger}_{+r}$ satisfies the requirements of fermion spin statistics, which can be written as the anticommutator

$$\{a^{\dagger}_{+r}, a^{\dagger}_{+r}\} = 0,$$

which implies

$$a^{\dagger}_{+r}a^{\dagger}_{+r}|0\rangle = 0, \tag{9.46}$$

and therefore two identical particles can not be produced in the same state. For the SU(2) isospin flavour symmetry $\hat{T}_-|\text{u}\uparrow\rangle = |\text{d}\uparrow\rangle$, which implies

$$\hat{T}_-a^{\dagger}_{+r}|0\rangle = a^{\dagger}_{-r}|0\rangle,$$

where $a^{\dagger}_{-r}$ is the creation operator for a spin-up down quark with $I_3 = -\frac{1}{2}$. Therefore one can write $\hat{T}_-a^{\dagger}_{+r} = a^{\dagger}_{-r}$. Applying the isospin lowering operator to (9.46) gives

$$\hat{T}_-(a^{\dagger}_{+r}a^{\dagger}_{+r})|0\rangle = a^{\dagger}_{-r}a^{\dagger}_{+r}|0\rangle + a^{\dagger}_{+r}a^{\dagger}_{-r}|0\rangle = 0,$$

and hence

$$\{a_{+r}^{\dagger}, a_{-r}^{\dagger}\} = 0.$$

Therefore, within the assumed SU(2) flavour symmetry, the creation operators for up- and down-quarks satisfy the same anticommutation relations as the creation operators for two up-quarks or two down-quarks. Consequently, within the SU(2) or SU(3) flavour symmetries, the requirement that the overall wavefunction is anti-symmetric applies equally to states where the flavours of the quarks are different.

## Problems

**9.1** By writing down the general term in the binomial expansion of

$$\left(1 + i\frac{1}{n}\alpha \cdot \hat{\mathbf{G}}\right)^n,$$

show that

$$\hat{U}(\alpha) = \lim_{n\to\infty}\left(1 + i\frac{1}{n}\alpha \cdot \hat{\mathbf{G}}\right)^n = \exp\left(i\alpha \cdot \mathbf{G}\right).$$

**9.2** For an infinitesimal rotation about the $z$-axis through an angle $\epsilon$ show that

$$\hat{U} = 1 - i\epsilon\hat{J}_z,$$

where $\hat{J}_z$ is the angular momentum operator $\hat{J}_z = x\hat{p}_y - y\hat{p}_x$.

**9.3** By considering the isospin states, show that the rates for the following strong interaction decays occur in the ratios

$$\Gamma(\Delta^- \to \pi^- n) : \Gamma(\Delta^0 \to \pi^- p) : \Gamma(\Delta^0 \to \pi^0 n) : \Gamma(\Delta^+ \to \pi^+ n) :$$
$$\Gamma(\Delta^+ \to \pi^0 p) : \Gamma(\Delta^{++} \to \pi^+ p) = 3:1:2:1:2:3.$$

**9.4** If quarks and antiquarks were spin-zero particles, what would be the multiplicity of the $L = 0$ multiplet(s). Remember that the overall wavefunction for bosons must be symmetric under particle exchange.

**9.5** The neutral vector mesons can decay leptonically through a virtual photon, for example by $V(q\bar{q}) \to \gamma \to e^+e^-$. The matrix element for this decay is proportional to $\langle\psi|\hat{Q}_q|\psi\rangle$, where $\psi$ is the meson flavour wavefunction and $\hat{Q}_q$ is an operator that is proportional to the quark charge. Neglecting the relatively small differences in phase space, show that

$$\Gamma(\rho^0 \to e^+e^-) : \Gamma(\omega \to e^+e^-) : \Gamma(\phi \to e^+e^-) \approx 9:1:2.$$

**9.6** Using the meson mass formulae of (9.37) and (9.38), obtain predictions for the masses of the $\pi^\pm$, $\pi^0$, $\eta$, $\eta'$, $\rho^0$, $\rho^\pm$, $\omega$ and $\phi$. Compare the values obtained to the experimental values listed in Table 9.1.

**9.7** Compare the experimentally measured values of the masses of the $J^P = \frac{3}{2}^+$ baryons, given in Table 9.2, with the predictions of (9.41). You will need to consider the combined spin of *any* two quarks in a spin-$\frac{3}{2}$ baryon state.

**9.8** Starting from the wavefunction for the $\Sigma^-$ baryon:

(a) obtain the wavefunction for the $\Sigma^0$ and therefore find the wavefunction for the $\Lambda$;
(b) using (9.41), obtain predictions for the masses of the $\Sigma^0$ and the $\Lambda$ baryons and compare these to the measured values.

**9.9** Show that the quark model predictions for the magnetic moments of the $\Sigma^+$, $\Sigma^-$ and $\Omega^-$ baryons are

$$\mu(\Sigma^+) = \tfrac{1}{3}\left(4\mu_u - \mu_s\right), \quad \mu(\Sigma^-) = \tfrac{1}{3}\left(4\mu_d - \mu_s\right) \quad \text{and} \quad \mu(\Omega^-) = 3\mu_s.$$

What values of the quark constituent masses are required to give the best agreement with the measured values of

$$\mu(\Sigma^+) = (2.46 \pm 0.01)\mu_N, \quad \mu(\Sigma^-) = (-1.16 \pm 0.03)\mu_N \quad \text{and}$$
$$\mu(\Omega^-) = (-2.02 \pm 0.06)\mu_N?$$

**9.10** If the colour did not exist, baryon wavefunctions would be constructed from

$$\psi = \phi_{\text{flavour}}\,\chi_{\text{spin}}\,\eta_{\text{space}}.$$

Taking $L = 0$ and using the flavour and spin wavefunctions derived in the text:

(a) show that it is still possible to construct a wavefunction for a spin-up proton for which $\phi_{\text{flavour}}\,\chi_{\text{spin}}$ is totally antisymmetric;
(b) predict the baryon multiplet structure for this model;
(c) for this colourless model, show that $\mu_p$ is negative and that the ratio of the neutron and proton magnetic moments would be

$$\frac{\mu_n}{\mu_p} = -2.$$

# 10 Quantum Chromodynamics (QCD)

This chapter provides an introduction to the theory of Quantum Chromodynamics (QCD). Firstly, the concept of a local gauge symmetry is described and then used to obtain the form of the QCD interaction. Superficially QCD appears like a stronger version of QED with eight gluons replacing the single photon, but because the gluons carry the charge of the interaction, QCD behaves very differently. A number of important topics are discussed including colour confinement, hadronisation, renormalisation, running coupling constants and colour factors. The last part of chapter provides an introduction to hadron–hadron collisions at the Tevatron and the LHC.

## 10.1 The local gauge principle

Gauge invariance is a familiar idea from electromagnetism, where the physical **E** and **B** fields, which are obtained from the scalar and vector potentials $\phi$ and **A**, do not change under the gauge transformation

$$\phi \to \phi' = \phi - \frac{\partial \chi}{\partial t} \quad \text{and} \quad \mathbf{A} \to \mathbf{A}' = \mathbf{A} + \boldsymbol{\nabla}\chi.$$

This gauge transformation can be written more succinctly as

$$A_\mu \to A'_\mu = A_\mu - \partial_\mu \chi, \tag{10.1}$$

where $A_\mu = (\phi, -\mathbf{A})$ and $\partial_\mu = (\partial_0, \boldsymbol{\nabla})$.

In relativistic quantum mechanics, the gauge invariance of electromagnetism can be related to a local gauge principle. Suppose there is a fundamental symmetry of the Universe that requires the invariance of physics under *local* phase transformations defined by

$$\psi(x) \to \psi'(x) = \hat{U}(x)\psi(x) = e^{iq\chi(x)}\psi(x). \tag{10.2}$$

This is similar to the U(1) global phase transformation of $\psi \to \psi' = e^{i\phi}\psi$ of (9.3), but here the phase $q\chi(x)$ can be different at all points in space-time. For this local U(1) phase transformation, the free-particle Dirac equation

$$i\gamma^\mu \partial_\mu \psi = m\psi, \tag{10.3}$$

becomes

$$\begin{aligned} i\gamma^\mu \partial_\mu (e^{iq\chi(x)}\psi) &= m e^{iq\chi(x)}\psi \\ \Rightarrow \quad e^{iq\chi}\, i\gamma^\mu \left[\partial_\mu \psi + iq(\partial_\mu \chi)\psi\right] &= e^{iq\chi} m\psi \\ i\gamma^\mu (\partial_\mu + iq\partial_\mu \chi)\psi &= m\psi, \end{aligned} \tag{10.4}$$

which differs from (10.3) by the term $-q\gamma^\mu(\partial_\mu \chi)\psi$. Hence, as it stands, the *free-particle* Dirac equation does not possess the hypothesised invariance under a U(1) local phase transformation. More generally, local phase invariance is not possible for a free theory, i.e. one without interactions. The required invariance can be established only by modifying the Dirac equation to include a new degree of freedom $A_\mu$ such that the original form of the Dirac equation of (10.3) becomes

$$i\gamma^\mu (\partial_\mu + iqA_\mu)\psi - m\psi = 0, \tag{10.5}$$

where $A_\mu$ will be interpreted as the field corresponding to a massless gauge boson. Equation (10.5) is invariant under the local phase transformation defined in (10.2) provided $A_\mu$ transforms as

$$A_\mu \to A'_\mu = A_\mu - \partial_\mu \chi,$$

in order to cancel the unwanted $-q\gamma^\mu(\partial_\mu \chi)\psi$ term in (10.4). Stating this another way, for physical predictions to remain unchanged under a local U(1) phase transformation, it is necessary to introduce a new field that exhibits the *observed* gauge invariance of classical electromagnetism, as given in (10.1). More significantly, the modified Dirac equation of (10.5) no longer corresponds to a wave equation for a free particle, there is now an interaction term of the form

$$q\gamma^\mu A_\mu \psi. \tag{10.6}$$

This is identical to the QED interaction term of (5.13) which was previously identified using minimal substitution.

The requirement that physics is invariant under local U(1) phase transformations implies the existence of a *gauge field* which couples to Dirac particles in exactly the same way as the photon. This is a profound statement; all of QED, including ultimately Maxwell's equations, can be derived by requiring the invariance of physics under local U(1) transformations of the form $\hat{U} = e^{iq\chi(x)}$.

### 10.1.1 From QED to QCD

Quantum Electrodynamics (QED) corresponds to a U(1) local gauge symmetry of the Universe. The underlying symmetry associated with Quantum Chromodynamics (QCD), which is the Quantum Field Theory of the strong interaction, is invariance under SU(3) local phase transformations,

$$\psi(x) \to \psi'(x) = \exp\left[ig_S\,\boldsymbol{\alpha}(x)\cdot\hat{\mathbf{T}}\right]\psi(x). \tag{10.7}$$

Here $\hat{\mathbf{T}} = \{T^a\}$ are the eight generators of the SU(3) symmetry group, which are related to the Gell-Mann matrices of (9.31) by

$$T^a = \tfrac{1}{2}\lambda^a,$$

and $\alpha^a(x)$ are eight functions of the space-time coordinate $x$. Because the generators of SU(3) are represented by $3\times3$ matrices, the wavefunction $\psi$ must now include three additional degrees of freedom that can be represented by a three component vector analogous to the representation of the u, d and s quarks in SU(3) flavour symmetry. This new degree of freedom is termed "colour" with red, blue and green labelling the states. The SU(3) local phase transformation corresponds to "rotating" states in this colour space about an axis whose direction is different at every point in space-time. For the local gauge transformation of (10.7), the Dirac equation becomes

$$i\gamma^\mu\left[\partial_\mu + ig_S(\partial_\mu\boldsymbol{\alpha})\cdot\hat{\mathbf{T}}\right]\psi = m\psi. \tag{10.8}$$

The required local gauge invariance can be asserted by introducing eight new fields $G^a_\mu(x)$, where the index $a = 1, \ldots, 8$, each corresponding to one of the eight generators of the SU(3) symmetry. The Dirac equation, including the interactions with the new gauge fields,

$$i\gamma^\mu\left[\partial_\mu + ig_S G^a_\mu T^a\right]\psi - m\psi = 0, \tag{10.9}$$

is invariant under local SU(3) phase transformations provided the new fields transform as

$$G^k_\mu \to G^{k\,\prime}_\mu = G^k_\mu - \partial_\mu\alpha_k - g_S f_{ijk}\alpha_i G^j_\mu. \tag{10.10}$$

The last term in (10.10) arises because the generators of the SU(3) symmetry do not commute and the $f_{ijk}$ are the *structure constants* of the SU(3) group, defined by the commutation relations $[\lambda_i, \lambda_j] = 2if_{ijk}\lambda_k$. Because the generators SU(3) do not commute, QCD is known as a non-Abelian gauge theory and the presence of the additional term in (10.10) gives rise to gluon self-interactions (see Appendix F). The mathematical forms of these triple and quartic gluon vertices, shown in Figure 10.1, are completely specified by the SU(3) gauge symmetry. Putting aside these self-interactions for now, the required SU(3) local gauge invariance necessitates the modification of the Dirac equation to include new interaction terms, one for each of the eight generators of the gauge symmetry. The eight new fields $G^a$ are

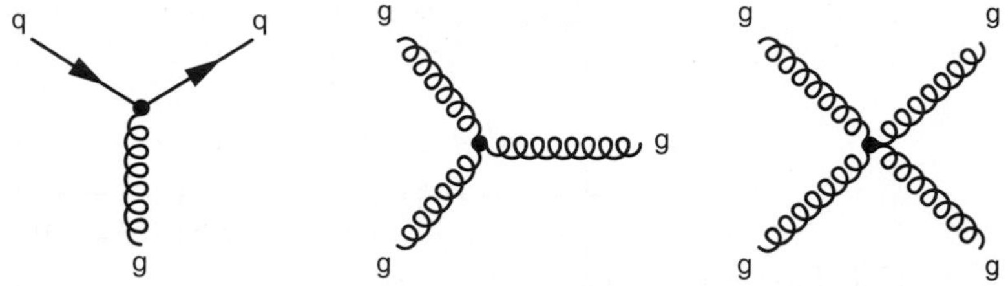


Fig. 10.1 The predicted QCD interaction vertices arising from the requirement of SU(3) local gauge invariance.

the gluons of QCD and from (10.9) it can be seen that the form of qqg interaction vertex is

$$g_S T^a \gamma^\mu G^a_\mu \psi = g_S \tfrac{1}{2}\lambda^a \gamma^\mu G^a_\mu \psi. \tag{10.11}$$

## 10.2 Colour and QCD

The underlying theory of quantum chromodynamics appears to be very similar to that of QED. The QED interaction is mediated by a massless photon corresponding to the single generator of the U(1) local gauge symmetry, whereas QCD is mediated by eight massless gluons corresponding to the eight generators of the SU(3) local gauge symmetry. The single charge of QED is replaced by three conserved "colour" charges, $r$, $b$ and $g$ (where colour is simply a label for the orthogonal states in the SU(3) colour space). Only particles that have non-zero colour charge couple to gluons. For this reason the leptons, which are colour neutral, do not feel the strong force. The quarks, which carry colour charge, exist in three orthogonal colour states. Unlike the approximate SU(3) flavour symmetry, discussed in Chapter 9, the SU(3) colour symmetry is exact and QCD is invariant under unitary transformations in colour space. Consequently, the strength of QCD interaction is independent of the colour charge of the quark. In QED the antiparticles have the opposite electric charge to the particles. Similarly, in QCD the antiquarks carry the opposite colour charge to the quarks, $\bar{r}$, $\bar{g}$ and $\bar{b}$.

The three colour states of QCD can be represented by colour wavefunctions,

$$r = \begin{pmatrix} 1 \\ 0 \\ 0 \end{pmatrix}, \quad g = \begin{pmatrix} 0 \\ 1 \\ 0 \end{pmatrix} \quad \text{and} \quad b = \begin{pmatrix} 0 \\ 0 \\ 1 \end{pmatrix}.$$

Following the discussion of SU(3) flavour symmetry in Chapter 9, the colour states of quarks and antiquarks can be labelled by two additive quantum numbers, the

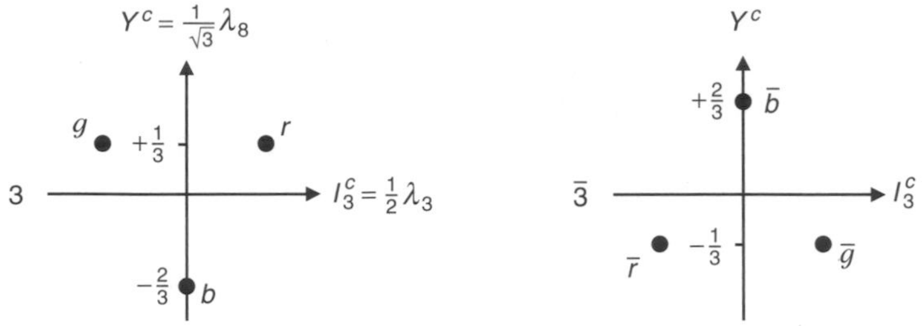


Fig. 10.2 The representations of the colour of quarks and the anticolour of antiquarks.

third component of colour isospin $I_3^c$ and colour hypercharge $Y^c$ as indicated in Figure 10.2.

### 10.2.1 The quark–gluon vertex

The SU(3) local gauge symmetry of QCD implies a conserved colour charge and an interaction between quarks and gluons of the form given by (10.11). By comparing the QCD interaction term to that for QED given in (10.6),

$$-iq\gamma^\mu A_\mu \psi \rightarrow -ig_S \tfrac{1}{2}\lambda^a \gamma^\mu G_\mu^a \psi,$$

the QCD vertex factor can be identified as

$$-iq\gamma^\mu \rightarrow -ig_S \gamma^\mu \tfrac{1}{2}\lambda^a.$$

Apart from the different coupling constant, the quark–gluon interaction only differs from the QED interaction in the appearance of the $3 \times 3$ Gell-Mann matrices that only act on the colour part of the quark wavefunction. The quark wavefunctions therefore need to include this colour degree of freedom. This can be achieved by writing

$$u(p) \rightarrow c_i u(p),$$

where $u(p)$ is a Dirac spinor and $c_i$ represents one of the possible colour states

$$c_1 = r = \begin{pmatrix} 1 \\ 0 \\ 0 \end{pmatrix}, \quad c_2 = g = \begin{pmatrix} 0 \\ 1 \\ 0 \end{pmatrix} \quad \text{and} \quad c_3 = b = \begin{pmatrix} 0 \\ 0 \\ 1 \end{pmatrix}.$$

Consequently, the quark current associated with the QCD vertex, shown in Figure 10.3, can be written

$$j_q^\mu = \bar{u}(p_3) c_j^\dagger \left\{-\tfrac{1}{2} i g_S \lambda^a \gamma^\mu\right\} c_i u(p_1), \tag{10.12}$$

where the $c_i$ and $c_j$ are the colour wavefunctions of the quarks and the index $a$ refers to gluon corresponding to the SU(3) generator $T^a$. (In other textbooks you may see the colour index appended to the spinor $c_i u(p) \rightarrow u_i(p)$.)

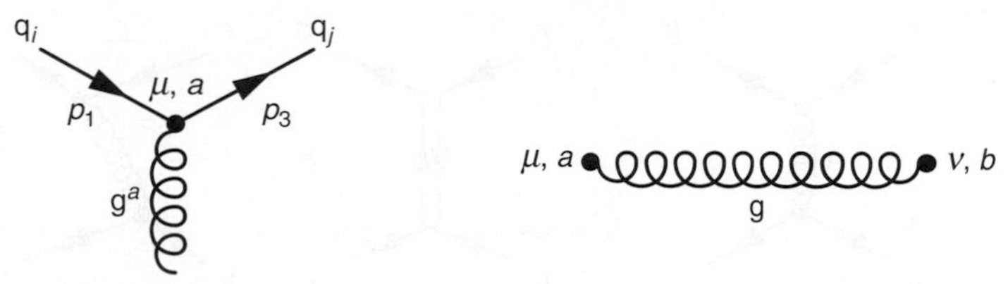


**Fig. 10.3** The QCD quark–gluon vertex representing the interaction of quarks with colours *i* and *j* with a gluon of type *a* and the gluon propagator.

In the quark current of (10.12), the $3 \times 3$ Gell-Mann matrix $\lambda^a$ acts on the three-component colour wavefunction, whereas the $4 \times 4$ $\gamma$-matrices act on the four components of the Dirac spinor. Therefore the colour part of the current factorises, allowing (10.12) to be written as

$$\overline{u}(p_3)c_j^{\dagger}\{-\tfrac{1}{2}ig_s\lambda^a\gamma^{\mu}\}c_iu(p_1) = -\tfrac{1}{2}ig_s\left[c_j^{\dagger}\lambda^a c_i\right] \times [\overline{u}(p_3)\gamma^{\mu}u(p_1)].$$

The factorised colour part of the interaction is

$$c_j^{\dagger}\lambda^a c_i = c_j^{\dagger}\begin{pmatrix}\lambda^a_{1i}\\ \lambda^a_{2i}\\ \lambda^a_{3i}\end{pmatrix} = \lambda^a_{ji}.$$

Hence the qqg vertex can be written as

$$-\tfrac{1}{2}ig_S\,\lambda^a_{ji}\,[\overline{u}(p_3)\gamma^{\mu}u(p_1)],$$

where $\lambda^a_{ji}$ is just a number, namely the $ji$th element of $\lambda^a$. Therefore, the Feynman rule associated with the QCD vertex is

$$-\tfrac{1}{2}ig_S\,\lambda^a_{ji}\gamma^{\mu}.$$

For lowest-order diagrams, the Feynman rule for the gluon propagator of Figure 10.3 is

$$-i\frac{g_{\mu\nu}}{q^2}\delta^{ab},$$

where the delta-function ensures that the gluon of type $a$ emitted at the vertex labelled $\mu$ is the same as that which is absorbed at vertex $\nu$.

## 10.3 Gluons

The QCD interaction vertex includes a factor $\lambda^a_{ji}$, where $i$ and $j$ label the colours of the quarks. Consequently, gluons corresponding to the non-diagonal Gell-Mann

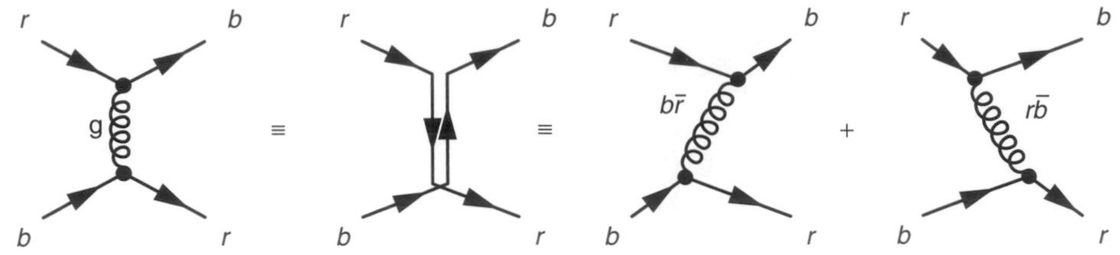


Fig. 10.4 Colour flow for the $t$-channel process $rb \to br$. Shown as the Feynman diagram, the colour flow and the two time-ordered diagrams.

matrices connect quark states of different colour. In order for colour to be conserved at the interaction vertex, the gluons must carry colour charge. For example, the gluon corresponding to $\lambda_4$, defined in (9.31), which has non-zero entries in the 13 and 31 positions, contributes to interactions involving the changes of colour $r \to b$ and $b \to r$. This is illustrated in Figure 10.4, which shows the QCD process of qq → qq scattering where the colour flow corresponds to $br \to rb$, illustrated both in terms the colour flow in the Feynman diagram and as the two corresponding time-ordered diagrams. Because colour is a conserved charge, the interaction involves the exchange of a $b\bar{r}$ gluon in the first time-ordering and a $r\bar{b}$ gluon in the second time-ordering. From this discussion, it is clear that gluons must carry simultaneously both colour charge and anticolour charge.

Since gluons carry a combination of colour and anticolour, there are six gluons with different colour and anticolour, $r\bar{g}$, $g\bar{r}$, $r\bar{b}$, $b\bar{r}$, $g\bar{b}$ and $b\bar{g}$. Naïvely one might expect three gluons corresponding to $r\bar{r}$, $g\bar{g}$ and $b\bar{b}$. However, the physical gluons correspond to the fields associated with the generators $\lambda_{1,..,8}$ of the SU(3) gauge symmetry. The gluons are therefore an octet of coloured states, analogous to the $q\bar{q}$ meson SU(3) flavour states. The colour assignments of the eight physical gluons can be written

$$r\bar{g},\ g\bar{r},\ r\bar{b},\ b\bar{r},\ g\bar{b},\ b\bar{g},\ \tfrac{1}{\sqrt{2}}(r\bar{r} - g\bar{g}) \quad \text{and} \quad \tfrac{1}{\sqrt{6}}(r\bar{r} + g\bar{g} - 2b\bar{b}).$$

Even though two of these gluon states have $I_3^c = Y^c = 0$, they are part of a colour octet and therefore still carry colour charge (unlike the colourless singlet state).

## 10.4 Colour confinement

There is a wealth of experimental evidence for the existence of quarks. However, despite many experimental attempts to detect free quarks, which would be observed as fractionally charged particles, they have never been seen directly. The non-observation of free quarks is explained by the hypothesis of *colour confinement*, which states that coloured objects are always confined to colour singlet states and

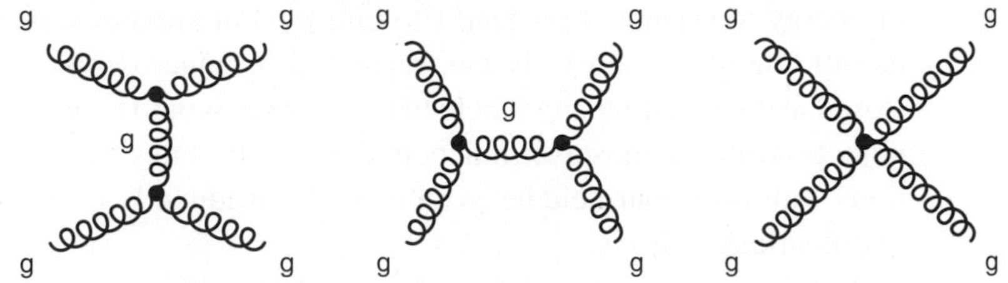


Fig. 10.5 Lowest-order Feynman diagrams for the process gg → gg, formed from the triple and quartic gluon vertices of Figure 10.1.

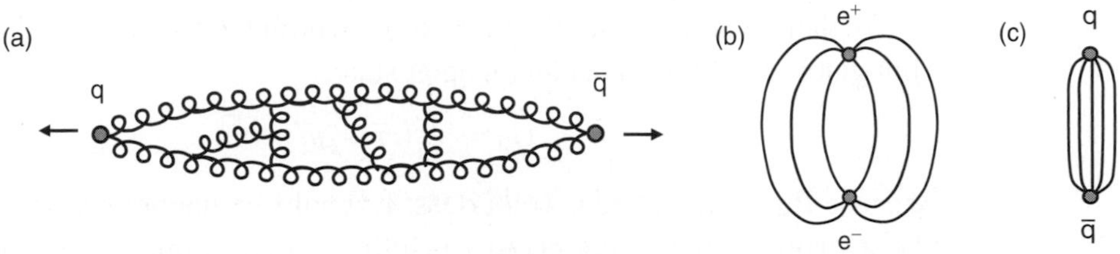


Fig. 10.6 Qualitative picture of the effect of gluon–gluon interactions on the long-range QCD force.

that no objects with non-zero colour charge can propagate as free particles. Colour confinement is believed to originate from the gluon–gluon self-interactions that arise because the gluons carry colour charge, allowing gluons to interact with other gluons through diagrams such as those shown in Figure 10.5.

There is currently no analytic proof of the concept of colour confinement, although there has been recent progress using the techniques of lattice QCD. Nevertheless, a qualitative understanding of the likely origin can be obtained by considering what happens when two free quarks are pulled apart. The interaction between the quarks can be thought of in terms of the exchange of virtual gluons. Because they carry colour charge, there are attractive interactions between these exchanged virtual gluons, as indicated in Figure 10.6a. The effect of these interactions is to squeeze the colour field between the quarks into a tube. Rather than the field lines spreading out as in QED (Figure 10.6b), they are confined to a tube between the quarks, as indicated in Figure 10.6c. At relatively large distances, the energy density in the tube between the quarks containing the gluon field is constant. Therefore the energy stored in the field is proportional the separation of the quarks, giving a term in the potential of the form

$$V(\mathbf{r}) \sim \kappa r, \tag{10.13}$$

where experimentally $\kappa \sim 1\,\mathrm{GeV/fm}$. This experimentally determined value for $\kappa$ (see Section 10.8) corresponds to a very large force of $\mathcal{O}(10^5)\,\mathrm{N}$ between *any* two unconfined quarks, regardless of separation! Because the energy stored in the colour field increases linearly with distance, it would require an infinite amount of energy to separate two quarks to infinity. Put another way, if there are two free colour charges in the Universe, separated by macroscopic distances, the energy stored in the resulting gluon field between them would be vast. As a result, coloured objects arrange themselves into bound hadronic states that are colourless combinations with no colour field between them. Consequently quarks are always confined to colourless hadrons.

Another consequence of the colour confinement hypothesis is that gluons, being coloured, are also confined to colourless objects. Therefore, unlike photons (the force carriers of QED), gluons do not propagate over macroscopic distances. It is interesting to note that had nature chosen a U(3) local gauge symmetry, rather than SU(3), there would be a ninth gluon corresponding to the additional U(1) generator. This gluon would be the colour singlet state,

$$G_9 = \frac{1}{\sqrt{3}}(r\bar{r} + g\bar{g} + b\bar{b}).$$

Because this gluon state is colourless, it would be unconfined and would behave like a strongly interacting photon, resulting in an infinite range strong force; the Universe would be a very different (and not very hospitable) place with long-range strong interactions between all quarks.

### 10.4.1 Colour confinement and hadronic states

Colour confinement implies that quarks are always observed to be confined to bound colourless states. To understand exactly what is meant by "colourless", it is worth recalling the states formed from the combination of spin for two spin-half particles. The four possible spin combinations give rise to a triplet of spin-1 states and a spin-0 singlet ($2 \otimes 2 = 3 \oplus 1$):

$$|1,+1\rangle = \uparrow\uparrow,\ \ |1,0\rangle = \frac{1}{\sqrt{2}}(\uparrow\downarrow + \downarrow\uparrow),\ \ |1,-1\rangle = \downarrow\downarrow \ \text{ and } \ |0,0\rangle = \frac{1}{\sqrt{2}}(\uparrow\downarrow - \downarrow\uparrow).$$

The singlet state is "spinless" in the sense that it carries no angular momentum. In a similar way, SU(3) colour singlet states are colourless combinations which have zero colour quantum numbers, $I_3^c = Y^c = 0$. It should be remembered that $I_3^c = Y^c = 0$ is a necessary but not sufficient condition for a state to be colourless. The action of any of the SU(3) colour ladder operators on a colour singlet state must yield zero, in which case the state is analogous to the spinless $|0,0\rangle$ singlet state.

The colour confinement hypothesis implies that only colour singlet states can exist as free particles. Consequently, all bound states of quarks and antiquarks must occur in colour singlets. This places a strong restriction on the structure of possible hadronic states; the allowed combinations of quarks and antiquarks are those where a colour singlet state can be formed. The algebra of the exact SU(3) colour symmetry was described in Chapter 9 in the context of SU(3) flavour symmetry and the results can be directly applied to colour with the replacements, u $\rightarrow r$, d $\rightarrow g$ and s $\rightarrow b$.

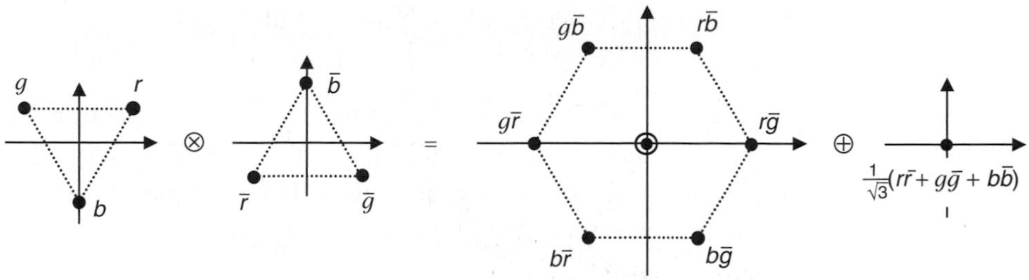


**Fig. 10.7** The colour combination of a quark and an antiquark, $3 \otimes \bar{3} = 8 \oplus 1$.

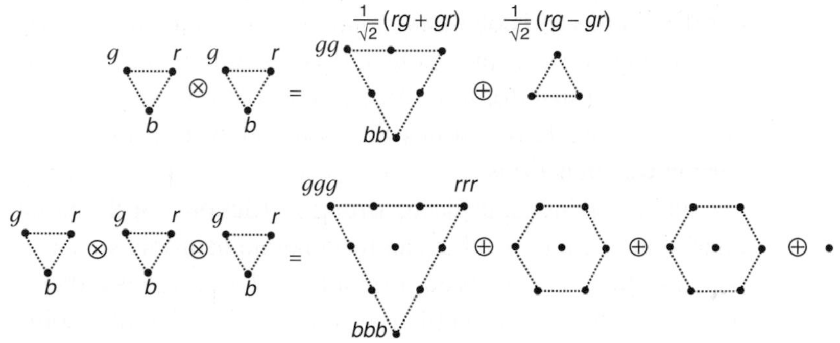


**Fig. 10.8** The multiplets from the colour combinations of two quarks, $3 \otimes 3 = 6 \oplus \bar{3}$, and three quarks, $3 \otimes 3 \otimes 3 = 10 \oplus 8 \oplus 8 \oplus 1$.

First consider the possible colour wavefunctions for a bound $q\bar{q}$ state. The combination of a colour with an anticolour is mathematically identical to the construction of meson *flavour* wavefunctions in SU(3) flavour symmetry. The resulting colour multiplets, shown in Figure 10.7, are a coloured octet and a colourless singlet. The colour confinement hypothesis implies that all hadrons must be colour singlets, and hence the colour wavefunction for mesons is

$$\psi^c(q\bar{q}) = \frac{1}{\sqrt{3}}(r\bar{r} + g\bar{g} + b\bar{b}).$$

The addition of another quark (or antiquark) to either the octet or singlet state in Figure 10.7 will not yield a state with $I_3^c = Y^c = 0$. Therefore, it can be concluded that bound states of $qq\bar{q}$ or $q\bar{q}\bar{q}$ do not exist in nature.

These arguments can be extended to the combinations of two and three quarks as shown in Figure 10.8. The combination of two colour triplets yields a colour sextet and a colour triplet ($\bar{3}$). The absence of a colour singlet state for the qq system, implies that bound states of two quarks are always coloured objects and therefore do not exist in nature. However, the combination of three colours yields a single singlet state with the colour wavefunction

$$\psi^c(\text{qqq}) = \frac{1}{\sqrt{6}}(rgb - rbg + gbr - grb + brg - bgr), \tag{10.14}$$

analogous to the SU(3) flavour singlet wavefunction of Section 9.6.4. This state clearly satisfies the requirement that $I_3^c = Y^c = 0$. The colour ladder operators can be used to confirm it is a colour singlet. For example, the action of the colour isospin raising operator $T_+^c$, for which $T_+^c g = r$, gives

$$T_+^c \psi^c(\text{qqq}) = \frac{1}{\sqrt{6}}(rrb - rbr + rbr - rrb + brr - brr) = 0,$$

as required. Hence a SU(3) colour singlet state can be formed from the combination of three quarks and colourless bound states of qqq are observed in nature. Since the colour singlet wavefunction of (10.14) is totally antisymmetric, and it is the only colour singlet state for three quarks, the colour wavefunction for baryons is always antisymmetric. This justifies the assumption used in Chapter 9 to construct the baryon wavefunctions.

Colour confinement places strong restrictions on the possible combinations of quarks and antiquarks that can form bound hadronic states. To date, all confirmed observed hadronic states correspond to colour singlets either in the form of mesons ($\text{q}\overline{\text{q}}$), baryons (qqq) or antibaryons ($\overline{\text{q}}\,\overline{\text{q}}\,\overline{\text{q}}$). In principle, combinations of ($\text{q}\overline{\text{q}}$) and (qqq) such as pentaquark states ($\text{qqqq}\overline{\text{q}}$) could exist, either as bound states in their own right or as hadronic molecules such as ($\text{q}\overline{\text{q}}$)-(qqq). In recent years there have been a number of claims for the existence of pentaquark states, but the evidence is (at best) far from convincing.

### 10.4.2 Hadronisation and jets

In processes such as $\text{e}^+\text{e}^- \to \text{q}\overline{\text{q}}$, the two (initially free) high-energy quarks are produced travelling back to back in the centre-of-mass frame. As a consequence of colour confinement, the quarks do not propagate freely and are observed as jets of colourless particles. The process by which high-energy quarks (and gluons) produce jets is known as hadronisation.

A qualitative description of the hadronisation process is shown in Figurc 10.9. The five stages correspond to: (i) the quark and antiquark produced in an interaction initially separate at high velocities; (ii) as they separate the colour field is restricted to a tube with energy density of approximately 1 GeV/fm; (iii) as the quarks separate further, the energy stored in the colour field is sufficient to provide the energy necessary to form new $\text{q}\overline{\text{q}}$ pairs and breaking the colour field into smaller "strings" is energetically favourable; (iv) this process continues and further $\text{q}\overline{\text{q}}$ pairs are produced until (v) all the quarks and antiquarks have sufficiently low energy to combine to form colourless hadrons. The hadronisation process results in two jets of hadrons, one following the initial quark direction and the other in the initial antiquark direction. Hence, in high-energy experiments, quarks and gluons

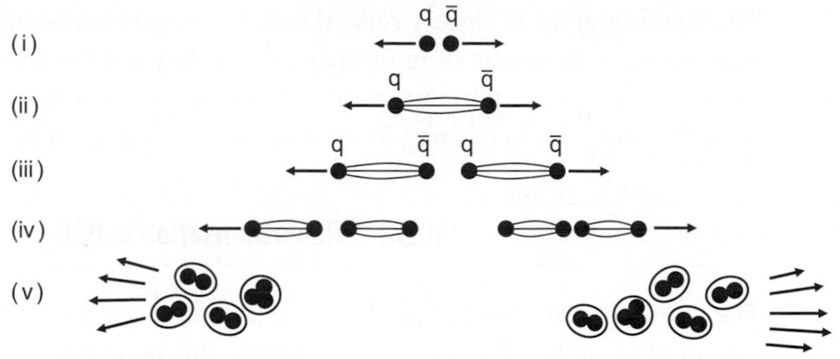


Fig. 10.9 Qualitative picture of the steps in the hadronisation process.

are always observed as jets of hadrons (see for example, Figures 10.19 and 10.30). The precise process of hadronisation is poorly understood. Nevertheless, there are a number of phenomenological models (often with many free parameters) that are able to provide a reasonable description of the experimental data. Whilst these models are motived by QCD, they are a long way from a first-principles theoretical description of the hadronisation process.

## 10.5 Running of $\alpha_S$ and asymptotic freedom

At low-energy scales, the coupling constant of QCD is large, $\alpha_S \sim O(1)$. Consequently, the perturbation expansion discussed in the context of QED in Section 6.1, does not converge rapidly. For this reason (low-energy) QCD processes are not calculable using traditional perturbation theory. Nevertheless, in recent years, there has been a significant progress with the computational technique of *lattice QCD*, where quantum-mechanical calculations are performed on a discrete lattice of space-time points. Such calculations are computationally intensive, with a single calculation often taking many months, even using specially adapted supercomputing facilities. With lattice QCD it has been possible to calculate the proton mass with a precision of a few per cent, thus providing a first principles test of the validity of QCD in the non-perturbative regime. Despite this success, most practical calculations in particle physics are based on perturbation theory. For this reason, it might seem problematic that perturbation theory cannot be applied in QCD processes because of the large value of $\alpha_S$. Fortunately, it turns out that $\alpha_S$ is not constant; its value depends on the energy scale of the interaction being considered. At high energies, $\alpha_S$ becomes sufficiently small that perturbation theory can again be used. In this way, QCD divides into a non-perturbative low-energy regime, where first-principles calculations are not currently possible, such as the hadronisation process, and a high-energy regime where perturbation theory can be used.

The running of $\alpha_S$ is closely related to the concept of renormalisation. A thorough mathematical treatment of renormalisation is beyond the level of this book. Nevertheless, it is necessary to introduce the basic ideas in order to provide a qualitative understanding of the running of the coupling constants of both QED and QCD.

### 10.5.1 *Renormalisation in QED

The strength of the interaction between a photon and an electron is determined by the coupling at the QED vertex, which up to this point has been taken to be constant with value $e$. The experimentally measured value of the electron charge $e$, which corresponds to $\alpha \approx 1/137$, is obtained from measurements of the strength of the static Coulomb potential in atomic physics. This is not the same as the strength of coupling between an electron and photon that appears in Feynman diagrams, which can be written as $e_0$ (often referred to as the bare electron charge); the experimentally measured value of $e$ is the effective strength of the interaction which results from the sum over all relevant QED higher-order diagrams.

Up to this point, only the lowest-order contribution to the QED coupling between a photon and a charged fermion, shown in Figure 10.10a, has been considered. However, for each QED vertex in a Feynman diagram, there is an infinite set of higher-order corrections; for example, the $\mathcal{O}(e^2)$ corrections to the QED $e^-\gamma e^-$ vertex are shown in Figures 10.10b–10.10e. The experimentally measured strength of the QED interaction is the effective strength from the sum over of all such diagrams. The diagram of Figure 10.10b represents correction to the propagator and the diagrams in Figures 10.10c–10.10e represent corrections to the electron four-vector current. In principle, both types of diagram will modify the strength of the interaction relative to the lowest-order diagram alone.

For each higher-order diagram, it is relatively straightforward to write down the matrix element using the Feynman rules for QED. Each loop in a Feynman diagram enters as an integral over the four momenta of the particles in the loop and such diagrams lead to divergent (infinite) results. Fortunately, the infinities associated with the loop corrections to the photon propagator can be absorbed into the definition of the electron charge (described below). However, the corrections to four-vector

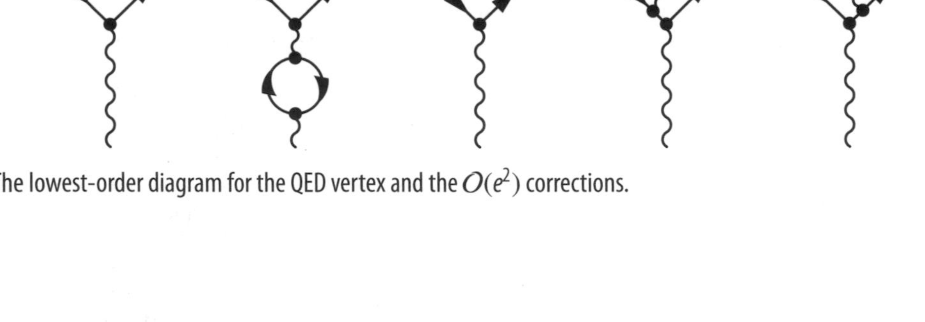

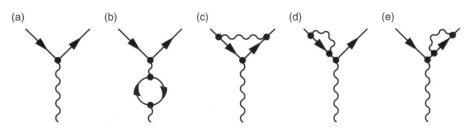

Fig. 10.10 The lowest-order diagram for the QED vertex and the $\mathcal{O}(e^2)$ corrections.

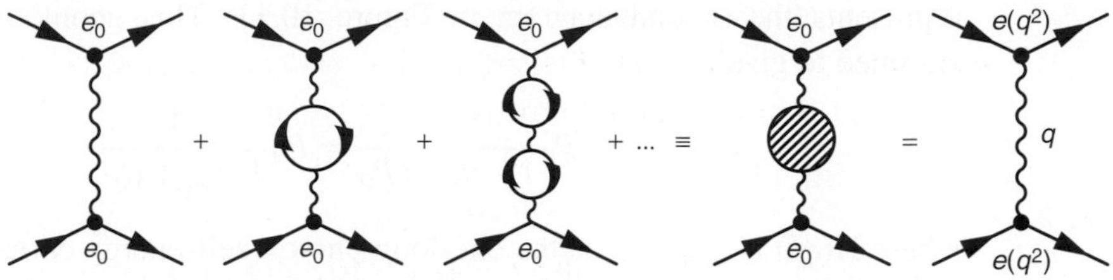


Fig. 10.11 Renormalisation in QED, relating the running charge $e(q^2)$ to the bare charge $e_0$.

current, Figures 10.10c–10.10e, are potentially more problematic as they involve loops that include virtual fermions. Consequently, the results of the corresponding loop integrals will depend on the fermion masses. In principle, this would result in the effective strength of the QED interaction being dependent on the mass of the particle involved, which is not the case. However, in a field theory with local gauge invariance such as QED, the effect of the diagram of Figure 10.10c is exactly cancelled by the effects of diagrams 10.10d–10.10e. This type of cancellation, that is known as a Ward identity, holds to all orders in perturbation theory. Consequently, here we only need to consider the higher-order corrections to the photon propagator.

The infinite series of corrections to the photon propagator, known as the photon self-energy terms, are accounted for by replacing the lowest-order photon exchange diagram by the infinite series of loop diagrams expressed in terms on the bare electron charge, $e_0$. As a result of the loop corrections, the photon propagator including the self-energy terms, will no longer have a simple $1/q^2$ form. The physical effects of the modification to the propagator can be accounted for by retaining the $1/q^2$ dependence for the effective propagator and absorbing the corrections into the definition of the charge, which now necessarily depends on $q^2$. This procedure is shown in Figure 10.11, where the infinite sum over the self-energy corrections to the photon with bare charge $e_0$, indicated by the blob, is replaced by a $1/q^2$ propagator with effective charge $e(q^2)$.

The effective photon propagator, here denoted as $P$, can be expressed in terms of the propagator with the bare charge,

$$P_0 = \frac{e_0^2}{q^2},$$

by inserting an infinite series of the fermion loops. Each loop introduces a correction factor $\pi(q^2)$, such that the effective propagator is given by

$$P = P_0 + P_0\,\pi(q^2)\,P_0 + P_0\,\pi(q^2)\,P_0\,\pi(q^2)\,P_0 + \cdots,$$

where, for example, the second term in the above sum corresponds to a single loop correction $\pi(q^2)$ inserted between two bare $P_0$ propagator terms, and therefore represents the second diagram in Figure 10.11. This geometric series can be summed to give

$$P = P_0 \frac{1}{1 - \pi(q^2)\, P_0} = P_0 \frac{1}{1 - e_0^2\, \Pi(q^2)},$$

where $\Pi(q^2) = \pi(q^2)/q^2$ is the one-loop photon self-energy correction. The effective propagator can then be expressed in terms of the running coupling $e(q^2)$ as

$$P \equiv \frac{e^2(q^2)}{q^2} = \frac{e_0^2}{q^2} \frac{1}{1 - e_0^2\, \Pi(q^2)}.$$

Since scattering cross sections are known to be finite, it is an experimentally established fact that $e(q^2)$ is finite, therefore

$$e^2(q^2) = \frac{e_0^2}{1 - e_0^2\, \Pi(q^2)}, \tag{10.15}$$

is finite, even though the denominator contains $\Pi(q^2)$ which is divergent. If the physical electron charge is known at some scale $q^2 = \mu^2$, then (10.15) can be rearranged to give an expression for the bare charge

$$e_0^2 = \frac{e^2(\mu^2)}{1 + e^2(\mu^2)\Pi(\mu^2)},$$

which can be substituted back into (10.15) to give the exact relation,

$$e^2(q^2) = \frac{e^2(\mu^2)}{1 - e^2(\mu^2) \cdot [\Pi(q^2) - \Pi(\mu^2)]}. \tag{10.16}$$

As a result of the loop integral for the photon self-energy, both $\Pi(q^2)$ and $\Pi(\mu^2)$ are separately divergent. However, the difference $\Pi(q^2) - \Pi(\mu^2)$ is finite and calculable. Although the infinities have been renormalised away, the finite difference between the effective strength of the interaction at different values of $q^2$ remains. Consequently, the coupling strength is no longer constant, it runs with the $q^2$ scale of thc virtual photon. For values of $q^2$ and $\mu^2$ larger than the electron mass squared, it can be shown that

$$\Pi(q^2) - \Pi(\mu^2) \approx \frac{1}{12\pi^2} \ln\left(\frac{q^2}{\mu^2}\right).$$

Substituting this into (10.16) and writing $\alpha(q^2) = e^2(q^2)/4\pi$ gives

$$\alpha(q^2) = \frac{\alpha(\mu^2)}{1 - \alpha(\mu^2)\frac{1}{3\pi}\ln\left(\frac{q^2}{\mu^2}\right)}, \tag{10.17}$$

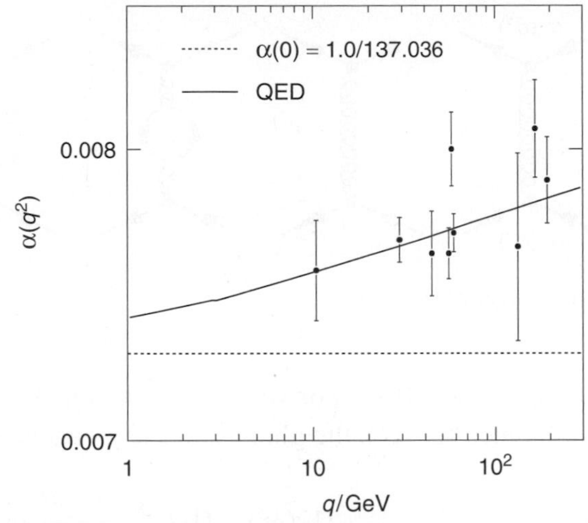


Fig. 10.12 Measurements of $\alpha(q^2)$ at different $q^2$ scales from $e^+e^- \to f\bar{f}$ with the OPAL experiment at LEP. The dotted line shows the low-$q^2$ limit of $\alpha \approx 1/137$. Adapted from Abbiendi *et al.* (2004).

and the coupling has acquired a dependence on the $q^2$ of the photon. Hence, the lowest-order QED diagram with a running coupling constant $\alpha(q^2)$ incorporates the effects of the virtual loop diagrams in the photon propagator. The above derivation applies equally to $s$-channel and $t$-channel processes and (10.17) holds in both cases. In a $t$-channel process both $q^2$ and $\mu^2$ are negative and the running of the coupling constant is often written as $\alpha(Q^2)$. It should be noted that $\alpha(q^2)$ should be read as $\alpha(|q^2|)$.

The minus sign in (10.17) implies that the coupling of QED increases with increasing $|q^2|$, although the evolution is rather slow. In measurements from atomic physics at $q^2 \approx 0$, the fine-structure constant is determined to be

$$\alpha(q^2 \approx 0) = \frac{1}{137.035\,999\,074(94)}.$$

The QED coupling $\alpha(q^2)$ has also been measured in $e^+e^-$ annihilation at LEP; the results from the highest $q^2$ measurements are shown in Figure 10.12. At a mean centre-of-mass energy of $\sqrt{s} = 193\,\text{GeV}$, it is found that

$$\alpha = \frac{1}{127.4 \pm 2.1},$$

providing a clear demonstration of the running of the coupling of QED.

### 10.5.2 Running of $\alpha_S$

The treatment of renormalisation in QCD is similar to that of QED. However, owing to the gluon–gluon self-interactions, there are additional loop diagrams, as

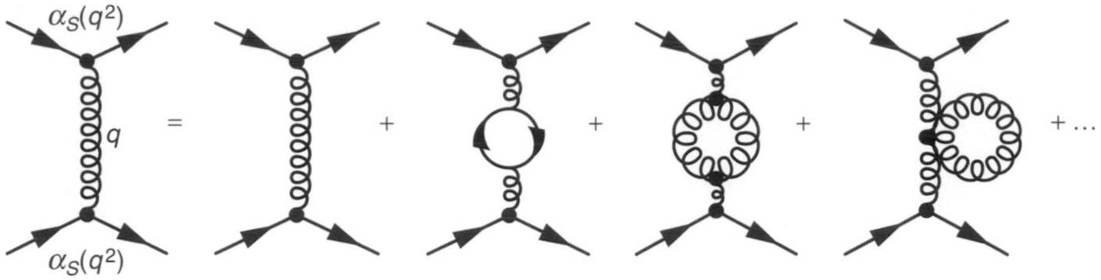


Fig. 10.13 Renormalisation in QCD.

shown in Figure 10.13. For values of $q^2$ and $\mu^2$ larger than the confinement scale, the difference between the gluon self-energy again grows logarithmically

$$\Pi_S(q^2) - \Pi_S(\mu^2) \approx -\frac{B}{4\pi}\ln\left(\frac{q^2}{\mu^2}\right),$$

where the $B$ depends to the numbers of fermionic (quark) and bosonic (gluon) loops. For $N_f$ quark flavours and $N_c$ colours,

$$B = \frac{11N_c - 2N_f}{12\pi}.$$

The effect of the bosonic loops enters the expression for the $q^2$ evolution of $\alpha_S$ with the opposite sign to the pure fermion loops, with the fermion loops leading to a negative contributions (which was also the case for QED) and the gluon loops leading to positive contributions. The corresponding evolution of $\alpha_S(q^2)$ is

$$\alpha_S(q^2) = \frac{\alpha_S(\mu^2)}{1 + B\alpha_S(\mu^2)\ln\left(\frac{q^2}{\mu^2}\right)}.$$

For $N_c = 3$ colours and $N_f \leq 6$ quarks, $B$ is greater than zero and hence $\alpha_S$ *decreases* with increasing $q^2$.

There are many ways in which $\alpha_S$ can be measured. These include studies of the hadronic decays of the tau-lepton, the observed spectra of bound states of heavy quarks ($c\bar{c}$ and $b\bar{b}$), measurements of deep inelastic scattering, and jet production rates in $e^+e^-$ annihilation. Figure 10.14 shows a summary of the most precise measurements of $\alpha_S$ which span $|q| = 2 - 200$ GeV. The predicted decrease in $\alpha_S$ with increasing $|q|$ is clearly observed and the data are consistent with the QCD predictions for the running of $\alpha_S$ with a value of $\alpha_S$ at $|q^2| = m_Z^2$ of

$$\alpha_S(m_Z^2) = 0.1184 \pm 0.0007.$$

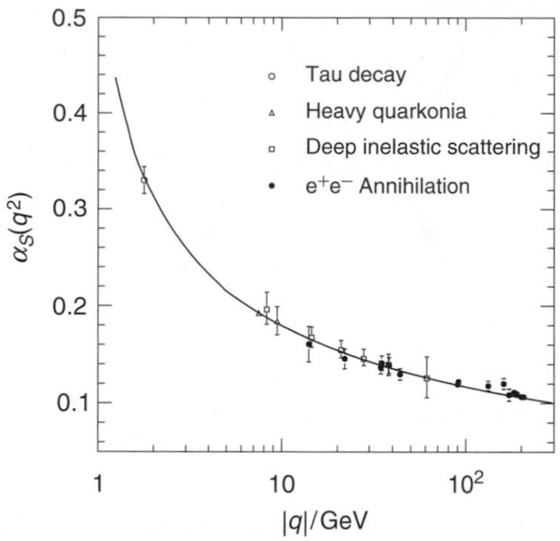


Fig. 10.14 Measurements of $\alpha_S$ at different $|q|$ scales. The barely noticeable kinks in the QCD prediction occur at the thresholds for producing $s\bar{s}$, $c\bar{c}$ and $b\bar{b}$; these affect the evolution of $\alpha_S$ as the number of effective fermion flavours $N_f$ changes. Adapted from Bethke (2009).

### Asymptotic freedom

The strength of the QCD coupling varies considerably over the range of energies relevant to particle physics. At $|q| \sim 1\,\text{GeV}$, $\alpha_S$ is of $\mathcal{O}(1)$ and perturbation theory cannot be used. This non-perturbative regime applies to the discussion of bound hadronic states and the latter stages in the hadronisation process. At $|q| > 100\,\text{GeV}$, which is the typical scale for modern high-energy collider experiments, $\alpha_S \sim 0.1$, which is sufficiently small that perturbation theory again can be used. This property of QCD is known as *asymptotic freedom*. It is the reason that, in the previous discussion of deep inelastic scattering at high $q^2$, the quarks could be treated as quasi-free particles, rather than being strongly bound within the proton. It should be noted that at high $q^2$, even though $\alpha_S \sim 0.1$ is sufficiently small for perturbation theory to be applicable, unlike QED, it is not so small that higher-order corrections can be neglected. For this reason, QCD calculations for processes at the LHC are almost always calculated beyond lowest order. These calculations, which often involve many Feynman diagrams, are extremely challenging.

## 10.6 QCD in electron–positron annihilation

A number of the properties of QCD can be studied at an electron–positron collider, primarily through the production of $q\bar{q}$ pairs in the annihilation process $e^+e^- \rightarrow q\bar{q}$, shown Figure 10.15. There are a number of advantages in studying QCD at an electron–positron collider compared to at a hadron collider. The QED production

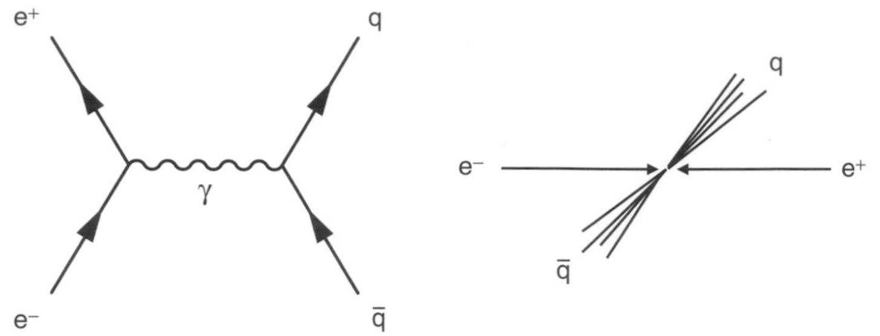


Fig. 10.15 The lowest-order QED Feynman diagram for $e^+e^- \to q\bar{q}$ production and the appearance of the interaction in a detector as a final state consisting of two jets of hadrons.

process of $e^+e^-$ annihilation is well understood and can be calculated to high precision; there are no uncertainties related to the parton distribution functions. In addition, the observed final state corresponds to the underlying hard interaction. This is not the case for hadron–hadron collisions, where the remnants of the colliding hadrons are also observed, typically as forward-going jets.

The differential cross-section for the process $e^+e^- \to \mu^+\mu^-$ was calculated in Chapter 6. Assuming that quarks are spin-half particles, the angular dependence of the differential cross section for $e^+e^- \to q\bar{q}$ is expected to be

$$\frac{d\sigma}{d\Omega} \propto (1 + \cos^2\theta),$$

where $\theta$ is the angle between the incoming $e^-$ and the final-state quark. Because the quark and antiquark will hadronise into jets of hadrons, it is not generally possible to identify experimentally which flavour of quark was produced. For this reason, the $e^+e^- \to q\bar{q}$ cross section is usually expressed as an inclusive sum over all quark flavours, $e^+e^- \to$ hadrons. Furthermore, it is also not usually possible to identify which jet came from the quark and which jet came from the antiquark. To reflect this ambiguity, the differential cross section is usually quoted in terms of $|\cos\theta|$. For example, Figure 10.16 shows the observed angular distribution of the jets in the process $e^+e^- \to$ hadrons in the centre-of-mass energy range $38.8 < \sqrt{s} < 46.5$ GeV. The angular distribution is consistent with expected $(1 + \cos^2\theta)$ form, demonstrating that quarks are indeed spin-half particles.

The total QED $e^+e^- \to \mu^+\mu^-$ cross section, was calculated previously

$$\sigma(e^+e^- \to \mu^+\mu^-) = \frac{4\pi\alpha^2}{3s}. \tag{10.18}$$

The corresponding cross section for the QED production of a $q\bar{q}$ pair is

$$\sigma(e^+e^- \to q\bar{q}) = 3 \times \frac{4\pi\alpha^2}{3s} Q_q^2, \tag{10.19}$$

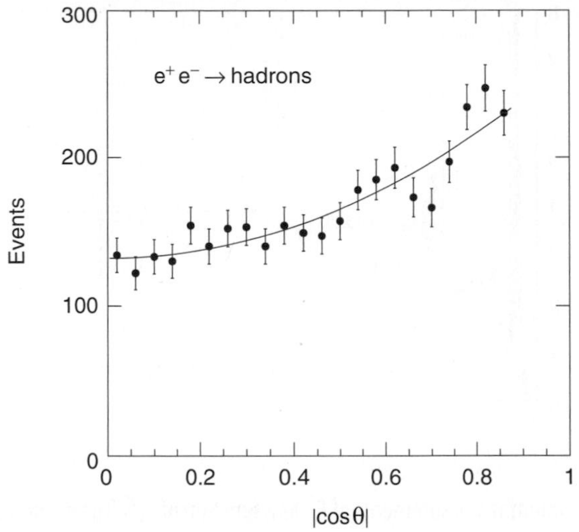


Fig. 10.16 The angular distribution of the jets produced in $e^+e^-$ annihilation at centre-of-mass energies between 38.8 GeV $< \sqrt{s} <$ 46.5 GeV as observed in the CELLO experiment at the PETRA $e^+e^-$ collider at DESY. The expected $(1+\cos^2\theta)$ distribution for the production of spin-half particles is also shown. Adapted from Behrend *et al.* (1987).

where the factor of three accounts for the sum over the three possible colour combinations of the final-state $q\overline{q}$ that can be produced as $g\overline{g}$, $r\overline{r}$ or $b\overline{b}$. The inclusive QED cross section for $\sigma(e^+e^- \to \text{hadrons})$ is the sum of the cross sections for the quark flavours that are kinematically accessible at a given centre-of-mass energy ($\sqrt{s} > 2m_q$),

$$\sigma(e^+e^- \to \text{hadrons}) = \frac{4\pi\alpha^2}{s} \times 3 \sum_{\text{flavours}} Q_q^2. \tag{10.20}$$

It is convenient to express the inclusive cross section of (10.20) in terms of a ratio relative to the $\mu^+\mu^-$ cross section of (10.18),

$$R_\mu \equiv \frac{\sigma(e^+e^- \to \text{hadrons})}{\sigma(e^+e^- \to \mu^+\mu^-)} = 3 \sum_{\text{flavours}} Q_q^2. \tag{10.21}$$

This has the advantage that a number of experimental systematic uncertainties cancel since $R_\mu$ is related to the ratio of the observed numbers of events. The expected value of $R_\mu$ depends on the sum of the squares of the charges of the quark flavours that can be produced at a particular centre-of-mass energy. For $\sqrt{s} \lesssim 3\,\text{GeV}$, only u, d and s quarks can be produced, giving the predicted value

$$R_\mu^{\text{d,u,s}} = 3 \times \left(\tfrac{4}{9} + \tfrac{1}{9} + \tfrac{1}{9}\right) = 2.$$

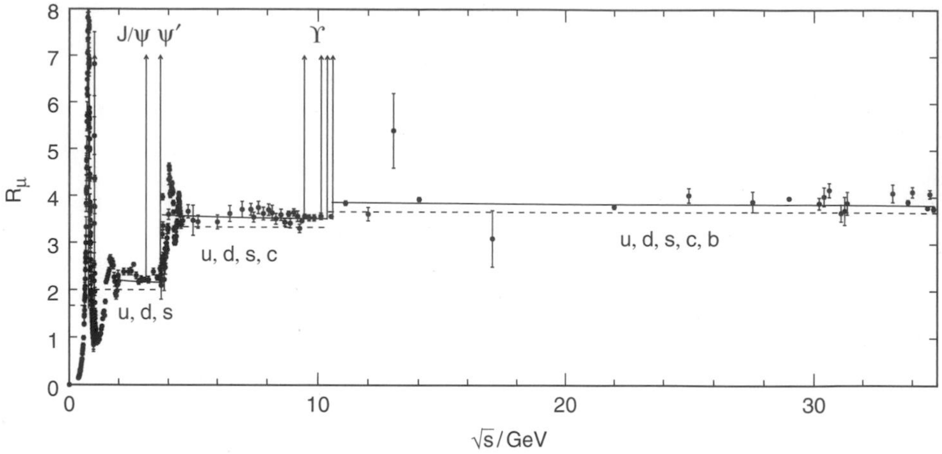


Fig. 10.17 Experimental measurements of $R_\mu$ as a function of $\sqrt{s}$. The dashed line is the lowest-order prediction assuming three colours. The solid line includes the first-order QCD correction of $(1 + \alpha_S(q^2)/\pi)$. Based on data compiled by the Particle Data Group, Beringer *et al.* (2012).

Above the thresholds for $c\bar{c}$ production (3.1 GeV) and $b\bar{b}$ production (9.5 GeV) the predictions for $R_\mu$ are respectively

$$R_\mu^{c} = 3 \times \left(\tfrac{1}{9} + \tfrac{4}{9} + \tfrac{1}{9} + \tfrac{4}{9}\right) = \tfrac{10}{3} \quad \text{and} \quad R_\mu^{b} = 3 \times \left(\tfrac{1}{9} + \tfrac{4}{9} + \tfrac{1}{9} + \tfrac{4}{9} + \tfrac{1}{9}\right) = \tfrac{11}{3}.$$

Figure 10.17 shows the measurements of $R_\mu$ over a wide range of centre-of-mass energies. At relatively low energies, there is significant structure due to resonant production of bound $q\bar{q}$ states with the same spin and parity as the virtual photon, $J^P = 1^-$. These resonances greatly enhance the $e^+e^- \to$ hadrons cross section when the centre-of-mass energy is close to the mass of the state being produced. At very low energies, the resonance structure is dominated by the $J^P = 1^-$ mesons introduced in Section 9.6.3, namely the $\rho^0$(770 MeV), $\omega$(782 MeV) and $\phi$(1020 MeV) mesons. At higher energies, charmonium ($c\bar{c}$) and bottomonium ($b\bar{b}$) states are produced, such as the J/$\psi$(3097 MeV), $\psi'$(3686 MeV) and $\Upsilon$ states. These heavy quark resonances are discussed further in Section 10.8.

In the continuum between the meson resonances, the data disagree with the predictions for $R_\mu$ given in (10.21) at the level of approximately 10%. The origin of the discrepancy is that the cross sections of (10.18) and (10.19) are only relevant for the lowest-order process, whereas the measured cross sections will include $\mu^+\mu^-\gamma$, $q\bar{q}\gamma$ and $q\bar{q}g$ final states, as shown in Figure 10.18. The cross sections for these processes will be suppressed relative to the lowest-order process by respective factors of $\alpha$, $\alpha$ and $\alpha_S$ due to the additional vertex. The QED corrections are relatively

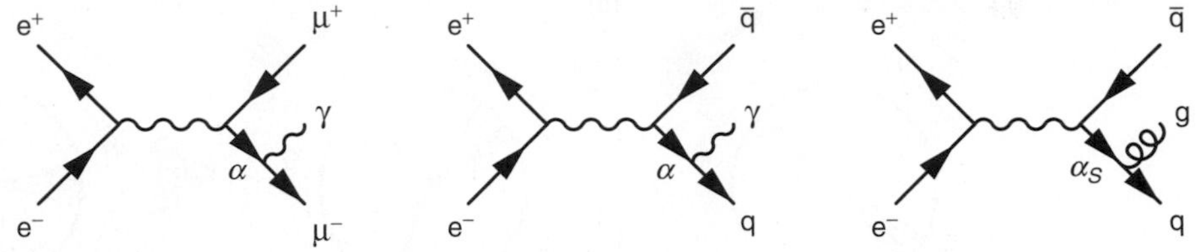


Fig. 10.18 Feynman diagrams for $e^+e^- \to \mu^+\mu^-\gamma, q\bar{q}\gamma, q\bar{q}g$.

small, but the $O(\alpha_S)$ correction cannot be neglected. If the first-order QCD correction from $e^+e^- \to q\bar{q}g$ is included, the prediction of (10.21) is modified to

$$R_\mu = \frac{\sigma(e^+e^- \to \text{hadrons})}{\sigma(e^+e^- \to \mu^+\mu^-)} = 3\left(1 + \frac{\alpha_S(q^2)}{\pi}\right)\sum_{\text{flavours}} Q_q^2. \tag{10.22}$$

With the QCD correction included, the prediction for $R_\mu$, shown by the solid line in Figure 10.17, is in excellent agreement with the experimental measurements away from the resonances. This agreement provides strong evidence for the existence of colour (which is never directly observed); without the additional colour degree of freedom, the prediction for $R_\mu$ would be a factor of three smaller and would be incompatible with the observed data.

## Gluon production in $e^+e^-$ annihilation

Jet production in high-energy electron–positron collisions also provides direct evidence for the existence of gluons. Figure 10.19 shows three examples of $e^+e^- \to$ hadrons events observed in the OPAL detector at LEP. Whilst the majority of the $e^+e^- \to$ hadrons events are produced with a clear two-jet topology, final states with three- or four-jets are also observed. The three-jet events originate from the process $e^+e^- \to q\bar{q}g$, where the gluon is radiated from either the final-state quark or antiquark, as shown in Figure 10.19b. The relative cross section for the production of three-jet events compared to the two-jet final states is proportional to $\alpha_S$. Hence the observed number of three-jet events relative to the number of two-jet events, provides one of the most precise measurements of $\alpha_S(q^2)$. Jet production in electron–positron collisions also provides a direct test of the SU(3) group structure of QCD. For example, one of the Feynman diagrams for four-jet production, shown in Figure 10.19c, involves the triple gluon vertex. The Feynman rules for this vertex are determined by the local gauge symmetry of QCD. By studying the angular distributions of the jets in four-jet events, it is possible to distinguish between an underlying SU(3) colour symmetry and alternative gauge symmetries. Needless to say, the experimental data are consistent with the predictions of SU(3).

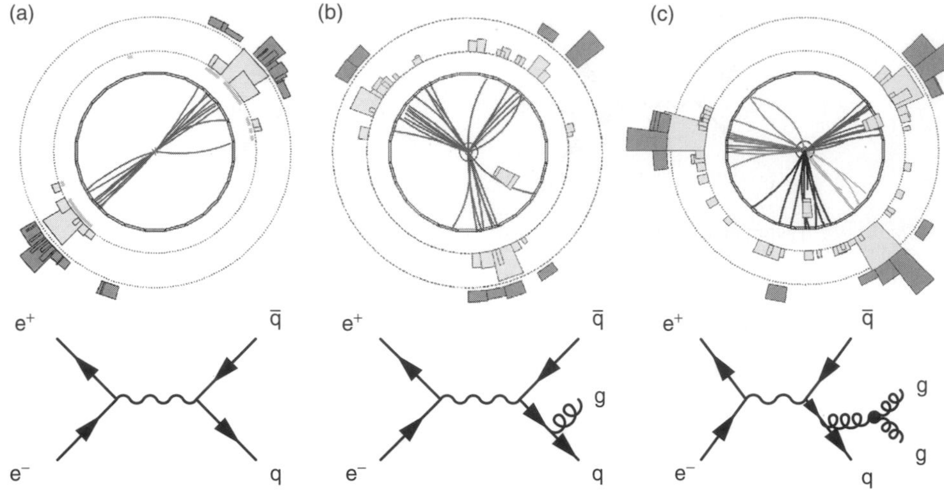


Fig. 10.19 Jet production in $e^+e^-$ annihilation. The example events were recorded at $\sqrt{s} = 91$ GeV by the OPAL experiment at LEP in the mid 1990s. They correspond to (a) $e^+e^- \to q\bar{q} \to$ two-jets, (b) $e^+e^- \to q\bar{q}g \to$ three-jets and (c) $e^+e^- \to q\bar{q}gg \to$ four-jets. Reproduced courtesy of the OPAL collaboration. Also shown are possible Feynman diagrams corresponding to the observed events. In the case of four-jet production there are also diagrams where both gluons are radiated from the quarks.

## 10.7 Colour factors

At hadron colliders, such as the LHC, the observed event rates are dominated by the QCD scattering of quarks and gluons. Figure 10.20 shows one of the parton-level processes contributing to the cross section for pp $\to$ two jets + $X$, where $X$ represents the remnants of the proton that are observed as forward jets in the direction of the incoming proton beams. The calculation of the corresponding matrix element needs to account for the different colours of the quarks and gluons that can contribute to the scattering process.

In the Feynman diagram of Figure 10.20, the incoming and outgoing quark colours are labelled by $i$, $j$, $k$ and $l$. The exchanged gluon is labelled by $a$ and $b$ at the two vertices, with the $\delta^{ab}$ term in the propagator ensuring that the gluon at vertex $\mu$ is the same as that at vertex $\nu$. The colour flow in the diagram corresponds to $ik \to jl$. There are $3^4$ possible colour combinations for the four quarks involved in this process. In addition, there and eight possible gluons that can be exchanged. Consequently, there are 648 distinct combinations of quark colours and gluons that potentially can contribute to the process. Fortunately, the effect of summing over all the colour and gluon combinations can be absorbed into a single *colour factor*.

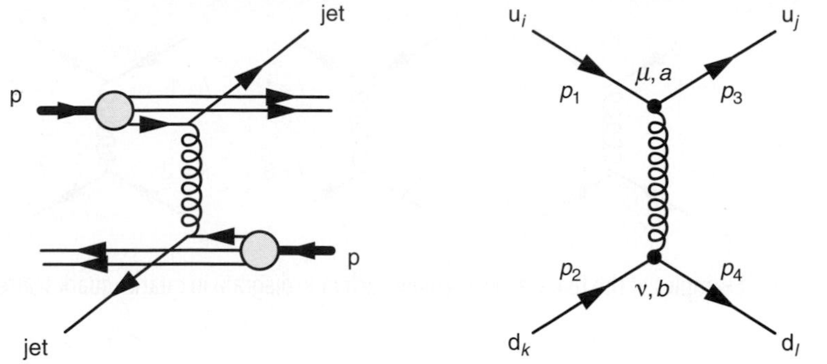


**Fig. 10.20** One of the processes contributing to two-jet production in pp collisions at the LHC, shown in terms of the colliding protons and one of the contributing Feynman diagram, with quark colours $ik \to jl$.

The matrix element for the Feynman diagram of Figure 10.20 can be written down using the Feynman rules for QCD:

$$-i\mathcal{M} = \left[\overline{u}(p_3)\{-\tfrac{1}{2}ig_S\lambda^a_{ji}\gamma^\mu\}u(p_1)\right]\frac{-ig_{\mu\nu}}{q^2}\delta^{ab}\left[\overline{u}(p_4)\{-\tfrac{1}{2}ig_S\lambda^b_{lk}\gamma^\nu\}u(p_2)\right].$$

This can be rearranged to give

$$\mathcal{M} = -\frac{g_S^2}{4}\lambda^a_{ji}\lambda^a_{lk}\frac{1}{q^2}g_{\mu\nu}[\overline{u}(p_3)\gamma^\mu u(p_1)][\overline{u}(p_4)\gamma^\nu u(p_2)]. \tag{10.23}$$

This matrix element resembles that for the QED process $e^-q \to e^-q$ given in (8.13),

$$\mathcal{M} = Q_q\frac{e^2}{q^2}g_{\mu\nu}[\overline{u}(p_3)\gamma^\mu u(p_1)][\overline{u}(p_4)\gamma^\nu u(p_2)].$$

The QCD matrix element for a particular combination of quark colours can be obtained from the calculated QED matrix element by making the replacements $-Q_q e^2 \to g_S^2$, or equivalently $-Q_q\alpha \to \alpha_S$, and multiplying by the colour factor $C(ik \to jl)$ that accounts for the sum over the eight possible exchanged gluons

$$C(ik \to jl) \equiv \frac{1}{4}\sum_{a=1}^{8}\lambda^a_{ji}\lambda^a_{lk}. \tag{10.24}$$

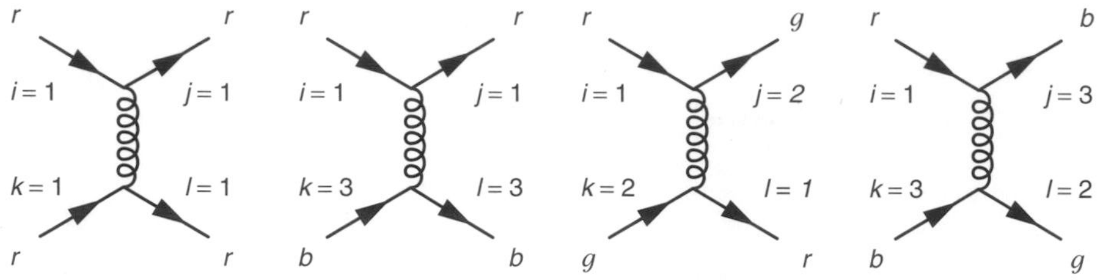


Fig. 10.21 Examples of the four classes of colour exchange diagram in quark–quark scattering.

The QCD colour factor $C(ik \to jl)$ can be evaluated using the Gell-Mann matrices

$$\lambda_1 = \begin{pmatrix} 0 & 1 & 0 \\ 1 & 0 & 0 \\ 0 & 0 & 0 \end{pmatrix}, \quad \lambda_4 = \begin{pmatrix} 0 & 0 & 1 \\ 0 & 0 & 0 \\ 1 & 0 & 0 \end{pmatrix}, \quad \lambda_6 = \begin{pmatrix} 0 & 0 & 0 \\ 0 & 0 & 1 \\ 0 & 1 & 0 \end{pmatrix}, \quad \lambda_3 = \begin{pmatrix} 1 & 0 & 0 \\ 0 & -1 & 0 \\ 0 & 0 & 0 \end{pmatrix},$$

$$\lambda_2 = \begin{pmatrix} 0 & -i & 0 \\ i & 0 & 0 \\ 0 & 0 & 0 \end{pmatrix}, \quad \lambda_5 = \begin{pmatrix} 0 & 0 & -i \\ 0 & 0 & 0 \\ i & 0 & 0 \end{pmatrix}, \quad \lambda_7 = \begin{pmatrix} 0 & 0 & 0 \\ 0 & 0 & -i \\ 0 & i & 0 \end{pmatrix}, \quad \lambda_8 = \frac{1}{\sqrt{3}}\begin{pmatrix} 1 & 0 & 0 \\ 0 & 1 & 0 \\ 0 & 0 & -2 \end{pmatrix},$$

where $\lambda_1$ and $\lambda_2$ correspond to the exchange of $r\overline{g}$ and $g\overline{r}$ gluons, $\lambda_4$ and $\lambda_5$ represent $r\overline{b}$ and $b\overline{r}$ gluons, $\lambda_6$ and $\lambda_7$ represent $g\overline{b}$ and $b\overline{g}$ gluons, and $\lambda_3$ and $\lambda_8$ represent the exchange of $\frac{1}{2}(r\overline{r} - g\overline{g})$ and $\frac{1}{\sqrt{6}}(r\overline{r} + g\overline{g} - 2b\overline{b})$ gluons. The $3^4$ possible combinations of the colours $i$, $j$, $k$ and $l$ can be categorised into the four classes of colour exchange, shown in Figure 10.21. These correspond to the following cases: all four colours are the same, e.g. $rr \to rr$; the two initial-state quarks have different colours but do not change colour, e.g. $rb \to rb$; the two initial-state quarks have different colours and exchange colour, e.g. $rb \to br$; and all three colours are involved. The different colour indices determine which elements of the $\lambda$-matrices are relevant to the scattering process, which in turn determines which gluons contribute.

From (10.24) the colour factor for $rr \to rr$ is

$$C(rr \to rr) \equiv \frac{1}{4}\sum_{a=1}^{8} \lambda_{11}^a \lambda_{11}^a.$$

Here the non-zero contributions arise from $\lambda_3$ and $\lambda_8$, which are the only Gell-Mann matrices with non-zero values in the 11-element. Hence

$$\begin{aligned} C(rr \to rr) &= \frac{1}{4}\sum_{a=1}^{8} \lambda_{11}^a \lambda_{11}^a = \frac{1}{4}(\lambda_{11}^3\lambda_{11}^3 + \lambda_{11}^8\lambda_{11}^8) \\ &= \frac{1}{4}\left(1 + \tfrac{1}{3}\right) = \tfrac{1}{3}. \end{aligned}$$

Because of the underlying exact SU(3) colour symmetry, there is no need to repeat the exercise for $gg \to gg$ or $bb \to bb$; the SU(3) colour symmetry guarantees that the same result will be obtained, thus

$$C(rr \to rr) = C(gg \to gg) = C(bb \to bb) = \tfrac{1}{3}. \tag{10.25}$$

For the second class of diagram of Figure 10.21, $rb \to rb$, the corresponding colour factor is

$$C(rb \to rb) = \frac{1}{4}\sum_{a=1}^{8} \lambda_{11}^{a}\lambda_{33}^{a}.$$

Hence only the gluons associated with the Gell-Mann matrices with non-zero entries in the 11 *and* 33 positions give a non-zero contribution, thus

$$C(rb \to rb) = \frac{1}{4}\sum_{a=1}^{8} \lambda_{11}^{a}\lambda_{33}^{a} = \frac{1}{4}\lambda_{11}^{8}\lambda_{33}^{8} = \frac{1}{4}\left(\tfrac{1}{\sqrt{3}} \cdot \tfrac{-2}{\sqrt{3}}\right) = -\frac{1}{6},$$

and, from the SU(3) colour symmetry,

$$\begin{aligned} &C(rb \to rb) = C(rg \to rg) = C(gr \to gr) = \\ &C(gb \to gb) = C(br \to br) = C(bg \to bg) = -\tfrac{1}{6}. \end{aligned} \tag{10.26}$$

For the third class of colour exchange of Figure 10.21, $rg \to gr$, the only non-zero contributions arise from the $\lambda$-matrices with non-zero entries in the 12 *and* 21 positions, therefore

$$C(rg \to gr) = \frac{1}{4}\sum_{a=1}^{8} \lambda_{21}^{a}\lambda_{12}^{a} = \frac{1}{4}\left(\lambda_{21}^{1}\lambda_{12}^{1} + \lambda_{21}^{2}\lambda_{12}^{2}\right) = \frac{1}{2},$$

and thus

$$\begin{aligned} &C(rb \to br) = C(rg \to gr) = C(gr \to rg) = \\ &C(gb \to bg) = C(br \to rb) = C(bg \to gb) = \tfrac{1}{2}. \end{aligned} \tag{10.27}$$

Finally, for the case where three different colours are involved, e.g. $rb \to bg$,

$$C(rb \to bg) = \frac{1}{4}\sum_{a=1}^{8} \lambda_{31}^{a}\lambda_{23}^{a}.$$

Because none of the $\lambda$-matrices has non-zero entries in both the 31 and 23 positions, the colour factor is zero. This should come as no surprise, colour is a conserved charge of the SU(3) colour symmetry and the process $rb \to bg$ would result in a net change of colour.

## Averaged colour factor

The colour factors calculated above account for the summation over the eight possible gluon intermediate states for a particular colour exchange $ik \to jl$. In the scattering process ud $\to$ ud, the colours of each of the initial-state quarks are equally likely to be $r$, $b$ or $g$. Therefore the nine possible initial-state colour combinations are equally probable. For a particular initial-state colour combination, the cross section will depend on the sum of the squared matrix elements for each of the nine possible orthogonal final-state colour combinations. The possible colour combinations are accounted for by the colour-averaged sum of squared matrix elements,

$$\langle|\mathcal{M}|^2\rangle = \frac{1}{9}\sum_{i,j,k,l=1}^{3} |\mathcal{M}(ij \to kl)|^2, \tag{10.28}$$

where the sum is over all possible colours in the initial- and final-state, and the factor of $\frac{1}{9}$ averages over the nine possible initial-state colour combinations. The colour part of (10.28),

$$\langle|C|^2\rangle = \frac{1}{9}\sum_{i,j,k,l=1}^{3} |C(ij \to kl)|^2, \tag{10.29}$$

can be evaluated using the expressions for the individual colour factors of (10.25)–(10.27). There are three colour combinations of the type $rr \to rr$ (i.e. $rr \to rr$, $bb \to bb$ and $gg \to gg$) each with an individual colour factor $\frac{1}{3}$, six combinations of the type $rb \to rb$ with colour factor $-\frac{1}{6}$ and six combinations of the type $rb \to br$ with colour factor $\frac{1}{2}$. Hence the overall colour factor is

$$\langle|C|^2\rangle = \frac{1}{9}\left[3\times\left(\frac{1}{3}\right)^2 + 6\times\left(-\frac{1}{6}\right)^2 + 6\times\left(\frac{1}{2}\right)^2\right] = \frac{2}{9}. \tag{10.30}$$

Hence, the entire effect of the 648 possible combinations of quark colours and types of gluons is encompassed into a single number.

The QCD cross section for the scattering process ud $\to$ ud can be obtained from the QED cross section for $e^-q \to e^-q$ of (8.19),

$$\frac{d\sigma}{dq^2} = \frac{2\pi Q_q^2\alpha^2}{q^4}\left[1 + \left(1 + \frac{q^2}{s}\right)^2\right],$$

by replacing $\alpha Q_q$ with $\alpha_S$ and by multiplying by the averaged colour factor of (10.30), to give

$$\frac{d\sigma}{dq^2} = \frac{4\pi\alpha_S^2}{9q^4}\left[1 + \left(1 + \frac{q^2}{\hat{s}}\right)^2\right], \tag{10.31}$$

where $\hat{s}$ is the centre-of-mass energy of the colliding ud system.

### 10.7.1 Colour in processes with antiquarks

Figure 10.22 shows the vertices for the basic QCD interaction between quarks and/or antiquarks. The quark current associated with the qqg vertex is given by (10.12)

$$j_{\mathrm{q}}^{\mu} = \overline{u}(p_3)c_j^{\dagger}\left\{-\tfrac{1}{2}ig_S\lambda^a\gamma^{\mu}\right\}c_i u(p_1),$$

where the outgoing quark enters as the adjoint spinor. In the equivalent expression for the $\overline{\mathrm{q}}\,\overline{\mathrm{q}}$g vertex, the incoming antiparticle is now represented by the adjoint spinor

$$j_{\overline{\mathrm{q}}}^{\mu} = \overline{v}(p_1)c_i^{\dagger}\left\{-\tfrac{1}{2}ig_S\lambda^a\gamma^{\mu}\right\}c_j v(p_3).$$

Consequently, the colour part of the expression is

$$c_i^{\dagger}\lambda^a c_j = c_i^{\dagger}\begin{pmatrix}\lambda^a_{1j}\\ \lambda^a_{2j}\\ \lambda^a_{3j}\end{pmatrix} = \lambda^a_{ij}.$$

The order of the indices $ij$ is swapped with respect to the quark case, $ji \to ij$. In general, the colour index associated with the adjoint spinor appears first, and thus the colour factor associated with the q$\overline{\mathrm{q}}$g annihilation vertex shown of Figure 10.22 is $\lambda^a_{ki}$.

Figure 10.23 shows the four possible combinations of two quarks/antiquarks interacting via the exchange of a single gluon. For the quark–antiquark and antiquark–antiquark scattering diagrams, the expressions for the colour factors are

$$C(\bar{i}\bar{k} \to \bar{j}\bar{l}) \equiv \frac{1}{4}\sum_{a=1}^{8}\lambda^a_{ij}\lambda^a_{kl}, \quad \text{and} \quad C(i\bar{k} \to j\bar{l}) \equiv \frac{1}{4}\sum_{a=1}^{8}\lambda^a_{ji}\lambda^a_{kl},$$

which can be compared to the expression for the colour factor for quark–quark scattering of (10.24). Because the Gell-Mann matrices have the property that either $\lambda^{\mathrm{T}} = \lambda$ or $\lambda^{\mathrm{T}} = -\lambda$, the same colour factors are obtained for qq and $\overline{\mathrm{q}}\,\overline{\mathrm{q}}$ scattering,

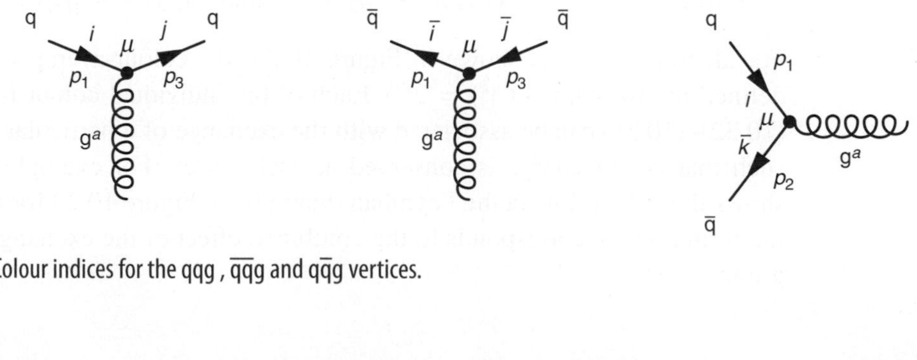


Fig. 10.22 Colour indices for the qqg, $\overline{\mathrm{q}}\overline{\mathrm{q}}$g and q$\overline{\mathrm{q}}$g vertices.

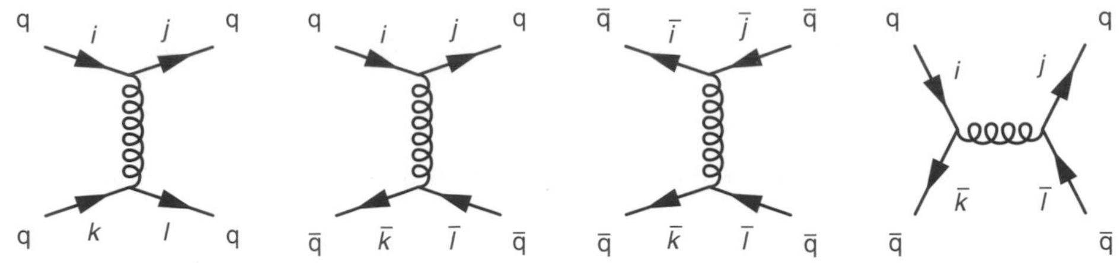


Fig. 10.23 The four diagrams involving the interaction of two quarks/antiquarks via a single gluon.

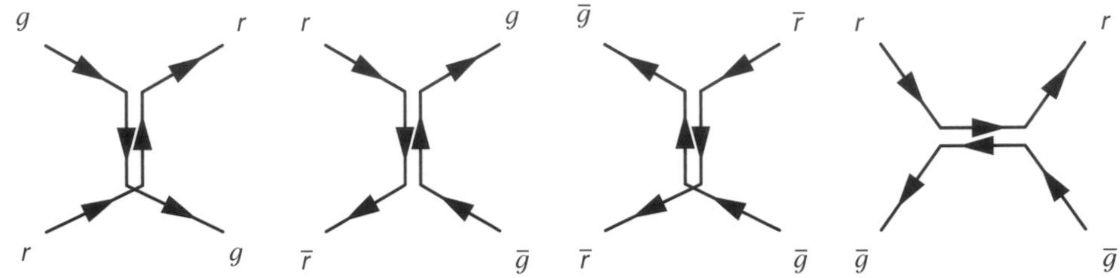


Fig. 10.24 Examples of colour flow in the $t$-channel scattering processes of ud $\rightarrow$ ud, u$\overline{\text{d}}$ $\rightarrow$ u$\overline{\text{d}}$ and $\overline{\text{u}}\overline{\text{d}}$ $\rightarrow$ $\overline{\text{u}}\overline{\text{d}}$ scattering, and the $s$-channel annihilation process q$\overline{\text{q}}$ $\rightarrow$ q$\overline{\text{q}}$. In each case the gluons exchanged are $r\overline{g}$ and $g\overline{r}$.

and it is straightforward to show that the non-zero colour factors for the $t$-channel scattering processes are

$$C(rr \rightarrow rr) = C(r\overline{r} \rightarrow r\overline{r}) = C(\overline{r}\overline{r} \rightarrow \overline{r}\overline{r}) = \tfrac{1}{3}, \tag{10.32}$$

$$C(rg \rightarrow rg) = C(r\overline{g} \rightarrow r\overline{g}) = C(\overline{r}\overline{g} \rightarrow \overline{r}\overline{g}) = -\tfrac{1}{6}, \tag{10.33}$$

$$C(rg \rightarrow gr) = C(r\overline{r} \rightarrow g\overline{g}) = C(\overline{r}\overline{g} \rightarrow \overline{g}\overline{r}) = \tfrac{1}{2}. \tag{10.34}$$

For the $s$-channel annihilation diagram, the expression for the individual colour factors is

$$C_s(i\overline{k} \rightarrow j\overline{l}) \equiv \frac{1}{4}\sum_{a=1}^{8} \lambda^a_{ki}\lambda^a_{jl},$$

from which it follows that

$$C_s(r\overline{r} \rightarrow r\overline{r}) = \tfrac{1}{3}, \quad C_s(r\overline{g} \rightarrow r\overline{g}) = \tfrac{1}{2} \quad \text{and} \quad C_s(r\overline{r} \rightarrow g\overline{g}) = -\tfrac{1}{6}. \tag{10.35}$$

For all four processes shown in Figure 10.23, the colour-averaged colour factor, defined in (10.29), is $\langle |C|^2 \rangle = 2/9$. Each of the individual colour factors given in (10.32)–(10.35) can be associated with the exchange of a particular type of gluon, such that colour charge is conserved at each vertex. For example, Figure 10.24 shows the colour flow in the Feynman diagrams of Figure 10.23 for the case where the virtual gluon corresponds to the combined effect of the exchange of $r\overline{g}$ and $g\overline{r}$ gluons.

### 10.7.2 *Colour sums revisited

The overall colour factor for quark (or antiquark) scattering via the exchange of a single gluon can be obtained directly by considering the factors that enter the expression for the matrix element squared. For example, the qq → qq matrix element $\mathcal{M}$ for colours $ik \to jl$ includes the colour factor

$$C = \frac{1}{4}\lambda^a_{ji}\lambda^a_{lk},$$

where summation over the repeated gluon indices is implied. The matrix element squared for this colour combination, $|\mathcal{M}|^2 = \mathcal{M}\mathcal{M}^\dagger$, is proportional to

$$\begin{aligned} CC^* &= \frac{1}{16}\lambda^a_{ji}\lambda^a_{lk}\cdot(\lambda^b_{ji})^*(\lambda^b_{lk})^* \\ &= \frac{1}{16}\lambda^a_{ji}\lambda^a_{lk}\lambda^b_{ij}\lambda^b_{kl}, \end{aligned} \qquad (10.36)$$

where the second line follows from the Gell-Mann matrices being Hermitian. The colour-averaged summed matrix element squared therefore can be written

$$\begin{aligned} \langle |C|^2\rangle = \langle CC^*\rangle &= \frac{1}{9}\frac{1}{16}\sum_{a,b=1}^{8}\sum_{ijkl=1}^{3}\lambda^a_{ji}\lambda^b_{ij}\lambda^a_{lk}\lambda^b_{kl} \\ &= \frac{1}{144}\sum_{a,b=1}^{8}\left[\mathrm{Tr}\left(\lambda^a\lambda^b\right)\right]^2. \end{aligned}$$

It is straightforward to show that $\mathrm{Tr}\left(\lambda^a\lambda^b\right) = 2\delta_{ab}$, and thus

$$\langle |C|^2\rangle = \frac{1}{144}\sum_{a,b=1}^{8}(2\delta_{ab})^2 = \frac{1}{144}\sum_{a=1}^{8}2^2 = \frac{2}{9}.$$

The same result will be obtained independent of the order in which the indices appear in the initial expression, and therefore the same colour-averaged colour factor is obtained for all four processes of Figure 10.23.

## 10.8 Heavy mesons and the QCD colour potential

Heavy quark $c\bar{c}$ (charmonium) and $b\bar{b}$ (bottomonium) bound states are observed as resonances in $e^+e^-$ annihilation, as seen previously in Figure 10.17. The multiple charmonium and bottomonium resonances correspond to eigenstates of the $q\bar{q}$ system in the QCD potential. Whilst only states with spin-parity $J^P = 1^-$ are produced in $e^+e^-$ annihilation, other states are observed in particle decays. Unlike the quarks

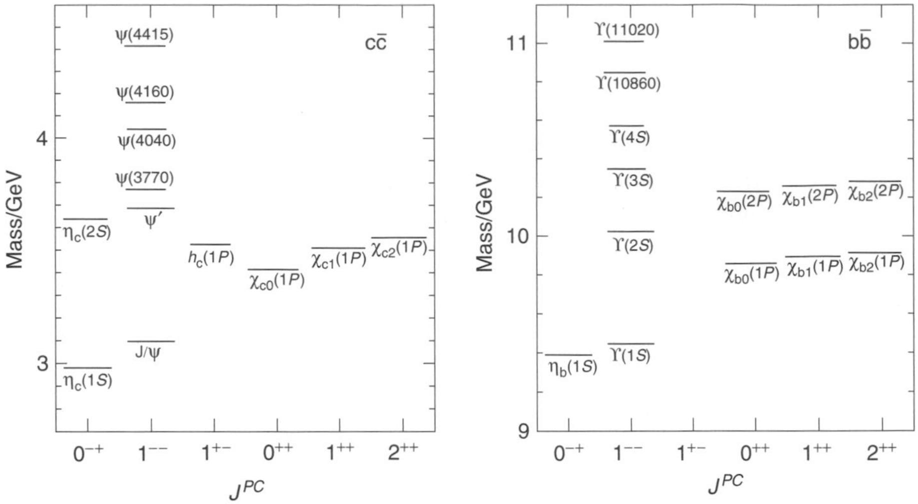


**Fig. 10.25** The masses and $J^{PC}$ assignments of the observed charmonium ($c\bar{c}$) and bottomonium ($b\bar{b}$) bound states, where $C$ is the charge conjugation quantum number discussed in Chapter 14.

in the light uds mesons, which are relativistic, the velocities of the heavy quarks in the charmonium and bottomonium states are relatively low, $\beta_c \sim 0.3$ and $\beta_b \sim 0.1$. In this case, the observed spectra of charmonium and bottomonium states, shown in Figure 10.25, provide a probe of the QCD potential in the non-relativistic limit.

In non-relativistic QCD (NRQCD), the interaction between two quarks (or between a quark and an antiquark) can be expressed as a static potential of the form $V(\mathbf{r})$. Owing to the gluon self-interactions (10.13), the potential at large distances is proportional to the separation of the quarks, $V(\mathbf{r}) \sim \kappa r$. The short-range component of the NRQCD potential can be obtained by considering the analogous situation for QED. The non-relativistic limit of QED gives rise to a repulsive Coulomb potential between two electrons (i.e. two particles), $V(\mathbf{r}) = \alpha/r$, and an attractive potential between an electron and its antiparticle, $V(\mathbf{r}) = -\alpha/r$. With the exception of the treatment of colour, which factorises from the spinor part, the fundamental QCD interaction has exactly the same $\overline{\phi}\gamma^\mu\psi$ form as QED. Therefore, the short-range NRQCD potential between two quarks must be

$$V_{qq}(\mathbf{r}) = +C\frac{\alpha_S}{r}, \tag{10.37}$$

and that between a quark and an antiquark is

$$V_{q\bar{q}}(\mathbf{r}) = -C\frac{\alpha_S}{r}, \tag{10.38}$$

where $C$ is the appropriate colour factor. Depending on the sign of this colour factor, which will depend on the colour wavefunction of the state, the short-range static potential for the $q\bar{q}$ system could be either attractive or repulsive.

From the colour confinement hypothesis, it is known that mesons are colour singlets, with a colour wavefunction

$$\psi = \frac{1}{\sqrt{3}}(r\bar{r} + g\bar{g} + b\bar{b}).$$

Thus, the expectation value of the NRQCD potential for a meson can be written

$$\langle V_{\mathrm{q\bar{q}}} \rangle = \langle \psi | V_{\mathrm{QCD}} | \psi \rangle = \tfrac{1}{3}\left( \langle r\bar{r} | V_{\mathrm{QCD}} | r\bar{r} \rangle + \cdots + \langle r\bar{r} | V_{\mathrm{QCD}} | b\bar{b} \rangle + \cdots \right), \tag{10.39}$$

where the dots indicate the other seven colour combinations. From the form of the NRQCD potential identified in (10.38),

$$\langle r\bar{r} | V_{\mathrm{QCD}} | r\bar{r} \rangle = -C(r\bar{r} \to r\bar{r}) \frac{\alpha_S}{\mathrm{r}} \quad \text{and} \quad \langle r\bar{r} | V_{\mathrm{QCD}} | b\bar{b} \rangle = -C(r\bar{r} \to b\bar{b}) \frac{\alpha_S}{\mathrm{r}},$$

and therefore the expectation value of the QCD potential of (10.39) can be written

$$\langle V_{\mathrm{q\bar{q}}} \rangle = -\frac{\alpha_S}{3\mathrm{r}} \left( C(r\bar{r} \to r\bar{r}) + \cdots + C(r\bar{r} \to b\bar{b}) + \cdots \right).$$

This expression contains three terms of the form $r\bar{r} \to r\bar{r}$ and six of the form $r\bar{r} \to g\bar{g}$, and therefore

$$\langle V_{\mathrm{q\bar{q}}} \rangle = -\frac{\alpha_S}{3\mathrm{r}} \left[ 3 \times C(r\bar{r} \to r\bar{r}) + 6 \times C(r\bar{r} \to g\bar{g}) \right].$$

Using the expressions for the colour factors for the $t$-channel exchange of a gluon between a quark and an antiquark, given in (10.32) and (10.34), the non-relativistic QCD potential can be written

$$\langle V_{\mathrm{q\bar{q}}} \rangle = -\frac{\alpha_S}{3\mathrm{r}} \left[ 3 \times \tfrac{1}{3} + 6 \times \tfrac{1}{2} \right] = -\frac{4}{3} \frac{\alpha_S}{\mathrm{r}}.$$

Hence, the short range NRQCD potential in the $\mathrm{q\bar{q}}$ colour singlet state is attractive. Adding in the long-range term of (10.13), gives the expression for the NRQCD potential

$$V_{\mathrm{q\bar{q}}}(\mathbf{r}) = -\frac{4}{3} \frac{\alpha_S}{\mathrm{r}} + \kappa \mathrm{r}. \tag{10.40}$$

The non-relativistic QCD potential of (10.40) can be used to obtain the predicted spectra for the $\mathrm{c\bar{c}}$ and $\mathrm{b\bar{b}}$ bound states. The more accurate predictions are obtained for the $\mathrm{b\bar{b}}$ system, where the non-relativistic treatment is a good approximation. Reasonable agreement with observed $l = 0$ and $l = 1$ charmonium and bottomonium states of Figure 10.25 is found assuming $\lambda \approx 1\,\mathrm{GeV/fm}$, providing further evidence for the presence of the linear term in the potential, which is believed to be responsible for colour confinement.

Figure 10.26 shows the non-relativistic QCD potential of (10.40) for $\alpha_S = 0.2$ and $\kappa = 1\,\mathrm{GeV/fm}$. The potential energy becomes positive at approximately 0.25 fm, with the linear term dominating at larger radii, setting the length scale for confinement for these heavy quark states.

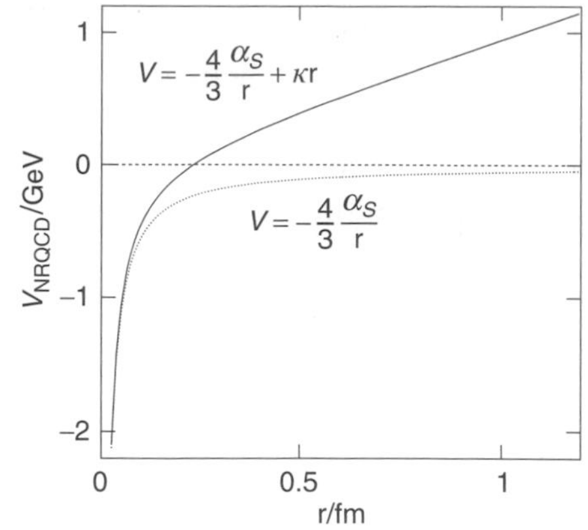


Fig. 10.26 The approximate form of the non-relativistic QCD potential for a bound $q\bar{q}$ state, assuming $\alpha_S = 0.2$ and $\kappa = 1\,\text{GeV}/\,\text{fm}$.

## 10.9 Hadron–hadron collisions

Hadron colliders, either proton–proton or proton–antiproton, provide a route to achieving higher centre-of-mass energies than is possible with circular $e^+e^-$ colliders, and are central to the search for the production of new particles at high-mass scales. The underlying process in hadron–hadron collisions is the interaction of two partons, which can be either quarks, antiquarks or gluons.

### 10.9.1 Hadron collider event kinematics

In electron–proton elastic scattering, a single variable was sufficient to describe the event kinematics. This was chosen to be the scattering angle of the electron. In electron–proton deep inelastic scattering two variables are required, reflecting the additional degree of freedom associated with the unknown momentum fraction $x$ of the struck quark. In hadron–hadron collisions, the momentum fractions $x_1$ and $x_2$ of the two interacting partons are unknown, and the event kinematics have to be described by three variables, for example $Q^2$, $x_1$ and $x_2$. These three independent kinematic variables can be related to three experimentally well-measured quantities. In hadron collider experiments, the scattered partons are observed as jets. In a process such as pp $\rightarrow$ two jets + $X$, the angles of the two-jets with respect to the beam axis are relatively well measured. Consequently, differential cross sections are usually described in terms of these two jet angles and the component of momentum of one of the jets in the plane transverse to the beam axis, referred to as the transverse momentum

$$p_T = \sqrt{p_x^2 + p_y^2},$$

where the $z$-axis defines the beam direction.

At a hadron–hadron collider, such as the LHC, the collisions take place in the centre-of-mass frame of the pp system, which is not the centre-of-mass frame of the colliding partons. The net longitudinal momentum of the colliding parton–parton system is given by $(x_1 - x_2)E_p$, where $E_p$ is the energy of the proton. Consequently, in a process such pp $\to$ 2 jets + $X$, the two final-state jets are boosted along the beam direction. For this reason, the jet angles are usually expressed in terms of the rapidity $y$, defined by

$$y = \frac{1}{2}\ln\left(\frac{E + p_z}{E - p_z}\right), \tag{10.41}$$

where $E$ and $p_z$ are the measured energy and $z$-component of momentum of a jet. The use of rapidity has the advantage that rapidity differences are invariant under boosts along the beam direction. This can be seen by considering the effect of a Lorentz transformation along the $z$-axis, where the rapidity $y$ in the boosted frame of reference is given by

$$\begin{aligned} y' &= \frac{1}{2}\ln\left[\frac{E' + p_z'}{E' - p_z'}\right] = \frac{1}{2}\ln\left[\frac{\gamma(E - \beta p_z) + \gamma(p_z - \beta E)}{\gamma(E - \beta p_z) - \gamma(p_z - \beta E)}\right] \\ &= \frac{1}{2}\ln\left[\frac{(1-\beta)(E + p_z)}{(1+\beta)(E - p_z)}\right] \\ &= y + \frac{1}{2}\ln\left(\frac{1-\beta}{1+\beta}\right). \end{aligned}$$

Hence, differences in rapidities are the same measured in any two frames, $\Delta y' = \Delta y$. Therefore, the *a priori* unknown longitudinal boost of the parton–parton system does not affect the distribution of rapidity differences.

The invariant mass of the system of particles forming a jet is referred to as the jet mass. The jet mass is not the same as the mass of the primary parton; it is mainly generated in the hadronisation process. For high-energy jets, the jet mass is usually small compared to the jet energy and $p_z \approx E\cos\theta$, where $\theta$ is the polar angle of the jet with respect to the beam axis. Hence the rapidity can be approximated by

$$y \approx \frac{1}{2}\ln\left(\frac{1+\cos\theta}{1-\cos\theta}\right) = \frac{1}{2}\ln\left(\cot^2\tfrac{\theta}{2}\right).$$

Therefore, the pseudorapidity $\eta$ defined as

$$\eta \equiv -\ln\left(\tan\tfrac{\theta}{2}\right)$$

can be used in place of rapidity $y$ when jet masses can be neglected. Figure 10.27 illustrates the polar angle ranges covered by different regions of pseudorapidity.

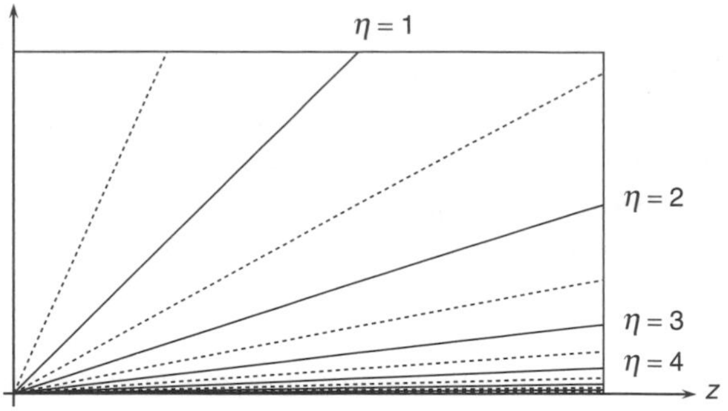


Fig. 10.27 Pseudorapidity regions relative to the beam axis ($z$).

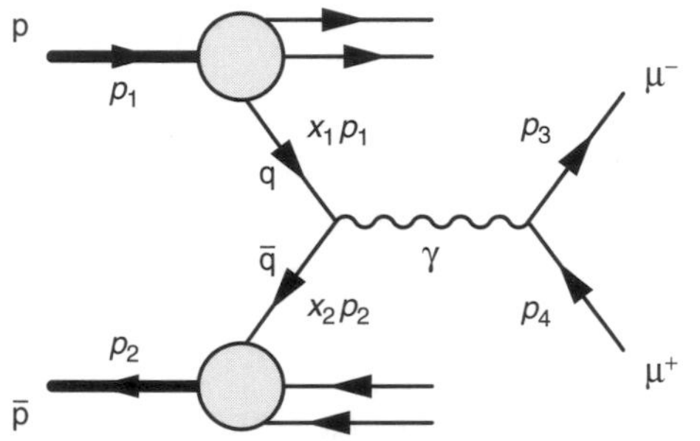


Fig. 10.28 The lowest-order Feynman diagram for the Drell–Yan process $p\overline{p} \to \mu^+\mu^- X$.

Broadly speaking, the differential cross sections for jet production in hadron–hadron collisions are approximately constant in pseudorapidity, implying that roughly equal numbers of jets are observed in each interval of pseudorapidity shown in Figure 10.27, reflecting the forward nature of jet production in pp and $p\overline{p}$ collisions.

### 10.9.2 The Drell–Yan process

The QED production of a pair of leptons in hadron–hadron collisions from the annihilation of an antiquark and a quark, shown in Figure 10.28, is known as the Drell–Yan process. It provides a useful example of a cross section calculation for hadron–hadron collisions, in this case $p\overline{p} \to \mu^+\mu^- X$, where $X$ represents the remnants of the colliding hadrons.

The QED annihilation cross section for $e^+e^- \to \mu^+\mu^-$ was calculated in Chapter 7. The corresponding cross section for $q\overline{q} \to \mu^+\mu^-$ annihilation is

$$\sigma(q\overline{q} \to \mu^+\mu^-) = \frac{1}{N_c} Q_q^2 \frac{4\pi\alpha^2}{3\hat{s}}, \tag{10.42}$$

where $Q_q$ is the quark/antiquark charge and $\hat{s}$ is the centre-of-mass energy of the colliding $q\overline{q}$ system. The factor $1/N_c$, where $N_c = 3$ is the number of colours, accounts for the conservation of colour charge, which implies that of the nine possible colour combinations of the $\mathrm{q\overline{q}}$ system, the annihilation process can only occur for three, $r\overline{r}$, $b\overline{b}$ and $g\overline{g}$. From the definition of the parton distribution functions, the contribution to the $\mathrm{p\overline{p}}$ Drell–Yan cross section from an up-quark within the proton with momentum fraction $x_1 \to x_1 + \delta x_1$ annihilating with an anti-up-quark within the antiproton with momentum fraction $x_2 \to x_2 + \delta x_2$ is

$$\mathrm{d}^2\sigma = Q_\mathrm{u}^2 \frac{4\pi\alpha^2}{9\hat{s}} u^\mathrm{p}(x_1)\mathrm{d}x_1\, \overline{u}^{\overline{\mathrm{p}}}(x_2)\,\mathrm{d}x_2, \tag{10.43}$$

where $\overline{u}^{\overline{\mathrm{p}}}(x_2)$ is the PDF for the anti-up-quark in the antiproton. Because the antiquark PDFs within the antiproton will be identical to the corresponding quark PDFs in the proton, $\overline{u}^{\overline{\mathrm{p}}}(x) = u^\mathrm{p}(x) \equiv u(x)$, Equation (10.43) can be written

$$\mathrm{d}^2\sigma = \frac{4}{9}\cdot\frac{4\pi\alpha^2}{9\hat{s}} u(x_1)u(x_2)\,\mathrm{d}x_1\mathrm{d}x_2. \tag{10.44}$$

The centre-of-mass energy of the $\mathrm{q\overline{q}}$ system can be expressed in terms of that of the proton–antiproton system using

$$\hat{s} = (x_1p_1 + x_2p_2)^2 = x_1^2p_1^2 + x_2^2p_2^2 + 2x_1x_2p_1\cdot p_2.$$

In the high-energy limit, where the proton mass squared can be neglected, $p_1^2 = p_2^2 \approx 0$ and

$$\hat{s} \approx x_1x_2(2p_1\cdot p_2) = x_1x_2s,$$

where $s$ is the centre-of-mass energy of the colliding $\mathrm{p\overline{p}}$ system. Hence (10.44), expressed in terms of $s$, becomes

$$\mathrm{d}^2\sigma = \frac{4}{9}\cdot\frac{4\pi\alpha^2}{9x_1x_2s} u(x_1)u(x_2)\,\mathrm{d}x_1\mathrm{d}x_2. \tag{10.45}$$

Accounting for the (smaller) contribution from the annihilation of a $\mathrm{\overline{u}}$ in the proton with a u in the antiproton and the contribution from $\mathrm{d\overline{d}}$ annihilation, leads to

$$\mathrm{d}^2\sigma = \frac{4\pi\alpha^2}{9x_1x_2s}\left[\tfrac{4}{9}\left\{u(x_1)u(x_2) + \overline{u}(x_1)\overline{u}(x_2)\right\} + \tfrac{1}{9}\left\{d(x_1)d(x_2) + \overline{d}(x_1)\overline{d}(x_2)\right\}\right]\mathrm{d}x_1\mathrm{d}x_2. \tag{10.46}$$

The Drell–Yan differential cross section is most usefully expressed in terms of the experimental observables. Here a suitable choice is the rapidity and the invariant mass of the $\mu^+\mu^-$ system, both of which can be determined from the momenta of the $\mu^+$ and $\mu^-$ as reconstructed in the tracking system of the detector. The coordinate transformation from $x_1$ and $x_2$ to these experimental observables is not entirely straightforward, but is shown to illustrate the general principle. The invariant mass of the $\mu^+\mu^-$ system is equal to the centre-of-mass energy of the colliding partons,

$$M^2 = x_1x_2s. \tag{10.47}$$

The rapidity of the $\mu^+\mu^-$ system is given by

$$y = \frac{1}{2}\ln\left(\frac{E_3 + E_4 + p_{3z} + p_{4z}}{E_3 + E_4 - p_{3z} - p_{4z}}\right) = \frac{1}{2}\ln\left(\frac{E_q + E_{\overline{q}} + p_{qz} + p_{\overline{q}z}}{E_q + E_{\overline{q}} - p_{qz} - p_{\overline{q}z}}\right),$$

where the equality of four-momenta of the $\mu^+\mu^-$ system and that of the colliding partons follows from energy and momentum conservation. The four-momenta of the colliding q and $\overline{q}$ are respectively given by

$$p_q = \frac{\sqrt{s}}{2}(x_1, 0, 0, x_1) \quad \text{and} \quad p_{\overline{q}} = \frac{\sqrt{s}}{2}(x_2, 0, 0, -x_2),$$

and hence

$$y = \frac{1}{2}\ln\left(\frac{(x_1 + x_2) + (x_1 - x_2)}{(x_1 + x_2) - (x_1 - x_2)}\right) = \frac{1}{2}\ln\frac{x_1}{x_2}. \tag{10.48}$$

From (10.47) and (10.48), $x_1$ and $x_2$ can be written in terms of $M$ and $y$,

$$x_1 = \frac{M}{\sqrt{s}}e^{y} \quad \text{and} \quad x_2 = \frac{M}{\sqrt{s}}e^{-y}. \tag{10.49}$$

The differential cross section in terms of $\mathrm{d}x_1\mathrm{d}x_2$ can be expressed in terms of $\mathrm{d}y\,\mathrm{d}M$ using the determinant of the Jacobian matrix for the coordinate transformation

$$\mathrm{d}y\,\mathrm{d}M = \frac{\partial(y, M)}{\partial(x_1, x_2)}\mathrm{d}x_1\mathrm{d}x_2 = \begin{vmatrix} \frac{\partial y}{\partial x_1} & \frac{\partial y}{\partial x_2} \\ \frac{\partial M}{\partial x_1} & \frac{\partial M}{\partial x_2} \end{vmatrix}\mathrm{d}x_1\,\mathrm{d}x_2,$$

where the partial derivatives obtained from (10.47) and (10.48) give

$$\mathrm{d}y\,\mathrm{d}M = \frac{s}{2M}\mathrm{d}x_1\mathrm{d}x_2.$$

Hence the differential cross section of (10.46) can be expressed as

$$\mathrm{d}^2\sigma = \frac{4\pi\alpha^2}{9M^2}f(x_1, x_2)\frac{2M}{s}\mathrm{d}y\,\mathrm{d}M,$$

where

$$f(x_1, x_2) = \left[\tfrac{4}{9}\left\{u(x_1)u(x_2) + \overline{u}(x_1)\overline{u}(x_2)\right\} + \tfrac{1}{9}\left\{d(x_1)d(x_2) + \overline{d}(x_1)\overline{d}(x_2)\right\}\right],$$

and thus, the Drell–Yan differential cross section, written in terms of the invariant mass and rapidity of the $\mu^+\mu^-$ system, is

$$\frac{\mathrm{d}^2\sigma}{\mathrm{d}y\,\mathrm{d}M} = \frac{8\pi\alpha^2}{9Ms}f(x_1, x_2),$$

where $x_1$ and $x_2$ are given by (10.49).

The above treatment of the Drell–Yan process considered only the QED photon-exchange diagram. However, any neutral particle which couples to both quarks and muons can contribute. For example, Figure 10.29 shows the measured differential cross section for $p\overline{p} \to \mu^+\mu^- X$ from the CDF experiment at the Tevatron,

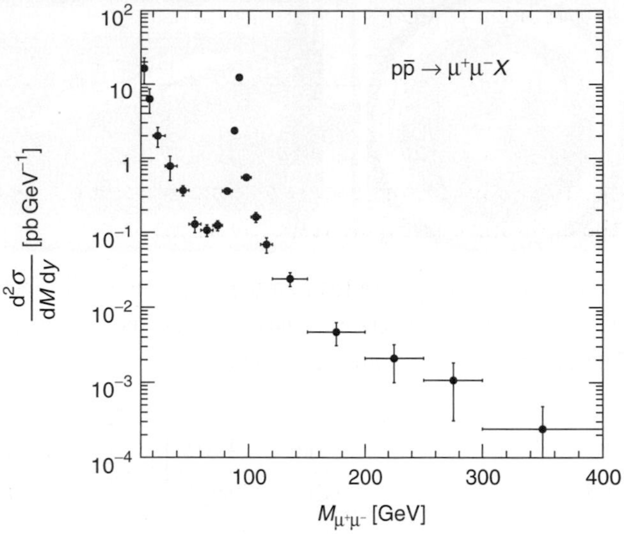


Fig. 10.29 The measured Drell–Yan cross section in $p\bar{p}$ collisions at $\sqrt{s} = 1.8$ TeV in the CDF detector at the Tevatron collider. Adapted from Abe *et al.* (1999).

which operated from 1989 to 2011. The strong enhancement in the cross section at $M_{\mu^+\mu^-} \sim 91$ GeV is due to the resonant production of the Z boson, through the annihilation process $q\bar{q} \to Z \to \mu^+\mu^-$. The Drell–Yan process also provides a way of searching for physics beyond the Standard Model through the production of new massive neutral particles that couple to both quarks and leptons, through the process $q\bar{q} \to X^0 \to \mu^+\mu^-$. To date, no such signals of physics beyond the Standard Model have been observed.

### 10.9.3 Jet production at the LHC

The Large Hadron Collider at CERN is the highest energy accelerator ever built. It is a pp collider designed to operate at an ultimate centre-of-mass energy of 14 TeV. The LHC commenced full operation at $\sqrt{s} = 7$ TeV in 2010 and ran at $\sqrt{s} = 8$ TeV in 2012. The most common, although not the most interesting, high-energy process at the LHC is the QCD production of two-jets. Figure 10.30 shows an example of a two-jet event recorded at $\sqrt{s} = 7$ TeV in the ATLAS experiment. Since the colliding partons have no momentum transverse to the beam axis, the jets are produce back to back in the transverse plane and have equal and opposite transverse momenta, $p_T$. In the other view, the jets are not back to back due to the boost of the final-state system from the net momentum of the colliding partons along the beam axis, $(x_1 - x_2)\sqrt{s}/2$.

The cross section for the production of two jets from the $t$-channel gluon exchange process $qq \to qq$ is given by (10.31),

Fig. 10.30 An example of a pp $\rightarrow$ two-jets $X$ event observed in the ATLAS detector a the LHC: (left) the transverse view (perpendicular to the beam direction) and (right) the $yz$-view with the $z$-axis along the beam direction. Reproduced courtesy of the ATLAS collaboration.

$$\frac{\mathrm{d}\sigma}{\mathrm{d}Q^2} = \frac{4\pi\alpha_S^2}{9Q^4}\left[1+\left(1-\frac{Q^2}{\hat{s}}\right)^2\right],$$

where $Q^2 = -q^2$ and $\hat{s} = x_1 x_2 s$ is the centre-of-mass energy of the colliding quarks. The contribution to the proton–proton cross section, expressed in terms of the parton density functions, is therefore

$$\frac{\mathrm{d}\sigma}{\mathrm{d}Q^2} = \frac{4\pi\alpha_S^2}{9Q^4}\left[1+\left(1-\frac{Q^2}{s x_1 x_2}\right)^2\right] g(x_1, x_2)\,\mathrm{d}x_1\mathrm{d}x_2.$$

where $g(x_1, x_2)$ is the sum over the products of the relevant parton distribution functions for the scattering process qq $\rightarrow$ qq, which for up- and down-quarks is

$$g(x_1, x_2) = [u(x_1)u(x_2) + u(x_1)d(x_2) + d(x_1)u(x_2) + d(x_1)d(x_2)]\,.$$

The differential cross section therefore can be written as

$$\frac{\mathrm{d}^3\sigma}{\mathrm{d}Q^2\,\mathrm{d}x_1\,\mathrm{d}x_2} = \frac{4\pi\alpha_S^2}{9Q^4}\left[1+\left(1-\frac{Q^2}{s x_1 x_2}\right)^2\right] g(x_1, x_2). \qquad (10.50)$$

This expression has three degrees of freedom; one from the underlying elastic scattering process, here written in terms of $Q^2$, and one from each of the parton momentum fractions, $x_1$ and $x_2$. In the process pp $\rightarrow$ two-jets $X$, thc experimentally well-measured quantities are the rapidities of the two final-state jets, $y_3$ and $y_4$, and the magnitude of the transverse momentum, $p_\mathrm{T}$ (which is the same for both jets). Equation (10.50) can be written in terms of these measured quantities using the determinant of the Jacobian for the coordinate transformation from $\{Q^2, x_1, x_2\}$ to $\{p_\mathrm{T}, y_3, y_4\}$ (see Problems 10.6 and 10.7). In principle, given knowledge of the PDFs, it would be possible to calculate the lowest-order QCD contribution to the LHC two-jet production cross section from the process qq $\rightarrow$ qq and express it in terms of these three experimental observables. However, qq $\rightarrow$ qq is just one of a number of parton-level processes that contribute to pp $\rightarrow$ two-jets $X$ at the LHC. For example, some of the other Feynman diagrams resulting in a two-jet

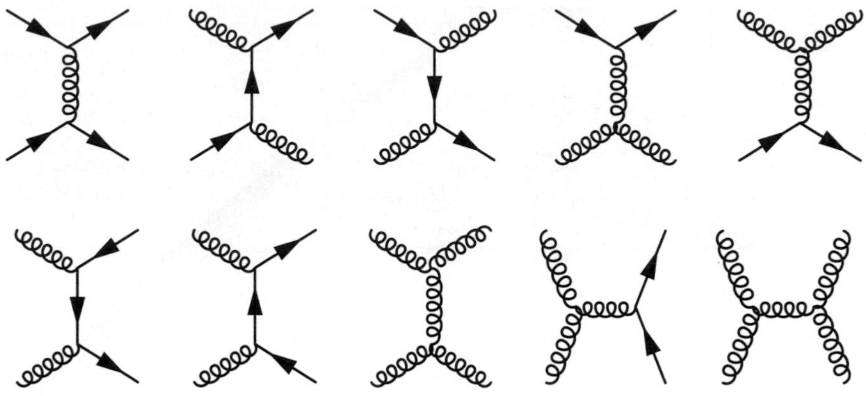

Fig. 10.31 Feynman diagrams for two-jet production in proton–proton collisions. There are also diagrams involving initial-state antiquarks.

final state are shown in Figure 10.31. The contributions from all processes, including the interference between diagrams with the same final-state partons, need to be summed to obtain the cross section for pp $\to$ two-jets $X$. Furthermore, for accurate predictions, the effects of higher-order QCD diagrams also need to be considered.

At this point, it should be clear that unlike the case of electron–positron annihilation, cross section calculations for the LHC are highly complex. Not only are the PDFs required, but multiple diagrams are involved and higher orders have to be included. In practice, such calculations are performed numerically in highly sophisticated computer programs. Nevertheless, the comparison of the predictions from these calculations with the experimental data from the LHC provides a powerful test of QCD. For example, Figure 10.32 shows early data from the CMS experiment. The plot shows the inclusive jet production cross section $\mathrm{d}^2\sigma/\mathrm{d}p_\mathrm{T}\mathrm{d}y$ in intervals of $\Delta y = 0.5$ of rapidity (which correspond to different ranges of polar angles in the detector). The $p_\mathrm{T}$ distribution is peaked towards zero, reflecting the $1/Q^4$ propagator term and the large values of the PDFs at low $x$. The measured cross sections for each interval of rapidity are similar, with roughly equal numbers of jets being observed in each of the (pseudo)rapidity intervals shown in Figure 10.27, demonstrating that jets are produced preferentially in the forward directions.

The data of Figure 10.32 are compared to next-to-leading-order (NLO) QCD predictions using the current knowledge of the PDFs. The predicted cross sections are in good agreement with the data that span a wide range of jet $p_\mathrm{T}$. In general, QCD is found to provide an excellent description of jet phenomena in hadron–hadron collisions. The success of QCD in describing the experimental results is an important achievement of modern particle physics and provides overwhelming evidence for the existence of the underlying SU(3) gauge symmetry.

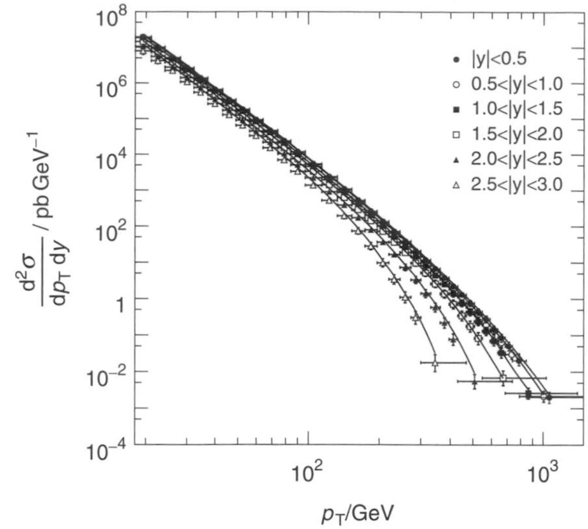


Fig. 10.32 The measurement of the inclusive differential cross section for jet production from data recorded at $\sqrt{s}$ = 7 TeV in the CMS experiment at the LHC. The curves are the predicted cross sections from NLO QCD. Adapted from Chatrchyan *et al.* (2011).

## Summary

Quantum Chromodynamics is the quantum field theory of the strong interaction. It corresponds to a non-Abelian SU(3) local gauge symmetry, with eight gluons associated with the eight generators. The interactions between the gluons and quarks are described by the qqg vertex factor

$$-\tfrac{1}{2} i g_S \lambda^a_{ji} \gamma^\mu,$$

where $i$ and $j$ are the colour charges of the quarks. The corresponding Feynman rule for the gluon propagator is

$$-i\frac{g_{\mu\nu}}{q^2}\delta^{ab}.$$

Whilst the Feynman rules for the QCD vertex and the gluon propagator resemble those of QED, the presence of gluon self-interactions leads to very different behaviour. For example, colour is confined and all freely propagating particles are colour singlet states; free quarks and gluons are not observed.

The running of $\alpha_S(Q^2)$ implies that the strength of the QCD interaction decreases with energy scale, a property known as asymptotic freedom. As a consequence, perturbative calculations can be used for high-energy QCD processes. Despite the practical difficulties of performing accurate calculations, QCD is found to provide an excellent description of hadron collider data and the SU(3) local gauge symmetry should be considered to be an experimentally established fact.

## Problems

**10.1** By considering the symmetry of the wavefunction, explain why the existence of the $\Omega^-(sss)$ $L = 0$ baryon provides evidence for a degree of freedom in addition to space $\times$ spin $\times$ flavour.

**10.2** From the expression for the running of $\alpha_S$ with $N_f = 3$, determine the value of $q^2$ at which $\alpha_S$ appears to become infinite. Comment on this result.

**10.3** Find the overall "colour factor" for $qq \to qq$ if QCD corresponded to a SU(2) colour symmetry.

**10.4** Calculate the non-relativistic QCD potential between quarks $q_1$ and $q_2$ in a $q_1q_2q_3$ baryon with colour wavefunction

$$\psi = \frac{1}{\sqrt{6}}(rgb - grb + gbr - bgr + brg - rbg).$$

**10.5** Draw the lowest-order QCD Feynman diagrams for the process $p\bar{p} \to$ two-jets $+ X$, where $X$ represents the remnants of the colliding hadrons.

**10.6** The observed events in the process $pp \to$ two-jets at the LHC can be described in terms of the jet $p_T$ and the jet rapidities $y_3$ and $y_4$.

(a) Assuming that the jets are massless, $E^2 = p_T^2 + p_z^2$, show that the four-momenta of the final-state jets can be written as

$$p_3 = (p_T \cosh y_3, +p_T \sin\phi, +p_T \cos\phi, p_T \sinh y_3),$$
$$p_4 = (p_T \cosh y_4, -p_T \sin\phi, -p_T \cos\phi, p_T \sinh y_4).$$

(b) By writing the four-momenta of the colliding partons in a pp collision as

$$p_1 = \frac{\sqrt{s}}{2}(x_1, 0, 0, x_1) \quad \text{and} \quad p_2 = \frac{\sqrt{s}}{2}(x_2, 0, 0, -x_1),$$

show that conservation of energy and momentum implies

$$x_1 = \frac{p_T}{\sqrt{s}}(e^{+y_3} + e^{+y_4}) \quad \text{and} \quad x_2 = \frac{p_T}{\sqrt{s}}(e^{-y_3} + e^{-y_4}).$$

(c) Hence show that

$$Q^2 = p_T^2(1 + e^{y_4 - y_3}).$$

**10.7** Using the results of the previous question show that the Jacobian

$$\frac{\partial(y_3, y_4, p_T^2)}{\partial(x_1, x_2, q^2)} = \frac{1}{x_1 x_2}.$$

**10.8** The total cross section for the Drell–Yan process $p\overline{p} \rightarrow \mu^+\mu^- X$ was shown to be

$$\sigma_{DY} = \frac{4\pi\alpha^2}{81s}\int_0^1\int_0^1 \frac{1}{x_1 x_2}\left[4u(x_1)u(x_2) + 4\overline{u}(x_1)\overline{u}(x_2) + d(x_1)d(x_2) + \overline{d}(x_1)\overline{d}(x_2)\right] dx_1 dx_2\,.$$

(a) Express this cross section in terms of the valence quark PDFs and a single PDF for the sea contribution, where $S(x) = \overline{u}(x) = \overline{d}(x)$.
(b) Obtain the corresponding expression for $pp \rightarrow \mu^+\mu^- X$.
(c) Sketch the region in the $x_1$–$x_2$ plane corresponding $s_{q\overline{q}} > s/4$. Comment on the expected ratio of the Drell–Yan cross sections in pp and $p\overline{p}$ collisions (at the same centre-of-mass energy) for the two cases: (i) $\hat{s} \ll s$ and (ii) $\hat{s} > s/4$, where $\hat{s}$ is the centre-of-mass energy of the colliding partons.

**10.9** Drell–Yan production of $\mu^+\mu^-$-pairs with an invariant mass $Q^2$ has been studied in $\pi^\pm$ interactions with carbon (which has equal numbers of protons and neutrons). Explain why the ratio

$$\frac{\sigma(\pi^+ C \rightarrow \mu^+\mu^- X)}{\sigma(\pi^- C \rightarrow \mu^+\mu^- X)}$$

tends to unity for small $Q^2$ and tends to $\frac{1}{4}$ as $Q^2$ approaches $s$.

# 11 The weak interaction

This chapter provides an introduction to the weak interaction, which is mediated by the massive $W^+$ and $W^-$ bosons. The main topics covered are: the origin of parity violation; the *V–A* form of the interaction vertex; and the connection to Fermi theory, which is the effective low-energy description of the weak charged current. The calculation of the decay rate of the charged pion is used to illustrate the rôle of helicity in weak decays. The purpose of this chapter is to describe the overall structure of the weak interaction; the applications are described in the following chapters on charged-current interactions, neutrino oscillations and CP violation in the weak decays of neutral mesons.

## 11.1 The weak charged-current interaction

At the fundamental level, QED and QCD share a number of common features. Both interactions are mediated by massless neutral spin-1 bosons and the spinor part of the QED and QCD interaction vertices have the same $\overline{u}(p')\gamma^\mu u(p)$ form. The *charged-current* weak interaction differs in almost all respects. It is mediated by massive charged $W^\pm$ bosons and consequently couples together fermions differing by one unit of electric charge. It is also the only place in the Standard Model where parity is not conserved. The parity violating nature of the interaction can be directly related to the form of the interaction vertex, which differs from that of QED and QCD.

## 11.2 Parity

The parity operation is equivalent to spatial inversion through the origin, $\mathbf{x} \to -\mathbf{x}$. In general, in quantum mechanics the parity transformation can be associated with the operator $\hat{P}$, defined by

$$\psi(\mathbf{x}, t) \to \psi'(\mathbf{x}, t) = \hat{P}\psi(\mathbf{x}, t) = \psi(-\mathbf{x}, t).$$

The original wavefunction is clearly recovered if the parity operator is applied twice,

$$\hat{P}\hat{P}\psi(\mathbf{x},t) = \hat{P}\psi(-\mathbf{x},t) = \psi(\mathbf{x},t),$$

and hence the parity operator is its own inverse,

$$\hat{P}\hat{P} = I. \tag{11.1}$$

If physics is invariant under parity transformations, then the parity operation must be unitary

$$\hat{P}^{\dagger}\hat{P} = I. \tag{11.2}$$

From (11.1) and (11.2), it can be inferred that

$$\hat{P}^{\dagger} = \hat{P},$$

and therefore $\hat{P}$ is a Hermitian operator that corresponds to an observable property of a quantum-mechanical system. Furthermore, *if* the interaction Hamiltonian commutes with $\hat{P}$, parity is an observable *conserved* quantity in the interaction. In this case, if $\psi(\mathbf{x},t)$ is an eigenstate of the Hamiltonian, it is also an eigenstate of the parity operator with an eigenvalue $P$,

$$\hat{P}\psi(\mathbf{x},t) = P\psi(\mathbf{x},t).$$

Acting on this eigenvalue equation with $\hat{P}$ gives

$$\hat{P}\hat{P}\psi(\mathbf{x},t) = P\hat{P}\psi(\mathbf{x},t) = P^2\psi(\mathbf{x},t),$$

which implies that $P^2 = 1$ since $\hat{P}\hat{P} = I$. Because $\hat{P}$ is Hermitian, its eigenvalues are real and are therefore equal to $\pm 1$.

### 11.2.1 Intrinsic parity

Fundamental particles, despite being point-like, possess an intrinsic parity. In Section 4.9, it was shown that the parity operator for Dirac spinors is $\gamma^0$, which in the Dirac–Pauli matrix representation is

$$\hat{P} = \gamma^0 = \begin{pmatrix} 1 & 0 & 0 & 0 \\ 0 & 1 & 0 & 0 \\ 0 & 0 & -1 & 0 \\ 0 & 0 & 0 & -1 \end{pmatrix}.$$

It was also shown that spin-half *particles*, which necessarily satisfy the Dirac equation, have the opposite parity to the corresponding *antiparticles*. By convention, the particle states are defined to have positive intrinsic parity; for example $P(e^-) = P(\nu_e) = P(q) = +1$, and therefore antiparticles have negative intrinsic parity, for example $P(\mathrm{e}^+) = P(\overline{\nu}_\mathrm{e}) = P(\overline{\mathrm{q}}) = -1$. From the Quantum Field Theory describing the force carrying particles, it can be shown that the vector bosons responsible for the electromagnetic, strong and weak forces all have negative intrinsic parity,

$$P(\gamma) = P(\mathrm{g}) = P(\mathrm{W}^\pm) = P(\mathrm{Z}) = -1.$$

### 11.2.2 Parity conservation in QED

Parity conservation in QED arises naturally from the form of the interaction. For example, the matrix element for the QED process of $\mathrm{e}^-\mathrm{q} \to \mathrm{e}^-\mathrm{q}$ scattering, shown in Figure 11.1, can be written as the four-vector scalar product

$$\mathcal{M} = \frac{Q_\mathrm{q} e^2}{q^2} j_\mathrm{e} \cdot j_\mathrm{q},$$

where the electron and quark currents are defined by

$$j_\mathrm{e}^\mu = \overline{u}(p_3)\gamma^\mu u(p_1) \quad \text{and} \quad j_\mathrm{q}^\nu = \overline{u}(p_4)\gamma^\nu u(p_2). \tag{11.3}$$

The equivalent matrix element for the parity transformed process, where the three-momenta of all the particles are reversed, can be obtained by applying the parity operator $\hat{P} = \gamma^0$ to the spinors of (11.3). Since Dirac spinors transform as

$$u \xrightarrow{\hat{P}} \hat{P}u = \gamma^0 u, \tag{11.4}$$

the adjoint spinors transform as

$$\overline{u} = u^\dagger \gamma^0 \xrightarrow{\hat{P}} (\hat{P}u)^\dagger \gamma^0 = u^\dagger \gamma^{0\dagger} \gamma^0 = u^\dagger \gamma^0 \gamma^0 = \overline{u}\gamma^0,$$

and hence

$$\overline{u} \xrightarrow{\hat{P}} \overline{u}\gamma^0. \tag{11.5}$$

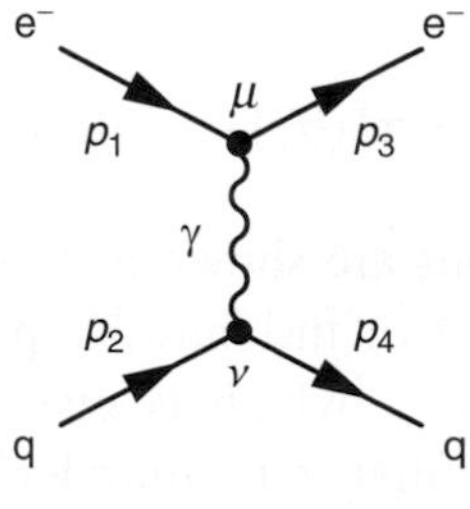


Fig. 11.1 The lowest-order Feynman diagram for the QED $t$-channel electron–quark scattering process.

From (11.4) and (11.5), it can be seen that the four-vector currents of (11.3) become

$$j_e^\mu = \overline{u}(p_3)\gamma^\mu u(p_1) \xrightarrow{\hat{P}} \overline{u}(p_3)\gamma^0\gamma^\mu\gamma^0 u(p_1).$$

Because $\gamma^0\gamma^0 = I$, the time-like component of the current is unchanged by the parity operation,

$$j_e^0 \xrightarrow{\hat{P}} \overline{u}\gamma^0\gamma^0\gamma^0 u = \overline{u}\gamma^0 u = j_e^0.$$

The space-like components of $j^\mu$, with indices $k = 1, 2, 3$, transform as

$$j_e^k \xrightarrow{\hat{P}} \overline{u}\gamma^0\gamma^k\gamma^0 u = -\overline{u}\gamma^k\gamma^0\gamma^0 u = -\overline{u}\gamma^k u = -j_e^k,$$

since $\gamma^0\gamma^k = -\gamma^k\gamma^0$. Therefore, as expected, the parity operation changes the signs of the space-like components of the four-vector current but the time-like component remains unchanged. Consequently, the four-vector scalar product in the QED matrix element, $j_e \cdot j_q = j_e^0 j_q^0 - j_e^k j_q^k$, transforms to

$$j_e \cdot j_q = j_e^0 j_q^0 - j_e^k j_q^k \xrightarrow{\hat{P}} j_e^0 j_q^0 - (-j_e^k)(-j_q^k) = j_e \cdot j_q, \tag{11.6}$$

and it can be concluded that the QED matrix element is invariant under the parity operation. Hence the terms in the Hamiltonian related to the QED interaction are invariant under parity transformations. This invariance implies that

parity is conserved in QED.

Apart from the colour factors, the QCD interaction has the same form as QED and consequently

parity is conserved in QCD.

The conservation of parity in strong and electromagnetic interactions needs to be taken into account when considering particle decays. For example, consider the two decays

$$\rho^0(1^-) \to \pi^+(0^-) + \pi^-(0^-) \quad \text{and} \quad \eta(0^-) \to \pi^+(0^-) + \pi^-(0^-),$$

where the $J^P$ values are shown in brackets. The total parity of the two-body final state is the product of the intrinsic parities of the particles and the parity of the orbital wavefunction, which is given by $(-1)^\ell$, where $\ell$ is the orbital angular momentum in the final state. In order to conserve angular momentum, the $\pi^+$ and $\pi^-$ in the $\rho^0 \to \pi^+\pi^-$ decay are produced with relative orbital angular moment

$\ell = 1$, whereas the $\pi^+\pi^-$ in the decay of the $\eta$ must have $\ell = 0$. Therefore, conservation of parity in the two decays can be expressed as follows:

$$P(\rho^0) = P(\pi^+) \cdot P(\pi^-) \cdot (-1)^{\ell=1} \quad \Rightarrow \quad -1 = (-1)(-1)(-1) \quad \checkmark$$
$$P(\eta) = P(\pi^+) \cdot P(\pi^-) \cdot (-1)^{\ell=0} \quad \Rightarrow \quad -1 = (-1)(-1)(+1) \quad \times$$

Hence the strong interaction decay process $\rho^0 \to \pi^+\pi^-$ is allowed, but the strong decay $\eta \to \pi^+\pi^-$ does not occur as it would violate the conservation of parity that is implicit in the strong interaction Hamiltonian. It can also be shown that the QED and QCD interactions are invariant under the charge conjugation operation $\hat{C}$, defined in Section 4.7.5, which changes particles into antiparticles and vice versa, and therefore there is a corresponding conserved quantity $C = \pm 1$.

### Scalars, pseudoscalars, vectors and axial vectors

Physical quantities can be classified according to their rank (dimensionality) and parity inversion properties. For example, single-valued *scalar* quantities, such as mass and temperature, are invariant under parity transformations. *Vector* quantities, such as position and momentum, change sign under parity transformations, $\mathbf{x} \to -\mathbf{x}$ and $\mathbf{p} \to -\mathbf{p}$. There is also a second class of vector quantity, known as an *axial vector*, which is somtimes referred to as a pseudovector. Axial vectors are formed from the cross product of two vector quantities, and therefore do not change sign under parity transformations. One example is angular momentum $\mathbf{L} = \mathbf{x} \times \mathbf{p}$. Because both $\mathbf{x}$ and $\mathbf{p}$ change sign under parity, the axial vector $\mathbf{L}$ is unchanged. Other examples of axial vectors include the magnetic moment and the magnetic flux density $\mathbf{B}$, which is related to the current density $\mathbf{j}$ by the Biot–Savart law, $\mathrm{d}\mathbf{B} \propto \mathbf{j} \times \mathrm{d}^3\mathbf{x}$. Scalar quantities can be formed out of scalar products of two vectors or two axial vectors, the simplest example being the magnitude squared of the momentum vector, $\mathrm{p}^2 = \mathbf{p} \cdot \mathbf{p}$. There is a second class of scalar quantity known as a *pseudoscalar*. Pseudoscalars are single-valued quantities formed from the product of a vector and an axial vector, and consequently change sign under the parity operation. One important example of a pseudoscalar is helicity, $h \propto \mathbf{S} \cdot \mathbf{p}$. The different scalar and vector quantities are listed in Table 11.1.

**Table 11.1** The parity properties of scalars, pseudoscalars, vectors and axial vectors.

| | Rank | Parity | Example |
|---|---|---|---|
| Scalar | 0 | + | Temperature, $T$ |
| Pseudoscalar | 0 | − | Helicity, $h$ |
| Vector | 1 | − | Momentum, $\mathbf{p}$ |
| Axial vector | 1 | + | Angular momentum, $\mathbf{L}$ |

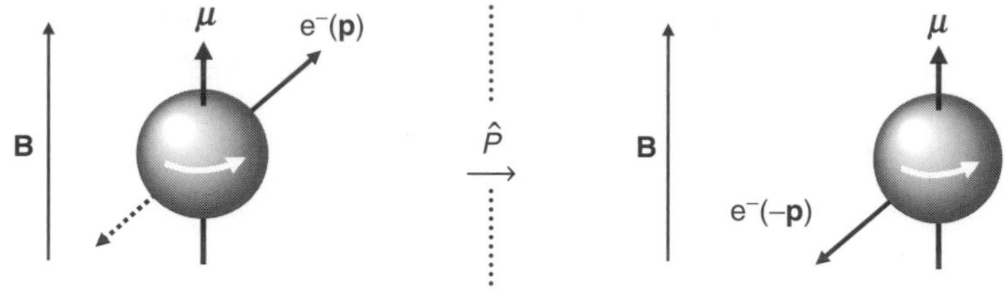


Fig. 11.2 The β-decay of polarised $^{60}$Co. On the left, an electron is emitted in a particular direction. On the right the parity inverted equivalent is shown.

### 11.2.3 Parity violation in nuclear β-decay

The parity inversion properties of the different types of physical quantity can be exploited to investigate whether parity is conserved in the weak interaction. In 1957, Wu and collaborators studied nuclear β-decay of polarised cobalt-60,

$$^{60}\mathrm{Co} \rightarrow {}^{60}\mathrm{Ni}^{*} + \mathrm{e}^{-} + \overline{\nu}_{\mathrm{e}}.$$

The $^{60}$Co nuclei, which possess a permanent nuclear magnetic moment $\boldsymbol{\mu}$, were aligned in a strong magnetic field **B** and the β-decay electrons were detected at different polar angles with respect to this axis, as shown in Figure 11.2. Because both **B** and $\boldsymbol{\mu}$ are axial vectors, they do not change sign under the parity transformation. Hence when viewed in the parity inverted "mirror", the only quantity that changes sign is the vector momentum of the emitted electron. Hence, if parity were conserved in the weak interaction, the rate at which electrons were emitted at a certain direction relative to the **B**-field would be identical to the rate in the opposite direction. Experimentally, it was observed that more electrons were emitted in the hemisphere opposite to the direction of the applied magnetic field than in the hemisphere in the direction of the applied field, thus providing a clear demonstration that

parity is NOT conserved in the weak interaction.

From this observation it can be concluded that, unlike QED and QCD, the weak interaction does not have four-vector currents of the form $j^{\mu} = \overline{u}(p')\gamma^{\mu}u(p)$.

## 11.3 $V-A$ structure of the weak interaction

QED and QCD are vector interactions with a current of the form $j^{\mu} = \overline{u}(p')\gamma^{\mu}u(p)$. This particular combination of spinors and $\gamma$-matrices transforms as a four vector (as shown in Appendix B.3). From the observation of parity violation, the

**Table 11.2** Lorentz-invariant bilinear covariant currents.

| Type | Form | Components | Boson spin |
|---|---|---|---|
| Scalar | $\overline{\psi}\phi$ | 1 | 0 |
| Pseudoscalar | $\overline{\psi}\gamma^5\phi$ | 1 | 0 |
| Vector | $\overline{\psi}\gamma^\mu\phi$ | 4 | 1 |
| Axial vector | $\overline{\psi}\gamma^\mu\gamma^5\phi$ | 4 | 1 |
| Tensor | $\overline{\psi}(\gamma^\mu\gamma^\nu - \gamma^\nu\gamma^\mu)\phi$ | 6 | 2 |

weak interaction vertex is required have a different form. However, the requirement of Lorentz invariance of the interaction matrix element severely restricts the possible forms of the interaction. The general bilinear combination of two spinors can be written $\overline{u}(p')\Gamma u(p)$, where $\Gamma$ is a $4\times 4$ matrix formed from products of the Dirac $\gamma$-matrices. It turns out that there are only five combinations of individual $\gamma$-matrices that have the correct Lorentz transformation properties, such that they can be combined into a Lorentz-invariant matrix element. These combinations are called *bilinear covariants* and give rise to the possible *scalar*, *pseudoscalar*, *vector*, *axial vector* and *tensor* currents listed in Table 11.2.

In QED, the factor $g_{\mu\nu}$ in the matrix element arises from the sum over the $(2J+1)+1$ polarisation states of the $J^P = 1^-$ virtual photon, which includes the time-like component of the polarisation four-vector. These four polarisation states correspond to the four degrees of freedom of the vector current $j^\mu = \overline{\psi}\gamma^\mu\phi$, labelled by the index $\mu = 0, 1, 2, 3$. The single component scalar and pseudoscalar interactions therefore can be associated with the exchange of a spin-0 boson ($J = 0$), which possesses just a single degree of freedom. Similarly, the six non-zero components of a tensor interaction can be associated with the exchange of a spin-2 boson ($J = 2$), with $(2J+1)+1 = 6$ polarisation states for the spin-2 virtual particle.

The most general Lorentz-invariant form for the interaction between a fermion and a boson is a linear combination of the bilinear covariants. If this is restricted to the exchange of a spin-1 (vector) boson, the most general form for the interaction is a linear combination of vector and axial vector currents,

$$j^\mu \propto \overline{u}(p')(g_V\gamma^\mu + g_A\gamma^\mu\gamma^5)u(p) = g_V j_V^\mu + g_A j_A^\mu,$$

where $g_V$ and $g_A$ are vector and axial vector coupling constants and the current has been decomposed into vector and axial vector components

$$j_V^\mu = \overline{u}(p')\gamma^\mu u(p) \quad \text{and} \quad j_A^\mu = \overline{u}(p')\gamma^\mu\gamma^5 u(p).$$

The parity transformation properties of $j^\mu_V$ were derived in Section 11.2.2. The parity transformation properties of the pure axial vector current can be obtained the same way

$$j^\mu_A = \overline{u}\gamma^\mu\gamma^5 u \xrightarrow{\hat{P}} \overline{u}\gamma^0\gamma^\mu\gamma^5\gamma^0 u = -\overline{u}\gamma^0\gamma^\mu\gamma^0\gamma^5 u,$$

which follows from $\gamma^5\gamma^0 = -\gamma^0\gamma^5$. Hence the time-like component of the axial vector current transforms as

$$j^0_A = \xrightarrow{\hat{P}} -\overline{u}\gamma^0\gamma^0\gamma^0\gamma^5 u = -\overline{u}\gamma^0\gamma^5 u = -j^0_A,$$

and the space-like components transform as

$$j^k_A = \xrightarrow{\hat{P}} -\overline{u}\gamma^0\gamma^k\gamma^0\gamma^5 u = +\overline{u}\gamma^k\gamma^5 u = +j^k_A.$$

Therefore, the scalar product of two axial vector currents is invariant under parity transformations

$$j_1 \cdot j_2 = j^0_1 j^0_2 - j^k_1 j^k_2 \xrightarrow{\hat{P}} (-j^0_1)(-j^0_2) - j^k_1 j^k_2 = j_1 \cdot j_2. \tag{11.7}$$

This should come as no surprise, the matrix element is a scalar quantity; if it is formed from the four-vector scalar product of either two vectors or two axial vectors it has to be invariant under the parity transformation.

To summarise, the parity transformation properties of the components of the vector and the axial vector currents are

$$j^0_V \xrightarrow{\hat{P}} +j^0_V, \quad j^k_V \xrightarrow{\hat{P}} -j^k_V, \quad \text{and} \quad j^0_A \xrightarrow{\hat{P}} -j^0_A, \quad j^k_A \xrightarrow{\hat{P}} +j^k_A.$$

Whilst the scalar products of two vector currents or two axial vector currents are unchanged in a parity transformation, the scalar product $j_V \cdot j_A$ transforms to $-j_V \cdot j_A$. Hence the combination of vector *and* axial vector currents provides a mechanism to explain the observed parity violation in the weak interaction.

Consider the (inverse-β-decay) charged-current weak interaction process $\nu_e d \rightarrow e^- u$, shown in Figure 11.3, with assumed currents of the form

$$j^\mu_{\nu e} = \overline{u}(p_3)(g_V\gamma^\mu + g_A\gamma^\mu\gamma^5)u(p_1) = g_V j^V_{\nu e} + g_A j^A_{\nu e},$$
$$j^\nu_{\mathrm{du}} = \overline{u}(p_4)(g_V\gamma^\nu + g_A\gamma^\nu\gamma^5)u(p_2) = g_V j^V_{\mathrm{du}} + g_A j^A_{\mathrm{du}}.$$

The matrix element is proportional to the four-vector scalar products of two currents

$$\mathcal{M}_{fi} \propto j_{\nu e} \cdot j_{\mathrm{du}} = g_V^2 j^V_{\nu e} \cdot j^V_{\mathrm{du}} + g_A^2 j^A_{\nu e} \cdot j^A_{\mathrm{du}} + g_V g_A (j^V_{\nu e} \cdot j^A_{\mathrm{du}} + j^A_{\nu e} \cdot j^V_{\mathrm{du}}).$$

The terms $j^V_{\nu e} \cdot j^V_{\nu e}$ and $j^A_{\nu e} \cdot j^A_{\nu e}$ do not change sign under a parity transformation, but the mixed $V$ and $A$ combinations do, and therefore

$$j_{\nu e} \cdot j_{\mathrm{du}} \xrightarrow{\hat{P}} g_V^2 j^V_{\nu e} \cdot j^V_{\mathrm{du}} + g_A^2 j^A_{\nu e} \cdot j^A_{\mathrm{du}} - g_V g_A (j^V_{\nu e} \cdot j^A_{\mathrm{du}} + j^A_{\nu e} \cdot j^V_{\mathrm{du}}).$$

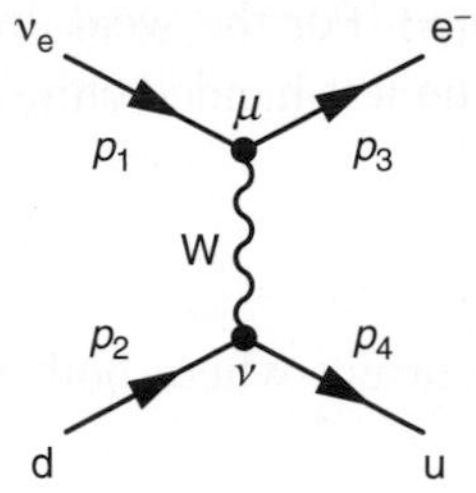


Fig. 11.3 The lowest-order Feynman diagram for the charged-current weak interaction $\nu_e d \to e^- u$.

Thus the relative strength of the parity violating part of the matrix element compared to the parity conserving part is given by

$$\frac{g_V g_A}{g_V^2 + g_A^2}.$$

Hence, if either $g_V$ or $g_A$ is zero, parity is conserved in the interaction. Furthermore, maximal parity violation occurs when $|g_V| = |g_A|$, corresponding to a pure $V-A$ or $V+A$ interaction. From experiment, it is known that the weak charged current due to the exchange of $W^{\pm}$ bosons is a vector minus axial vector ($V-A$) interaction of the form $\gamma^\mu - \gamma^\mu\gamma^5$, with a vertex factor of

$$\frac{-ig_W}{\sqrt{2}}\frac{1}{2}\gamma^\mu(1-\gamma^5). \tag{11.8}$$

Here $g_W$ is the weak coupling constant (which is often written simply as $g$). The origin of the additional numerical factors will be explained in Chapter 15. The corresponding four-vector current is given by

$$j^\mu = \frac{g_W}{\sqrt{2}}\overline{u}(p')\frac{1}{2}\gamma^\mu(1-\gamma^5)u(p).$$

## 11.4 Chiral structure of the weak interaction

In Chapter 6, the left- and right-handed *chiral* projection operators,

$$P_R = \tfrac{1}{2}(1+\gamma^5) \quad \text{and} \quad P_L = \tfrac{1}{2}(1-\gamma^5),$$

were introduced. Any spinor can be decomposed into left- and right-handed chiral components,

$$u = \tfrac{1}{2}(1+\gamma^5)u + \tfrac{1}{2}(1-\gamma^5)u = P_R u + P_L u = a_R u_R + a_L u_L,$$

with coefficients $a_R$ and $a_L$. In Section 6.4.1, it was shown that only two combinations of chiral spinors (RR and LL) gave non-zero values for the QED *vector* current, $\overline{u}(p')\gamma^\mu u(p)$. For the weak interaction, the $V - A$ vertex factor of (11.8) already includes the left-handed chiral projection operator,

$$\tfrac{1}{2}(1-\gamma^5).$$

In this case, the current where both the spinors are right-handed chiral states is also zero

$$\begin{aligned} j^\mu_{RR} &= \frac{g_W}{\sqrt{2}}\,\overline{u}_R(p')\gamma^\mu \tfrac{1}{2}(1-\gamma^5)u_R(p) \\ &= \frac{g_W}{\sqrt{2}}\,\overline{u}_R(p')\gamma^\mu P_L u_R(p) = 0, \end{aligned}$$

and the only non-zero current for particle spinors involves only left-handed chiral states. Hence only left-handed chiral *particle* states participate in the charged-current weak interaction. For antiparticle spinors $P_L$ projects out right-handed chiral states,

$$\tfrac{1}{2}(1-\gamma^5)v = v_R,$$

and therefore only right-handed chiral *antiparticle* states participate in the charged-current weak interaction. In the limit $E \gg m$, where the chiral and helicity states are the same, the $V - A$ term in the weak interaction vertex projects out left-handed helicity particle states and right-handed helicity antiparticles states. Hence, in this ultra-relativistic limit, the only allowed helicity combinations for the weak interaction vertices involving electrons/positrons and electron neutrinos/antineutrinos are those shown in Figure 11.4.

The maximally different coupling of the weak charged-current interaction to left-handed and right-handed chiral states is the origin of parity violation. For example, the left-hand plot of Figure 11.5 shows the helicity configuration of the allowed weak interaction of a high-energy left-handed $e^-$ and a right-handed $\overline{\nu}_e$. In the parity mirror, the vector quantities are reversed, $\mathbf{p} \to -\mathbf{p}$, but the axial vector spins of the particles remain unchanged, giving a RH particle and a LH antiparticle. Hence the parity operation transforms an allowed weak interaction into one that is not allowed, maximally violating the conservation of parity.

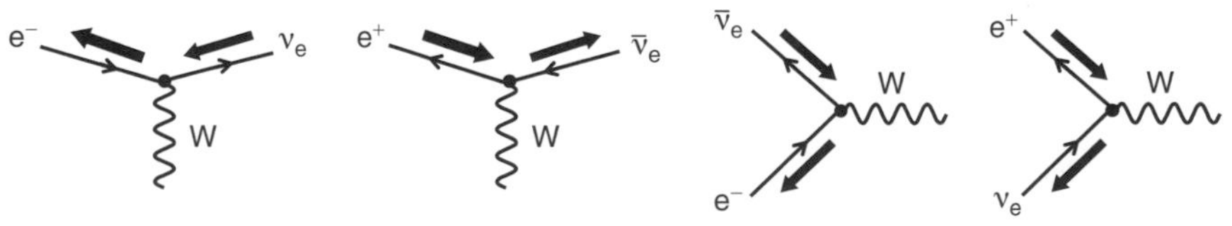


Fig. 11.4 The allowed helicity combinations in weak interaction vertices involving the $e^+$, $e^-$, $\nu_e$ and $\overline{\nu}_e$, in the limit where $E \gg m$ (where the helicity states are effectively the same as the chiral states).

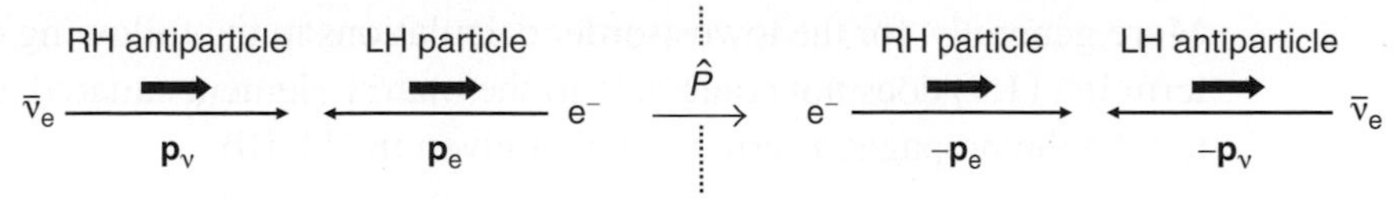


Fig. 11.5 The allowed helicity combination for a $e^-\bar{\nu}_e$ weak interaction, and (right) its parity transformed equivalent.

## 11.5 The W-boson propagator

The Feynman rule for the propagator of QED, corresponding to the exchange of the massless spin-1 photon, is

$$\frac{-ig_{\mu\nu}}{q^2}.$$

The weak interaction not only differs from QED and QCD in the form of the interaction vertex, but it is mediated by the massive W bosons, with $m_W \sim 80\,\text{GeV}$. Consequently, the $q^2$-dependence of the W-boson propagator is given by (5.7),

$$\frac{1}{q^2 - m_W^2}.$$

The $g_{\mu\nu}$ term in the Feynman rule for QED propagator is associated with the sum over the polarisation states of the virtual photon,

$$\sum_\lambda \epsilon_\mu^{\lambda *}\epsilon_\nu^\lambda = -g_{\mu\nu}.$$

Massive spin-1 particles differ from massless spin-1 particles in having the additional degree of freedom of a longitudinal polarisation state. In Appendix D, it is shown that the corresponding sum over the polarisation states of the exchanged virtual massive spin-1 boson gives

$$\sum_\lambda \epsilon_\mu^{\lambda *}\epsilon_\nu^\lambda = -g_{\mu\nu} + \frac{q_\mu q_\nu}{m_W^2}.$$

Therefore, the Feynman rule associated with the exchange of a virtual W boson is

$$\frac{-i}{q^2 - m_W^2}\left(g_{\mu\nu} - \frac{q_\mu q_\nu}{m_W^2}\right). \tag{11.9}$$

In the limit where $q^2 \ll m_W^2$, the $q_\mu q_\nu$ term is small and the propagator can be taken to be

$$\frac{-ig_{\mu\nu}}{q^2 - m_W^2}. \tag{11.10}$$

More generally, for the lowest-order calculations in the following chapters the $q_\mu q_\nu$ term in (11.9) does not contribute to the matrix element squared and it is sufficient to take the propagator term to be that given in (11.10).

### 11.5.1 Fermi theory

For most low-energy weak interactions, such as the majority of particle decays, $|q^2| \ll m_W^2$ and the W-boson propagator of (11.10) can be approximated by

$$i\frac{g_{\mu\nu}}{m_W^2}, \tag{11.11}$$

and the effective interaction no longer has any $q^2$ dependence. Physically this corresponds to replacing the propagator with an interaction which occurs at a single point in space-time, as indicated in Figure 11.6. Hence, in the low-energy limit, the weak charged-current can be expressed in terms of this four-fermion contact interaction.

The original description of the weak interaction, due to Fermi (1934), was formulated before the discovery of the parity violation and the matrix element for β-decay was expressed in terms of a contact interaction

$$\mathcal{M}_{fi} = G_F\, g_{\mu\nu}[\overline{\psi}_3\gamma^\mu\psi_1][\overline{\psi}_4\gamma^\nu\psi_2], \tag{11.12}$$

where the strength of the weak interaction is given by the Fermi constant $G_F$. After the discovery of parity violation by Wu *et al.* (1957), this expression was modified to

$$\mathcal{M}_{fi} = \frac{1}{\sqrt{2}}G_F\, g_{\mu\nu}[\overline{\psi}_3\gamma^\mu(1-\gamma^5)\psi_1][\overline{\psi}_4\gamma^\nu(1-\gamma^5)\psi_2], \tag{11.13}$$

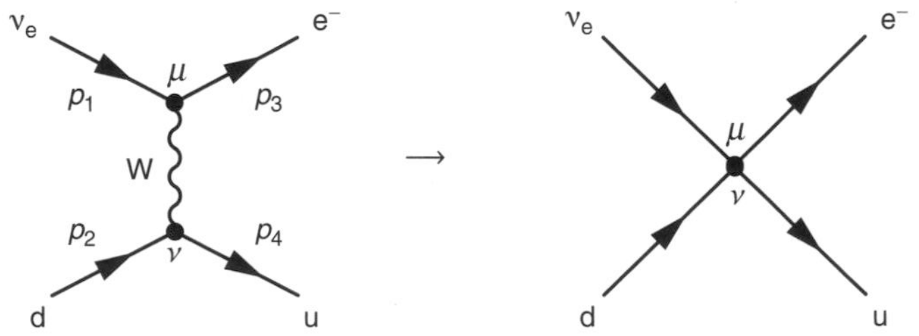


Fig. 11.6 The weak interaction Feynman diagram and the $q^2 \ll m_W^2$ limit of an effective contact interaction.

where the factor of $\sqrt{2}$ was introduced so that the numerical value of $G_F$ did not need to be changed. The expression of (11.13) can be compared to the full expression obtained using the Feynman rules for the weak interaction,

$$\mathcal{M}_{fi} = -\left[\frac{g_W}{\sqrt{2}}\overline{\psi}_3\frac{1}{2}\gamma^\mu(1-\gamma^5)\psi_1\right]\cdot\left[\frac{g_{\mu\nu} - q_\mu q_\nu/m_W^2}{q^2 - m_W^2}\right]\cdot\left[\frac{g_W}{\sqrt{2}}\overline{\psi}_4\frac{1}{2}\gamma^\nu(1-\gamma^5)\psi_2\right],$$

which in the limit of $q^2 \ll m_W^2$ reduces to

$$M_{fi} = \frac{g_W^2}{8m_W^2}g_{\mu\nu}[\overline{\psi}_3\gamma^\mu(1-\gamma^5)\psi_1][\overline{\psi}_4\gamma^\nu(1-\gamma^5)\psi_2]. \tag{11.14}$$

Hence, by comparing (11.13) and (11.14), it can be seen that the Feynman rules in the low-$q^2$ limit, give the same expression for the matrix element as obtained from Fermi theory and therefore the Fermi constant is related to the weak coupling strength by

$$\frac{G_F}{\sqrt{2}} = \frac{g_W^2}{8m_W^2}. \tag{11.15}$$

## Strength of the weak interaction

The strength of the weak interaction is most precisely determined from low-energy measurements, and in particular from the muon lifetime. For these low-energy measurements, where for example $m_\mu \ll m_W$, Fermi theory can be used. The calculation of the decay rate for $\mu^- \to e^-\nu_\mu\overline{\nu}_e$ includes a fairly involved integration over the three-body phase space of the final state and the results are simply quoted here. The muon lifetime $\tau_\mu$ is related to its mass by

$$\Gamma(\mu^- \to e^-\nu_\mu\overline{\nu}_e) = \frac{1}{\tau_\mu} = \frac{G_F^2 m_\mu^5}{192\pi^3}. \tag{11.16}$$

The precise measurements of the muon lifetime and mass,

$$m_\mu = 0.105\,658\,371\,5(35)\,\text{GeV} \quad \text{and} \quad \tau_\mu = 2.196\,981\,1(22)\times 10^{-6}\,\text{s},$$

provide a precise determination of the Fermi constant,

$$G_F = 1.166\,38 \times 10^{-5}\,\text{GeV}^{-2}.$$

However, $G_F$ does not express the fundamental strength of the weak interaction, it is related to the ratio of the coupling strength $g_W$ and the W-boson mass by (11.15). Nevertheless, $G_F$ is the quantity that is precisely measured in muon decay and it is still used parameterise the strength of weak interaction.

The value of fundamental coupling constant $g_W$ can be obtained from $G_F$ using the precise measurement of $m_W = 80.385 \pm 0.015\,\text{GeV}$ (see Chapter 16). From the relation of (11.15) and the measured values of $G_F$ and $m_W$, the dimensionless coupling constant of the weak interaction is

$$\alpha_W = \frac{g_W^2}{4\pi} = \frac{8m_W^2 G_F}{4\sqrt{2}\pi} \approx \frac{1}{30}.$$

Hence the weak interaction is in fact intrinsically stronger than the electromagnetic interaction, $\alpha_W > \alpha$. It is only the presence of the large mass of the W boson in the propagator that is responsible for the weakness of the low-energy weak interaction compared to that of QED. For a process where the exchanged boson carries four-momentum $q$, where $|q^2| \ll m_W^2$, the QED and weak interactions propagators are respectively

$$P_{QED} \sim \frac{1}{q^2} \quad \text{and} \quad P_W \sim \frac{1}{q^2 - m_W^2} \approx -\frac{1}{m_W^2}.$$

Therefore weak interaction decay rates, which are proportional to $|\mathcal{M}|^2$, are suppressed by a factor $q^4/m_W^4$ relative to QED decay rates. In contrast, in the high-energy limit where $|q^2| > m_W^2$, the $m_W^2$ term in the weak propagator is relatively unimportant and the electromagnetic and weak interactions have similar strength, as will be seen directly in the results from high-$Q^2$ electron–proton interactions, described in Section 12.5.

## 11.6 Helicity in pion decay

The charged pions ($\pi^\pm$) are the $J^P = 0^-$ meson states formed from $u\overline{d}$ and $d\overline{u}$. They are the lightest mesons with $m(\pi^\pm) \sim 140$ MeV and therefore cannot decay via the strong interaction; they can only decay through the weak interaction to final states with lighter fundamental fermions. Hence charged pions can only decay to final states with either electrons or muons. The three main decay modes of the $\pi^-$ are the charged-current weak processes $\pi^- \to e^-\overline{\nu}_e$, $\pi^- \to \mu^-\overline{\nu}_\mu$ and $\pi^- \to \mu^-\overline{\nu}_\mu\gamma$, with decays to $\mu^-\overline{\nu}_\mu$ dominating.

The Feynman diagrams for the decays $\pi^- \to e^-\overline{\nu}_e$ and $\pi^- \to \mu^-\overline{\nu}_\mu$ are shown in Figure 11.7. Because the strength of the weak interaction for the different lepton generations is found to be the same (see Chapter 12), it might be expected that the matrix elements for the decays $\pi^- \to e^-\overline{\nu}_e$ and $\pi^- \to \mu^-\overline{\nu}_\mu$ would be similar. For a two-body decay, the phase space factor is proportional the momentum of the decay products in the centre-of-mass frame, see (3.49). On this basis, the decay rate to $e^-\overline{\nu}_e$ would be expected to be greater than that to $\mu^-\overline{\nu}_\mu$. However, the opposite is found to be true; charged pions decay almost entirely by $\pi^- \to \mu^-\overline{\nu}_\mu$

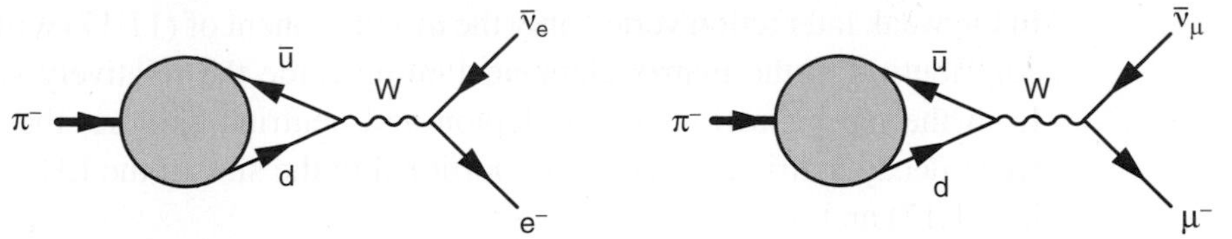


Fig. 11.7 Two of the three main decay modes for the $\pi^-$. The decay $\pi^- \to \mu^-\overline{\nu}_\mu\gamma$ (not shown) has a comparable branching ratio to that for $\pi^- \to e^-\overline{\nu}_e$.

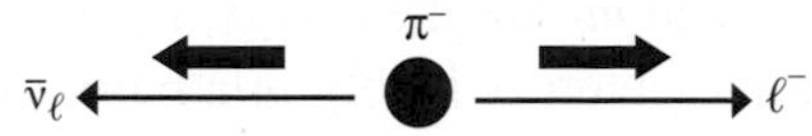


Fig. 11.8 The helicity configuration in $\pi^- \to \ell^-\overline{\nu}_\ell$ decay, where $\ell$ = e or $\mu$.

(or equivalently $\pi^+ \to \mu^+\nu_\mu$) with a branching ratio of 99.988% and the measured ratio of the decay rates to electrons and muons is

$$\frac{\Gamma(\pi^- \to e^-\overline{\nu}_e)}{\Gamma(\pi^- \to \mu^-\overline{\nu}_\mu)} = 1.230(4) \times 10^{-4}.$$

This counterintuitive result is a manifestation of the chiral structure of the weak interaction and provides a clear illustration of the difference between helicity, defined by $\boldsymbol{\sigma} \cdot \mathbf{p}/|\mathbf{p}|$, and chirality defined by the action of the chiral projection operators.

The weak interaction only couples to LH chiral particle states and RH chiral antiparticle states. Because neutrinos are effectively massless, $m_\nu \ll E$, the neutrino chiral states are, in all practical circumstances, equivalent to the helicity states. Therefore, the antineutrino from a $\pi^-$ decay is always produced in a RH helicity state. Because the pion is a spin-0 particle, the lepton–neutrino system must be produced in the spin-0 singlet state, with the charged lepton and neutrino spins in opposite directions. Therefore, because the neutrino is RH, conservation of angular momentum implies that the charged lepton is also produced in a RH helicity state, and the only allowed spin configuration is that of Figure 11.8. Since the weak interaction vertex is non-zero only for LH chiral particle states, the charged lepton has, in some sense, the "wrong helicity" for the weak interaction. If the charged leptons were also massless, the decay would not occur. However, chiral and helicity states are not equivalent and the weak decay to a RH *helicity* particle state can occur, although it may be highly suppressed.

In general the RH helicity spinor $u_\uparrow$ can be decomposed into RH and LH chiral components, $u_R$ and $u_L$, given by (6.38),

$$u_\uparrow \equiv \tfrac{1}{2}\left(1 + \frac{p}{E+m}\right)u_R + \tfrac{1}{2}\left(1 - \frac{p}{E+m}\right)u_L. \quad (11.17)$$

In the weak interaction vertex only the $u_L$ component of (11.17) will give a non-zero contribution to the matrix element. Putting aside the relatively small differences from the normalisations of the lepton and neutrino spinors, the charged-current weak decay matrix element is proportional to the size of the LH chiral component in (11.17) and

$$\mathcal{M} \sim \frac{1}{2}\left(1 - \frac{\mathrm{p}_\ell}{E_\ell + m_\ell}\right), \tag{11.18}$$

where $E_\ell$, $\mathrm{p}_\ell$ and $m_\ell$ are the energy, momentum and mass of the charged lepton. If the charged lepton is highly relativistic, the left-handed chiral component of the right-handed helicity state will be very small, resulting in a suppression of the decay rate.

Taking the mass of the neutrino to be zero, it is straightforward to show that

$$E_\ell = \frac{m_\pi^2 + m_\ell^2}{2m_\pi} \quad \text{and} \quad \mathrm{p}_\ell = \frac{m_\pi^2 - m_\ell^2}{2m_\pi}, \tag{11.19}$$

giving

$$\frac{\mathrm{p}_\ell}{E_\ell + m_\ell} = \frac{m_\pi - m_\ell}{m_\pi + m_\ell},$$

which when substituted into (11.18), demonstrates that

$$\mathcal{M} \sim \frac{m_\ell}{m_\pi + m_\ell}.$$

Because $m_\mu/m_e \approx 200$, pion decays to electrons are strongly suppressed with respect to those to muons. This helicity suppression reflects the fact that the electrons produced in pion decay are highly relativistic, $\beta = 0.999\,97$, and therefore the chiral states almost correspond to the helicity states. For the decay to muons, $\beta = 0.27$, and the $u_L$ coefficient in (11.17) is significant. The above discussion gives a qualitative explanation of why charged pions predominantly decay to muons rather than electrons. The full calculation, which is interesting in its own right, is given below.

### 11.6.1 Pion decay rate

Consider the $\pi^- \to \ell^-\overline{\nu}_\ell$ decay in its rest frame, where the direction of the charged lepton defines the $z$-axis, as shown in Figure 11.9. In this case, the four-momenta of the $\pi^-$, $\ell^-$ and $\overline{\nu}_\ell$ are respectively,

$$p_\pi = (m_\pi, 0, 0, 0), \quad p_\ell = p_3 = (E_\ell, 0, 0, \mathrm{p}) \quad \text{and} \quad p_{\overline{\nu}} = p_4 = (\mathrm{p}, 0, 0, -\mathrm{p}),$$

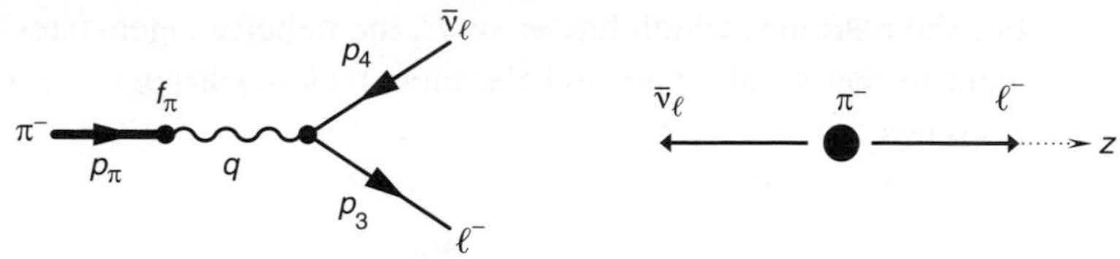


Fig. 11.9 The definition of the four-momenta in the process $\pi^- \to \ell^- \overline{\nu}_\ell$.

where p is the magnitude of the momentum of both the charged lepton and antineutrino in the centre-of-mass frame.

The weak leptonic current associated with the $\ell^- \overline{\nu}_\ell$ vertex is

$$j_\ell^\nu = \frac{g_W}{\sqrt{2}} \overline{u}(p_3) \tfrac{1}{2} \gamma^\nu (1 - \gamma^5) v(p_4).$$

Because the pion is a bound $q\overline{q}$ state, the corresponding hadronic current cannot be expressed in terms of free particle Dirac spinors. However, the pion current has to be a four-vector such that the four-vector scalar product with the leptonic current gives a Lorentz-invariant expression for the matrix element. Since the pion is a spin-0 particle, the only four-vector quantity that can be used is the pion four-momentum. Hence, the most general expression for the pion current is obtained by replacing $\overline{v}\gamma^\mu(1-\gamma^5)u$ with $f_\pi p_\pi^\mu$, where $f_\pi$ is a constant associated with the decay. The matrix element for the decay $\pi^- \to \ell^- \overline{\nu}_\ell$ therefore can be written as

$$\begin{aligned} \mathcal{M}_{fi} &= \left[ \frac{g_W}{\sqrt{2}} \frac{1}{2} f_\pi p_\pi^\mu \right] \times \left[ \frac{g_{\mu\nu}}{m_W^2} \right] \times \left[ \frac{g_W}{\sqrt{2}} \overline{u}(p_3) \gamma^\nu \tfrac{1}{2} (1 - \gamma^5) v(p_4) \right] \\ &= \frac{g_W^2}{4 m_W^2} g_{\mu\nu} f_\pi p_\pi^\mu \overline{u}(p_3) \gamma^\nu \tfrac{1}{2} (1 - \gamma^5) v(p_4), \end{aligned}$$

where the propagator has been approximated by the Fermi contact interaction (which is an extremely good approximation because $q^2 = m_\pi^2 \ll m_W^2$). In the pion rest frame, only the time-like component of the pion four-momentum is non-zero, $p_\pi^0 = m_\pi$, and hence

$$\mathcal{M}_{fi} = \frac{g_W^2}{4 m_W^2} f_\pi m_\pi \overline{u}(p_3) \gamma^0 \tfrac{1}{2} (1 - \gamma^5) v(p_4).$$

Because $\overline{u}\gamma^0 = u^\dagger \gamma^0 \gamma^0 = u^\dagger$, this can be written as

$$\mathcal{M}_{fi} = \frac{g_W^2}{4 m_W^2} f_\pi m_\pi u^\dagger(p_3) \tfrac{1}{2} (1 - \gamma^5) v(p_4). \tag{11.20}$$

For the neutrino, which has $m \ll E$, the helicity eigenstates are essentially equivalent to the chiral states and therefore $\frac{1}{2}(1-\gamma^5)v(p_4) = v_\uparrow(p_4)$, and thus (11.20) becomes

$$\mathcal{M}_{fi} = \frac{g_W^2}{4m_W^2} f_\pi m_\pi u^\dagger(p_3) v_\uparrow(p_4). \tag{11.21}$$

The spinors corresponding to the two possible helicity states of the charged lepton spinor are obtained from (4.65) with $(\theta = 0, \phi = 0)$,

$$u_\uparrow(p_3) = \sqrt{E_\ell + m_\ell}\begin{pmatrix} 1 \\ 0 \\ \frac{\mathrm{p}}{E_\ell + m_\ell} \\ 0 \end{pmatrix} \quad \text{and} \quad u_\downarrow(p_3) = \sqrt{E_\ell + m_\ell}\begin{pmatrix} 0 \\ 1 \\ 0 \\ -\frac{\mathrm{p}}{E_\ell + m_\ell} \end{pmatrix}, \tag{11.22}$$

and the right-handed antineutrino spinor is given by (4.66) with $(\theta = \pi, \phi = \pi)$,

$$v_\uparrow(p_4) = \sqrt{\mathrm{p}}\begin{pmatrix} 1 \\ 0 \\ -1 \\ 0 \end{pmatrix}. \tag{11.23}$$

From (11.22) and (11.23) it is immediately clear that $u_\downarrow^\dagger(p_3)v_\uparrow(p_4) = 0$. Therefore, as anticipated, of the four possible helicity combinations, the only non-zero matrix element corresponds to the case where both the charged lepton and the antineutrino are in RH helicity states. Using the explicit forms for the spinors, the matrix element of (11.21) is

$$\mathcal{M}_{fi} = \frac{g_W^2}{4m_W^2} f_\pi m_\pi \sqrt{E_\ell + m_\ell}\,\sqrt{\mathrm{p}}\left(1 - \frac{\mathrm{p}}{E_\ell + m_\ell}\right). \tag{11.24}$$

Equation (11.24) can be simplified using the expressions for $E_\ell$ and p given in (11.19), such that

$$\begin{aligned}\mathcal{M}_{fi} &= \frac{g_W^2}{4m_W^2} f_\pi m_\pi \cdot \frac{m_\pi + m_\ell}{\sqrt{2m_\pi}} \cdot \left(\frac{m_\pi^2 - m_\ell^2}{2m_\pi}\right)^{\frac{1}{2}} \cdot \frac{2m_\ell}{m_\pi + m_\ell} \\ &= \left(\frac{g_W}{2m_W}\right)^2 f_\pi m_\ell (m_\pi^2 - m_\ell^2)^{\frac{1}{2}}.\end{aligned}$$

Since the pion is a spin-0 particle, there is no need to average over the initial-state spins, and the matrix element squared is given by

$$\langle|\mathcal{M}_{fi}|^2\rangle \equiv |\mathcal{M}_{fi}|^2 = 2G_F^2 f_\pi^2 m_\ell^2(m_\pi^2 - m_\ell^2),$$

where $g_W$ has been expressed in terms of $G_F$ using (11.15). Finally, the decay rate can be determined from the expression for the two-body decay rate given by (3.49), where the integral over solid angle introduces a factor of $4\pi$ as there is no angular dependence in $\langle|\mathcal{M}_{fi}|^2\rangle$. Hence

$$\Gamma = \frac{4\pi}{32\pi^2 m_\pi^2}\,\mathrm{p}\,\langle|\mathcal{M}_{fi}|^2\rangle = \frac{G_F^2}{8\pi m_\pi^3} f_\pi^2\left[m_\ell(m_\pi^2 - m_\ell^2)\right]^2, \tag{11.25}$$

where p is given by (11.19). Therefore, to lowest order, the predicted ratio of the $\pi^- \to e^-\overline{\nu}_e$ to $\pi^- \to \mu^-\overline{\nu}_\mu$ decay rates is

$$\frac{\Gamma(\pi^- \to e^-\overline{\nu}_e)}{\Gamma(\pi^- \to \mu^-\overline{\nu}_\mu)} = \left[\frac{m_e(m_\pi^2 - m_e^2)}{m_\mu(m_\pi^2 - m_\mu^2)}\right]^2 = 1.26 \times 10^{-4},$$

which is in reasonable agreement with the measured value of $1.230(4) \times 10^{-4}$.

## 11.7 Experimental evidence for *V* − *A*

The $V - A$ nature of the weak interaction is an experimentally established fact. For example, if the weak interaction was a scalar $(\overline{\psi}\phi)$ or pseudoscalar $(\overline{\psi}\gamma^5\phi)$ interaction, the predicted ratio of the charged pion leptonic decay rates would be $\Gamma(\pi^- \to e^-\overline{\nu}_e)/\Gamma(\pi^- \to \mu^-\overline{\nu}_\mu) = 5.5$, in clear contradiction with the experimental observations. In general, any weak decay can be expressed in terms of a linear combination of the five bilinear covariants, scalar ($S$), pseudoscalar ($P$), vector ($V$), axial vector ($A$) and tensor ($T$):

$$g_S\overline{\psi}\phi,\ g_P\overline{\psi}\gamma^5\phi,\ g_V\overline{\psi}\gamma^\mu\phi,\ g_A\overline{\psi}\gamma^\mu\gamma^5\phi \text{ and } g_T\overline{\psi}(\gamma^\mu\gamma^\nu - \gamma^\nu\gamma^\mu)\phi.$$

By comparing these predictions with the experimental measurements, limits can be placed on the possible sizes of the different contributions. The most precise test of the $V - A$ structure of the weak interaction is based on measurements of the angular distribution of decays of approximately $10^{10}$ polarised muons by the TWIST experiment: see Bayes *et al.* (2011). The measurements are expressed in terms of the Michel parameters which parameterise the general combination of the possible $S+P+V+A+T$ interaction terms. For example, the Michel parameter $\rho$, which for a pure $V-A$ interaction should be 0.75, is measured to be $\rho = 0.749\,97 \pm 0.000\,26$. All such tests indicate that the charged-current weak interaction is described by a $V - A$ vertex factor.

## Summary

In this chapter, the general structure of the weak charged-current interaction was introduced. Unlike QED and QCD, the weak interaction does not conserve parity. Parity violation in the weak charged-current interaction is a direct consequence of the $V-A$ form of the weak charged-current, which treats left-handed and right-handed particles differently. The weak charged-current vertex factor was found to be

$$\frac{-ig_W}{\sqrt{2}}\frac{1}{2}\gamma^{\mu}(1-\gamma^5),$$

and the propagator associated with the exchange of the massive W bosons is

$$\frac{-i}{q^2-m_W^2}\left(g^{\mu\nu}-\frac{q^{\mu}q^{\nu}}{m_W^2}\right).$$

Because of the $V-A$ interaction only

LH chiral particle states and RH chiral antiparticle states

participate in the weak charged-current.

## Problems

**11.1** Explain why the strong decay $\rho^0 \to \pi^-\pi^+$ is observed, but the strong decay $\rho^0 \to \pi^0\pi^0$ is not.

Hint: you will need to consider conservation of angular momentum, parity and the symmetry of the $\pi^0\pi^0$ wavefunction.

**11.2** When $\pi^-$ mesons are stopped in a deuterium target they can form a bound $(\pi^- - \mathrm{D})$ state with zero orbital angular momentum, $\ell = 0$. The bound state decays by the strong interaction

$$\pi^-\mathrm{D} \to \mathrm{nn}.$$

By considering the possible spin and orbital angular momentum states of the nn system, and the required symmetry of the wavefunction, show that the pion has negative intrinsic parity.

Note: the deuteron has $J^P = 1^+$ and the pion is a spin-0 particle.

**11.3** Classify the following quantities as either scalars (S), pseudoscalars (P), vectors (V) or axial-vectors (A):

(a) mechanical power, $P = \mathbf{F}\cdot\mathbf{v}$;
(b) force, $\mathbf{F}$;
(c) torque, $\mathbf{G} = \mathbf{r}\times\mathbf{F}$;
(d) vorticity, $\boldsymbol{\Omega} = \boldsymbol{\nabla}\times\mathbf{v}$;

(e) magnetic flux, $\phi = \int \mathbf{B} \cdot d\mathbf{S}$;
(f) divergence of the electric field strength, $\boldsymbol{\nabla} \cdot \mathbf{E}$.

**11.4** In the annihilation process $e^+e^- \to q\bar{q}$, the QED vector interaction leads to non-zero matrix elements only for the chiral combinations $LR \to LR$, $LR \to RL$, $RL \to RL$, $RL \to LR$. What are the corresponding allowed chiral combinations for $S$, $P$ and $S-P$ interactions?

**11.5** Consider the decay at rest $\tau^- \to \pi^-\nu_\tau$, where the spin of the tau is in the positive $z$-direction and the $\nu_\tau$ and $\pi^-$ travel in the $\pm z$-directions. Sketch the allowed spin configurations assuming that the form of the weak charged-current interaction is (i) $V-A$ and (ii) $V+A$.

**11.6** Repeat the pion decay calculation for a pure scalar interaction and show that the predicted ratio of decay rates is

$$\frac{\Gamma(\pi^- \to e^-\bar{\nu}_e)}{\Gamma(\pi^- \to \mu^-\bar{\nu}_\mu)} \approx 5.5.$$

**11.7** Predict the ratio of the $K^- \to e^-\bar{\nu}_e$ and $K^- \to \mu^-\bar{\nu}_\mu$ weak interaction decay rates and compare your answer to the measured value of

$$\frac{\Gamma(K^- \to e^-\bar{\nu}_e)}{\Gamma(K^- \to \mu^-\bar{\nu}_\mu)} = (2.488 \pm 0.012) \times 10^{-5}.$$

**11.8** Charged kaons have several weak interaction decay modes, the largest of which are

$$K^+(u\bar{s}) \to \mu^+\nu_\mu, \quad K^+ \to \pi^+\pi^0 \quad \text{and} \quad K^+ \to \pi^+\pi^+\pi^-.$$

(a) Draw the Feynman diagrams for these three weak decays.
(b) Using the measured branching ratio

$$Br(K^+ \to \mu^+\nu_\mu) = 63.55 \pm 0.11\,\%,$$

*estimate* the lifetime of the charged kaon.

Note: charged pions decay almost 100% of the time by the weak interaction $\pi^+ \to \mu^+\nu_\mu$ and have a lifetime of $(2.6033 \pm 0.0005) \times 10^{-8}$ s.

**11.9** From the prediction of (11.25) and the above measured value of the charged pion lifetime, obtain a value for $f_\pi$.

**11.10** Calculate the partial decay width for the decay $\tau^- \to \pi^-\nu_\tau$ in the following steps.

(a) Draw the Feynman diagram and show that the corresponding matrix element is

$$\mathcal{M} \approx \sqrt{2}G_F f_\pi \bar{u}(p_\nu)\gamma^\mu \tfrac{1}{2}(1-\gamma^5)u(p_\tau)g_{\mu\nu}p_\pi^\nu.$$

(b) Taking the $\tau^-$ spin to be in the $z$-direction and the four-momentum of the neutrino to be

$$p_\nu = p^*(1, \sin\theta, 0, \cos\theta),$$

show that the leptonic current is

$$j^\mu = \sqrt{2m_\tau p^*}\,(-s, -c, -ic, s)\,,$$

where $s = \sin\left(\frac{\theta}{2}\right)$ and $c = \cos\left(\frac{\theta}{2}\right)$. Note that, for this configuration, the spinor for the $\tau^-$ can be taken to be $u_1$ for a particle at rest.

(c) Write down the four-momentum of the $\pi^-$ and show that

$$|\mathcal{M}|^2 = 4G_F^2 f_\pi^2 m_\tau^3 p^* \sin^2\left(\frac{\theta}{2}\right).$$

(d) Hence show that

$$\Gamma(\tau^- \to \pi^- \nu_\tau) = \frac{G_F^2 f_\pi^2}{16\pi} m_\tau^3 \left(\frac{m_\tau^2 - m_\pi^2}{m_\tau^2}\right)^2.$$

(e) Using the value of $f_\pi$ obtained in the previous problem, find a numerical value for $\Gamma(\tau^- \to \pi^- \nu_\tau)$.

(f) Given that the lifetime of the $\tau$-lepton is measured to be $\tau_\tau = 2.906 \times 10^{-13}$ s, find an approximate value for the $\tau^- \to \pi^- \nu_\tau$ branching ratio.

# 12 The weak interactions of leptons

In the previous chapter, the general structure of the charged-current weak interaction was introduced. In this chapter, these ideas are applied to the weak interactions of charged leptons and neutrinos. The scattering cross sections for neutrinos on nucleons are calculated from first principles and the experimental measurements are related to the nucleon parton distribution functions. In the final section, the high-energy charged-current process $e^-p \to \nu_e X$ is used as an example of the weak interaction in the limit $Q^2 > m_W^2$.

## 12.1 Lepton universality

From the observed decay rates of muons and tau leptons, it is found that the strength of the weak interaction is the same for all lepton flavours. For example, Figure 12.1 shows the Feynman diagram for muon decay, $\mu^- \to e^-\overline{\nu}_e\nu_\mu$. It involves two weak-interaction vertices, $\mu^-\nu_\mu W$ and $We^-\overline{\nu}_e$. In principle, the coupling at these two vertices could be different. Allowing for this possibility, the muon decay rate of (11.16) can be written

$$\Gamma(\mu^- \to e^-\overline{\nu}_e\nu_\mu) \equiv \frac{1}{\tau_\mu} = \frac{G_F^{(e)}G_F^{(\mu)}m_\mu^5}{192\pi^3}, \tag{12.1}$$

where the weak couplings to the electron and muon are respectively $G_F^{(e)}$ and $G_F^{(\mu)}$.

The same calculation for the decay rate $\tau^- \to e^-\overline{\nu}_e\nu_\tau$ gives

$$\Gamma(\tau^- \to e^-\overline{\nu}_e\nu_\tau) = \frac{G_F^{(e)}G_F^{(\tau)}m_\tau^5}{192\pi^3}. \tag{12.2}$$

The tau-lepton is sufficiently massive that it can also decay into a muon or to mesons formed from light quarks, as shown in Figure 12.2. Therefore the tau lifetime needs to expressed in terms of the total decay rate

$$\frac{1}{\tau_\tau} = \Gamma = \sum_i \Gamma_i,$$

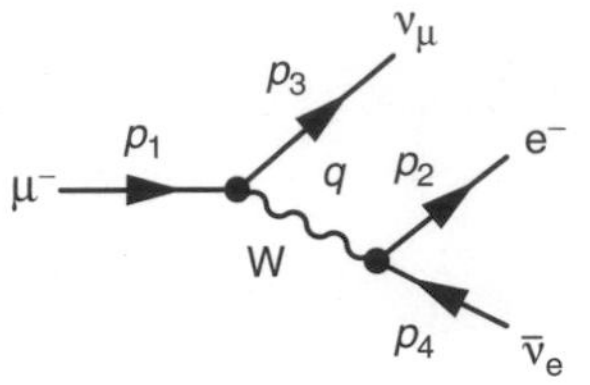


Fig. 12.1 The lowest-order Feynman diagram for muon decay.

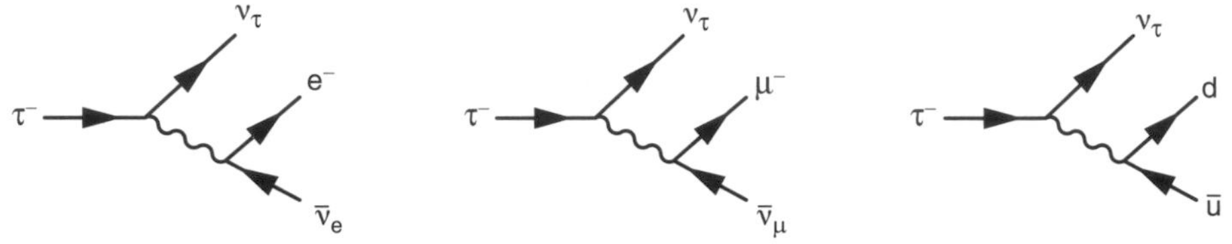


Fig. 12.2 The lowest-order Feynman diagrams for tau decay.

where $\Gamma_i$ are the partial decay rates for the individual decay modes. The ratio of the partial width $\Gamma(\tau^- \to e^-\overline{\nu}_e\nu_\tau)$ to the total decay rate gives the branching ratio

$$Br(\tau^- \to e^-\overline{\nu}_e\nu_\tau) = \frac{\Gamma(\tau^- \to e^-\overline{\nu}_e\nu_\tau)}{\Gamma} = \Gamma(\tau^- \to e^-\overline{\nu}_e\nu_\tau) \times \tau_\tau. \tag{12.3}$$

From (12.2) and (12.3), the tau lifetime can be expressed as

$$\tau_\tau = \frac{192\pi^3}{G_F^{(e)} G_F^{(\tau)} m_\tau^5} Br(\tau^- \to e^-\overline{\nu}_e\nu_\tau). \tag{12.4}$$

Comparing the expressions for the muon and tau-lepton lifetimes given in (12.1) and (12.4), gives the ratio

$$\frac{G_F^{(\tau)}}{G_F^{(\mu)}} = \frac{m_\mu^5 \tau_\mu}{m_\tau^5 \tau_\tau} Br(\tau^- \to e^-\overline{\nu}_e\nu_\tau). \tag{12.5}$$

The ratios of the couplings can be obtained from the measured branching ratios for the leptonic decays of the tau-lepton, which are

$$Br(\tau^- \to e^-\overline{\nu}_e\nu_\tau) = 0.1783(5) \quad \text{and} \quad Br(\tau^- \to \mu^-\overline{\nu}_\mu\nu_\tau) = 0.1741(4),$$

and the measured masses and lifetimes of the muon and tau-lepton,

$$m_\mu = 0.1056583715(35)\,\text{GeV} \quad \text{and} \quad \tau_\mu = 2.1969811(22) \times 10^{-6}\,\text{s},$$
$$m_\tau = 1.77682(16)\,\text{GeV} \quad \text{and} \quad \tau_\tau = 0.2906(10) \times 10^{-12}\,\text{s}.$$

From these measured values and the relation of (12.5), the ratio of the muon and tau weak charged-current coupling strengths is determined to be

$$\frac{G_F^{(\tau)}}{G_F^{(\mu)}} = 1.0023 \pm 0.0033.$$

Similarly, by comparing $Br(\tau^- \to e^-\overline{\nu}_e\nu_\tau)$ to $Br(\tau^- \to \mu^-\overline{\nu}_\mu\nu_\tau)$, taking into account the expected small difference due to phase space, gives

$$\frac{G_F^{(e)}}{G_F^{(\mu)}} = 1.000 \pm 0.004.$$

Therefore, within the accuracy of the experimental measurements, it can be concluded that $G_F^{(e)} = G_F^{(\mu)} = G_F^{(\tau)}$, providing strong experimental evidence for the lepton universality of the weak charged current; there is a universal coupling strength at the $We\nu_e$, $W\mu\nu_\mu$ and $W\tau\nu_\tau$ interaction vertices.

## 12.2 Neutrino scattering

Although neutrinos interact only weakly in matter, precise measurements of their properties can be made using sufficiently intense neutrino beams. The general scheme for producing a collimated beam of neutrinos is shown schematically in Figure 12.3. The neutrino beam is produced by firing an intense beam of high-energy protons at a target, resulting in a large flux of high-energy hadrons from the hard QCD interaction and subsequent hadronisation process. A significant fraction of the produced hadrons are charged pions, both $\pi^+$ and $\pi^-$. The charged pions of a particular charge sign can be focussed in the magnetic field generated by one or more "neutrino horns"; the other charge will be defocussed. In this way it is possible to produce a collimated beam of pions with a particular charge sign. The pion

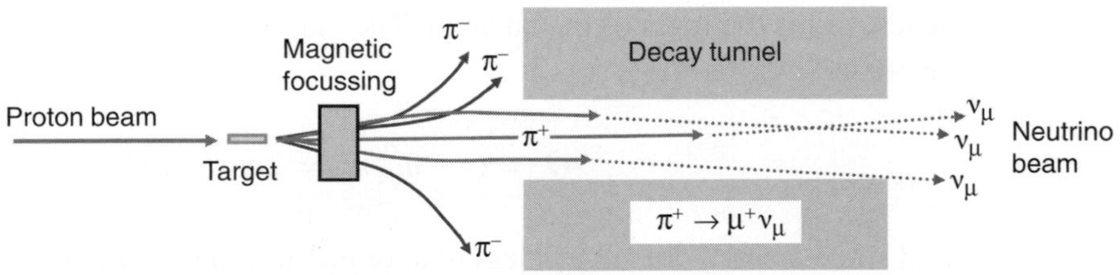


Fig. 12.3 The general scheme for producing a neutrino beam from a proton beam. The focussed pions produced from the target are allowed to decay in a long decay tunnel.

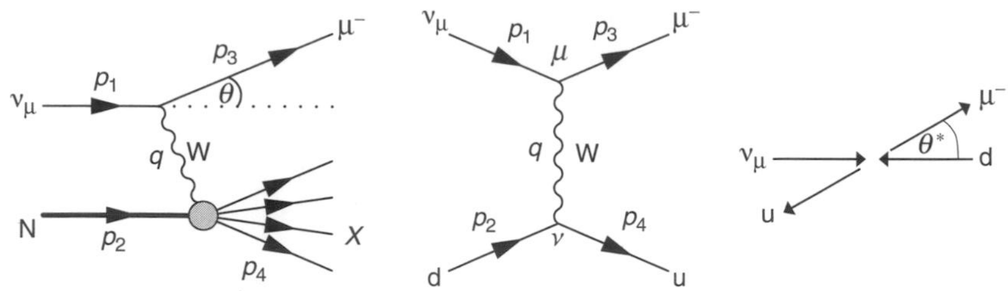


Fig. 12.4 Neutrino deep inelastic scattering and the Feynman diagram for the corresponding $\nu_\mu d$ process. In the centre-of-mass frame the $\mu^-$ is produced at an angle $\theta^*$ to the incoming neutrino direction.

beam is then allowed to decay in flight over a distance of order $\gamma c\tau_\pi$ in an evacuated decay tunnel, which is typically a few hundred metres long. Because the mass of the pion is small compared to the typical pion energy, the neutrinos produced in the $\pi^+ \to \mu^+\nu_\mu$ or $\pi^- \to \mu^-\overline{\nu}_\mu$ decays are boosted along the direction of motion of the pions. The choice of the sense of the magnetic field in the neutrino horns enables either $\pi^+$ or $\pi^-$ to be focussed, and therefore either a $\nu_\mu$ or $\overline{\nu}_\mu$ beam can be produced. The muons produced from the pion decays are stopped in the rock at the end of the decay tunnel, before they decay themselves. The result is a collimated beam of almost entirely muon neutrinos or muon antineutrinos.

The phenomenology of neutrino scattering closely follows that of electron–proton scattering, discussed in Chapter 8. At low $Q^2$, there is a quasi-elastic process, $\nu_\mu n \to \mu^- p$, which is quasi-elastic in the sense that the nucleon changes type but does not break up. At slightly higher neutrino energies (a few GeV), resonant inelastic processes such as $\nu_\mu n \to \mu^-\Delta^+ \to \mu^- p\pi^0$ are observed. At higher energies still, neutrino interactions are dominated by the neutrino deep inelastic scattering process, shown in Figure 12.4.

In the neutrino scattering experiments considered in this chapter, the neutrino energy is sufficiently high that only the deep inelastic process is of relevance. For a neutrino interacting with a nucleon at rest, the centre-of-mass energy squared is

$$s = (p_1 + p_2)^2 = (E_\nu + m_N)^2 - E_\nu^2 = 2m_N E_\nu + m_N^2, \tag{12.6}$$

where $m_N$ is the mass of the nucleon. The maximum $Q^2$ in the scattering process is restricted by

$$Q^2 = (s - m_N^2)xy = 2m_N E_\nu xy,$$

and since $x$ and $y$ are always less than or equal to one, for a given neutrino energy,

$$Q^2 \leq 2m_N E_\nu.$$

The highest-energy neutrinos produced in an accelerator-based neutrino beam had $E_\nu = 400\,\text{GeV}$, for which $Q^2 \lesssim 750\,\text{GeV}^2$. This is sufficiently small that, for all practical purposes, the weak interaction propagator can be approximated by

$$\frac{-ig_{\mu\nu}}{q^2 - m_W^2} \rightarrow \frac{ig_{\mu\nu}}{m_W^2}.$$

Furthermore, for high-energy neutrino interactions where the $m_N^2$ term in (12.6) can be neglected, the centre-of-mass energy squared is proportional to the neutrino energy,

$$s \approx 2m_N E_\nu. \tag{12.7}$$

The underlying interactions in neutrino–nucleon scattering are the parton-level processes, $\nu_\mu d \to \mu^- u$ and $\nu_\mu \bar{u} \to \mu^- \bar{d}$. The cross sections for these processes are calculated below. Because only left-handed chiral particle states and right-handed chiral antiparticle states participate in the weak charged-current, only one helicity combination needs to be considered in each case.

### 12.2.1 Neutrino–quark scattering cross section

Neglecting the $q^2$ dependence of the propagator, the matrix element for the Feynman diagram of Figure 12.4 is

$$-i\mathcal{M}_{fi} = \left[-i\frac{g_W}{\sqrt{2}}\bar{u}(p_3)\gamma^\mu \tfrac{1}{2}(1-\gamma^5)u(p_1)\right]\frac{ig_{\mu\nu}}{m_W^2}\left[-i\frac{g_W}{\sqrt{2}}\bar{u}(p_4)\gamma^\nu \tfrac{1}{2}(1-\gamma^5)u(p_2)\right]$$

$$\mathcal{M}_{fi} = \frac{g_W^2}{2m_W^2}g_{\mu\nu}\left[\bar{u}(p_3)\gamma^\mu \tfrac{1}{2}(1-\gamma^5)u(p_1)\right]\left[\bar{u}(p_4)\gamma^\nu \tfrac{1}{2}(1-\gamma^5)u(p_2)\right]. \tag{12.8}$$

For high-energy neutrino scattering, both the masses of the neutrinos and quarks are sufficiently small that the LH chiral states are effectively identical to the LH helicity states and (12.8) can be written

$$\mathcal{M}_{fi} = \frac{g_W^2}{2m_W^2}g_{\mu\nu}\left[\bar{u}_\downarrow(p_3)\gamma^\mu u_\downarrow(p_1)\right]\left[\bar{u}_\downarrow(p_4)\gamma^\nu u_\downarrow(p_2)\right] = \frac{g_W^2}{2m_W^2}\,j_\ell \cdot j_q, \tag{12.9}$$

where $j_\ell^\mu = \bar{u}_\downarrow(p_3)\gamma^\mu u_\downarrow(p_1)$ and $j_q^\nu = \bar{u}_\downarrow(p_4)\gamma^\nu u_\downarrow(p_2)$ are respectively the lepton and quark currents.

The matrix element is most easily evaluated in the centre-of-mass frame. Taking the initial neutrino direction to define the $z$-axis and $\theta^*$ to be the polar angle of the

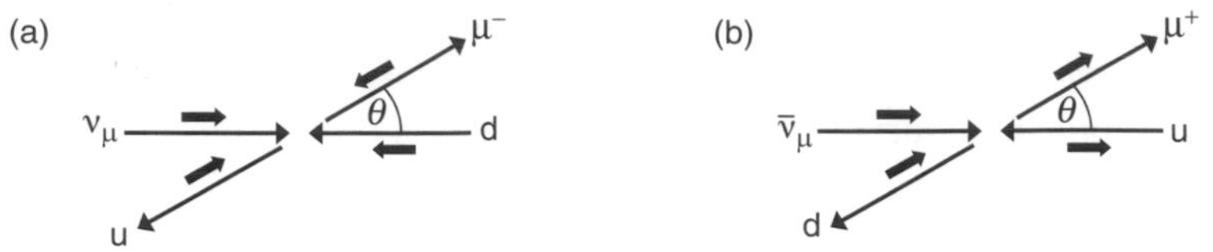


Fig. 12.5 The spin orientations of the particles in (a) charged-current $\nu_\mu d \to \mu^- u$ and (b) charged-current $\bar{\nu}_\mu u \to \mu^+ d$ weak interactions.

final-state $\mu^-$, then the spherical polar angles of the four particles, as indicated in Figure 12.4, are

$$(\theta_1,\phi_1) = (0,0),\quad (\theta_2,\phi_2) = (\pi,\pi),\quad (\theta_3,\phi_3) = (\theta^*,0) \text{ and } (\theta_4,\phi_4) = (\pi-\theta^*,\pi).$$

The corresponding LH spinors are given by (4.67),

$$u_\downarrow(p_1) = \sqrt{E}\begin{pmatrix}0\\1\\0\\-1\end{pmatrix},\; u_\downarrow(p_2) = \sqrt{E}\begin{pmatrix}-1\\0\\1\\0\end{pmatrix},\; u_\downarrow(p_3) = \sqrt{E}\begin{pmatrix}-s\\c\\s\\-c\end{pmatrix}\; u_\downarrow(p_4) = \sqrt{E}\begin{pmatrix}-c\\-s\\c\\s\end{pmatrix}, \tag{12.10}$$

where $c = \cos\frac{\theta^*}{2}$, $s = \sin\frac{\theta^*}{2}$ and $E$ is the energy of each of the four particles in the centre-of-mass frame. The lepton and quark currents then can be evaluated by using the relations of (6.12)–(6.15), giving

$$j_\ell^\mu = \bar{u}_\downarrow(p_3)\gamma^\mu u_\downarrow(p_1) = 2E(c,s,-is,c),$$
$$j_q^\nu = \bar{u}_\downarrow(p_4)\gamma^\nu u_\downarrow(p_2) = 2E(c,-s,-is,-c),$$

and hence

$$\mathcal{M}_{fi} = \frac{g_W^2}{2m_W^2}\, j_\ell \cdot j_q = \frac{g_W^2}{2m_W^2} 4E^2(c^2+s^2+s^2+c^2).$$

Therefore, the matrix element for $\nu_\mu d \to \mu^- u$ scattering is simply

$$\mathcal{M}_{fi} = \frac{g_W^2}{m_W^2}\hat{s}, \tag{12.11}$$

where $\hat{s} = (2E)^2$ is the $\nu_\mu d$ centre-of-mass energy. The matrix element of (12.11) does not depend on the polar angle $\theta^*$ and therefore represents an isotropic distribution of the final-state particles in the centre-of-mass frame. This can be understood in terms of the helicities of the colliding particles, shown in Figure 12.5a. Because both the quark and neutrino are left-handed, the interaction occurs in an $S_z = 0$ state and thus there is no preferred polar angle in the centre-of-mass frame.

In the limit where the particle masses can be neglected, the centre-of-mass frame differential cross section is given by (3.50),

$$\frac{d\sigma}{d\Omega^*} = \frac{1}{64\pi^2 \hat{s}} \langle |\mathcal{M}_{fi}|^2 \rangle,$$

where $\langle |\mathcal{M}_{fi}|^2 \rangle$ is the spin-averaged matrix element squared. In this weak charged-current process, the only non-zero matrix element is that for the helicity combination $LL \to LL$, calculated above. Previously the average over the spins of the two initial-state fermions gave rise to a factor 1/4 in the spin-averaged matrix element. Here, because it was produced in a weak decay, the neutrino will always be left-handed and it is only necessary to average over the two spin states of the quark. Hence the spin-averaged matrix element squared is

$$\langle |\mathcal{M}_{fi}|^2 \rangle = \frac{1}{2} \left( \frac{g_W^2}{m_W^2} \hat{s} \right)^2, \tag{12.12}$$

and the differential cross section is

$$\frac{d\sigma}{d\Omega^*} = \frac{1}{64\pi^2 \hat{s}} \langle |\mathcal{M}_{fi}|^2 \rangle = \left( \frac{g_W^2}{8\sqrt{2}\pi m_W^2} \right)^2 \hat{s}.$$

Using $G_F = \sqrt{2} g_W^2 / 8 m_W^2$, this can be written as

$$\frac{d\sigma_{\nu q}}{d\Omega^*} = \frac{G_F^2}{4\pi^2} \hat{s}, \tag{12.13}$$

and the total cross section, obtained by integrating over $d\Omega^*$, is

$$\sigma_{\nu q} = \frac{G_F^2 \hat{s}}{\pi}. \tag{12.14}$$

### 12.2.2 Antineutrino–quark scattering

Figure 12.6 shows the Feynman diagram for antineutrino–quark scattering. In order to conserve electric charge, the antineutrino can interact with an up-quark, but not a down-quark. The corresponding matrix element is

$$\mathcal{M}_{fi} = \frac{g_W^2}{2m_W^2} g_{\mu\nu} \left[ \overline{v}(p_1) \gamma^\mu \tfrac{1}{2}(1-\gamma^5) v(p_3) \right] \left[ \overline{u}(p_4) \gamma^\nu \tfrac{1}{2}(1-\gamma^5) u(p_2) \right].$$

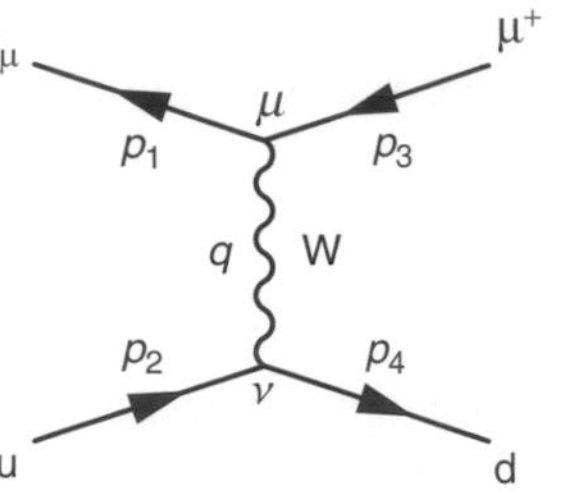


Fig. 12.6 The Feynman diagram for $\overline{\nu}_\mu \text{u} \to \mu^+ \text{d}$.

In the high-energy limit, only LH helicity particles and RH helicity antiparticles participate in the charged-current weak interaction and thus the only non-zero matrix element is given by

$$\mathcal{M}_{fi} = \frac{g_W^2}{2m_W^2} g_{\mu\nu} \left[\overline{v}_\uparrow(p_1)\gamma^\mu v_\uparrow(p_3)\right]\left[\overline{u}_\downarrow(p_4)\gamma^\nu u_\downarrow(p_2)\right].$$

Proceeding as before, it is straightforward to show that this leads to

$$\mathcal{M}_{\overline{\nu}\text{q}} = \tfrac{1}{2}(1+\cos\theta^*)\frac{g_W^2}{m_W^2}\hat{s}, \tag{12.15}$$

where $\theta^*$ is the polar angle of the $\mu^+$ in the centre-of-mass frame. This matrix element differs from the corresponding matrix element for neutrino–quark scattering (12.11) by the factor $\frac{1}{2}(1+\cos\theta^*)$. The origin of this difference can be understood in terms of spins of the particles, as shown in Figure 12.5b. The $V-A$ nature of the weak interaction means that the $\overline{\nu}\text{q}$ interaction occurs in an $S_z = 1$ state and, from the discussion of Section 6.3, this results in an angular dependence of the matrix element of $\frac{1}{2}(1+\cos\theta^*)$. Therefore, the $\overline{\nu}_\mu\text{u}$ differential cross section is related to that for $\nu_\mu\text{d}$ by

$$\frac{\text{d}\sigma_{\overline{\nu}\text{q}}}{\text{d}\Omega^*} = \tfrac{1}{4}(1+\cos\theta^*)^2\,\frac{\text{d}\sigma_{\nu\text{q}}}{\text{d}\Omega^*} = \frac{G_F^2}{16\pi^2}(1+\cos\theta^*)^2\,\hat{s},$$

and the total cross section is obtained by integrating over solid angle with

$$\int (1+\cos\theta^*)^2\text{d}\Omega^* = \int_0^{2\pi}\text{d}\phi^*\int_{-1}^{+1}(1+x)^2\text{d}x = \frac{16\pi}{3},$$

where the substitution $x = \cos\theta^*$ was used. Hence, the total antineutrino–quark cross section is

$$\sigma_{\overline{\nu}\text{q}} = \frac{G_F^2\hat{s}}{3\pi},$$

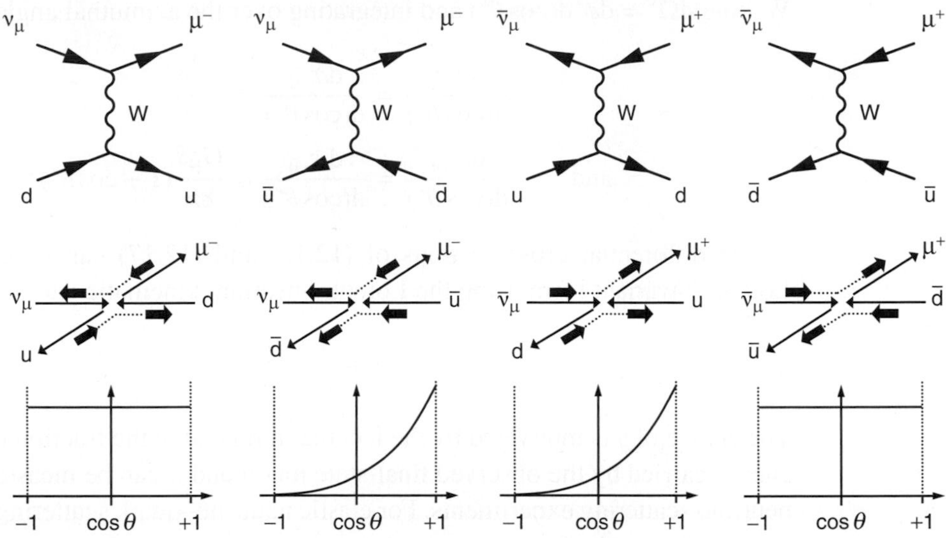


Fig. 12.7 The four possible Feynman diagrams and corresponding helicity combinations for the weak charged-current interactions involving $\nu_\mu$, $\overline{\nu}_\mu$ and any light (ud) quark or antiquark. Also shown are the corresponding angular distributions in the centre-of-mass frame.

which is a factor three smaller than the neutrino–quark cross section

$$\frac{\sigma_{\overline{\nu}q}}{\sigma_{\nu q}} = \frac{1}{3}.$$

## Neutrino–nucleon differential cross sections

Neutrino interactions in matter are described by their interactions with the constituent quarks of nucleons. From the conservation of electric charge, there are only four possible interactions between a $\nu_\mu$ or $\overline{\nu}_\mu$ and the light constituents (u, d, $\overline{u}$ and $\overline{d}$) of the nucleon. These are $\nu_\mu d \to \mu^- u$, $\nu_\mu \overline{u} \to \mu^- \overline{d}$, $\overline{\nu}_\mu u \to \mu^- d$ and $\overline{\nu}_\mu \overline{d} \to \mu^+ \overline{u}$. For each process only one helicity combination is involved. The differential cross sections for scattering from the antiquarks can be obtained directly from those derived above by considering the spin state in which the interaction occurs, as shown in Figure 12.7. The differential cross section for neutrino/antineutrino scattering from antiquarks can be equated to the corresponding scattering cross section for quarks that occurs in the same spin state. Hence

$$\frac{d\sigma_{\overline{\nu}\overline{d}}}{d\Omega^*} = \frac{d\sigma_{\nu d}}{d\Omega^*} = \frac{G_F^2 \hat{s}}{4\pi^2} \quad \text{and} \quad \frac{d\sigma_{\overline{\nu}u}}{d\Omega^*} = \frac{d\sigma_{\nu\overline{u}}}{d\Omega^*} = \frac{G_F^2 \hat{s}}{16\pi^2}(1+\cos\theta^*)^2.$$

Writing $d\Omega^* = d\phi^* d(\cos\theta^*)$ and integrating over the azimuthal angle gives

$$\frac{d\sigma_{\overline{\nu}\overline{d}}}{d(\cos\theta^*)} = \frac{d\sigma_{\nu d}}{d(\cos\theta^*)} = \frac{G_F^2 \hat{s}}{2\pi} \tag{12.16}$$

and

$$\frac{d\sigma_{\overline{\nu}u}}{d(\cos\theta^*)} = \frac{d\sigma_{\nu\overline{u}}}{d(\cos\theta^*)} = \frac{G_F^2 \hat{s}}{8\pi}(1+\cos\theta^*)^2. \tag{12.17}$$

The differential cross sections of (12.16) and (12.17) can be expressed in a Lorentz-invariant form using the Lorentz-invariant kinematic variable

$$y \equiv \frac{p_2 \cdot q}{p_2 \cdot p_1}.$$

The choice of $y$ is motivated by the fact that it related to the fraction of the neutrino energy carried by the observed final-state muon and it can be measured directly in neutrino scattering experiments. For elastic neutrino–quark scattering in the centre-of-mass frame, the four-momenta appearing in the expression for $y$ are

$$p_1 = (E,0,0,E), \quad p_2 = (E,0,0,-E) \quad \text{and} \quad p_3 = (E,0,E\sin\theta^*, E\cos\theta^*),$$

and therefore $y$ can be written as

$$y = \frac{p_2 \cdot q}{p_2 \cdot p_1} = \frac{p_2 \cdot (p_1 - p_3)}{p_2 \cdot p_1} = \tfrac{1}{2}(1 - \cos\theta^*).$$

Differentiating $y$ with respect to $\cos\theta^*$ gives

$$\frac{dy}{d(\cos\theta^*)} = -\frac{1}{2},$$

and thus

$$\frac{d\sigma}{dy} = \left|\frac{d(\cos\theta^*)}{dy}\right| \frac{d\sigma}{d(\cos\theta^*)} = 2\frac{d\sigma}{d(\cos\theta^*)},$$

and the differential cross section of (12.16) can be expressed as

$$\frac{d\sigma_{\nu q}}{dy} = \frac{d\sigma_{\overline{\nu}\overline{q}}}{dy} = \frac{G_F^2}{\pi}\hat{s}. \tag{12.18}$$

Furthermore, using $(1-y) = \frac{1}{2}(1+\cos\theta^*)$, the Lorentz-invariant forms of the $\overline{\nu}q$ and $\nu\overline{q}$ differential cross sections of (12.17) are

$$\frac{d\sigma_{\overline{\nu}q}}{dy} = \frac{d\sigma_{\nu\overline{q}}}{dy} = \frac{G_F^2}{\pi}(1-y)^2\hat{s}. \tag{12.19}$$

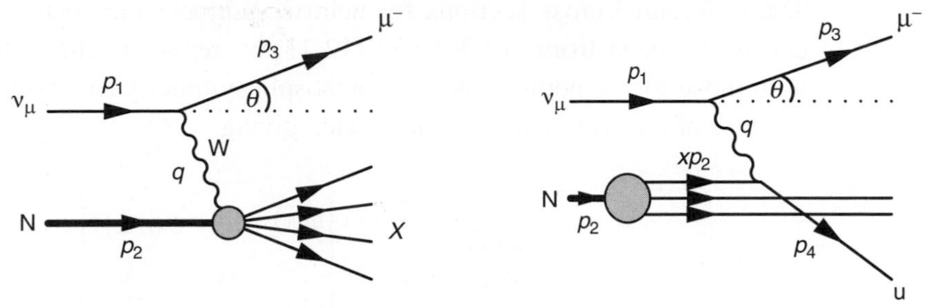


Fig. 12.8 The parton model picture of the charged-current scattering of a neutrino from a d quark in a nucleon.

### 12.2.3 Neutrino deep inelastic scattering

The interaction of high-energy neutrinos and antineutrinos with matter can be expressed in terms of deep inelastic scattering with protons and neutrons. The underlying interaction, shown in Figure 12.8, is between a neutrino/antineutrino and a quark/antiquark carrying a fraction $x$ of the nucleon momentum. Neglecting the contributions from the strange quark sea, the underlying physical processes for neutrino–proton deep inelastic scattering are $\nu_\mu d \to \mu^- u$ and $\nu_\mu \overline{u} \to \mu^- \overline{d}$ with differential cross sections given by (12.18) and (12.19). The number of down-quarks within a proton in the momentum fraction range $x \to x+\mathrm{d}x$ is $d(x)\,\mathrm{d}x$, where $d(x)$ is the down-quark parton distribution function for the proton. The equivalent expression for the anti-up-quarks is $\overline{u}(x)\,\mathrm{d}x$. The contribution to the total neutrino–proton scattering cross section from $\nu d$ and $\nu\overline{u}$ scattering is therefore

$$\frac{\mathrm{d}\sigma^{\nu p}}{\mathrm{d}\hat{y}} = \frac{G_F^2}{\pi}\hat{s}\left[d(x) + (1-\hat{y})^2\overline{u}(x)\right]\mathrm{d}x,$$

where $\hat{s}$ and $\hat{y}$ refer to the neutrino–quark system. The kinematic variables $\hat{s}$ and $\hat{y}$ can be expressed in terms of the neutrino–proton system using

$$\hat{s} = (p_1 + xp_2)^2 \approx 2xp_1 \cdot p_2 = xs \quad \text{and} \quad \hat{y} = \frac{p_q \cdot q}{p_q \cdot p_1} = \frac{xp_2 \cdot q}{xp_2 \cdot p_1} = y.$$

Hence, the parton model differential cross section for neutrino–proton scattering is

$$\frac{\mathrm{d}^2\sigma^{\nu p}}{\mathrm{d}x\,\mathrm{d}y} = \frac{G_F^2}{\pi}sx\left[d(x) + (1-y)^2\overline{u}(x)\right]. \tag{12.20}$$

The underlying processes for antineutrino–proton scattering are $\overline{\nu}u$ and $\overline{\nu}\overline{d}$ scattering, and the corresponding differential cross section is

$$\frac{\mathrm{d}^2\sigma^{\overline{\nu}p}}{\mathrm{d}x\,\mathrm{d}y} = \frac{G_F^2}{\pi}sx\left[(1-y)^2u(x) + \overline{d}(x)\right]. \tag{12.21}$$

The differential cross sections for neutrino/antineutrino scattering with neutrons can be obtained from (12.20) and (12.21) by replacing the PDFs for the proton with those for the neutron, and using isospin symmetry to relate the neutron PDFs to those of the proton, $d^{\mathrm{n}}(x) = u(x)$, etc., giving

$$\frac{\mathrm{d}^2\sigma^{\nu\mathrm{n}}}{\mathrm{d}x\,\mathrm{d}y} = \frac{G_\mathrm{F}^2}{\pi} sx\left[u(x) + (1-y)^2\overline{d}(x)\right], \tag{12.22}$$

$$\frac{\mathrm{d}^2\sigma^{\overline{\nu}\mathrm{n}}}{\mathrm{d}x\,\mathrm{d}y} = \frac{G_\mathrm{F}^2}{\pi} sx\left[(1-y)^2 d(x) + \overline{u}(x)\right]. \tag{12.23}$$

In practice, because neutrino cross sections are so small, massive detectors are required. Consequently, the majority of the experiments that have studied high-energy neutrino deep inelastic scattering have employed detectors constructed from a dense material, usually steel (which is predominantly iron). Therefore, the measured neutrino cross sections are a combination of the cross sections for protons and neutrons. For an isoscalar target, with an equal number of protons and neutrons, the average neutrino scattering cross section per nucleon is

$$\frac{\mathrm{d}^2\sigma^{\nu\mathrm{N}}}{\mathrm{d}x\,\mathrm{d}y} = \frac{1}{2}\left(\frac{\mathrm{d}^2\sigma^{\nu\mathrm{p}}}{\mathrm{d}x\,\mathrm{d}y} + \frac{\mathrm{d}^2\sigma^{\nu\mathrm{n}}}{\mathrm{d}x\,\mathrm{d}y}\right),$$

which, from (12.20) and (12.22), can be written in terms of the PDFs as

$$\frac{\mathrm{d}^2\sigma^{\nu\mathrm{N}}}{\mathrm{d}x\,\mathrm{d}y} = \frac{G_\mathrm{F}^2 m_\mathrm{N}}{\pi} E_\nu\, x\left[d(x) + u(x) + (1-y)^2\left\{\overline{u}(x) + \overline{d}(x)\right\}\right], \tag{12.24}$$

where the centre-of-mass energy has been expressed in terms of the neutrino energy, $s \approx 2m_\mathrm{N}E_\nu$. The integral of (12.24) over the momentum fraction $x$ of the struck quark (or antiquark) gives the differential cross section in terms of $y$,

$$\frac{\mathrm{d}\sigma^{\nu\mathrm{N}}}{\mathrm{d}y} = \frac{G_\mathrm{F}^2 m_\mathrm{N}}{\pi} E_\nu\left[f_\mathrm{q} + (1-y)^2 f_{\overline{\mathrm{q}}}\right], \tag{12.25}$$

where $f_\mathrm{q}$ and $f_{\overline{\mathrm{q}}}$ are the fractions of the nucleon momentum respectively carried by the quarks and the antiquarks,

$$f_\mathrm{q} = \int_0^1 x\left[u(x) + d(x)\right]\mathrm{d}x \quad \text{and} \quad f_{\overline{\mathrm{q}}} = \int_0^1 x\left[\overline{u}(x) + \overline{d}(x)\right]\mathrm{d}x.$$

Likewise, the average antineutrino–nucleon scattering cross section, obtained from (12.21) and (12.23), is

$$\frac{\mathrm{d}^2\sigma^{\overline{\nu}\mathrm{N}}}{\mathrm{d}x\mathrm{d}y} = \frac{G_\mathrm{F}^2 m_\mathrm{N}}{\pi} E_\nu\, x\left[(1-y)^2\left\{u(x) + d(x)\right\} + \overline{u}(x) + \overline{d}(x)\right], \tag{12.26}$$

which when integrated over $x$ gives

$$\frac{\mathrm{d}\sigma^{\overline{\nu}\mathrm{N}}}{\mathrm{d}y} = \frac{G_{\mathrm{F}}^2 m_{\mathrm{N}}}{\pi} E_\nu \left[(1-y)^2 f_{\mathrm{q}} + f_{\overline{\mathrm{q}}}\right]. \tag{12.27}$$

Here the factor $(1-y)^2$ appears in front of the quark term rather than the antiquark term as in (12.25). The $y$-dependence of the differential cross sections for neutrino/antineutrino nucleon scattering can be utilised to provide a direct measurement of the antiquark content of the proton and neutron.

## 12.3 Neutrino scattering experiments

Over the past few decades there have been several high-energy neutrino beam experiments, such as the CDHS experiment at CERN, which took data from 1976 to 1984. The CDHS experiment used a collimated neutrino beam of either $\nu_\mu$ or $\overline{\nu}_\mu$ in the energy range 30 GeV–200 GeV, created from a 400 GeV proton beam. The neutrino interactions were observed in a detector with a mass of 1250 tons, which consisted of 19 magnetised iron modules, separated by wire drift chambers that provided a precise measurement of the position of the muon produced in charged-current neutrino interactions. The iron modules were built up of several iron plates with planes of plastic scintillator detectors in between them. The muon momentum was reconstructed from its curvature in the magnetic field of the detector and the energy of the final-state hadronic system was determined from the total energy deposited in the scintillator detectors between the iron plates.

An example of a charged-current $\nu_\mu$ interaction in the CDHS detector is shown in Figure 12.9. The upper half of the event display shows the deposited energy, as a function of distance along the detector. The energy deposition at the start of the interaction provides a measurement of the energy of the hadronic system, $E_X$.

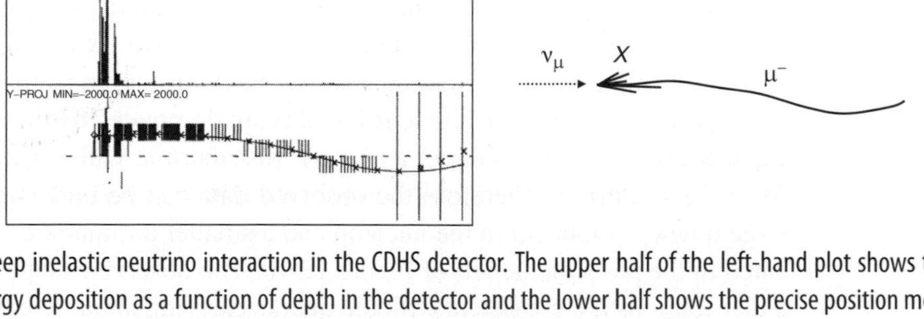


Fig. 12.9 A deep inelastic neutrino interaction in the CDHS detector. The upper half of the left-hand plot shows the energy deposition as a function of depth in the detector and the lower half shows the precise position measurements provided by the drift chambers. The right-hand figure shows the interpretation of the event.

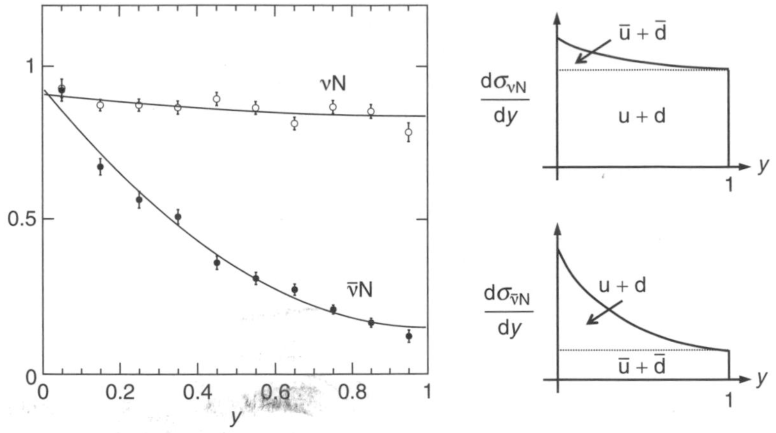


Fig. 12.10 The measured shape (arbitrary normalisation) of the νN and $\overline{\nu}$N differential cross sections from the CDHS experiment. Adapted from de Groot *et al.* (1979). The relative normalisation of the ν and $\overline{\nu}$ data have been corrected to correspond to the same flux. The plots on the right show the expected shapes of the measured differential cross sections in terms of quark and antiquark components.

The lower half of the event display shows the hit positions, from which the muon momentum can be reconstructed from the measured curvature. Thus, for each observed neutrino interaction, the neutrino energy can be reconstructed as

$$E_\nu = E_\mu + E_X,$$

and the value of $y$ for each interaction can be determined from (8.3),

$$y = \left(1 - \frac{E_\mu}{E_\nu}\right) = \frac{E_X}{E_X + E_\mu}.$$

The measured $y$-distributions for νN and $\overline{\nu}$N scattering from the CDHS experiment are shown in Figure 12.10. The data can be compared to expected distributions given in (12.25) and (12.27). The ν beam data show a clear contribution from νq scattering with a flat $y$ distribution and smaller contribution from ν$\overline{\text{q}}$ scattering with a $(1-y)^2$ distribution. The antineutrino data are dominated by the contribution from $\overline{\nu}$q scattering, which results in a $(1 - y)^2$ distribution, with a smaller contribution from $\overline{\nu}\,\overline{\text{q}}$ scattering. Therefore the observed data can be understood in terms of a large quark component in the nucleon and a smaller antiquark component, as indicated in the plots on the left of Figure 12.10. The observed shapes of the measured $y$ distributions are consistent with the quarks carrying about five times the momentum fraction of the proton as the antiquarks.

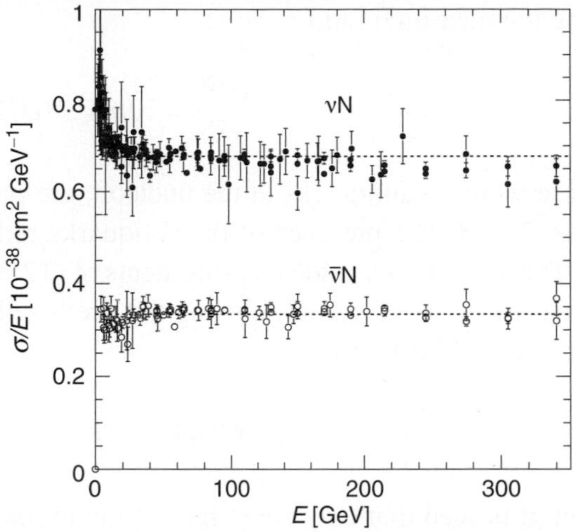


Fig. 12.11 A summary of the measurements of the total νN and $\overline{\nu}$N cross sections divided by the laboratory frame neutrino energy. The data are taken from references given in Beringer *et al.* (2012). The lines show the average values of the cross sections measured by a subset of the experiments in the range 30–200 GeV.

The total expected neutrino–nucleon cross section can be obtained by integrating (12.25) over $y$ to give

$$\sigma^{\nu N} = \frac{G_F^2 m_N E_\nu}{\pi}\left[f_q + \frac{1}{3}f_{\overline{q}}\right]. \tag{12.28}$$

Similarly, the integral of (12.27) gives the total antineutrino–nucleon cross section,

$$\sigma^{\overline{\nu}N} = \frac{G_F^2 m_N E_{\overline{\nu}}}{\pi}\left[\frac{1}{3}f_q + f_{\overline{q}}\right]. \tag{12.29}$$

The total neutrino deep inelastic cross section is proportional to the laboratory frame neutrino energy $E_\nu$, and the ratio of the νN to the $\overline{\nu}$N total cross section is

$$\frac{\sigma^{\nu N}}{\sigma^{\overline{\nu}N}} = \frac{3f_q + f_{\overline{q}}}{f_q + 3f_{\overline{q}}}. \tag{12.30}$$

Figure 12.11 summarises the experimental measurements of $\sigma^{\nu N}/E_\nu$ and $\sigma^{\overline{\nu}N}/E_{\overline{\nu}}$ for neutrino energies of up to 350 GeV. In the deep inelastic region, above about 30 GeV, the cross sections are approximately proportional to the neutrino energy and the measurements in the range 30–150 GeV are consistent with

$$\sigma^{\nu N}/E_\nu = 0.677 \pm 0.014 \times 10^{-38}\,\mathrm{cm^2\,GeV^{-1}}, \tag{12.31}$$

$$\sigma^{\overline{\nu}N}/E_{\overline{\nu}} = 0.334 \pm 0.008 \times 10^{-38}\,\mathrm{cm^2\,GeV^{-1}}, \tag{12.32}$$

giving the measured ratio

$$\frac{\sigma^{\nu\mathrm{N}}}{\sigma^{\overline{\nu}\mathrm{N}}} = 1.984 \pm 0.012.$$

If there were no antiquarks in the nucleon, the parton model prediction would be $\sigma^{\nu\mathrm{N}}/\sigma^{\overline{\nu}\mathrm{N}} = 3$. The presence of the antiquarks reduces this ratio to approximately two. The total cross section measurements of (12.31) and (12.32) can be interpreted as measurements of $f_\mathrm{q}$ and $f_{\overline{\mathrm{q}}}$ using the predicted deep inelastic cross sections of (12.28) and (12.29), giving

$$f_\mathrm{q} \approx 0.41 \quad \text{and} \quad f_{\overline{\mathrm{q}}} \approx 0.08.$$

Again, it is seen that only about half of the momentum of the proton is carried by the quarks/antiquarks, the remaining 50% is due to the gluons that do not interact with the W bosons of the weak interaction. Just under a tenth of momentum of the proton is carried by the antiquarks.

## 12.4 Structure functions in neutrino interactions

In the limit where particle masses can be neglected, the $\mathrm{e}^\pm\mathrm{p}$ deep inelastic cross section of (8.11) can be written in terms of $y$ as

$$\frac{\mathrm{d}^2\sigma^{\mathrm{e}^\pm\mathrm{p}}}{\mathrm{d}x\,\mathrm{d}y} = \frac{4\pi\alpha^2 s}{Q^4}\left[(1-y)\,F_2(x,Q^2) + y^2 x F_1(x,Q^2)\right], \tag{12.33}$$

where $F_1(x,Q^2)$ and $F_2(x,Q^2)$ are the structure functions described in Chapter 8. Equation (12.33) is the most general Lorentz-invariant *parity conserving* expression for the $\mathrm{e}^\pm\mathrm{p}$ cross section mediated by single photon exchange. The corresponding general expressions for neutrino/antineutrino deep inelastic scattering, which are modified to allow parity violation, are

$$\begin{aligned}
\frac{\mathrm{d}^2\sigma^{\nu\mathrm{p}}}{\mathrm{d}x\,\mathrm{d}y} &= \frac{G_\mathrm{F}^2 s}{2\pi}\left[(1-y)\,F_2^{\nu\mathrm{p}}(x,Q^2) + y^2 x F_1^{\nu\mathrm{p}}(x,Q^2) + y\left(1-\frac{y}{2}\right)xF_3^{\nu\mathrm{p}}(x,Q^2)\right],\\
\frac{\mathrm{d}^2\sigma^{\overline{\nu}\mathrm{p}}}{\mathrm{d}x\,\mathrm{d}y} &= \frac{G_\mathrm{F}^2 s}{2\pi}\left[(1-y)\,F_2^{\overline{\nu}\mathrm{p}}(x,Q^2) + y^2 x F_1^{\overline{\nu}\mathrm{p}}(x,Q^2) - y\left(1-\frac{y}{2}\right)xF_3^{\overline{\nu}\mathrm{p}}(x,Q^2)\right].
\end{aligned} \tag{12.34}$$

The structure functions are the experimental observables of deep inelastic scattering experiments that can then be interpreted in the parton model. By equating the

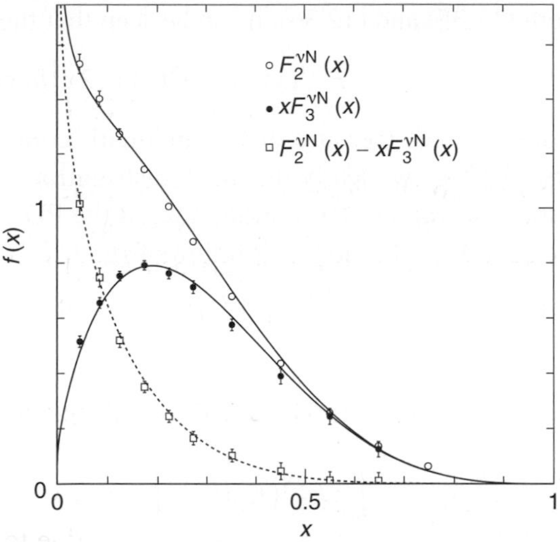


Fig. 12.12 Measurements of $F_2^{\nu N}(x)$ and $xF_3^{\nu N}(x)$ in neutrino/antineutrino nucleon deep inelastic scattering in the NuTeV experiment at Fermilab for 7.5 GeV$^2$ $< Q^2 <$ 13.0 GeV$^2$, compared to the expected distributions from the parton distribution functions at $Q^2 = 10$ GeV$^2$ shown in Figure 8.17. Data from Tzanov *et al.* (2006).

powers of $y$ in (12.34) with those in the parton model prediction of (12.20), the structure functions can be expressed as

$$F_2^{\nu p} = 2xF_1^{\nu p} = 2x\left[d(x) + \overline{u}(x)\right],$$
$$xF_3^{\nu p} = 2x\left[d(x) - \overline{u}(x)\right].$$

The equivalent expressions for neutrino–nucleon scattering are

$$F_2^{\nu N} = 2xF_1^{\nu N} = \tfrac{1}{2}(F_2^{\nu p} + F_2^{\nu n}) = x\left[u(x) + d(x) + \overline{u}(x) + \overline{d}(x)\right], \quad (12.35)$$
$$xF_3^{\nu N} = \tfrac{1}{2}(xF_3^{\nu p} + xF_3^{\nu n}) = x\left[u(x) + d(x) - \overline{u}(x) - \overline{d}(x)\right]. \quad (12.36)$$

Using the parton model prediction that $F_2^{\nu N} = 2xF_1^{\nu N}$, the $y$-dependence of the cross sections can be expressed in terms of $F_2(x, Q^2)$ and $F_3(x, Q^2)$,

$$\frac{\mathrm{d}^2\sigma^{\nu N}}{\mathrm{d}x\,\mathrm{d}y} = \frac{G_F^2 s}{2\pi}\left[\left(1 - y + \frac{y^2}{2}\right)F_2^{\nu N}(x, Q^2) + y\left(1 - \frac{y}{2}\right)xF_3^{\nu N}(x, Q^2)\right]. \quad (12.37)$$

The structure functions, $F_2^{\nu N}(x, Q^2)$ and $F_3^{\nu N}(x, Q^2)$, can be obtained from the experimental measurements of the $y$-dependence of the measured neutrino cross sections at a particular value of $x$. For example, Figure 12.12 shows experimental measurements of $F_2^{\nu N}$ and $xF_3^{\nu N}$ at $Q^2 \sim 10\,\mathrm{GeV}^2$ from the NuTeV experiment at Fermilab, which had a similar sandwich structure of steel plates and active detectors to the CDHS experiment described in Section 12.3. The data are compared to the predictions obtained using the PDFs shown in Figure 8.17, including the contribution from the strange quarks.

From (12.35) and (12.36), it can be seen that the difference

$$F_2^{\nu N}(x) - xF_3^{\nu N}(x) = 2x\left[\overline{u}(x) + \overline{d}(x)\right],$$

provides a direct measure of the antiquark content of the nucleon. The data of Figure 12.12 show clearly that the $x$ distribution of the antiquarks is, as expected, largest at low values of $x$. Furthermore, if the PDFs are written in terms of valence and sea quark contributions, it follows from (12.36) that

$$F_3^{\nu N}(x) = u_V(x) + d_V(x).$$

Therefore $F_3^{\nu N}(x)$ provides a direct measurement of the sum of the PDFs for the valence quarks alone. If there are three valence quarks within the nucleon, then

$$\int_0^1 F_3^{\nu N}(x)\,dx = \int_0^1 u_V(x) + d_V(x)\,dx = 3,$$

which is known as the Gross–Llewellyn-Smith sum rule. The measurement of $xF_3^{\nu N}(x)$, shown in Figure 12.12, is consistent with this prediction.

## 12.5 Charged-current electron–proton scattering

Section 8.5 described electron–proton scattering at a centre-of-mass energy of $\sqrt{s} = 300\,\text{GeV}$ at the HERA collider. In addition to electron–proton collisions, which constituted the largest part of HERA operation, HERA also collided positrons and protons. For collisions at $Q^2 < m_W^2$, the $e^-p$ and $e^+p$ interactions are dominated by neutral-current QED $t$-channel deep inelastic scattering, mediated by photon exchange. However, since $Q^2 \approx sxy$, the $Q^2$ values obtainable at HERA extend above $m_W^2$. Therefore, for the interactions at the highest $Q^2$ values, charged-current scattering processes mediated by $W^\pm$ exchange, become increasingly important. Figure 12.13 shows the parton-level Feynman diagrams for the weak charged-current $e^-p \to \nu_e X$ and $e^+p \to \overline{\nu}_e X$ scattering processes. The experimental signature for high-$Q^2$ weak charged-current interactions is similar to the event shown in Figure 8.13, except that there is no final-state charged lepton. As

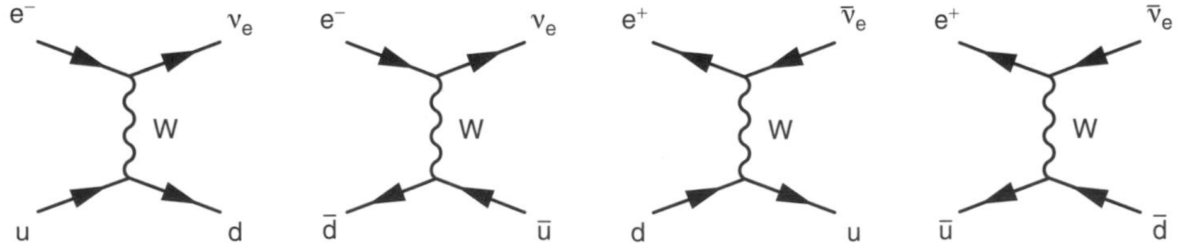


Fig. 12.13 Lowest-order diagrams contributing to charged-current $e^-p$ and $e^+p$ scattering.

discussed in Section 1.3, although the neutrino is not observed, its presence can be inferred from the momentum imbalance in the plane transverse to the beam axis. Consequently, the charged-current interactions mediated by the exchange of a W boson are clearly identifiable from the more common QED neutral-current interactions.

The calculations of the differential cross sections for charged current $e^-p$ and $e^+p$ scattering follow closely that for $\nu p$ and $\overline{\nu}p$ scattering described above. Consequently, the results can be obtained directly from the neutrino scattering cross sections by replacing the approximate low-$Q^2$ form of the W-boson propagator by

$$\frac{1}{m_W^2} \rightarrow \frac{1}{Q^2 + m_W^2},$$

and by making the replacement $u(x) \leftrightarrow d(x)$ and $u(x) \leftrightarrow d(x)$ to reflect the different quarks involved in the electron scattering interaction, which can be seen by comparing the Feynman diagrams of Figure 12.7 to those of Figure 12.13. With these replacements, the neutrino–proton cross section of (12.20) can be adapted to give the electron–proton weak charged-current scattering cross section,

$$\frac{d^2\sigma_{CC}^{e^-p}}{dx\,dy} = \frac{1}{2}\frac{G_F^2}{\pi}\frac{m_W^4}{(Q^2 + m_W^2)^2} sx\left(u(x) + (1-y)^2\overline{d}(x)\right), \tag{12.38}$$

where the additional factor of one-half comes from the need to average over the two spin states of the electron and quark, rather than just the quark spin in neutrino scattering. Using $Q^2 = sxy$, (12.38) can be written

$$\frac{d^2\sigma_{CC}^{e^-p}}{dx\,dQ^2} = \frac{d^2\sigma_{CC}^{e^-p}}{dx\,dy}\left|\frac{\partial y}{\partial Q^2}\right| = \frac{1}{sx}\frac{d^2\sigma_{CC}^{e^-p}}{dx\,dy}.$$

Hence the differential cross section for charged-current $e^-p \rightarrow \nu_e X$ scattering is

$$\frac{d^2\sigma_{CC}^{e^-p}}{dx\,dQ^2} = \frac{G_F^2 m_W^4}{2\pi(Q^2 + m_W^2)^2}\left(u(x) + (1-y)^2\overline{d}(x)\right), \tag{12.39}$$

and the corresponding expression for charged-current $e^+p \rightarrow \overline{\nu}_e X$ scattering is

$$\frac{d^2\sigma_{CC}^{e^+p}}{dx\,dQ^2} = \frac{G_F^2 m_W^4}{2\pi(Q^2 + m_W^2)^2}\left((1-y)^2 d(x) + \overline{u}(x)\right). \tag{12.40}$$

Figure 12.14 shows the measured charged-current and neutral-current differential cross sections for electron–proton and positron–proton scattering obtained by the H1 experiment at HERA. The main features of this plot can be understood in terms of the $Q^2$ dependence of the photon and W-boson propagators. For $Q^2 \ll m_W^2$, the W-boson propagator is approximately independent of $Q^2$,

$$1/(m_W^2 + Q^2) \approx 1/m_W^2,$$

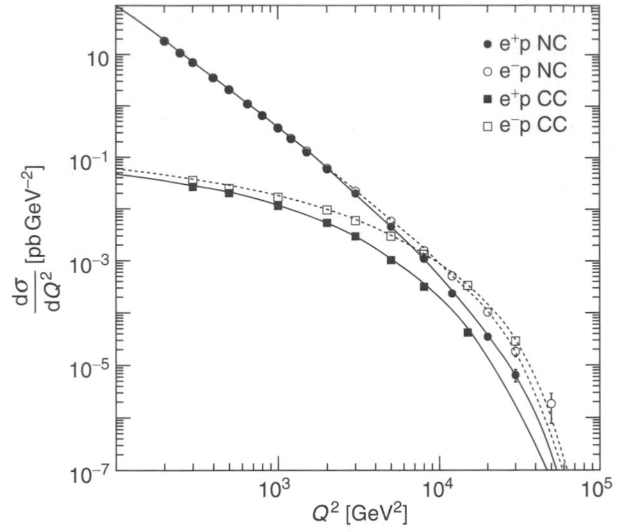


Fig. 12.14 The double-differential charged-current and neutral-current current cross sections for electron–proton and positron–proton scattering measured by the H1 experiment at $\sqrt{s} = 319$ GeV at the HERA collider. Adapted from data in Aaron *et al.* (2012).

which explains the relative flatness of the charged-current cross section at low $Q^2$. At higher values of $Q^2$, the charged-current cross section decreases more rapidly with $Q^2$ due to the $1/(m_W^2 + Q^2)$ dependence of the propagator.

The observed differences in the $e^-p$ and $e^+p$ charged-current cross sections can be understood in terms of the PDFs involved. At low $Q^2$, the cross sections are dominated by interactions with low-$x$ sea quarks and antiquarks. Because the sea quark PDFs are approximately the same for up- and down-quarks, the $e^-p$ and $e^+p$ cross sections given in (12.39) and (12.40) are roughly equal. Whereas at high $Q^2$, interactions with valence quarks become more important ($Q^2 = xys$) and therefore the $e^-p$ cross section is greater than that for $e^+p$ because $u_V(x) > d_V(x)$ and due to the presence of the factor of $(1-y)^2$ multiplying $d(x)$ in (12.40).

From the $Q^2$ dependence of the $\gamma$ and W-boson propagators and the size of the QED and weak coupling constants,

$$\frac{\sigma_{\rm NC}^{QED}}{\sigma_{\rm CC}^{\rm W}} \sim \frac{\alpha^2}{\alpha_W^2}\cdot\frac{(Q^2+m_{\rm W}^2)^2}{Q^4}.$$

Consequently, at low $Q^2$ the neutral-current cross section, which is dominated by photon exchange, is larger than weak charged-current cross section. However, for $Q^2 \gtrsim m_{\rm W}^2$, the photon and W-boson propagator terms are approximately equal and the observed convergence of the neutral-current and charged-current cross sections reflects the similar intrinsic strengths of the electromagnetic and weak interactions, $\alpha \sim 1/137$ and $\alpha_W \sim 1/30$. Hence, the high-energy $e^{\pm}p$ scattering data from HERA provide direct experimental evidence that the coupling constants of QED and the weak interaction are not very different.

Finally, it is worth noting that for $Q^2 \gtrsim m_Z^2$, the neutral-current cross section includes significant contributions from both $\gamma$- and Z-exchange diagrams. The contribution to the scattering amplitude from the Z boson leads to the observed small differences between the $e^-p$ and $e^+p$ neutral-current cross sections at high $Q^2$.

## Summary

The main aim of this chapter was to provide an introduction to cross section calculations for the weak charged-current. In the limit where the masses of the particles can be neglected, these calculations are relatively straightforward, since for each Feynman diagram only one helicity combination needs to be considered. In the limit where $Q^2 \ll m_W^2$, the W-boson propagator has little $Q^2$ dependence and the neutrino interaction cross sections are proportional to the laboratory frame neutrino energy. The total neutrino and antineutrino interaction cross sections per nucleon, assuming equal numbers of protons and neutrons, were found to be

$$\sigma^{\nu N} = \frac{G_F^2 m_N E_\nu}{\pi}\left[f_q + \frac{1}{3}f_{\bar{q}}\right] \quad \text{and} \quad \sigma^{\bar{\nu}N} = \frac{G_F^2 m_N E_{\bar{\nu}}}{\pi}\left[\frac{1}{3}f_q + f_{\bar{q}}\right].$$

The $1/(Q^2 + m_W^2)$ form of the W-boson propagator was evident in the discussion of $e^-p$ and $e^+p$ scattering, where both charged-current (W exchange) and neutral current ($\gamma$ and Z exchange) processes contribute. The convergence of the charged-current and neutral-current cross sections provides a direct demonstration that the QED and weak coupling constants might be related, hinting at the unified electroweak theory described in Chapter 15.

## Problems

**12.1** Explain why the tau-lepton branching ratios are observed to be approximately

$$Br(\tau^- \to e^-\nu_\tau\bar{\nu}_e) : Br(\tau^- \to e^-\nu_\tau\bar{\nu}_\mu) : Br(\tau^- \to \nu_\tau + \text{hadrons}) \approx 1 : 1 : 3.$$

**12.2** Assuming that the process $\nu_e e^- \to e^- \nu_e$ occurs only by the weak charged-current interaction (i.e. ignoring the Z-exchange neutral-current process), show that

$$\sigma_{CC}^{\nu_e e^-} \approx \frac{2m_e E_\nu G_F^2}{\pi},$$

where $E_\nu$ is neutrino energy in the laboratory frame in which the struck $e^-$ is at rest.

**12.3** Using the above result, estimate the probability that a 10 MeV Solar $\nu_e$ will undergo a charged-current weak interaction with an electron in the Earth if it travels along a trajectory passing through the centre of the Earth. Take the Earth to be a sphere of radius 6400 km and uniform density $\rho = 5520\ \mathrm{kg\,m^{-3}}$.

**12.4** By equating the powers of $y$ in (12.34) with those in the parton model prediction of (12.20), show that the structure functions can be expressed as

$$F_2^{\nu p} = 2xF_1^{\nu p} = 2x\left[d(x) + \overline{u}(x)\right] \quad \text{and} \quad xF_3^{\nu p} = 2x\left[d(x) - \overline{u}(x)\right].$$

**12.5** In the quark–parton model, show that $F_2^{eN} = \frac{1}{2}(Q_u^2 + Q_d^2)F_2^{\nu N}$.
Hence show that the measured value of

$$F_2^{eN}/F_2^{\nu N} = 0.29 \pm 0.02,$$

is consistent with the up- and down-quarks having respective charges of $+\frac{2}{3}$ and $-\frac{1}{3}$.

**12.6** Including the contributions from strange quarks, the neutrino–nucleon scattering structure functions can be expressed as

$$F_2^{\nu p} = 2x[d(x) + s(x) + \overline{u}(x)] \quad \text{and} \quad F_2^{\nu n} = 2x[u(x) + \overline{d}(x) + \overline{s}(x)],$$

where $s(x)$ and $\overline{s}(x)$ are respectively the strange and anti-strange quark PDFs of the nucleon. Assuming $s(x) = \overline{s}(x)$, obtain an expression for $xs(x)$ in terms of the structure functions for neutrino–nucleon and electron–nucleon scattering

$$F_2^{\nu N} = \frac{1}{2}\left(F_2^{\nu p}(x) + F_2^{\nu n}(x)\right) \quad \text{and} \quad F_2^{eN} = \frac{1}{2}\left(F_2^{ep}(x) + F_2^{en}(x)\right).$$

**12.7** The H1 and ZEUS experiments at HERA measured the cross sections for the charged-current processes $e^- p \to \nu_e X$ and $e^+ p \to \overline{\nu}_e X$ for different degrees of electron longitudinal polarisation. For example, the ZEUS measurements of the total $e^+ p \to \overline{\nu}_e X$ cross section at $\sqrt{s} = 318$ GeV and $Q^2 > 200\ \mathrm{GeV}^2$ for positron polarisations of $P_e = -36\%, 0\%$ and $+33\%$ are:

$$\sigma(-0.36) = 22.9 \pm 1.1\,\mathrm{pb}, \quad \sigma(0) = 34.8 \pm 1.34\,\mathrm{pb} \quad \text{and} \quad \sigma(+0.33) = 48.0 \pm 1.8\,\mathrm{pb},$$

see Abramowicz *et al.* (2010) and references therein. Plot these data and predict the corresponding cross section for $P_e = -1.0$, i.e. when the positrons are all left-handed. What does this tell you about the nature of the weak charged-current interaction?

# 13 Neutrinos and neutrino oscillations

This chapter focusses on the properties of neutrinos and in particular the phenomenon of neutrino oscillations, whereby neutrinos undergo flavour transitions as they propagate over large distances. Neutrino oscillations are a quantum-mechanical phenomenon and can be described in terms of the relationship between the eigenstates of the weak interaction $\nu_e$, $\nu_\mu$ and $\nu_\tau$, and the eigenstates of the free-particle Hamiltonian, known as the mass eigenstates, $\nu_1$, $\nu_2$ and $\nu_3$. The mathematical description of neutrino oscillations is first introduced for two flavours and then extended to three flavours. The predictions are compared to the recent experimental data from reactor and long-baseline neutrino oscillation experiments.

## 13.1 Neutrino flavours

Unlike the charged leptons, which can be detected from the continuous track defined by the ionisation of atoms as they traverse matter, neutrinos are never directly observed; they are only detected through their weak interactions. Different neutrino flavours can only be distinguished by the flavours of charged lepton produced in charged-current weak interactions. Consequently, the electron neutrino $\nu_e$, is *defined* as the neutrino state produced in a charged-current weak interaction along with an electron. Similarly, by definition, the weak charged-current interactions of a $\nu_e$ will produce an electron. For many years it was assumed that the $\nu_e$, $\nu_\mu$ and $\nu_\tau$ were massless fundamental particles. This assumption was based, at least in part, on experimental evidence. For example, it was observed that the interactions of the neutrino/antineutrino produced along with a positron/electron in a nuclear $\beta$-decay, would produce an electron/positron as indicated in Figure 13.1. This naturally led to the idea that the electron neutrino carried some property related to the electron that is conserved in weak interactions, which was referred to as electron number. Similarly, in beam neutrino experiments, such as those described in Chapter 12, it was observed that the neutrinos produced from $\pi^+ \to \mu^+\nu_\mu$ decays always produced a muon in charged-current weak interactions.

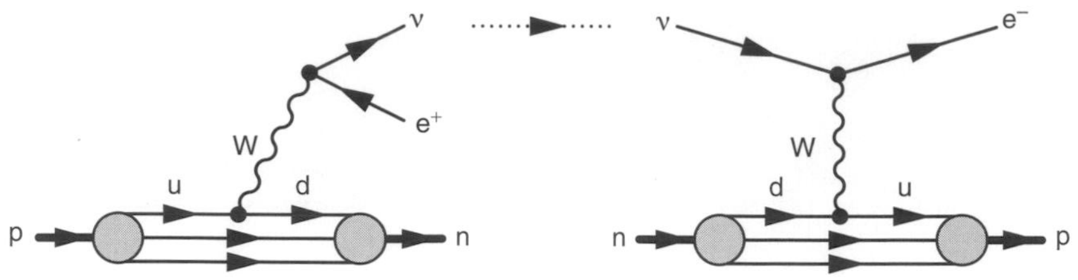


Fig. 13.1 Neutrino production and subsequent detection where the $\nu_e$ state is associated with positrons/electrons.

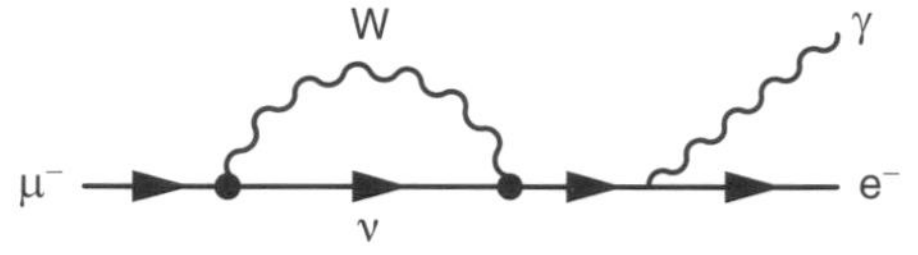


Fig. 13.2 A possible Feynman diagram for $\mu^- \to e^- \gamma$ for the case where the $\nu_e$ and $\nu_\mu$ are not distinct.

Further evidence for the distinct nature of the electron and muon neutrinos was provided by the non-observation of the decay $\mu^- \to e^- \gamma$, which is known to have a very small branching ratio,

$$BR(\mu^- \to e^- \gamma) < 10^{-11}.$$

In principle, this decay could occur via the Feynman diagram shown in Figure 13.2. The absence of the decay suggests that the neutrino associated with $W\mu^-\nu$ vertex is distinct from the neutrino associated with the $We^-\nu$ vertex.

Until the late 1990s, relatively little was known about neutrinos beyond that there are three distinct flavours and that they are extremely light (and possibly massless). However, even at that time several experiments had reported possible anomalies in the observed interaction rates of atmospheric and solar neutrinos. This picture changed with the publication of the solar and atmospheric neutrino data from the Super-Kamiokande detector, which provided compelling experimental evidence for the phenomenon of neutrino flavour oscillations over very large distances. The subsequent study of neutrino oscillations has been one of the highlights of particle physics in recent years.

## 13.2 Solar neutrinos

Nuclear fusion in the Sun produces a large flux of electron neutrinos, $2 \times 10^{38}\ \nu_e\ \mathrm{s}^{-1}$. Despite the smallness of neutrino interaction cross sections and the large distance to the Sun, solar neutrinos can be observed with a sufficiently massive detector. Nuclear fusion in the Sun proceeds through a number of distinct processes, each of which has several stages. The resulting solar neutrino energy spectrum is shown in

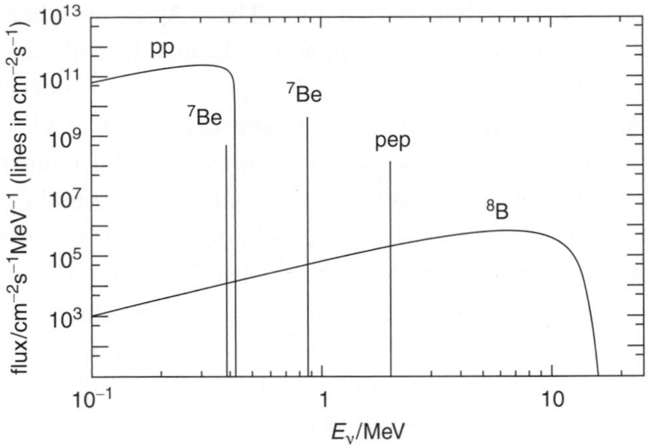


Fig. 13.3 The flux of solar neutrinos from the main processes in the Sun. The two $^7$Be lines are from the electron capture reactions $^7\text{Be} + e^- \rightarrow {}^7\text{Li} + \nu_e$. The pep line is from $p + e^- + p \rightarrow {}^2\text{H} + \nu_e$. Adapted from Bahcall and Pinsonneault (2004).

Figure 13.3. The main hydrogen burning process, known as the pp cycle, proceeds through three steps:

$$\begin{aligned} p + p &\rightarrow D + e^+ + \nu_e, \\ D + p &\rightarrow {}^3_2\text{He} + \gamma, \\ {}^3_2\text{He} + {}^3_2\text{He} &\rightarrow {}^4_2\text{He} + p + p. \end{aligned}$$

Because the binding energy of the deuteron ${}^2_1$D is only 2.2 MeV, the neutrinos produced in the process $p + p \rightarrow D + e^+ + \nu_e$ have low energies, $E_\nu < 0.5$ MeV. Consequently, they are difficult to detect. For this reason, the majority of experiments have focussed on the detection of the higher-energy solar neutrinos from rarer fusion processes. The highest energy solar neutrinos originate from the β-decay of boron-8 ($^8$B) that is produced from the fusion of two helium nuclei,

$$\begin{aligned} {}^4_2\text{He} + {}^3_2\text{He} &\rightarrow {}^7_4\text{Be} + \gamma, \\ {}^7_4\text{Be} + p &\rightarrow {}^8_5\text{B} + \gamma, \end{aligned}$$

with the subsequent β-decay,

$$ {}^8_5\text{B} \rightarrow {}^8_4\text{Be}^* + e^+ + \nu_e, $$

giving neutrinos with energies up to 15 MeV.

A number of experimental techniques have been used to detect solar neutrinos. The earliest experiment, based in the Homestake Mine in South Dakota, USA, used a radiochemical technique to measure the flux of solar neutrinos. It consisted of a tank of 615 tons of dry-cleaning fluid, $C_2Cl_4$. The solar neutrino flux was measured by counting the number of $^{37}$Ar atoms produced in the inverse β-decay process, $\nu_e + {}^{37}_{17}\text{Cl} \rightarrow {}^{37}_{18}\text{Ar} + e^-$. The $^{37}$Ar atoms where extracted from the tank and counted through their radioactive decays. Despite the huge flux of neutrinos, only 1.7 interactions per day were expected. The observed rate was only $0.48 \pm 0.04$ neutrino interactions per day; see Cleveland *et al.* (1998). This apparent deficit of solar neutrinos became known as the solar neutrino problem. The Homestake experiment was sensitive to the relatively high-energy $^8$B neutrinos. Subsequently, the SAGE and GALLEX radiochemical experiments used gallium as a target, and were sensitive to the low-energy neutrinos from the first step of the pp chain. These experiments also observed a deficit of solar neutrinos.

Radiochemical experiments played an important role in demonstrating the existence of the solar neutrino deficit; Ray Davis, who conceived the Homestake experiment, was awarded the Nobel prize for its discovery. However, it was the results from the large water Čerenkov detectors that firmly established the origin of the deficit of solar neutrinos.

### 13.2.1 The Super-Kamiokande experiment

The 50 000 ton Super-Kamiokande water Čerenkov detector, shown schematically in Figure 13.4a, was designed to detect Čerenkov radiation (see Section 1.2.1) from relativistic particles produced within the volume of the detector. In essence, Super-Kamiokande is a large vessel of water surrounded by photo-multiplier tubes (PMTs) that are capable of detecting single photons.

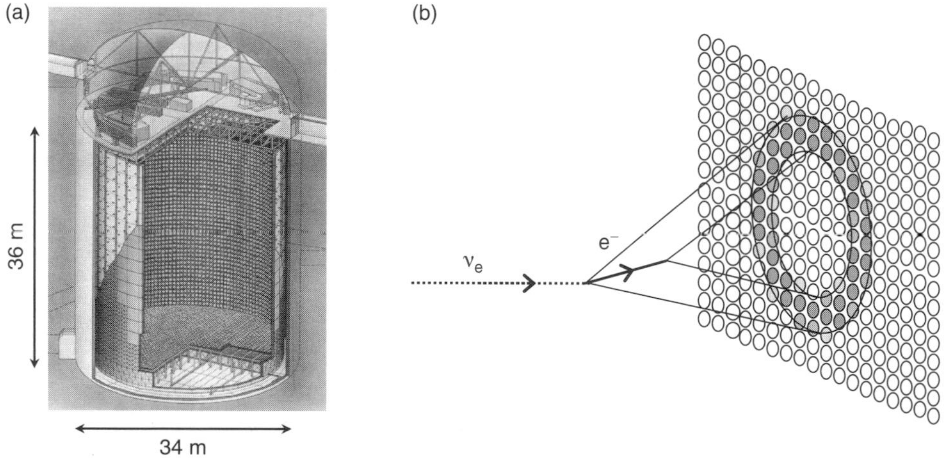


Fig. 13.4 (a) The Super-Kamiokande experiment comprising a tank of 50 000 tons of water viewed by 11 146 PMTs. (b) A neutrino interaction in the Super-Kamiokande experiment showing the ring of Čerenkov light produced by the relativistic e$^-$ with $v > c/n$ as detected as signals in the PMTs on the walls of the detector. Left-hand diagram courtesy of the Super-Kamiokande collaboration.

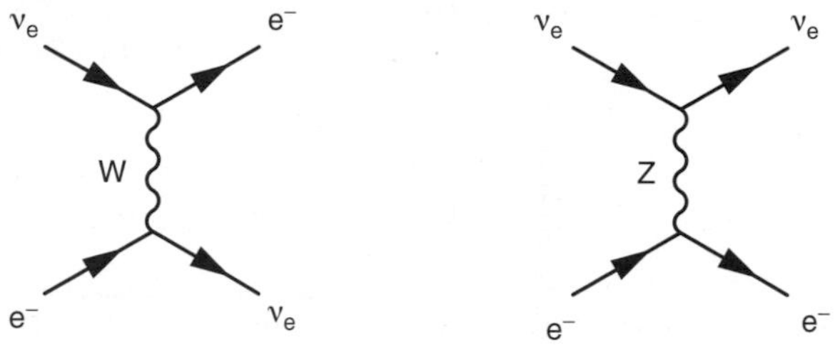


Fig. 13.5 The two Feynman diagrams contributing to $\nu_e e^- \to \nu_e e^-$ elastic scattering.

Because oxygen is a particularly stable nucleus, the charged-current process $\nu_e + {}^{16}_{8}\mathrm{O} \to {}^{16}_{9}\mathrm{F} + e^-$ is kinematically forbidden for the neutrino energies being considered here. Consequently, solar neutrinos are detected by the elastic scattering process $\nu_e e^- \to \nu_e e^-$, shown in Figure 13.5. The final-state electron is relativistic and can be detected from the Čerenkov radiation photons that are emitted at a fixed angle to its direction of motion as it travels through water. The photons form a ring of hits in the PMTs on the sides of the detector, as shown in Figure 13.4b. The number of detected photons provides a measure of the neutrino energy and the direction of the electron can be determined from the orientation of the Čerenkov ring. In this way, Super-Kamiokande is able to detect electron neutrino elastic scattering interactions down to neutrino energies of about 5 MeV. Below this energy, background from the β-decays of radioisotopes dominates. Because of this effective threshold, the Super-Kamiokande detector is sensitive primarily to the flux of $^8$B neutrinos.

The angular distribution of the scattered electron with respect to the incoming neutrino direction is isotropic in the centre-of-mass frame, as was the case for neutrino–quark scattering cross section of (12.13). Because the centre-of-mass frame is boosted in the direction of the neutrino, in the laboratory frame the scattered electron tends to follow the direction of the solar neutrino. Consequently, the directional correlation with the Sun is retained.

Figure 13.6 shows the reconstructed electron direction with respect to the direction of the Sun for neutrino interactions with $E_\nu \gtrsim 5\,\mathrm{MeV}$. The peak towards $\cos\theta_{\mathrm{sun}} = 1$ provides clear evidence for a flux of neutrinos from the Sun. The flat background arises from the β-decay of radioisotopes. Whilst Super-Kamiokande observes clear evidence of solar neutrinos from the Sun, the flux of electron neutrinos is measured to be about half that expected.

### 13.2.2 The SNO experiment

Results from Super-Kamiokande and other solar neutrino experiments demonstrated a clear deficit of electron neutrinos from the Sun. The Sudbury Neutrino Observatory (SNO) experiment in Canada was designed to measure both the $\nu_e$ and *total*

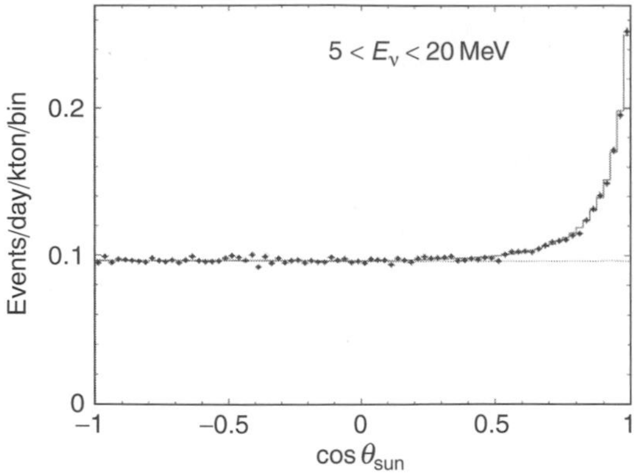


Fig. 13.6 The Super-Kamiokande solar neutrino data plotted as a function of the cosine of the polar angle of the electron with respect to the direction of the Sun. From Fukada *et al.* (2001).

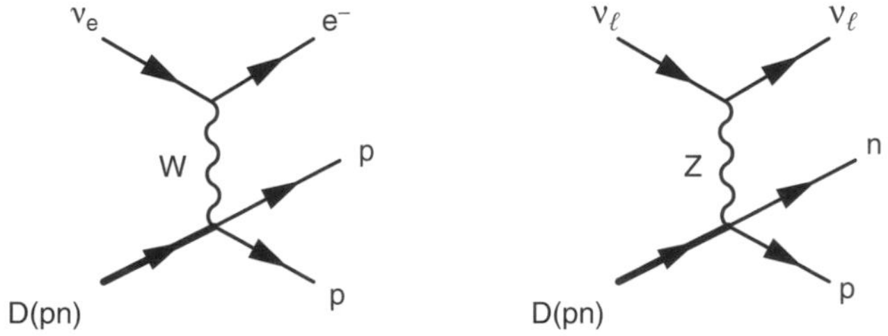


Fig. 13.7 The $\nu_e$ charged-current (CC) and neutral-current (NC) weak interactions with the deuteron.

neutrino flux from the Sun. SNO consisted of 1000 tons of heavy water, $D_2O$, inside a 12 m diameter vessel, viewed by 9,600 PMTs. Heavy water was used because the deuteron, the bound state of a proton and a neutron, has a binding energy of just 2.2 MeV, which is relatively small compared to the energies of the $^8$B solar neutrinos. For this reason, solar neutrinos can be detected in SNO through three different physical processes. Crucially, the different processes have different sensitivities to the fluxes of electron, muon and tau neutrinos, $\phi(\nu_e)$, $\phi(\nu_\mu)$ and $\phi(\nu_\tau)$.

Because of the low binding energy of the deuteron, the charged-current (CC) interaction of electron neutrinos, $\nu_e + D \rightarrow e + p + p$, shown in Figure 13.7 (left), is kinematically allowed. The final-state electron can be detected from the resulting Čerenkov ring. From the discussion of Section 12.2.2, it can be appreciated that in the centre-of-mass frame the angular distribution of the electron relative to the incoming neutrino is almost isotropic. Because $E_\nu \ll m_D$, the laboratory frame is almost equivalent to the centre-of-mass frame and therefore the final-state electron does not correlate strongly with the direction of the Sun. The charged-current interaction with the deuteron is only sensitive to the $\nu_e$ flux and therefore

$$\text{CC rate} \propto \phi(\nu_e). \tag{13.1}$$

All flavours of neutrinos can interact with the deuteron via the neutral-current (NC) interaction of Figure 13.7 (right), where the momentum imparted to the deuteron is sufficient to break up this loosely bound state. The neutron produced in the final state will (eventually) be captured in the reaction $n + {}^2_1H \rightarrow {}^3_1H + \gamma$, releasing a 6.25 MeV photon. Through its subsequent interactions, this photon will produce relativistic electrons that give a detectable Čerenkov signal. The neutral-current process is equally sensitive to all neutrino flavours, thus

$$\text{NC rate} \propto \phi(\nu_e) + \phi(\nu_\mu) + \phi(\nu_\tau). \tag{13.2}$$

Finally, neutrinos can interact with the atomic electrons through the elastic scattering (ES) processes of Figure 13.5. For electron neutrinos, both the charged-current process and the neutral-current process contribute to the cross section, whereas for $\nu_\mu$ and $\nu_\tau$ only the neutral-current process, which has a smaller cross section, contributes. The observed elastic scattering rate is therefore sensitive to all flavours of neutrinos but has greater sensitivity to $\nu_e$,

$$\text{ES rate} \propto \phi(\nu_e) + 0.154\left[\phi(\nu_\mu) + \phi(\nu_\tau)\right]. \tag{13.3}$$

The electrons from the ES scattering process point back to the Sun and can therefore be distinguished from those from the CC process.

The different angular and energy distributions of the Čerenkov rings from CC, NC and ES interactions allows the rates for each individual process to be determined separately. Using the knowledge of the interaction cross sections, the measured rates can be interpreted in terms of the neutrino fluxes using (13.1)–(13.3), with the CC process providing a measure of the $\nu_e$ flux and the NC process providing a measure of the total neutrino flux ($\nu_e + \nu_\mu + \nu_\tau$). The observed CC rate was consistent with a flux of $\nu_e$ of $1.8 \times 10^{-6}\,\text{cm}^{-2}\,\text{s}^{-1}$ and the observed NC rate was consistent with a total neutrino flux of $5.1 \times 10^{-6}\,\text{cm}^{-2}\,\text{s}^{-1}$, providing clear evidence for an unexpected $\nu_\mu/\nu_\tau$ flux from the Sun.

The observed neutrino rates in SNO from the CC, NC and ES processes can be combined to place constraints on the separate $\nu_e$ and $\nu_\mu + \nu_\tau$ fluxes, as shown in Figure 13.8, giving the overall result

$$\phi(\nu_e) = (1.76 \pm 0.10) \times 10^{-6}\,\text{cm}^{-2}\,\text{s}^{-1},$$
$$\phi(\nu_\mu) + \phi(\nu_\tau) = (3.41 \pm 0.63) \times 10^{-6}\,\text{cm}^{-2}\,\text{s}^{-1}.$$

The total neutrino flux, obtained from the NC process is consistent with the expectation from theoretical modelling of the Sun that predicts a $\nu_e$ flux of

$$\phi(\nu_e)_{pred} = (5.1 \pm 0.9) \times 10^{-6}\,\text{cm}^{-2}\,\text{s}^{-1}.$$

The SNO data therefore demonstrate that the total flux of neutrinos from the Sun is consistent with the theoretical expectation, but rather than consisting of only $\nu_e$, there is a large $\nu_\mu$ and/or $\nu_\tau$ component. Since $\nu_\mu/\nu_\tau$ cannot be produced in

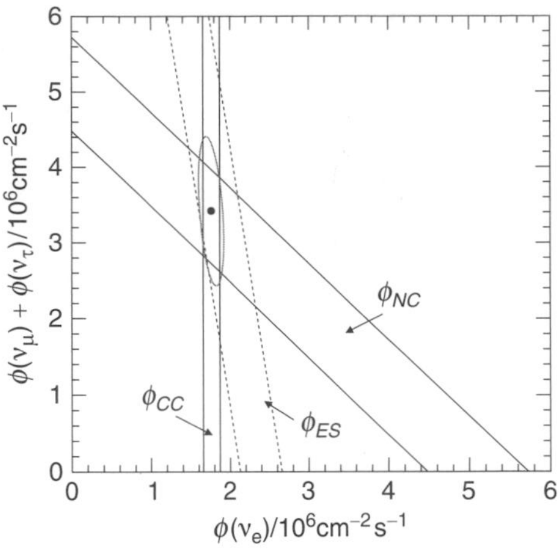


Fig. 13.8 The constraints on the $\nu_e$ and $\nu_\mu + \nu_\tau$ fluxes from the Sun from SNO data. The bands indicate the one standard deviation errors from the different processes. The CC process is sensitive only to $\nu_e$. The NC process is sensitive to all neutrinos. The ES rate is proportional to $\phi(\nu_e) + 0.154\left[\phi(\nu_\mu) + \phi(\nu_\tau)\right]$. The ellipse is the resulting 68% confidence limit from combining the three measurements. Adapted from Ahmad *et al.* (2002).

the fusion processes in the Sun, SNO provides clear evidence of neutrino flavour transformations over large distances.

## 13.3 Mass and weak eigenstates

The neutrino flavour transformations observed by SNO and other experiments can be explained by the phenomenon of neutrino oscillations. The physical states of particle physics, termed the mass eigenstates, are the stationary states of the free-particle Hamiltonian and satisfy

$$\hat{H}\psi = i\frac{\partial\psi}{\partial t} = E\psi.$$

The time evolution of a mass eigenstate takes the form of (2.25),

$$\psi(\mathbf{x}, t) = \phi(\mathbf{x})e^{-iEt}.$$

The neutrino mass eigenstates (the fundamental particles) are labelled $\nu_1$, $\nu_2$ and $\nu_3$. There is no reason to believe that the mass eigenstates should correspond to the weak eigenstates, $\nu_e$, $\nu_\mu$ and $\nu_\tau$, which are produced along with the respective flavour of charged lepton in a weak interaction. This important distinction between mass and weak eigenstates is illustrated in Figure 13.9. Here any one of the three mass eigenstates can be produced in conjunction with the electron in the initial

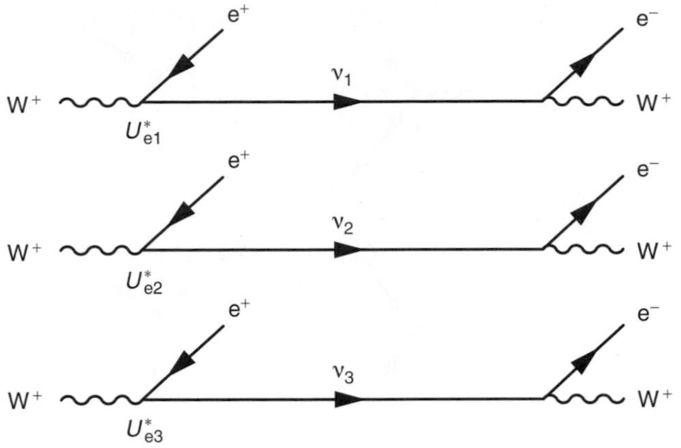


Fig. 13.9 The $We^+\nu_e$ vertex in terms of the mass eigenstates.

weak interaction. Since it is not possible to know which mass eigenstate was produced, the system has to be described by a coherent linear superposition of $\nu_1$, $\nu_2$ and $\nu_3$ states. In quantum mechanics, the basis of weak eigenstates can be related to the basis of mass eigenstates by a unitary matrix $U$,

$$\begin{pmatrix} \nu_e \\ \nu_\mu \\ \nu_\tau \end{pmatrix} = \begin{pmatrix} U_{e1} & U_{e2} & U_{e3} \\ U_{\mu 1} & U_{\mu 2} & U_{\mu 3} \\ U_{\tau 1} & U_{\tau 2} & U_{\tau 3} \end{pmatrix} \begin{pmatrix} \nu_1 \\ \nu_2 \\ \nu_3 \end{pmatrix}. \tag{13.4}$$

Hence the electron neutrino, which is the quantum state produced along with a positron in a charged-current weak interaction, is the linear combination of the mass eigenstates defined by the relative charged-current weak interaction couplings of the $\nu_1$, $\nu_2$ and $\nu_3$ at the $W^+ \to e^+\nu$ vertex

$$|\psi\rangle = U^*_{e1}|\nu_1\rangle + U^*_{e2}|\nu_2\rangle + U^*_{e3}|\nu_3\rangle. \tag{13.5}$$

The neutrino state subsequently propagates as a coherent linear superposition of the three mass eigenstates until it interacts and the wavefunction collapses into a weak eigenstate, producing an observable charged lepton of a particular flavour. If the masses of the $\nu_1$, $\nu_2$ and $\nu_3$ are not the same, phase differences arise between the different components of the wavefunction and the phenomenon of neutrino oscillations occurs. In this way, a neutrino produced along with one flavour of charged lepton can interact to produce a charged lepton of a different flavour.

### 13.3.1 The leptonic charged-current vertex revisited

In Chapter 12, the charged-current interaction between a charged lepton and a neutrino was described in terms of the neutrino weak eigenstates. For example, the weak charged-current vertex has the form

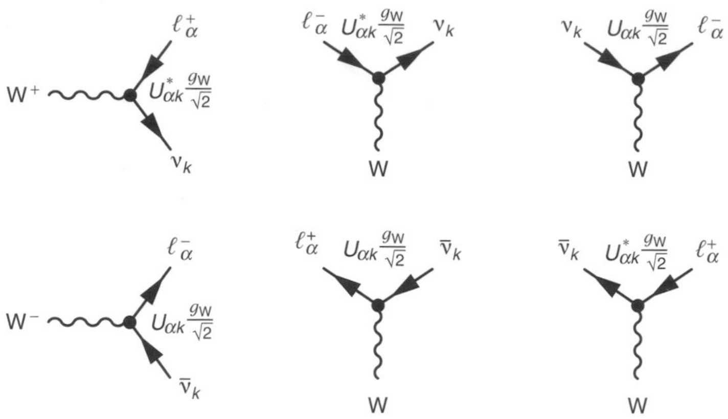


Fig. 13.10 The charged-current weak interaction vertices for charged lepton of flavour $\alpha = \mathrm{e}, \mu, \tau$ and a neutrino of type $k = 1, 2, 3$.

$$-i\frac{g_W}{\sqrt{2}}\overline{\mathrm{e}}\gamma^\mu\tfrac{1}{2}(1-\gamma^5)\nu_\mathrm{e},$$

where here $\nu_\mathrm{e}$ and $\overline{\mathrm{e}}$ denote the electron neutrino spinor and the electron adjoint spinor. In terms of the neutrino mass eigenstates, the weak charged-current for a lepton of flavour $\alpha = \mathrm{e}, \mu, \tau$ and a neutrino of type $k = 1, 2, 3$ takes the form

$$-i\frac{g_W}{\sqrt{2}}\overline{\ell}_\alpha\gamma^\mu\tfrac{1}{2}(1-\gamma^5)U_{\alpha k}\nu_k.$$

Defining the neutrino state produced in a weak interaction using the matrix $U$ of (13.4) implies that, when the neutrino appears as the adjoint spinor, the factor $U^*_{\alpha k}$ appears in the weak interaction vertex. Consequently, the couplings between neutrinos or antineutrinos and the charged leptons are those shown in Figure 13.10.

## 13.4 Neutrino oscillations of two flavours

The full treatment of neutrino oscillations for three flavours is developed in Section 13.5. However, the main features can be readily understood by considering just two flavours. For example, consider the weak eigenstates $\nu_\mathrm{e}$ and $\nu_\mu$, which here are taken to be coherent linear superpositions of the mass eigenstates $\nu_1$ and $\nu_2$. The mass eigenstates propagate as plane waves of the form

$$|\nu_1(t)\rangle = |\nu_1\rangle e^{i(\mathbf{p}_1\cdot\mathbf{x}-E_1t)} = e^{-ip_1\cdot x},$$
$$|\nu_2(t)\rangle = |\nu_2\rangle e^{i(\mathbf{p}_2\cdot\mathbf{x}-E_2t)} = e^{-ip_2\cdot x},$$

where $(E_1, \mathbf{p}_1)$ and $(E_2, \mathbf{p}_2)$ are the respective energy and three-momenta of the $\nu_1$ and $\nu_2$, and $p\cdot x = Et - \mathbf{p}\cdot\mathbf{x}$ is the (Lorentz-invariant) phase. In the two-flavour

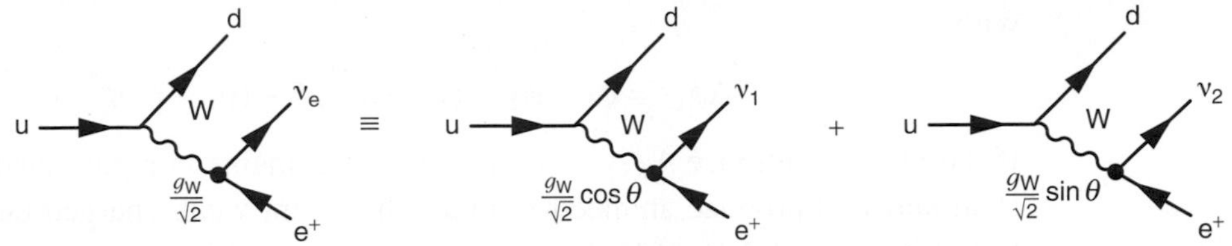


Fig. 13.11 Nuclear β-decay and its relation to the mass eigenstates for two flavours.

treatment of neutrino oscillations, the weak eigenstates are related to the mass eigenstates by a $2 \times 2$ unitary matrix that can be expressed in terms of a single mixing angle $\theta$,

$$\begin{pmatrix} \nu_e \\ \nu_\mu \end{pmatrix} = \begin{pmatrix} \cos\theta & \sin\theta \\ -\sin\theta & \cos\theta \end{pmatrix} \begin{pmatrix} \nu_1 \\ \nu_2 \end{pmatrix}. \tag{13.6}$$

Now suppose at time $t = 0$, a neutrino is produced in the process $u \to de^+\nu_e$, as shown in Figure 13.11. The wavefunction at time $t = 0$ is the coherent linear superposition of $\nu_1$ and $\nu_2$, corresponding to the $\nu_e$ state

$$|\psi(0)\rangle = |\nu_e\rangle \equiv \cos\theta\,|\nu_1\rangle + \sin\theta\,|\nu_2\rangle.$$

The state subsequently evolves according to the time dependence of the mass eigenstates

$$|\psi(\mathbf{x}, t)\rangle = \cos\theta\,|\nu_1\rangle e^{-ip_1\cdot x} + \sin\theta\,|\nu_2\rangle e^{-ip_2\cdot x},$$

where $p_1$ and $p_2$ are the four-momenta associated with the mass eigenstates $\nu_1$ and $\nu_2$. If the neutrino then interacts at a time $T$ and at a distance $L$ along its direction of flight, the neutrino state at this space-time point is

$$|\psi(L, T)\rangle = \cos\theta\,|\nu_1\rangle e^{-i\phi_1} + \sin\theta\,|\nu_2\rangle e^{-i\phi_2}, \tag{13.7}$$

where the phases of the two mass eigenstates are written as

$$\phi_i = p_i\cdot x = E_i T - \mathrm{p}_i L.$$

Equation (13.7) can be written in terms of the weak eigenstates using the inverse of (13.6),

$$\begin{pmatrix} \nu_1 \\ \nu_2 \end{pmatrix} = \begin{pmatrix} \cos\theta & -\sin\theta \\ \sin\theta & \cos\theta \end{pmatrix} \begin{pmatrix} \nu_e \\ \nu_\mu \end{pmatrix},$$

leading to

$$\begin{aligned} |\psi(L, T)\rangle &= \cos\theta\left(\cos\theta\,|\nu_e\rangle - \sin\theta\,|\nu_\mu\rangle\right)e^{-i\phi_1} + \sin\theta\left(\sin\theta\,|\nu_e\rangle + \cos\theta\,|\nu_\mu\rangle\right)e^{-i\phi_2} \\ &= (e^{-i\phi_1}\cos^2\theta + e^{-i\phi_2}\sin^2\theta)\,|\nu_e\rangle - (e^{-i\phi_1} - e^{-i\phi_2})\cos\theta\sin\theta\,|\nu_\mu\rangle \\ &= e^{-i\phi_1}\left[(\cos^2\theta + e^{i\Delta\phi_{12}}\sin^2\theta)\,|\nu_e\rangle - (1 - e^{i\Delta\phi_{12}})\cos\theta\sin\theta\,|\nu_\mu\rangle\right], \end{aligned} \tag{13.8}$$

with

$$\Delta\phi_{12} = \phi_1 - \phi_2 = (E_1 - E_2)T - (\mathrm{p}_1 - \mathrm{p}_2)L. \tag{13.9}$$

If the phase difference $\Delta\phi_{12} = 0$, the neutrino remains in a pure electron neutrino state and will produce an electron in a subsequent weak charged-current interaction. However, if $\Delta\phi_{12} \neq 0$, there is now a muon neutrino component to the wavefunction. The relative sizes of the electron and muon neutrino components of the wavefunction can be obtained by writing (13.8) as

$$|\psi(L,T)\rangle = c_e|\nu_e\rangle + c_\mu|\nu_\mu\rangle,$$

where $c_e = \langle\nu_e|\psi\rangle$ and $c_\mu = \langle\nu_\mu|\psi\rangle$. The probability that the neutrino, which was produced as a $\nu_e$, will interact to produce a muon is $P(\nu_e \to \nu_\mu) = c_\mu c_\mu^*$. Comparison with (13.8) gives

$$\begin{aligned} P(\nu_e \to \nu_\mu) = c_\mu c_\mu^* &= (1 - e^{i\Delta\phi_{12}})(1 - e^{-i\Delta\phi_{12}})\cos^2\theta\sin^2\theta \\ &= \tfrac{1}{4}(2 - 2\cos\Delta\phi_{12})\sin^2(2\theta) \\ &= \sin^2(2\theta)\sin^2\left(\frac{\Delta\phi_{12}}{2}\right). \end{aligned} \tag{13.10}$$

Hence, the $\nu_e \to \nu_\mu$ oscillation probability depends on the mixing angle $\theta$ and the phase difference between the mass eigenstates, $\Delta\phi_{12}$. The derivation of the phase difference, $\Delta\phi_{12} = (E_1 - E_2)T - (\mathrm{p}_1 - \mathrm{p}_2)L$, in terms of the masses of $\nu_1$ and $\nu_2$ requires care. One could assume, without any real justification, that the momenta of the two mass eigenstates are equal, $\mathrm{p}_1 = \mathrm{p}_2 = \mathrm{p}$, in which case

$$\Delta\phi_{12} = (E_1 - E_2)T = \left[\mathrm{p}\left(1 + \frac{m_1^2}{\mathrm{p}^2}\right)^{\frac{1}{2}} - \mathrm{p}\left(1 + \frac{m_2^2}{\mathrm{p}^2}\right)^{\frac{1}{2}}\right]T. \tag{13.11}$$

Because $m \ll E$, the square roots in (13.11) are approximately,

$$\left(1 + \frac{m^2}{\mathrm{p}^2}\right)^{\frac{1}{2}} \approx 1 + \frac{m^2}{2\mathrm{p}^2},$$

and therefore

$$\Delta\phi_{12} \approx \frac{m_1^2 - m_2^2}{2\mathrm{p}}L, \tag{13.12}$$

where it has been assumed that $T \approx L$ (in natural units), which follows since the neutrino velocity $\beta \approx 1$. At first sight, this treatment appears perfectly reasonable. However, it overlooks that fact that the different mass eigenstates will propagate with different velocities, and therefore will travel the distance $L$ in different times. This objection only can be overcome with a proper wave-packet treatment of the propagation of the coherent state, which yields the same expression as given in

(13.12). However, it is worth noting that the expression of (13.12), which was obtained assuming $p_1 = p_2$, also can be obtained by assuming either $E_1 = E_2$ or $\beta_1 = \beta_2$. This can be seen by writing the phase difference of (13.9) as

$$\begin{aligned}\Delta\phi_{12} &= (E_1 - E_2)T - \left(\frac{p_1^2 - p_2^2}{p_1 + p_2}\right)L \\ &= (E_1 - E_2)T - \left(\frac{E_1^2 - m_1^2 - E_2^2 + m_2^2}{p_1 + p_2}\right)L \\ &= (E_1 - E_2)\left[T - \left(\frac{E_1 + E_2}{p_1 + p_1}\right)L\right] + \left(\frac{m_1^2 - m_2^2}{p_1 + p_2}\right)L. \end{aligned} \tag{13.13}$$

The first term on the RHS of (13.13) clearly vanishes if it is assumed that $E_1 = E_2$. This term also vanishes if a common velocity is assumed, $\beta_1 = \beta_2 = \beta$ (see Problem 13.1). Hence, although a wave-packet treatment of the neutrino oscillation phenomenon is desirable, it is comforting to see that the same result for the phase difference $\Delta\phi$ is obtained from the assumption of either $p_1 = p_2$, $E_1 = E_2$ or $\beta_1 = \beta_2$.

Combining the results of (13.10) and (13.12) and writing $p = E_\nu$, gives the two-flavour neutrino oscillation probability

$$P(\nu_e \to \nu_\mu) = \sin^2(2\theta)\sin^2\left(\frac{(m_1^2 - m_2^2)L}{4E_\nu}\right). \tag{13.14}$$

It is convenient to express the oscillation probability in units more suited to the length and energy scales encountered in practice. Writing $L$ in km, $\Delta m^2$ in $\mathrm{eV}^2$ and the neutrino energy in GeV, (13.14) can be written

$$P(\nu_e \to \nu_\mu) = \sin^2(2\theta)\sin^2\left(1.27\frac{\Delta m^2[\mathrm{eV}^2]L[\mathrm{km}]}{E_\nu[\mathrm{GeV}]}\right). \tag{13.15}$$

The corresponding electron neutrino survival probability, $P(\nu_e \to \nu_e)$, either can be obtained from $c_e^* c_e$ or from the conservation of probability, $P(\nu_e \to \nu_e) = 1 - P(\nu_e \to \nu_\mu)$,

$$P(\nu_e \to \nu_e) = 1 - \sin^2(2\theta)\sin^2\left(\frac{(m_1^2 - m_2^2)L}{4E_\nu}\right). \tag{13.16}$$

Figure 13.12 shows an illustrative example of the oscillation probability as function of distance for $E_\nu = 1\,\mathrm{GeV}$, $\Delta m^2 = 0.002\,\mathrm{eV}^2$ and $\sin^2(2\theta) = 0.8$. The wavelength of the oscillations is given by

$$\lambda_{\mathrm{osc}}[\mathrm{km}] = \frac{\pi E_\nu[\mathrm{GeV}]}{1.27\,\Delta m^2[\mathrm{eV}^2]}.$$

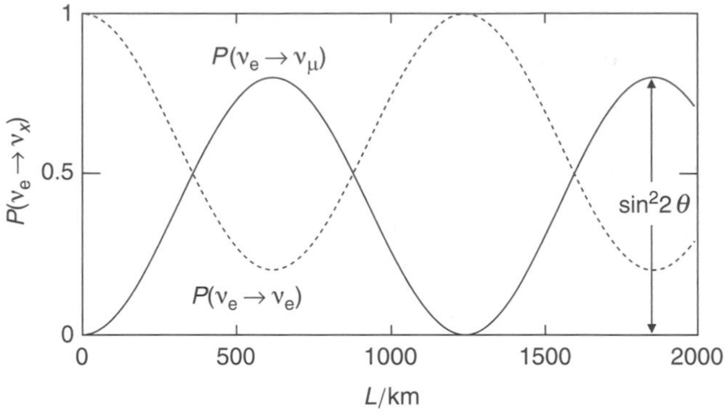


Fig. 13.12 The two-flavour oscillation probability $P(\nu_e \to \nu_\mu)$ and the survival probability $P(\nu_e \to \nu_e)$ plotted as function of $L$ for $E_\nu = 1\,\text{GeV}$, $\Delta m^2 = 0.002\,\text{eV}^2$ and $\sin^2(2\theta) = 0.8$.

For small values of $\Delta m^2$, neutrino flavour oscillations only develop over very large distances. This explains why neutrino flavour appeared to be conserved in earlier neutrino experiments. Finally, it should be noted that the amplitude of the oscillations is determined by $\sin^2(2\theta)$, with $\sin^2(2\theta) = 1$ corresponding to maximal mixing.

## 13.5 Neutrino oscillations of three flavours

The derivation of the neutrino oscillation probability for two flavours contains nearly all of the essential physics, namely the relationship between the weak and mass eigenstates and that the oscillations originate from the phase difference between the mass eigenstates in the time-dependent wavefunction. The full three-flavour derivation of the neutrino oscillation probabilities follows closely the steps above, although the algebra is more involved.

In the three-flavour treatment of neutrino oscillations, the three weak eigenstates are related to the mass eigenstates by the $3 \times 3$ unitary Pontecorvo–Maki–Nakagawa–Sakata (PMNS) matrix,

$$\begin{pmatrix} \nu_e \\ \nu_\mu \\ \nu_\tau \end{pmatrix} = \begin{pmatrix} U_{e1} & U_{e2} & U_{e3} \\ U_{\mu 1} & U_{\mu 2} & U_{\mu 3} \\ U_{\tau 1} & U_{\tau 2} & U_{\tau 3} \end{pmatrix} \begin{pmatrix} \nu_1 \\ \nu_2 \\ \nu_3 \end{pmatrix}. \tag{13.17}$$

The elements of the PMNS matrix are fundamental parameters of the lepton flavour sector of the Standard Model. The mass eigenstates can be expressed in terms of the weak eigenstates using the unitarity of the PMNS matrix that implies $U^{-1} = U^\dagger \equiv (U^*)^T$ and hence

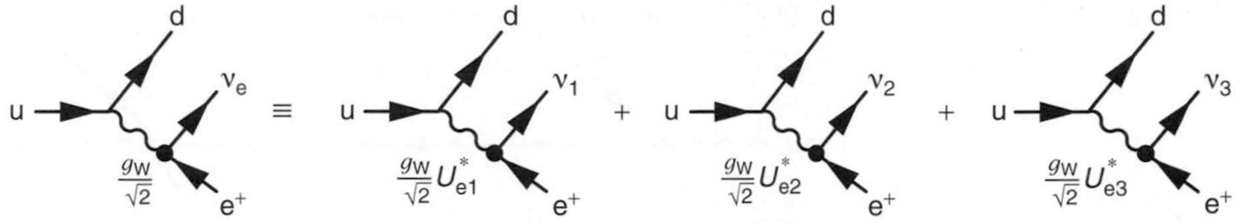


Fig. 13.13 The Feynman diagram for $\beta^+$-decay broken down into the contributions from the different mass eigenstates.

$$\begin{pmatrix} \nu_1 \\ \nu_2 \\ \nu_3 \end{pmatrix} = \begin{pmatrix} U^*_{e1} & U^*_{\mu 1} & U^*_{\tau 1} \\ U^*_{e2} & U^*_{\mu 2} & U^*_{\tau 2} \\ U^*_{e3} & U^*_{\mu 3} & U^*_{\tau 3} \end{pmatrix} \begin{pmatrix} \nu_e \\ \nu_\mu \\ \nu_\tau \end{pmatrix}.$$

The unitarity condition, $UU^\dagger = I$, also implies that

$$\begin{pmatrix} U_{e1} & U_{e2} & U_{e3} \\ U_{\mu 1} & U_{\mu 2} & U_{\mu 3} \\ U_{\tau 1} & U_{\tau 2} & U_{\tau 3} \end{pmatrix} \begin{pmatrix} U^*_{e1} & U^*_{\mu 1} & U^*_{\tau 1} \\ U^*_{e2} & U^*_{\mu 2} & U^*_{\tau 2} \\ U^*_{e3} & U^*_{\mu 3} & U^*_{\tau 3} \end{pmatrix} = \begin{pmatrix} 1 & 0 & 0 \\ 0 & 1 & 0 \\ 0 & 0 & 1 \end{pmatrix},$$

which gives nine relations between the elements of the PMNS matrix, for example

$$U_{e1}U^*_{e1} + U_{e2}U^*_{e2} + U_{e3}U^*_{e3} = 1, \tag{13.18}$$

$$U_{e1}U^*_{\mu 1} + U_{e2}U^*_{\mu 2} + U_{e3}U^*_{\mu 3} = 0. \tag{13.19}$$

Now consider the neutrino state that is produced in a charged-current weak interaction along with an electron, as indicated in Figure 13.13. The neutrino, which enters the weak interaction vertex as the adjoint spinor, corresponds to a coherent linear superposition of mass eigenstates with a wavefunction at time $t = 0$ of

$$|\psi(0)\rangle = |\nu_e\rangle \equiv U^*_{e1}|\nu_1\rangle + U^*_{e2}|\nu_2\rangle + U^*_{e3}|\nu_3\rangle.$$

The time evolution of the wavefunction is determined by the time evolution of the mass eigenstates and can be written as

$$|\psi(\mathbf{x}, t)\rangle = U^*_{e1}|\nu_1\rangle e^{-i\phi_1} + U^*_{e2}|\nu_2\rangle e^{-i\phi_2} + U^*_{e3}|\nu_3\rangle e^{-i\phi_3},$$

where as before $\phi_i = p_i{\cdot}x_i = (E_i t - \mathbf{p}_i{\cdot}\mathbf{x})$ is the phase of the plane wave representing each mass eigenstate. The subsequent *charged-current weak interactions* of the neutrino can be described in terms of its weak eigenstates by writing

$$\begin{aligned} |\psi(\mathbf{x}, t)\rangle = {} & U^*_{e1}(U_{e1}|\nu_e\rangle + U_{\mu 1}|\nu_\mu\rangle + U_{\tau 1}|\nu_\tau\rangle)e^{-i\phi_1} \\ & + U^*_{e2}(U_{e2}|\nu_e\rangle + U_{\mu 2}|\nu_\mu\rangle + U_{\tau 2}|\nu_\tau\rangle)e^{-i\phi_2} \\ & + U^*_{e3}(U_{e3}|\nu_e\rangle + U_{\mu 3}|\nu_\mu\rangle + U_{\tau 3}|\nu_\tau\rangle)e^{-i\phi_3}. \end{aligned} \tag{13.20}$$

Because the neutrino appears as the spinor in the weak interaction vertex producing a charged lepton, the mass eigenstates are expressed in terms of the elements of the

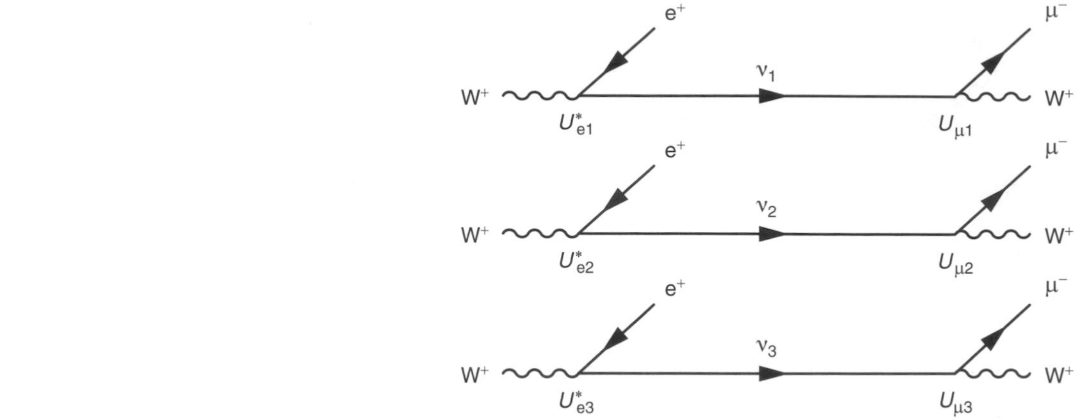


Fig. 13.14 The processes and elements of the PMNS that contribute to $\nu_e \to \nu_\mu$ oscillations.

PMNS matrix and not its complex conjugate. It should be noted that the weak states ($\nu_e$, $\nu_\mu$ and $\nu_\tau$) in (13.20) really refer to the flavour of the lepton produced in a subsequent charged-current weak interaction of the neutrino. Gathering up the terms for each weak eigenstate, Equation (13.20) can be written

$$\begin{aligned} |\psi(\mathbf{x},t)\rangle = \; & (U_{e1}^* U_{e1} e^{-i\phi_1} + U_{e2}^* U_{e2} e^{-i\phi_2} + U_{e3}^* U_{e3} e^{-i\phi_3})|\nu_e\rangle \\ & (U_{e1}^* U_{\mu 1} e^{-i\phi_1} + U_{e2}^* U_{\mu 2} e^{-i\phi_2} + U_{e3}^* U_{\mu 3} e^{-i\phi_3})|\nu_\mu\rangle \\ & (U_{e1}^* U_{\tau 1} e^{-i\phi_1} + U_{e2}^* U_{\tau 2} e^{-i\phi_2} + U_{e3}^* U_{\tau 3} e^{-i\phi_3})|\nu_\tau\rangle. \end{aligned} \tag{13.21}$$

This can be expressed in the form $|\psi(\mathbf{x},t)\rangle = c_e|\nu_e\rangle + c_\mu|\nu_\mu\rangle + c_\tau|\nu_\tau\rangle$, from which the oscillation probabilities can be obtained, for example

$$\begin{aligned} P(\nu_e \to \nu_\mu) &= |\langle \nu_\mu|\psi(\mathbf{x},t)\rangle|^2 = c_\mu c_\mu^* \\ &= |U_{e1}^* U_{\mu 1} e^{-i\phi_1} + U_{e2}^* U_{\mu 2} e^{-i\phi_2} + U_{e3}^* U_{\mu 3} e^{-i\phi_3}|^2. \end{aligned} \tag{13.22}$$

This expression can be understood as the magnitude squared of the sum of the diagrams shown in Figure 13.14, taking into account the relative phase differences that develop over the propagation distance. The oscillation probabilities are defined in terms of the flavours of the *charged* leptons produced in the weak interactions and the relevant PMNS matrix elements. If the phases were all the same, then the complex conjugate of the unitarity relation of (13.19), $U_{e1}^* U_{\mu 1} + U_{e2}^* U_{\mu 2} + U_{e3}^* U_{\mu 3} = 0$, would imply $P(\nu_e \to \nu_\mu) = 0$ and, as before, neutrino flavour oscillations only occur if the neutrinos have mass, and the masses are not all the same.

Equation (13.22) can be simplified using the complex number identity,

$$|z_1 + z_2 + z_3|^2 \equiv |z_1|^2 + |z_2|^2 + |z_3|^2 + 2\,\Re\mathrm{e}\{z_1 z_2^* + z_1 z_3^* + z_2 z_3^*\}, \tag{13.23}$$

giving

$$P(\nu_e \to \nu_\mu) = |U_{e1}^* U_{\mu 1}|^2 + |U_{e2}^* U_{\mu 2}|^2 + |U_{e3}^* U_{\mu 3}|^2 + 2\,\Re\{U_{e1}^* U_{\mu 1} U_{e2} U_{\mu 2}^* e^{-i(\phi_1-\phi_2)}\}$$
$$+ 2\,\Re\{U_{e1}^* U_{\mu 1} U_{e3} U_{\mu 3}^* e^{-i(\phi_1-\phi_3)}\} + 2\,\Re\{U_{e2}^* U_{\mu 2} U_{e3} U_{\mu 3}^* e^{-i(\phi_2-\phi_3)}\}. \quad (13.24)$$

This can be simplified further by applying the identity (13.23) to the modulus squared of the complex conjugate of the unitarity relation of (13.19), which gives

$$|U_{e1}^* U_{\mu 1}|^2 + |U_{e2}^* U_{\mu 2}|^2 + |U_{e3}^* U_{\mu 3}|^2 +$$
$$2\,\Re\{U_{e1}^* U_{\mu 1} U_{e2} U_{\mu 2}^* + U_{e1}^* U_{\mu 1} U_{e3} U_{\mu 3}^* + U_{e2}^* U_{\mu 2} U_{e3} U_{\mu 3}^*\} = 0,$$

and thus, (13.24) can be written as

$$P(\nu_e \to \nu_\mu) = 2\,\Re\{U_{e1}^* U_{\mu 1} U_{e2} U_{\mu 2}^* [e^{i(\phi_2-\phi_1)} - 1]\}$$
$$2\,\Re\{U_{e1}^* U_{\mu 1} U_{e3} U_{\mu 3}^* [e^{i(\phi_3-\phi_1)} - 1]\}$$
$$2\,\Re\{U_{e2}^* U_{\mu 2} U_{e3} U_{\mu 3}^* [e^{i(\phi_3-\phi_2)} - 1]\}. \quad (13.25)$$

The electron neutrino survival probability $P(\nu_e \to \nu_e)$ can be obtained in a similar manner starting from (13.24) and using the unitarity relation of (13.18). In this case, each element of the PMNS matrix is paired with the corresponding complex conjugate, e.g. $U_{e1} U_{e1}^*$, and the combinations of PMNS matrix elements give real numbers. Therefore, the electron neutrino survival probability is

$$P(\nu_e \to \nu_e) = 1 + 2|U_{e1}|^2|U_{e2}|^2\,\Re\{[e^{i(\phi_2-\phi_1)} - 1]\}$$
$$+ 2|U_{e1}|^2|U_{e3}|^2\,\Re\{[e^{i(\phi_3-\phi_1)} - 1]\}$$
$$+ 2|U_{e2}|^2|U_{e3}|^2\,\Re\{[e^{i(\phi_3-\phi_2)} - 1]\}. \quad (13.26)$$

Equation (13.26) can be simplified by noting

$$\Re\{e^{i(\phi_j-\phi_i)} - 1\} = \cos(\phi_j - \phi_i) - 1 = -2\sin^2\left(\frac{\phi_j - \phi_i}{2}\right) = -2\sin^2\Delta_{ji},$$

where $\Delta_{ji}$ is defined as

$$\Delta_{ji} = \frac{\phi_j - \phi_i}{2} = \frac{(m_j^2 - m_i^2)L}{4E_\nu}.$$

Hence, (13.26) can be written

$$P(\nu_e \to \nu_e) = 1 - 4|U_{e1}|^2|U_{e2}|^2 \sin^2\Delta_{21}$$
$$- 4|U_{e1}|^2|U_{e3}|^2 \sin^2\Delta_{31} - 4|U_{e2}|^2|U_{e3}|^2 \sin^2\Delta_{32}. \quad (13.27)$$

The electron neutrino survival probability depends on three differences of squared masses, $\Delta m_{21}^2 = m_2^2 - m_1^2$, $\Delta m_{31}^2 = m_3^2 - m_1^2$ and $\Delta m_{32}^2 = m_3^2 - m_2^2$. Only two of these differences are independent and $\Delta_{31}$ can be expressed as

$$\Delta_{31} = \Delta_{32} + \Delta_{21}. \tag{13.28}$$

Before using the above formulae to describe the experimental data, it is worth discussing the current knowledge of neutrino masses, the nature of the PMNS matrix and the discrete symmetries related to the neutrino oscillation phenomena, including the possibility of CP violation.

### 13.5.1 Neutrino masses and the neutrino mass hierarchy

Since the neutrino oscillation probabilities depend on differences of the squared neutrino masses, the experimental measurements of neutrino oscillations do not constrain the overall neutrino mass scale. To date, there are no direct measurements of neutrino masses, only upper limits. From studies of the end point of the electron energy distribution in the nuclear $\beta$-decay of tritium, it is known that the mass of the lightest neutrino is $\lesssim 2\,\mathrm{eV}$. Tighter, albeit model-dependent, limits can be obtained from cosmology. The density of low-energy relic neutrinos from the Big Bang is large, $\mathcal{O}(100)\,\mathrm{cm}^{-3}$ for each flavour. Consequently, neutrino masses potentially impact the evolution of the Universe. From recent cosmological measurements of the large-scale structure of the Universe, it can be deduced that

$$\sum_{i=1}^{3} m_{\nu_i} \lesssim 1\,\mathrm{eV}.$$

Whilst the neutrino masses are not known, it is clear that they are much smaller than those of either the charged leptons or the quarks. Even with neutrino masses at the eV scale, they are smaller by a factor of at least $10^6$ than the mass of the electron and smaller by a factor of at least $10^9$ than the mass of the tau-lepton. The current hypothesis for this large difference, known as the seesaw mechanism, is discussed in Chapter 17.

The results of recent neutrino oscillation experiments, which are described in Sections 13.7 and 13.8, provide determinations of differences of the squares of the neutrino masses

$$\Delta m_{21}^2 = m_2^2 - m_1^2 \approx 8 \times 10^{-5}\,\mathrm{eV}^2,$$
$$|\Delta m_{32}^2| = |m_3^2 - m_2^2| \approx 2 \times 10^{-3}\,\mathrm{eV}^2.$$

Regardless of the absolute mass scale of the lightest neutrino, there are two possible hierarchies for the neutrino masses, shown in Figure 13.15. In the *normal* hierarchy $m_3 > m_2$ and in the *inverted* mass hierarchy $m_3 < m_2$. Current experiments

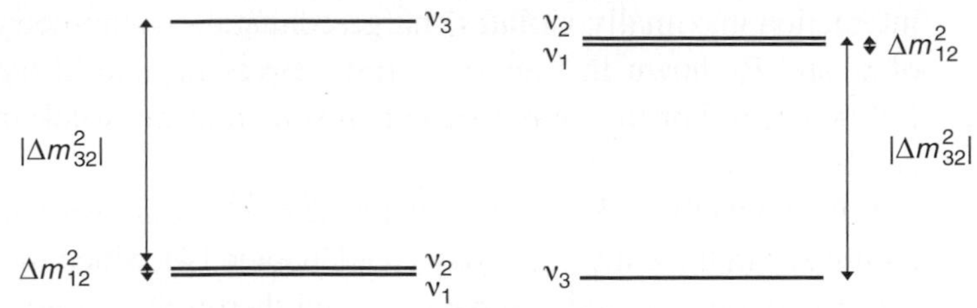


Fig. 13.15 The two possible neutrino mass hierarchies, normal where $m_3 > m_2$ and inverted where $m_2 > m_3$.

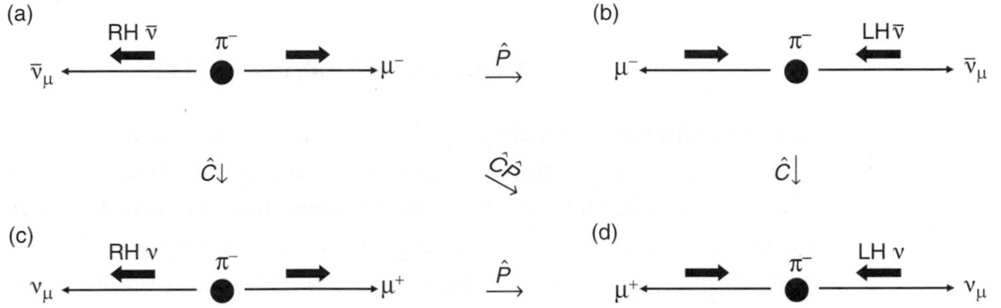


Fig. 13.16 The effect of parity, charge-conjugation and combined CP operation on pion decay.

are not sensitive enough to distinguish between these two possibilities. However, regardless of the hierarchy, because $\Delta m^2_{21} \ll |\Delta m^2_{32}|$ in most circumstances it is reasonable to make the approximation

$$|\Delta m^2_{31}| \approx |\Delta m^2_{32}|.$$

### 13.5.2 CP violation in neutrino oscillations

The $V - A$ chiral structure of the weak charged-current implies that parity is maximally violated. It also implies that charge-conjugation symmetry is maximally violated. This can be seen by considering the weak decay $\pi^- \to \mu^-\overline{\nu}_\mu$. Because neutrino masses are extremely small compared to the energies involved, the antineutrino is effectively always emitted in a RH helicity state, as shown in Figure 13.16a. The effect of the parity operator, shown in Figure 13.16b, is to reverse the particle momenta leaving the particle spins (axial-vectors) unchanged. The result of the parity transformation is a final state with a LH antineutrino, for which the weak charged-current matrix element is zero.

The effect of the charge conjugation operator $\hat{C}$ is to replace particles by their antiparticles and *vice versa*, is shown in Figure 13.16c. Charge conjugation results in a RH neutrino in the final state. Since only LH particle states participate in the weak interaction, the matrix element for this process is also zero. Thus the weak interaction maximally violates charge-conjugation symmetry. The combined effect of C and P, shown in Figure 13.16d, results in a valid weak decay involving a LH neutrino. For this reason it is plausible that the weak interaction respects the combined CP symmetry.

It is known that CP violation is needed to account for the excess of matter over antimatter in the Universe today (see Chapter 14). Since the QED and QCD interactions conserve C and P separately, and therefore conserve CP, the only possible place in the Standard Model where CP-violating effects can occur is in the weak interaction.

## Time reversal symmetry and CPT

Parity is a discrete symmetry operation corresponding to $\mathbf{x} \to -\mathbf{x}$. Similarly, time reversal is a discrete symmetry operation that has the effect $t \to -t$. Following the arguments of Chapter 11, it should be clear that the vector nature of the QED and QCD interactions, implies that the matrix elements of QED and QCD are invariant under time reversal. More generally, all local Lorentz-invariant Quantum Field Theories can be shown to be invariant under the combined operation of C, P and T. One consequence of this CPT symmetry is that particles and antiparticles have identical masses, magnetic moments, etc. The best experimental limit on CPT invariance comes from the equality of the masses of the flavour eigenstates of the neutral kaons, $\mathrm{K^0(d\bar{s})}$ and $\mathrm{\overline{K}^0(s\bar{d})}$, where

$$\frac{|m(\mathrm{K}^0) - m(\overline{\mathrm{K}}^0)|}{m(\mathrm{K}^0)} < 10^{-18}.$$

CPT is believed to be an exact symmetry of the Universe. This implies that, if physics is unchanged by the combined operation of C and P, then time reversal symmetry also holds. The corollary is that CP violation implies that T reversal symmetry is also violated and vice versa.

## CP and T violation in neutrino oscillations

It is instructive to consider the effects of the discrete symmetry transformations, CP, T and CPT, in the context of neutrino oscillations. If time reversal symmetry applies, then the oscillation probability for $P(\nu_e \to \nu_\mu)$ will be equal to $P(\nu_\mu \to \nu_e)$. The oscillation probability $P(\nu_e \to \nu_\mu)$ is given by (13.25)

$$P(\nu_e \to \nu_\mu) = 2\,\Re\mathrm{e}\{U_{e1}^* U_{\mu 1} U_{e2} U_{\mu 2}^* [e^{i(\phi_2-\phi_1)} - 1]\} + \cdots . \qquad (13.29)$$

The corresponding expression for the oscillation probability $P(\nu_\mu \to \nu_e)$ is obtained by swapping the e and $\mu$ labels

$$P(\nu_\mu \to \nu_e) = 2\,\Re\mathrm{e}\{U_{\mu 1}^* U_{e1} U_{\mu 2} U_{e2}^* [e^{i(\phi_2-\phi_1)} - 1]\} + \cdots .$$

The elements of the PMNS matrix that appear in the expression for $P(\nu_\mu \to \nu_e)$ are the complex conjugates of those in the expression for $P(\nu_e \to \nu_\mu)$. Hence, unless all elements $U_{ei}$ and $U_{\mu j}$ are real, time reversal symmetry does not necessarily hold in neutrino oscillations, which in turn implies the possibility of CP violation.

The effect of the CP operation on $\nu_e \to \nu_\mu$ flavour transformations is

$$\nu_e \to \nu_\mu \quad \xrightarrow{\hat{C}\hat{P}} \quad \overline{\nu}_e \to \overline{\nu}_\mu,$$

where $C$ transforms particles in to antiparticles and $P$ ensures that the LH neutrinos transform to RH antineutrinos. The oscillation probability $P(\overline{\nu}_e \to \overline{\nu}_\mu)$ can be obtained from that for $P(\nu_e \to \nu_\mu)$ by noting that whether the element of the PMNS matrix appears as $U$ or $U^*$ depends on whether the neutrino appears as the spinor or adjoint spinor in the weak interaction vertex (see Section 13.3.1). Consequently

$$P(\overline{\nu}_e \to \overline{\nu}_\mu) = 2\,\Re\mathrm{e}\{U_{e1}U^*_{\mu 1}U^*_{e2}U_{\mu 2}[e^{i(\phi_2-\phi_1)} - 1]\} + \cdots .$$

Again, unless all the elements $U_{ei}$ and $U_{\mu j}$ are real, $P(\nu_e \to \nu_\mu) \neq P(\overline{\nu}_e \to \overline{\nu}_\mu)$, and CP can be violated in neutrino oscillations. Finally, consider the combined CPT operation

$$\nu_e \to \nu_\mu \quad \xrightarrow{\hat{C}\hat{P}\hat{T}} \quad \overline{\nu}_\mu \to \overline{\nu}_e,$$

where the effect of time reversal swaps the e and $\mu$ labels and the effect CP is to exchange $U \leftrightarrow U^*$ and therefore

$$P(\overline{\nu}_\mu \to \overline{\nu}_e) = 2\,\Re\mathrm{e}\{U_{\mu 1}U^*_{e1}U^*_{\mu 2}U_{e2}[e^{i(\phi_2-\phi_1)} - 1]\} + \cdots = P(\nu_e \to \nu_\mu).$$

As expected, neutrino oscillations are invariant under the combined action of CPT.

The imaginary components of the PMNS matrix, provide a possible source of CP violation in the Standard Model. The relative magnitude of the CP violation in neutrino oscillations is given by $P(\nu_e \to \nu_\mu) - P(\overline{\nu}_e \to \overline{\nu}_\mu)$. This can be shown to be (see Problem 13.4)

$$P(\nu_e \to \nu_\mu) - P(\overline{\nu}_e \to \overline{\nu}_\mu) = 16\,\Im\mathrm{m}\{U^*_{e1}U_{\mu 1}U_{e2}U^*_{\mu 2}\}\sin\Delta_{12}\sin\Delta_{13}\sin\Delta_{23}. \tag{13.30}$$

With the current experimental knowledge of the PMNS matrix elements, it is known that the difference $P(\nu_e \to \nu_\mu) - P(\overline{\nu}_e \to \overline{\nu}_\mu)$ is at most a few percent. CP violating effects in neutrino oscillations are small and are beyond the sensitivity of the current generation of experiments.

### The PMNS matrix

In the Standard Model, the unitarity PMNS matrix can be described in terms of three real parameters and a single phase. The reasoning is subtle. A general $3 \times 3$ matrix can be described by nine complex numbers. The unitarity of the PMNS matrix, $UU^\dagger = I$, provides nine constraints, leaving nine independent parameters. If the PMNS matrix were real, it would be correspond to the orthogonal rotation matrix $R$ and could be described by three rotation angles, $\theta_{12}$, $\theta_{13}$ and $\theta_{23}$

$$R = \begin{pmatrix} 1 & 0 & 0 \\ 0 & c_{23} & s_{23} \\ 0 & -s_{23} & c_{23} \end{pmatrix} \times \begin{pmatrix} c_{13} & 0 & s_{13} \\ 0 & 1 & 0 \\ -s_{13} & 0 & c_{13} \end{pmatrix} \times \begin{pmatrix} c_{12} & s_{12} & 0 \\ -s_{12} & c_{12} & 0 \\ 0 & 0 & 1 \end{pmatrix}, \tag{13.31}$$

where $s_{ij} = \sin\theta_{ij}$ and $c_{ij} = \cos\theta_{ij}$. In this form $\theta_{12}$ is the angle of rotation about the three-axis, $\theta_{13}$ is the angle of rotation about the new two-axis, and $\theta_{23}$ is a rotation about the resulting one-axis.

Since the PMNS matrix is unitary, not real, there are six additional degrees of freedom that appear as complex phases of the form $\exp(i\delta)$. It turns out that not all of these phases are physically relevant. This can be seen by writing the currents for the possible leptonic weak interaction charged-current vertices as

$$-i\frac{g_W}{\sqrt{2}}(\overline{\mathrm{e}}, \overline{\mu}, \overline{\tau})\gamma^\mu \tfrac{1}{2}(1-\gamma^5) \begin{pmatrix} U_{\mathrm{e}1} & U_{\mathrm{e}2} & U_{\mathrm{e}3} \\ U_{\mu 1} & U_{\mu 2} & U_{\mu 3} \\ U_{\tau 1} & U_{\tau 2} & U_{\tau 3} \end{pmatrix} \begin{pmatrix} \nu_1 \\ \nu_2 \\ \nu_3 \end{pmatrix}.$$

These four-vector currents are unchanged by the transformation,

$$\ell_\alpha \to \ell_\alpha e^{i\theta_\alpha}, \quad \nu_k \to \nu_k e^{i\theta_k} \quad \text{and} \quad U_{\alpha k} \to U_{\alpha k} e^{i(\theta_\alpha - \theta_k)}, \tag{13.32}$$

where $\ell_\alpha$ is the charged lepton of type $\alpha = \mathrm{e}, \mu, \tau$. Hence, it might appear that the six complex phases in the PMNS matrix can be absorbed into the definitions of the phases of the neutrino and charged leptons without any physical consequences. This is not the case because an overall phase factor in the PMNS matrix multiplying all elements has no physical consequence. For this reason, it is possible to pull out a common phase $U \to Ue^{i\theta}$. In this way all phases can be defined relative to, for example, the phase of the electron $\theta_\mathrm{e}$ such that $\theta_k = \theta_\mathrm{e} + \theta'_k$. In this case the transformation of (13.32) becomes

$$\ell_\alpha \to \ell_\alpha e^{i(\theta_\mathrm{e} + \theta'_\alpha)}, \quad \nu_k \to \nu_k e^{i(\theta_\mathrm{e} + \theta'_k)} \quad \text{and} \quad U_{\alpha k} \to U_{\alpha k} e^{i(\theta'_\alpha - \theta'_k)},$$

from which it can be seen that only five phases of the PMNS matrix can be absorbed into the definition of the particles since $\theta'_\mathrm{e} = 0$ and the common phase $e^{i\theta_\mathrm{e}}$ has no physical consequences. Hence the PMNS matrix can be expressed in terms of three mixing angles, $\theta_{12}$, $\theta_{23}$ and $\theta_{13}$ and a single complex phase $\delta$.

The PMNS matrix is usually written as

$$U_{\mathrm{PMNS}} = \begin{pmatrix} U_{\mathrm{e}1} & U_{\mathrm{e}2} & U_{\mathrm{e}3} \\ U_{\mu 1} & U_{\mu 2} & U_{\mu 3} \\ U_{\tau 1} & U_{\tau 2} & U_{\tau 3} \end{pmatrix} = \begin{pmatrix} 1 & 0 & 0 \\ 0 & c_{23} & s_{23} \\ 0 & -s_{23} & c_{23} \end{pmatrix} \begin{pmatrix} c_{13} & 0 & s_{13}e^{-i\delta} \\ 0 & 1 & 0 \\ -s_{13}e^{i\delta} & 0 & c_{13} \end{pmatrix} \begin{pmatrix} c_{12} & s_{12} & 0 \\ -s_{12} & c_{12} & 0 \\ 0 & 0 & 1 \end{pmatrix}.$$

This form is particularly convenient because $\theta_{13}$ is known to be relatively small and thus the central matrix is almost diagonal. The individual elements of the PMNS matrix, obtained from the matrix multiplication, are

$$\begin{pmatrix} U_{e1} & U_{e2} & U_{e3} \\ U_{\mu 1} & U_{\mu 2} & U_{\mu 3} \\ U_{\tau 1} & U_{\tau 2} & U_{\tau 3} \end{pmatrix} = \begin{pmatrix} c_{12}c_{13} & s_{12}c_{13} & s_{13}e^{-i\delta} \\ -s_{12}c_{23} - c_{12}s_{23}s_{13}e^{i\delta} & c_{12}c_{23} - s_{12}s_{23}s_{13}e^{i\delta} & s_{23}c_{13} \\ s_{12}s_{23} - c_{12}c_{23}s_{13}e^{i\delta} & -c_{12}s_{23} - s_{12}c_{23}s_{13}e^{i\delta} & c_{23}c_{13} \end{pmatrix}. \tag{13.33}$$

It is worth noting that, in the two-flavour treatment of neutrino oscillations, the general form of the unitary transformation between weak and mass eigenstates has four parameters, a rotation angle and three complex phases. But, all three complex phases can be absorbed into the definitions of the particles, and the resulting matrix depends on a single angle, as assumed in (13.6). In this case the matrix is entirely real and therefore cannot accommodate the CP violation. Hence CP violation originating from the PMNS matrix occurs only for three or more generations of leptons.

## 13.6 Neutrino oscillation experiments

Early experimental results on neutrino oscillations were obtained from studies of solar neutrinos and the neutrinos produced in cosmic-ray-induced cascades in the atmosphere. More recent results have been obtained from long-baseline neutrino oscillation beam experiments and from the study of electron antineutrinos from nuclear fission reactors. There are two possible signatures for neutrino oscillations. Firstly neutrino oscillations can result in the *appearance* of "wrong" flavour charged leptons, for example the observation of $e^-$ and/or $\tau^-$ from an initially pure beam of $\nu_\mu$. Alternatively, neutrino oscillations can be observed as the *disappearance* of the "right" flavour charged lepton, where fewer than expected $\mu^-$ are produced from an initially pure $\nu_\mu$ beam.

### 13.6.1 Neutrino interaction thresholds

The observable experimental effects resulting from neutrino oscillations depend on the type of neutrino interactions that are detectable. Neutrinos can be detected in matter through their charged-current and neutral-current weak interactions, either with atomic electrons or with nucleons, as shown in Figure 13.17. Unless kinematically forbidden, interactions with nucleons will dominate, since the neutrino interaction cross sections are proportional to the centre-of-mass energy squared, $s \approx 2mE_\nu$, where $m$ is the mass of the target particle (see for example, Section 12.3).

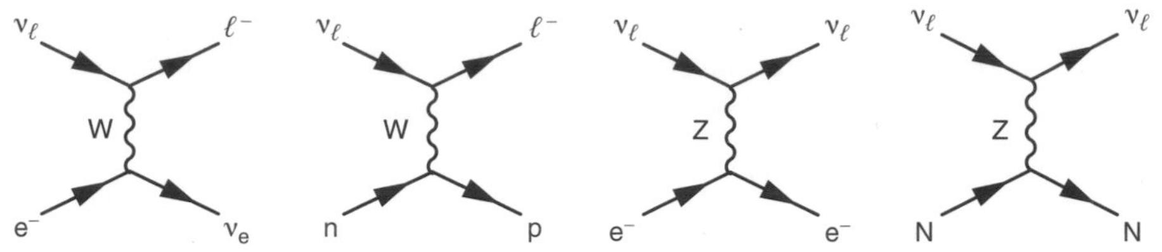


Fig. 13.17 The Feynman diagrams for CC and NC neutrino interactions in matter. For $\overline{\nu}_e$ there is also an $s$-channel process, $\overline{\nu}_e e^- \to \overline{\nu}_e e^-$.

Whether an appearance signal can be observed depends on whether the interaction is kinematically allowed. Charged-current neutrino interactions are allowed if the centre-of-mass energy is sufficient to produce a charged lepton and the final-state hadronic system. The threshold is determined by the lowest $W^2$ process, $\nu_\ell n \to \ell^- p$. In the laboratory frame, where the neutron is at rest, the centre-of-mass energy squared is given by

$$s = (p_\nu + p_n)^2 = (E_\nu + m_n)^2 - E_\nu^2 = 2E_\nu m_n + m_n^2.$$

The $\nu_\ell n \to \ell^- p$ interaction is only kinematically allowed if $s > (m_\ell + m_p)^2$,

$$E_\nu > \frac{(m_p^2 - m_n^2) + m_\ell^2 + 2m_p m_\ell}{2m_n}.$$

From this expression, the laboratory frame neutrino threshold energies for charged-current interactions with a nucleon are

$$E_{\nu_e} > 0, \quad E_{\nu_\mu} > 110\,\text{MeV} \quad \text{and} \quad E_{\nu_\tau} > 3.5\,\text{GeV}.$$

For electron neutrinos with energies of order a few MeV, the nuclear binding energy also has to be taken into account.

Charged-current interactions with an atomic electron $\nu_\ell e^- \to \nu_e \ell^-$ are kinematically allowed if $s > m_\ell^2$, where $m_\ell$ is the mass of the final-state charged lepton. In the laboratory frame

$$s = (p_\nu + p_e)^2 = (E_\nu + m_e)^2 - E_\nu^2 = 2E_\nu m_e + m_e^2,$$

and hence

$$E_\nu > \frac{m_\ell^2 - m_e^2}{2m_e},$$

leading to laboratory frame thresholds for charged-current $\nu e^-$ scattering of

$$E_{\nu_e} > 0, \quad E_{\nu_\mu} > 11\,\text{GeV} \quad \text{and} \quad E_{\nu_\tau} > 3090\,\text{GeV}.$$

Consequently, for the neutrino energies encountered in most experiments, interactions with atomic electrons are relevant only for electron neutrinos/antineutrinos.

## 13.7 Reactor experiments

Nuclear fission reactors produce a large flux of electron antineutrinos from the β-decays of radioisotopes such as $^{235}$U, $^{238}$U, $^{239}$Pu and $^{241}$Pu, which are produced in nuclear fission. The mean energy of the reactor antineutrinos is about 3 MeV and the flux is known precisely from the power produced by the reactor (which is closely monitored). The $\overline{\nu}_e$ can be detected through the inverse β-decay process,

$$\overline{\nu}_e + p \to e^+ + n.$$

If the $\overline{\nu}_e$ oscillate to other neutrino flavours, they will not be detected since the neutrino energy is well below threshold to produce a muon or tau-lepton the final state. Hence it is only possible to observe the disappearance of reactor $\overline{\nu}_e$. The $\overline{\nu}_e$ survival probability is given by (13.27), which with the approximation $\Delta_{31} \approx \Delta_{32}$ becomes

$$P(\overline{\nu}_e \to \overline{\nu}_e) \approx 1 - 4|U_{e1}|^2|U_{e2}|^2 \sin^2\Delta_{21} - 4|U_{e3}|^2\left[|U_{e1}|^2 + |U_{e2}|^2\right]\sin^2\Delta_{32}.$$

Using the unitarity relation of (13.18), this can be written as

$$P(\overline{\nu}_e \to \overline{\nu}_e) \approx 1 - 4|U_{e1}|^2|U_{e2}|^2 \sin^2\Delta_{21} - 4|U_{e3}|^2\left[1 - |U_{e3}|^2\right]\sin^2\Delta_{32}, \quad (13.34)$$

which can be expressed in terms of the PMNS matrix elements of (13.33) as

$$\begin{aligned} P(\overline{\nu}_e \to \overline{\nu}_e) &= 1 - 4(c_{12}c_{13})^2(s_{12}c_{13})^2\sin^2\Delta_{21} - 4s_{13}^2(1 - s_{13}^2)\sin^2\Delta_{32} \\ &= 1 - \cos^4(\theta_{13})\sin^2(2\theta_{12})\sin^2\left(\frac{\Delta m_{21}^2 L}{4E_{\overline{\nu}}}\right) - \sin^2(2\theta_{13})\sin^2\left(\frac{\Delta m_{32}^2 L}{4E_{\overline{\nu}}}\right). \end{aligned} \quad (13.35)$$

Figure 13.18 shows the expected $\overline{\nu}_e$ survival probability assuming $\theta_{12} = 30°$, $\theta_{23} = 45°$, $\theta_{13} = 10°$ and

$$\Delta m_{21}^2 = 8 \times 10^{-5}\,\text{eV}^2 \quad \text{and} \quad \Delta m_{32}^2 = 2.5 \times 10^{-3}\,\text{eV}^2.$$

The oscillations occur on two different length scales. The short wavelength component, which depends on $\Delta m_{32}^2$, oscillates with an amplitude of $\sin^2(2\theta_{13})$ about the longer wavelength component, with wavelength determined by $\Delta m_{21}^2$. Hence, measurements of the $\overline{\nu}_e$ survival probability at distances of $\mathcal{O}(1)$ km are sensitive to $\theta_{13}$ and measurements at distances of $\mathcal{O}(100)$ km are sensitive to $\Delta m_{21}^2$ and $\theta_{12}$.

### 13.7.1 The short-baseline reactor experiments

Close to a fission reactor, where the long wavelength contribution to neutrino oscillations has yet to develop, the electron antineutrino survival probability of (13.35) can be approximated by

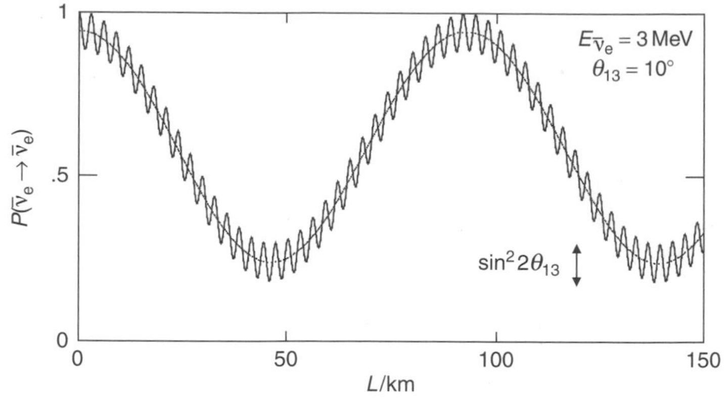


Fig. 13.18 The $P(\overline{\nu}_e \to \overline{\nu}_e)$ survival probability as a function of distance for 3 MeV $\overline{\nu}_e$ assuming $\theta_{13} = 10°$.

$$P(\overline{\nu}_e \to \overline{\nu}_e) \approx 1 - \sin^2(2\theta_{13}) \sin^2\left(\frac{\Delta m_{32}^2 L}{4E_{\overline{\nu}}}\right). \tag{13.36}$$

Until recently, such short-baseline neutrino oscillations had not been observed, and $\theta_{13}$ only was known to be small. The first conclusive observations of a non-zero value of $\theta_{13}$ were published in 2012.

The Daya Bay experiment in China detects neutrinos from six reactor cores each producing 2.9 GW of power. The experiment consists of six detectors, two at a mean flux-weighted distance of 470 m from the reactors, one at 576 m and three at 1.65 km. Each detector consists of a large vessel containing 20 tons of liquid scintillator loaded with gadolinium. The vessels are viewed by arrays of photomultiplier tubes. Electron antineutrinos are detected by the inverse β-decay reaction $\overline{\nu}_e + p \to e^+ + n$. The subsequent annihilation of the positron with an electron gives two *prompt* photons. The low-energy neutron scatters in the liquid scintillator until it is captured by a gadolinium nucleus. The neutron capture, which occurs on a timescale of 100 μs, produces photons from $n + Gd \to Gd^* \to Gd + \gamma$. The photons from both the annihilation process and neutron capture produce Compton scattered electrons. These electrons then ionise the liquid scintillator producing scintillation light. The signature for a $\overline{\nu}_e$ interaction is therefore the coincidence of a prompt pulse of scintillation light from the annihilation and a delayed pulse from the neutron capture 10–100 μs later. The observed amount of prompt light provides a measure of the neutrino energy.

The signal for neutrino oscillations at Daya Bay is a deficit of antineutrinos that depends on the distance from the reactors and a distortion of the observed $e^+$ energy spectrum. By comparing the data recorded in the three *far* detectors at 1.65 km from the reactors, with the data from the three *near* detectors, many systematic uncertainties cancel. In the absence of neutrino oscillations, the rates in the near and far detectors will be compatible and the same energy distribution

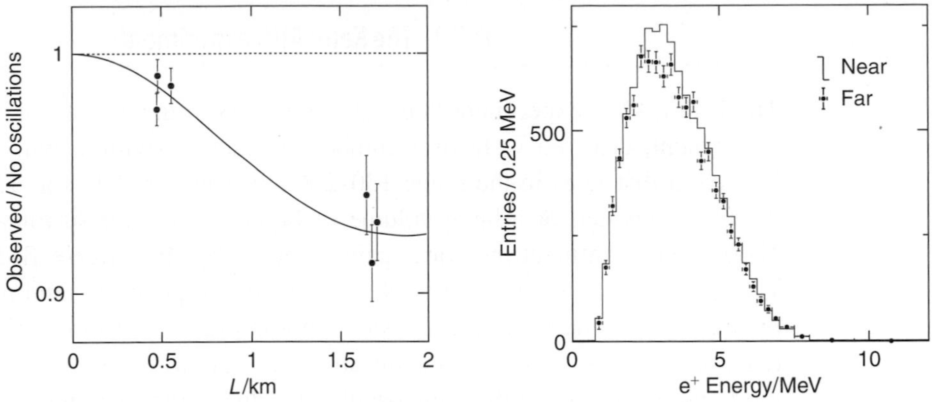


Fig. 13.19 Left: the observed antineutrino rates in the Daya Bay experiment scaled to the expectation for no oscillations, plotted as a function of the flux-weighted distance to the reactors. Right: the observed background-subtracted $e^+$ energy spectrum in the far detectors compared to the corresponding scaled distributions from the near detectors. Adapted from An *et al.* (2012).

will be observed in all detectors. The left-hand plot of Figure 13.19 shows the observed background-subtracted rates in the near and far detectors relative to the unoscillated expectation. The results show a clear deficit of events compared to the unoscillated expectation and this deficit increases with the distance from the reactors. Accounting for scaling of the fluxes with distance, the observed ratio of far/near rates is

$$N_{\text{far}}/N_{\text{near}} = 0.940 \pm 0.012.$$

The right-hand plot of Figure 13.19 shows the observed $e^+$ energy spectrum in the far detectors compared to that in the near detectors, scaled to the same integrated neutrino flux. A clear difference is observed, with the maximum deficit in the far detectors occurring in the 2–4 MeV range, consistent with neutrino oscillations with the known value of $\Delta m^2_{32} = 2.3 \times 10^{-3}\,\text{eV}^2$ (see Section 13.8). The observed ratio of far-to-near event rates gives $\sin^2(2\theta_{13}) = 0.092 \pm 0.017$.

Recent results from the RENO reactor experiment in South Korea, which is similar in design to the Daya Bay experiment, also show a deficit of electron antineutrinos, compatible with $\sin^2(2\theta_{13}) = 0.113 \pm 0.023$, see Ahn *et al.* (2012). Based on the initial Daya Bay and RENO results, it can be concluded that

$$\sin^2(2\theta_{13}) \simeq 0.10 \pm 0.01.$$

Further, albeit less significant, evidence for a non-zero value of $\theta_{13}$ has been provided by the Double-Chooz, MINOS and T2K experiments.

### 13.7.2 The KamLAND experiment

The KamLAND experiment, located in the same mine as the Super-Kamiokande experiment, detected $\overline{\nu}_e$ from a number of reactors (with a total power 70 GW) located at distances in the range 130–240 km from the detector. The KamLAND detector consisted of a large volume of liquid scintillator surrounded by almost 1800 PMTs. Antineutrinos are again detected by the inverse β-decay reaction, $\overline{\nu}_e + p \to e^+ + n$, giving a prompt signal from the positron annihilation followed by a delayed signal from the 2.2 MeV photon produced from the neutron capture reaction, $n + p \to D + \gamma$. At the distances relevant to the KamLAND experiment, the $L/E$ dependence of the rapid oscillations due to the $\Delta m^2_{32}$ term in (13.35) is not resolved because the neutrino sources (the reactors) are not at a single distance $L$ and also because the energy resolution is insufficient to resolve the rapid neutrino energy dependence. Consequently, only the average value of

$$\langle \sin^2 \Delta_{32} \rangle = \tfrac{1}{2},$$

is relevant. Therefore the survival probability of (13.35) can be written

$$\begin{aligned} P(\overline{\nu}_e \to \overline{\nu}_e) &= 1 - \cos^4(\theta_{13}) \sin^2(2\theta_{12}) \sin^2 \Delta_{21} - \tfrac{1}{2} \sin^2(2\theta_{13}) \\ &= \cos^4(\theta_{13}) + \sin^4(\theta_{13}) - \cos^4(\theta_{13}) \sin^2(2\theta_{12}) \sin^2 \Delta_{21}. \end{aligned}$$

Neglecting the $\sin^4(\theta_{13})$ term, which is small ($< 0.001$), gives

$$P(\overline{\nu}_e \to \overline{\nu}_e) \approx \cos^4(\theta_{13}) \left[ 1 - \sin^2(2\theta_{12}) \sin^2 \left( \frac{\Delta m^2_{21} L}{4E_{\overline{\nu}}} \right) \right]. \tag{13.37}$$

Hence, the effective survival probability for reactor neutrinos at large distances has the same form as the two-flavour oscillation formula multiplied by $\cos^4(\theta_{13}) \approx 0.95$.

The KamLAND experiment observed 1609 reactor $\overline{\nu}_e$ interactions compared to the expectation of 2179±89 in the absence of neutrino oscillations. For each event, a measurement of the neutrino energy was obtained from the amount of light associated with the prompt scintillation signal from the positron annihilation. By comparing the energy distribution of the observed data with the expected distribution, the survival probability can be plotted as a function of $L_0/E_{\overline{\nu}}$, where $L_0 = 180$ km is the flux-weighted average distance to the reactors contributing to $\overline{\nu}_e$ interactions in KamLAND, as shown in Figure 13.20. The range of $L/E$ sampled is determined by the energies of the neutrinos produced in nuclear reactors, ~2–7 MeV. The data show a clear oscillation signal with a decrease and subsequent rise in the mean oscillation probability. The measured distribution can be compared to the expectation of (13.37) after accounting for the experimental energy resolution and range of distances sampled, which smears out the effect of the oscillations.

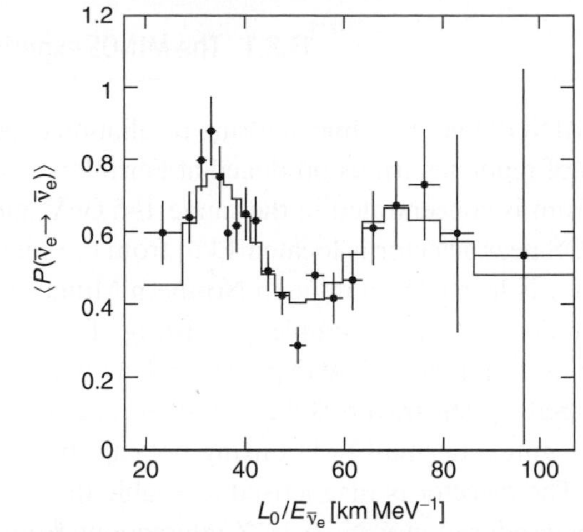


Fig. 13.20 KamLAND data showing the measured mean survival probability as a function of the measured neutrino energy divided by the flux-weighted mean distance to the reactors, $L_0$. The histogram shows the expected distribution for the oscillation parameters that best describe the data. Adapted from Abe *et al.* (2008).

The location of the minimum at $L_0/E_{\overline{\nu}_e} \sim 50\,\text{km MeV}^{-1}$ provides a tight constraint on $\Delta m^2_{21} = m^2_2 - m^2_1$,

$$\Delta m^2_{21} = (7.6 \pm 0.2) \times 10^{-5}\,\text{eV}^2.$$

A measurement of the mixing angle $\theta_{12}$ can also be obtained, which when combined with the more precise determination from the solar neutrino data of SNO (see Section 13.2) gives

$$\sin^2(2\theta_{12}) = 0.87 \pm 0.04.$$

## 13.8 Long-baseline neutrino experiments

In recent years, intense accelerator-based neutrino beams, produced in a similar manner to that described in Section 12.2, have been used to study neutrino oscillations. One advantage of a neutrino beam experiment is that the energy spectrum can be tailored to a specific measurement. Long-baseline neutrino oscillation experiments typically use two detectors, one sufficiently close to the source of the beam to allow a measurement of the unoscillated neutrino energy spectrum, and one far from the source to measure the oscillated spectrum. The use of a near and far detector means that many systematic uncertainties cancel, allowing precise measurements to be made.

### 13.8.1 The MINOS experiment

The MINOS long-baseline neutrino oscillation experiment uses an intense 0.3 MW beam of muon neutrinos produced at Fermilab near Chicago. The neutrino energy spectrum is concentrated in the range 1–5 GeV and peaks at 3 GeV. The 1000 ton MINOS near detector is located 1 km from the source and the 5400 ton MINOS far detector is located in a mine in Northern Minnesota, 735 km from the source. The detectors are relatively simple, consisting of planes of iron, which provide the bulk of the mass, interleaved with planes of 4 cm wide strips of plastic scintillator. When a charged particle traverses the scintillator, light is produced. This scintillation light is transmitted to small PMTs using optical fibres that are embedded in the scintillator. The detector is magnetised to enable the measurement of the momentum of muons produced in $\nu_\mu N \to \mu^- X$ interactions from their curvature. The amount of scintillation light gives a measure of the energy of the hadronic final state $X$ produced in the interaction. Hence, on an event-by-event basis, the neutrino energy is reconstructed, $E_\nu = E_\mu + E_X$. An example of a neutrino interaction in the MINOS detector is shown in Figure 13.21.

MINOS studied the neutrino oscillations of an almost pure $\nu_\mu$ beam. Because $\theta_{13}$ is relatively small, $\nu_\mu \to \nu_\tau$ oscillations dominate. Since $L$ is fixed, the oscillations are observed as a distortion of the energy spectrum. It is found that the first maximum of the oscillation probability occurs at 1.3 GeV. Despite the fact that the oscillations are dominated by $\nu_\mu \to \nu_\tau$, most of the oscillated $\nu_\tau$ are below threshold for producing a tau-lepton and therefore MINOS makes a disappearance measurement of $|\Delta m^2_{32}|$ and $\theta_{32}$. With the approximation $\Delta_{31} \approx \Delta_{32}$, the $\nu_\mu \to \nu_\mu$ survival probability is given by (13.34) with the $U_{ei}$ replaced by $U_{\mu i}$,

$$P(\nu_\mu \to \nu_\mu) \approx 1 - 4|U_{\mu 1}|^2 |U_{\mu 2}|^2 \sin^2 \Delta_{21} - 4|U_{\mu 3}|^2 (1 - |U_{\mu 3}|^2) \sin^2 \Delta_{32}.$$

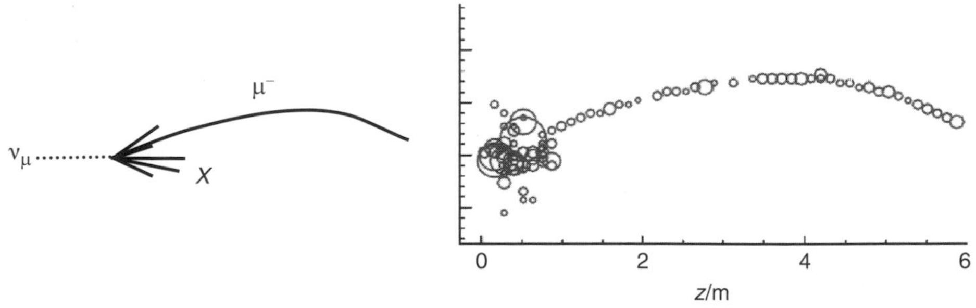


Fig. 13.21 A $\nu_\mu$ charged-current weak interaction, $\nu_\mu N \to \mu^- X$, in the MINOS detector. The sizes of the circles indicate the amount of light recorded in the scintillator strips. The muon momentum is determined from the curvature in the magnetic field and the energy of the hadronic system from the amount of light close to the interaction vertex.

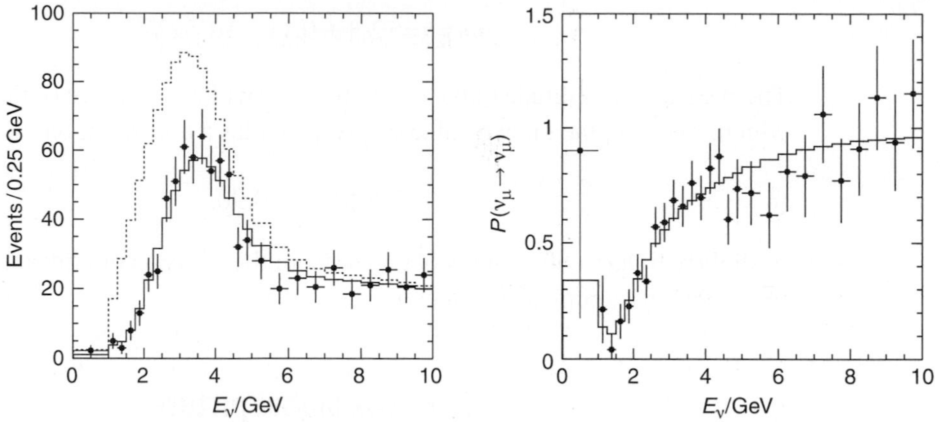


Fig. 13.22 Left: the MINOS far detector energy spectrum compared to the unoscillated prediction (dashed). Right: the oscillation probability as measured from the ratio of the far detector data to the unoscillated prediction. Adapted from Adamson *et al.* (2011).

For the MINOS experiment, with $L = 735\,\text{km}$ and $E_\nu > 1\,\text{GeV}$, the contribution to the oscillation probability from the long wavelength component associated with $\Delta_{21}$ can be neglected and therefore

$$P(\nu_\mu \to \nu_\mu) \approx 1 - 4|U_{\mu3}|^2(1 - |U_{\mu3}|^2)\sin^2\Delta_{32}.$$

Using the parameterisation of the PMNS matrix given in (13.33), this can be written

$$\begin{aligned} P(\nu_\mu \to \nu_\mu) &= 1 - 4\sin^2(\theta_{23})\cos^2(\theta_{13})\left[1 - \sin^2(\theta_{23})\cos^2(\theta_{13})\right]\sin^2\Delta_{32} \\ &= 1 - \left[\sin^2(2\theta_{23})\cos^4(\theta_{13}) + \sin^2(2\theta_{13})\sin^2(\theta_{23})\right]\sin^2\Delta_{32} \\ &\approx 1 - A\sin^2\left(\frac{\Delta m^2_{32}L}{4E_\nu}\right), \end{aligned} \tag{13.38}$$

where $A = \sin^2(2\theta_{23})\cos^4(\theta_{13}) + \sin^2(2\theta_{13})\sin^2(\theta_{23})$. Because $\theta_{13}$ is relatively small the dominant term in the amplitude of the oscillations is from $\sin^2(2\theta_{23})\cos^4(\theta_{13})$.

By comparing the energy spectrum of charged-current neutrino interactions in the near and far detectors, MINOS directly measures the oscillation probability as a function of $E_\nu$. Figure 13.22 shows the measured far detector energy spectrum compared to the expected spectrum for no oscillations, determined from the unoscillated near detector data. A clear deficit of neutrinos is observed at low energies, where the oscillation probability is highest. The right plot of Figure 13.22 shows the ratio of the measured far detector energy spectrum to the expectation without neutrino oscillations that is obtained from the near detector data. This provides a direct measurement of the survival probability $P(\nu_\mu \to \nu_\mu)$, albeit slightly smeared out by the experimental energy resolution. The position of the minimum in the measured oscillation curve at $E_\nu \sim 1.5\,\text{GeV}$ determines

$$|\Delta m_{32}^2| = (2.3 \pm 0.1) \times 10^{-3}\,\mathrm{eV}^2.$$

The measured amplitude of the oscillations provides a measure of the parameter $A$, which, by using the known value of $\theta_{13}$, provides a constraint on the mixing angle

$$\sin^2(2\theta_{23}) \gtrsim 0.90.$$

A slightly tighter constraint is obtained from the analysis of atmospheric neutrinos in the Super-Kamiokande detector.

## 13.9 The global picture

For reasons of space, it has only been possible to describe a few notable experiments that provide an illustration of the main experimental techniques used to study neutrino oscillations; there are other experiments. For example, the CERN to Gran Sasso neutrino experiment (CNGS) is searching for $\nu_\mu \to \nu_\tau$ appearance. At the time of writing, two candidate $\nu_\tau$ interactions have been observed in the OPERA detector of the CNGS experiment. Furthermore, the T2K experiment in Japan is studying $\nu_\mu$ disappearance and $\nu_\mu \to \nu_e$ appearance in an intense beam.

When the results from all experiments are taken together, a detailed picture of the properties of neutrinos emerges. The existence of neutrino oscillations implies that the neutrinos have mass, even if the masses are very small. The differences of the squares of the neutrino masses have been measured to better than 5% by the KamLAND and MINOS experiments,

$$m_2^2 - m_1^2 = (7.6 \pm 0.2) \times 10^{-5}\,\mathrm{eV}^2,$$
$$|m_3^2 - m_2^2| = (2.3 \pm 0.1) \times 10^{-3}\,\mathrm{eV}^2.$$

Three of the four parameters of the PMNS matrix, describing the lepton flavour sector of the Standard Model, have been determined. From the recent results of the SNO, KamLAND, Super-Kamiokande, MINOS, Daya Bay, RENO and Double-Chooz experiments it is known that

$$\sin^2(2\theta_{12}) = 0.87 \pm 0.04,$$
$$\sin^2(2\theta_{23}) > 0.92,$$
$$\sin^2(2\theta_{13}) \approx 0.10 \pm 0.01.$$

From the above measurements, the magnitudes of the elements of PMNS matrix are determined to be approximately

$$\begin{pmatrix} |U_{e1}| & |U_{e2}| & |U_{e3}| \\ |U_{\mu 1}| & |U_{\mu 2}| & |U_{\mu 3}| \\ |U_{\tau 1}| & |U_{\tau 2}| & |U_{\tau 3}| \end{pmatrix} \sim \begin{pmatrix} 0.85 & 0.50 & 0.17 \\ 0.35 & 0.60 & 0.70 \\ 0.35 & 0.60 & 0.70 \end{pmatrix}. \tag{13.39}$$

The final parameter of the PMNS matrix, the phase $\delta$, is not yet known. The focus of the next generation of experiments will be to measure this phase and thus establish whether CP is violated in leptonic weak interactions.

## Summary

The $\nu_e$, $\nu_\mu$ and $\nu_\tau$ are not fundamental particle states, but are mixtures of the mass eigenstates, $\nu_1$, $\nu_2$ and $\nu_3$. The relationship between the weak and mass eigenstates is determined by the unitary PMNS matrix

$$\begin{pmatrix} \nu_e \\ \nu_\mu \\ \nu_\tau \end{pmatrix} = \begin{pmatrix} U_{e1} & U_{e2} & U_{e3} \\ U_{\mu 1} & U_{\mu 2} & U_{\mu 3} \\ U_{\tau 1} & U_{\tau 2} & U_{\tau 3} \end{pmatrix} \begin{pmatrix} \nu_1 \\ \nu_2 \\ \nu_3 \end{pmatrix}.$$

The PMNS matrix can be expressed in terms of four fundamental parameters of the Standard Model; three rotation angles, $\theta_{12}$, $\theta_{13}$ and $\theta_{23}$, and a complex phase $\delta$ that admits the possibility of CP violation in the leptonic sector.

Neutrinos propagate as coherent linear superpositions of the mass eigenstates, for example

$$|\nu_e\rangle = U_{e1}|\nu_1\rangle e^{-i\phi_1} + U_{e2}|\nu_2\rangle e^{-i\phi_2} + U_{e3}|\nu_3\rangle e^{-i\phi_3}.$$

If $m(\nu_1) \neq m(\nu_2) \neq m(\nu_3)$, phase differences develop between the different components, giving rise to the observable effect of neutrino oscillations, with oscillation probabilities of the form

$$P(\nu_e \to \nu_\mu) = \sin^2(2\theta)\sin^2\left(1.27\frac{\Delta m^2[\mathrm{eV}^2]L[\mathrm{km}]}{E_\nu[\mathrm{GeV}]}\right).$$

The study of neutrino oscillations provides a determination of the differences in the squares of the neutrino masses

$$m_2^2 - m_1^2 \approx 7.6 \times 10^{-5}\,\mathrm{eV}^2 \quad \text{and} \quad |m_3^2 - m_2^2| \approx 2.3 \times 10^{-3}\,\mathrm{eV}^2,$$

and measurements of the mixing angles of the PMNS matrix

$$\theta_{12} \approx 35^\circ,\ \ \theta_{23} \approx 45^\circ \ \text{ and } \ \theta_{13} \approx 10^\circ.$$

## Problems

**13.1** By writing $p_1 = \beta E_1$ and $p_2 = \beta E_2$, and assuming $\beta_1 = \beta_2 = \beta$, show that Equation (13.13) reduces to (13.12), i.e.

$$\Delta\phi_{12} = (E_1 - E_2)\left[T - \left(\frac{E_1 + E_2}{p_1 + p_1}\right)L\right] + \left(\frac{m_1^2 - m_2^2}{p_1 + p_2}\right)L \approx \frac{m_1^2 - m_2^2}{2p}L,$$

where $p = p_1 \approx p_2$ and it is assumed that $p_1 \gg m_1$ and $p_2 \gg m_2$.

**13.2** Show that when $L$ is given in km and $\Delta m^2$ is given in $eV^2$, the two-flavour oscillation probability expressed in natural units becomes

$$\sin^2(2\theta)\sin^2\left(\frac{\Delta m^2[\mathrm{GeV}^2]L[\mathrm{GeV}^{-1}]}{4E_\nu[\mathrm{GeV}]}\right) \rightarrow \sin^2(2\theta)\sin^2\left(1.27\frac{\Delta m^2[\mathrm{eV}^2]L[\mathrm{km}]}{4E_\nu[\mathrm{GeV}]}\right).$$

**13.3** From Equation (13.24) and the unitarity relation of (13.18), show that

$$\begin{aligned} P(\nu_e \rightarrow \nu_e) = 1 &+ 2|U_{e1}|^2|U_{e2}|^2\,\Re e\{[e^{-i(\phi_1-\phi_2)} - 1]\} \\ &+ 2|U_{e1}|^2|U_{e3}|^2\,\Re e\{[e^{-i(\phi_1-\phi_3)} - 1]\} \\ &+ 2|U_{e2}|^2|U_{e3}|^2\,\Re e\{[e^{-i(\phi_2-\phi_3)} - 1]\}. \end{aligned}$$

**13.4** Derive Equation (13.30) in the following three steps.

(a) By writing the oscillation probability $P(\nu_e \rightarrow \nu_\mu)$ as

$$P(\nu_e \rightarrow \nu_\mu) = 2\sum_{i<j}\Re e\left\{U_{ei}^*U_{\mu i}U_{ej}U_{\mu j}^*\left[e^{i(\phi_j-\phi_i)} - 1\right]\right\},$$

and writing $\Delta_{ij} = (\phi_i - \phi_j)/2$, show that

$$\begin{aligned} P(\nu_e \rightarrow \nu_\mu) = &-4\sum_{i<j}\Re e\{U_{ei}^*U_{\mu i}U_{ej}U_{\mu j}^*\}\sin^2\Delta_{ij} \\ &+ 2\sum_{i<j}\Im m\{U_{ei}^*U_{\mu i}U_{ej}U_{\mu j}^*\}\sin 2\Delta_{ij}. \end{aligned}$$

(b) Defining $-J \equiv \Im m\{U_{e1}^*U_{\mu 1}U_{e3}U_{\mu 3}^*\}$, use the unitarity of the PMNS matrix to show that

$$\Im m\{U_{e1}^*U_{\mu 1}U_{e3}U_{\mu 3}^*\} = -\Im m\{U_{e2}^*U_{\mu 2}U_{e3}U_{\mu 3}^*\} = -\Im m\{U_{e1}^*U_{\mu 1}U_{e2}U_{\mu 2}^*\} = -J.$$

(c) Hence, using the identity

$$\sin A + \sin B - \sin(A+B) = 4\sin\left(\frac{A}{2}\right)\sin\left(\frac{B}{2}\right)\sin\left(\frac{A+B}{2}\right),$$

show that

$$P(\nu_e \rightarrow \nu_\mu) = -4\sum_{i<j}\Re e\{U_{ei}^*U_{\mu i}U_{ej}U_{\mu j}^*\}\sin^2\Delta_{ij} + 8J\sin\Delta_{12}\sin\Delta_{13}\sin\Delta_{23}.$$

(d) Hence show that

$$P(\nu_e \rightarrow \nu_\mu) - P(\overline{\nu}_e \rightarrow \overline{\nu}_\mu) = 16\,\Im m\{U_{e1}^*U_{\mu 1}U_{e2}U_{\mu 2}^*\}\sin\Delta_{12}\sin\Delta_{13}\sin\Delta_{23}.$$

(e) Finally, using the current knowledge of the PMNS matrix determine the maximum possible value of $P(\nu_e \to \nu_\mu) - P(\overline{\nu}_e \to \overline{\nu}_\mu)$.

**13.5** The general unitary transformation between mass and weak eigenstates for two flavours can be written as

$$\begin{pmatrix} \nu_e \\ \nu_\mu \end{pmatrix} = \begin{pmatrix} \cos\theta \exp(i\delta_1) & \sin\theta \exp\left(i\left[\frac{\delta_1+\delta_2}{2} - \delta\right]\right) \\ -\sin\theta \exp\left(i\left[\frac{\delta_1+\delta_2}{2} + \delta\right]\right) & \cos\theta \exp(i\delta_2) \end{pmatrix} \begin{pmatrix} \nu_1 \\ \nu_2 \end{pmatrix}.$$

(a) Show that the matrix in the above expression is indeed unitary.
(b) Show that the three complex phases $\delta_1$, $\delta_2$ and $\delta$ can be eliminated from the above expression by the transformation

$$\ell_\alpha \to \ell_\alpha e^{i(\theta_e+\theta'_\alpha)}, \quad \nu_k \to \nu_k e^{i(\theta_e+\theta'_k)} \quad \text{and} \quad U_{\alpha k} \to U_{\alpha k} e^{i(\theta'_\alpha - \theta'_k)},$$

without changing the physical form of the two-flavour weak charged current

$$-i\frac{g_W}{\sqrt{2}}(\overline{e}, \overline{\mu})\gamma^\mu \tfrac{1}{2}(1-\gamma^5)\begin{pmatrix} U_{e1} & U_{e2} \\ U_{\mu 1} & U_{\mu 2} \end{pmatrix}\begin{pmatrix} \nu_1 \\ \nu_2 \end{pmatrix}.$$

**13.6** The derivations of (13.37) and (13.38) used the trigonometric relations

$$1 - \tfrac{1}{2}\sin^2(2\theta_{13}) = \cos^4(\theta_{13}) + \sin^4(\theta_{13}),$$

and

$$4\sin^2\theta_{23}\cos^2\theta_{13}(1 - \sin^2\theta_{23}\cos^2\theta_{13}) = (\sin^2 2\theta_{23}\cos^4\theta_{13} + \sin^2 2\theta_{13}\sin^2\theta_{23}).$$

Convince yourself these relations hold.

**13.7** Use the data of Figure 13.20 to obtain estimates of $\sin^2(2\theta_{12})$ and $|\Delta m^2_{21}|$.

**13.8** Use the data of Figure 13.22 to obtain estimates of $\sin^2(2\theta_{23})$ and $|\Delta m^2_{32}|$.

**13.9** The T2K experiment uses an off-axis $\nu_\mu$ beam produced from $\pi^+ \to \mu^+\nu_\mu$ decays. Consider the case where the pion has velocity $\beta$ along the $z$-direction in the laboratory frame and a neutrino with energy $E^*$ is produced at an angle $\theta^*$ with respect to the $z'$-axis in the $\pi^+$ rest frame.

(a) Show that the neutrino energy in the pion rest frame is $p^* = (m_\pi^2 - m_\mu^2)/2m_\pi$.
(b) Using a Lorentz transformation, show that the energy $E$ and angle of production $\theta$ of the neutrino in the laboratory frame are

$$E = \gamma E^*(1 + \beta\cos\theta^*) \quad \text{and} \quad E\cos\theta = \gamma E^*(\cos\theta^* + \beta),$$

where $\gamma = E_\pi/m_\pi$.
(c) Using the expressions for $E^*$ and $\theta^*$ in terms of $E$ and $\theta$, show that

$$\gamma^2(1 - \beta\cos\theta)(1 + \beta\cos\theta^*) = 1.$$

(d) Show that maximum value of $\theta$ in the laboratory frame is $\theta_{max} = 1/\gamma$.
(e) In the limit $\theta \ll 1$, show that

$$E \approx 0.43 E_\pi \frac{1}{1 + \beta\gamma^2\theta^2},$$

and therefore on-axis ($\theta = 0$) the neutrino energy spectrum follows that of the pions.
(f) Assuming that the pions have a flat energy spectrum in the range 1–5 GeV, sketch the form of the resulting neutrino energy spectrum at the T2K far detector (Super-Kamiokande), which is off-axis at $\theta = 2.5°$. Given that the Super-Kamiokande detector is 295 km from the beam, explain why this angle was chosen.

# 14 CP violation and weak hadronic interactions

CP violation is an essential part of our understanding of both particle physics and the evolution of the early Universe. It is required to explain the observed dominance of matter over antimatter in the Universe. In the Standard Model, the only place where CP violating effects can be accommodated is in the weak interactions of quarks and leptons. This chapter describes the weak charged-current interactions of the quarks and concentrates on the observations of CP violation in the neutral kaon and B-meson systems. This is not an easy topic and it is developed in several distinct stages. The detailed quantum mechanical derivations of the mixing of neutral meson states are given in two starred sections.

## 14.1 CP violation in the early Universe

The atoms in our local region of the Universe are formed from electrons, protons and neutrons rather than their equivalent antiparticles. The possibility that there are galaxies and/or regions of space dominated by antimatter can be excluded by the astronomical searches for photons from the $e^+e^-$ annihilation process that would occur at the interfaces between matter and antimatter dominated regions of the Universe. The predominance of matter is believed to have arisen in the early evolution of the Universe.

In the early Universe, when the thermal energy $k_B T$ was large compared to the masses of the hadrons, there were an equal number of baryons and antibaryons. The baryons and antibaryons were initially in thermal equilibrium with the soup of relatively high-energy photons that pervaded the early Universe, through processes such as

$$\gamma + \gamma \rightleftharpoons \mathrm{p} + \bar{\mathrm{p}}. \tag{14.1}$$

As the Universe expanded, its temperature decreased as did the mean energy of the photons. At some point, the forward reaction of (14.1) effectively ceased. Furthermore, with the expansion, the number density of baryons and antibaryons also decreased and eventually became sufficiently low that annihilation processes such as the backward reaction of (14.1) became very rare. At this point in time, the number of baryons and antibaryons in the Universe was effectively fixed. This process is known as Big Bang baryogenesis. The calculations of the thermal freeze out of the baryons *without* CP violation predict equal number densities of baryons and antibaryons, $n_B = n_{\overline{B}}$, and a baryon to photon number density ratio of

$$n_B = n_{\overline{B}} \sim 10^{-18} n_\gamma.$$

This prediction is in contradiction with the observed matter-dominated Universe, where the baryon–antibaryon asymmetry, which can be inferred from the relative abundances of light isotopes formed in the process of Big Bang nucleosynthesis, is

$$\frac{n_B - n_{\overline{B}}}{n_\gamma} \sim 10^{-9}.$$

Broadly speaking, to generate this asymmetry, for every $10^9$ antibaryons in the early Universe there must have been $10^9 + 1$ baryons, which annihilated to give $\mathcal{O}(10^9)$ photons, leaving 1 baryon.

To explain the observed matter–antimatter asymmetry in the Universe, three conditions, originally formulated by Sakharov (1967), must be satisfied. In the early Universe there must have been: (i) baryon number violation such that $n_B - n_{\overline{B}}$ is not constant; (ii) C and CP violation, because if CP is conserved, for every reaction that creates a net number of baryons over antibaryons there would be a CP conjugate reaction generating a net number of antibaryons over baryons; and (iii) departure from thermal equilibrium, since in thermal equilibrium any baryon number violating process will be balanced by the inverse reaction. The Standard Model of particle physics provides the possibility of CP violation in the weak interactions of quarks and leptons. To date, CP violation has only been observed in the quark sector, where many detailed measurements have been made. Despite the clear observations of CP violating effects in the weak interactions of quarks, this is not sufficient to explain the matter–antimatter asymmetry in the Universe and ultimately another source needs to be identified.

## 14.2 The weak interactions of quarks

In Section 12.1, it was shown that there is a universal coupling strength of the weak interaction to charged leptons and the corresponding neutrino weak eigenstates; $G_F^{(e)} = G_F^{(\mu)} = G_F^{(\tau)}$. The strength of the weak interaction for quarks can be determined from the study of nuclear β-decay, where $|\mathcal{M}|^2 \propto G_F^{(e)} G_F^{(\beta)}$ and $G_F^{(\beta)}$ gives the coupling at the weak interaction vertex of the quarks in Figure 14.1. From the observed β-decay rates for superallowed nuclear transitions, the strength of the

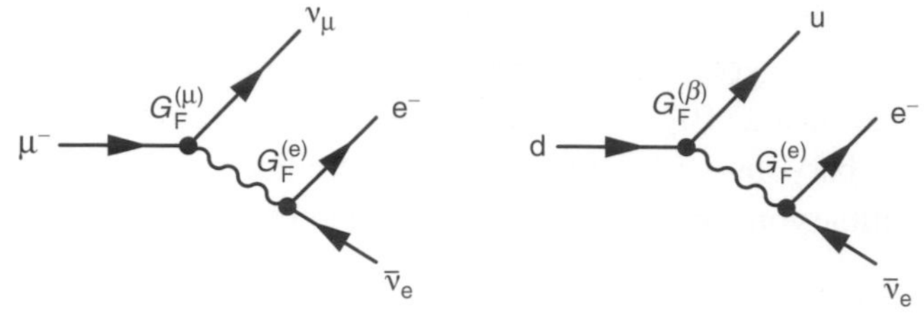


Fig. 14.1 The lowest-order Feynman diagrams for $\mu^-$-decay and the underlying quark-level process in nuclear $\beta^-$-decay.

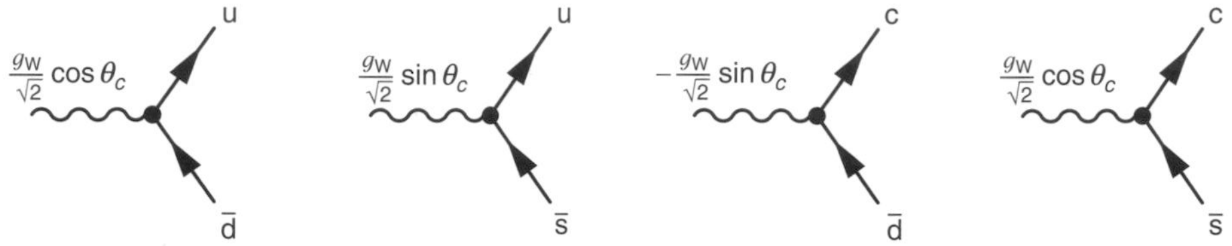


Fig. 14.2 The weak interaction couplings of the d, s, u and c in terms of the Cabibbo angle, $\theta_c$.

coupling at the ud quark weak interaction vertex is found to be 5% smaller than that at the $\mu^-\nu_\mu$ vertex,

$$G_F^{(\mu)} = (1.166\,3787 \pm 0.000\,0006) \times 10^{-5}\,\text{GeV}^{-2},$$
$$G_F^{(\beta)} = (1.1066 \pm 0.0011) \times 10^{-5}\,\text{GeV}^{-2}.$$

Furthermore, different coupling strengths are found for the ud and us weak charged-current vertices. For example, the measured decay rate for $K^-(u\overline{s}) \to \mu^-\overline{\nu}_\mu$ compared to that of $\pi^-(u\overline{d}) \to \mu^-\overline{\nu}_\mu$ is approximately a factor 20 smaller than would be expected for a universal weak coupling to the quarks. These observations were originally explained by the Cabibbo hypothesis. In the Cabibbo hypothesis, the weak interactions of quarks have the same strength as the leptons, but the weak eigenstates of quarks differ from the mass eigenstates. The weak eigenstates, labelled d′ and s′, are related to the mass eigenstates, d and s, by the unitary matrix,

$$\begin{pmatrix} d' \\ s' \end{pmatrix} = \begin{pmatrix} \cos\theta_c & \sin\theta_c \\ -\sin\theta_c & \cos\theta_c \end{pmatrix} \begin{pmatrix} d \\ s \end{pmatrix}, \tag{14.2}$$

where $\theta_c$ is known as the Cabibbo angle. This idea is very similar to the two-flavour mixing of the neutrino mass and weak eigenstates encountered in Section 13.4. In the Cabibbo model, the weak interactions of quarks are described by ud′ and cs′ couplings, shown in Figure 14.2.

Nuclear β-decay involves the weak coupling between u and d quarks. Therefore, with the Cabibbo hypothesis, β-decay matrix elements are proportional to $g_W \cos\theta_c$ and decay rates are proportional to $G_F \cos^2\theta_c$. Similarly, the matrix

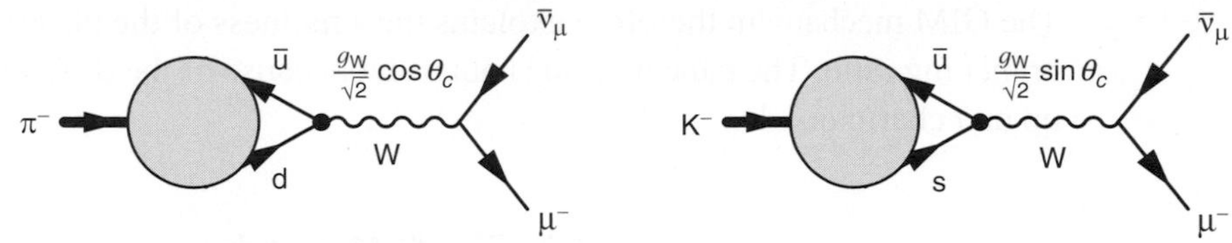


Fig. 14.3 The main decay modes of the $\pi^-$ and $K^-$.

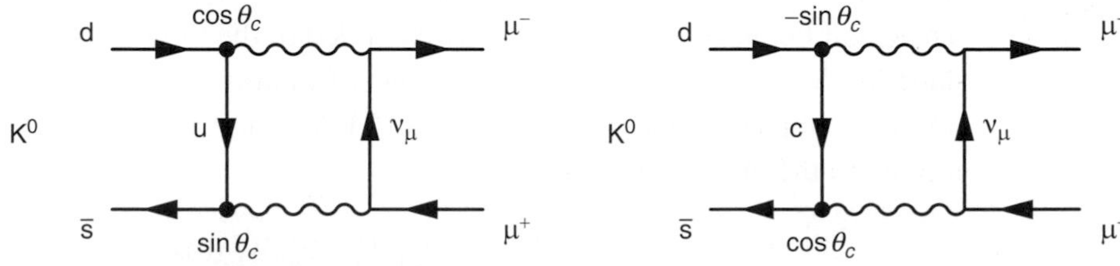


Fig. 14.4 Two box diagrams for the decay $K_L \rightarrow \mu^+\mu^-$. The distinction between the $K_L$ and the $K^0$ is described in Section 14.4.

elements for the decays $K^- \rightarrow \mu^-\overline{\nu}_\mu$ and $\pi^- \rightarrow \mu^-\overline{\nu}_\mu$, shown in Figure 14.3, respectively include factors of $\cos\theta_c$ and $\sin\theta_c$. Consequently, after accounting for the difference in phase space, the $K^-$ decay rate is suppressed by a factor of $\tan^2\theta_c$ relative to that for the $\pi^-$. The observed β-decay rates and the measured ratio of $\Gamma(K^-(u\overline{s}) \rightarrow \mu^-\overline{\nu}_\mu)$ to $\Gamma(\pi^-(ud) \rightarrow \mu^-\overline{\nu}_\mu)$ can be explained by the Cabibbo hypothesis with $\theta_c \simeq 13°$.

When the Cabibbo mechanism was first proposed, the charm quark had not been discovered. Since the Cabibbo mechanism allows for ud and us couplings, the flavour changing neutral-current (FCNC) decay of the neutral kaon $K_L \rightarrow \mu^+\mu^-$ can occur via the exchange of a virtual up-quark, as shown in the first box diagram of Figure 14.4. However, the observed branching ratio,

$$BR(K_L \rightarrow \mu^+\mu^-) = (6.84 \pm 0.11) \times 10^{-9},$$

is much smaller than expected from this diagram alone. This observation was explained by the GIM mechanism; see Glashow, Iliopoulos and Maiani (1970). In the GIM mechanism, which was formulated before the discovery of the charm quark, a postulated fourth quark coupled to the s′ weak eigenstate. In this case, the decay $K_L \rightarrow \mu^+\mu^-$ can also proceed via the exchange of a virtual charm quark, as shown in the second box diagram of Figure 14.4. The matrix elements for the two $K_L \rightarrow \mu^+\mu^-$ box diagrams are respectively

$$\mathcal{M}_u \propto g_W^4 \cos\theta_c \sin\theta_c \quad \text{and} \quad \mathcal{M}_c \propto -g_W^4 \cos\theta_c \sin\theta_c.$$

Because both diagrams give the same final state, the amplitudes must be summed

$$|\mathcal{M}|^2 = |\mathcal{M}_u + \mathcal{M}_c|^2 \approx 0.$$

The GIM mechanism therefore explains the smallness of the observed $K_L \to \mu^+\mu^-$ branching ratio. The cancellation is not exact because of the different masses of the up and charm quarks.

## 14.3 The CKM matrix

The Cabibbo mechanism is naturally extended to the three generations of the Standard Model, where the weak interactions of quarks are described in terms of the unitary Cabibbo–Kobayashi–Maskawa (CKM) matrix. The weak eigenstates are related to the mass eigenstates by

$$\begin{pmatrix} \mathrm{d}' \\ \mathrm{s}' \\ \mathrm{b}' \end{pmatrix} = \begin{pmatrix} V_{\mathrm{ud}} & V_{\mathrm{us}} & V_{\mathrm{ub}} \\ V_{\mathrm{cd}} & V_{\mathrm{cs}} & V_{\mathrm{cb}} \\ V_{\mathrm{td}} & V_{\mathrm{ts}} & V_{\mathrm{tb}} \end{pmatrix} \begin{pmatrix} \mathrm{d} \\ \mathrm{s} \\ \mathrm{b} \end{pmatrix}. \tag{14.3}$$

Consequently, the weak charged-current vertices involving quarks are given by

$$-i\frac{g_W}{\sqrt{2}}(\bar{\mathrm{u}}, \bar{\mathrm{c}}, \bar{\mathrm{t}})\gamma^\mu \tfrac{1}{2}(1-\gamma^5) \begin{pmatrix} V_{\mathrm{ud}} & V_{\mathrm{us}} & V_{\mathrm{ub}} \\ V_{\mathrm{cd}} & V_{\mathrm{cs}} & V_{\mathrm{cb}} \\ V_{\mathrm{td}} & V_{\mathrm{ts}} & V_{\mathrm{tb}} \end{pmatrix} \begin{pmatrix} \mathrm{d} \\ \mathrm{s} \\ \mathrm{b} \end{pmatrix},$$

where, for example, d is a down-quark spinor and $\bar{\mathrm{u}}$ is the adjoint spinor for an up-quark. The relative strength of the interaction is defined by the relevant element of the CKM matrix. For example, the weak charged-current associated with the duW vertex shown in the top left plot of Figure 14.5 is

$$j^\mu_{\mathrm{du}} = -i\tfrac{g_W}{\sqrt{2}} V_{\mathrm{ud}}\, \bar{\mathrm{u}}\gamma^\mu \tfrac{1}{2}(1-\gamma^5)\mathrm{d}.$$

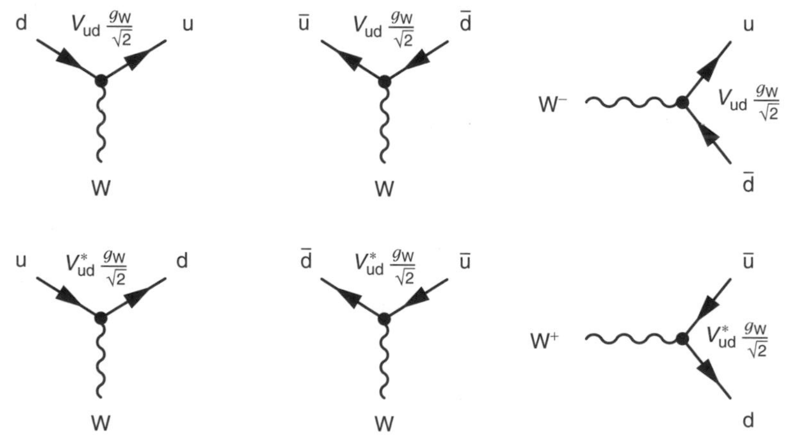


Fig. 14.5 The charged-current weak interaction vertices involving u and d quarks.

The CKM matrix is defined such that the associated vertex factor contains $V_{\rm ud}$ when the charge $-\frac{1}{3}$ quark enters the weak current as the spinor. If the charge $-\frac{1}{3}$ quark is represented by an adjoint spinor, $\overline{\rm d} = {\rm d}^\dagger\gamma^0$, the vertex factor from the CKM matrix is $V^*_{\rm ud}$. For example, the current associated with the vertex in the bottom left plot of Figure 14.5 is

$$j^\mu_{\rm ud} = -i\frac{g_{\rm W}}{\sqrt{2}}\,V^*_{\rm ud}\,\overline{\rm d}\gamma^\mu\tfrac{1}{2}(1-\gamma^5){\rm u}.$$

The CKM matrix, which is the analogous to the PMNS matrix for the weak interactions of leptons, is unitary and can be described by three rotation angles and a complex phase,

$$V_{\rm CKM} = \begin{pmatrix} V_{\rm ud} & V_{\rm us} & V_{\rm ub} \\ V_{\rm cd} & V_{\rm cs} & V_{\rm cb} \\ V_{\rm td} & V_{\rm ts} & V_{\rm tb} \end{pmatrix} = \begin{pmatrix} 1 & 0 & 0 \\ 0 & c_{23} & s_{23} \\ 0 & -s_{23} & c_{23} \end{pmatrix} \times \begin{pmatrix} c_{13} & 0 & s_{13}e^{-i\delta'} \\ 0 & 1 & 0 \\ -s_{13}e^{i\delta'} & 0 & c_{13} \end{pmatrix} \times \begin{pmatrix} c_{12} & s_{12} & 0 \\ -s_{12} & c_{12} & 0 \\ 0 & 0 & 1 \end{pmatrix}, \tag{14.4}$$

where $s_{ij} = \sin\phi_{ij}$ and $c_{ij} = \cos\phi_{ij}$.

Whilst the structure of the weak interactions of quarks and leptons is the same, the phenomenology is very different. Quarks do not propagate as free particles, but hadronise on a length scale of $10^{-15}$ m. Consequently, the final states of weak interactions involving quarks have to be described in terms of mesons or baryons. The observed hadronic states are composed of particular quark flavours and, therefore, it is the quark mass (flavour) eigenstates that form the observable quantities in hadronic weak interactions. Consequently, the nine individual elements of the CKM matrix can be measured separately. For example, $V_{\rm ud}$ is determined from superallowed nuclear β-decays,

$$|V_{\rm ud}| = \cos\theta_c = 0.974\,25(22).$$

The weak coupling between the u and s quarks can be determined from the measured branching ratio of the $\rm K^0 \to \pi^- e^+ \nu_e$ decay shown in Figure 14.6a,

$$|V_{\rm us}| = 0.225\,2(9).$$

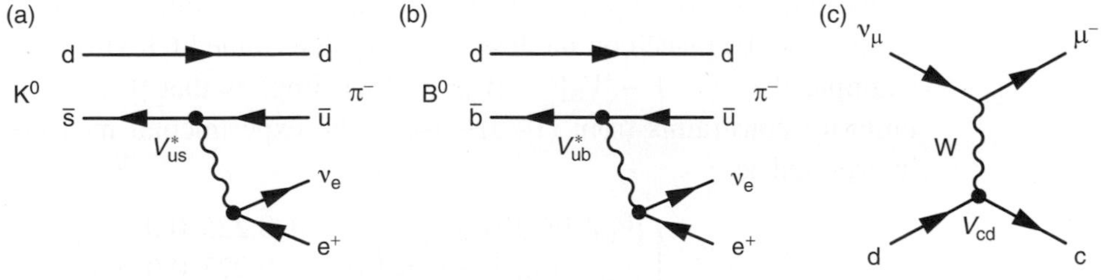


Fig. 14.6 The Feynman diagrams for a) $\rm K^0 \to \pi^- e^+ \nu_e$, b) $\rm B^0 \to \pi^- e^+ \nu_e$ and c) $\nu_\mu {\rm d} \to \mu^- {\rm c}$.

The large numbers of $B^0(d\bar{b})$ and $\overline{B}^0(b\bar{d})$ mesons produced at the BaBar and Belle experiments, described in Section 14.6.3, allow precise measurements of the branching ratios for decays such as $B^0 \to \pi^- e^+ \nu_e$, shown in Figure 14.6b. The measurements of the inclusive and exclusive branching ratios of the B-mesons imply

$$|V_{ub}| = (4.15 \pm 0.49) \times 10^{-3}.$$

The CKM matrix element $V_{cs}$ can be determined from the leptonic decays of the $D_s^+(u\bar{s})$ meson, for example $D_s^+ \to \mu^+\nu_\mu$, and $V_{cb}$ can be determined from the semi-leptonic decay modes of B-mesons to final states with charm quarks, giving

$$|V_{cs}| = 1.006 \pm 0.023 \quad \text{and} \quad |V_{cb}| = (40.9 \pm 1.1) \times 10^{-3}.$$

The CKM matrix element $V_{cd}$ is most precisely measured in neutrino–nucleon scattering, $\nu_\mu d \to \mu^- c$, shown in Figure 14.6c. The final-state charm quark can be identified from its semi-leptonic decay $c \to s\mu^+\nu_\mu$, which gives an experimental signature of a pair of oppositely charged muons, one from the charm production process and one from its decay. The observed production rate of opposite sign muons in neutrino deep inelastic scattering gives

$$|V_{cd}| = 0.230(11).$$

The experimental situation for the CKM matrix elements involving top quarks is somewhat less clear. The observations of $B^0 \leftrightarrow \overline{B}^0$ oscillations, described in Section 14.6, can be interpreted in the Standard Model as measurements of

$$|V_{td}| = (8.4 \pm 0.6) \times 10^{-3} \quad \text{and} \quad |V_{ts}| = (42.9 \pm 2.6) \times 10^{-3}.$$

From the observed decay modes of the top quark at CDF and D0, it is known that the top quark decays predominantly via $t \to bW$ and therefore $|V_{tb}|$ is close to unity, although the current experimental error is at the 10% level.

In the Standard Model, the CKM matrix is unitary, $V^\dagger V = I$, which implies that

$$|V_{ud}|^2 + |V_{us}|^2 + |V_{ub}|^2 = 1, \tag{14.5}$$

$$|V_{cd}|^2 + |V_{cs}|^2 + |V_{cb}|^2 = 1, \tag{14.6}$$

$$|V_{td}|^2 + |V_{ts}|^2 + |V_{tb}|^2 = 1. \tag{14.7}$$

The measurements of the individual CKM matrix elements, described above, are consistent with these three unitarity relations. Assuming unitarity, further constraints can be placed on the less precisely determined CKM matrix elements, for example $|V_{tb}|^2 = 1 - |V_{ts}|^2 - |V_{tb}|^2$, which implies that $|V_{tb}| = 0.999$. With the unitarity constraints from (14.5)–(14.7), the experimental measurements can be interpreted as

$$\begin{pmatrix} |V_{ud}| & |V_{us}| & |V_{ub}| \\ |V_{cd}| & |V_{cs}| & |V_{cb}| \\ |V_{td}| & |V_{ts}| & |V_{tb}| \end{pmatrix} \approx \begin{pmatrix} 0.974 & 0.225 & 0.004 \\ 0.225 & 0.973 & 0.041 \\ 0.009 & 0.040 & 0.999 \end{pmatrix}. \tag{14.8}$$

Unlike the PMNS matrix of the lepton flavour sector, the off-diagonal terms in the CKM matrix are relatively small. This implies that the rotation angles between the quark mass and weak eigenstates in (14.4) are also small, $\phi_{12} = 13°$, $\phi_{23} = 2.3°$ and $\phi_{13} = 0.2°$. The smallness of these angles leads to the near diagonal form of the CKM matrix. Consequently, the weak interactions of quarks of different generations are suppressed relative to those of the same generation, ud, cs and tb. The suppression is largest for the couplings between first and third generation quarks, ub and td.

Because of the near diagonal nature of the CKM matrix, it is convenient to express it as an expansion in the relatively small parameter $\lambda = \sin\theta_c = 0.225$. In the widely used Wolfenstein parameterisation, the CKM matrix is written in terms of four real parameters, $\lambda$, $A$, $\rho$ and $\eta$. To $O(\lambda^4)$ the CKM matrix then can be parameterised as

$$\begin{pmatrix} V_{ud} & V_{us} & V_{ub} \\ V_{cd} & V_{cs} & V_{cb} \\ V_{td} & V_{ts} & V_{tb} \end{pmatrix} = \begin{pmatrix} 1-\lambda^2/2 & \lambda & A\lambda^3(\rho - i\eta) \\ -\lambda & 1-\lambda^2/2 & A\lambda^2 \\ A\lambda^3(1-\rho-i\eta) & -A\lambda^2 & 1 \end{pmatrix} + O(\lambda^4). \quad (14.9)$$

In the Wolfenstein parameterisation, the complex components of the CKM matrix reside only in $V_{ub}$ and $V_{td}$ (although if higher-order terms are included, $V_{cd}$ and $V_{ts}$ also have a small complex components that are proportional to $\lambda^5$). For CP to be violated in the quark sector, the CKM matrix must contain an irreducible complex phase and this corresponds to $\eta$ being non-zero. The experimental measurements of branching ratios only constrain the magnitudes of the individual CKM matrix elements, and do not provide any information about this complex phase. To constrain $\eta$ and $\rho$ separately, measurements that are sensitive to the amplitudes, rather than amplitudes squared are required. Such measurements can be made in the neutral kaon and neutral B-mesons systems.

## 14.4 The neutral kaon system

The first experimental observation of CP violation was made in the neutral kaon system. The $K^0(d\bar{s})$ and $\overline{K}^0(s\bar{d})$ are the lightest mesons containing strange quarks. They are produced copiously in strong interactions, for example in processes

$$\pi^-(d\bar{u}) + p(uud) \rightarrow \Lambda(uds) + K^0(d\bar{s}),$$

$$p(uud) + \bar{p}(\bar{u}\bar{u}\bar{d}) \rightarrow K^+(u\bar{s}) + \overline{K}^0(s\bar{d}) + \pi^-(d\bar{u}).$$

The $K^0$ and $\overline{K}^0$ are the eigenstates of the strong interaction and are referred to as the flavour states. Since they are the lightest hadrons containing strange quarks, the $K^0$ and $\overline{K}^0$ can decay only by the weak interaction. Because the neutral kaons

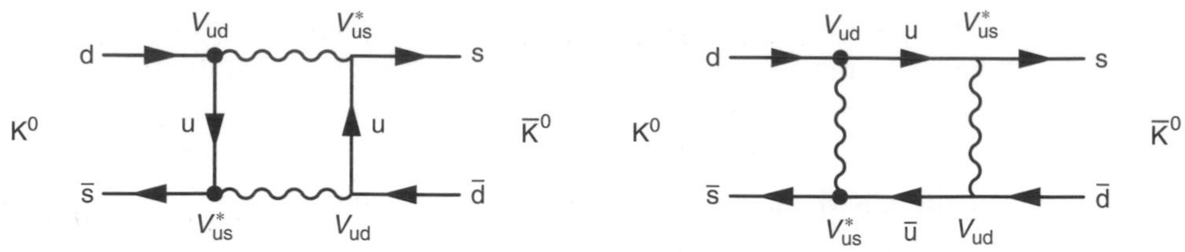


Fig. 14.7 Two box diagrams for $K^0 \leftrightarrow \overline{K}^0$ mixing. There are corresponding diagrams involving all nine combinations of virtual up, charm and top quarks.

are relatively light, $m(K) = 498$ MeV, only decays to final states with either leptons (e/μ) or pions are kinematically allowed. The weak interaction also provides a mechanism whereby the neutral kaons can mix through the $K^0 \leftrightarrow \overline{K}^0$ box diagrams shown in Figure 14.7.

In quantum mechanics, the physical states are the eigenstates of the free-particle Hamiltonian. These are the stationary states introduced in Section 2.3.3. Until now, independent stationary states have been used to describe each type of particle. Here however, because of the $K^0 \leftrightarrow \overline{K}^0$ mixing process, a neutral kaon that is produced as a $K^0$ will develop a $\overline{K}^0$ component. For this reason, the $K^0$–$\overline{K}^0$ system has to be considered as a whole. The physical neutral kaon states are the stationary states of the *combined* Hamiltonian of the $K^0$–$\overline{K}^0$ system, including the weak interaction mixing Hamiltonian. Consequently, the neutral kaons propagate as linear combinations of the $K^0$ and $\overline{K}^0$. These physical states are known as the K-short ($K_S$) and the K-long ($K_L$). The $K_S$ and $K_L$ are observed to have very similar masses, $m(K_S) \approx m(K_L) \approx 498$ MeV, but quite different lifetimes,

$$\tau(K_S) = 0.9 \times 10^{-10}\,\text{s} \quad \text{and} \quad \tau(K_L) = 0.5 \times 10^{-7}\,\text{s}.$$

If CP were an exact symmetry of the weak interaction, the $K_S$ and $K_L$ would be equivalent to the CP eigenstates of the neutral kaon system (the proof of this statement is given in Section 14.4.3). The CP states can be identified by considering the action of the parity and charge conjugation operators on the neutral kaons. The flavour eigenstates, $K^0(d\bar{s})$ and $\overline{K}^0(s\bar{d})$, have spin-parity $J^P = 0^-$ and therefore

$$\hat{P}|K^0\rangle = -|K^0\rangle \quad \text{and} \quad \hat{P}|\overline{K}^0\rangle = -|\overline{K}^0\rangle.$$

The $K^0$ and $\overline{K}^0$ are not eigenstates of the charge conjugation operator $\hat{C}$ that has the effect of replacing particles with antiparticles and vice versa. However, since they are neutral particles with opposite flavour content, one can write

$$\hat{C}|K^0(d\bar{s})\rangle = e^{i\zeta}|\overline{K}^0(\bar{d}s)\rangle \quad \text{and} \quad \hat{C}|\overline{K}^0(\bar{d}s)\rangle = e^{-i\zeta}|K^0(d\bar{s})\rangle,$$

where $\zeta$ is an unobservable phase factor, which is conventionally[1] chosen to be $\zeta = \pi$ such that

$$\hat{C}|\mathrm{K}^0(\mathrm{d\bar{s}})\rangle = -|\overline{\mathrm{K}}^0(\bar{\mathrm{d}}\mathrm{s})\rangle \quad \text{and} \quad \hat{C}|\overline{\mathrm{K}}^0(\bar{\mathrm{d}}\mathrm{s})\rangle = -|\mathrm{K}^0(\mathrm{d\bar{s}})\rangle.$$

With this choice, the combined action of $\hat{C}\hat{P}$ on the neutral kaon flavour eigenstates are

$$\hat{C}\hat{P}|\mathrm{K}^0\rangle = +|\overline{\mathrm{K}}^0\rangle \quad \text{and} \quad \hat{C}\hat{P}|\overline{\mathrm{K}}^0\rangle = +|\mathrm{K}^0\rangle.$$

Consequently, the orthogonal linear combinations

$$\mathrm{K}_1 = \tfrac{1}{\sqrt{2}}(\mathrm{K}^0 + \overline{\mathrm{K}}^0) \quad \text{and} \quad \mathrm{K}_2 = \tfrac{1}{\sqrt{2}}(\mathrm{K}^0 - \overline{\mathrm{K}}^0), \tag{14.10}$$

are CP eigenstates with

$$\hat{C}\hat{P}|\mathrm{K}_1\rangle = +|\mathrm{K}_1\rangle \quad \text{and} \quad \hat{C}\hat{P}|\mathrm{K}_2\rangle = -|\mathrm{K}_2\rangle.$$

If CP were conserved in the weak interaction, these states would correspond to the physical $\mathrm{K}_S$ and $\mathrm{K}_L$ particles. In practice, CP is observed to be violated but at a relatively low level, and to a reasonable approximation it is found that

$$|\mathrm{K}_S\rangle \approx |\mathrm{K}_1\rangle \quad \text{and} \quad |\mathrm{K}_L\rangle \approx |\mathrm{K}_2\rangle.$$

### 14.4.1 Kaon decays to pions

Neutral kaons propagate as the physical particles $\mathrm{K}_S$ and $\mathrm{K}_L$, which have well-defined masses and lifetimes. The $\mathrm{K}_S$ and $\mathrm{K}_L$ mainly decay to hadronic final states of either two/three pions or to semi-leptonic final states with electrons or muons. For the hadronic decays, the $\mathrm{K}_S$ decays mostly to $\pi\pi$ final states, whereas the main hadronic decays of the $\mathrm{K}_L$ are to $\pi\pi\pi$ final states,

$$\Gamma(\mathrm{K}_S \to \pi\pi) \gg \Gamma(\mathrm{K}_S \to \pi\pi\pi) \quad \text{and} \quad \Gamma(\mathrm{K}_L \to \pi\pi\pi) \gg \Gamma(\mathrm{K}_L \to \pi\pi).$$

The differences in the lifetimes of the $\mathrm{K}_S$ and $\mathrm{K}_L$ can be attributed to the different hadronic decay modes that are a consequence of the (near) conservation of CP in kaon decays, as discussed below

First consider the decays to two pions. The two pions can be produced with relative orbital angular momentum $\ell$, as indicated in Figure 14.8a. Because kaons and pions both have $J^P = 0^-$, the pions produced in the decay $\mathrm{K} \to \pi^0\pi^0$ must be in an $\ell = 0$ state in order to conserve angular momentum. The overall parity of the $\pi^0\pi^0$ system, which is given by the symmetry of the spatial wavefunction and the intrinsic parity of the pion, is therefore

$$P(\pi^0\pi^0) = (-1)^\ell P(\pi^0)P(\pi^0) = (+1)\times(-1)\times(-1) = +1.$$

[1] Sometimes, the convention $\zeta = 0$ is used, leading to a different definition of the $\mathrm{K}_1$ and $\mathrm{K}_2$ in terms of the flavour eigenstates. However, provided this weak phase is treated consistently, there are no physical consequences in the choice.

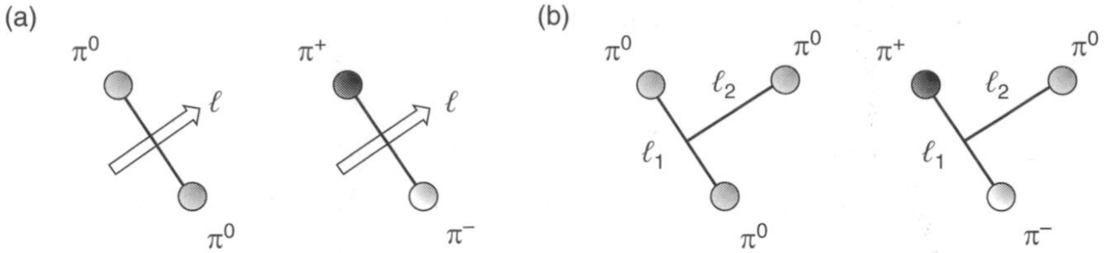


Fig. 14.8 Angular momentum in the two- and three-pion systems.

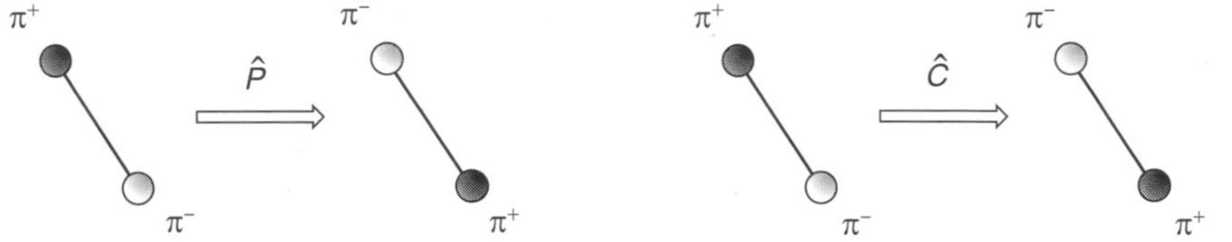


Fig. 14.9 The effect of the parity and charge conjugation operations on the $\pi^+\pi^-$ system in an $\ell = 0$ angular momentum state.

The flavour wavefunction of the $\pi^0$ is

$$|\pi^0\rangle = \frac{1}{\sqrt{2}}(u\bar{u} - d\bar{d}),$$

and consequently the $\pi^0$ is an eigenstate of $\hat{C}$ with eigenvalue +1. Therefore

$$C(\pi^0\pi^0) = C(\pi^0)C(\pi^0) = +1,$$

and since $P(\pi^0\pi^0) = +1$, the $\pi^0\pi^0$ system must be produced in a CP-even state,

$$CP(\pi^0\pi^0) = +1.$$

The angular momentum arguments given above apply equally to the $\pi^+\pi^-$ system, and therefore $P(\pi^+\pi^-) = +1$. The effect of the parity operation on the $\pi^+\pi^-$ system is to swap the positions of the two particles, with no change in sign. Because the charge conjugation operation turns a $\pi^+$ into a $\pi^-$ and vice versa, the effect of the charge conjugation on the $\pi^+\pi^-$ system is also to swap the positions of the particles, with no change in sign. Hence, here the parity and charge conjugation operations have the same effect, as shown in Figure 14.9, and thus $C(\pi^+\pi^-) = P(\pi^+\pi^-) = +1$. Therefore, the decay of a neutral kaon into two pions always produces a CP-even final state,

$$CP(\pi^0\pi^0) = +1 \quad \text{and} \quad CP(\pi^+\pi^-) = +1.$$

If CP is conserved in kaon decay (which it is to a very good approximation), the decay $K \to \pi\pi$ can only occur if the neutral kaon state has CP = +1.

The corresponding arguments for the decays $K \to \pi^0\pi^0\pi^0$ and $K \to \pi^+\pi^-\pi^0$ are slightly more involved. Here, the orbital angular momentum has to be decomposed into two components; the relative angular momentum of the first two particles, $\mathbf{L}_1$, and the relative angular momentum of the third with respect to the centre of mass of the first two, $\mathbf{L}_2$, as indicated in Figure 14.8b. Because both kaons and pions are spin-0 particles, the total orbital angular momentum in the decay $K \to \pi\pi\pi$ must be zero, $\mathbf{L} = \mathbf{L}_1 + \mathbf{L}_2 = 0$. This only can be the case if $\ell_1 = \ell_2 = \ell$. The overall parity of the final state in a $K \to \pi\pi\pi$ decay is therefore

$$P(\pi\pi\pi) = (-1)^{\ell_1}(-1)^{\ell_2}(P(\pi))^3 = (-1)^{2\ell}(-1)^3 = -1.$$

For the $\pi^0\pi^0\pi^0$ final state, the effect of the charge conjugation operator is

$$C(\pi^0\pi^0\pi^0) = C(\pi^0)C(\pi^0)C(\pi^0) = (+1)^3 = +1,$$

and therefore $CP(\pi^0\pi^0\pi^0) = -1$. The effect of the charge conjugation operator on the $\pi^+\pi^-\pi^0$ system follows from the arguments given previously,

$$C(\pi^+\pi^-\pi^0) = C(\pi^+\pi^-)C(\pi^0) = +C(\pi^+\pi^-) = P(\pi^+\pi^-) = (-1)^{\ell_1},$$

where again the effect of $\hat{C}(\pi^+\pi^-)$ is the same as that of $\hat{P}(\pi^+\pi^-)$. Because $m(K) - 3m(\pi) \approx 80\,\text{MeV}$, the kinetic energy of the three-pion system is relatively small, and the decays where $\ell_1 = \ell_2 > 0$ are suppressed to the point where the contribution is negligible. For this reason $\ell_1$ can be taken to be zero and thus

$$CP(\pi^0\pi^0\pi^0) = -1 \quad \text{and} \quad CP(\pi^+\pi^-\pi^0) = -1.$$

Therefore, the $K \to \pi\pi\pi$ decay modes of neutral kaons always result in a CP-odd final state.

If CP were conserved in the decays of neutral kaons, the hadronic decays of the CP-eigenstates $|K_1\rangle$ and $|K_2\rangle$ would be exclusively $K_1 \to \pi\pi$ and $K_2 \to \pi\pi\pi$. Because the phase space available for decays to two and three pions is very different, $m(K) - 2m(\pi) \approx 220\,\text{MeV}$ compared to $m(K) - 3m(\pi) \approx 80\,\text{MeV}$, the decay rate to two pions is much larger than that to three pions. Hence, the short-lived $K_S$, which decays mostly to two pions, can be identified as being a close approximation to the CP-even state

$$K_S \approx K_1 = \frac{1}{\sqrt{2}}(K^0 + \overline{K}^0), \tag{14.11}$$

and the longer lived $K_L$ as

$$K_L \approx K_2 = \frac{1}{\sqrt{2}}(K^0 - \overline{K}^0). \tag{14.12}$$

If CP were *exactly* conserved in the weak interaction, then $K_S \equiv K_1$ and $K_L \equiv K_2$.

### CP violation in hadronic kaon decays

The decays of neutral kaons have been extensively studied using kaon beams produced from hadronic interactions. If a neutral kaon is produced in the strong interaction $p\overline{p} \to K^-\pi^+K^0$, at the time of production, the kaon is the flavour eigenstate,

$$|K(0)\rangle = |K^0\rangle.$$

In the absence of CP violation, where $K_S \equiv K_1$ and $K_L \equiv K_2$, the $|K^0\rangle$ flavour state can be written in terms of the CP eigenstates using (14.11) and (14.12),

$$|K(0)\rangle = |K^0\rangle = \tfrac{1}{\sqrt{2}}[\,|K_1\rangle + |K_2\rangle] = \tfrac{1}{\sqrt{2}}[\,|K_S\rangle + |K_L\rangle]. \tag{14.13}$$

The subsequent time evolution is described in terms of the $K_S$ and $K_L$, which are the observed physical neutral kaons with well-defined masses and lifetimes. In the rest frame of the kaon, the time-evolution of the $K_S$ and $K_L$ states are given by

$$|K_S(t)\rangle = |K_S\rangle \exp[-im_S t - \Gamma_S t/2], \tag{14.14}$$
$$|K_L(t)\rangle = |K_L\rangle \exp[-im_L t - \Gamma_L t/2], \tag{14.15}$$

where the $\exp[-\Gamma t/2]$ terms ensure that the probability densities decay exponentially. For example

$$\langle K_S(t)|K_S(t)\rangle \propto e^{-\Gamma_S t} = e^{-t/\tau_S}.$$

Hence the time evolution of the state of (14.13) is

$$|K(t)\rangle = \tfrac{1}{\sqrt{2}}\left[|K_S\rangle e^{-(im_S+\Gamma_S/2)t} + |K_L\rangle e^{-(im_L+\Gamma_L/2)t}\right],$$

which can be written as

$$|K(t)\rangle = \tfrac{1}{\sqrt{2}}\left[\theta_S(t)|K_S\rangle + \theta_L(t)K_L\right], \tag{14.16}$$

with

$$\theta_S(t) = \exp[-(im_S + \Gamma_S/2)\,t] \quad \text{and} \quad \theta_L(t) = \exp[-(im_L + \Gamma_L/2)\,t]. \tag{14.17}$$

The decay rate to the CP-even two-pion final state is proportional to the $K_1$ component of the wavefunction, which in the limit where CP is conserved is equivalent to the $K_S$ component. Therefore, if CP is conserved, the decay rate to two pions from a beam that was initially in a pure $|K^0\rangle$ state is

$$\Gamma(K^0_{t=0} \to \pi\pi) \propto |\langle K_S|K(t)\rangle|^2 \propto |\theta_S(t)|^2 = e^{-\Gamma_S t} = e^{-t/\tau_S},$$

and similarly

$$\Gamma(K^0_{t=0} \to \pi\pi\pi) \propto |\langle K_L|\psi(t)\rangle|^2 \propto e^{-t/\tau_L}.$$

If a kaon beam, which originally consisted of $K^0(d\overline{s})$, propagates over a large distance ($L \gg c\tau_S$), the $K_S$ component will decay away leaving a pure $K_L$ beam, as indicated in Figure 14.10. The same would be true for an initial $\overline{K}^0$ beam.

If CP were conserved in the weak interactions of quarks, the $K_L$ would correspond exactly to the CP-odd $K_2$ state and at large distances from the production

**Table 14.1** The main decay modes of the $K_S$ and $K_L$.

| $K_S$ Decays | BR | $K_L$ Decays | BR |
|---|---|---|---|
| $K_S \to \pi^+\pi^-$ | 69.2% | $K_L \to \pi^+\pi^-$ | 0.20% |
| $K_S \to \pi^0\pi^0$ | 30.7% | $K_L \to \pi^0\pi^0$ | 0.09% |
| $K_S \to \pi^+\pi^-\pi^0$ | $\sim 3 \times 10^{-5}$% | $K_L \to \pi^+\pi^-\pi^0$ | 12.5% |
| $K_S \to \pi^0\pi^0\pi^0$ | – | $K_L \to \pi^0\pi^0\pi^0$ | 19.5% |
| $K_S \to \pi^-e^+\nu_e$ | 0.03% | $K_L \to \pi^-e^+\nu_e$ | 20.3% |
| $K_S \to \pi^+e^-\overline{\nu}_e$ | 0.03% | $K_L \to \pi^+e^-\overline{\nu}_e$ | 20.3% |
| $K_S \to \pi^-\mu^+\nu_\mu$ | 0.02% | $K_L \to \pi^-\mu^+\nu_\mu$ | 13.5% |
| $K_S \to \pi^+\mu^-\overline{\nu}_\mu$ | 0.02% | $K_L \to \pi^+\mu^-\overline{\nu}_\mu$ | 13.5% |

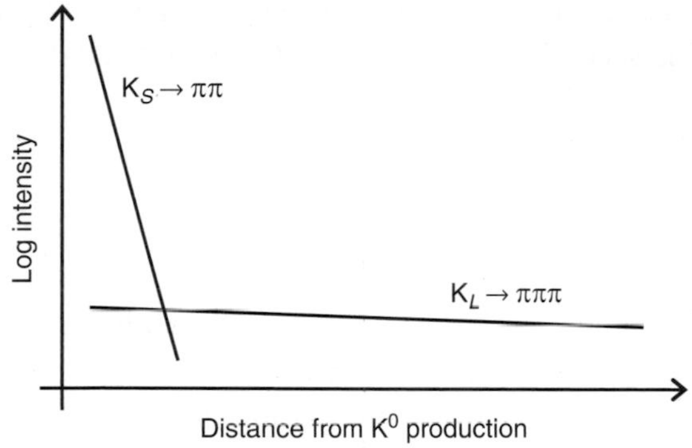


Fig. 14.10 Expected decay rates to pions from an initially pure $K^0$ beam, assuming no CP violation.

of a kaon beam, the hadronic decays to two pions would never be detected. The first experimental evidence for CP violation was the observation of 45 $K_L \to \pi^+\pi^-$ decays out of a total of 22 700 $K_L$ decays at a large distance from the production of the neutral kaon beam; see Christenson *et al.* (1964). This provided the first direct evidence for CP violation in the neutral kaon system, albeit only at the level of 0.2%, for which Cronin and Fitch were awarded the Nobel prize.

The branching ratios for the main decay modes of the $K_S$ and $K_L$ are listed in Table 14.1, including the relatively rare CP violating hadronic decays. The smallness of the semi-leptonic *branching ratios* of the $K_S$ compared to the $K_L$, reflects the relatively large $K_S \to \pi\pi$ decay rate; the semi-leptonic partial decay *rates* of the $K_S$ and $K_L$ are almost identical (see Section 14.5.4).

## 14.4.2 The origin of CP violation

There are two main ways of introducing CP violation into the neutral kaon system. If CP is violated in the $K^0 \leftrightarrow \overline{K}^0$ mixing process (see Section 14.4.3), then the

$K_S$ and $K_L$ will not correspond to the CP eigenstates, $K_1$ and $K_2$. Given that the observed level of CP violation is relatively small, the $K_S$ and $K_L$ can be related to the CP eigenstates by the small (complex) parameter $\varepsilon$,

$$|K_S\rangle = \frac{1}{\sqrt{1+|\varepsilon|^2}}\left(|K_1\rangle + \varepsilon|K_2\rangle\right) \quad \text{and} \quad |K_L\rangle = \frac{1}{\sqrt{1+|\varepsilon|^2}}\left(|K_2\rangle + \varepsilon|K_1\rangle\right),$$

such that $K_S \approx K_1$ and $K_L \approx K_2$. In this case, the observed $K_L \to \pi\pi$ decays are accounted for by

$$|K_L\rangle = \frac{1}{\sqrt{1+|\varepsilon|^2}}\left(|K_2\rangle + \varepsilon|K_1\rangle\right)$$

$$\hookrightarrow \pi\pi$$

$$\hookrightarrow \pi\pi\pi$$

and the relative rate of decays to two pions will be depend on $\varepsilon$.

The second possibility is that CP is violated directly in the decay of a CP eigenstate,

$$|K_L\rangle = |K_2\rangle$$

$$\hookrightarrow \pi\pi$$

$$\hookrightarrow \pi\pi\pi$$

The relative strength of this direct CP violation in neutral kaon decay is parameterised by $\varepsilon'$ with $\Gamma(K_2 \to \pi\pi)/\Gamma(K_2 \to \pi\pi\pi) = \varepsilon'$. Experimentally, it is known that CP is violated in both mixing and directly in the decay. The results of the NA48 experiment at CERN and the KTeV experiment at Fermilab, demonstrate that direct CP violation is a relatively small effect,

$$\Re e\left(\frac{\varepsilon'}{\varepsilon}\right) = (1.65 \pm 0.26) \times 10^{-3},$$

and $\varepsilon$ is already a small parameter. Therefore, the main contribution to CP violation in the neutral kaon system is from $K^0 \leftrightarrow \overline{K}^0$ mixing. The quantum mechanics of mixing in the neutral kaon system is described in detail in the following starred section.

### 14.4.3 *The quantum mechanics of kaon mixing

To fully understand the physics of the neutral kaon system, it is necessary to consider the quantum mechanical time evolution of the combined $K^0$–$\overline{K}^0$ system. This is not an easy topic, but the results are important.

In the absence of neutral kaon mixing, the time dependence of the wavefunction of the $K^0$ would be

$$|K^0(t)\rangle = |K^0\rangle e^{-\Gamma t/2} e^{-imt}, \tag{14.18}$$

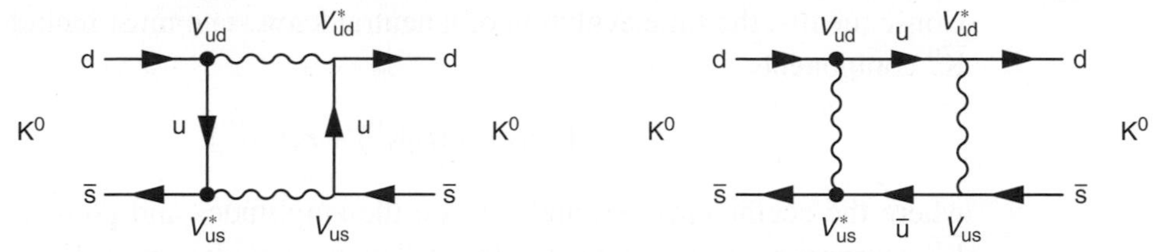


**Fig. 14.11** Two box diagrams for $K^0 \leftrightarrow K^0$. There are corresponding diagrams involving all nine combinations of virtual up, charm and top quarks.

where $m$ is the mass of the particle and the term $\Gamma = 1/\tau$ ensures the probability density decays away exponentially. The time-dependent wavefunction of (14.18) clearly satisfies the differential equation

$$i\frac{\partial}{\partial t}|K^0(t)\rangle = (m - \tfrac{i}{2}\Gamma)|K^0(t)\rangle,$$

and therefore the *effective* Hamiltonian $\mathcal{H}$ can be identified as

$$\mathcal{H}|K^0(t)\rangle = (m - \tfrac{i}{2}\Gamma)|K^0(t)\rangle. \tag{14.19}$$

Because of the inclusion of the exponential decay term in the wavefunction, the effective Hamiltonian is not Hermitian and also the expectation values of operators corresponding to physical observable will not be constant. The mass $m$ in the effective Hamiltonian of (14.19) includes contributions from the masses of the constituent quarks and from the potential energy of the system. The potential energy includes contributions from the strong interaction potential (which is the dominant term), the coulomb interaction and the weak interaction. The interaction terms can be expressed as expectation values of the corresponding interaction Hamiltonians. Therefore the mass of the $K^0$, when taken in isolation, can be written as

$$m = m_d + m_{\bar{s}} + \langle K^0|\hat{H}_{QCD} + \hat{H}_{EM} + \hat{H}_W|K^0\rangle + \sum_j \frac{\langle K^0|\hat{H}_W|j\rangle\langle j|\hat{H}_W|K^0\rangle}{E_j - m_K}. \tag{14.20}$$

The last term in this expression comes from the small second-order $O(G_F^2)$ contribution to the weak interaction potential from the $K^0 \leftrightarrow K^0$ box diagrams of Figure 14.11. The decay rate $\Gamma$ that appears in (14.19) is given by Fermi's golden rule

$$\Gamma = 2\pi \sum_f |\langle f|\hat{H}_W|K^0\rangle|^2 \rho_f,$$

where the sum is taken over all possible final states, labelled $f$, and $\rho_f$ is the density of states for that decay mode.

Up to this point, the $K^0$ has been considered in isolation. However, a $K^0$ will develop a $\overline{K}^0$ component through the $K^0 \leftrightarrow \overline{K}^0$ mixing diagrams of Figure 14.7.

Consequently, the time evolution of a neutral kaon state must include both $K^0$ and $\overline{K}^0$ components,

$$|K(t)\rangle = a(t)|K^0\rangle + b(t)|\overline{K}^0\rangle, \tag{14.21}$$

where the coefficients $a(t)$ and $b(t)$ are the amplitudes and phases of the $K^0$ and $\overline{K}^0$ components of the state at a time $t$. The time evolution of $|K(t)\rangle$, analogous to (14.19), now has to be written as the coupled equations

$$\begin{pmatrix} M_{11} - \frac{i}{2}\Gamma_{11} & M_{12} - \frac{i}{2}\Gamma_{12} \\ M_{21} - \frac{i}{2}\Gamma_{21} & M_{22} - \frac{i}{2}\Gamma_{22} \end{pmatrix} \begin{pmatrix} a(t)\,|K^0\rangle \\ b(t)\,|\overline{K}^0\rangle \end{pmatrix} = i\frac{\partial}{\partial t} \begin{pmatrix} a(t)\,|K^0\rangle \\ b(t)\,|\overline{K}^0\rangle \end{pmatrix}, \tag{14.22}$$

and the effective Hamiltonian becomes

$$\mathcal{H} = \mathbf{M} - \tfrac{i}{2}\mathbf{\Gamma} = \begin{pmatrix} M_{11} & M_{12} \\ M_{21} & M_{22} \end{pmatrix} - \frac{i}{2}\begin{pmatrix} \Gamma_{11} & \Gamma_{12} \\ \Gamma_{21} & \Gamma_{22} \end{pmatrix}. \tag{14.23}$$

It is important to understand the physical meaning of the terms in (14.23). First consider the decay matrix $\mathbf{\Gamma}$ that accounts for the decay of the state $|K(t)\rangle$. Here the total decay rate is given by Fermi's golden rule, which to lowest order is

$$\Gamma = 2\pi \sum_f |\langle f|\hat{H}_W|K(t)\rangle|^2 \rho_f \equiv 2\pi \sum_f \langle K(t)|\hat{H}_W|f\rangle\langle f|\hat{H}_W|K(t)\rangle\,\rho_f.$$

By writing $|K(t)\rangle$ in terms of $K^0$ and $\overline{K}^0$, the matrix element squared for the decay to a final state $f$ becomes

$$\begin{aligned} |\langle f|\hat{H}_W|K(t)\rangle|^2 &= |a(t)|^2\,|\langle f|\hat{H}_W|K^0\rangle|^2 + |b(t)|^2\,|\langle f|\hat{H}_W|\overline{K}^0\rangle|^2 \\ &\quad + a(t)b(t)^*\,\langle \overline{K}^0|\hat{H}_W|f\rangle\langle f|\hat{H}_W|K^0\rangle + a(t)^*b(t)\,\langle K^0|\hat{H}_W|f\rangle\langle f|\hat{H}_W|\overline{K}^0\rangle. \end{aligned}$$

The diagonal elements of $\mathbf{\Gamma}$ are therefore given by the decay rates

$$\Gamma_{11} = 2\pi \sum_f |\langle f|\hat{H}_W|K^0\rangle|^2 \rho_f \quad \text{and} \quad \Gamma_{22} = 2\pi \sum_f |\langle f|\hat{H}_W|\overline{K}^0\rangle|^2 \rho_f,$$

and are therefore real numbers. The off-diagonal terms of $\mathbf{\Gamma}$ account for the interference between the decays of the $K^0$ and $\overline{K}^0$ components of $K(t)$. Because the two interference terms are the Hermitian conjugates of each other, $\Gamma_{12} = \Gamma_{21}^*$, and the matrix $\mathbf{\Gamma}$ is itself Hermitian.

Now consider the mass matrix $\mathbf{M}$. The diagonal elements are the mass terms for the $K^0$ and $\overline{K}^0$ *flavour* eigenstates, with $M_{11}$ given by (14.20) and

$$M_{22} = m_s + m_{\overline{d}} + \langle \overline{K}^0|\hat{H}_{QCD} + \hat{H}_{EM} + \hat{H}_W|\overline{K}^0\rangle + \sum_j \frac{\langle \overline{K}^0|\hat{H}_W|j\rangle\langle j|\hat{H}_W|\overline{K}^0\rangle}{E_j - m_K}. \tag{14.24}$$

The off-diagonal terms of **M** are due to the $K^0 \leftrightarrow \overline{K}^0$ mixing diagrams of Figure 14.7, and can be written

$$M_{12} = M_{21}^* = \sum_j \frac{\langle \overline{K}^0 | \hat{H}_W | j \rangle \langle j | \hat{H}_W | K^0 \rangle}{E_j - m_K}.$$

There is no off-diagonal term of the form $\langle \overline{K}^0 | \hat{H}_W | K^0 \rangle$ because there is no Feynman diagram for $K^0 \leftrightarrow \overline{K}^0$ mixing involving the exchange of a single W boson. Since $M_{12} = M_{21}^*$ and the diagonal terms of **M** are real, the mass matrix is Hermitian. If there were no mixing in the neutral kaon system $M_{12}$, $M_{21}$, $\Gamma_{12}$ and $\Gamma_{21}$ would all be zero, and the time evolution equation of (14.22) would decouple into two independent equations of the form of (14.19), describing the independent time evolution of the $K^0$ and $\overline{K}^0$.

From the required CPT symmetry of the Standard Model, the masses and decay rates of the flavour states $K^0$ and $\overline{K}^0$ must be equal, $M_{11} = M_{22} = M$ and $\Gamma_{11} = \Gamma_{22} = \Gamma$. Therefore the effective Hamiltonian of (14.23) can be written as

$$\mathcal{H} = \mathbf{M} - \tfrac{i}{2}\mathbf{\Gamma} = \begin{pmatrix} M & M_{12} \\ M_{12}^* & M \end{pmatrix} - \frac{i}{2}\begin{pmatrix} \Gamma & \Gamma_{12} \\ \Gamma_{12}^* & \Gamma \end{pmatrix}. \tag{14.25}$$

Because the off-diagonal elements of **M** arise from second-order weak interaction box diagrams, they are much smaller than the diagonal elements that include the fermion masses and the strong interaction Hamiltonian. The off-diagonal terms of $\mathbf{\Gamma}$, which can be of the same order of magnitude as the diagonal terms, are either positive or negative. Because of the presence of the non-zero off-diagonal terms in $\mathcal{H}$, the flavour eigenstates $K^0$ and $\overline{K}^0$ are no longer the eigenstates of the Hamiltonian.

The neutral kaon state of (14.21), evolves in time according to

$$\begin{pmatrix} M - \frac{i}{2}\Gamma & M_{12} - \frac{i}{2}\Gamma_{12} \\ M_{12}^* - \frac{i}{2}\Gamma_{12}^* & M - \frac{i}{2}\Gamma \end{pmatrix}\begin{pmatrix} a(t)\,|K^0\rangle \\ b(t)\,|\overline{K}^0\rangle \end{pmatrix} = i\frac{\partial}{\partial t}\begin{pmatrix} a(t)\,|K^0\rangle \\ b(t)|\overline{K}^0\rangle \end{pmatrix}. \tag{14.26}$$

The eigenstates of this effective Hamiltonian can be found by transforming (14.26) into the basis where $\mathcal{H}$ is diagonal. The required transformation can be found by first solving the eigenvalue equation

$$\begin{pmatrix} M - \frac{i}{2}\Gamma & M_{12} - \frac{i}{2}\Gamma_{12} \\ M_{12}^* - \frac{i}{2}\Gamma_{12}^* & M - \frac{i}{2}\Gamma \end{pmatrix}\begin{pmatrix} p \\ q \end{pmatrix} = \lambda\begin{pmatrix} p \\ q \end{pmatrix}. \tag{14.27}$$

The non-trivial solutions to (14.27) can be obtained from the characteristic equation, $\det(\mathcal{H} - \lambda I) = 0$, which gives

$$(M - \tfrac{i}{2}\Gamma - \lambda)^2 - (M_{12}^* - \tfrac{i}{2}\Gamma_{12}^*)(M_{12} - \tfrac{i}{2}\Gamma_{12}) = 0.$$

Solving this quadratic equation for $\lambda$ gives the two eigenvalues

$$\lambda_\pm = M - \tfrac{i}{2}\Gamma \pm \left[(M_{12}^* - \tfrac{i}{2}\Gamma_{12}^*)(M_{12} - \tfrac{i}{2}\Gamma_{12})\right]^{\frac{1}{2}}. \tag{14.28}$$

The corresponding eigenstates, found by substituting these two eigenvalues back into (14.27), have

$$\frac{q}{p} = \pm\xi \equiv \pm\left(\frac{M^*_{12} - \frac{i}{2}\Gamma^*_{12}}{M_{12} - \frac{i}{2}\Gamma_{12}}\right)^{\frac{1}{2}}. \tag{14.29}$$

The normalised eigenstates, here denoted $K_+$ and $K_-$, which ultimately will be identified as the $K_S$ and $K_L$, are therefore

$$\begin{pmatrix} |K_+\rangle \\ |K_-\rangle \end{pmatrix} = \frac{1}{\sqrt{1+|\xi|^2}}\begin{pmatrix} 1 & \xi \\ 1 & -\xi \end{pmatrix}\begin{pmatrix} |K^0\rangle \\ |\overline{K}^0\rangle \end{pmatrix} = \frac{1}{\sqrt{1+|\xi|^2}}\begin{pmatrix} |K^0\rangle + \xi\overline{K}^0 \\ |K^0\rangle - \xi|\overline{K}^0\rangle \end{pmatrix}.$$

Equation (14.26), which has the form $\mathcal{H}\mathbf{K} = i\partial\mathbf{K}/\partial t$, can be written in the diagonal basis using the matrix $\mathbf{S}$ formed from the eigenvectors of $\mathcal{H}$, such that $\mathcal{H}' = \mathbf{S}^{-1}\mathcal{H}\mathbf{S}$ is diagonal,

$$\mathcal{H}' = \mathbf{S}^{-1}\mathcal{H}\mathbf{S} = \begin{pmatrix} \lambda_+ & 0 \\ 0 & \lambda_- \end{pmatrix}.$$

In the diagonal basis (14.26) becomes

$$i\frac{\partial}{\partial t}\begin{pmatrix} |K_+(t)\rangle \\ |K_-(t)\rangle \end{pmatrix} = \begin{pmatrix} \lambda_+ & 0 \\ 0 & \lambda_- \end{pmatrix}\begin{pmatrix} |K_+(t)\rangle \\ |K_-(t)\rangle \end{pmatrix}. \tag{14.30}$$

Hence the states $K_+$ and $K_-$ propagate as independent particles and therefore can be identified as the physical mass eigenstates of the neutral kaon system. The time dependences of the $K_+$ and $K_-$ states are given by the solutions of (14.30),

$$|K_+(t)\rangle = \frac{1}{\sqrt{1+|\xi|^2}}\left(|K^0\rangle + \xi|\overline{K}^0\rangle\right)e^{-i\lambda_+ t}$$
$$|K_-(t)\rangle = \frac{1}{\sqrt{1+|\xi|^2}}\left(|K^0\rangle - \xi|\overline{K}^0\rangle\right)e^{-i\lambda_- t},$$

with the real and imaginary parts of $\lambda_\pm$ determining respectively the masses and decay rates of the two physical states. From (14.28),

$$\lambda_+ - \lambda_- = 2\left[(M^*_{12} - \tfrac{i}{2}\Gamma^*_{12})(M_{12} - \tfrac{i}{2}\Gamma_{12})\right]^{\frac{1}{2}}, \tag{14.31}$$

and therefore $\lambda_+$ and $\lambda_-$ can be written as

$$\lambda_\pm = M - \tfrac{i}{2}\Gamma \pm \tfrac{1}{2}(\lambda_+ - \lambda_-) = M \pm \Re e\left(\frac{\lambda_+ - \lambda_-}{2}\right) - \tfrac{i}{2}\left(\Gamma \mp \Im m\{\lambda_+ - \lambda_-\}\right).$$

It is not *a priori* clear which of the two eigenvalues, $\lambda_+$ and $\lambda_-$, is associated with the $K_S$ and which is associated with the $K_L$, but both can be written in the form

$$\lambda = [M \pm \Delta m/2] - \tfrac{i}{2}[\Gamma \pm \Delta\Gamma/2],$$

with

$$\Delta m = |\,\mathfrak{Re}\,(\lambda_+ - \lambda_-)| \quad \text{and} \quad \Delta\Gamma = \pm|\Delta\Gamma| = \pm 2|\,\mathfrak{Im}\,(\lambda_+ - \lambda_-)|.$$

Here $\Delta m$ is *defined* to be positive and the sign of $\Delta\Gamma$ depends on the relative signs of the real and imaginary parts of (14.31), which in turn depends on the off-diagonal terms of the effective Hamiltonian. For the neutral kaon system it turns out that $\Delta\Gamma < 0$, and therefore the heavier state has the smaller decay rate. Consequently, the physical eigenstates of the neutral kaon system consist of a heavier state of mass $M + \Delta m/2$ that can be identified as the longer-lived $K_L$ state and a lighter state with a larger decay rate and mass $M - \Delta m/2$ that can be identified as the $K_S$,

$$\lambda_S = m_S - \tfrac{i}{2}\Gamma_S \quad \text{with} \quad m_S = M - \Delta m/2 \quad \text{and} \quad \Gamma_S = \Gamma + |\Delta\Gamma|/2,$$
$$\lambda_L = m_L - \tfrac{i}{2}\Gamma_L \quad \text{with} \quad m_L = M + \Delta m/2 \quad \text{and} \quad \Gamma_L = \Gamma - |\Delta\Gamma|/2.$$

Because the off-diagonal terms in the effective Hamiltonian arise from the weak interaction alone, $\Delta m \ll M$, and the mass difference between the $K_L$ and $K_S$ is very small.

If the CKM matrix were entirely real, which would imply that $M_{12} = M_{12}^*$ and $\Gamma_{12} = \Gamma_{12}^*$, the parameter $\xi$ defined in (14.29) would be unity. In this case, the physical states would be

$$K_S \equiv K_1 = \tfrac{1}{\sqrt{2}}\left(K^0 + \overline{K}^0\right) \quad \text{and} \quad K_L \equiv K_2 = \tfrac{1}{\sqrt{2}}\left(K^0 - \overline{K}^0\right). \tag{14.32}$$

Hence, *if* the CKM matrix were entirely real, in which case the weak interactions of quarks would conserve CP, the physical states of the neutral kaon system would be the CP eigenstates, $K_1$ and $K_2$. In practice, CP violation is observed in the neutral kaon system, albeit at a very low level and therefore $\xi \neq 1$.

Because CP-violating effects are observed to be relatively small, it is convenient to rewrite $\xi$ in terms of the (small) complex parameter $\varepsilon$ defined by

$$\xi = \frac{1-\varepsilon}{1+\varepsilon},$$

such that the physical $K_S$ and $K_L$ states are

$$|K_S(t)\rangle = \frac{1}{\sqrt{2(1+|\varepsilon|^2)}}\left[(1+\varepsilon)|K^0\rangle + (1-\varepsilon)|\overline{K}^0\rangle\right]e^{-i\lambda_S t}, \tag{14.33}$$

$$|K_L(t)\rangle = \frac{1}{\sqrt{2(1+|\varepsilon|^2)}}\left[(1+\varepsilon)|K^0\rangle - (1-\varepsilon)|\overline{K}^0\rangle\right]e^{-i\lambda_L t}. \tag{14.34}$$

Using (14.10), the physical states also can be expressed in terms of the CP eigenstates $K_1$ and $K_2$,

$$|K_S(t)\rangle = \frac{1}{\sqrt{1+|\varepsilon|^2}}\left[|K_1\rangle + \varepsilon|K_2\rangle\right] e^{-i\lambda_S t}, \tag{14.35}$$

$$|K_L(t)\rangle = \frac{1}{\sqrt{1+|\varepsilon|^2}}\left[|K_2\rangle + \varepsilon|K_1\rangle\right] e^{-i\lambda_L t}. \tag{14.36}$$

## 14.5 Strangeness oscillations

The previous chapter described how neutrino oscillations arise because neutrinos are created and interact as weak eigenstates but propagate as mass eigenstates. A similar phenomenon occurs in the neutral kaon system. The physical mass eigenstates are the $K_S$ and $K_L$. However, the hadronic decays to $\pi\pi$ or $\pi\pi\pi$ have to be described in terms of the CP eigenstates and the semi-leptonic decays of the $K_S$ and $K_L$ have to be described in terms of the flavour eigenstates, $K^0$ and $\overline{K}^0$. For example, Figure 14.12 shows the Feynman diagrams for the allowed decays $K^0 \to \pi^- e^+ \nu_e$ and $\overline{K}^0 \to \pi^+ e^- \overline{\nu}_e$. There are no corresponding Feynman diagrams for $K^0 \to \pi^+ e^- \overline{\nu}_e$ and $\overline{K}^0 \to \pi^- e^+ \nu_e$ because the charge of the lepton depends on whether $s \to u$ or $\overline{s} \to \overline{u}$ decay is involved:

$$K^0 \to \pi^- e^+ \nu_e \quad \text{and} \quad \overline{K}^0 \to \pi^+ e^- \overline{\nu}_e,$$
$$K^0 \nrightarrow \pi^+ e^- \overline{\nu}_e \quad \text{and} \quad \overline{K}^0 \nrightarrow \pi^- e^+ \nu_e.$$

Hence neutral kaons are produced and decay as flavour and/or CP eigenstates, but propagate as the $K_S$ and $K_L$ mass eigenstates. The result is the phenomenon of strangeness oscillations, which occurs regardless of whether CP is violated or not.

### 14.5.1 Strangeness oscillations neglecting CP violation

Consider a neutral kaon that is produced as the flavour eigenstate $K^0$. The time evolution of the wavefunction is described in terms of the $K_S$ and $K_L$ mass eigenstates,

$$|K(t)\rangle = \tfrac{1}{\sqrt{2}}\left[\,\theta_S(t)|K_S\rangle + \theta_L(t)|K_L\rangle\right], \tag{14.37}$$

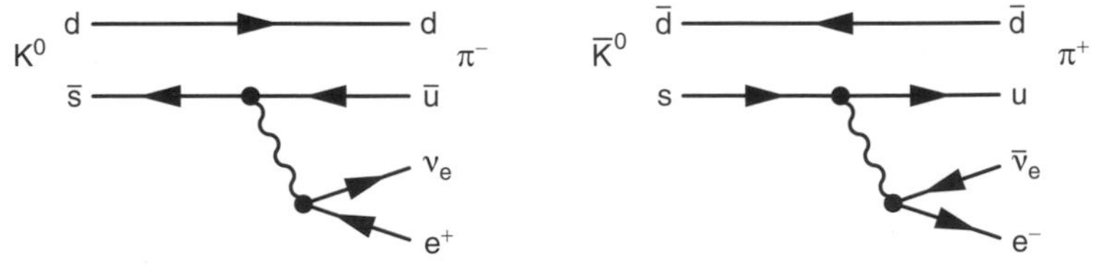


Fig. 14.12 The Feynman diagrams for $K^0 \to \pi^- e^+ \nu_e$ and $\overline{K}^0 \to \pi^+ e^- \overline{\nu}_e$.

where $\theta_S(t)$ and $\theta_L(t)$ are given by (14.17). In the limit where CP violation is neglected, in which case $K_S = K_1$ and $K_L = K_2$, this can be expressed in terms of the flavour eigenstates using (14.11) and (14.12),

$$\begin{aligned}|K(t)\rangle &\approx \tfrac{1}{2}\left(\theta_S\left[|K^0\rangle + |\overline{K}^0\rangle\right] + \theta_L\left[|K^0\rangle - |\overline{K}^0\rangle\right]\right)\\ &= \tfrac{1}{2}(\theta_S + \theta_L)|K^0\rangle + \tfrac{1}{2}(\theta_S - \theta_L)|\overline{K}^0\rangle.\end{aligned}$$

Because the masses of the $K_S$ and $K_L$ are slightly different, the oscillatory parts of $\theta_S(t)$ and $\theta_L(t)$ differ, and the initially pure $K^0$ beam will develop a $\overline{K}^0$ component. The corresponding strangeness oscillation probabilities are

$$P(K^0_{t=0} \to K^0) = |\langle K^0|K(t)\rangle|^2 = \tfrac{1}{4}|\theta_S + \theta_L|^2, \tag{14.38}$$

$$P(K^0_{t=0} \to \overline{K}^0) = |\langle \overline{K}^0|K(t)\rangle|^2 = \tfrac{1}{4}|\theta_S - \theta_L|^2. \tag{14.39}$$

This can be simplified by using the identity, $|\theta_S \pm \theta_L|^2 = |\theta_S|^2 + |\theta_L|^2 \pm 2\,\Re\mathrm{e}(\theta_S\theta_L^*)$,

$$\begin{aligned}|\theta_S(t) \pm \theta_L(t)|^2 &= e^{-\Gamma_S t} + e^{-\Gamma_L t} \pm 2\,\Re\mathrm{e}\left\{e^{-im_S t}e^{-\frac{1}{2}\Gamma_S t}\cdot e^{+im_L t}e^{-\frac{1}{2}\Gamma_L t}\right\}\\ &= e^{-\Gamma_S t} + e^{-\Gamma_L t} \pm 2e^{-\frac{1}{2}(\Gamma_S+\Gamma_L)t}\,\Re\mathrm{e}\left\{e^{i(m_L - m_S)t}\right\}\\ &= e^{-\Gamma_S t} + e^{-\Gamma_L t} \pm 2e^{-\frac{1}{2}(\Gamma_S+\Gamma_L)t}\cos(\Delta m\, t),\end{aligned}$$

where $\Delta m = m(K_L) - m(K_S)$. Substituting the above expression into (14.38) and (14.39) leads to

$$P(K^0_{t=0} \to K^0) = \tfrac{1}{4}\left[e^{-\Gamma_S t} + e^{-\Gamma_L t} + 2e^{-\frac{1}{2}(\Gamma_S+\Gamma_L)t}\cos(\Delta m\, t)\right], \tag{14.40}$$

$$P(K^0_{t=0} \to \overline{K}^0) = \tfrac{1}{4}\left[e^{-\Gamma_S t} + e^{-\Gamma_L t} - 2e^{-\frac{1}{2}(\Gamma_S+\Gamma_L)t}\cos(\Delta m\, t)\right]. \tag{14.41}$$

The above equations are reminiscent of the two-flavour neutrino oscillation probabilities, except here the amplitudes of the oscillations decay at a rate given by the arithmetic mean of the $K_S$ and $K_L$ decay rates.

Using the measured value of $\Delta m$ (see Section 14.5.2), the corresponding period of the strangeness oscillations is

$$T_{osc} = \frac{2\pi\hbar}{\Delta m} \approx 1.2 \times 10^{-9}\,\mathrm{s},$$

which turns out to be greater than the $K_S$ lifetime, $\tau(K_S) = 0.9 \times 10^{-10}$ s. Consequently, after one oscillation period, the $K_S$ and oscillatory components of (14.40) and (14.41) will have decayed away leaving an essentially pure $K_L$ beam. The resulting oscillation probabilities are plotted in Figure 14.13. Because of the relatively rapid decay of the $K_S$ component, the oscillatory structure is not particularly pronounced. Nevertheless, the observation of strangeness oscillations provides a method to measure $\Delta m$.

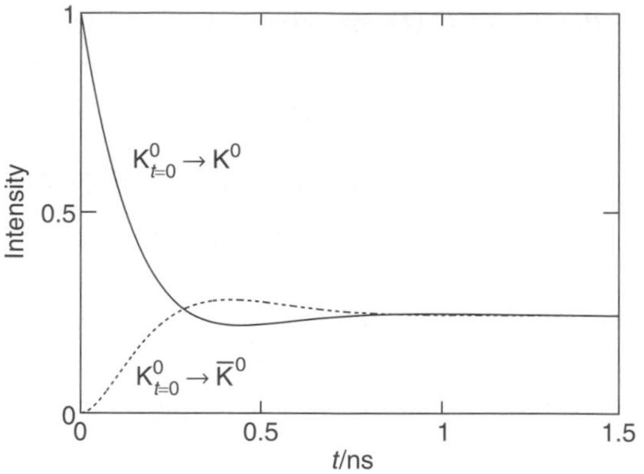


Fig. 14.13 The effect of strangeness oscillations, showing the relative $K^0$ and $\overline{K}^0$ components in a beam that was produced as a $K^0$ plotted against time.

### 14.5.2 The CPLEAR experiment

Strangeness oscillations can be studied by using the semi-leptonic decays of the neutral kaon system. Because the decays $K^0 \rightarrow \pi^+\ell^-\overline{\nu}_\ell$ and $\overline{K}^0 \rightarrow \pi^-\ell^+\nu_\ell$ ($\ell =$ e, μ) do not occur, the charge of the observed lepton in the semi-leptonic decays $K^0 \rightarrow \pi^-\ell^+\nu_\ell$ and $\overline{K}^0 \rightarrow \pi^+\ell^-\overline{\nu}_\ell$ uniquely tags the flavour eigenstate from which the decay originated.

The CPLEAR experiment, which operated from 1990 to 1996 at CERN, studied strangeness oscillations and CP violation in the neutral kaon system. It used a low-energy antiproton beam to produce kaons through the strong interaction processes

$$\overline{p}p \rightarrow K^-\pi^+K^0 \quad \text{and} \quad \overline{p}p \rightarrow K^+\pi^-\overline{K}^0.$$

The energy of the beam was sufficiently low that the particles were produced almost at rest. This enabled the production and decay to be observed in the same detector. The charge of the observed $K^\pm\pi^\mp$ identifies the flavour state of the neutral kaon produced in the $\overline{p}p$ interaction as being either a $\overline{K}^0$ or $K^0$. The neutral kaon then propagates at a low velocity as the linear combinations of the $K_S$ and $K_L$ with the time dependence given by (14.37). The charge of the observed lepton in the semi-leptonic decay then identifies the decay as coming from either a $K^0$ or $\overline{K}^0$, thus tagging the flavour component of the wavefunction at the time of decay. For example, Figure 14.14 shows an event in the CPLEAR detector where a $K^0$ is produced at the origin along with a $K^-\pi^+$, where the $K^-$ is distinguished from a $\pi^-$ by the absence of an associated signal in the Čerenkov detectors used for particle identification, see Section 1.2.1. The neutral kaon state subsequently decays as a $\overline{K}^0$, identified by its leptonic decay $\overline{K}^0 \rightarrow \pi^+e^-\overline{\nu}_e$. The relative rates of decays

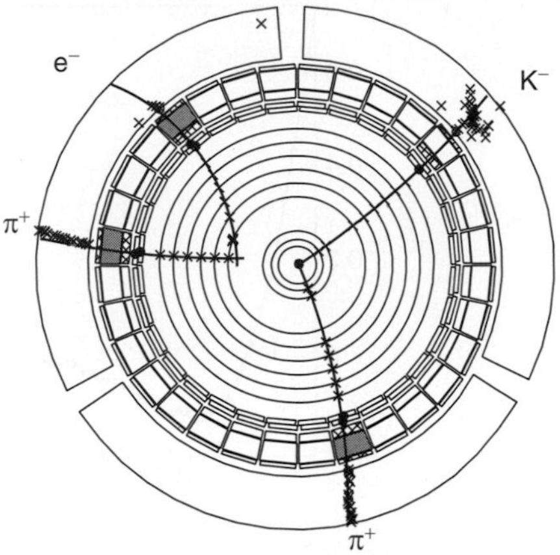


Fig. 14.14 An event in the CPLEAR detector where a $K^0$ is produced in $\overline{p}p \rightarrow K^-\pi^+K^0$ and decays as $\overline{K}^0 \rightarrow \pi^+e^-\overline{\nu}_e$. The grey boxes indicate signals from relativistic particles in the Čerenkov detectors. Courtesy of the CPLEAR Collaboration.

from $K^0$ and $\overline{K}^0$ as a function of the distance between the production point and the decay vertex, provides a direct measure of the relative $K^0$ and $\overline{K}^0$ components of the neutral kaon wavefunction as a function of time.

For a kaon initially produced as a $K^0$, the decay rates to $\pi^-e^+\nu_e$ and $\pi^+e^-\overline{\nu}_e$, denoted $R_+$ and $R_-$ respectively, are given by (14.40) and (14.41),

$$R_+ \propto P(K^0_{t=0} \rightarrow K^0) = N\tfrac{1}{4}\left[e^{-\Gamma_S t} + e^{-\Gamma_L t} + 2e^{-(\Gamma_S+\Gamma_L)t/2}\cos(\Delta m\, t)\right],$$
$$R_- \propto P(K^0_{t=0} \rightarrow \overline{K}^0) = N\tfrac{1}{4}\left[e^{-\Gamma_S t} + e^{-\Gamma_L t} - 2e^{-(\Gamma_S+\Gamma_L)t/2}\cos(\Delta m\, t)\right],$$

where $N$ is an overall normalisation factor related to the number of $p\overline{p}$ interactions. The corresponding expressions for the decays of neutral kaons that were produced as the $\overline{K}^0$ flavour state are

$$\overline{R}_+ \propto P(\overline{K}^0_{t=0} \rightarrow K^0) = N\tfrac{1}{4}\left[e^{-\Gamma_S t} + e^{-\Gamma_L t} - 2e^{-(\Gamma_S+\Gamma_L)t/2}\cos(\Delta m\, t)\right],$$
$$\overline{R}_- \propto P(\overline{K}^0_{t=0} \rightarrow \overline{K}^0) = N\tfrac{1}{4}\left[e^{-\Gamma_S t} + e^{-\Gamma_L t} + 2e^{-(\Gamma_S+\Gamma_L)t/2}\cos(\Delta m\, t)\right].$$

Because the QCD interaction is charge conjugation symmetric, equal numbers of $K^0$ and $\overline{K}^0$ are produced in the $\overline{p}p$ strong interaction and the same normalisation factor applies to $R_\pm$ and $\overline{R}_\pm$. The dependence on the overall normalisation can be removed by expressing the experimental measurements in terms of the asymmetry,

$$A_{\Delta m}(t) = \frac{(R_+ + \overline{R}_-) - (R_- + \overline{R}_+)}{(R_+ + \overline{R}_-) + (R_- + \overline{R}_+)},$$

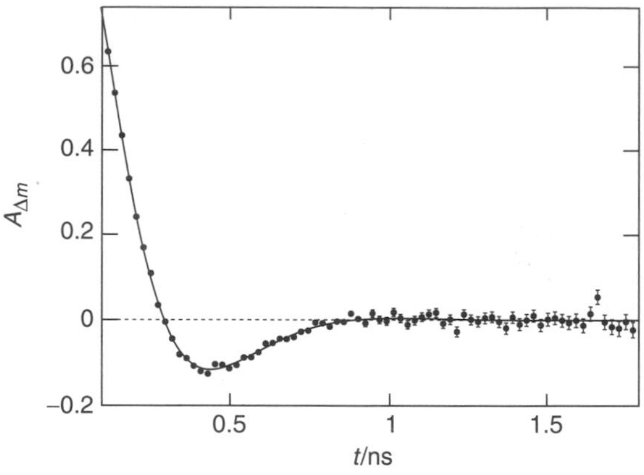


Fig. 14.15 The CPLEAR measurement of $A_{\Delta m}$ as a function of time. The curve shows expression of (14.42) for $\Delta m = 3.485 \times 10^{-15}$ GeV, modified to include the effects of the experimental timing resolution. Adapted from Angelopoulos *et al.* (2001).

which has the advantage that a number of potential systematic biases cancel. This asymmetry can be expressed as a function of time using the above expressions for $R_\pm$ and $\overline{R}_\pm$,

$$A_{\Delta m}(t) = \frac{2e^{-(\Gamma_S+\Gamma_L)t/2}\cos(\Delta m t)}{e^{-\Gamma_S t} + e^{-\Gamma_L t}}. \tag{14.42}$$

The experimental measurements of $A_{\Delta m}(t)$ from the CPLEAR experiment are shown in Figure 14.15. The effects of strangeness oscillations are clearly seen and the position of the minimum provides a precise measurement of $\Delta m$. The combined results from several experiments, including the CPLEAR experiment and the KTeV experiment at Fermilab, give

$$\Delta m = m(\mathrm{K}_L) - m(\mathrm{K}_S) = (3.483 \pm 0.006) \times 10^{-15}\ \mathrm{GeV}.$$

### 14.5.3 CP violation in the neutral kaon system

CP violation in the neutral kaon system has been studied by a number of experiments, including CPLEAR. If there is CP violation in $\mathrm{K}^0 \leftrightarrow \overline{\mathrm{K}}{}^0$ mixing process, the physical states of the neutral hadron system are

$$|\mathrm{K}_S\rangle = \frac{1}{\sqrt{1+|\varepsilon|^2}}\left(|\mathrm{K}_1\rangle + \varepsilon|\mathrm{K}_2\rangle\right) \quad \text{and} \quad |\mathrm{K}_L\rangle = \frac{1}{\sqrt{1+|\varepsilon|^2}}\left(|\mathrm{K}_2\rangle + \varepsilon|\mathrm{K}_1\rangle\right), \tag{14.43}$$

which can be expressed in terms of the flavour eigenstates as

$$|K_S\rangle = \frac{1}{\sqrt{2(1+|\varepsilon|^2)}}\left[(1+\varepsilon)|K^0\rangle + (1-\varepsilon)|\overline{K}^0\rangle\right],$$

$$|K_L\rangle = \frac{1}{\sqrt{2(1+|\varepsilon|^2)}}\left[(1+\varepsilon)|K^0\rangle - (1-\varepsilon)|\overline{K}^0\rangle\right].$$

The corresponding expressions for the flavour eigenstates in terms of the physical $K_S$ and $K_L$ are

$$|K^0\rangle = \tfrac{1}{1+\varepsilon}\sqrt{\tfrac{1+|\varepsilon|^2}{2}}\,(|K_S\rangle + |K_L\rangle) \quad \text{and} \quad |\overline{K}^0\rangle = \tfrac{1}{1-\varepsilon}\sqrt{\tfrac{1+|\varepsilon|^2}{2}}\,(|K_S\rangle - |K_L\rangle).$$

Therefore, accounting for the possibility of CP violation in neutral kaon mixing, a neutral kaon state that was produced as a $K^0$ evolves as

$$|K(t)\rangle = \frac{1}{1+\varepsilon}\sqrt{\frac{1+|\varepsilon|^2}{2}}\,[\theta_S(t)|K_S\rangle + \theta_L(t)|K_L\rangle], \tag{14.44}$$

where as before $\theta_S(t)$ and $\theta_L(t)$ are given by (14.17). Direct CP violation in kaon decay is a relatively small effect ($\varepsilon'/\varepsilon \sim 10^{-3}$) and decays to the $\pi\pi$ final state can be taken to originate almost exclusively from the CP-even $K_1$ component of the wavefunction. The time evolution of (14.44) can be expressed in terms of the $K_1$ and $K_2$ states using (14.43)

$$\begin{aligned}|K(t)\rangle &= \tfrac{1}{\sqrt{2}}\frac{1}{(1+\varepsilon)}\,[\theta_S(|K_1\rangle + \varepsilon|K_2\rangle) + \theta_L(|K_2\rangle + \varepsilon|K_1\rangle)] \\ &= \tfrac{1}{\sqrt{2}}\frac{1}{(1+\varepsilon)}\,[(\theta_S + \varepsilon\theta_L)|K_1\rangle + (\theta_L + \varepsilon\theta_S)|K_2\rangle].\end{aligned}$$

The decay rate to two pions is therefore given by

$$\Gamma(K^0_{t=0} \to \pi\pi) \propto |\langle K_1|K(t)\rangle|^2 = \tfrac{1}{2}\left|\frac{1}{1+\varepsilon}\right|^2 |\theta_S + \varepsilon\theta_L|^2. \tag{14.45}$$

Because $|\varepsilon| \ll 1$,

$$\left|\frac{1}{1+\varepsilon}\right|^2 = \frac{1}{(1+\varepsilon^*)(1+\varepsilon)} \approx \frac{1}{1+2\,\Re\mathrm{e}\{\varepsilon\}} \approx 1 - 2\,\Re\mathrm{e}\{\varepsilon\}.$$

The term $|\theta_S + \varepsilon\theta_L|^2$ can be simplified using $|\theta_S \pm \varepsilon\theta_L|^2 = |\theta_S|^2 + |\theta_L|^2 \pm 2\,\Re\mathrm{e}(\theta_S\varepsilon^*\theta_L^*)$ and by writing $\varepsilon = |\varepsilon|e^{i\phi}$,

$$\begin{aligned}|\theta_S + \varepsilon\theta_L|^2 &= \left|e^{-im_S t - \Gamma_S t/2} + |\varepsilon|e^{i\phi}e^{-im_L t - \Gamma_L t/2}\right|^2 \\ &= e^{-\Gamma_S t} + |\varepsilon|^2 e^{-\Gamma_L t} + 2|\varepsilon|e^{-(\Gamma_S+\Gamma_L)t/2}\cos(\Delta m\, t - \phi).\end{aligned}$$

Therefore (14.45) can be written as

$$\Gamma(K^0_{t=0} \to \pi\pi) = \frac{N}{2}\,(1 - 2\,\Re\mathrm{e}\{\varepsilon\})\left[e^{-\Gamma_S t} + |\varepsilon|^2 e^{-\Gamma_L t} + 2|\varepsilon|e^{-(\Gamma_S+\Gamma_L)t/2}\cos(\Delta m\, t - \phi)\right], \tag{14.46}$$

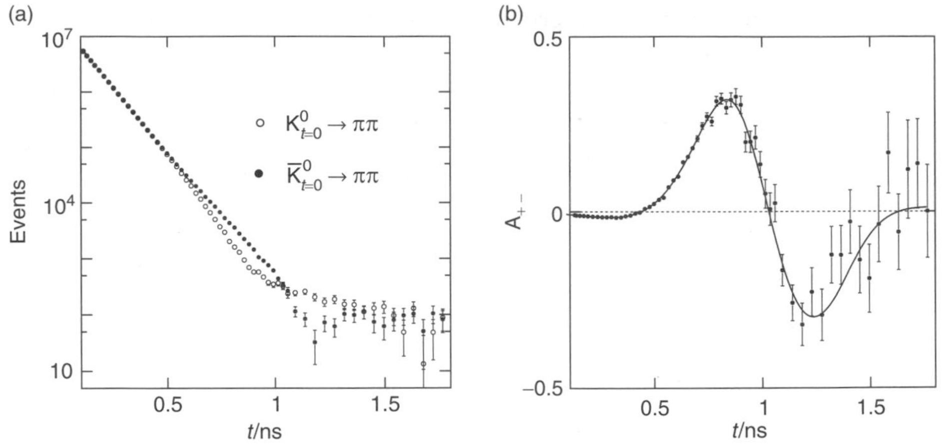


Fig. 14.16 The CPLEAR measurement of $A_{+-}$. Adapted from Angelopoulos *et al.* (2000).

where $N$ is a normalisation factor. The first term in the square brackets corresponds to the contribution from $K_S$ decays. The second term is the contribution from $K_L$ decays, which is small since $|\varepsilon|^2 \ll 1$. The final term is the interference between the $K_S$ and $K_L$ components of the wavefunction. The corresponding expression for the decay rate to two pions from a state that was initially a $\overline{K}^0$ is

$$\Gamma(\overline{K}^0_{t=0} \to \pi\pi) = \frac{N}{2}(1 + 2\,\Re e\{\varepsilon\})\left[e^{-\Gamma_S t} + |\varepsilon|^2 e^{-\Gamma_L t} - 2|\varepsilon|e^{-(\Gamma_S+\Gamma_L)t/2}\cos(\Delta m\, t - \phi)\right]. \tag{14.47}$$

Here the interference term has the opposite sign to that of (14.46). For $t \ll \tau_S$ and $t \gg \tau_L$ the expressions of (14.46) and (14.47) are approximately equal, but at intermediate times, the interference term results in a significant difference in the $\pi\pi$ decay rates. Figure 14.16a shows the numbers of $K \to \pi^+\pi^-$ decays observed in the CPLEAR experiment, plotted as a function of the neutral kaon decay time for events that were initially tagged as either a $K^0$ or $\overline{K}^0$. The difference in the region of $t \sim 1$ ns is the result of this interference term and the magnitude of the difference is proportional to $|\varepsilon|$.

In practice, the experimental measurement of $\varepsilon$ at CPLEAR was obtained from the asymmetry $A_{+-}$, defined as

$$A_{+-} = \frac{\Gamma\left(\overline{K}^0_{t=0} \to \pi^+\pi^-\right) - \Gamma\left(K^0_{t=0} \to \pi^+\pi^-\right)}{\Gamma\left(\overline{K}^0_{t=0} \to \pi^+\pi^-\right) + \Gamma\left(K^0_{t=0} \to \pi^+\pi^-\right)}. \tag{14.48}$$

From (14.46) and (14.47), this can be expressed as

$$A_{+-} = \frac{4\,\Re e\{\varepsilon\}\left[e^{-\Gamma_S t} + |\varepsilon|^2 e^{-\Gamma_L t}\right] - 4|\varepsilon|e^{-(\Gamma_L+\Gamma_S)t/2}\cos(\Delta m\, t - \phi)}{2\left[e^{-\Gamma_S t} + |\varepsilon|^2 e^{-\Gamma_L t}\right] - 8\,\Re e\{\varepsilon\}|\varepsilon|e^{-(\Gamma_L+\Gamma_S)t/2}\cos(\Delta m\, t - \phi)}.$$

Since $\varepsilon$ is small, the term in the denominator that is proportional to $|\varepsilon|\,\Re e\{\varepsilon\}$ can be neglected at all times, giving

$$\begin{aligned} A_{+-} &\approx 2\,\Re e\{\varepsilon\} - \frac{2|\varepsilon| e^{-(\Gamma_L+\Gamma_S)t/2}\cos(\Delta m\, t - \phi)}{e^{-\Gamma_S t} + |\varepsilon|^2 e^{-\Gamma_L t}} \\ &= 2\,\Re e\{\varepsilon\} - \frac{2|\varepsilon| e^{(\Gamma_S-\Gamma_L)t/2}\cos(\Delta m\, t - \phi)}{1 + |\varepsilon|^2 e^{(\Gamma_S-\Gamma_L)t}}. \end{aligned} \tag{14.49}$$

Hence both the phase and magnitude of $\varepsilon$ can be cleanly extracted from the experimental measurement of $A_{+-}$. Figure 14.16b shows the asymmetry $A_{+-}$ obtained from the CPLEAR data of Figure 14.16a. The measured asymmetry is well described by (14.49) with the measured parameters

$$|\varepsilon| = (2.264 \pm 0.035) \times 10^{-3} \quad \text{and} \quad \phi = (43.19 \pm 0.73)^\circ. \tag{14.50}$$

The non-zero value of $|\varepsilon|$ provides clear evidence for CP violation in the weak interaction. Because $\phi$ is close to 45°, the real and imaginary parts of $\varepsilon$ are roughly the same size, $\Re e\{\varepsilon\} \approx \Im m\{\varepsilon\}$.

### 14.5.4 CP violation in leptonic decays

We can also observe CP violation in the semi-leptonic decays of the $K_L$ from measurements at a large distance from the production of the $K^0/\overline{K}^0$. Since the semi-leptonic decays occur from a particular kaon flavour eigenstate, the relative decay rates can be obtained from the $K_L$ wavefunction expressed in terms of its $K^0$ and $\overline{K}^0$ components,

$$|K_L\rangle = \frac{1}{\sqrt{2(1+|\varepsilon^2|)}}\left[(1+\varepsilon)|K^0\rangle - (1-\varepsilon)|\overline{K}^0\rangle\right].$$

(with $(1+\varepsilon)|K^0\rangle \to \pi^- e^+ \nu_e$ and $(1-\varepsilon)|\overline{K}^0\rangle \to \pi^+ e^- \overline{\nu}_e$)

Hence the decay rates are

$$\Gamma(K_L \to \pi^+ e^- \overline{\nu}_e) \propto |\langle \overline{K}^0 | K_L\rangle|^2 \propto |1-\varepsilon|^2 \approx 1 - 2\,\Re e\,(\varepsilon),$$
$$\Gamma(K_L \to \pi^- e^+ \nu_e) \propto |\langle K^0 | K_L\rangle|^2 \propto |1+\varepsilon|^2 \approx 1 + 2\,\Re e\,(\varepsilon).$$

The experimental measurements are conveniently expressed in terms of the charge asymmetry $\delta$ defined as

$$\delta = \frac{\Gamma(K_L \to \pi^- e^+ \nu_e) - \Gamma(K_L \to \pi^+ e^- \overline{\nu}_e)}{\Gamma(K_L \to \pi^- e^+ \nu_e) + \Gamma(K_L \to \pi^+ e^- \overline{\nu}_e)} \approx 2\,\Re e\,(\varepsilon) = 2|\varepsilon|\cos\phi.$$

Experimentally, the number of observed $K_L \to \pi^- e^+ \nu_e$ decays is found to be 0.66% larger than the number of $K_L \to \pi^+ e^- \overline{\nu}_e$ decays, giving

$$\delta = 0.327 \pm 0.012\%. \tag{14.51}$$

This is consistent with the expectation from the measured values of $|\varepsilon|$ and $\phi$ given in (14.50), which predict a charge asymmetry of $\delta = 0.33\%$.

Interestingly, the small difference in the $K_L \to \pi^- e^+ \nu_e$ and $K_L \to \pi^- e^- \overline{\nu}_e$ decay rates can be used to provide an unambiguous definition of what we mean by matter as opposed to antimatter, which in principle, could be communicated to aliens in a distant galaxy; the electrons in the atoms in our region of the Universe have the same charge sign as those emitted least often in the decays of the long-lived neutral kaons. Interesting, but perhaps of little practical use.

### 14.5.5 Interpretation of the neutral kaon data

The size of the mass splitting $\Delta m = m(K_L) - m(K_S)$ and magnitude of the CP violating parameter $\varepsilon$ can be related to the elements of CKM matrix and how they enter the matrix elements for $K^0 \leftrightarrow \overline{K}^0$ mixing. In the box diagrams responsible for neutral kaon mixing, shown in Figure 14.17, there are nine possible combinations of u, c and t flavours for the two virtual quarks. The matrix element for each box diagram has the dependence

$$\mathcal{M}_{qq'} \propto V_{qd} V^*_{qs} V^*_{q's} V_{q'd}.$$

For reasons that are explained below, to first order, the value of $\varepsilon$ is determined by the matrix elements for box diagrams involving at least one top quark, whereas the dominant contributions to the $K_L$ and $K_S$ mass splitting arises from box diagrams with combinations of virtual up- and charm quarks. A full treatment of these calculations is beyond the scope of this book, but the essential physical concepts can be readily understood.

The mass splitting $\Delta m$ can be related to the magnitude of the matrix elements for $K^0 \leftrightarrow \overline{K}^0$ mixing. Owing to the smallness of $|V_{td}|$ and $|V_{ts}|$, the diagrams involving the top quark can, to a first approximation, be neglected (see Problem 14.8). Hence the overall matrix element for $K^0 \leftrightarrow \overline{K}^0$ mixing can be written

$$\mathcal{M} \approx \mathcal{M}_{uu} + \mathcal{M}_{uc} + \mathcal{M}_{cu} + \mathcal{M}_{cc}.$$

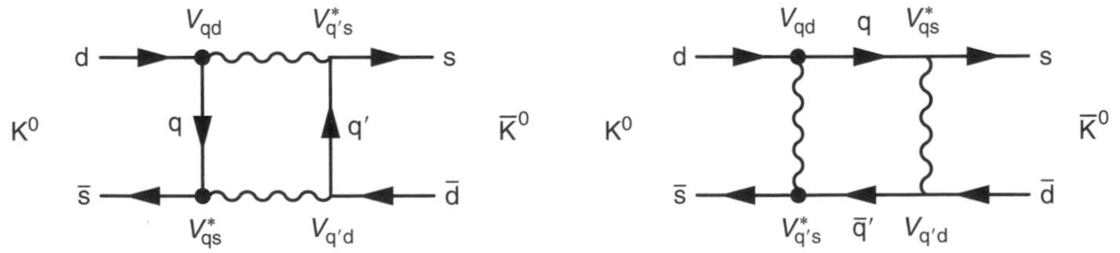


Fig. 14.17 The box diagrams for $K^0 \leftrightarrow \overline{K}^0$ mixing, where the virtual quarks can be any of the nine combinations of $q, q' = \{u, c, t\}$.

The individual matrix elements will be proportional to $G_F^2$ and will include propagator terms for the two virtual quarks involved and hence

$$\mathcal{M} \sim G_F^2 \left[ \frac{V_{ud}V_{us}^*V_{ud}V_{us}^*}{(k^2 - m_u^2)^2} + 2\frac{V_{ud}V_{us}^*V_{cd}V_{cs}^*}{(k^2 - m_u^2)(k^2 - m_c^2)} + \frac{V_{cd}V_{cs}^*V_{cd}V_{cs}^*}{(k^2 - m_c^2)^2} \right],$$

where $k$ is the four-momentum appearing in the box of virtual particles. Writing $V_{ud} \approx V_{cs} \approx \cos\theta_c$ and $V_{us} \approx -V_{cd} \approx \sin\theta_c$, this can be expressed as

$$\mathcal{M} \sim G_F^2 \left[ \frac{\sin^2\theta_c\cos^2\theta_c}{(k^2 - m_u^2)^2} - 2\frac{\sin^2\theta_c\cos^2\theta_c}{(k^2 - m_u^2)(k^2 - m_c^2)} + \frac{\sin^2\theta_c\cos^2\theta_c}{(k^2 - m_c^2)^2} \right]$$
$$\sim G_F^2 \sin^2\theta_c \cos^2\theta_c \frac{(m_c^2 - m_u^2)^2}{(k^2 - m_u^2)^2(k^2 - m_c^2)^2}.$$

If the masses of the up- and charm quarks were identical, this contribution to the matrix element for $K^0 \leftrightarrow \overline{K}^0$ mixing would vanish. The evaluation of the matrix element, which involves the integration over the four-momentum $k$, is non-trivial and the resulting expression for $\Delta m$ is simply quoted here

$$\Delta m \approx \frac{G_F^2}{3\pi^2} \sin^2\theta_c \cos^2\theta_c f_K^2 m_K \frac{(m_c^2 - m_u^2)^2}{m_c^2}. \tag{14.52}$$

In this expression $f_K \sim 170\,\text{MeV}$ is the kaon decay factor, analogous to that introduced in Section 11.6.1 in the context of $\pi^\pm$ decay. Although the above analysis is rather simplistic, it gives a reasonable estimate of the magnitude of $\Delta m$. Taking the charm quark mass to be 1.3 GeV, Equation (14.52) gives the predicted value of $\Delta m \sim 5 \times 10^{-15}\,\text{GeV}$, which is within a factor of two of the observed value. The smallness of $\Delta m$ is due to the presence of the $G_F^2$ term from the two exchanged W bosons in the box diagram.

### The Standard Model interpretation of $\varepsilon$

CP violation in $K^0 \leftrightarrow \overline{K}^0$ mixing arises because the matrix element for $K^0 \rightarrow \overline{K}^0$ is not the same as that for $\overline{K}^0 \rightarrow K^0$. For example, the matrix elements for $K^0 \rightarrow \overline{K}^0$ and $\overline{K}^0 \rightarrow K^0$, arising from the exchange of a charm and a top quark, shown in Figure 14.18, are respectively proportional to

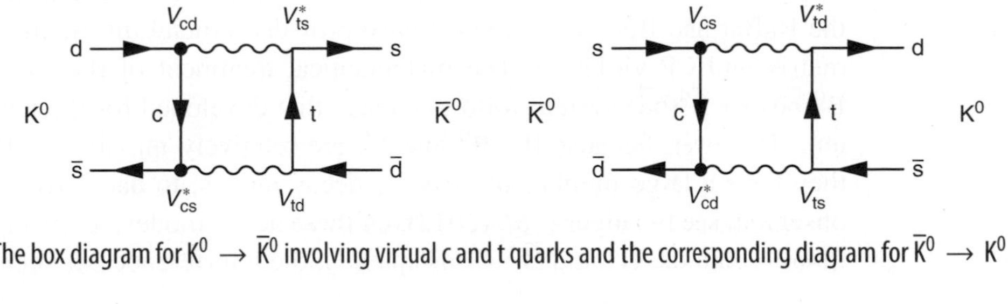


Fig. 14.18 The box diagram for $K^0 \rightarrow \overline{K}^0$ involving virtual c and t quarks and the corresponding diagram for $\overline{K}^0 \rightarrow K^0$.

$$\mathcal{M}_{12} \propto V_{cd}V_{cs}^*V_{td}V_{ts}^* \quad \text{and} \quad \mathcal{M}_{21} \propto V_{cd}^*V_{cs}V_{td}^*V_{ts} = \mathcal{M}_{12}^*.$$

CP violation in mixing occurs if $\mathcal{M}_{12} \neq \mathcal{M}_{12}^*$. It can be shown (see Problem 14.9) that

$$|\varepsilon| \approx \frac{\Im\mathfrak{m}\{\mathcal{M}_{12}\}}{\sqrt{2}\,\Delta m}.$$

The imaginary part of $\mathcal{M}_{12}$ can be expressed in terms of the possible combinations of exchanged u, c, t quarks in the box diagrams,

$$\Im\mathfrak{m}\{\mathcal{M}_{12}\} = \sum_{q,q'} \mathcal{A}_{qq'}\,\Im\mathfrak{m}\,(V_{qd}V_{qs}^*V_{q'd}V_{q's}^*),$$

where the parameters $\mathcal{A}_{qq'}$ are constants that depend on the masses of the exchanged quarks. In the Wolfenstein parameterisation of the CKM matrix given in (14.9), the imaginary elements of the CKM matrix are $V_{td}$ and $V_{ub}$. Since $V_{ub}$ is not relevant for kaon mixing, CP violation in neutral kaon mixing is associated with box diagrams involving at least one top quark, and therefore

$$|\varepsilon| \propto \mathcal{A}_{ut}\,\Im\mathfrak{m}\,(V_{ud}V_{us}^*V_{td}V_{ts}^*) + \mathcal{A}_{ct}\,\Im\mathfrak{m}\,(V_{cd}V_{cs}^*V_{td}V_{ts}^*) + \mathcal{A}_{tt}\,\Im\mathfrak{m}\,(V_{td}V_{ts}^*V_{td}V_{ts}^*). \tag{14.53}$$

Writing the elements of the CKM matrix in terms of $A$, $\lambda$, $\rho$ and $\eta$ of the Wolfenstein parametrisation (14.9), it can be shown that (see Problem 14.10)

$$|\varepsilon| \propto \eta(1 - \rho + \text{constant}).$$

Hence the measurement of a non-zero value of $|\varepsilon|$ implies that $\eta \neq 0$ and provides an experimental constraint on the possible values of the parameters $\eta$ and $\rho$.

## 14.6 B-meson physics

The oscillations of neutral mesons are not confined to kaons, they have also been observed for the heavy neutral meson systems,

$$B^0(\overline{b}d) \leftrightarrow \overline{B}^0(b\overline{d}), \quad B_s^0(\overline{b}s) \leftrightarrow \overline{B}_s^0(b\overline{s}) \quad \text{and} \quad D^0(\overline{c}u) \leftrightarrow \overline{D}^0(c\overline{u}).$$

In particular, the results from the studies of the $B^0(\overline{b}d)$ and $\overline{B}^0(b\overline{d})$ mesons by the BaBar and Belle experiments have provided crucial information on the CKM matrix and CP violation. The mathematical treatment of the oscillations of the $B^0(\overline{b}d) \leftrightarrow \overline{B}^0(b\overline{d})$ system follows closely that developed for the neutral kaon system. However, because the $B^0$ and $\overline{B}^0$ are relatively massive, $m(B) \sim 5.3\,\text{GeV}$, they have a large number of possible decay modes; to date, over 400 have been observed; see Beringer *et al.* (2012). Of these decay modes, relatively few are common to both the $B^0$ and $\overline{B}^0$. Consequently, the interference between the decays of the $B^0$ and $\overline{B}^0$ is small. Because of this, it can be shown (see the following starred section) that $B^0 \leftrightarrow \overline{B}^0$ oscillations can be described by a single angle $\beta$ and that the physical eigenstates of the neutral B-meson system are

$$|B_L\rangle = \frac{1}{\sqrt{2}}\left[|B^0\rangle + e^{-i2\beta}|\overline{B}^0\rangle\right] \quad \text{and} \quad |B_H\rangle = \frac{1}{\sqrt{2}}\left[|B^0\rangle - e^{-i2\beta}|\overline{B}^0\rangle\right]. \tag{14.54}$$

The $B_L$ and $B_H$ are respectively a lighter and heavier state with almost identical lifetimes; again the mass difference $m(B_L) - m(B_H)$ is very small.

### 14.6.1 *B-meson mixing

The treatment of B-mixing given here, makes a number of approximations to simplify the discussion in order to focus on the main physical concepts. The physical neutral B-meson states are the eigenstates of the overall Hamiltonian of the $B^0$ and $\overline{B}^0$ system, analogous to the kaon states discussed in Section 14.4.3. There are a large number of B-meson decay modes, of which only a few are common to both the $B^0$ and $\overline{B}^0$, and the contribution to the effective Hamiltonian of (14.25) from the interference between the decays of the $B^0$ and $\overline{B}^0$ can be neglected, $\Gamma_{12} = \Gamma_{21}^* \approx 0$. In this case

$$\mathcal{H} \approx \begin{pmatrix} M - \frac{i}{2}\Gamma & M_{12} \\ M_{12}^* & M - \frac{i}{2}\Gamma \end{pmatrix}, \tag{14.55}$$

where $M_{12}$ is due to the box diagrams for $B^0 \leftrightarrow \overline{B}^0$ mixing. The eigenvalues of (14.55), which determine the masses and lifetimes of the physical states, are

$$\lambda_H = m_H + \tfrac{1}{2}i\Gamma_H \approx M + |M_{12}| - \tfrac{1}{2}i\Gamma,$$
$$\lambda_L = m_L + \tfrac{1}{2}i\Gamma_L \approx M - |M_{12}| - \tfrac{1}{2}i\Gamma.$$

leading to a heavier state $B_H$ and a lighter state $B_L$ with masses

$$m_H = M + |M_{12}| \quad \text{and} \quad m_L = M - |M_{12}|. \tag{14.56}$$

Because the interference term $\Gamma_{12}$ is sufficiently small that it can be neglected, the imaginary parts of $\lambda_H$ and $\lambda_L$ are the same. Consequently, the $B_H$ and $B_L$ have approximately the same lifetime, which is measured to be

$$\Gamma_H \approx \Gamma_L \approx \Gamma \approx 4.3 \times 10^{-13}\,\text{GeV}.$$

The corresponding physical eigenstates of the effective Hamiltonian are

$$|B_L\rangle = \frac{1}{\sqrt{1+|\xi|^2}}(|B^0\rangle + \xi|\overline{B}^0\rangle) \quad \text{and} \quad |B_H\rangle = \frac{1}{\sqrt{1+|\xi|^2}}(|B^0\rangle - \xi|\overline{B}^0\rangle), \tag{14.57}$$

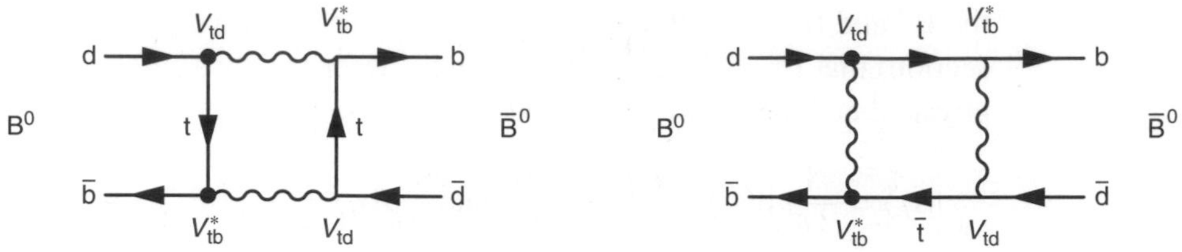


Fig. 14.19 The dominant box diagrams for $B^0 \leftrightarrow \overline{B}^0$ mixing.

where $\xi$ is given by (14.29),

$$\xi = \left(\frac{M_{12}^* - \frac{i}{2}\Gamma_{12}^*}{M_{12} - \frac{i}{2}\Gamma_{12}}\right)^{\frac{1}{2}} \approx \frac{M_{12}^*}{|M_{12}|}, \tag{14.58}$$

from which it follows that $|\xi| \approx 1$.

In $K^0 \leftrightarrow \overline{K}^0$ mixing, the contributions from different flavours of virtual quarks in the box diagrams are of a similar order of magnitude. Here, because $|V_{tb}| \gg |V_{ts}| > |V_{td}|$, only the box diagrams involving two top quarks, shown in Figure 14.19, contribute significantly to the mixing process and

$$M_{12}^* \propto (V_{td}V_{tb}^*)^2.$$

In the Wolfenstein parametrisation of the CKM matrix (14.9), $V_{tb}$ is real and thus

$$\xi = \frac{M_{12}^*}{|M_{12}|} = \frac{(V_{td}V_{tb}^*)^2}{|(V_{td}V_{tb}^*)^2|} = \frac{V_{td}^2}{|V_{td}^2|}. \tag{14.59}$$

By writing $V_{td}$ as

$$V_{td} = |V_{td}|e^{-i\beta},$$

the expression for $\xi$ given in (14.59) is simply

$$\xi = e^{-i2\beta}.$$

Hence, the physical neutral B-meson states of (14.57) are

$$|B_L\rangle = \frac{1}{\sqrt{2}}\left(|B^0\rangle + e^{-i2\beta}|\overline{B}^0\rangle\right) \quad \text{and} \quad |B_H\rangle = \frac{1}{\sqrt{2}}\left(|B^0\rangle - e^{-i2\beta}|\overline{B}^0\rangle\right). \tag{14.60}$$

From (14.56), it can be seen that the mass difference

$$\Delta m_d = m(B_H) - m(B_L) = 2|M_{12}| \propto |(V_{td}V_{tb}^*)^2|. \tag{14.61}$$

Because $V_{tb} \approx 1$, it follows that the $B_H - B_L$ mass difference is proportional to $|V_{td}^2|$. Consequently, the measurement of $\Delta m_d$ in $B^0 \leftrightarrow \overline{B}^0$ mixing provides a way of determining $|V_{td}|$.

### 14.6.2 Neutral B-meson oscillations

The mathematical description of the phenomenon of B-meson oscillations follows that developed for the kaon system. Suppose a $B^0(\overline{b}d)$ is produced at a time $t = 0$, such that $|B(0)\rangle = |B^0\rangle$. Then from (14.60), the flavour state $B^0$ can be expressed in terms of the physical $B_H$ and $B_L$ mass eigenstates

$$|B^0\rangle = \frac{1}{\sqrt{2}}\left(|B_L\rangle + |B_H\rangle\right).$$

The wavefunction evolves according to the time dependence of the physical states,

$$|B(t)\rangle = \frac{1}{\sqrt{2}}\left[\theta_L(t)|B_L\rangle + \theta_H(t)|B_H\rangle\right], \tag{14.62}$$

where the time dependencies of the physical states are

$$\theta_L = e^{-\Gamma t/2}e^{-im_L t} \quad \text{and} \quad \theta_H = e^{-\Gamma t/2}e^{-im_H t}.$$

Equation (14.62) can be expressed in terms of the flavour eigenstates using (14.60),

$$|B(t)\rangle = \tfrac{1}{2}\left[(\theta_L + \theta_H)|B^0\rangle + e^{-i2\beta}(\theta_L - \theta_H)|\overline{B}^0\rangle\right] = \tfrac{1}{2}\left[\theta_+|B^0\rangle + \xi\theta_-|\overline{B}^0\rangle\right], \tag{14.63}$$

where $\theta_\pm = \theta_L \pm \theta_H$ and $\xi = e^{-2i\beta}$. By writing $m_L = M - \Delta m_d/2$ and $m_H = M + \Delta m_d/2$,

$$\theta_\pm(t) = e^{-\Gamma t/2}e^{-iMt} \times \left[e^{i\Delta m_d t/2} \pm e^{-i\Delta m_d t/2}\right], \tag{14.64}$$

from which it follows that $\theta_+$ and $\theta_-$ are given by

$$\theta_+ = 2e^{-\Gamma t/2}e^{-iMt}\cos\left(\frac{\Delta m_d t}{2}\right) \quad \text{and} \quad \theta_- = 2ie^{-\Gamma t/2}e^{-iMt}\sin\left(\frac{\Delta m_d t}{2}\right).$$

The probabilities of the state decaying as a $|B^0\rangle$ or a $|\overline{B}^0\rangle$ are therefore

$$P(B^0_{t=0} \to B^0) = |\langle B(t)|B^0\rangle|^2 = \tfrac{1}{4}e^{-\Gamma t}|\theta_+|^2 = e^{-\Gamma t}\cos^2\left(\tfrac{1}{2}\Delta m_d t\right),$$
$$P(B^0_{t=0} \to \overline{B}^0) = |\langle B(t)|\overline{B}^0\rangle|^2 = \tfrac{1}{4}e^{-\Gamma t}|\xi\theta_-|^2 = |\xi|^2 e^{-\Gamma t}\sin^2\left(\tfrac{1}{2}\Delta m_d t\right). \tag{14.65}$$

The corresponding expressions for a state that was produced as a $\overline{B}^0$ are

$$P(\overline{B}^0_{t=0} \to \overline{B}^0) = e^{-\Gamma t}\cos^2\left(\tfrac{1}{2}\Delta m_d t\right) \text{ and } P(\overline{B}^0_{t=0} \to B^0) = \left|\frac{1}{\xi}\right|^2 e^{-\Gamma t}\sin^2\left(\tfrac{1}{2}\Delta m_d t\right).$$

Because the contribution to the effective Hamiltonian for the neutral B-meson system from the interference between $B^0$ and $\overline{B}^0$ decays can be neglected, $|\xi| = |e^{-i2\beta}| = 1$ and therefore

$$P(\overline{B}^0_{t=0} \to \overline{B}^0) \approx P(B^0_{t=0} \to B^0) \quad \text{and} \quad P(\overline{B}^0_{t=0} \to B^0) \approx P(B^0_{t=0} \to \overline{B}^0).$$

Consequently, it is very hard to observe CP violation in neutral B-meson *mixing*. Nevertheless, $B^0 \leftrightarrow \overline{B}^0$ oscillations can be utilised to measure $\Delta m_d$, which from (14.61) provides a measurement of $|V_{td}|$.

### 14.6.3 The BaBar and Belle experiments

The BaBar (1999–2008) and Belle (1999–2010) experiments were designed to provide precise measurements of CP violation in the neutral B-meson system. The experiments utilised the high-luminosity PEP2 and KEKB $e^+e^-$ colliders at SLAC in California and KEK in Japan. To produce very large numbers of $B^0\overline{B}^0$ pairs, the colliders operated at a centre-of-mass energy of 10.58 GeV, which corresponds to the mass of the $\Upsilon(4S)$ $b\overline{b}$ resonance. The $\Upsilon(4S)$ predominantly decays by either $\Upsilon(4S) \to B^+B^-$ or $\Upsilon(4S) \to B^0\overline{B}^0$, with roughly equal branching ratios. The masses of the charged and neutral B-mesons are 5.279 GeV, and therefore they are produced almost at rest in the centre-of-mass frame of the $\Upsilon(4S)$. Because the lifetimes of the neutral B-mesons are short ($\tau = 1.519\times10^{-12}$ s) and they are produced with relatively low velocities, they travel only a short distance in the centre-of-mass frame before decaying. Consequently, in the centre-of-mass frame it would be hard to separate the decays of two B-mesons produced in $e^+e^- \to \Upsilon(4S) \to B^0\overline{B}^0$. For this reason, the PEP2 and KEKB colliders operated as asymmetric b-factories, where the electron beam energy was higher than that of the positron beam. For example, PEP2 collided a 9 GeV electron beam with a 3.1 GeV positron beam. Owing to the asymmetric beam energies, the $\Upsilon(4S)$ is boosted along the beam axis; at the PEP2 collider the $\Upsilon(4S)$ is produced with $\beta\gamma = 0.56$. As a result of this boost, the mean distance between the two B-meson decay vertices in the beam direction is increased to $\Delta z \sim 200\,\mu m$. This separation is large enough for the two B-meson decay vertices to be resolved using a high-precision silicon vertex detector, as described in Section 1.3.1.

The oscillations of B-mesons can be studied through their leptonic decays,

$$B^0(\overline{b}d) \to D^-(\overline{c}d)\,\mu^+\,\nu_\mu \quad \text{and} \quad \overline{B}^0(b\overline{d}) \to D^+(c\overline{d})\,\mu^-\,\overline{\nu}_\mu.$$

The sign of the lepton identifies the B-meson flavour state, since the decays $B^0 \to D^+\mu^-\overline{\nu}_\mu$ and $\overline{B}^0 \to D^-\mu^+\nu_\mu$ do not occur. After production in $e^+e^- \to B^0\overline{B}^0$, the two B-mesons propagate as a coherent state. When one of the B-mesons decays into a particular flavour eigenstate, the overall wavefunction collapses, fixing the flavour state of the other B-meson. For example, Figure 14.20 illustrates the case where at $t = 0$ one of the B-mesons is observed to decay to $D^+\mu^-\overline{\nu}_\mu$, tagging it as a $\overline{B}^0$. At this instant in time, the second B-meson corresponds to a pure $B^0$ state, $|B(0)\rangle = |B^0\rangle$. The wavefunction of the second B-meson then evolves according to (14.63). When the second B-meson decays $\Delta t$ later, the charge sign of the observed lepton tags the flavour eigenstate in which the decay occurred. Thus $B^0 \leftrightarrow \overline{B}^0$ oscillations can be studied by measuring the rates where the two B-meson decays are the same flavour, $B^0B^0$ and $\overline{B}^0\overline{B}^0$, or are of opposite flavour, $B^0\overline{B}^0$. The same flavour (SF) decays give like-sign leptons, $\mu^+\mu^+$ or $\mu^-\mu^-$, and the opposite flavour (OF) decays give opposite-sign leptons, $\mu^+\mu^-$. The relative rates depend on the

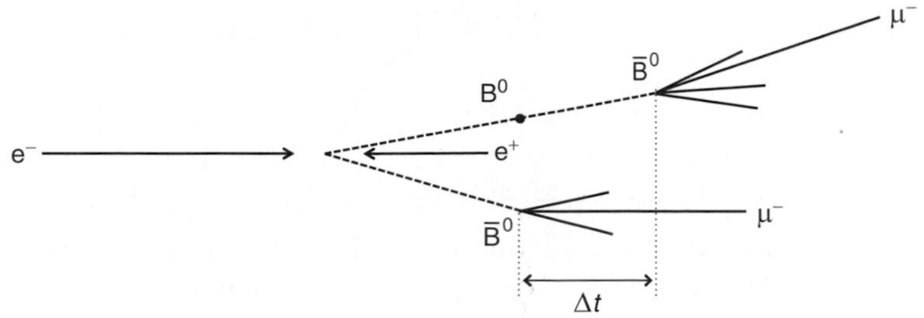


Fig. 14.20 The process $e^+e^- \to B^0\overline{B}^0$ followed by two same-flavour (SF) leptonic $\overline{B}^0$ decays, following $B^0 \to \overline{B}^0$ oscillation.

time between the two decays $\Delta t$. Because the B-mesons are produced almost at rest in the centre-of-mass frame, the proper time between the two decays is given by $\Delta t = \Delta z/\beta\gamma c$, where $\beta$ and $\gamma$ are determined from the known velocity of the $\Upsilon$.

The mass difference $\Delta m_d = m(B_H) - m(B_L)$ is determined from the lepton flavour asymmetry $A(\Delta t)$ defined as

$$A(\Delta t) = \frac{N_{OF} - N_{SF}}{N_{SF} + N_{OF}},$$

where $N_{OF}$ is the number of observed opposite flavour decays and $N_{SF}$ is the corresponding number of same flavour decays. The observed asymmetry can be expressed in terms of the oscillation probabilities as

$$A(\Delta t) = \frac{[P(B^0_{t=0} \to B^0) + P(\overline{B}^0_{t=0} \to \overline{B}^0)] - [P(B^0_{t=0} \to \overline{B}^0) + P(\overline{B}^0_{t=0} \to B^0)]}{[P(B^0_{t=0} \to B^0) + P(\overline{B}^0_{t=0} \to \overline{B}^0)] + [P(B^0_{t=0} \to \overline{B}^0) + P(\overline{B}^0_{t=0} \to B^0)]},$$

which, using (14.65) and the subsequent relations, gives

$$A(\Delta t) = \cos^2\left(\tfrac{1}{2}\Delta m_d t\right) - \sin^2\left(\tfrac{1}{2}\Delta m_d t\right) = \cos\left(\Delta m_d t\right). \tag{14.66}$$

Figure 14.21 shows the measurement of $A(\Delta t)$ from the Belle experiment. The data do not follow the pure cosine form of (14.66) due to a number of experimental effects, including the presence of background, the misidentification of the lepton charge and the experimental $\Delta t$ resolution. Nevertheless, the effects of $B^0 \leftrightarrow \overline{B}^0$ oscillations are clearly observed. When combined, the results from the BaBar and Belle experiments give

$$\Delta m_d = (0.507 \pm 0.005)\,\text{ps}^{-1} \equiv (3.34 \pm 0.03) \times 10^{-13}\,\text{GeV}.$$

From (14.61) and the knowledge that $V_{tb} \approx 1$, the measured value of $\Delta m_d$ can be interpreted as a measurement of

$$|V_{td}| = (8.4 \pm 0.6) \times 10^{-3}.$$

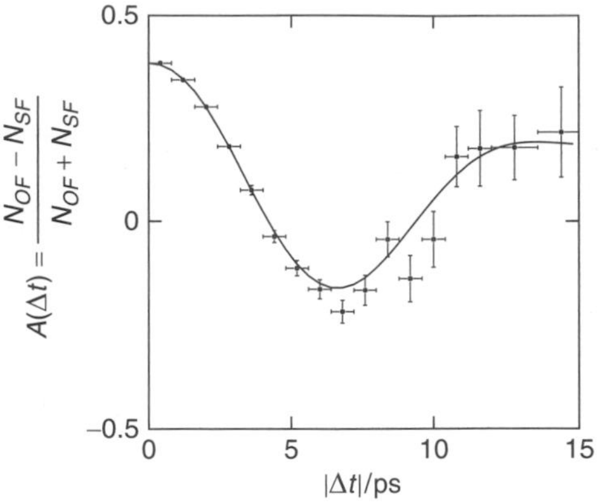


Fig. 14.21 The Belle measurement of $A_{\Delta t}$. The line represents a fit to the data including contributions from background and the effects of experimental resolution. Adapted from Abe *et al.* (2005).

In a similar manner, $|V_{ts}|$ can be extracted from the measurements of oscillations in the $B^0_s(\overline{b}s) \leftrightarrow \overline{B}^0_s(b\overline{s})$ system from the CDF and LHCb experiments, see Abulencia *et al.* (2006) and Aaij *et al.* (2012). When the results from these two experiments are averaged they give $\Delta m_s = 17.72 \pm 0.04\,\text{ps}^{-1}$. Taking $V_{tb} \approx 1$ this result leads to

$$|V_{ts}| = (4.3 \pm 0.3) \times 10^{-2}.$$

### 14.6.4 CP violation in the B-meson system

In general, CP violation can be observed as three distinct effects:

(i) direct CP violation in decay such that $\Gamma(A \to X) \neq \Gamma(\overline{A} \to \overline{X})$, as parameterised by $\varepsilon'$ in the neutral kaon system;
(ii) CP violation in the mixing of neutral mesons as parameterised by $\varepsilon$ in the kaon system;
(iii) CP violation in the interference between decays to a common final state $f$ with and without mixing, for example $B^0 \to f$ and $B^0 \to \overline{B}^0 \to f$.

In the Standard Model, the effects of CP violation in $B^0 \leftrightarrow \overline{B}^0$ mixing is small. Nevertheless, CP-violating effects in the interference between decays $B^0 \to f$ and $B^0 \to \overline{B}^0 \to f$ can be relatively large and have been studied extensively by the BaBar and Belle experiments in a number of final states; here the decay $B \to J/\psi\, K_S$ is used to illustrate the main ideas. To simplify the notation, the $J/\psi$ meson is written simply as $\psi$.

The $\psi$ charmonium ($c\overline{c}$) state has $J^P = 1^-$ and is a CP eigenstate with $CP = +1$. Neglecting CP violation in neutral kaon mixing, the $K_S$ is to a good approximation,

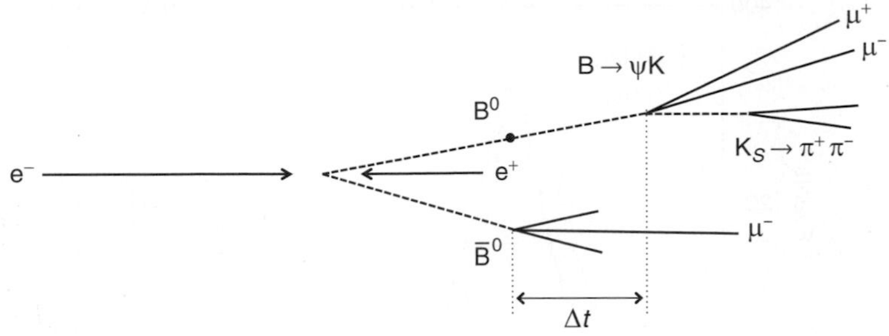


Fig. 14.22 The process $e^+e^- \to B^0\overline{B}^0$ where the leptonic decay tags the flavour of the other B-meson as being a $B^0$ that subsequently decays to a $\psi\,K_S$. In this illustrative example, $\psi \to \mu^+\mu^-$ and $K_S \to \pi^+\pi^-$.

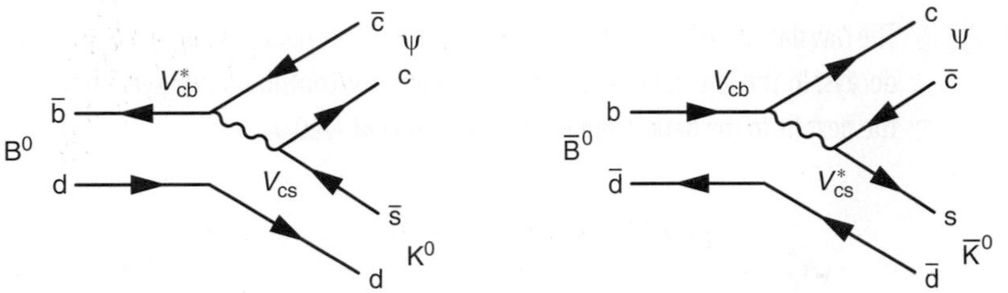


Fig. 14.23 The Feynman diagrams for $B^0 \to \psi\,K^0$ and $\overline{B}^0 \to \psi\,\overline{K}^0$.

the $CP = +1$ eigenstate of the neutral kaon system with $J^P = 0^-$. Since the $B^0$ and $\overline{B}^0$ are spin-0 mesons, the decays $B^0 \to \psi\,K_S$ and $\overline{B}^0 \to \psi\,K_S$ must result in an $\ell = 1$ orbital angular momentum state. Therefore the $CP$ state of the combined $\psi\,K_S$ system is

$$CP(\psi\,K_S) = CP(\psi) \times CP(K_S) \times (-1)^{\ell} = (+1)(+1)(-1) = -1.$$

Similarly, the decay $B \to \psi\,K_L$ occurs in a CP-even state, $CP(\psi\,K_L) = +1$.

Figure 14.22 shows the topology of a typical neutral B-meson decay to $\psi\,K_S$. In this example, the charge of the muon in the leptonic $\overline{B}^0 \to D^+\mu^-\overline{\nu}_\mu$ decay tags it as $\overline{B}^0$ and hence at time $t = 0$, the other B-meson is in a $B^0$ flavour state, $|B(0)\rangle = |B^0\rangle$. The decay to $\psi\,K_S$ can either occur directly by $B^0 \to \psi\,K_S$ or after mixing, $B^0 \to \overline{B}^0 \to \psi\,K_S$. It is the interference between the two amplitudes for these processes, which have different phases, that provides the sensitivity to the CP violating angle $\beta$. The $B \to \psi\,K_S$ decays can be identified from the clear experimental signatures, for example $\psi \to \mu^+\mu^-$ and $K_S \to \pi^+\pi^-$.

The $B^0/\overline{B}^0 \to \psi\,K_S$ decays proceed in two stages. First the $B^0/\overline{B}^0$ decays to the corresponding *flavour* eigenstate, $B^0 \to \psi K^0$ and $\overline{B}^0 \to \psi\overline{K}^0$, as shown in Figure 14.23. Subsequently, the neutral kaon system evolves as as a linear combination of the physical $K_S$ and $K_L$ states and then decays to the CP states $K_S\psi$ and $K_L\psi$. CP violation in the interference between $B^0 \to \psi\,K_S$ and $B^0 \to \overline{B}^0 \to \psi\,K_S$ is measurable through the asymmetry,

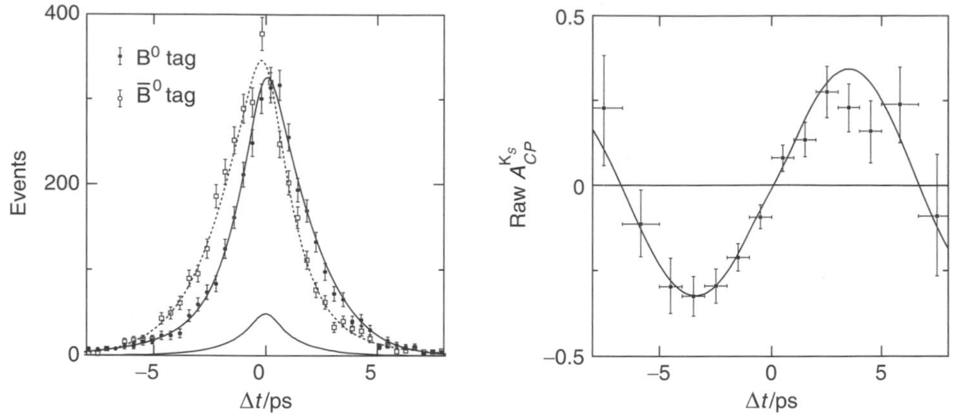


Fig. 14.24 The raw data and distribution of $A_{CP}^{K_S}$ from the BaBar experiment based on a sample of $4\times10^8$ $\Upsilon(4S) \to B^0\overline{B}^0$ decays. In the left-hand plot, the small symmetric contribution arises from the background. The lines show the best fit to the data. Adapted from Aubert *et al.* (2007).

$$A_{CP}^{K_S} = \frac{\Gamma(\overline{B}^0_{t=0} \to \psi\, K_S) - \Gamma(B^0_{t=0} \to \psi\, K_S)}{\Gamma(\overline{B}^0_{t=0} \to \psi\, K_S) + \Gamma(B^0_{t=0} \to \psi\, K_S)} = \sin(\Delta m_d t)\,\sin(2\beta). \qquad (14.67)$$

Figure 14.24 shows the experimental data from the BaBar experiment. The left-hand plot shows the raw numbers of observed decays to $\psi\, K_S$, from both events that were tagged as $B^0$ or $\overline{B}^0$, plotted as a function of $\Delta t$. Here $\Delta t$ is the difference in the proper time of the tagged $B^0/\overline{B}^0$ semi-leptonic decay and the observed $B \to \psi\, K_S$ decay. The curves show the expected distributions including a symmetric background contribution from other B-meson decays. The right-hand plot of Figure 14.24 shows the raw asymmetry obtained from these data. This has the expected sinusoidal form of (14.67) and the amplitude provides a measurement of $\sin(2\beta)$,

$$\sin(2\beta) = 0.685 \pm 0.032.$$

This observation of a non-zero value of $\sin(2\beta)$ is a direct manifestation of CP violation in the B-meson system. The Belle experiment (see Adachi *et al.* (2012)) measured $\sin(2\beta) = 0.670 \pm 0.032$.

## 14.7 CP violation in the Standard Model

There is now a wealth of experimental data on CP violation associated with the weak interactions of quarks. This chapter has focussed on the observations of CP violation in $K^0 - \overline{K}^0$ mixing and in the interference between the amplitudes for $B^0 \to J/\psi\, K_S$ and $B^0 \to \overline{B}^0 \to J/\psi\, K_S$ decays. Direct CP violation in the decays of kaons and B-mesons has also been observed, for example, as a difference between the rates $\Gamma(\overline{B}^0 \to K^-\pi^+)$ and $\Gamma(B^0 \to K^+\pi^-)$.

In the Standard Model, CP violation in the weak interactions of hadrons is described by the single irreducible complex phase in the CKM matrix. In the Wolfenstein parametrisation of (14.9), CP violation is associated with the parameter $\eta$. To $O(\lambda^4)$, the parameter $\eta$ appears only in $V_{ub}$ and $V_{td}$, with

$$V_{ub} \approx A\lambda^3(\rho - i\eta) \quad \text{and} \quad V_{td} \approx A\lambda^3(1 - \rho - i\eta).$$

The measurements of non-zero values of $|\varepsilon|$ and $\sin(2\beta)$ separately imply that $\eta \neq 0$. However, it is only when the experimental measurements are combined, that the values of $\rho$ and $\eta$ can be determined.

In the Standard Model, the CKM matrix is unitary, $V^\dagger V = I$. This property places constraints on the possible values of the different elements of the CKM matrix. These constraints are usually expressed in terms of unitarity triangles. For example, the unitarity of the CKM matrix implies that

$$V_{ud}V_{ub}^* + V_{cd}V_{cb}^* + V_{td}V_{tb}^* = 0. \tag{14.68}$$

In the Wolfenstein parametrisation, of these six CKM matrix elements, $V_{ud}$, $V_{tb}$, $V_{cd}$ and $V_{cb}$ are all real and only $V_{cd}$ is negative. Hence (14.68) can be divided by $V_{cd}V_{cb}$ to give

$$1 - \frac{|V_{ud}|}{|V_{cd}||V_{cb}|}V_{ub}^* - \frac{|V_{tb}|}{|V_{cd}||V_{cb}|}V_{td} = 0. \tag{14.69}$$

Since $V_{ub}^*$ and $V_{td}$ are complex, $V_{ub}^* = A\lambda^3(\rho + i\eta)$ and $V_{td} = A\lambda^3(1 - \rho - i\eta)$, the unitarity relation of (14.69) is a vector equation in the complex $\rho$–$\eta$ plane, with the three vectors forming the closed triangle, as shown in Figure 14.25a.

From $V_{td} = |V_{td}|e^{-i\beta} = A\lambda^3(1 - \rho - i\eta)$, it can be seen that

$$\beta = \arg(1 - \rho + i\eta) \quad \text{or equivalently} \quad \tan\beta = \frac{\eta}{1-\rho}.$$

Consequently, the angle $\beta$ corresponds to the internal angle of the unitarity triangle shown in Figure 14.25a. Therefore, the measurement of $\sin(2\beta)$ described previously constrains the angle between two of the sides of the unitarity triangle as shown in Figure 14.25b, which also shows the constraint in the $\rho$–$\eta$ plane obtained from the measurement of $|\varepsilon|$ in neutral kaon mixing,

$$|\varepsilon| \propto \eta(1 - \rho + \text{constant}).$$

The measurement of $\Delta m_d$ determines $|V_{td}|$. When this is combined with the knowledge that $|V_{tb}| \approx 1$ and the measurements of $|V_{cd}|$ and $|V_{cb}|$ described in Section 14.3, it constrains of the length of the upper side of the unitarity triangle, as shown in Figure 14.25b.

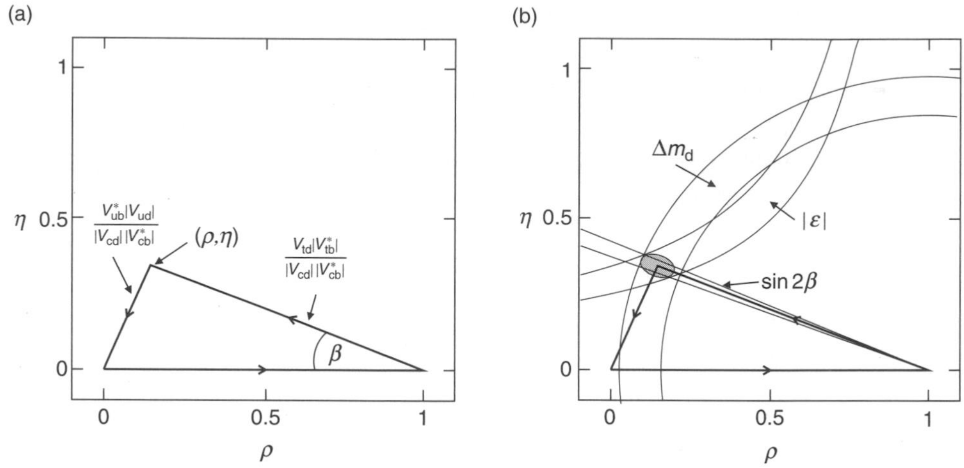


Fig. 14.25 (a) The unitarity relation, $V_{ud}V^*_{ub} + V_{cd}V^*_{cb} + V_{td}V^*_{tb} = 0$, shown in the $\rho$–$\eta$ plane. (b) The constraints from the measurements of $|\varepsilon|$, $\Delta m_d$ and $\sin 2\beta$. The shaded ellipse shows the combination of these constraints that give a measurement of $\eta$ and $\rho$.

The experimental constraints from the measurements of $|\varepsilon|$, $\sin(2\beta)$ and $\Delta m_d$ are consistent with a common point in the $\rho$–$\eta$ plane, as indicated by the ellipse in Figure 14.25b, thus providing experimental confirmation of the unitarity relation $V_{ud}V^*_{ub} + V_{cd}V^*_{cb} + V_{td}V^*_{tb} = 0$. From a global fit to these and other results (see Beringer *et al.* (2012)) the Wolfenstein parameters are determined to be

$$\lambda = 0.2253 \pm 0.0007, \ A = 0.811^{+0.022}_{-0.012}, \ \rho = 0.13 \pm 0.02, \ \eta = 0.345 \pm 0.014.$$

The experimental measurements described in this chapter provide a strong test of the Standard Model prediction that the unitarity triangle of (14.69) is closed. Any deviation from this prediction would indicate physics beyond the Standard Model. To date, all measurements in the quark flavour sector are consistent a unitary CKM matrix, where the observed CP violation is described by a single complex phase.

Whilst the Standard Model provides an explanation of the observed CP violation in the quark sector, this is not sufficient to explain the matter–antimatter asymmetry in the Universe. There are suggestions that CP violation in the lepton sector during the early evolution of the Universe might account for the observed matter-antimatter asymmetry. However, it is also possible that there are as yet undiscovered CP violating processes beyond the Standard Model. In the coming years the LHCb experiment at the LHC and the Belle II experiment at KEK will probe CP violation in the quark sector with ever increasing precision and may shed further light on this important question.

## Summary

CP violation is an essential part of our understanding of particle physics. In the Standard Model it can be accommodated in the irreducible complex phases in the PMNS and CKM matrices. In the decays of hadrons, CP violation has been observed in three ways: (i) direct CP violation in decay; (ii) indirect CP violation in the mixing of neutral mesons; and (iii) CP violation in the interference between decays with and without oscillations.

This chapter concentrated on the measurements of oscillations and CP violation in the neutral kaon and neutral B-meson systems. Many of the effects arise from the distinction between the different neutral meson states. For example, neutral kaons are produced in the strong interaction as the flavour eigenstates, $K^0(\bar{s}d)$ and $\overline{K}^0(s\bar{d})$, but the physical particles with definite masses and lifetimes are the eigenstates of the overall Hamiltonian of the $K^0$–$\overline{K}^0$ system are

$$|K_S\rangle \propto (1+\varepsilon)|K^0\rangle + (1-\varepsilon)|\overline{K}^0\rangle \quad \text{and} \quad |K_L\rangle \propto (1+\varepsilon)|K^0\rangle - (1-\varepsilon)|\overline{K}^0\rangle,$$

where the parameter $\varepsilon$ is non-zero only if CP is violated. If CP were conserved in the weak interaction, the physical states would correspond to the CP eigenstates

$$|K_S\rangle \propto |K^0\rangle + |\overline{K}^0\rangle \quad \text{and} \quad |K_L\rangle \propto |K^0\rangle - |\overline{K}^0\rangle.$$

Oscillations arise because neutral mesons are produced as flavour eigenstates and decay as either flavour or CP eigenstates, but propagate as the physical mass eigenstates.

The studies of the neutral mesons and their oscillations, provide constraints on the values of the elements of the CKM matrix and allow CP violation to be studied in the quark sector. To date, all such experimental measurements are consistent with the Standard Model predictions from the single complex phase in the unitary CKM matrix.

## Problems

**14.1** Draw the lowest-order Feynman diagrams for the decays

$$K^0 \to \pi^+\pi^-, \quad K^0 \to \pi^0\pi^0, \quad \overline{K}^0 \to \pi^+\pi^- \quad \text{and} \quad \overline{K}^0 \to \pi^0\pi^0,$$

and state how the corresponding matrix elements depend on the Cabibbo angle $\theta_c$.

**14.2** Draw the lowest-order Feynman diagrams for the decays

$$\mathrm{B}^0 \to \mathrm{D}^- \pi^+, \quad \mathrm{B}^0 \to \pi^+ \pi^- \quad \text{and} \quad \mathrm{B}^0 \to \mathrm{J}/\psi\, \mathrm{K}^0,$$

and place them in order of decreasing decay rate.

The flavour content of the above mesons is $\mathrm{B}^0(\mathrm{d\bar{b}})$, $\mathrm{D}^-(\mathrm{d\bar{c}})$, $\mathrm{J}/\psi(\mathrm{c\bar{c}})$ and $\mathrm{K}^0(\mathrm{d\bar{s}})$.

**14.3** Draw the lowest-order Feynman diagrams for the weak decays

$$\mathrm{D}^0(\mathrm{c\bar{u}}) \to \mathrm{K}^-(\mathrm{s\bar{u}}) + \pi^+(\mathrm{u\bar{d}}) \quad \text{and} \quad \mathrm{D}^0(\mathrm{c\bar{u}}) \to \mathrm{K}^+(\mathrm{u\bar{s}}) + \pi^-(\mathrm{d\bar{u}}),$$

and explain the observation that

$$\frac{\Gamma(\mathrm{D}^0 \to \mathrm{K}^+\pi^-)}{\Gamma(\mathrm{D}^0 \to \mathrm{K}^-\pi^+)} \approx 4 \times 10^{-3}.$$

**14.4** A hypothetical $\overline{\mathrm{T}}^0(\mathrm{t\bar{u}})$ meson decays by the weak charged-current decay chain,

$$\overline{\mathrm{T}}^0 \to W\pi \to (X\pi)\,\pi \to (Y\pi)\,\pi\,\pi \to (Z\pi)\,\pi\,\pi\,\pi.$$

Suggest the most likely identification of the $W$, $X$, $Y$ and $Z$ mesons and state why this decay chain would be preferred over the direct decay $\overline{\mathrm{T}}^0 \to Z\,\pi$.

**14.5** For the cases of two, three and four generations, state:

**(a)** the number of free parameters in the corresponding $n \times n$ unitary matrix relating the quark flavour and weak states;
**(b)** how many of these parameters are real and how many are complex phases;
**(c)** how many of the complex phases can be absorbed into the definitions of phases of the fermions without any physical consequences;
**(d)** whether CP violation can be accommodated in quark mixing.

**14.6** Draw the lowest-order Feynman diagrams for the strong interaction processes

$$\bar{\mathrm{p}}\mathrm{p} \to \mathrm{K}^-\pi^+\mathrm{K}^0 \quad \text{and} \quad \bar{\mathrm{p}}\mathrm{p} \to \mathrm{K}^+\pi^-\overline{\mathrm{K}}^0.$$

**14.7** In the neutral kaon system, time-reversal violation can be expressed in terms of the asymmetry

$$A_T = \frac{\Gamma(\overline{\mathrm{K}}^0 \to \mathrm{K}^0) - \Gamma(\mathrm{K}^0 \to \overline{\mathrm{K}}^0)}{\Gamma(\overline{\mathrm{K}}^0 \to \mathrm{K}^0) + \Gamma(\mathrm{K}^0 \to \overline{\mathrm{K}}^0)}.$$

Show that this is equivalent to

$$A_T = \frac{\Gamma(\overline{\mathrm{K}}^0_{t=0} \to \pi^-\mathrm{e}^+\nu_\mathrm{e}) - \Gamma(\mathrm{K}^0_{t=0} \to \pi^+\mathrm{e}^-\overline{\nu}_\mathrm{e})}{\Gamma(\overline{\mathrm{K}}^0_{t=0} \to \pi^-\mathrm{e}^+\nu_\mathrm{e}) + \Gamma(\mathrm{K}^0_{t=0} \to \pi^+\mathrm{e}^-\overline{\nu}_\mathrm{e})},$$

and therefore

$$A_T \approx 4|\varepsilon| \cos\phi.$$

**14.8** The $\mathrm{K}_S - \mathrm{K}_L$ mass difference can be expressed as

$$\Delta m = m(\mathrm{K}_L) - m(\mathrm{K}_S) \approx \sum_{q,q'} \frac{G_\mathrm{F}^2}{3\pi^2} f_\mathrm{K}^2 m_\mathrm{K} |V_{qd} V_{qs}^* V_{q'd} V_{q's}^*|\, m_q m_{q'},$$

where q and q′ are the quark flavours appearing in the box diagram. Using the values for the CKM matrix elements given in (14.8), obtain expressions for the relative contributions to $\Delta m$ arising from the different combinations of quarks in the box diagrams.

**14.9** Indirect CP violation in the neutral kaon system is expressed in terms of $\varepsilon = |\varepsilon| e^{i\phi}$. Writing

$$\xi = \frac{1-\varepsilon}{1+\varepsilon} \approx 1 - 2\varepsilon = \left( \frac{M_{12}^* - \frac{i}{2}\Gamma_{12}^*}{M_{12} - \frac{i}{2}\Gamma_{12}} \right)^{\frac{1}{2}},$$

show that

$$\varepsilon \approx \frac{1}{2} \times \left( \frac{(\Im m\{M_{12}\} - \frac{i}{2}\Im m\{\Gamma_{12}\})}{M_{12} - \frac{i}{2}\Gamma_{12}} \right)^{\frac{1}{2}} \approx \frac{\Im m\{M_{12}\} - i\Im m\{\Gamma_{12}/2\}}{\Delta m - i\Delta\Gamma/2}.$$

Using the knowledge that $\phi \approx 45°$ and the measurements of $\Delta m$ and $\Delta\Gamma$, deduce that $\Im m\{M_{12}\} \gg \Im m\{\Gamma_{12}\}$ and therefore

$$|\varepsilon| \sim \frac{1}{\sqrt{2}} \frac{\Im m\{M_{12}\}}{\Delta m}.$$

**14.10** Using (14.53) and the explicit form of Wolfenstein parametrisation of the CKM matrix, show that

$$|\varepsilon| \propto \eta(1 - \rho + \text{constant}).$$

**14.11** Show that the $B^0 - \overline{B}^0$ mass difference is dominated by the exchange of two top quarks in the box diagram.

**14.12** Calculate the velocities of the B-mesons produced in the decay at rest of the $\Upsilon(4S) \to B^0\overline{B}^0$.

**14.13** Given the lifetimes of the neutral B-mesons are $\tau = 1.53$ ps, calculate the mean distance they travel when produced at the KEKB collider in collisions of 8 GeV electrons and 3.5 GeV positrons.

**14.14** From the measured values

$$|V_{ud}| = 0.974\,25 \pm 0.000\,22 \quad \text{and} \quad |V_{ub}| = (4.15 \pm 0.49) \times 10^{-3},$$
$$|V_{cd}| = 0.230 \pm 0.011 \quad \text{and} \quad |V_{cb}| = 0.041 \pm 0.001,$$

calculate the length of the corresponding side of the unitarity triangle in Figure 14.25 and its uncertainty. By sketching this constraint and that from the measured value of $\beta$, obtain approximate constraints on the values of $\rho$ and $\eta$.

# 15 Electroweak unification

One of the main goals of particle physics is to provide a unified picture of the fundamental particles and their interactions. In the nineteenth century, Maxwell provided a description of electricity and magnetism as different aspects of a unified electromagnetic theory. In the 1960s, Glashow, Salam and Weinberg (GSW) developed a unified picture of the electromagnetic and weak interactions. One consequence of the GSW electroweak model is the prediction of a weak neutral-current mediated by the neutral Z boson with well-defined properties. This short chapter describes electroweak unification and the properties of the W and Z bosons.

## 15.1 Properties of the W bosons

The W boson is a spin-1 particle with a mass of approximately 80 GeV. Its wave-function can be written in terms of a plane wave and a polarisation four-vector,

$$W^{\mu} = \epsilon^{\mu}_{\lambda} e^{-ip\cdot x} = \epsilon^{\mu}_{\lambda} e^{i(\mathbf{p}\cdot\mathbf{x}-Et)}.$$

For a massive spin-1 particle the polarisation four-vector $\epsilon^{\mu}_{\lambda}$ is restricted to one of three possible polarisation states (see Appendix D). For a W boson travelling in the $z$-direction, the three orthogonal polarisation states $\lambda$ can be written as

$$\epsilon^{\mu}_{-} = \frac{1}{\sqrt{2}}(0,1,-i,0), \quad \epsilon^{\mu}_{L} = \frac{1}{m_{\mathrm{W}}}(p_z,0,0,E) \quad \text{and} \quad \epsilon^{\mu}_{+} = -\frac{1}{\sqrt{2}}(0,1,i,0). \tag{15.1}$$

These states represent two transverse polarisation modes $\epsilon_{\pm}$, corresponding to circularly polarised spin-1 states with $S_z = \pm 1$, and a longitudinal $S_z = 0$ state.

### 15.1.1 W-boson decay

The calculation of the W-boson decay rate provides a good illustration of the use of polarisation four-vectors in matrix element calculations. The lowest-order Feynman diagram for the $\mathrm{W}^- \to \mathrm{e}^-\overline{\nu}_{\mathrm{e}}$ decay is shown in Figure 15.1. The matrix element for the decay is obtained using the appropriate Feynman rules. The final-state electron and antineutrino are written respectively as the adjoint particle spinor

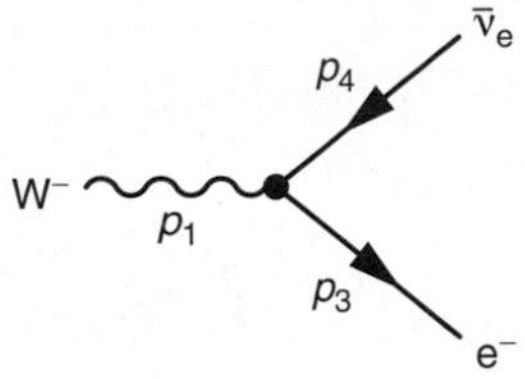


Fig. 15.1 The lowest-order Feynman diagram for $W^- \to e^-\overline{\nu}_e$.

$\overline{u}(p_3)$ and the antiparticle spinor $v(p_4)$. The initial-state $W^-$ is written as $\epsilon_\mu^\lambda(p_1)$, where $\lambda$ indicates one of the three possible polarisation states. Finally, the vertex factor for the weak charged-current is the usual $V - A$ interaction

$$-i\frac{g_W}{\sqrt{2}}\frac{1}{2}\gamma^\mu(1-\gamma^5).$$

Using these Feynman rules, the matrix element for $W^- \to e^-\overline{\nu}_e$ is given by

$$-i\mathcal{M}_{fi} = \epsilon_\mu^\lambda(p_1)\,\overline{u}(p_3)\left[-i\frac{g_W}{\sqrt{2}}\gamma^\mu\frac{1}{2}(1-\gamma^5)\right]v(p_4),$$

and therefore

$$\mathcal{M}_{fi} = \frac{g_W}{\sqrt{2}}\,\epsilon_\mu^\lambda(p_1)\,\overline{u}(p_3)\gamma^\mu\frac{1}{2}(1-\gamma^5)v(p_4). \tag{15.2}$$

This expression can be written as the four-vector scalar product of the W-boson four-vector polarisation and the lepton current,

$$\mathcal{M}_{fi} = \frac{g_W}{\sqrt{2}}\,\epsilon_\mu^\lambda(p_1)j^\mu, \tag{15.3}$$

where the leptonic weak charged-current $j^\mu$ is given by

$$j^\mu = \overline{u}(p_3)\gamma^\mu\frac{1}{2}(1-\gamma^5)v(p_4). \tag{15.4}$$

It is convenient to consider the $W^- \to e^-\overline{\nu}_e$ decay in the rest frame of the W boson, as illustrated in Figure 15.2. Given that $m_W \gg m_e$, the mass of the electron can be neglected and the four-vectors of the $W^-$, $e^-$ and $\overline{\nu}_e$ can be taken to be

$$\begin{aligned}
p_1 &= (m_W, 0, 0, 0),\\
p_3 &= (E, E\sin\theta, 0, E\cos\theta),\\
p_4 &= (E, -E\sin\theta, 0, -E\cos\theta),
\end{aligned}$$

with $E = m_W/2$. In the ultra-relativistic limit, where the helicity states are the same as the chiral states, only left-handed helicity particle states and right-handed helicity antiparticle states contribute to the weak interaction. In this case, the leptonic current of (15.4) can be written

$$j^\mu = \overline{u}_\downarrow(p_3)\gamma^\mu v_\uparrow(p_4), \tag{15.5}$$

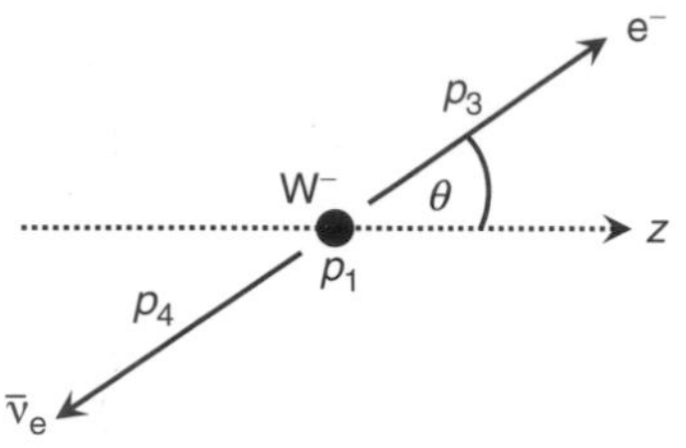


Fig. 15.2 The decay $W^- \to e^-\overline{\nu}_e$ in the W-boson rest frame.

where $\overline{u}_\downarrow(p_3)$ and $v_\uparrow(p_4)$ are respectively the left-handed particle and right-handed antiparticle helicity spinors for the electron and electron antineutrino. The leptonic current of (15.5) is identical to that encountered for the $\mu^+\mu^-$ current in the $s$-channel process $e^+e^- \to \mu^+\mu^-$ and is given by (6.17) with $E = m_W/2$,

$$j^\mu = m_W(0, -\cos\theta, -i, \sin\theta).$$

For a W boson at rest, the three possible polarisation states of (15.1) are

$$\epsilon_-^\mu = \tfrac{1}{\sqrt{2}}(0,1,-i,0), \quad \epsilon_L^\mu = (0,0,0,1) \quad \text{and} \quad \epsilon_+^\mu = -\tfrac{1}{\sqrt{2}}(0,1,i,0).$$

Therefore, from (15.3), the matrix elements for the decay $W^- \to e^-\overline{\nu}_e$ in the three possible W-boson polarisation states are

$$\begin{aligned}
\mathcal{M}_- &= \tfrac{g_W m_W}{2}(0,1,-i,0)\cdot(0,-\cos\theta,-i,\sin\theta) = \tfrac{1}{2}g_W m_W(1+\cos\theta),\\
\mathcal{M}_L &= \tfrac{g_W m_W}{\sqrt{2}}(0,0,0,1)\cdot m_W(0,-\cos\theta,-i,\sin\theta) = -\tfrac{1}{\sqrt{2}}g_W m_W \sin\theta,\\
\mathcal{M}_+ &= -\tfrac{g_W m_W}{2}(0,1,i,0)\cdot m_W(0,-\cos\theta,-i,\sin\theta) = \tfrac{1}{2}g_W m_W(1-\cos\theta).
\end{aligned}$$

Hence, for the three possible W-boson polarisations

$$\begin{aligned}
|\mathcal{M}_-|^2 &= g_W^2 m_W^2 \tfrac{1}{4}(1+\cos\theta)^2,\\
|\mathcal{M}_L|^2 &= g_W^2 m_W^2 \tfrac{1}{2}\sin^2\theta,\\
|\mathcal{M}_+|^2 &= g_W^2 m_W^2 \tfrac{1}{4}(1-\cos\theta)^2.
\end{aligned}$$

The resulting angular distributions of the decay products for each of the different W-boson polarisations can be understood by noting that the LH and RH helicities of the electron and antineutrino imply that they are produced in a spin-1 state aligned with the direction of the neutrino, as shown in Figure 15.3. The angular

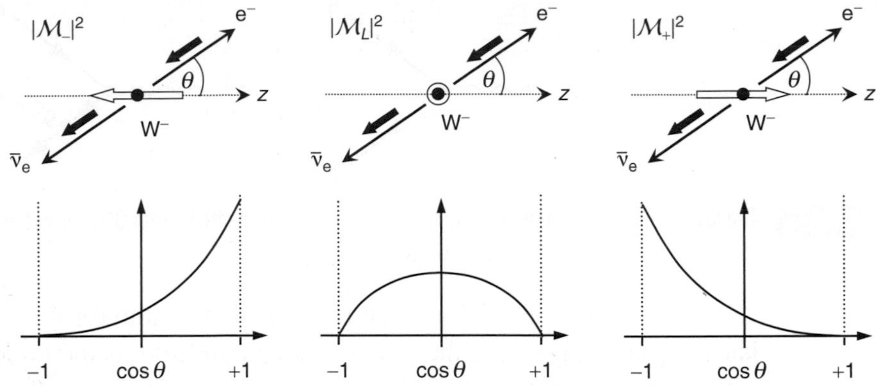


Fig. 15.3 The angular distributions of the electron and electron antineutrino in the decay $W^- \to e^-\overline{\nu}_e$ for the three possible W-boson polarisation states.

distributions then follow from the quantum mechanical properties of spin-1, as discussed in Section 6.3.

The total decay rate is determined by the spin-averaged matrix element squared, which (for unpolarised W decays) is given by

$$\begin{aligned}\left\langle |\mathcal{M}_{fi}|^2 \right\rangle &= \tfrac{1}{3}\left(|\mathcal{M}_-|^2 + |\mathcal{M}_L|^2 + |\mathcal{M}_+|^2\right)\\ &= \tfrac{1}{3} g_W^2 m_W^2 \left[\tfrac{1}{4}(1+\cos\theta)^2 + \tfrac{1}{2}\sin^2\theta + \tfrac{1}{4}(1-\cos\theta)^2\right]\\ &= \tfrac{1}{3} g_W^2 m_W^2. \end{aligned} \tag{15.6}$$

Hence, after averaging over the three polarisation states of the W boson, there is no preferred direction for the final-state particles that are, as expected, produced isotropically in the W-boson rest frame. The $W^- \to e^-\overline{\nu}_e$ decay rate is obtained by substituting the expression for the spin-averaged matrix element of (15.6) into the decay rate formula of (3.49),

$$\Gamma = \frac{p^*}{32\pi^2 m_W^2}\int \left\langle |\mathcal{M}_{fi}|^2 \right\rangle d\Omega^* = \frac{p^*}{8\pi m_W^2}\left\langle |\mathcal{M}_{fi}|^2 \right\rangle,$$

where $p^*$ is the momentum of the electron (or antineutrino) in the centre-of-mass frame. If the masses of the final-state particles are neglected, $p^* = m_W/2$, and therefore the $W^- \to e^-\overline{\nu}_e$ decay rate is given by

$$\Gamma(W^- \to e^-\overline{\nu}_e) = \frac{g_W^2 m_W}{48\pi}. \tag{15.7}$$

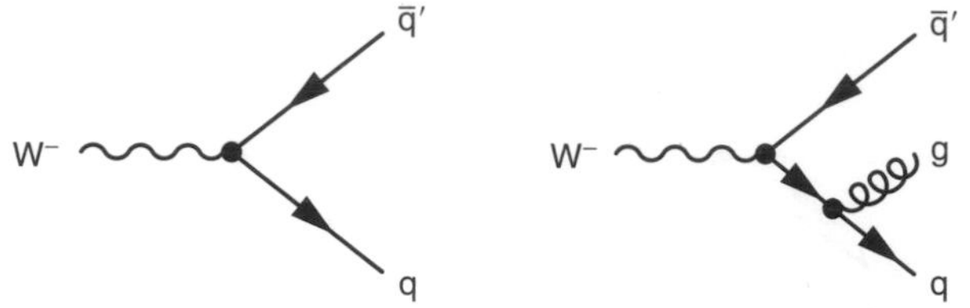


Fig. 15.4 The lowest-order Feynman diagram for $W^- \to q\overline{q}'$ and the first-order QCD correction from $W^- \to q\overline{q}'g$.

The expression of (15.7) gives the partial decay width for $W^- \to e^-\overline{\nu}_e$. To calculate the total decay rate of the W boson, all possible decay modes have to be considered. From the lepton universality of the weak charged-current (and neglecting the very small differences due to the lepton masses), the three leptonic decay modes have the same partial decay rates,

$$\Gamma(W^- \to e^-\overline{\nu}_e) = \Gamma(W^- \to \mu^-\overline{\nu}_\mu) = \Gamma(W^- \to \tau^-\overline{\nu}_\tau).$$

The W boson also can decay to all flavours of quarks with the exception of the top quark, which is too massive ($m_t > m_W$). The decay rate of the W boson to a particular quark flavour needs to account for the elements of the CKM matrix and the three possible colours of the final-state quarks, therefore the decay rates relative to $\Gamma_{e\nu} = \Gamma(W^- \to e^-\overline{\nu}_e)$ are

$$\Gamma(W^- \to d\overline{u}) = 3|V_{ud}|^2\,\Gamma_{e\nu}, \quad \Gamma(W^- \to d\overline{c}) = 3|V_{cd}|^2\,\Gamma_{e\nu},$$
$$\Gamma(W^- \to s\overline{u}) = 3|V_{us}|^2\,\Gamma_{e\nu}, \quad \Gamma(W^- \to s\overline{c}) = 3|V_{cs}|^2\,\Gamma_{e\nu},$$
$$\Gamma(W^- \to b\overline{u}) = 3|V_{ub}|^2\,\Gamma_{e\nu}, \quad \Gamma(W^- \to b\overline{c}) = 3|V_{cb}|^2\,\Gamma_{e\nu}.$$

From the unitarity of the CKM matrix,

$$|V_{ud}|^2 + |V_{us}|^2 + |V_{ub}|^2 = 1 \quad \text{and} \quad |V_{cd}|^2 + |V_{cs}|^2 + |V_{cb}|^2 = 1,$$

and the lowest-order prediction for the W-boson decay rate to quarks is

$$\Gamma(W^- \to q\overline{q}') = 6\,\Gamma(W^- \to e^-\overline{\nu}_e).$$

In addition to the lowest-order $W \to q\overline{q}'$ process, the QCD correction from the process $W \to q\overline{q}'g$, shown in Figure 15.4, enhances the decay rate to hadronic final states by a factor

$$\kappa_{QCD} = \left[1 + \frac{\alpha_S(m_W)}{\pi}\right] \approx 1.038. \tag{15.8}$$

Thus the total decay rate of the W boson to either quarks or to the three possible leptonic final states is

$$\Gamma_W = (3 + 6\,\kappa_{QCD})\,\Gamma(W^- \to e^-\overline{\nu}_e) \approx 9.2 \times \frac{g_W^2 m_W}{48\pi} = 2.1\,\text{GeV},$$

and the branching ratio of the W boson to hadronic final states is

$$BR(W \to q\overline{q}') = \frac{6\,\kappa_{QCD}}{3 + 6\,\kappa_{QCD}} = 67.5\%. \tag{15.9}$$

The prediction of $\Gamma_W = 2.1\,\text{GeV}$ is in good agreement with the measured value of $\Gamma_W = 2.085 \pm 0.042\,\text{GeV}$ (see Chapter 16). Because the mass of the W boson is large, so is the total decay width, and the lifetime of the W boson is only $\mathcal{O}(10^{-25}\,\text{s})$.

### 15.1.2 W-pair production

The fact that the force carrying particles of the weak interaction possess the charge of the electromagnetic interaction is already suggestive that the weak and electromagnetic forces are somehow related. Further hints of electroweak unification are provided by the observation that the coupling constants of the electromagnetic and weak interactions are of the same order of magnitude (see Section 11.5.1). However, there are also strong theoretical arguments for why a theory with just the weak charged current must be incomplete.

Pairs of W bosons can be produced in $e^+e^-$ annihilation at an electron–positron collider or in $q\overline{q}$ annihilation at a hadron collider. The three lowest-order Feynman diagrams for the process $e^+e^- \to W^+W^-$ are shown in Figure 15.5. The *t*-channel neutrino exchange diagram represents a purely weak charged-current process. The *s*-channel photon exchange diagram is an electromagnetic process, which arises because the $W^+$ and $W^-$ carry electromagnetic charge. With the first two diagrams of Figure 15.5 alone, the calculated $e^+e^- \to W^+W^-$ cross section is found to increase with centre-of-mass energy without limit, as shown in Figure 15.6.

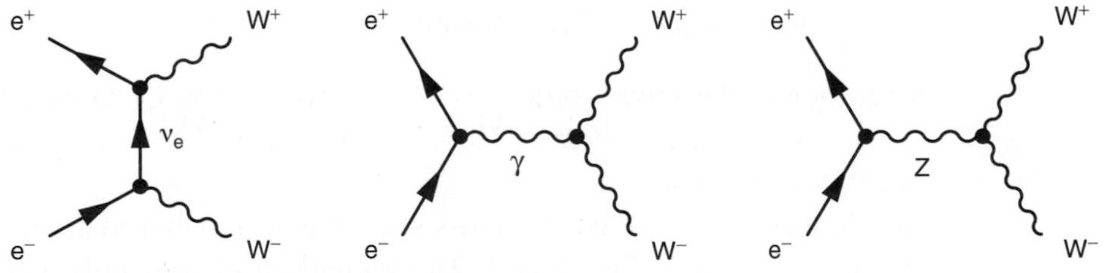


Fig. 15.5 The three lowest-order Feynman diagram for $e^+e^- \to W^+W^-$.

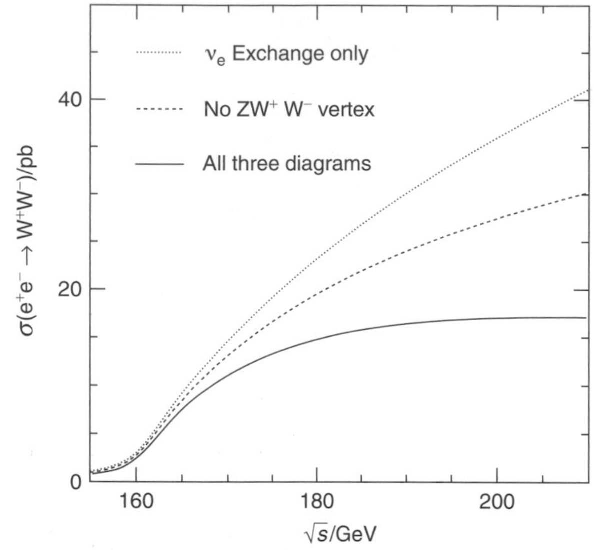


Fig. 15.6 The predicted $e^+e^- \to W^+W^-$ production cross section for three cases: only the $\nu_e$-exchange diagram; the $\nu_e$-exchange and $\gamma$-exchange diagrams; and all three Feynman diagrams of Figure 15.5.

At some relatively high centre-of-mass energy, the cross section violates quantum mechanical unitarity, whereby particle probability is no longer conserved; the calculated number of W-pairs produced in the interaction exceeds the incident $e^+e^-$ flux. This problematic high-energy behaviour of the $e^+e^- \to W^+W^-$ cross section indicates that the theory with just the first two diagrams of Figure 15.5 is incomplete. Because the *s*- and *t*-channel diagrams interfere negatively, the problem would be even worse with the neutrino exchange diagram alone,

$$|\mathcal{M}_\nu + \mathcal{M}_\gamma|^2 < |\mathcal{M}_\nu|^2.$$

The problem of unitarity violation in $e^+e^- \to W^+W^-$ production is resolved naturally in the electroweak theory, which predicts an additional gauge boson, the neutral Z. Because the contribution to the $e^+e^- \to W^+W^-$ cross section from the Z-exchange diagram interferes negatively,

$$|\mathcal{M}_\nu + \mathcal{M}_\gamma + \mathcal{M}_Z|^2 < |\mathcal{M}_\nu + \mathcal{M}_\gamma|^2,$$

the calculated $e^+e^- \to W^+W^-$ cross section is well behaved at all centre-of-mass energies, as shown in Figure 15.6. This partial cancellation only works because the couplings of the $\gamma$, $W^\pm$ and the new Z boson are related to each other in the unified electroweak model.

## 15.2 The weak interaction gauge group

In Section 10.1 it was shown that QED and QCD are associated with respective U(1) and SU(3) local gauge symmetries. The charged-current weak interaction is associated with invariance under SU(2) local phase transformations,

$$\varphi(x) \to \varphi'(x) = \exp\left[ig_W\,\boldsymbol{\alpha}(x)\cdot\mathbf{T}\right]\varphi(x). \tag{15.10}$$

Here $\mathbf{T}$ are the three generators of the SU(2) group that can be written in terms of the Pauli spin matrices,

$$\mathbf{T} = \tfrac{1}{2}\boldsymbol{\sigma},$$

and $\boldsymbol{\alpha}(x)$ are three functions which specify the local phase at each point in space-time. The required local gauge invariance can only be satisfied by the introduction of three gauge fields, $W_\mu^k$ with $k = 1, 2, 3$, corresponding to three gauge bosons $W^{(1)}$, $W^{(2)}$ and $W^{(3)}$. Because the generators of the SU(2) gauge transformation are the $2 \times 2$ Pauli spin-matrices, the wavefunction $\varphi(x)$ in (15.10) must be written in terms of two components. In analogy with the definition of isospin, $\varphi(x)$ is termed a weak isospin doublet. Since the weak charged-current interaction associated with the $W^\pm$ couples together different fermions, the weak isospin doublets must contain flavours differing by one unit of electric charge, for example

$$\varphi(x) = \begin{pmatrix} \nu_e(x) \\ e^-(x) \end{pmatrix}.$$

In this weak isospin space, the $\nu_e$ and $e^-$ have total weak isospin $I_W = \frac{1}{2}$ and third component of weak isospin $I_W^{(3)}(\nu_e) = +\frac{1}{2}$ and $I_W^{(3)}(e^-) = -\frac{1}{2}$. Since the observed form of the weak charged-current interaction couples only to left-handed chiral particle states and right-handed chiral antiparticle states, the gauge transformation of (15.10) can affect only LH particles and RH antiparticles. To achieve this, RH particle and LH antiparticle chiral states are placed in weak isospin singlets with $I_W = 0$ and are therefore unaffected by the SU(2) local gauge transformation. The weak isospin doublets are composed only of LH chiral particle states and RH chiral antiparticle states and, for this reason, the symmetry group of the weak interaction is referred to as $SU(2)_L$.

The weak isospin doublets are constructed from the weak eigenstates and therefore account for the mixing in the CKM and PMNS matrices. For example, the u quark appears in a doublet with the weak eigenstate d′, as defined in (14.3). The upper member of the doublet, with $I_W^{(3)} = +1/2$, is always the particle state which differs by plus one unit in electric charge relative to the lower member of the doublet,

$$\begin{pmatrix} \nu_e \\ e^- \end{pmatrix}_L, \quad \begin{pmatrix} \nu_\mu \\ \mu^- \end{pmatrix}_L, \quad \begin{pmatrix} \nu_\tau \\ \tau^- \end{pmatrix}_L, \quad \begin{pmatrix} u \\ d' \end{pmatrix}_L, \quad \begin{pmatrix} c \\ s' \end{pmatrix}_L, \quad \begin{pmatrix} t \\ b' \end{pmatrix}_L.$$

This common ordering within the doublets is necessary for a consistent definition of the physical $W^\pm$ bosons. The right-handed particle chiral states are placed in weak isospin singlets with $I_W = I_W^{(3)} = 0$,

$$e_R^-, \quad \mu_R^-, \quad \tau_R^-, \quad u_R, \quad c_R, \quad t_R, \quad d_R, \quad s_R, \quad b_R.$$

Because the weak isospin singlets are unaffected by the $SU(2)_L$ local gauge transformation of the weak interaction, they do not couple to the gauge bosons of the symmetry.

The requirement of local gauge invariance implies the modification of the Dirac equation to include a new interaction term, analogous to (10.11),

$$ig_W T_k \gamma^\mu W_\mu^k \varphi_L = ig_W \tfrac{1}{2}\sigma_k \gamma^\mu W_\mu^k \varphi_L, \tag{15.11}$$

where $\varphi_L$ represents a weak isospin doublet of left-handed chiral particles. This form of the interaction gives rise to three weak currents, one for each of the three gauge fields $W^k$. In the case of the weak isospin doublet formed from the left-handed electron and the electron neutrino,

$$\varphi_L = \begin{pmatrix} \nu_L \\ e_L \end{pmatrix},$$

the three weak currents, one for each of the Pauli spin-matrices, are

$$j_1^\mu = \frac{g_W}{2}\overline{\varphi}_L\gamma^\mu\sigma_1\varphi_L, \quad j_2^\mu = \frac{g_W}{2}\overline{\varphi}_L\gamma^\mu\sigma_2\varphi_L \quad \text{and} \quad j_3^\mu = \frac{g_W}{2}\overline{\varphi}_L\gamma^\mu\sigma_3\varphi_L,$$

where $\overline{\varphi}_L = \begin{pmatrix} \overline{\nu}_L & \overline{e}_L \end{pmatrix}$ contains the left-handed chiral adjoint spinors, $\overline{\nu}_L$ and $\overline{e}_L$. The weak charged-currents are related to the weak isospin raising and lowering operators, $\sigma_\pm = \frac{1}{2}(\sigma_1 \pm i\sigma_2)$, which step between the two states within a weak isospin doublet. The four-vector currents corresponding to the exchange of the physical $W^\pm$ bosons are

$$\begin{aligned} j_\pm^\mu = \tfrac{1}{\sqrt{2}}\left(j_1^\mu \pm i j_2^\mu\right) &= \frac{g_W}{\sqrt{2}}\overline{\varphi}_L\gamma^\mu\tfrac{1}{2}(\sigma_1 \pm i\sigma_2)\varphi_L, \\ &= \frac{g_W}{\sqrt{2}}\overline{\varphi}_L\gamma^\mu\sigma_\pm\varphi_L. \end{aligned}$$

The physical W bosons can be identified as the linear combinations

$$W_\mu^\pm = \tfrac{1}{\sqrt{2}}\left(W_\mu^{(1)} \mp iW_\mu^{(2)}\right), \tag{15.12}$$

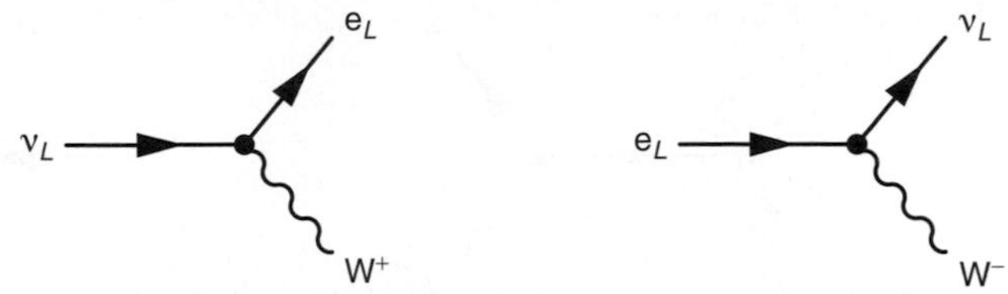


Fig. 15.7 The weak charged-current interaction vertices between the $e^-$ and the $\nu_e$ weak eigenstate.

such that the weak currents can be written

$$\mathbf{j}^\mu \cdot \mathbf{W}_\mu = j_1^\mu W_\mu^{(1)} + j_2^\mu W_\mu^{(2)} + j_3^\mu W_\mu^{(3)} \equiv j_+^\mu W_\mu^+ + j_-^\mu W_\mu^- + j_3^\mu W_\mu^{(3)}.$$

The current $j_+^\mu$, which corresponds to the exchange of a $W^+$ boson, can be expressed as

$$\begin{aligned} j_+^\mu &= \frac{g_W}{\sqrt{2}}\overline{\varphi}_L\gamma^\mu\sigma_+\varphi_L = \frac{g_W}{\sqrt{2}}\begin{pmatrix}\overline{\nu}_L & \overline{e}_L\end{pmatrix}\gamma^\mu\begin{pmatrix}0 & 1\\ 0 & 0\end{pmatrix}\begin{pmatrix}\nu_L\\ e_L\end{pmatrix}\\ &= \frac{g_W}{\sqrt{2}}\overline{\nu}_L\gamma^\mu e_L \equiv \frac{g_W}{\sqrt{2}}\overline{\nu}\gamma^\mu\tfrac{1}{2}(1-\gamma^5)e. \end{aligned}$$

Similarly, the current corresponding to the $W^-$ vertex is

$$\begin{aligned} j_-^\mu &= \frac{g_W}{\sqrt{2}}\overline{\varphi}_L\gamma^\mu\sigma_-\varphi_L = \frac{g_W}{\sqrt{2}}\begin{pmatrix}\overline{\nu}_L & \overline{e}_L\end{pmatrix}\gamma^\mu\begin{pmatrix}0 & 0\\ 1 & 0\end{pmatrix}\begin{pmatrix}\nu_L\\ e_L\end{pmatrix}\\ &= \frac{g_W}{\sqrt{2}}\overline{e}_L\gamma^\mu \nu_L \equiv \frac{g_W}{\sqrt{2}}\overline{e}\gamma^\mu\tfrac{1}{2}(1-\gamma^5)\nu. \end{aligned}$$

Thus, the $SU(2)_L$ symmetry of the weak interaction results in the now familiar weak charged-currents

$$j_+^\mu = \frac{g_W}{\sqrt{2}}\overline{\nu}\gamma^\mu\tfrac{1}{2}(1-\gamma^5)e \quad\text{and}\quad j_-^\mu = \frac{g_W}{\sqrt{2}}\overline{e}\gamma^\mu\tfrac{1}{2}(1-\gamma^5)\nu,$$

corresponding to the $W^+$ and $W^-$ vertices shown in Figure 15.7.

In addition to the two weak charged-currents, $j_+$ and $j_-$, which arise from linear combinations of the $W^{(1)}$ and $W^{(2)}$, the $SU(2)_L$ gauge symmetry implies the existence a weak neutral-current given by

$$j_3^\mu = g_W\overline{\varphi}_L\gamma^\mu\tfrac{1}{2}\sigma_3\varphi_L.$$

The weak neutral-current, written in terms of the component fermions, is

$$\begin{aligned} j_3^\mu &= g_W\tfrac{1}{2}\begin{pmatrix}\overline{\nu}_L & \overline{e}_L\end{pmatrix}\gamma^\mu\begin{pmatrix}1 & 0\\ 0 & -1\end{pmatrix}\begin{pmatrix}\nu_L\\ e_L\end{pmatrix}\\ &= g_W\tfrac{1}{2}\overline{\nu}_L\gamma^\mu\nu_L - g_W\tfrac{1}{2}\overline{e}_L\gamma^\mu e_L. \end{aligned} \tag{15.13}$$

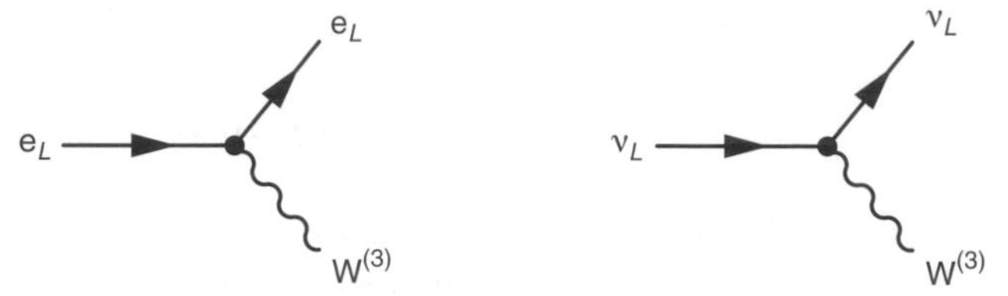


Fig. 15.8 The weak neutral-current interaction of the $e^-$ and $\nu_e$.

Hence the $SU(2)_L$ symmetry of the weak interaction implies the existence of the weak neutral-current corresponding to the vertices shown in Figure 15.8, with

$$j_3^\mu = I_W^{(3)} g_W \overline{f} \gamma^\mu \tfrac{1}{2}(1-\gamma^5) f, \tag{15.14}$$

where f denotes the fermion spinor. The sign in this expression is determined by the third component of weak isospin $I_W^{(3)} = \pm 1/2$. Because RH particles/LH antiparticles have $I_W^{(3)} = 0$, they do not couple to the weak neutral-current corresponding to the $W^{(3)}$.

## 15.3 Electroweak unification

It is tempting to identify the neutral-current of (15.14) as that due to the exchange of the Z boson, in which case the Z boson would correspond to the $W^{(3)}$ of the $SU(2)_L$ local gauge symmetry. This would imply that the weak neutral-current coupled only to left-handed particles and right-handed antiparticles. This is in contradiction with experiment, which shows that the physical Z boson couples to both left- and right-handed chiral states (although not equally).

Of the four observed bosons of QED and the weak interaction, the photon and the Z boson, with the corresponding fields $A_\mu$ and $Z_\mu$, are both neutral. Consequently, it is plausible that they can be expressed in terms of quantum state formed from two neutral bosons, one of which is the $W^{(3)}$ associated with the $SU(2)_L$ local gauge symmetry. In the electroweak model of Glashow, Salam and Weinberg (GSW), the U(1) gauge symmetry of electromagnetism is replaced with a new $U(1)_Y$ local gauge symmetry

$$\psi(x) \to \psi'(x) = \hat{U}(x)\psi(x) = \exp\left[ig'\frac{Y}{2}\zeta(x)\right]\psi(x), \tag{15.15}$$

giving rise to a new gauge field $B_\mu$ that couples to a new kind of charge, termed weak hypercharge $Y$. The resulting interaction term is

$$g'\frac{Y}{2}\gamma^\mu B_\mu \psi, \tag{15.16}$$

which has the same form as the interaction from the U(1) symmetry of QED,

$$Qe\gamma^{\mu}A_{\mu}\psi,$$

with $Qe$ replaced by $Yg'/2$. In the unified electroweak model, the photon and Z boson are written as linear combinations of the $B_{\mu}$ and neutral $W_{\mu}^{(3)}$ of the weak interaction,

$$A_{\mu} = +B_{\mu}\cos\theta_{W} + W_{\mu}^{(3)}\sin\theta_{W}, \tag{15.17}$$

$$Z_{\mu} = -B_{\mu}\sin\theta_{W} + W_{\mu}^{(3)}\cos\theta_{W}, \tag{15.18}$$

where $\theta_W$ is the weak mixing angle. This mixing of the neutral fields of the $U(1)_Y$ and $SU(2)_L$ gauge symmetries might seem contrived; however, it arises naturally in the Higgs mechanism (see Chapter 17). From (15.17) and (15.18), the physical currents of QED and the weak neutral current are

$$j_{em}^{\mu} = j_{Y}^{\mu}\cos\theta_{W} + j_{3}^{\mu}\sin\theta_{W}, \tag{15.19}$$

$$j_{Z}^{\mu} = -j_{Y}^{\mu}\sin\theta_{W} + j_{3}^{\mu}\cos\theta_{W}. \tag{15.20}$$

The GSW model of electroweak unification implies that the couplings of the weak and electromagnetic interactions are related. This can be seen by considering the interactions of the electron and the electron neutrino. The weak neutral-current associated with the $W^{(3)}$ is given by (15.13) and involves only left-handed particles,

$$j_{3}^{\mu} = \tfrac{1}{2}g_{W}\,\overline{\nu}_{L}\gamma^{\mu}\nu_{L} - \tfrac{1}{2}g_{W}\,\overline{e}_{L}\gamma^{\mu}e_{L}. \tag{15.21}$$

The currents from the interaction term of (15.16), which treats left- and right-handed states equally, are

$$j_{Y}^{\mu} = \tfrac{1}{2}g'Y_{e_L}\,\overline{e}_{L}\gamma^{\mu}e_{L} + \tfrac{1}{2}g'Y_{e_R}\,\overline{e}_{R}\gamma^{\mu}e_{R} + \tfrac{1}{2}g'Y_{\nu_L}\,\overline{\nu}_{L}\gamma^{\mu}\nu_{L} + \tfrac{1}{2}g'Y_{\nu_R}\,\overline{\nu}_{R}\gamma^{\mu}\nu_{R}, \tag{15.22}$$

where, for example, $Y_{e_L}$ is the weak hypercharge of the left-handed electron. The current for the electromagnetic interaction, written in terms of the chiral components of the electron, is

$$j_{em}^{\mu} = Q_{e}e\,\overline{e}_{L}\gamma^{\mu}e_{L} + Q_{e}e\,\overline{e}_{R}\gamma^{\mu}e_{R}.$$

Since the neutrino is a neutral particle its electromagnetic current is zero. For the GSW model to work, it must reproduce the observed couplings of QED. From (15.19) the electromagnetic current can be written

$$j_{em}^{\mu} = Q_{e}e\,\overline{e}_{L}\gamma^{\mu}e_{L} + Q_{e}e\,\overline{e}_{R}\gamma^{\mu}e_{R} = j_{Y}^{\mu}\cos\theta_{W} + j_{3}^{\mu}\sin\theta_{W},$$

where $j_Y^\mu$ and $j_3^\mu$, which include terms for the electron neutrino, are given by (15.21) and (15.22). Hence the terms in electromagnetic current $j_{em}^\mu$, including those for the neutrinos which are zero, can be equated to

$$\overline{e}_L\gamma^\mu e_L: \qquad Q_e e = \tfrac{1}{2}g' Y_{e_L}\cos\theta_W - \tfrac{1}{2}g_W\sin\theta_W, \tag{15.23}$$

$$\overline{\nu}_L\gamma^\mu \nu_L: \qquad 0 = \tfrac{1}{2}g' Y_{\nu_L}\cos\theta_W + \tfrac{1}{2}g_W\sin\theta_W, \tag{15.24}$$

$$\overline{e}_R\gamma^\mu e_R: \qquad Q_e e = \tfrac{1}{2}g' Y_{e_R}\cos\theta_W, \tag{15.25}$$

$$\overline{\nu}_R\gamma^\mu \nu_R: \qquad 0 = \tfrac{1}{2}g' Y_{\nu_R}\cos\theta_W. \tag{15.26}$$

Equations (15.23)–(15.26) relate the couplings of electromagnetism to those of the weak interaction and the couplings associated with the $U(1)_Y$ symmetry.

In the GSW model, the underlying gauge symmetry of the electroweak sector of the Standard Model is the $U(1)_Y$ of weak hypercharge and the $SU(2)_L$ of the weak interaction, written as $U(1)_Y \times SU(2)_L$. For invariance under $U(1)_Y$ *and* $SU(2)_L$ local gauge transformations, the weak hypercharges of the particles in a weak isospin doublet must be the same, for example $Y_{e_L} = Y_{\nu_L}$. If this were not the case, a $U(1)_Y$ local gauge transformation would introduce a phase difference between the two components of a weak isospin doublet, breaking the $SU(2)_L$ symmetry. The weak hypercharge assignments of the fermions can be expressed as a linear combination of the electromagnetic charge $Q$ and the third component of weak isospin $I_W^{(3)}$,

$$Y = \alpha Q + \beta I_W^{(3)}.$$

The charges and third component of weak isospin for the left-handed electron and the left-handed electron neutrino are respectively $\left(Q = -1, I_W^{(3)} = -\frac{1}{2}\right)$ and $\left(Q = 0, I_W^{(3)} = +\frac{1}{2}\right)$, and therefore

$$Y_{\nu_L} = +\tfrac{1}{2}\beta \quad \text{and} \quad Y_{e_L} = -\alpha - \tfrac{1}{2}\beta.$$

From the requirement that $Y_{e_L} = Y_{\nu_L}$, it follows that $\beta = -\alpha$ and the weak hypercharge can be identified as

$$Y = 2\left(Q - I_W^{(3)}\right). \tag{15.27}$$

The factor of two in (15.27) is purely conventional; a different choice could be absorbed into the definition of $g'$ without modifying the actual couplings. The weak hypercharges of the $e_L$ and $\nu_L$ are therefore

$$Y_{e_L} = Y_{\nu_L} = -1.$$

Since $Y_{e_L} = Y_{\nu_L}$, subtracting (15.23) from (15.24) gives the relationship between the weak and electromagnetic couplings in terms of the weak mixing angle,

$$e = g_W \sin\theta_W. \tag{15.28}$$

The sum of (15.23) and (15.24) gives

$$Q_e e = \tfrac{1}{2}g'(Y_{e_L} + Y_{\nu_L})\cos\theta_W.$$

Since $Q_{e_L} = -1$ and $Y_{e_L} = Y_{\nu_L} = -1$, the coupling $g'$ is related to the electron charge by

$$e = g'\cos\theta_W. \tag{15.29}$$

Finally, from (15.27), the weak hypercharge assignments of the $I_W^{(3)} = 0$ right-handed states are

$$Y_{e_R} = -2 \quad \text{and} \quad Y_{\nu_R} = 0,$$

which when substituted into (15.25) and (15.26) give the correct electromagnetic charges of $Q = -1$ and $Q = 0$ for the $e_R$ and $\nu_R$.

The unified electroweak model is able to provide a consistent picture of the electromagnetic interactions of the fermions with the relation,

$$e = g_W \sin\theta_W = g'\cos\theta_W, \tag{15.30}$$

and where the weak hypercharge is given by

$$Y = 2\left(Q - I_W^{(3)}\right).$$

The weak mixing angle has been measured in a number of different ways, including the studies of $e^+e^- \to Z \to f\bar{f}$, described in Chapter 16. The average of the measurements of $\sin^2\theta_W$ gives

$$\sin^2\theta_W = 0.23146 \pm 0.00012. \tag{15.31}$$

From (15.28) and the measured value of $\sin^2\theta_W$, the expected ratio of the weak to electromagnetic coupling constants is

$$\frac{\alpha}{\alpha_W} = \frac{e^2}{g_W^2} = \sin^2\theta_W \sim 0.23,$$

consistent with the measured values discussed previously in Section 11.5.1.

### 15.3.1 The couplings of the Z

At this point it might be tempting to think that the procedure for electroweak unification has just replaced two independent couplings, $e$ and $g_W$, by a single unified coupling and the weak mixing angle. However, once the couplings in the electroweak model are chosen to reproduce the observed electromagnetic couplings,

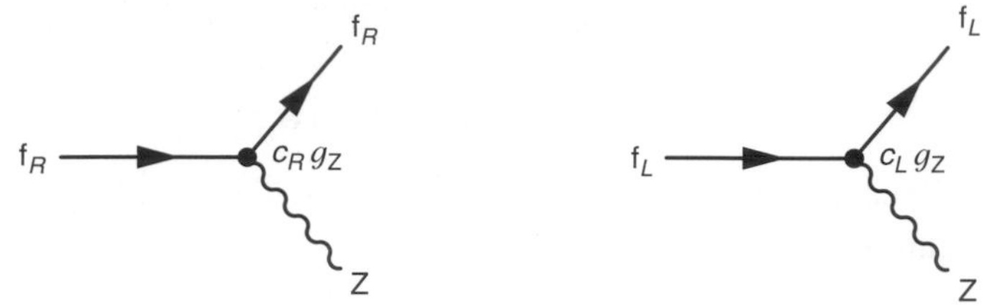


Fig. 15.9 The weak neutral-current interaction vertices for the physical Z boson and the chiral states of a fermion f.

the couplings of the Z boson to all the fermions are completely specified. The current from the interaction between the Z boson and a fermion of flavour f is given by (15.20),

$$j_Z^\mu = -\tfrac{1}{2}g' \sin\theta_W[Y_{f_L}\overline{u}_L\gamma^\mu u_L + Y_{f_R}\overline{u}_R\gamma^\mu u_R] + I_W^{(3)} g_W \cos\theta_W[\overline{u}_L\gamma^\mu u_L],$$

where $u_{L/R}$ and $\overline{u}_{L/R}$ are the spinors and adjoints spinors for LH and RH chiral states. Using (15.27) to express the weak hypercharge in terms of $Q$ and $I_W^{(3)}$ implies

$$j_Z^\mu = -g' \sin\theta_W\left[\left(Q_f - I_W^{(3)}\right)\overline{u}_L\gamma^\mu u_L + Q_f\overline{u}_R\gamma^\mu u_R\right] + I_W^{(3)} g_W \cos\theta_W\left[\overline{u}_L\gamma^\mu u_L\right].$$

Collecting up the factors in front of the left- and right-handed currents gives

$$j_Z^\mu = \left[-g'\left(Q_f - I_W^{(3)}\right)\sin\theta_W + I_W^{(3)} g_W \cos\theta_W\right]\overline{u}_L\gamma^\mu u_L - \left[g' \sin\theta_W Q_f\right]\overline{u}_R\gamma^\mu u_R.$$

From (15.30) it can be seen that $g' = g_W \tan\theta_W$ and therefore

$$j_Z^\mu = g_W\left[-\left(Q_f - I_W^{(3)}\right)\frac{\sin^2\theta_W}{\cos\theta_W} + I_W^{(3)}\cos\theta_W\right]\overline{u}_L\gamma^\mu u_L - g_W\left[\frac{\sin^2\theta_W}{\cos\theta_W}Q_f\right]\overline{u}_R\gamma^\mu u_R. \tag{15.32}$$

Defining the coupling to the physical Z boson as

$$g_Z = \frac{g_W}{\cos\theta_W} \equiv \frac{e}{\sin\theta_W\cos\theta_W},$$

allows the neutral-current due to the Z boson to be written as

$$j_Z^\mu = g_Z\left(I_W^{(3)} - Q_f\sin^2\theta_W\right)\overline{u}_L\gamma^\mu u_L - g_Z\left(Q_f\sin^2\theta_W\right)\overline{u}_R\gamma^\mu u_R.$$

Hence the couplings of the Z boson to left- and right-handed chiral states, shown in Figure 15.9, are

$$j_Z^\mu = g_Z\left(c_L\overline{u}_L\gamma^\mu u_L + c_R\overline{u}_R\gamma^\mu u_R\right), \tag{15.33}$$

with

$$c_L = I_W^{(3)} - Q_f\sin^2\theta_W \quad \text{and} \quad c_R = -Q_f\sin^2\theta_W. \tag{15.34}$$

Thus, the Z boson couples to both left- and right-handed chiral states, but not equally. This should come as no surprise; the current associated with the Z boson

**Table 15.1** The charge, $I_W^{(3)}$ and weak hypercharge assignments of the fundamental fermions and their couplings to the Z assuming $\sin^2\theta_W = 0.231\,46$.

| fermion | $Q_f$ | $I_W^{(3)}$ | $Y_L$ | $Y_R$ | $c_L$ | $c_R$ | $c_V$ | $c_A$ |
|---|---|---|---|---|---|---|---|---|
| $\nu_e, \nu_\mu, \nu_\tau$ | 0 | $+\frac{1}{2}$ | −1 | 0 | $+\frac{1}{2}$ | 0 | $+\frac{1}{2}$ | $+\frac{1}{2}$ |
| $e^-, \mu^-, \tau^-$ | −1 | $-\frac{1}{2}$ | −1 | −2 | −0.27 | +0.23 | −0.04 | $-\frac{1}{2}$ |
| u, c, t | $+\frac{2}{3}$ | $+\frac{1}{2}$ | $+\frac{1}{3}$ | $+\frac{4}{3}$ | +0.35 | −0.15 | +0.19 | $+\frac{1}{2}$ |
| d, s, b | $-\frac{1}{3}$ | $-\frac{1}{2}$ | $+\frac{1}{3}$ | $-\frac{2}{3}$ | −0.42 | +0.08 | −0.35 | $-\frac{1}{2}$ |

has contributions from the weak interaction, which couples only to left-handed particles, and from the $B_\mu$ field associated with the $U(1)_Y$ local gauge symmetry, which treats left- and right-handed states equally.

The couplings of the Z boson to fermions also can be expressed in terms of vector and axial-vector couplings using the chiral projection operators of (6.33),

$$\overline{u}_L\gamma^\mu u_L = \overline{u}\gamma^\mu\tfrac{1}{2}(1-\gamma^5)u \quad \text{and} \quad \overline{u}_R\gamma^\mu u_R = \overline{u}\gamma^\mu\tfrac{1}{2}(1+\gamma^5)u,$$

such that the current $j_Z^\mu$ of (15.33) becomes

$$\begin{aligned} j_Z^\mu &= g_Z\overline{u}\gamma^\mu\left[c_L\tfrac{1}{2}(1-\gamma^5)+c_R\tfrac{1}{2}(1+\gamma^5)\right]u \\ &= g_Z\overline{u}\gamma^\mu\,\tfrac{1}{2}\left[(c_L+c_R)-(c_L-c_R)\gamma^5\right]u. \end{aligned}$$

Therefore the weak neutral-current can be written as

$$j_Z^\mu = \tfrac{1}{2}g_Z\overline{u}\left(c_V\gamma^\mu - c_A\gamma^\mu\gamma^5\right)u, \tag{15.35}$$

where the vector and axial-vector couplings of the Z boson are

$$c_V = (c_L + c_R) = I_W^{(3)} - 2Q\sin^2\theta_W, \tag{15.36}$$

$$c_A = (c_L - c_R) = I_W^{(3)}. \tag{15.37}$$

In terms of these vector and axial-vector couplings, the Feynman rule associated with the Z-boson interaction vertex is

$$-i\tfrac{1}{2}g_Z\gamma^\mu\left[c_V - c_A\gamma^5\right]. \tag{15.38}$$

Because the weak neutral-current contains both vector and axial-vector couplings, it does not conserve parity (see Section 11.3); this also immediately follows from its different couplings to left- and right-handed chiral states.

In the Standard Model, once $\sin^2\theta_W$ is known, the couplings of the Z boson to the fermions are predicted exactly. For $\sin^2\theta_W = 0.23146$, the predicted couplings of the fermions to the Z boson are listed in Table 15.1, both in terms of the vector and axial-vector couplings $(c_V, c_A)$ and the couplings to left- and right-handed chiral states, $(c_L, c_R)$.

## 15.4 Decays of the Z

The calculation of the Z-boson total decay width and branching ratios follows closely that for the decay of the W boson, given in Section 15.1.1. However, whereas the W boson couples only to left-handed chiral particle states, the Z boson couples to both left- and right-handed states. Nevertheless, because the weak neutral-current is a vector/axial-vector interaction, the currents due to certain chiral combinations are still zero. For example, for the decay $Z \to f\bar{f}$, the weak neutral-current where both the fermion and antifermion are right-handed is zero, which can be seen from

$$\begin{aligned}\overline{u}_R\gamma^\mu(c_V - c_A\gamma^5)v_R &= u^\dagger \tfrac{1}{2}(1+\gamma^5)\gamma^0\gamma^\mu(c_V - c_A\gamma^5)\tfrac{1}{2}(1-\gamma^5)v \\ &= \tfrac{1}{4}u^\dagger\gamma^0(1-\gamma^5)(1+\gamma^5)\gamma^\mu(c_V - c_A\gamma^5)v \\ &= \tfrac{1}{4}\overline{u}\gamma^\mu P_L P_R(c_V - c_A\gamma^5)v = 0.\end{aligned}$$

Consequently, in the limit where the masses of the fermions in the decay $Z \to f\bar{f}$ can be neglected, only the two helicity combinations shown in Figure 15.10 give non-zero matrix elements for the decay of the Z boson.

The Z-boson decay rate either can be calculated from first principles (see Problem 15.3) or can be obtained from the spin-averaged matrix element of (15.6), derived previously for W-boson decay. For the helicity combination where the decay of the Z boson gives a LH particle and RH antiparticle, the spin-averaged matrix element is the same as that for W-boson decay, but with

$$\tfrac{1}{2}g_W^2 \to g_Z^2 c_L^2, \quad \Rightarrow \quad \langle|\mathcal{M}_L|^2\rangle = \tfrac{2}{3}c_L^2 g_Z^2 m_Z^2.$$

The corresponding matrix element for the Z decay to a RH particle and LH antiparticle will be proportional to $c_R$ rather than $c_L$. After averaging over the spin states and decay angle, all other factors will be the same. Therefore the spin-averaged matrix element squared for $Z \to f\bar{f}$ is

$$\langle|\mathcal{M}|^2\rangle = \langle|\mathcal{M}_L|^2 + |\mathcal{M}_R|^2\rangle = \tfrac{2}{3}(c_L^2 + c_R^2)g_Z^2 m_Z^2. \tag{15.39}$$

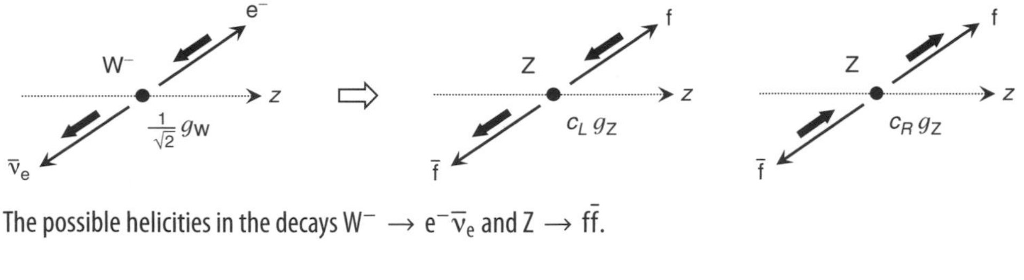


Fig. 15.10 The possible helicities in the decays $W^- \to e^-\overline{\nu}_e$ and $Z \to f\bar{f}$.

This can be expressed in terms of the vector and axial-vector couplings of the Z boson using $c_V = c_L + c_R$ and $c_A = c_L - c_R$, which implies that $c_V^2 + c_A^2 = 2(c_L^2 + c_R^2)$. Hence

$$\langle|\mathcal{M}|^2\rangle = \tfrac{1}{3}(c_V^2 + c_A^2)g_Z^2 m_Z^2, \tag{15.40}$$

from which it follows that

$$\Gamma(\mathrm{Z} \to \mathrm{f\bar{f}}) = \frac{g_Z^2 m_Z}{48\pi}(c_V^2 + c_A^2). \tag{15.41}$$

### 15.4.1 Z width and branching ratios

The Z-boson partial decay rates depend on $g_Z$ and $m_Z$. The measured value of mass of the Z boson is $m_Z = 91.2\,\mathrm{GeV}$ (see Section 16.2.1). The numerical value of $g_Z$ can be obtained from the measured values of the Fermi constant and $\sin^2\theta_W$,

$$g_Z^2 = \frac{g_W^2}{\cos^2\theta_W} = \frac{8m_W^2}{\sqrt{2}\cos^2\theta_W}G_F \approx 0.55.$$

The partial decay rate to a particular fermion flavour can be calculated from (15.41) using the appropriate vector and axial-vector couplings. For example, in the case of the decay $\mathrm{Z} \to \nu_e\bar{\nu}_e$, the neutrino vector and axial-vector couplings are $c_V = c_A = \frac{1}{2}$, and therefore

$$\Gamma(\mathrm{Z} \to \nu_e\bar{\nu}_e) = \frac{g_Z^2 m_Z}{48\pi}\left(\frac{1}{4} + \frac{1}{4}\right) = 167\,\mathrm{MeV}. \tag{15.42}$$

Because the Z boson couples to all fermions, it can decay to all flavours with the exception of the top quark ($m_t > m_Z$). The total decay width $\Gamma_Z$ is given by the sum of the partial decay widths

$$\Gamma_Z = \sum_f \Gamma(\mathrm{Z} \to \mathrm{f\bar{f}}).$$

The Z-boson couplings, listed in Table 15.1, are the same for all three generations, and thus the total decay width can be written

$$\Gamma_Z = 3\,\Gamma(\mathrm{Z} \to \nu_e\bar{\nu}_e) + 3\,\Gamma(\mathrm{Z} \to \mathrm{e^+e^-}) + 3 \times 2\,\Gamma(\mathrm{Z} \to \mathrm{u\bar{u}}) + 3 \times 3\,\Gamma(\mathrm{Z} \to \mathrm{d\bar{d}}),$$

where the additional factors of three multiplying the decays to quarks account for colour, and only two decays to up-type quarks are possible since $m_t > m_Z$. Using the couplings in Table 15.1, and multiplying the hadronic decay widths by $[1 + \alpha_S(Q^2)/\pi]$ to account for the gluon radiation in the decay, the total decay width of the Z is predicted to be

$$\Gamma_Z \approx 2.5\,\mathrm{GeV},$$

and the branching ratios of the Z boson, given by $Br(\mathrm{Z} \to \mathrm{f\overline{f}}) = \Gamma(\mathrm{Z} \to \mathrm{f\overline{f}})/\Gamma_\mathrm{Z}$, are

$$\begin{aligned}
Br(\mathrm{Z} \to \nu_\mathrm{e}\overline{\nu}_\mathrm{e}) = Br(\mathrm{Z} \to \nu_\mu\overline{\nu}_\mu) = Br(\mathrm{Z} \to \nu_\tau\overline{\nu}_\tau) &\approx 6.9\%,\\
Br(\mathrm{Z} \to \mathrm{e^+e^-}) = Br(\mathrm{Z} \to \mu^+\mu^-) = Br(\mathrm{Z} \to \tau^+\tau^-) &\approx 3.5\%,\\
Br(\mathrm{Z} \to \mathrm{u\overline{u}}) = Br(\mathrm{Z} \to \mathrm{c\overline{c}}) &\approx 12\%,\\
Br(\mathrm{Z} \to \mathrm{d\overline{d}}) = Br(\mathrm{Z} \to \mathrm{s\overline{s}}) = Br(\mathrm{Z} \to \mathrm{b\overline{b}}) &\approx 15\%.
\end{aligned}$$

Grouping together the decays to neutrinos, charged leptons, and quarks gives

$$Br(\mathrm{Z} \to \nu\overline{\nu}) \approx 21\%, \quad Br(\mathrm{Z} \to \ell^+\ell^-) \approx 10\% \ \text{ and } \ Br(\mathrm{Z} \to \text{hadrons}) \approx 69\%,$$

and thus almost 70% of Z decays result in final states with jets.

## Summary

In the Standard Model, the weak charged-current is associated with an $\mathrm{SU}(2)_L$ local gauge symmetry. This gives rise to the $\mathrm{W}^+$ and $\mathrm{W}^-$ bosons and a neutral gauge field, $W^{(3)}$. In the GSW model of electroweak unification, this neutral field mixes with a photon-like field of the $\mathrm{U}(1)_Y$ gauge symmetry to give the physical photon and Z-boson fields

$$\begin{aligned}
A_\mu &= +B_\mu \cos\theta_\mathrm{W} + W^{(3)}_\mu \sin\theta_\mathrm{W}\\
Z_\mu &= -B_\mu \sin\theta_\mathrm{W} + W^{(3)}_\mu \cos\theta_\mathrm{W},
\end{aligned}$$

where $\theta_\mathrm{W}$ is the weak mixing angle. Within this unified model, the couplings of the $\gamma$, W and Z are related by

$$e = g_\mathrm{W} \sin\theta_\mathrm{W} = g_\mathrm{Z} \sin\theta_\mathrm{W} \cos\theta_\mathrm{W}.$$

Within the unified electroweak model, once $\theta_\mathrm{W}$ is known, the properties of the Z boson are completely specified. The precise tests of these predictions are main the subject of the next chapter.

## Problems

**15.1** Draw all possible lowest-order Feynman diagrams for the processes:

$$\mathrm{e^+e^-} \to \mu^+\mu^-, \quad \mathrm{e^+e^-} \to \nu_\mu\overline{\nu}_\mu, \quad \nu_\mu\mathrm{e^-} \to \nu_\mu\mathrm{e^-} \quad \text{and} \quad \overline{\nu}_\mathrm{e}\mathrm{e^-} \to \overline{\nu}_\mathrm{e}\mathrm{e^-}.$$

**15.2** Draw the lowest-order Feynman diagram for the decay $\pi^0 \to \nu_\mu\overline{\nu}_\mu$ and explain why this decay is effectively forbidden.

**15.3** Starting from the matrix element, work through the calculation of the $Z \to f\bar{f}$ partial decay rate, expressing the answer in terms of the vector and axial-vector couplings of Z. Taking $\sin^2\theta_W = 0.2315$, show that

$$R_\mu = \frac{\Gamma(Z \to \mu^+\mu^-)}{\Gamma(Z \to \text{hadrons})} \approx \frac{1}{20}.$$

**15.4** Consider the purely neutral-current (NC) process $\nu_\mu e^- \to \nu_\mu e^-$.

(a) Show that in the limit where the electron mass can be neglected, the spin-averaged matrix element for $\nu_\mu e^- \to \nu_\mu e^-$ can be written

$$\langle|\mathcal{M}|^2\rangle = \frac{1}{2}\left(\left|\mathcal{M}_{LL}^{NC}\right|^2 + \left|\mathcal{M}_{LR}^{NC}\right|^2\right),$$

where

$$\mathcal{M}_{LL}^{NC} = 2c_L^{(\nu)}c_L^{(e)}\frac{g_Z^2 s}{m_Z^2} \quad \text{and} \quad \mathcal{M}_{RR}^{NC} = 2c_L^{(\nu)}c_R^{(e)}\frac{g_Z^2 s}{m_Z^2}\frac{1}{2}(1+\cos\theta^*),$$

and $\theta^*$ is the angle between the directions of the incoming and scattered neutrino in the centre-of-mass frame.

(b) Hence find an expression for the $\nu_\mu e^-$ neutral-current cross section in terms of the laboratory frame neutrino energy.

**15.5** The two lowest-order Feynman diagrams for $\nu_e e^- \to \nu_e e^-$ are shown in Figure 13.5. Because both diagrams produce the same final state, the amplitudes have to be added before the matrix element is squared. The matrix element for the charged-current (CC) process is

$$\mathcal{M}_{LL}^{CC} = \frac{g_W^2 s}{m_W^2}.$$

(a) In the limit where the lepton masses and the $q^2$ term in the W-boson propagator can be neglected, write down expressions for spin-averaged matrix elements for the processes

$$\nu_\mu e^- \to \nu_\mu e^-, \quad \nu_e e^- \to \nu_e e^- \quad \text{and} \quad \nu_\mu e^- \to \nu_e \mu^-.$$

(b) Using the relation $g_Z/m_Z = g_W/m_W$, show that

$$\sigma(\nu_\mu e^- \to \nu_\mu e^-) : \sigma(\nu_e e^- \to \nu_e e^-) : \sigma(\nu_\mu e^- \to \nu_e \mu^-) = c_L^2 + \tfrac{1}{3}c_R^2 : (1 + c_L)^2 + \tfrac{1}{3}c_R^2 : 1,$$

where $c_L$ and $c_R$ refer to the couplings of the left- and right-handed charged leptons to the Z.

(c) Find numerical values for these ratios of NC + CC : NC : CC cross sections and comment on the sign of the interference between the NC and CC diagrams.

# 16 Tests of the Standard Model

Over the course of the previous 15 chapters, the main elements of the Standard Model of particle physics have been described. There are 12 fundamental spin-half fermions, which satisfy the Dirac equation, and 12 corresponding antiparticles. The interactions between particles are described by the exchange of spin-1 gauge bosons where the form of the interaction is determined by the local gauge principle. The underlying gauge symmetry of the Standard Model is $U(1)_Y \times SU(2)_L \times SU(3)$, with the electromagnetic and weak interactions described by the unified electroweak theory. The precise predictions of the electroweak theory were confronted with equally precise experimental measurements of the properties of the W and Z bosons at the LEP and Tevatron colliders. These precision tests of the Standard Model are the main subject of this chapter.

## 16.1 The Z resonance

The unified electroweak model introduced in Chapter 15 provides precise predictions for the properties of the Z boson. These predictions were tested with high precision at the Large Electron–Positron (LEP) collider at CERN, where large numbers of Z bosons were produced in $e^+e^-$ annihilation at the Z resonance.

Because the neutral Z boson couples to all flavours of fermions, the photon in any QED process can be replaced by a Z. For example, Figure 16.1 shows the two lowest-order Feynman diagrams for the annihilation process $e^+e^- \to \mu^+\mu^-$. The respective couplings and propagator terms that enter the matrix elements for the photon and Z exchange diagrams are

$$\mathcal{M}_\gamma \propto \frac{e^2}{q^2} \quad \text{and} \quad \mathcal{M}_Z \propto \frac{g_Z^2}{q^2 - m_Z^2}. \tag{16.1}$$

In the $s$-channel annihilation process, the four-momentum of the virtual particle is equal to the centre-of-mass energy squared, $q^2 = s$. Owing to the presence of the $m_Z^2$ term in the Z-boson propagator, the QED process dominates at low

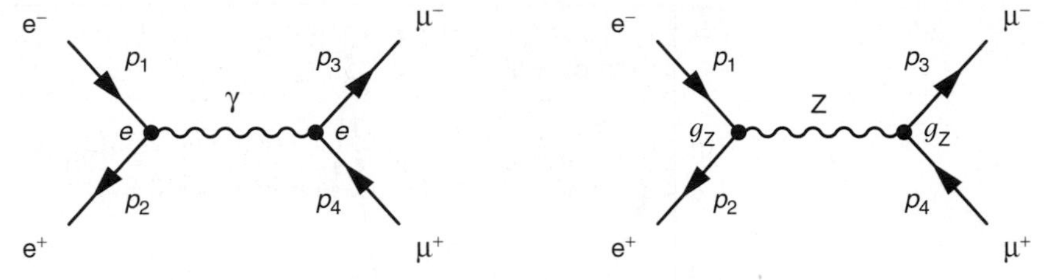


Fig. 16.1 The lowest-order Feynman diagrams for the annihilation process $e^+e^- \to \mu^+\mu^-$.

centre-of-mass energies, $\sqrt{s} \ll m_Z$. This is why the Z-boson diagram could be neglected in the discussion of electron–positron annihilation in Chapter 6. At very high centre-of-mass energies, $\sqrt{s} \gg m_Z$, the QED and Z exchange processes are both important because the strengths of the couplings of the photon and the Z boson are comparable. In the region $\sqrt{s} \sim m_Z$, the Z-boson process dominates. Indeed, from (16.1) it would appear that the matrix element diverges at $\sqrt{s} = m_Z$. This apparent problem arises because the Z-boson propagator of (16.1) does not account for the Z boson being an unstable particle.

There are a number of ways of deriving the propagator for a decaying state. Here, the form of the Z-boson propagator is obtained from the time evolution of the wavefunction for a decaying state. The time dependence of the quantum mechanical wavefunction for a stable particle, as measured in its rest frame, is given by $e^{-imt}$. For an unstable particle, with total decay rate $\Gamma = 1/\tau$, this must be modified to

$$\psi \propto e^{-imt} \quad \to \quad \psi \propto e^{-imt}e^{-\Gamma t/2}, \tag{16.2}$$

such that the probability density decays away as $\psi\psi^* \propto e^{-\Gamma t} = e^{-t/\tau}$. The introduction of the exponential decay term in (16.2) can be obtained from the replacement

$$m \to m - i\Gamma/2.$$

This suggests that the finite lifetime of the Z boson can be accounted for in the propagator of (16.1) by making the replacement

$$m_Z^2 \to (m_Z - i\Gamma_Z/2)^2 = m_Z^2 - im_Z\Gamma_Z - \tfrac{1}{4}\Gamma_Z^2.$$

For the Z boson, the total decay width $\Gamma_Z \ll m_Z$, and to a good approximation the $\frac{1}{4}\Gamma_Z^2$ term can be neglected. In this case, the Z-boson propagator of (16.1) becomes

$$\frac{1}{q^2 - m_Z^2} \to \frac{1}{q^2 - m_Z^2 + im_Z\Gamma_Z}. \tag{16.3}$$

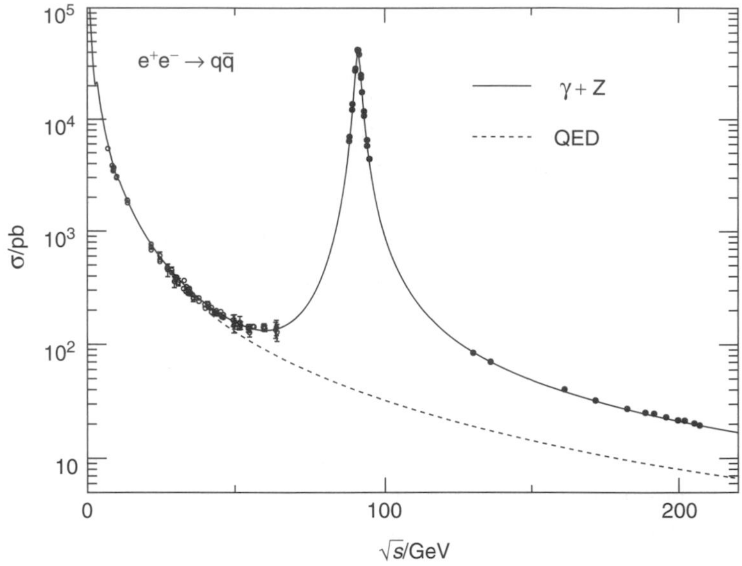


Fig. 16.2 The measurements of the $e^+e^- \to q\bar{q}$ cross section from LEP close to and above Z resonance. Also shown are the lower-energy measurements from earlier experiments. The dashed line shows the contribution to the cross section from the QED process alone. Adapted from LEP and SLD Collaborations (2006).

The cross section for $e^+e^- \to Z \to \mu^+\mu^-$, with $q^2 = s$, is therefore proportional to

$$\sigma \propto |\mathcal{M}|^2 \propto \left|\frac{1}{s - m_Z^2 + im_Z\Gamma_Z}\right|^2 = \frac{1}{(s - m_Z^2)^2 + m_Z^2\Gamma_Z^2}.$$

Hence the $e^+e^- \to Z$ annihilation cross section peaks sharply at $\sqrt{s} = m_Z$, and the resulting Lorentzian dependence on the centre-of-mass energy is referred to as a Breit–Wigner resonance.

The experimental measurements of the $e^+e^- \to q\bar{q}$ cross section over a wide range of centre-of-mass energies are shown in Figure 16.2. The data are compared to the prediction from the $s$-channel γ- and Z-exchange Feynman diagrams, including the interference between the two processes,

$$|\mathcal{M}|^2 = |\mathcal{M}_\gamma + \mathcal{M}_Z|^2.$$

The predicted cross section from the QED process alone is also shown. For centre-of-mass energies below 40 GeV, the cross section is dominated by the QED photon exchange diagram. In the region $\sqrt{s} = 50 - 80\,\text{GeV}$, both γ and Z processes are important. Close to the Z resonance, the Z-boson exchange diagram dominates; at the peak of the resonance it is about three orders of magnitude greater than pure QED contribution. For $\sqrt{s} \gg m_Z$, the contributions from the photon and Z-exchange diagrams are of the same order of magnitude, reflecting the unified description of QED and the weak interaction where $g_Z \sim e$.

### 16.1.1 Z production cross section

In principle, the cross section for $e^+e^-$ annihilation close to the Z resonance, in for example the process $e^+e^- \to \mu^+\mu^-$, needs to account for the two Feynman diagrams of Figure 16.1. However, for $\sqrt{s} \sim m_Z$ the QED contribution to the total cross section can be neglected. In this case, only the matrix element for $e^+e^- \to Z \to \mu^+\mu^-$ needs to be considered. This matrix element can be obtained by using the propagator of (16.3) and the weak neutral-current vertex factor of (15.38),

$$\mathcal{M}_{fi} = -\frac{g_Z^2}{(s - m_Z^2 + im_Z\Gamma_Z)} g_{\mu\nu} \left[\overline{v}(p_2)\gamma^\mu \tfrac{1}{2}\left(c_V^e - c_A^e\gamma^5\right) u(p_1)\right] \times \left[\overline{u}(p_3)\gamma^\nu \tfrac{1}{2}\left(c_V^\mu - c_A^\mu\gamma^5\right) v(p_4)\right],$$

where $c_V^e$, $c_A^e$, $c_V^\mu$ and $c_A^\mu$ are the vector and axial-vector couplings of the Z to the electron and muon. Given the chiral nature of vector boson interactions, it is convenient to re-express this matrix element in terms of the couplings of the Z boson to left- and right-handed chiral states by writing $c_V = (c_L + c_R)$ and $c_A = (c_L - c_R)$. In this case, the matrix element can be written

$$\mathcal{M}_{fi} = -\frac{g_Z^2}{(s - m_Z^2 + im_Z\Gamma_Z)} g_{\mu\nu} \left[c_L^e\overline{v}(p_2)\gamma^\mu P_L u(p_1) + c_R^e\overline{v}(p_2)\gamma^\mu P_R u(p_1)\right] \times \left[c_L^\mu\overline{u}(p_3)\gamma^\nu P_L v(p_4) + c_R^\mu\overline{u}(p_3)\gamma^\nu P_R v(p_4)\right], \quad (16.4)$$

where $P_L = \frac{1}{2}(1-\gamma^5)$ and $P_R = \frac{1}{2}(1+\gamma^5)$ are the chiral projection operators. Given that $m_Z \gg m_\mu$, the fermions in the process $e^+e^- \to Z \to \mu^+\mu^-$ are ultra-relativistic and the helicity and chiral states are essentially the same. The chiral projection operators in (16.4) therefore have the effect

$$P_L u = u_\downarrow, \quad P_R u = u_\uparrow, \quad P_L v = v_\uparrow \quad \text{and} \quad P_R v = v_\downarrow.$$

Furthermore, as described in Section 15.4, helicity combinations such as $\overline{u}_\uparrow\gamma^\mu v_\uparrow$ give zero matrix elements. Consequently the matrix element of (16.4) is only non-zero for the four helicity combinations shown in Figure 16.3, with the corresponding matrix elements

$$\mathcal{M}_{RR} = -P_Z(s)\, g_Z^2\, c_R^e c_R^\mu\, g_{\mu\nu}\, [\overline{v}_\downarrow(p_2)\gamma^\mu u_\uparrow(p_1)]\, [\overline{u}_\uparrow(p_3)\gamma^\nu v_\downarrow(p_4)], \quad (16.5)$$

$$\mathcal{M}_{RL} = -P_Z(s)\, g_Z^2\, c_R^e c_L^\mu\, g_{\mu\nu}\, [\overline{v}_\downarrow(p_2)\gamma^\mu u_\uparrow(p_1)]\, [\overline{u}_\downarrow(p_3)\gamma^\nu v_\uparrow(p_4)], \quad (16.6)$$

$$\mathcal{M}_{LR} = -P_Z(s)\, g_Z^2\, c_L^e c_R^\mu\, g_{\mu\nu}\, [\overline{v}_\uparrow(p_2)\gamma^\mu u_\downarrow(p_1)]\, [\overline{u}_\uparrow(p_3)\gamma^\nu v_\downarrow(p_4)], \quad (16.7)$$

$$\mathcal{M}_{LL} = -P_Z(s)\, g_Z^2\, c_L^e c_L^\mu\, g_{\mu\nu}\, [\overline{v}_\uparrow(p_2)\gamma^\mu u_\downarrow(p_1)]\, [\overline{u}_\downarrow(p_3)\gamma^\nu v_\uparrow(p_4)], \quad (16.8)$$

where $P_Z(s) = 1/(s - m_Z^2 + im_Z\Gamma_Z)$ is the Z propagator and the labels on the different matrix elements $\mathcal{M}$ denote the helicity states of the $e^-$ and $\mu^-$.

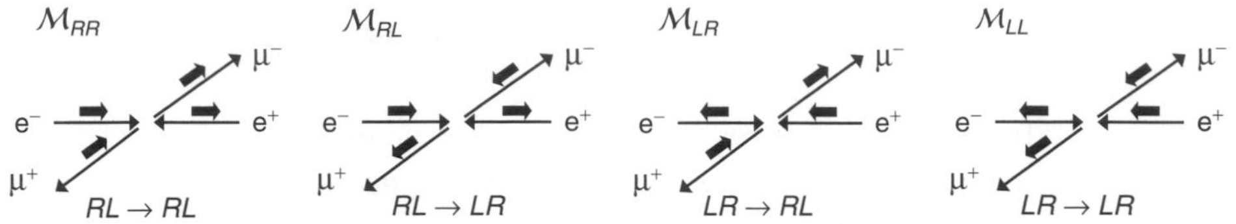


Fig. 16.3 The four possible helicity combinations contributing to $e^+e^- \to Z \to \mu^+\mu^-$. The corresponding matrix elements are labelled by the helicity states of the $e^-$ and $\mu^-$.

The combinations of four-vector currents in (16.5)–(16.8) are identical to those encountered in Chapter 6 for the pure QED process $e^+e^- \to \mu^+\mu^-$, where for example

$$g_{\mu\nu}[\overline{v}_{\downarrow}(p_2)\gamma^{\mu}u_{\uparrow}(p_1)][\overline{u}_{\uparrow}(p_3)\gamma^{\nu}v_{\downarrow}(p_4)] = s(1+\cos\theta).$$

Using the previously derived results of (6.20) and (6.21), the matrix elements squared for the four helicity combinations in the process $e^+e^- \to Z \to \mu^+\mu^-$ are

$$|\mathcal{M}_{RR}|^2 = |P_Z(s)|^2\, g_Z^4 s^2 (c_R^e)^2 (c_R^\mu)^2 (1+\cos\theta)^2, \tag{16.9}$$
$$|\mathcal{M}_{RL}|^2 = |P_Z(s)|^2\, g_Z^4 s^2 (c_R^e)^2 (c_L^\mu)^2 (1-\cos\theta)^2, \tag{16.10}$$
$$|\mathcal{M}_{LR}|^2 = |P_Z(s)|^2\, g_Z^4 s^2 (c_L^e)^2 (c_R^\mu)^2 (1-\cos\theta)^2, \tag{16.11}$$
$$|\mathcal{M}_{LL}|^2 = |P_Z(s)|^2\, g_Z^4 s^2 (c_L^e)^2 (c_L^\mu)^2 (1+\cos\theta)^2, \tag{16.12}$$

where $|P_Z(s)|^2 = 1/[(s-m_Z^2)^2 + m_Z^2\Gamma_Z^2]$.

For unpolarised $e^-$ and $e^+$ beams, the spin-averaged matrix element is given by

$$\langle|\mathcal{M}|^2\rangle = \frac{1}{4}\left(|\mathcal{M}_{RR}|^2 + |\mathcal{M}_{LL}|^2 + |\mathcal{M}_{LR}|^2 + |\mathcal{M}_{RL}|^2\right),$$

where the factor of one quarter arises from averaging over the two possible helicity states for each of the electron and the positron, and therefore from (16.9)–(16.12),

$$\langle|\mathcal{M}|^2\rangle = \frac{1}{4}\frac{g_Z^4 s^2}{(s-m_Z^2)^2 + m_Z^2\Gamma_Z^2} \times \left\{\left[(c_R^e)^2(c_R^\mu)^2 + (c_L^e)^2(c_L^\mu)^2\right](1+\cos\theta)^2 \right.$$
$$\left. + \left[(c_R^e)^2(c_L^\mu)^2 + (c_L^e)^2(c_R^\mu)^2\right](1-\cos\theta)^2\right\}. \tag{16.13}$$

The terms in the braces can be grouped into

$$\{\cdots\} = \left[(c_R^e)^2 + (c_L^e)^2\right]\left[(c_R^\mu)^2 + (c_L^\mu)^2\right]\left(1+\cos^2\theta\right)$$
$$+ 2\left[(c_R^e)^2 - (c_L^e)^2\right]\left[(c_R^\mu)^2 - (c_L^\mu)^2\right]\cos\theta, \tag{16.14}$$

which can then be expressed back in terms of the vector and axial-vector couplings of the electron and muon to the Z boson using

$$c_V^2 + c_A^2 = 2(c_L^2 + c_R^2) \quad \text{and} \quad c_V c_A = c_L^2 - c_R^2,$$

giving

$$\{\ldots\} = \tfrac{1}{4}\left[(c_V^{\mathrm{e}})^2 + (c_A^{\mathrm{e}})^2\right]\left[(c_V^{\mu})^2 + (c_A^{\mu})^2\right]\left(1+\cos^2\theta\right) + 2c_V^{\mathrm{e}}c_A^{\mathrm{e}}c_V^{\mu}c_A^{\mu}\cos\theta.$$

Finally, the $\mathrm{e^+e^-} \to \mathrm{Z} \to \mu^+\mu^-$ differential cross section is obtained by substituting the spin-averaged matrix element squared into (3.50),

$$\frac{\mathrm{d}\sigma}{\mathrm{d}\Omega} = \frac{1}{256\pi^2 s}\cdot\frac{g_Z^4 s^2}{(s-m_Z^2)^2 + m_Z^2\Gamma_Z^2}\times$$
$$\left\{\tfrac{1}{4}\left[(c_V^{\mathrm{e}})^2 + (c_A^{\mathrm{e}})^2\right]\left[(c_V^{\mu})^2 + (c_A^{\mu})^2\right]\left(1+\cos^2\theta\right) + 2c_V^{\mathrm{e}}c_A^{\mathrm{e}}c_V^{\mu}c_A^{\mu}\cos\theta\right\}. \quad (16.15)$$

The total cross section is determined by integrating over the solid angle $\mathrm{d}\Omega$. This is most easily performed by writing $\mathrm{d}\Omega = \mathrm{d}\phi\,\mathrm{d}(\cos\theta)$ and making the substitution $x = \cos\theta$, such that

$$\int (1+\cos^2\theta)\,\mathrm{d}\Omega = \int_0^{2\pi}\mathrm{d}\phi\int_{-1}^{+1}(1+x^2)\,\mathrm{d}x = \frac{16\pi}{3} \quad \text{and} \quad \int \cos\theta\,\mathrm{d}\Omega = 0.$$

The resulting cross section for the process $\mathrm{e^+e^-} \to \mathrm{Z} \to \mu^+\mu^-$ is

$$\sigma(\mathrm{e^+e^-}\to\mathrm{Z}\to\mu^+\mu^-) = \frac{1}{192\pi}\frac{g_Z^4 s}{(s-m_Z^2)^2 + m_Z^2\Gamma_Z^2}\left[(c_V^{\mathrm{e}})^2 + (c_A^{\mathrm{e}})^2\right]\left[(c_V^{\mu})^2 + (c_A^{\mu})^2\right].$$

Thus, the total $\mathrm{e^+e^-} \to \mathrm{Z} \to \mu^+\mu^-$ cross section is proportional to the product of the sum of the squares of the vector and axial-vector couplings of the initial-state electrons and the final-state muons. Using the expression for the partial decay widths of the Z boson, given in (15.41), the sums $c_V^2 + c_A^2$ for the electron and muon can be related to $\Gamma_{\mathrm{ee}} = \Gamma(\mathrm{Z}\to\mathrm{e^+e^-})$ and $\Gamma_{\mu\mu} = \Gamma(\mathrm{Z}\to\mu^+\mu^-)$,

$$\Gamma_{\mathrm{ee}} = \frac{g_Z^2 m_Z}{48\pi}\left[(c_V^{\mathrm{e}})^2 + (c_A^{\mathrm{e}})^2\right] \quad \text{and} \quad \Gamma_{\mu\mu} = \frac{g_Z^2 m_Z}{48\pi}\left[(c_V^{\mu})^2 + (c_A^{\mu})^2\right].$$

Hence, the total cross section, expressed in terms of the partial decay widths, is

$$\sigma(\mathrm{e^+e^-}\to\mathrm{Z}\to\mu^+\mu^-) = \frac{12\pi s}{m_Z^2}\frac{\Gamma_{\mathrm{ee}}\Gamma_{\mu\mu}}{(s-m_Z^2)^2 + m_Z^2\Gamma_Z^2}. \quad (16.16)$$

The cross sections for other final-state fermions are given by simply replacing $\Gamma_{\mu\mu}$ by the partial width $\Gamma_{\mathrm{ff}} = \Gamma(\mathrm{e^+e^-}\to\mathrm{f\bar{f}})$.

The properties of the Z resonance are described by (16.16). The maximum value of the cross section, which occurs at the centre-of-mass energy $\sqrt{s} = m_Z$, is

$$\sigma_{\mathrm{ff}}^0 = \frac{12\pi}{m_Z^2}\frac{\Gamma_{\mathrm{ee}}\Gamma_{\mathrm{ff}}}{\Gamma_Z^2}. \quad (16.17)$$

From (16.16) it is straightforward to show that the cross section falls to half of its peak value at

$$\sqrt{s} = m_Z \pm \Gamma_Z/2.$$

Therefore $\Gamma_Z$ is not only the total decay rate of the Z boson, it is also the full-width-at-half-maximum (FWHM) of the cross section as a function of centre-of-mass energy. Hence the mass and total width of the Z boson can be determined directly from measurements of the centre-of-mass energy dependence of the cross section for $e^+e^- \to Z \to f\bar{f}$. Furthermore, once $m_Z$ and $\Gamma_Z$ are known, the measured value of peak cross section for a particular final-state fermion $\sigma^0_{f\bar{f}}$ can be related to the product of the partial decay widths using (16.17),

$$\Gamma_{ee}\Gamma_{f\bar{f}} = \frac{\sigma^0_{f\bar{f}}\Gamma_Z^2 m_Z^2}{12\pi}.$$

Hence, the observed peak cross sections can be used to determine the partial decay widths of the Z boson for the different visible final states.

## 16.2 The Large Electron–Positron collider

The LEP collider, which operated at CERN from 1989 to 2000, is the highest energy electron–positron collider ever built. The circular accelerator was located in the 26 km circumference underground tunnel that is now home to the LHC. The electrons and positrons circulated in opposite directions and collided at four interaction points, spaced equally around the ring, accommodating four large general purpose detectors, ALEPH, DELPHI, L3 and OPAL. From 1989 to 1995, LEP operated at centre-of-mass energies close to the Z mass and the four experiments accumulated over 17 million Z events between them, allowing its properties to be determined with high precision. From 1996 to 2000, LEP operated above the threshold for $W^+W^-$ production and the LEP experiments accumulated a total of more than 30 000 $e^+e^- \to W^+W^-$ events over the centre-of-mass energy range 161–208 GeV, allowing the properties of the W boson to be studied in detail.

### 16.2.1 Measurement of the mass and width of the Z

At LEP, the mass and width of the Z boson were determined from the centre-of-mass energy dependence of the measured $e^+e^- \to Z \to q\bar{q}$ cross section. In principle, the cross section is described by the Breit–Wigner resonance of (16.16), with the maximum occurring at $\sqrt{s} = m_Z$ and the FWHM giving $\Gamma_Z$. In practice, this is not quite the case. In addition to the lowest-order Feynman diagram, there are two higher-order QED diagrams where a photon is radiated from either the

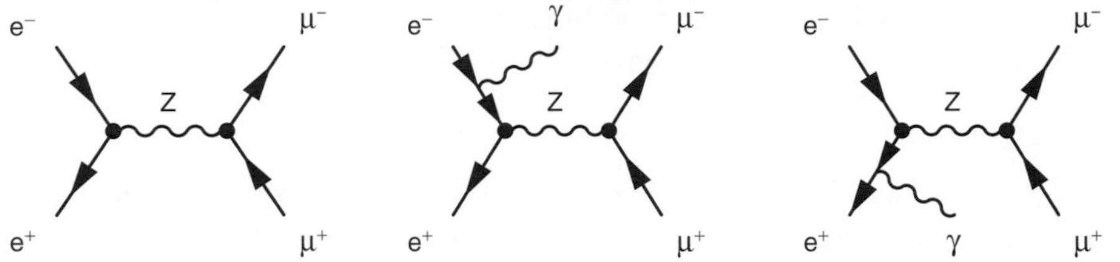


Fig. 16.4 The lowest-order Feynman diagram for the annihilation process $e^+e^- \to \mu^+\mu^-$ and the diagrams including initial-state radiation.

initial-state electron or positron, as shown in Figure 16.4. The effect of initial-state radiation (ISR) is to distort the shape of the Z resonance curve. If an ISR photon with energy $E_\gamma$ is radiated, the energy of the $e^+$ or $e^-$ is reduced from $E$ to $E' = E - E_\gamma$, where $E$ is the nominal energy of the electron and positron beams. In the limit where the photon is emitted collinear with the incoming electron/positron (which is usually the case), the four-momenta of the electron and positron at the Z production vertex are $p_1 = (E - E_\gamma, 0, 0, E - E_\gamma)$ and $p_2 = (E, 0, 0, -E)$. For collisions at a nominal centre-of-mass energy of $\sqrt{s}$, the effect of ISR is to produce a distribution of the four-momentum $q_Z$ of the virtual Z bosons. This can be expressed as the effective centre-of-mass energy squared at the $e^+e^-$ annihilation vertex $s' = q_Z^2$, given by the square of the sum of four-momenta of the $e^+$ and $e^-$ after ISR,

$$s' = (p_1 + p_2)^2 = (2E - E_\gamma)^2 - E_\gamma^2 = 4E^2\left(1 - \frac{E_\gamma}{E}\right) = s\left(1 - \frac{E_\gamma}{E}\right).$$

The impact of ISR is to reduce the effective centre-of-mass energy for the collisions where ISR photons are emitted; even if the accelerator is operated at a nominal centre-of-mass energy equal to $m_Z$, some fraction of the Z bosons will be produced with $q_Z^2 < m_Z^2$.

The distribution of $\sqrt{s'}$ can be written in terms of the normalised probability distribution $f(s', s)$. The measured cross section is the convolution of the $s'$ distribution with the cross section $\sigma(s')$ obtained from (16.16),

$$\sigma_{\text{meas}}(s) = \int \sigma(s') f(s', s)\, \mathrm{d}s'.$$

The effect of ISR is to distort the measured Z resonance. However, because ISR is a QED process, the function $f(s', s)$ can be calculated to high precision. Consequently, the measured cross section can be corrected back to the underlying Breit–Wigner distribution. Figure 16.5 shows the measured $e^+e^- \to Z \to q\bar{q}$ cross section as a function of centre-of-mass energy. The data are compared to the expected distribution including ISR. The dashed curve shows the reconstructed shape of the Z resonance after the deconvolution of the effects of ISR. Close to and below the peak of the resonance, ISR results in a reduction in the measured cross section because the centre-of-mass energy at the $e^+e^-$ vertex is moved further from the peak of

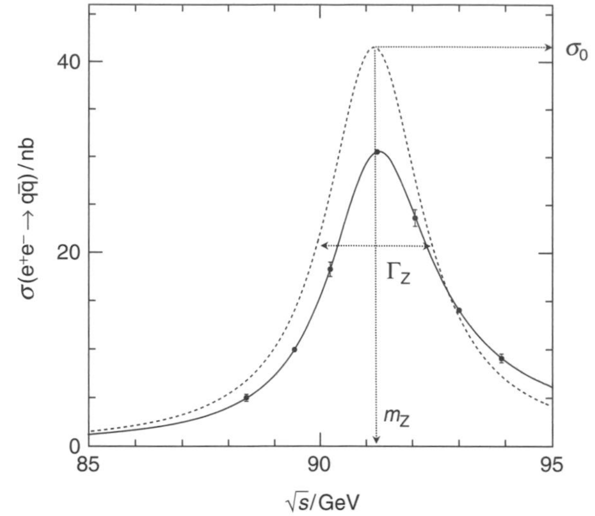


Fig. 16.5 The measurements of the $e^+e^- \to q\bar{q}$ cross section from LEP close to the Z resonance. The expected resonance curves, before and after correcting for ISR are shown by the solid and dashed lines respectively. The measurement errors have been multiplied by a factor of 10 to improve visibility. Adapted from LEP and SLD Collaborations (2006).

the resonance. Above the peak of the resonance, ISR increases the cross section because the average centre-of-mass energy is moved closer to the peak.

From the measurements of the Z resonance at LEP, shown in Figure 16.5, the mass of the Z boson is determined to be

$$m_Z = 91.1875 \pm 0.0021\,\text{GeV}.$$

Owing to the large numbers of Z bosons produced at LEP, $m_Z$ could be measured with a precision of 0.002%, making it one of the more precisely known fundamental parameters. To achieve this high level of precision, the average centre-of-mass energy of the LEP collider had to be known to 2 MeV. This required a detailed understanding of a number of potential systematic biases. For example, tidal effects due to the gravitational pull of the Moon distort the rock surrounding the LEP accelerator by a small amount, resulting in ±0.15 mm variations in the 4.3 km radius of the accelerator. These variations are sufficient to change the beam energy by approximately ±10 MeV. Nevertheless, the position of the Moon is known and the effect of these tidal variations could be accounted for. A more subtle and unexpected effect was the observation of apparent jumps in the beam energies at specific times of the day. After much investigation, the origin was identified as leakage currents from the local high-speed railway. These leakage currents followed the path of least resistance in a circuit formed from the rails, a local river and the LEP ring. The small currents that ran along the LEP ring, were sufficient to modify the magnetic field in the accelerator, leading to small changes in the beam energy. Once understood, the affected data could be treated appropriately.

### The width of the Z-boson

The total width of the Z boson, determined from the FWHM of the Breit–Wigner resonance curve after unfolding the effects of ISR (shown in Figure 16.5) is

$$\Gamma_Z = 2.4952 \pm 0.0023\,\text{GeV},$$

corresponding to a lifetime of just $2.6 \times 10^{-25}$ s. The total width of the Z is the sum of the partial decay widths for all its decay modes,

$$\Gamma_Z = \Gamma_{ee} + \Gamma_{\mu\mu} + \Gamma_{\tau\tau} + \Gamma_{\text{hadrons}} + \Gamma_{\nu_e\nu_e} + \Gamma_{\nu_\mu\nu_\mu} + \Gamma_{\nu_\tau\nu_\tau}, \tag{16.18}$$

where $\Gamma_{\text{hadrons}}$ is the partial decay width to all final states with quarks. Assuming the lepton universality of the weak neutral-current, (16.18) can be written

$$\Gamma_Z = 3\Gamma_{\ell\ell} + \Gamma_{\text{hadrons}} + 3\Gamma_{\nu\nu},$$

where $\Gamma_{\ell\ell}$ and $\Gamma_{\nu\nu}$ are respectively the partial decay widths to a single flavour of charged lepton or neutrino. Although the decays to neutrinos are not observed, they still affect the observable total width of the Z resonance.

To date only three generations of fermions have been observed. This in itself does not preclude the possibility of a fourth generation, provided the fourth-generation particles are sufficiently massive to have avoided detection. However, if there were a fourth-generation neutrino, with similar properties to the three known generations, the neutrino would be sufficiently light for the decay $Z \to \nu_4\overline{\nu}_4$ to occur. This possibility can be tested through its observable effect on $\Gamma_Z$. For $N_\nu$ light neutrino generations, the expected width of the Z boson is

$$\Gamma_Z = 3\Gamma_{\ell\ell} + \Gamma_{\text{hadrons}} + N_\nu\Gamma_{\nu\nu}. \tag{16.19}$$

Hence the number of light neutrino generations that exist in nature can be obtained from the measured values of $\Gamma_Z$, $\Gamma_{\ell\ell}$ and $\Gamma_{\text{hadrons}}$ using

$$N_\nu = \frac{(\Gamma_Z - 3\Gamma_{\ell\ell} - \Gamma_{\text{hadrons}})}{\Gamma_{\nu\nu}^{\text{SM}}}, \tag{16.20}$$

where $\Gamma_{\nu\nu}^{\text{SM}}$ is the Standard Model prediction of (15.42). The individual partial decay widths to particles other than neutrinos, can be determined from the measured cross sections at the peak of the Z resonance using

$$\sigma^0(e^+e^- \to Z \to f\overline{f}) = \frac{12\pi}{m_Z^2}\frac{\Gamma_{ee}\Gamma_{ff}}{\Gamma_Z^2}. \tag{16.21}$$

Given that $m_Z$ and $\Gamma_Z$ are known precisely, the measured peak cross section for $e^+e^- \to Z \to e^+e^-$ determines $\Gamma_{ee}^2$. Once $\Gamma_{ee}$ is known, the partial decay widths of

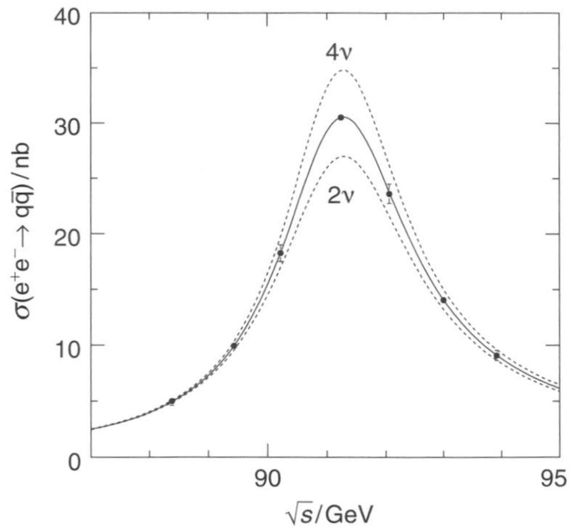


Fig. 16.6 The measurements of the $e^+e^- \to q\bar{q}$ cross section from LEP close to the Z resonance compared to the Standard Model expectation for two, three and four light neutrino generations. Adapted from LEP and SLD Collaborations (2006).

the Z boson to the other visible final states can be determined from the respective peak cross sections, again using (16.21). Using the measured values of the partial decay widths and the relation of (16.20), the number of light neutrino generations is determined to be

$$N_\nu = 2.9840 \pm 0.0082. \tag{16.22}$$

Figure 16.6 compares the measured $e^+e^- \to q\bar{q}$ cross section, close to the Z resonance, with the expected cross sections for two, three and four neutrino generations. The consistency of the data with the predictions for three neutrino generations provides strong evidence that there are exactly three generations of light neutrinos (assuming Standard Model couplings) from which it can be inferred that there are probably only three generations of fermions.

### 16.2.2 Measurements of the weak mixing angle

The weak mixing angle $\theta_W$ is one of the fundamental parameters of the Standard Model. The Standard Model vector and axial-vector couplings of the fermions to the Z boson are given by (15.36) and (15.37),

$$c_V = I_W^{(3)} - 2Q\sin^2\theta_W \quad \text{and} \quad c_A = I_W^{(3)},$$

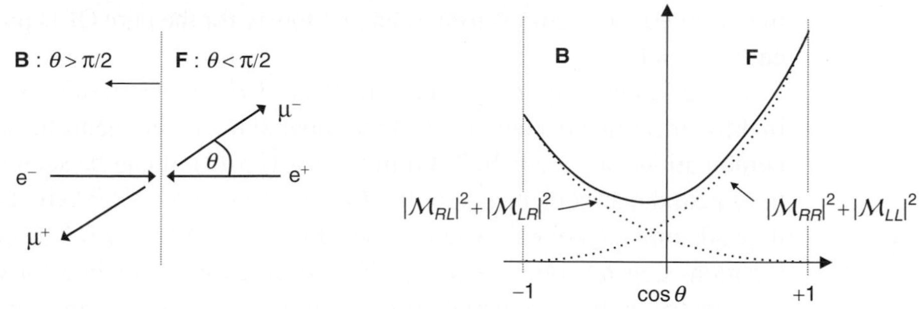


Fig. 16.7 The definition of the forward ($\theta_{\mu^-} < \pi/2$) and backward ($\theta_{\mu^-} > \pi/2$) hemispheres and the contributions to the differential cross section from the different helicity combinations.

and thus $\sin^2\theta_W$ can be obtained from measurements of $c_V$. In practice, the relevant experimental observables depend on the ratio of couplings,

$$\frac{c_V}{c_A} = 1 - \frac{2Q\sin^2\theta_W}{I_W^{(3)}}.$$

For the charged leptons, with $Q = -1$ and $I_W^{(3)} = -1/2$,

$$\frac{c_V^\ell}{c_A^\ell} = 1 - 4\sin^2\theta_W. \tag{16.23}$$

There are a number of ways in which the ratio $c_V/c_A$ can be obtained from measurements at LEP. The simplest conceptually is the measurement of the forward–backward asymmetry $A_{FB}$ of the leptons produced in $e^+e^- \to Z \to \ell^+\ell^-$. $A_{FB}$ reflects the asymmetry in angular distribution of the final-state leptons and is defined as

$$A_{FB}^\ell = \frac{\sigma_F - \sigma_B}{\sigma_F + \sigma_B}. \tag{16.24}$$

Here $\sigma_F$ and $\sigma_B$ are the respective cross sections for the negatively charged lepton being produced in the forward ($\theta_{\ell^-} < \pi/2$) and backward ($\theta_{\ell^-} > \pi/2$) hemispheres, as indicated in Figure 16.7.

The differential cross section for $e^+e^- \to Z \to \mu^+\mu^-$ of (16.14) has the form

$$\frac{d\sigma}{d\Omega} \propto a(1+\cos^2\theta) + 2b\cos\theta, \tag{16.25}$$

where the coefficients $a$ and $b$ are given by

$$a = \left[(c_L^e)^2 + (c_R^e)^2\right]\left[(c_L^\mu)^2 + (c_R^\mu)^2\right] \quad \text{and} \quad b = \left[(c_L^e)^2 - (c_R^e)^2\right]\left[(c_L^\mu)^2 - (c_R^\mu)^2\right].$$

If the couplings of the Z boson to left-handed (LH) and right-handed (RH) fermions were the same, $b$ would be equal to zero and the angular distribution would have the symmetric $(1+\cos^2\theta)$ form seen previously for the pure QED process, in which case $A_{\mathrm{FB}} = 0$.

The different couplings of the Z boson to LH and RH fermions manifests itself in differences in the magnitudes of the squared matrix elements for the four helicity combinations of Figure 16.3. From (16.9)–(16.12), it can be seen that the sum of the squared matrix elements for the $RL \to RL$ and $LR \to LR$ helicity combinations depends on $(1+\cos\theta)^2$, whereas the sum for the $RL \to LR$ and $LR \to RL$ combinations depends on $(1-\cos\theta)^2$. This difference results in a forward–backward asymmetry in the differential cross section, as indicated in the right-hand plot of Figure 16.7.

The forward and backward cross sections, $\sigma_F$ and $\sigma_B$, can be obtained by integrating the differential cross section of (16.25) over the two different polar angle ranges, $0 < \theta < \pi/2$ and $\pi/2 < \theta < \pi$. Writing $\mathrm{d}\Omega = \mathrm{d}\phi\,\mathrm{d}(\cos\theta)$,

$$\sigma_F \equiv 2\pi \int_0^1 \frac{\mathrm{d}\sigma}{\mathrm{d}\Omega}\mathrm{d}(\cos\theta) \quad \text{and} \quad \sigma_B \equiv 2\pi \int_{-1}^0 \frac{\mathrm{d}\sigma}{\mathrm{d}\Omega}\mathrm{d}(\cos\theta).$$

From the form of the differential cross section of (16.25), $\sigma_F$ and $\sigma_B$ are

$$\sigma_F \propto \int_0^1 \left[a(1+\cos^2\theta) + 2b\cos\theta\right]\mathrm{d}(\cos\theta) = \int_0^1 \left[a(1+x^2) + 2bx\right]\mathrm{d}x = \left(\tfrac{4}{3}a + b\right),$$

$$\sigma_B \propto \int_{-1}^0 \left[a(1+\cos^2\theta) + 2b\cos\theta\right]\mathrm{d}(\cos\theta) = \int_{-1}^0 \left[a(1+x^2) + 2bx\right]\mathrm{d}x = \left(\tfrac{4}{3}a - b\right).$$

Thus the forward–backward asymmetry is given by

$$A_{\mathrm{FB}} = \frac{\sigma_F - \sigma_B}{\sigma_F + \sigma_B} = \frac{3b}{4a}.$$

From the expressions for the coefficients $a$ and $b$, the forward–backward asymmetry is related to the left- and right-handed couplings of the fermions by

$$A_{\mathrm{FB}} = \frac{3}{4}\left[\frac{(c_L^{\mathrm{e}})^2 - (c_R^{\mathrm{e}})^2}{(c_L^{\mathrm{e}})^2 + (c_R^{\mathrm{e}})^2}\right]\cdot\left[\frac{(c_L^{\mu})^2 - (c_R^{\mu})^2}{(c_L^{\mu})^2 + (c_R^{\mu})^2}\right].$$

This can be written in the form

$$A_{\mathrm{FB}} = \frac{3}{4}\mathcal{A}_{\mathrm{e}}\mathcal{A}_{\mu},$$

where the asymmetry parameter $\mathcal{A}_{\mathrm{f}}$ for a particular flavour f is defined by

$$\mathcal{A}_{\mathrm{f}} = \frac{(c_L^{\mathrm{f}})^2 - (c_R^{\mathrm{f}})^2}{(c_L^{\mathrm{f}})^2 + (c_R^{\mathrm{f}})^2} \equiv \frac{2c_V^{\mathrm{f}}c_A^{\mathrm{f}}}{(c_V^{\mathrm{f}})^2 + (c_A^{\mathrm{f}})^2}. \tag{16.26}$$

At LEP, $A_{\mathrm{FB}}$ is most cleanly measured using the $\mathrm{e}^+\mathrm{e}^-$, $\mu^+\mu^-$ and $\tau^+\tau^-$ final states, where the charges of the leptons are determined from the sense of the curvature of the measured particle track in the magnetic field of the detector. By counting the

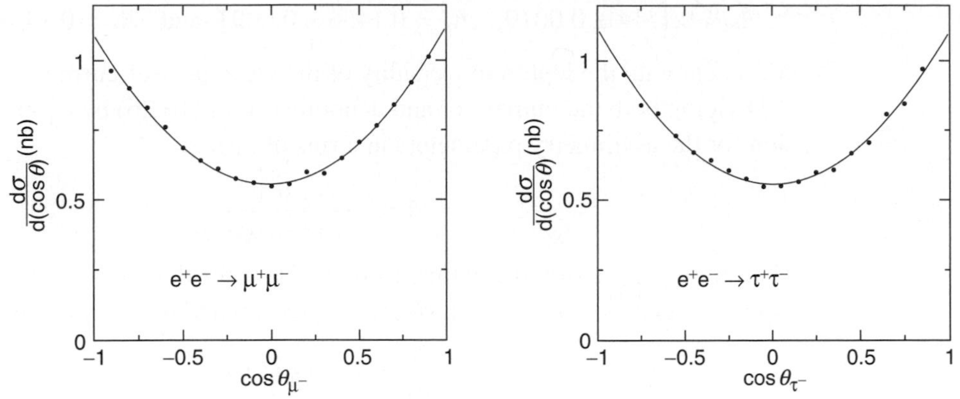


Fig. 16.8 Measurements of the differential cross sections for $e^+e^- \to \mu^+\mu^-$ and $e^+e^- \to \tau^+\tau^-$ at $\sqrt{s} = m_Z$. Adapted from the OPAL Collaboration, Abbiendi *et al.* (2001).

numbers of events where the $\ell^-$ is produced in either the forward or backward hemispheres, $N_F$ and $N_B$, the measured forward–backward asymmetry is simply

$$A_{\rm FB} = \frac{N_F - N_B}{N_F + N_B}.$$

In the measurement of $A_{\rm FB}$, many potential systematic biases cancel because they tend to affect both forward and backward events in the same manner, consequently a precise measurement can be made. In practice, $A_{\rm FB}$ is obtained from the observed angular distribution, rather than simply counting events. For example, Figure 16.8 shows the OPAL measurements of the $e^+e^- \to \mu^+\mu^-$ and $e^+e^- \to \tau^+\tau^-$ differential cross sections at $\sqrt{s} = m_Z$. The observed asymmetry is small, but non-zero.

Combining the results from the four LEP experiments gives

$$A^{\rm e}_{FB} = 0.0145 \pm 0.0025,\quad A^{\mu}_{FB} = 0.0169 \pm 0.0013 \text{ and } A^{\tau}_{FB} = 0.0188 \pm 0.0017.$$

These measurements can be expressed in terms of the asymmetry parameter defined in (16.26), giving

$$A^{\rm e}_{FB} = \frac{3}{4}\mathcal{A}_{\rm e}^2,\quad A^{\mu}_{FB} = \frac{3}{4}\mathcal{A}_{\rm e}\mathcal{A}_{\mu} \quad\text{and}\quad A^{\tau}_{FB} = \frac{3}{4}\mathcal{A}_{\rm e}\mathcal{A}_{\tau}.$$

Hence the measurements of the forward–backward asymmetries can be interpreted as measurements of the asymmetry parameters for the individual lepton flavours, with the $e^+e^- \to Z \to e^+e^-$ process uniquely determining $\mathcal{A}_{\rm e}$.

There are a number of other ways of measuring the lepton asymmetry parameters at LEP and elsewhere. For example, the results from the left-right asymmetry measured in the SLD detector at the Stanford Linear Collider (SLC) provides a precise measurement of $\mathcal{A}_{\rm e}$ alone (see Problem 16.4). The combined results from LEP and SLC give

$$\mathcal{A}_e = 0.1514 \pm 0.0019, \quad \mathcal{A}_\mu = 0.1456 \pm 0.0091 \quad \text{and} \quad \mathcal{A}_\tau = 0.1449 \pm 0.0040,$$

consistent with the lepton universality of the weak neutral-current.

Dividing both the numerator and denominator of (16.26) by $c_A^2$ gives the expression for the asymmetry parameters in terms of $c_V/c_A$,

$$\mathcal{A} = \frac{2c_V/c_A}{1 + (c_V/c_A)^2}.$$

Therefore, the measured asymmetry parameters for the leptons can be interpreted as measurements of $c_V/c_A$, which then can be related to $\sin^2\theta_W$ using (16.23),

$$\frac{c_V}{c_A} = 1 - 4\sin^2\theta_W.$$

When the various measurements of $\sin^2\theta_W$ from the Z resonance and elsewhere are combined, the weak mixing angle is determined to be

$$\sin^2\theta_W = 0.23146 \pm 0.00012.$$

The lepton forward–backward asymmetries are small because $\sin^2\theta_W$ is nearly 1/4.

## 16.3 Properties of the W boson

The studies of the Z boson provide a number of important results, including the precise measurements of $m_Z$, $\Gamma_Z$ and $\sin^2\theta_W$. Further constraints on the electroweak sector of the Standard Model can be obtained from studies of the W boson. From 1996 to 2000, the LEP collider operated at $\sqrt{s} > 161\,\text{GeV}$, above the threshold for production of $e^+e^- \to W^+W^-$. In $e^+e^- \to W^+W^-$ events, each W boson can decay either leptonically, for example $W^- \to \mu^-\overline{\nu}_\mu$, or hadronically, for example $W^- \to d\overline{u}$. Consequently, $e^+e^- \to W^+W^-$ interactions at LEP are observed in the three distinct event topologies shown in Figure 16.9. Events where both W bosons decay leptonically are observed as two charged leptons and an imbalance of momentum in the transverse plane due to the two unseen neutrinos. Events where one W decays leptonically and the other decays hadronically are observed as two jets, a single charged lepton and an imbalance of momentum from the neutrino. Finally, events where both W bosons decay hadronically produce four jets. The distinctive event topologies enable $e^+e^- \to W^+W^-$ events to be identified with high efficiency and little ambiguity.

The observed numbers of events in each of the three $W^+W^-$ topologies can be related to branching ratio for $W \to q\overline{q}'$. For example, the numbers of fully hadronic decays and fully leptonic decays are respectively proportional to

$$N_{qqqq} \propto \left[BR(W \to q\overline{q}')\right]^2 \quad \text{and} \quad N_{\ell\nu\ell\nu} \propto \left[1 - BR(W \to q\overline{q}')\right]^2.$$

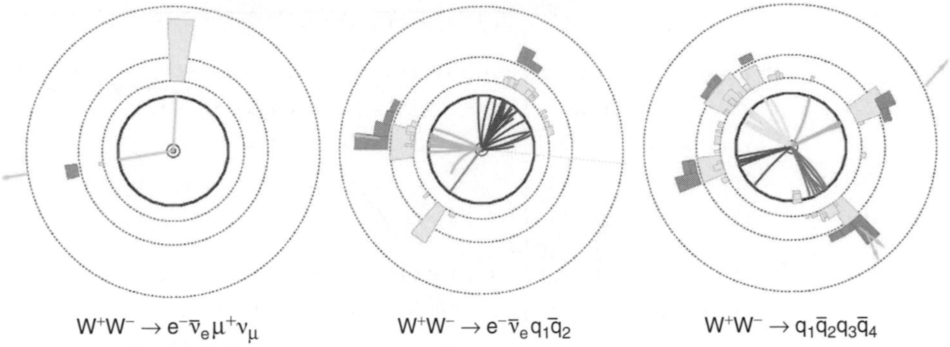


Fig. 16.9 The three possible event topologies for the decays of $W^+W^-$ in $e^+e^- \to W^+W^-$ at LEP. Reproduced courtesy of the OPAL Collaboration.

Consequently, the relative numbers of observed events in the three $W^+W^-$ topologies gives a precise measurement of the W-boson branching ratio to hadrons,

$$BR(\mathrm{W} \to \mathrm{q}\overline{\mathrm{q}}') = 67.41 \pm 0.27\%.$$

This is consistent with the Standard Model expectation of 67.5% obtained from (15.9). Furthermore, the decays $\mathrm{W} \to \mathrm{e}\nu$, $\mathrm{W} \to \mu\nu$ and $\mathrm{W} \to \tau\nu$ are observed to occur with equal frequencies, consistent with the expectation from the lepton universality of the charged-current weak interaction.

Figure 16.10 shows the combined measurements of the $e^+e^- \to W^+W^-$ cross section from the four LEP experiments. The data are consistent with the Standard Model expectation determined from the three Feynman diagrams of Figure 15.5. The contribution to the total cross section from the *s*-channel Z-exchange diagram, shown in Figure 16.11, depends on the strength of the $W^+W^-Z$ coupling, which in the Standard Model is fixed by the local gauge symmetry and the electroweak unification mechanism. The predicted cross section without the contribution from the Z-exchange diagram, also shown in Figure 16.10, clearly does not reproduce the data. The $e^+e^- \to W^+W^-$ cross section measurements therefore provide a test the Standard Model prediction of the strength of coupling at the $W^+W^-Z$ vertex. Yet again, the Standard Model provides an excellent description of the data.

### 16.3.1 Measurement of the W boson mass and width

The mass and width of the Z boson are determined from the shape of the resonance in the Z production cross section in $e^+e^-$ collisions. The production of W-pairs at LEP is not a resonant process; for $\sqrt{s} > 2m_\mathrm{W}$, the Z boson in the *s*-channel Feynman diagram of Figure 16.11 is far from being on-mass shell. Consequently different techniques are required to measure the mass and width of the W boson. In principle, it is possible to measure the W boson mass and width from the shape

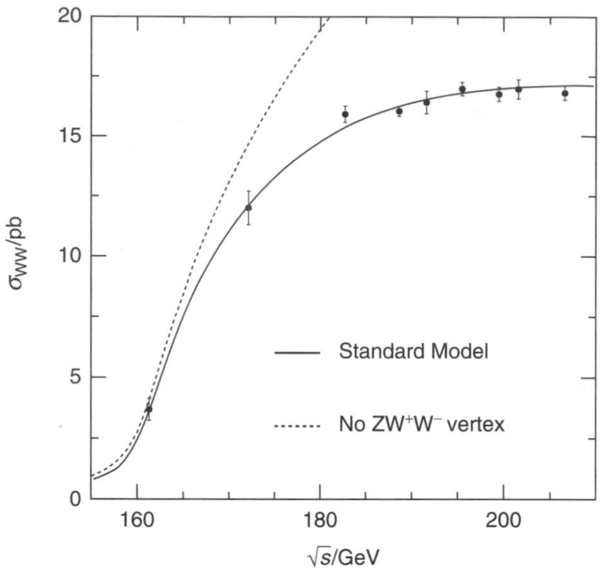


Fig. 16.10 The measurements of the $e^+e^- \to W^+W^-$ cross section at LEP. The curve shows the Standard Model prediction from the three Feynman diagrams of Figure 15.5. Also shown is the prediction without the Z-exchange diagram. The points show the combined results from the ALEPH, DELPHI, L3 and OPAL Collaborations.

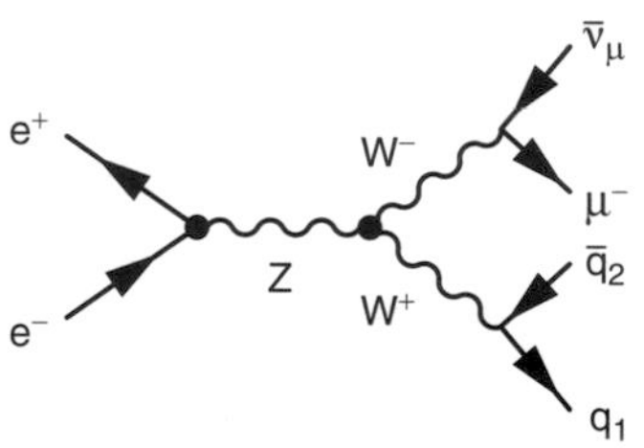


Fig. 16.11 One of the three Feynman diagrams for $e^+e^- \to W^+W^- \to \mu^-\overline{\nu}_\mu q_1\overline{q}_2$.

of the $e^+e^- \to W^+W^-$ cross section close to threshold at $\sqrt{s} \sim 2m_W$, where the position and sharpness of the turn-on of the cross section depend on $m_W$ and $\Gamma_W$. Significantly above threshold, where the majority of the LEP data were recorded, $m_W$ and $\Gamma_W$ are determined from the *direct reconstruction* of the invariant masses of the W-decay products.

Up to this point, the production and decay of the W bosons in the process $e^+e^- \to W^+W^-$ have been discussed as independent processes. This distinction, which effectively treats the W bosons as real on-shell particles, is not strictly correct. The W bosons should be considered as virtual particles. For example, Figure 16.11 shows one of the three Feynman diagrams for $e^+e^- \to \mu^-\overline{\nu}_\mu q_1\overline{q}_2$, which proceeds via the production and decay of two virtual W bosons. In this

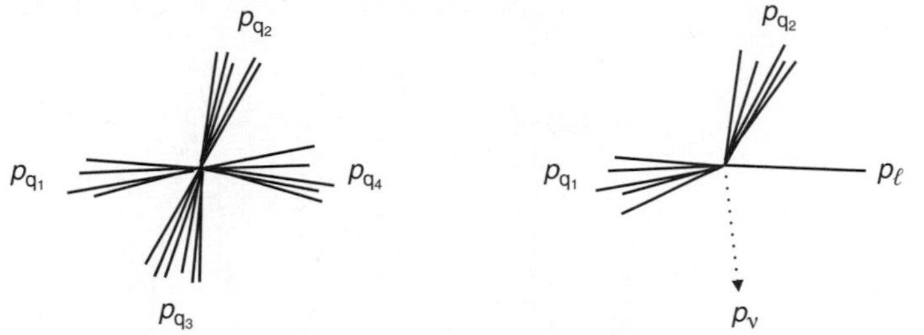


Fig. 16.12 The WW $\to$ qqqq and WW $\to$ $q\overline{q}\ell\nu$ event topologies used at LEP to determine the W-boson mass from the direct reconstruction of the decay products.

diagram, there are three propagators, one for the Z boson and one for each of the two W bosons. The propagators for the virtual W bosons have the form

$$\frac{1}{q^2 - m_W^2 + im_W\Gamma_W},$$

where $q$ is the four-momentum of the W boson and the imaginary term accounts for its finite lifetime. The contribution of the two W-boson propagators to the matrix element squared is therefore

$$|\mathcal{M}|^2 \propto \frac{1}{(q_1^2 - m_W^2)^2 + m_W^2\Gamma_W^2} \times \frac{1}{(q_2^2 - m_W^2)^2 + m_W^2\Gamma_W^2}, \tag{16.27}$$

where $q_1$ and $q_2$ are the four-momenta of the two W bosons. Hence, the invariant mass of the two fermions from each W-boson decay is not fixed to be exactly $m_W$, but will distributed as a Lorentzian centred on $m_W$ with width $\Gamma_W$. A precise determination of the W-boson mass and width can be obtained from the direct reconstruction of the four-momenta of the four fermions in the $W^+W^- \to \ell\nu q_1\overline{q}_2$ and $W^+W^- \to q_1\overline{q}_2q_3\overline{q}_4$ decay topologies, shown in Figure 16.12.

For $W^+W^- \to q_1\overline{q}_2q_3\overline{q}_4$ decays, the measured four-momenta of the four jets can be used directly to reconstruct the invariant masses of the two W bosons,

$$m_1^2 = q_1^2 = (p_{q1} + p_{q2})^2 \quad \text{and} \quad m_2^2 = q_2^2 = (p_{q3} + p_{q4})^2.$$

The masses of the two W bosons produced in $W^+W^- \to \ell\nu q_1\overline{q}_2$ decays can be determined by reconstructing the momentum of the neutrino. Because the $e^+e^-$ collisions occur in the centre-of-mass frame, the total four-momentum of the final-state particles is constrained to

$$P_{\text{tot}} = (\sqrt{s}, \mathbf{0}).$$

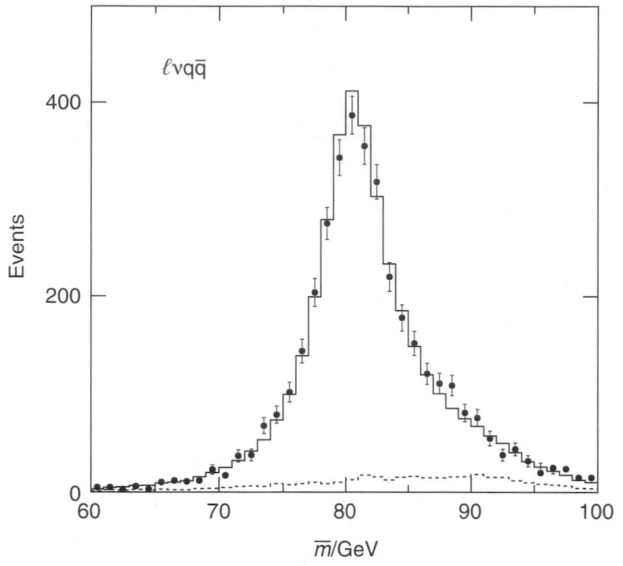


Fig. 16.13 The reconstructed invariant mass distribution of W bosons in $e^+e^- \to W^+W^- \to \ell\nu q\bar{q}$ events in the L3 experiment at LEP. The dashed line indicates the background from events other than $W^+W^- \to \ell\nu q\bar{q}$. Adapted from Achard *et al.* (2006).

Consequently, the neutrino four-momentum can be obtained from the measured four-momenta of the charged lepton and the two jets,

$$p_\nu = P_{\text{tot}} - p_\ell - p_{q_1} - p_{q_2}.$$

Thus, the masses of the two W bosons in the $W^+W^- \to \ell\nu q_1\bar{q}_2$ decays can be measured from

$$m_1^2 = q_1^2 = (p_{q_1} + p_{q_2})^2 \quad \text{and} \quad m_2^2 = q_2^2 = (p_\ell + p_\nu)^2.$$

Whilst jet angles are generally well reconstructed, the experimental jet energy resolution is relatively poor. The resolution on the reconstructed invariant masses in each observed event can be improved by using the constraint $P_{\text{tot}} = (\sqrt{s}, \mathbf{0})$, which implies that both W bosons will have half the centre-of-mass energy and will have equal and opposite three-momenta.

Figure 16.13 shows the reconstructed W-mass distribution for $\ell\nu q_1\bar{q}_2$ events observed in the L3 experiment. For each observed $W^+W^-$ event, the average reconstructed W-boson mass $\overline{m} = \frac{1}{2}(m_1 + m_2)$ is plotted. The position of the peak determines $m_W$ and the width of the distribution, after accounting for the experimental resolution, determines $\Gamma_W$. The results from the four LEP experiments give

$$m_W = 80.376 \pm 0.033\,\text{GeV} \quad \text{and} \quad \Gamma_W = 2.195 \pm 0.083\,\text{GeV}.$$

It is worth noting that, owing to the Lorentzian form of the propagator, the virtual W bosons in the Feynman diagram of Figure 16.11 tend to be produced in the range $q^2 \sim (m_W \pm \Gamma_W)^2$ and are therefore usually close to being on-shell, as can be seen from (16.13). For this reason, the process $e^+e^- \to W^+W^- \to f_1\bar{f}_2f_3\bar{f}_4$ can be approximated as the production of two real W bosons, each of which subsequently decays to two fermions. For more accurate calculations, such as the Standard Model prediction shown in Figure 16.10, this approximation is not sufficient to match the experimental precision and the process has to be treated as the production of four fermions through two virtual W bosons.

### 16.3.2 Measurement of the W mass at the Tevatron

The study of W-boson pair production at LEP provides precise measurements of $m_W$, $\Gamma_W$ and the W-boson branching ratios. Precision measurements can also be made at hadron colliders. For example, $m_W$ has been measured precisely at the Tevatron in the process $p\bar{p} \to WX$, where $X$ is the hadronic system from initial-state QCD radiation and the remnants of the colliding hadrons. In $p\bar{p}$ collisions, the W boson is produced in parton-level processes such as $u\bar{d} \to W^+ \to \mu^+\bar{\nu}_\mu$. In order to reconstruct the mass of the W boson, the momentum of the neutrino needs to be determined.

At a hadron collider, the centre-of-mass energy of the underlying $q\bar{q}'$ annihilation process is not known on an event-by-event basis. If $x_1$ and $x_2$ are the momentum fractions of the proton and antiproton carried by the annihilating q and $\bar{q}'$, the four-momentum of the final state is

$$P_{\text{tot}} = \left[(x_1 + x_2)\frac{\sqrt{s}}{2},\, 0,\, 0,\, (x_1 - x_2)\frac{\sqrt{s}}{2}\right].$$

Consequently, the final-state W boson will be boosted along the beam ($z$) axis. Because the momentum fractions $x_1$ and $x_2$ are unknown, the components of the momentum of the final-state system only balance in the transverse ($xy$) plane. The typical $W \to \mu\nu$ event topology, as seen in the plane transverse to the beam axis, is illustrated in the left plot of Figure 16.14. The transverse components of the momentum of the neutrino can be reconstructed from the transverse momentum of the muon, $\mathbf{p}_T^\mu = p_x\hat{\mathbf{x}} + p_y\hat{\mathbf{y}}$, and the (usually small) transverse momentum $\mathbf{u}_T$ of the hadronic system $X$,

$$\mathbf{p}_T^\nu = -\mathbf{p}_T^\mu - \mathbf{u}_T.$$

Owing to the unknown momentum fractions of the colliding partons, the $z$-component of the neutrino momentum cannot be determined.

Because the $z$-component of the momentum of the neutrino is unknown, the invariant mass of the products from the decaying W boson can not be determined on an event-by-event-basis. However, it is possible to define the transverse mass

$$m_T \equiv \left[2\left(p_T^\mu p_T^\nu - \mathbf{p}_T^\mu \cdot \mathbf{p}_T^\nu\right)\right]^{\frac{1}{2}}.$$

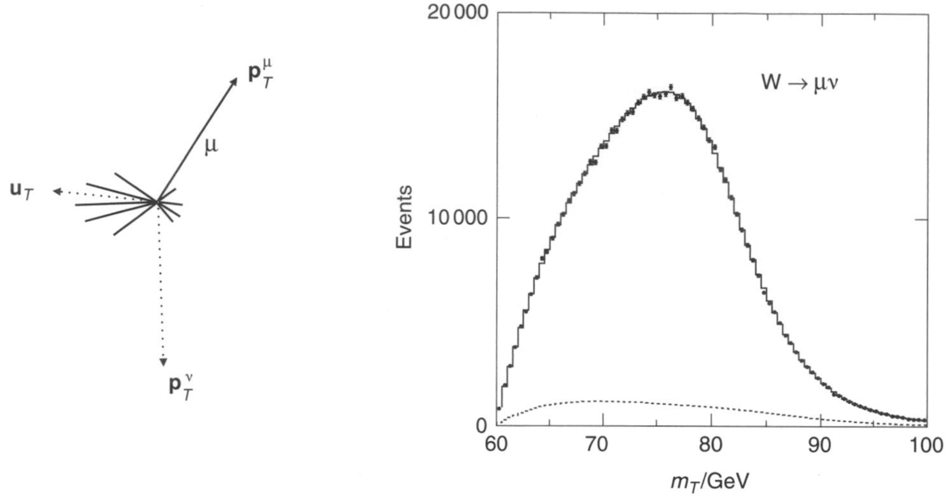


Fig. 16.14 The W $\rightarrow \mu\nu$ event topology for W bosons produced in $p\bar{p}$ collisions at the Tevatron and the reconstructed transverse mass distribution from the CDF experiment, adapted from Aaltonen *et al.* (2012).

Figure 16.14 shows the reconstructed $m_T$ distribution from over 600 000 W $\rightarrow \mu\nu$ decays observed in the CDF detector. Because the longitudinal components of the momentum are not included, $m_T$ does not peak at $m_W$ and the distribution of $m_T$ is relatively broad. Nevertheless, these disadvantages are outweighed by the very large W-production cross section at a hadron–hadron collider. Because of the large numbers of events, $m_W$ can be measured even more precisely than at LEP. The sensitivity to $m_W$ comes from the shape of the $m_T$ distribution and the position of the broad peak. The combined results from the CDF and D0 experiments at the Tevatron and the four LEP experiments give

$$m_W = 80.385 \pm 0.015\,\text{GeV} \quad \text{and} \quad \Gamma_W = 2.085 \pm 0.042\,\text{GeV}.$$

## 16.4 Quantum loop corrections

The data from LEP, the Tevatron and elsewhere provide precise measurements of the fundamental parameters of the electroweak model. The masses of the weak gauge bosons are determined to be

$$m_Z = 91.1875 \pm 0.0021\,\text{GeV} \quad \text{and} \quad m_W = 80.385 \pm 0.015\,\text{GeV}.$$

The weak mixing angle is determined to be

$$\sin^2\theta_W = 0.23146 \pm 0.00012,$$

and the strengths of the weak and electromagnetic interaction are

$$G_F = 1.166\,378\,7(6) \times 10^{-5}\,\text{GeV}^{-2} \quad \text{and} \quad \alpha(m_Z^2) = \frac{1}{128.91 \pm 0.02},$$

where $\alpha$ is given at the electroweak scale of $q^2 = m_Z^2$. In the Standard Model, the masses of the W and Z bosons are not free parameters, they are determined by the Higgs mechanism, described in Chapter 17. Consequently, if any three of the parameters $m_Z$, $m_W$, $G_F$, $\alpha$ and $\sin^2\theta_W$ are known, the other two are determined by exact relations from the electroweak unification mechanism of the Standard Model. For example, the mass of the W boson is related to $\alpha$, $G_F$ and $\theta_W$ by

$$m_W = \left(\frac{\pi\alpha}{\sqrt{2}G_F}\right)^{\frac{1}{2}} \frac{1}{\sin\theta_W},$$

and the masses of the W and Z bosons are related by

$$m_W = m_Z \cos\theta_W.$$

These constraints, coupled with the precise measurements described above, allow the electroweak sector of the Standard Model to be tested to high precision. For example, using the measurements of $m_Z$ and $\sin^2\theta_W$, the predicted mass of the W boson obtained from $m_W = m_Z\cos\theta_W$ is

$$m_W^{\text{pred}} = 79.937 \pm 0.009\,\text{GeV}.$$

Despite being of the right order, this prediction is thirty standard deviations smaller than the measured value of $m_W = 80.385 \pm 0.015$ GeV. This apparent discrepancy does not represent a failure of the Standard Model; it can be explained by higher-order contributions from virtual quantum loop corrections. For example, the mass of the W boson includes contributions from virtual loops, of which the two largest are shown in Figure 16.15. As a result of these quantum loops, the physical W-boson mass differs from the lowest-order prediction $m_W^0$ by

$$m_W = m_W^0 + a\,m_t^2 + b\ln\left(\frac{m_H}{m_W}\right) + \cdots, \tag{16.28}$$

where $a$ and $b$ are calculable constants, and $m_t$ and $m_H$ are the masses of the top quark and Higgs boson.

The difference between the predicted lowest-order W-boson mass and the measured value effectively measures the size of these quantum loop corrections. Since the dependence on the Higgs mass is only logarithmic, the dominant correction in (16.28) comes from the top quark mass. In 1994 the measurements of the electroweak parameters at LEP implied a top quark mass of $175 \pm 11$ GeV. Shortly afterwards, the top quark was discovered at the Tevatron with a mass consistent

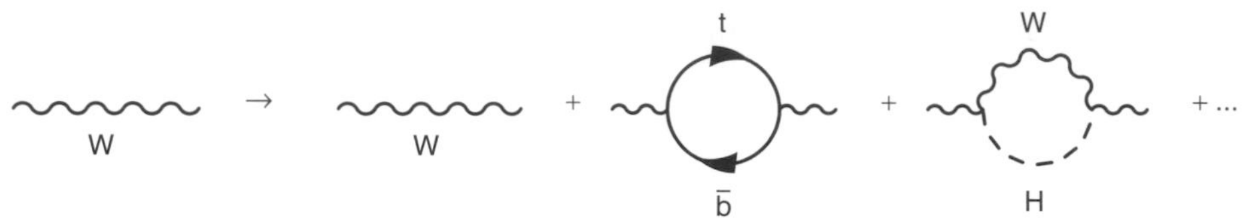


Fig. 16.15 The two largest loop corrections to the W mass.

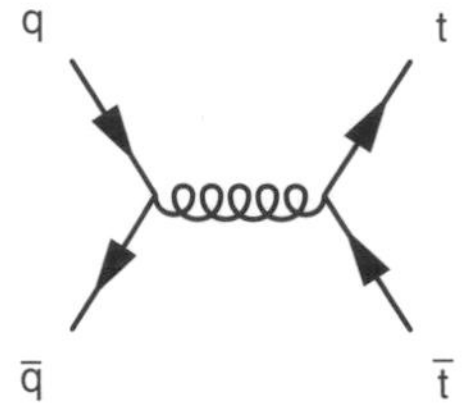


Fig. 16.16 The lowest-order Feynman diagram for $t\bar{t}$ production in $p\bar{p}$ collisions at the Tevatron.

with this prediction. This direct observation of the effects of quantum loop corrections provides an impressive validation of the electroweak sector of the Standard Model.

Because the electroweak measurements are sufficiently precise to be sensitive to quantum loop corrections, they strongly constrain possible models for physics beyond the Standard Model; any new particle or interaction that gives rise to a significant contribution to the quantum loop corrections to the W-boson mass will not be consistent with the experimental data.

## 16.5 The top quark

The top quark is by some way the most massive of the quarks. In fact, with a mass of approximately 175 GeV, it is the most massive fundamental particle in the Standard Model, $m(\mathrm{t}) > m(\mathrm{H}) > m(\mathrm{Z}) > m(\mathrm{W})$. Because of its mass, the top quark could not be observed directly at LEP and was only discovered in 1994 in $\mathrm{p\bar{p}}$ collisions at the Tevatron. In $\mathrm{p\bar{p}}$ collisions, top quarks are predominantly produced in pairs in the QCD process $\mathrm{q\bar{q}} \to \mathrm{t\bar{t}}$, shown in Figure 16.16.

Owing to its large mass, the lifetime of the top quark is very short. Consequently, the top pairs produced in the process $\mathrm{q\bar{q}} \to \mathrm{t\bar{t}}$ do not have time to form bound states, such as those observed in the resonant production of charmonium ($\mathrm{c\bar{c}}$) and bottomonium ($\mathrm{b\bar{b}}$) states (discussed in Section 10.8). Because $|V_{\mathrm{tb}}| \gg |V_{\mathrm{ts}}| > |V_{\mathrm{td}}|$, the top quark decays almost entirely by $\mathrm{t} \to \mathrm{bW^+}$. Hence top quark pair production and decay at the Tevatron (and at the LHC) proceeds mostly by

$$\mathrm{q\bar{q}} \to \mathrm{t\bar{t}} \to \mathrm{bW^+\,\bar{b}W^-}.$$

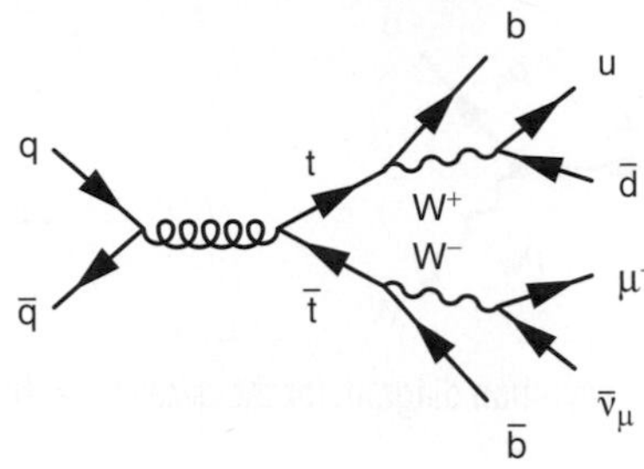


Fig. 16.17 The lowest-order Feynman diagram for $t\bar{t}$ production in $p\bar{p}$ collisions, where the W bosons decay to give a semi-leptonic final state.

The Feynman diagram for the production of the $t\bar{t}$ semi-leptonic final state, where one W boson decays leptonically and the other decays hadronically, is shown in Figure 16.17. The corresponding matrix element contains four propagators for massive particles, two for the top quarks and two for the W bosons. Because $\Gamma_W \ll m_W$, the largest contributions to the matrix element will be when the W bosons are produced almost on-shell with $q^2 \sim m_W^2$. Similarly, the presence of the propagators for the two virtual top quarks implies that

$$|\mathcal{M}|^2 \propto \frac{1}{(q_1^2 - m_t^2)^2 + m_t^2\Gamma_t^2} \times \frac{1}{(q_2^2 - m_t^2)^2 + m_t^2\Gamma_t^2}.$$

As a result, the invariant masses of each of the $W^+b$ and $W^-\bar{b}$ systems, produced in the top decays, will be distributed according Lorentzian centred on $m_t$ with width $\Gamma_t$.

### 16.5.1 Decay of the top quark

Because the W boson from the decay of a top quark is close to being on-shell, $q^2 \sim m_W^2$, the top decay width can be calculated from the Feynman diagram for $t \to bW^+$ shown in Figure 16.18, where the W boson is treated as a real on-shell final-state particle. The corresponding matrix element is obtained from the quark spinors, the weak charged-current vertex factor and the term $\epsilon^*(p_W)$ for the polarisation state of the W boson,

$$-i\mathcal{M} = \left[\overline{u}(p_b)\,\frac{-ig_W}{\sqrt{2}}\gamma^\mu\tfrac{1}{2}(1-\gamma^5)\,u(p_t)\right] \times \epsilon^*_\mu(p_W),$$

and thus

$$\mathcal{M} = \tfrac{g_W}{\sqrt{2}}\,\epsilon^*_\mu(p_W)\,\overline{u}(p_b)\gamma^\mu\tfrac{1}{2}(1-\gamma^5)u(p_t). \tag{16.29}$$

It is convenient to consider the decay in the rest frame of the top quark and to take the final-state b-quark direction to define the $z$-axis. Neglecting the mass of the b-quark, the four-momenta of the t, b and $W^+$ can be written

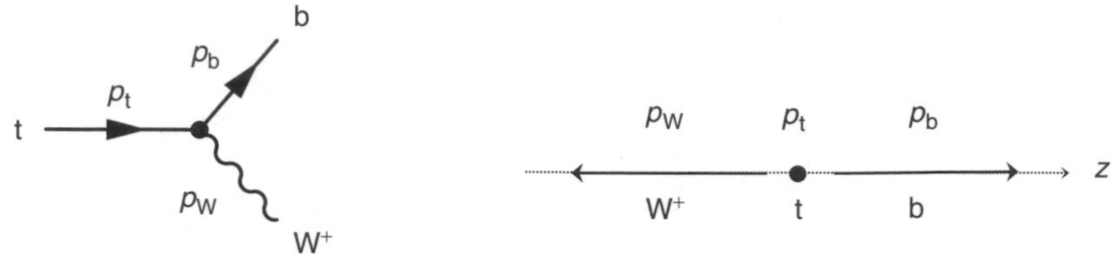


Fig. 16.18 The lowest-order Feynman diagram for the decay $t \to bW^+$ and coordinates used for the calculation of the decay rate.

$$p_t = (m_t, 0, 0, 0), \quad p_b = (p^*, 0, 0, p^*) \quad \text{and} \quad p_W = (E^*, 0, 0, -p^*),$$

where $p^*$ is the magnitude of the momentum of the final-state particles in the centre-of-mass frame and $E^*$ is the energy of the W boson, $E^{*2} = (p^*)^2 + m_W^2$. The weak interaction couples only to left-handed chiral particle states. Here, in the limit $p^* \gg m_b$, the chiral states are equivalent to the helicity states and, consequently, the b-quark can only be produced in a left-handed helicity state such that for the configuration of Figure 16.18 its spin points in the negative $z$-direction. Hence, the matrix element of (16.29) can be written as

$$\mathcal{M} = \frac{g_W}{\sqrt{2}} \epsilon^*_\mu(p_W)\, \overline{u}_\downarrow(p_b)\gamma^\mu u(p_t). \tag{16.30}$$

From (4.67), the LH helicity spinor for the b-quark is

$$u_\downarrow(p_b) \approx \sqrt{p^*}\begin{pmatrix} 0 \\ 1 \\ 0 \\ -1 \end{pmatrix}.$$

The two possible spin states of the top quark can be written using the $u_1$ and $u_2$ spinors, which for a top quark at rest are (4.48),

$$u_1(p_t) = \sqrt{2m_t}\begin{pmatrix} 1 \\ 0 \\ 0 \\ 0 \end{pmatrix} \quad \text{and} \quad u_2(p_t) = \sqrt{2m_t}\begin{pmatrix} 0 \\ 1 \\ 0 \\ 0 \end{pmatrix},$$

respectively representing $S_z = +\frac{1}{2}$ and $S_z = -\frac{1}{2}$ states. The four-vector quark currents for the two possible spin states, calculated using the relations of (6.12)–(6.15), are

$$j_1^\mu = \overline{u}_\downarrow(p_b)\gamma^\mu u_1(p_t) = \sqrt{2m_t p^*}(0, -1, -i, 0),$$
$$j_2^\mu = \overline{u}_\downarrow(p_b)\gamma^\mu u_2(p_t) = \sqrt{2m_t p^*}(1, 0, 0, 1).$$

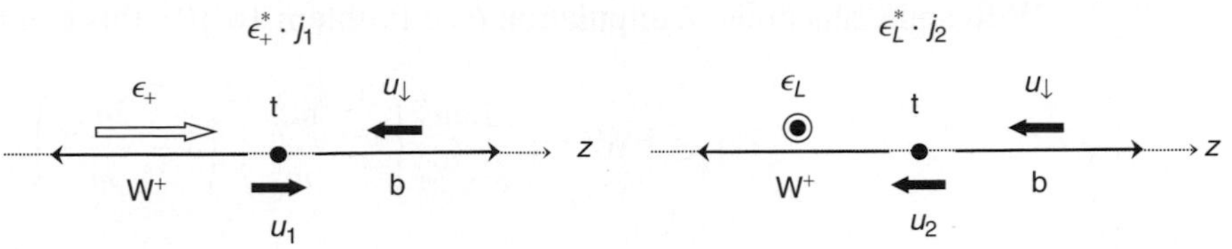


Fig. 16.19 The two allowed spin combinations in the process $t \to bW^+$.

The three possible polarisation states of the W boson, given by (15.1), are

$$\epsilon_+^*(p_W) = -\tfrac{1}{\sqrt{2}}(0,1,-i,0),$$
$$\epsilon_-^*(p_W) = \tfrac{1}{\sqrt{2}}(0,+1,+i,0),$$
$$\epsilon_L^*(p_W) = \tfrac{1}{m_W}(-p^*,0,0,E^*),$$

which correspond to the $S_z = \pm 1$ and the longitudinal polarisation states. For a particular top quark spin and W polarisation state, the matrix element of (16.30) is given by the four-vector scalar product

$$\mathcal{M} = \tfrac{g_W}{\sqrt{2}} j_i \cdot \epsilon_\lambda^*.$$

The only combinations of the two possible quark currents and the three possible W-boson polarisations for which the matrix element is non-zero are $\epsilon_+^* \cdot j_1$ and $\epsilon_L^* \cdot j_2$. These two combinations correspond to the spin states shown in Figure 16.19, which (unsurprisingly) are the only configurations that conserve angular momentum.

The matrix elements for these two allowed spin configurations are

$$\mathcal{M}_1 = \frac{g_W}{\sqrt{2}}\,\epsilon_+^* \cdot j_1 = -\tfrac{g_W}{\sqrt{2}}\,\sqrt{m_t p^*}\,(0,1,i,0)\cdot(0,-1,-i,0) = -g_W\,\sqrt{2m_t p^*},$$
$$\mathcal{M}_2 = \frac{g_W}{\sqrt{2}}\,\epsilon_L^* \cdot j_2 = \tfrac{g_W}{m_W}\,\sqrt{m_t p^*}\,(-p^*,0,0,E^*)\cdot(1,0,0,1) = -\frac{g_W}{m_W}\,\sqrt{m_t p^*}(E^* + p^*).$$

From conservation of energy, $E^* + p^* = m_t$, and therefore the spin-averaged matrix element squared for the decay $t \to bW^+$ is

$$\langle|\mathcal{M}^2|\rangle = \tfrac{1}{2}\left(|\mathcal{M}_1^2| + |\mathcal{M}_2^2|\right) = \tfrac{1}{2} g_W^2 m_t p^* \left(2 + \frac{m_t^2}{m_W^2}\right),$$

where the factor of one half averages over the two spin states of the t-quark. The total decay rate is obtained by substituting the spin-averaged matrix element into the formula of (3.49) which, after integrating over the $4\pi$ of solid angle, gives

$$\Gamma(t \to bW^+) = \frac{p^*}{8\pi m_t^2}\langle|\mathcal{M}^2|\rangle = \frac{g_W^2 p^{*2}}{16\pi m_t}\left(2 + \frac{m_t^2}{m_W^2}\right). \qquad (16.31)$$

With some algebraic manipulation (see Problem 16.10), this can be written as

$$\Gamma(\mathrm{t} \to \mathrm{bW}^+) = \frac{G_\mathrm{F} m_\mathrm{t}^3}{8\sqrt{2}\pi}\left(1 - \frac{m_\mathrm{W}^2}{m_\mathrm{t}^2}\right)^2\left(1 + \frac{2m_\mathrm{W}^2}{m_\mathrm{t}^2}\right), \tag{16.32}$$

where $g_\mathrm{W}^2$ is given in terms of the Fermi constant, $G_\mathrm{F} = \sqrt{2}g_\mathrm{W}^2/(8m_\mathrm{W}^2)$. For the measured values of $m_\mathrm{t} = 173\,\mathrm{GeV}$, $m_\mathrm{W} = 80.4\,\mathrm{GeV}$ and $G_\mathrm{F} = 1.166 \times 10^{-5}\,\mathrm{GeV}^{-2}$, the lowest-order calculation of the total decay width of the top quark gives

$$\Gamma_\mathrm{t} = 1.5\,\mathrm{GeV}. \tag{16.33}$$

Hence the top quark lifetime is of order $\tau_\mathrm{t} = 1/\Gamma_\mathrm{t} \approx 5 \times 10^{-25}\,\mathrm{s}$. This is sufficiently short that the top quarks produced at the Tevatron decay in a distance of order $10^{-16}\,\mathrm{m}$. This is small compared to the typical length scale for the hadronisation process, and therefore the $\mathrm{t\bar{t}}$ pairs produced at the Tevatron not only decay before forming a bound state, but also decay before hadronising.

### 16.5.2 Measurement of the top quark mass

The mass of the top quark has been measured in the process $\mathrm{p\bar{p}} \to \mathrm{t\bar{t}}$ by direct reconstruction of the top quark decay products, similar to the procedure used to measure the W-boson mass at LEP. Since both top quarks decay to a b-quark and a W boson there are three distinct final-state topologies:

$$\mathrm{t\bar{t}} \to (\mathrm{bW^+})(\mathrm{\bar{b}W^-}) \to (\mathrm{b\,q_1\bar{q}_2})(\mathrm{\bar{b}\,q_3\bar{q}_4}) \to 6\text{ jets},$$

$$\mathrm{t\bar{t}} \to (\mathrm{bW^+})(\mathrm{\bar{b}W^-}) \to (\mathrm{b\,q_1\bar{q}_2})(\mathrm{\bar{b}}\,\ell^-\bar{\nu}_\ell) \to 4\text{ jets} + 1,\ \text{charged lepton} + 1\nu$$

$$\mathrm{t\bar{t}} \to (\mathrm{bW^+})(\mathrm{\bar{b}W^-}) \to (\mathrm{b}\,\ell^+\nu_\ell)(\mathrm{\bar{b}}\,\ell'^-\bar{\nu}_{\ell'}) \to 2\text{ jets} + 2\text{ charged leptons} + 2\nu\text{s}.$$

The measurement of the top quark mass is more complicated than the corresponding measurement of the W-mass at LEP, but the principle is the same. The b-quark jets are identified from the tagging of secondary vertices (see Section 1.3.1) and the remaining jets have to be associated to the W-boson decay(s), as indicated in Figure 16.20. Because the momentum of the $\mathrm{t\bar{t}}$ system in the beam ($z$) direction is not known (see Section 16.3.2), it might appear that there is insufficient information to fully reconstruct the neutrino momentum in observed $\mathrm{t\bar{t}} \to \text{four jets} + \ell + \nu$ events. However, the invariant mass of the two jets associated with a W boson and the invariant mass of the lepton and neutrino both can be constrained to $m_\mathrm{W}$ within $\pm\Gamma_\mathrm{W}$. Furthermore, the invariant masses of the particles forming the two reconstructed top quarks can be required to be equal. These additional constraints provide a system of equations that allow the momentum of the neutrino to be determined from the technique of kinematic fitting. In both the fully hadronic and $\mathrm{t\bar{t}} \to \text{four jets} + \ell + \nu$ decay topologies, these constraints improve the event-by-event mass resolution.

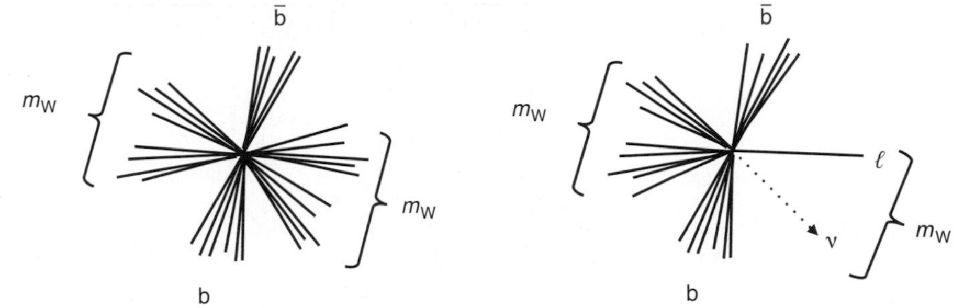


Fig. 16.20 The two main $t\bar{t}$ event topologies used at the Tevatron to determine the top quark mass from the direct reconstruction of the decay products.

The top quark mass has been determined by both the CDF and D0 collaborations using the measured four-momenta of the jets and leptons in observed $t\bar{t} \to$ six jets and $t\bar{t} \to$ four jets $+ \ell + \nu$ events. As an example, Figure 16.21 shows the reconstructed top mass distribution from an analysis of data recorded by the CDF experiment. Whilst the reconstructed mass peak is relatively broad due to experimental resolution, a clear peak is observed allowing the top mass to be determined with a precision of $\mathcal{O}(1\%)$. The current average of the top quark mass measurements from the CDF and D0 experiments is

$$m_t = 173.5 \pm 1.0\,\text{GeV}.$$

The total width of the top quark is measured to be $\Gamma_t = 2.0 \pm 0.6\,\text{GeV}$. The top width is determined much less precisely than the top quark mass because the width of the distribution in Figure 16.21 is dominated by the experimental resolution. Nevertheless, the current measurement is consistent with the result of the lowest-order calculation presented in Section 16.5.1.

### A window on the Higgs boson

Just as the electroweak measurements at LEP provided a prediction of the top quark mass through its quantum loop corrections to the W-boson mass, the precise determination of the top quark mass at the Tevatron provides a window on the Higgs boson. Its measurement determines the size of the largest loop correction to the W-boson mass, which arises from the virtual $t\bar{b}$ loop in Figure 16.15. The next largest correction arises from the WH loop that leads to the logarithmic term in (16.28). The electroweak measurements at LEP and the Tevatron, when combined with the direct measurement of the top quark mass, constrain the mass of the Standard Model Higgs boson to be in the range

$$50\,\text{GeV} \lesssim m_H \lesssim 150\,\text{GeV}.$$

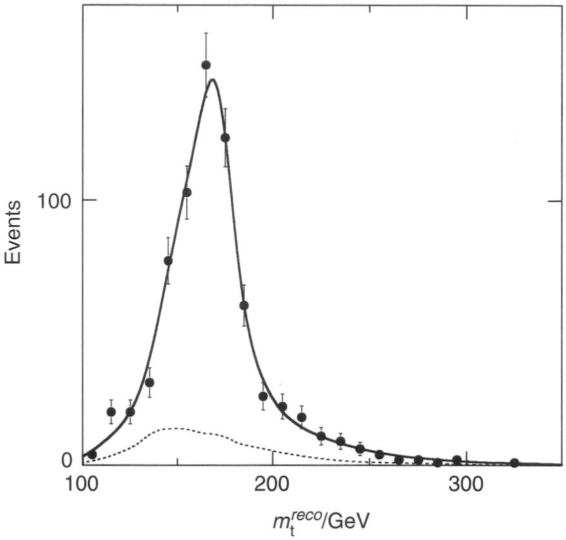


Fig. 16.21 The distribution of the reconstructed top mass in selected $t\bar{t} \to$ four jets $+ \ell + \nu$ events in the CDF detector at the Tevatron. The dashed curve indicates the expected contribution from processes other than $t\bar{t}$. Adapted from Aaltonen *et al.* (2011).

Direct searches for the Higgs boson at LEP placed a lower bound on its mass of

$$m_{\mathrm{H}} > 115\,\mathrm{GeV}.$$

Hence, prior to the turn-on of the LHC, the window for the Standard Model Higgs boson was already quite narrow.

## Summary

The Z boson was studied with very high precision at the LEP $e^+e^-$ collider. The resulting measurements provided a stringent test of the predictions of the GSW model for electroweak unification. The mass of the Z boson, which is a fundamental parameter of the Standard Model, was determined to be

$$m_{\mathrm{Z}} = 91.1875 \pm 0.0021\,\mathrm{GeV}.$$

The observed couplings of the Z boson are consistent with the Standard Model expectations with

$$\sin^2\theta_{\mathrm{W}} = 0.23146 \pm 0.00012.$$

The measurements of $\Gamma_Z$ and the $e^+e^- \to Z \to f\bar{f}$ cross sections, demonstrate that there are only three generations of light neutrinos (assuming standard couplings), which is strongly suggestive that there are only three generations of fundamental fermions. Furthermore, studies of the W boson at LEP and the Tevatron and the studies of the top quark at the Tevatron, show that

$$m_W = 80.385 \pm 0.015\,\text{GeV} \quad \text{and} \quad m_t = 173.5 \pm 1.0\,\text{GeV}.$$

Remarkably, when combined, the above measurements of the electroweak sector of the Standard Model are sufficiently precise to reveal effects at the quantum loop level. These precision measurements and the ability of the Standard Model to describe the electroweak data, represent one of the highlights of modern physics. However, there is a serious problem with the Standard Model as it has been presented so far; the fact the W and Z bosons have mass breaks the required gauge symmetry of the Standard Model. The solution to this apparent contradiction is the Higgs mechanism.

## Problems

**16.1** After correcting for QED effects, including initial-state radiation, the measured $e^+e^- \to \mu^+\mu^-$ and $e^+e^- \to$ hadrons cross sections at the peak of the Z resonance give

$$\sigma^0(e^+e^- \to Z \to \mu^+\mu^-) = 1.9993\,\text{nb} \quad \text{and} \quad \sigma^0(e^+e^- \to Z \to \text{hadrons}) = 41.476\,\text{nb}.$$

(a) Assuming lepton universality, determine $\Gamma_{\ell\ell}$ and $\Gamma_{\text{hadrons}}$.
(b) Hence, using the measured value of $\Gamma_Z = 2.4952 \pm 0.0023$ GeV and the theoretical value of $\Gamma_{\nu\nu}$ given by Equation (15.41), obtain an estimate of the number of light neutrino flavours.

**16.2** Show that the $e^+e^- \to Z \to \mu^+\mu^-$ differential cross section can be written as

$$\frac{d\sigma}{d\Omega} \propto (1+\cos^2\theta) + \tfrac{8}{3}A_{FB}\cos\theta.$$

**16.3** From the measurement of the muon asymmetry parameter,

$$\mathcal{A}_\mu = 0.1456 \pm 0.0091,$$

determine the corresponding value of $\sin^2\theta_W$.

**16.4** The $e^+e^-$ Stanford Linear Collider (SLC), operated at $\sqrt{s} = m_Z$ with left- and right-handed longitudinally polarised beams. This enabled the $e^+e^- \to Z \to f\bar{f}$ cross section to be measured separately for left-handed and right-handed electrons.

Assuming that the electron beam is 100% polarised and that the positron beam is unpolarised, show that the left–right asymmetry $A_{LR}$ is given by

$$A_{LR} = \frac{\sigma_L - \sigma_R}{\sigma_L + \sigma_R} = \frac{(c_L^e)^2 - (c_R^e)^2}{(c_L^e)^2 + (c_R^e)^2} = \mathcal{A}_e,$$

where $\sigma_L$ and $\sigma_R$ are respectively the measured cross sections at the Z resonance for LH and RH electron beams.

**16.5** From the expressions for the matrix elements given in (16.8), show that:

(a) the average polarisation of the tau leptons produced in the process $e^+e^- \to Z \to \tau^+\tau^-$ is

$$\langle P_{\tau^-} \rangle = \frac{N_\uparrow - N_\downarrow}{N_\uparrow + N_\downarrow} = -\mathcal{A}_\tau,$$

where $N_\uparrow$ and $N_\downarrow$ are the respective numbers of $\tau^-$ produced in RH and LH helicity states;

(b) the tau polarisation where the $\tau^-$ is produced at an angle $\theta$ with respect to the initial-state $e^-$ is

$$P_{\tau^-}(\cos\theta) = \frac{N_\uparrow(\cos\theta) - N_\downarrow(\cos\theta)}{N_\uparrow(\cos\theta) + N_\downarrow(\cos\theta)} = -\frac{\mathcal{A}_\tau(1+\cos^2\theta) + 2\mathcal{A}_e \cos\theta}{(1+\cos^2\theta) + \frac{8}{3}A_{FB}\cos\theta}.$$

**16.6** The average tau polarisation in the process $e^+e^- \to Z \to \tau^+\tau^-$ can be determined from the energy distribution of $\pi^-$ in the decay $\tau^- \to \pi^-\nu_\tau$. In the $\tau^-$ rest frame, the $\pi^-$ four-momentum can be written $p = (E^*, p^*\sin\theta^*, 0, p^*\cos\theta^*)$, where $\theta^*$ is the angle with respect to the $\tau^-$ spin, and the differential partial decay width is

$$\frac{d\Gamma}{d\cos\theta^*} \propto \frac{(p^*)^2}{m_\tau}(1+\cos\theta^*).$$

(a) Without explicit calculation, explain this angular dependence.

(b) For the case where the $\tau^-$ is right-handed, show that the observed energy distribution of the $\pi^-$ *in the laboratory frame* is

$$\frac{d\Gamma_{\tau_\uparrow^-}}{dE_{\pi^-}} \propto x,$$

where $x = E_\pi/E_\tau$.

(c) What is the corresponding $\pi^-$ energy distribution for the decay of a LH helicity $\tau^-$.

(d) If the observed pion energy distribution is consistent with

$$\frac{d\Gamma}{dx} = 1.14 - 0.28x \equiv 0.86x + 1.14(1-x),$$

determine $\mathcal{A}_\tau$ and the corresponding value of $\sin^2\theta_W$.

**16.7** There are ten possible lowest-order Feynman diagrams for the process $e^+e^- \to \mu^-\bar{\nu}_\mu u\bar{d}$, of which only three involve a $W^+W^-$ intermediate state. Draw the other seven diagrams (they are all *s*-channel processes involving a single virtual W).

**16.8** Draw the two lowest-order Feynman diagrams for $e^+e^- \to ZZ$.

**16.9** In the OPAL experiment at LEP, the efficiencies for selecting $W^+W^- \to \ell\nu q_1\bar{q}_2$ and $W^+W^- \to q_1\bar{q}_2q_3\bar{q}_4$ events were 83.8% and 85.9% respectively. After correcting for background, the observed numbers of $\ell\nu q_1\bar{q}_2$ and $q_1\bar{q}_2q_3\bar{q}_4$ events were respectively 4192 and 4592. Determine the measured value of the W-boson hadronic branching ratio $BR(W \to q\bar{q}')$ and its statistical uncertainty.

**16.10** Suppose the four jets in an identified $e^+e^- \to W^+W^-$ event at LEP are measured to have momenta,

$$p_1 = 82.4 \pm 5\,\text{GeV},\ \ p_2 = 59.8 \pm 5\,\text{GeV},\ \ p_3 = 23.7 \pm 5\,\text{GeV and } p_4 = 42.6 \pm 5\,\text{GeV},$$

and directions given by the Cartesian unit vectors,

$$\hat{\mathbf{n}}_1 = (0.72, 0.33, 0.61), \quad \hat{\mathbf{n}}_2 = (-0.61, 0.58, -0.53),$$
$$\hat{\mathbf{n}}_3 = (-0.63, -0.72, -0.25), \quad \hat{\mathbf{n}}_4 = (-0.14, -0.96, -0.25).$$

Assuming that the jets can be treated as massless particles, find the most likely association of the four jets to the two W bosons and obtain values for the invariant masses of the (off-shell) W bosons in this event. Optionally, calculate the uncertainties on the reconstructed masses assuming that the jet directions are perfectly measured.

**16.11** Show that the momenta of the final-state particles in the decay $t \rightarrow W^+b$ are

$$p^* = \frac{m_t^2 - m_W^2}{2m_t},$$

and show that the decay rate of (16.31) leads to the expression for $\Gamma_t$ given in (16.32).

# 17 The Higgs boson

The Higgs mechanism and the associated Higgs boson are essential parts of the Standard Model. The Higgs mechanism is the way that the W and Z bosons acquire mass without breaking the local gauge symmetry of the Standard Model. It also gives mass to the fundamental fermions. This chapter describes the Higgs mechanism and the discovery of the Higgs boson at the LHC. The Higgs mechanism is subtle and to gain a full understanding requires the additional theoretical background material covered in the sections on Lagrangians and local gauge invariance in quantum field theory.

## 17.1 The need for the Higgs boson

The apparent violation of unitarity in the $e^+e^- \to W^+W^-$ cross section was resolved by the introduction of the Z boson. A similar issue arises in the $W^+W^- \to W^+W^-$ scattering process, where the cross section calculated from the Feynman diagrams shown in Figure 17.1 violates unitarity at a centre-of-mass energy of about 1 TeV. The unitarity violating amplitudes originate from $W_LW_L \to W_LW_L$ scattering, where the W bosons are longitudinally polarised. Consequently, unitary violation in WW scattering can be associated with the W bosons being massive, since longitudinal polarisation states do not exist for massless particles. The unitarity violation of the $W_LW_L \to W_LW_L$ cross section can be cancelled by the diagrams involving the exchange of a scalar particle, shown in Figure 17.2. In the Standard Model this scalar is the Higgs boson. This cancellation can work only if the couplings of the scalar particle are related to the electroweak couplings, which naturally occurs in the Higgs mechanism.

The Higgs mechanism is an integral part of the Standard Model. Without it, the Standard Model is not a consistent theory; the underlying gauge symmetry of the electroweak interaction is broken by the masses of the associated gauge bosons. As shown by 't Hooft, only theories with local gauge invariance are renormalisable, such that the cancellation of all infinities takes place among only a finite number of interaction terms. Consequently, the breaking of the local gauge invariance of the electroweak theory by the gauge boson masses can not simply be dismissed.

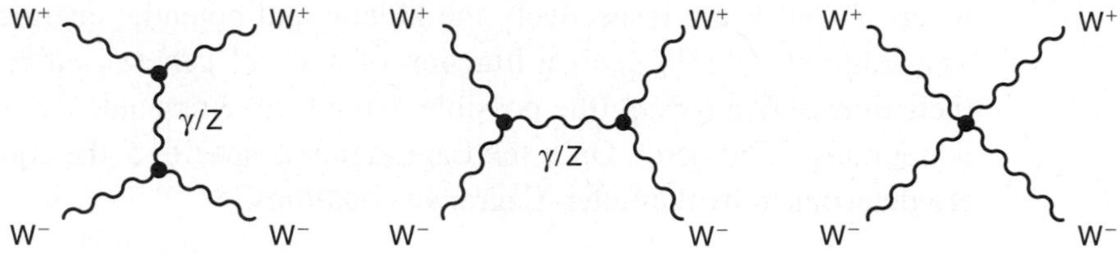


Fig. 17.1 The lowest-order Feynman diagrams for $W^+W^- \to W^+W^-$. The final diagram, corresponds to the quartic coupling of four W bosons.

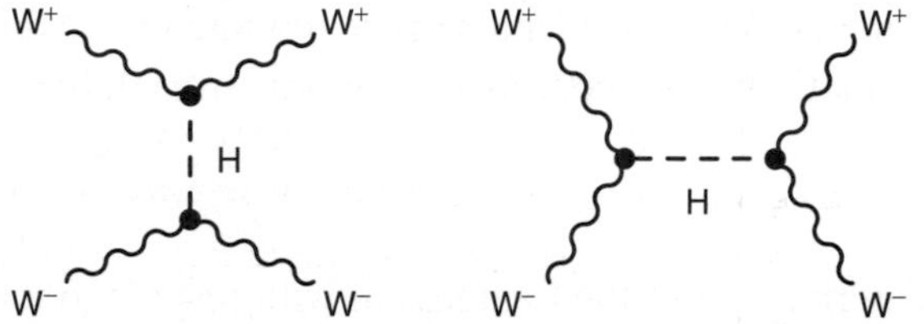


Fig. 17.2 Higgs boson exchange diagrams for $W^+W^- \to W^+W^-$.

The Higgs mechanism generates the masses of the electroweak gauge bosons in a manner that preserves the local gauge invariance of the Standard Model.

## 17.2 Lagrangians in Quantum Field Theory

The Higgs mechanism is described in terms of the Lagrangian of the Standard Model. In quantum mechanics, single particles are described by wavefunctions that satisfy the appropriate wave equation. In Quantum Field Theory (QFT), particles are described by excitations of a quantum field that satisfies the appropriate quantum mechanical field equations. The dynamics of a quantum field theory can be expressed in terms of the Lagrangian density. Whilst the development of QFT is outside the scope of this book, an understanding of the Lagrangian formalism is necessary for the discussion of the Higgs mechanism. The purpose of this section is to provide a pedagogical introduction to the Lagrangian of the Standard Model, which ultimately contains all of the fundamental particle physics.

### 17.2.1 Classical fields

In classical dynamics, the motion of a system can be described in terms of forces and the resulting accelerations using Newton's second law, $\mathbf{F} = m\ddot{\mathbf{x}}$. The same equations of motion can be obtained from the Lagrangian $L$ defined as

$$L = T - V, \tag{17.1}$$

where $T$ and $V$ are respectively the kinetic and potential energies of the system. The Lagrangian $L(q_i, \dot{q}_i)$ is a function of a set of generalised coordinates $q_i$ and their time derivatives $\dot{q}_i$ (the possible explicit time dependence of the Lagrangian is not considered here). Once the Lagrangian is specified, the equations of motion are determined by the Euler–Lagrange equations,

$$\frac{\mathrm{d}}{\mathrm{d}t}\left(\frac{\partial L}{\partial \dot{q}_i}\right) - \frac{\partial L}{\partial q_i} = 0. \tag{17.2}$$

For example, consider a particle moving in one dimension where the Lagrangian is a function of the coordinate $x$ and its time derivative $\dot{x}$, with

$$L = T - V = \tfrac{1}{2}m\dot{x}^2 - V(x).$$

The derivatives of the Lagrangian with respect to $x$ and $\dot{x}$ are

$$\frac{\partial L}{\partial \dot{x}} = m\dot{x} \quad \text{and} \quad \frac{\partial L}{\partial x} = -\frac{\partial V}{\partial x},$$

and the Euler–Lagrange equation (17.2) for the coordinate $q_i = x$ is simply

$$m\ddot{x} = -\frac{\partial V(x)}{\partial x}.$$

Since the derivative of the potential gives the force, this is equivalent to $F = m\ddot{x}$ and Newton's second law of motion is recovered.

The Lagrangian treatment of a discrete system of particles, described by $n$ generalised coordinates $q_i$, can be extended to a continuous system by replacing the Lagrangian of (17.1) with the Lagrangian *density* $\mathcal{L}$,

$$L\left(q_i, \frac{\mathrm{d}q_i}{\mathrm{d}t}\right) \rightarrow \mathcal{L}\left(\phi_i, \partial_\mu \phi_i\right).$$

In the Lagrangian density, the generalised coordinates $q_i$ are replaced by the *fields* $\phi_i(t, x, y, z)$, and the time derivatives of the generalised coordinates $\dot{q}_i$ are replaced by the derivatives of the fields with respect to each of the four space-time coordinates,

$$\partial_\mu \phi_i \equiv \frac{\partial \phi_i}{\partial x^\mu}.$$

The fields are continuous functions of the space-time coordinates $x^\mu$ and the Lagrangian $L$ itself is given by

$$L = \int \mathcal{L}\, \mathrm{d}^3\mathbf{x}.$$

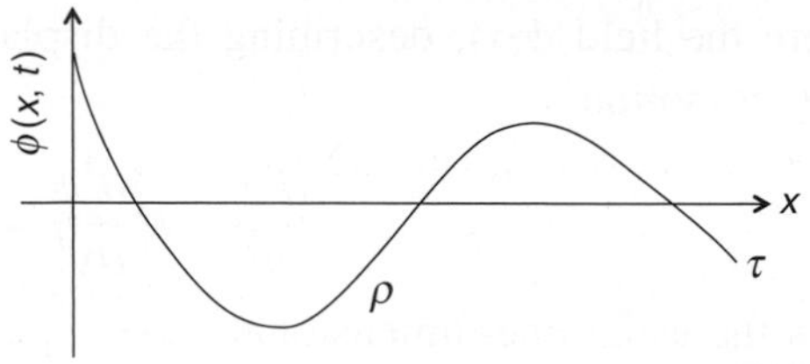


Fig. 17.3 The scalar field $\phi(x,t)$ representing the transverse displacement of a string of mass per unit length $\rho$ under tension $\tau$.

Using the principle of least action,[1] the equivalent of the Euler–Lagrange equation for the fields $\phi_i$ can be shown to be

$$\partial_\mu\left(\frac{\partial\mathcal{L}}{\partial(\partial_\mu\phi_i)}\right)-\frac{\partial\mathcal{L}}{\partial\phi_i}=0. \tag{17.3}$$

The field $\phi_i(x^\mu)$ represents a continuous quantity with a value at each point in space-time. It can be a scalar such as temperature $T(\mathbf{x},t)$, a vector such as the electric field strength $\mathbf{E}(\mathbf{x},t)$, or a tensor.

To illustrate the application of classical field theory, consider the relatively simple example of a string of mass per unit length $\rho$ under tension $\tau$, as indicated in Figure 17.3. Here the scalar field $\phi(x,t)$ represents the transverse displacement of the string as a function of $x$ and $t$. The kinetic and potential energies of the string, written in terms of the derivatives of the field, are

$$T=\int\tfrac{1}{2}\rho\left(\frac{\partial\phi}{\partial t}\right)^2\mathrm{d}x \quad\text{and}\quad V=\int\tfrac{1}{2}\tau\left(\frac{\partial\phi}{\partial x}\right)^2\mathrm{d}x.$$

Hence the Lagrangian density is

$$\mathcal{L}=\tfrac{1}{2}\rho\left[\left(\frac{\partial\phi}{\partial t}\right)^2-v^2\left(\frac{\partial\phi}{\partial x}\right)^2\right]\equiv\tfrac{1}{2}\rho\left[(\partial_0\phi)^2-v^2(\partial_1\phi)^2\right], \tag{17.4}$$

where $v=\sqrt{\tau/\rho}$. Once the Lagrangian density has been specified, the equations of motion follow from the Euler–Lagrange equation of (17.3). For the Lagrangian density of (17.4), the relevant partial derivatives are

$$\frac{\partial\mathcal{L}}{\partial(\partial_0\phi)}=\rho\,\partial_0\phi\equiv\rho\frac{\partial\phi}{\partial t},\quad \frac{\partial\mathcal{L}}{\partial(\partial_1\phi)}=-\rho v^2\,\partial_1\phi\equiv-\rho v^2\frac{\partial\phi}{\partial x}\quad\text{and}\quad\frac{\partial\mathcal{L}}{\partial\phi}=0,$$

and the Euler–Lagrange equation gives

$$\rho\partial_0(\partial_0\phi)-\rho v^2\partial_1(\partial_1\phi)=0\quad\text{or equivalently}\quad\rho\frac{\partial}{\partial t}\left(\frac{\partial\phi}{\partial t}\right)-\rho v^2\frac{\partial}{\partial x}\left(\frac{\partial\phi}{\partial x}\right)=0.$$

[1] The derivation can be found in any standard text on classical or Quantum Field Theory.

Therefore the field $\phi(x)$, describing the displacement of the string, satisfies the equation of motion

$$\frac{\partial^2\phi}{\partial t^2} - v^2\frac{\partial^2\phi}{\partial x^2} = 0,$$

which is the usual one-dimensional wave equation with phase velocity given by $v = \sqrt{\tau/\rho}$. Hence, it can be seen that Lagrangian density determines the wave equation for the field.

### 17.2.2 Relativistic fields

In Quantum Field Theory, the single particle wavefunctions of quantum mechanics are replaced by (multi-particle) excitations of a quantum field, which itself satisfies the appropriate field equation. The field equation is determined by the form of the Lagrangian density, which henceforth will be referred to simply as the Lagrangian. In the above example of a string under tension, it was shown that the Lagrangian,

$$\mathcal{L} = \tfrac{1}{2}\rho\left[(\partial_0\phi)^2 - v^2(\partial_1\phi)^2\right],$$

gives the usual wave equation for the displacement of the string. Similarly the dynamics of the quantum mechanical fields describing spin-0, spin-half and spin-1 particles are determined by the appropriate Lagrangian densities.

#### Relativistic scalar fields

In QFT, spin-0 scalar particles are described by excitations of the scalar field $\phi(x)$ satisfying the Klein–Gordon equation, first encountered in Section 4.1. The Lagrangian for a free non-interacting scalar field can be identified as

$$\mathcal{L}_S = \tfrac{1}{2}(\partial_\mu\phi)(\partial^\mu\phi) - \tfrac{1}{2}m^2\phi^2. \tag{17.5}$$

To see that this Lagrangian corresponds to the Klein–Gordon equation, it is helpful to write (17.5) in full,

$$\mathcal{L}_S = \tfrac{1}{2}\left[(\partial_0\phi)(\partial_0\phi) - (\partial_1\phi)(\partial_1\phi) - (\partial_2\phi)(\partial_2\phi) - (\partial_3\phi)(\partial_3\phi)\right] - \tfrac{1}{2}m^2\phi^2,$$

from which the partial derivatives appearing in the Euler–Lagrange equation are

$$\frac{\partial\mathcal{L}}{\partial\phi} = -m^2\phi, \quad \frac{\partial\mathcal{L}}{\partial(\partial_0\phi)} = \partial_0\phi \equiv \partial^0\phi \quad\text{and}\quad \frac{\partial\mathcal{L}}{\partial(\partial_k\phi)} = -\partial_k\phi \equiv \partial^k\phi,$$

where $k = 1, 2, 3$. Substituting these partial derivatives into the Euler–Lagrange equation of (17.3) gives

$$\partial_\mu\partial^\mu\phi + m^2\phi = 0,$$

which is the Klein–Gordon equation for a free scalar field $\phi(x)$.

### Relativistic spin-half fields

The Lagrangian for the spinor field $\psi(x)$ satisfying the free-particle Dirac equation is

$$\mathcal{L}_D = i\overline{\psi}\gamma^\mu\partial_\mu\psi - m\overline{\psi}\psi. \tag{17.6}$$

Here the field $\psi(x)$ is a four-component complex spinor, which can be expressed in terms of eight independent real fields,

$$\psi(x) = \begin{pmatrix} \psi_1 \\ \psi_2 \\ \psi_3 \\ \psi_4 \end{pmatrix} = \begin{pmatrix} \Psi_1 + i\Phi_1 \\ \Psi_2 + i\Phi_2 \\ \Psi_3 + i\Phi_3 \\ \Psi_4 + i\Phi_4 \end{pmatrix}.$$

In principle, the Euler–Lagrange equation can be solved in terms of these eight fields. However, the eight independent components of the complex Dirac spinor $\psi$ also can be expressed as linear combinations of $\psi$ and the adjoint spinor $\overline{\psi}$. Hence, the *independent* fields can be taken to be the four components the spinor and the four components of the adjoint spinor. The partial derivatives with respect to one of the components of the adjoint spinor are

$$\frac{\partial\mathcal{L}}{\partial(\partial_\mu\overline{\psi}_i)} = 0 \quad \text{and} \quad \frac{\partial\mathcal{L}}{\partial\overline{\psi}_i} = i\gamma^\mu\partial_\mu\psi - m\psi,$$

which when substituted into the Euler–Lagrange equation give

$$-\frac{\partial\mathcal{L}}{\partial\overline{\psi}_i} = 0,$$

and consequently, the spinor field $\psi$ satisfies the Dirac equation,

$$i\gamma^\mu(\partial_\mu\psi) - m\psi = 0.$$

### Relativistic vector fields

Maxwell's equations for the electromagnetic field $A^\mu = (\phi, \mathbf{A})$ can be expressed in a covariant form (see Appendix D.1) as

$$\partial_\mu F^{\mu\nu} = j^\nu,$$

where $F^{\mu\nu}$ is the field-strength tensor,

$$F^{\mu\nu} = \partial^\mu A^\nu - \partial^\nu A^\mu = \begin{pmatrix} 0 & -E_x & -E_y & -E_z \\ E_x & 0 & -B_z & B_y \\ E_y & B_z & 0 & -B_x \\ E_z & -B_y & B_x & 0 \end{pmatrix}, \tag{17.7}$$

and $j = (\rho, \mathbf{J})$ is the four-vector current associated with the charge and current densities $\rho$ and $\mathbf{J}$. The corresponding Lagrangian (see Problem 17.4) is

$$\mathcal{L}_{EM} = -\tfrac{1}{4}F^{\mu\nu}F_{\mu\nu} - j^{\mu}A_{\mu}.$$

In the absence of sources $j^{\mu} = 0$, and the Lagrangian for the free photon field is

$$\mathcal{L}_{EM} = -\tfrac{1}{4}F^{\mu\nu}F_{\mu\nu}. \tag{17.8}$$

Using the form of the field strength tensor of (17.7), this is equivalent to $\mathcal{L}_{EM} = \frac{1}{2}(\mathbf{E}^2 - \mathbf{B}^2)$, from which the corresponding Hamiltonian density $\mathcal{H}_{EM} = \frac{1}{2}(\mathbf{E}^2 + \mathbf{B}^2)$ gives the normal expression for the energy density of an electromagnetic field (in Heaviside–Lorentz units with $\epsilon_0 = \mu_0 = 1$). If the photon had mass, the free-particle Lagrangian of (17.8) would be modified to

$$\mathcal{L}_{\text{Proca}} = -\tfrac{1}{4}F^{\mu\nu}F_{\mu\nu} + \tfrac{1}{2}m_{\gamma}^2 A^{\mu}A_{\mu}, \tag{17.9}$$

which is known as the Proca Lagrangian, from which the field equations for a massive spin-1 particle can be obtained.

### 17.2.3 Noether's theorem

In the following section, the ideas of local gauge invariance are considered in the context of the symmetries of the Lagrangian. Here a simple example is used to illustrate the connection between a symmetry of the Lagrangian and a conservation law. The Lagrangian for a mass $m$ orbiting in the gravitational potential of a fixed body of mass $M$ is

$$\begin{aligned} L = T - V &= \tfrac{1}{2}m\mathrm{v}^2 + \frac{GMm}{r} \\ &= \tfrac{1}{2}m\dot{r}^2 + \tfrac{1}{2}mr^2\dot{\phi}^2 + \frac{GMm}{r}, \end{aligned}$$

where $r$ and $\phi$ are the polar coordinates of the mass $m$ in the plane of the orbit. The Lagrangian does not depend on the polar angle $\phi$ and therefore is invariant under the infinitesimal transformation, $\phi \to \phi' = \phi + \delta\phi$. Since the Lagrangian does not depend on $\phi$, the corresponding Euler–Lagrange equation implies

$$\frac{\mathrm{d}}{\mathrm{d}t}\left(\frac{\partial L}{\partial \dot{\phi}}\right) = 0,$$

and consequently

$$J = \frac{\partial L}{\partial \dot{\phi}} = mr^2\dot{\phi},$$

is a constant of the motion. The rotational symmetry of the Lagrangian therefore implies the existence of a conserved quantity, which in this example is the angular momentum of the orbiting body $m$.

In field theory, Noether's theorem relates a symmetry of the Lagrangian to a conserved current. For example, the Lagrangian for the free Dirac field

$$\mathcal{L} = i\overline{\psi}\gamma^\mu\partial_\mu\psi - m\overline{\psi}\psi, \tag{17.10}$$

is unchanged by the global U(1) phase transformation,

$$\psi \rightarrow \psi' = e^{i\theta}\psi.$$

In Appendix E it is shown that the corresponding conserved current is the usual four-vector current

$$j^\mu = \overline{\psi}\gamma^\mu\psi,$$

which automatically satisfies the continuity equation $\partial_\mu j^\mu = 0$.

## 17.3 Local gauge invariance

In Section 10.1, the electromagnetic interaction was introduced by requiring the Dirac equation to be invariant under a U(1) *local* phase transformation. The required local gauge symmetry is expressed naturally as the invariance of the Lagrangian under a local phase transformation of the fields,

$$\psi(x) \rightarrow \psi'(x) = e^{iq\chi(x)}\psi(x). \tag{17.11}$$

The local nature of the gauge transformation means that the derivatives acting on the field also act on the local phase $\chi(x)$. With this transformation, the Lagrangian for a free spin-half particle of (17.6),

$$\mathcal{L} = i\overline{\psi}\gamma^\mu\partial_\mu\psi - m\overline{\psi}\psi, \tag{17.12}$$

becomes

$$\begin{aligned}\mathcal{L} \rightarrow \mathcal{L}' &= ie^{-iq\chi}\overline{\psi}\gamma^\mu\left[e^{iq\chi}\partial_\mu\psi + iq\left(\partial_\mu\chi\right)e^{iq\chi}\psi\right] - me^{-iq\chi}\overline{\psi}e^{iq\chi}\psi \\ &= \mathcal{L} - q\overline{\psi}\gamma^\mu\left(\partial_\mu\chi\right)\psi.\end{aligned} \tag{17.13}$$

Hence, as it stands, the free-particle Lagrangian for a Dirac field is not invariant under U(1) local phase trasformations. The required gauge invariance can be restored by replacing the derivative $\partial_\mu$ in (17.12) with the *covariant derivative* $D_\mu$,

$$\partial_\mu \rightarrow D_\mu = \partial_\mu + iqA_\mu,$$

where $A_\mu$ is a new field. The desired cancellation of the unwanted $q\overline{\psi}\gamma^\mu(\partial_\mu\chi)\psi$ term in (17.13) is achieved provided the new field transforms as

$$A_\mu \rightarrow A'_\mu = A_\mu - \partial_\mu\chi. \tag{17.14}$$

The required U(1) local gauge invariance of the Lagrangian corresponding to the Dirac equation can be achieved only by the introduction of the field $A_\mu$ with well-defined gauge transformation properties. Hence the gauge-invariant Lagrangian for a spin-half fermion

$$\mathcal{L} = \overline{\psi}(i\gamma^\mu\partial_\mu - m)\psi - q\overline{\psi}\gamma^\mu A_\mu\psi,$$

now contains a term describing the interaction of the fermion with the new field $A_\mu$, which can be identified as the photon. Hence the Lagrangian of QED, describing the fields for the electron (with $q = -e$), the massless photon and the interactions between them can be written as

$$\mathcal{L}_{QED} = \overline{\psi}(i\gamma^\mu\partial_\mu - m_e)\psi + e\overline{\psi}\gamma^\mu\psi A_\mu - \tfrac{1}{4}F_{\mu\nu}F^{\mu\nu}. \tag{17.15}$$

The kinetic term for the massless spin-1 field $F_{\mu\nu}F^{\mu\nu}$ is already invariant under U(1) local phase transformations (see Problem 17.3).

The connection to Maxwell's equations can be made apparent by writing the QED Lagrangian of (17.15) in terms of the four-vector current, $j^\mu = -e\overline{\psi}\gamma^\mu\psi$,

$$\mathcal{L} = \overline{\psi}(i\gamma^\mu\partial_\mu - m_e)\psi - j^\mu A_\mu - \tfrac{1}{4}F_{\mu\nu}F^{\mu\nu}.$$

The Euler–Lagrange equation for the derivatives with respect to the photon field $A^\mu$ gives (see Problem 17.4)

$$\partial_\mu F^{\mu\nu} = j^\nu,$$

which is the covariant form of Maxwell's equations. Hence the whole of electromagnetism can be derived by requiring a local U(1) gauge symmetry of the Lagrangian for a particle satisfying the Dirac equation.

The weak interaction and QCD are respectively obtained by extending the local gauge principle to require that the Lagrangian is invariant under $SU(2)_L$ and SU(3) local phase transformations. The prescription for achieving the required gauge invariance is to replace the four-derivative $\partial_\mu$ with the covariant derivative $D_\mu$ defined in terms of the generators of the group. For example, for the $SU(2)_L$ symmetry of the weak interaction

$$\partial_\mu \to D_\mu = \partial_\mu + ig_W\mathbf{T}\cdot\mathbf{W}_\mu(x),$$

where the $\mathbf{T} = \frac{1}{2}\sigma$ are the three generators of SU(2) and $\mathbf{W}(x)$ are the three new gauge fields. The generators of the SU(2) and SU(3) symmetry groups do not commute and the corresponding local gauge theories are termed non-Abelian. In a non-Abelian gauge theory, the transformation properties of the gauge fields are not independent and additional gauge boson self-interaction terms have to be added to the field-strength tensor for it to be gauge invariant. The focus of this chapter is the Higgs mechanism and therefore the more detailed discussion of non-Abelian gauge theories is deferred to Appendix F.

## 17.4 Particle masses

The local gauge principle provides an elegant description of the interactions in the Standard Model. The success of the Standard Model in describing the experimental data, including the high-precision electroweak measurements, places the local gauge principle on a solid experimental basis. However, the required local gauge invariance of the Standard Model is broken by the terms in the Lagrangian corresponding to particle masses. For example, if the photon were massive, the Lagrangian of QED would contain an additional term $\frac{1}{2}m_\gamma^2 A_\mu A^\mu$,

$$\mathcal{L}_{\text{QED}} \rightarrow \overline{\psi}(i\gamma^\mu\partial_\mu - m_e)\psi + e\overline{\psi}\gamma^\mu A_\mu\psi - \tfrac{1}{4}F_{\mu\nu}F^{\mu\nu} + \tfrac{1}{2}m_\gamma^2 A_\mu A^\mu.$$

For the U(1) local gauge transformation of (17.11), the photon field transforms as

$$A_\mu \rightarrow A'_\mu = A_\mu - \partial_\mu\chi,$$

and the new mass term becomes

$$\tfrac{1}{2}m_\gamma^2 A_\mu A^\mu \rightarrow \tfrac{1}{2}m_\gamma^2 \left(A_\mu - \partial_\mu\chi\right)(A^\mu - \partial^\mu\chi) \neq \tfrac{1}{2}m_\gamma^2 A_\mu A^\mu,$$

from which it is clear that the photon mass term is not gauge invariant. Hence the required U(1) local gauge symmetry can only be satisfied if the gauge boson of an interaction is *massless*. This restriction is not limited to the U(1) local gauge symmetry of QED, it also applies to the $SU(2)_L$ and SU(3) gauge symmetries of the weak interaction and QCD. Whilst the local gauge principle provides an elegant route to describing the nature of the observed interactions, it works only for massless gauge bosons. This is not a problem for QED and QCD where the gauge bosons are massless, but it is in apparent contradiction with the observation of the large masses of W and Z bosons.

The problem with particle masses is not restricted to the gauge bosons. Writing the electron spinor field as $\psi = \text{e}$, the electron mass term in QED Lagrangian can be written in terms of the chiral particle states as

$$\begin{aligned} -m_e\bar{\text{e}}\text{e} &= -m_e\bar{\text{e}}\left[\tfrac{1}{2}(1-\gamma^5) + \tfrac{1}{2}(1+\gamma^5)\right]\text{e} \\ &= -m_e\bar{\text{e}}\left[\tfrac{1}{2}(1-\gamma^5)\text{e}_L + \tfrac{1}{2}(1+\gamma^5)\text{e}_R\right] \\ &= -m_e(\bar{\text{e}}_R\text{e}_L + \bar{\text{e}}_L\text{e}_R). \end{aligned} \tag{17.16}$$

In the $SU(2)_L$ gauge transformation of the weak interaction, left-handed particles transform as weak isospin doublets and right-handed particles as singlets, and therefore the mass term of (17.16) breaks the required gauge invariance.

## 17.5 The Higgs mechanism

In the Standard Model, particles acquire masses through their interactions with the Higgs field. In this section, the Higgs mechanism is developed in three distinct stages. First it is shown how mass terms for a scalar field can arise from a broken symmetry. This mechanism is then extended to show how the mass of a gauge boson can be generated from a broken U(1) local gauge symmetry. Finally, the full Higgs mechanism is developed by breaking the $SU(2)_L \times U(1)_Y$ local gauge symmetry of the electroweak sector of the Standard Model.

### 17.5.1 Interacting scalar fields

A Lagrangian consists of two parts, a kinetic term involving the derivatives of the fields and a potential term expressed in terms of the fields themselves. For example, in the Lagrangian of QED (17.15), the kinetic terms for the electron and photon are

$$i\overline{\psi}\gamma^\mu\partial_\mu\psi \quad \text{and} \quad -\tfrac{1}{4}F_{\mu\nu}F^{\mu\nu}.$$

The potential term, which represents the interactions between the electron and photon fields, is

$$e\overline{\psi}\gamma^\mu\psi A_\mu.$$

This can be associated with the normal three-point interaction vertex of QED, shown on the left of Figure 17.4. In general, the nature of the interactions between the fields and the strength of the coupling is determined by the terms in the Lagrangian involving the combinations of the fields, here $\overline{\psi}\psi A$.

Now, consider a scalar field $\phi$ with the potential

$$V(\phi) = \tfrac{1}{2}\mu^2\phi^2 + \tfrac{1}{4}\lambda\phi^4. \tag{17.17}$$

The corresponding Lagrangian is given by

$$\begin{aligned}\mathcal{L} &= \tfrac{1}{2}(\partial_\mu\phi)(\partial^\mu\phi) - V(\phi)\\ &= \tfrac{1}{2}(\partial_\mu\phi)(\partial^\mu\phi) - \tfrac{1}{2}\mu^2\phi^2 - \tfrac{1}{4}\lambda\phi^4.\end{aligned} \tag{17.18}$$

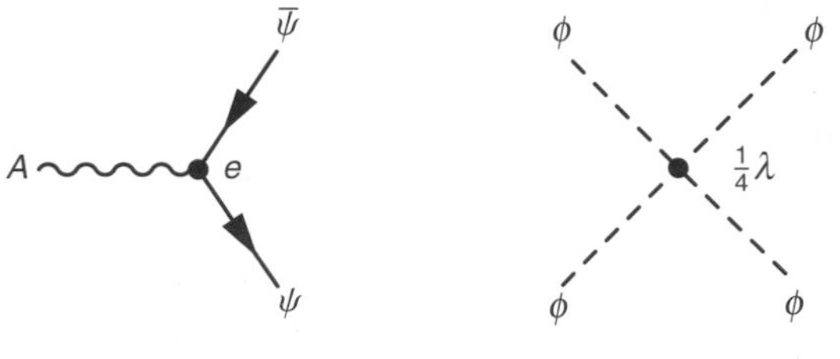


Fig. 17.4 The three-point interaction of QED and the four-point interaction for a scalar field with the potential $\lambda\phi^4/4$.

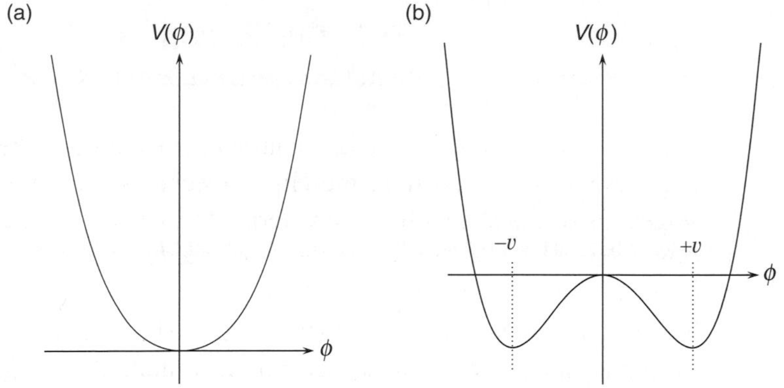


Fig. 17.5 The one-dimensional potential $V(\phi) = \mu^2\phi^2/2 + \lambda\phi^4/4$ for $\lambda > 0$ and the cases where (a) $\mu^2 > 0$ and (b) $\mu^2 < 0$.

The term proportional to $(\partial_\mu\phi)(\partial^\mu\phi)$ can be associated with the kinetic energy of the scalar particle and $\frac{1}{2}\mu^2\phi^2$ can represent the mass of the particle. The $\phi^4$ term can be identified as self-interactions of the scalar field, corresponding to the four-point interaction vertex shown in the right-hand plot of Figure 17.4.

The vacuum state is the lowest energy state of the field $\phi$ and corresponds to the minimum of the potential of (17.17). For the potential to have a finite minimum, $\lambda$ must be positive. If $\mu^2$ is also chosen to be positive, the resulting potential, shown in Figure 17.5a, has a minimum at $\phi = 0$. In this case, the vacuum state corresponds to the field $\phi$ being zero and the Lagrangian of (17.18) represents a scalar particle with mass $\mu$ and a four-point self-interaction term proportional to $\phi^4$. However, whilst $\lambda$ must be greater than zero for there to be a finite minimum, there is no such restriction for $\mu^2$. If $\mu^2 < 0$, the associated term in the Lagrangian can no longer be interpreted as a mass and the potential of (17.17) has minima at

$$\phi = \pm v = \pm \left| \sqrt{\frac{-\mu^2}{\lambda}} \right|,$$

as shown in Figure 17.5b. For $\mu^2 < 0$, the lowest energy state does not occur at $\phi = 0$ and the field is said to have a non-zero vacuum expectation value $v$. Since the potential is symmetric, there are two degenerate possible vacuum states. The actual vacuum state of the field either will be $\phi = +v$ or $\phi = -v$. The choice of the vacuum state breaks the symmetry of the Lagrangian, a process known as *spontaneous symmetry breaking*. A familiar example of spontaneous symmetry breaking is a ferromagnet with magnetisation $\mathbf{M}$. The Lagrangian (or Hamiltonian) depends on $\mathbf{M}^2$ and has no preferred direction. However, below the Curie temperature, the spins will be aligned in a particular direction, spontaneously breaking the underlying rotational symmetry of the Lagrangian.

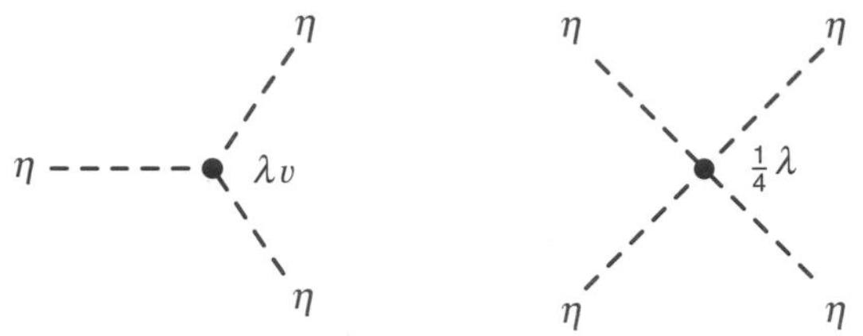


Fig. 17.6 The self-interactions of the field $\eta$ that lead to Feynman diagrams for the processes $\eta \to \eta\eta$ and $\eta\eta \to \eta\eta$.

If the vacuum state of the scalar field is chosen to be at $\phi = +v$, the excitations of the field, which describe the particle states, can be obtained by considering perturbations of the field $\phi$ around the vacuum state by writing $\phi(x) = v + \eta(x)$. Since the vacuum expectation value $v$ is a constant, $\partial_\mu \phi = \partial_\mu \eta$ and the Lagrangian of (17.18), expressed in terms of the field $\eta$, is

$$\begin{aligned}\mathcal{L}(\eta) &= \tfrac{1}{2}(\partial_\mu \eta)(\partial^\mu \eta) - V(\eta) \\ &= \tfrac{1}{2}(\partial_\mu \eta)(\partial^\mu \eta) - \tfrac{1}{2}\mu^2(v+\eta)^2 - \tfrac{1}{4}\lambda(v+\eta)^4.\end{aligned}$$

Since the minimum of the potential is given by $\mu^2 = -\lambda v^2$, this expression can be written as

$$\mathcal{L}(\eta) = \tfrac{1}{2}(\partial_\mu \eta)(\partial^\mu \eta) - \lambda v^2 \eta^2 - \lambda v \eta^3 - \tfrac{1}{4}\lambda \eta^4 + \tfrac{1}{4}\lambda v^4. \tag{17.19}$$

From the comparison with the Lagrangian for a free scalar field of (17.5), it can be seen that the term proportional to $\eta^2$ can be interpreted as a mass

$$m_\eta = \sqrt{2\lambda v^2} = \sqrt{-2\mu^2},$$

and therefore the Lagrangian of (17.19) describes a massive scalar field. The terms proportional to $\eta^3$ and $\eta^4$ can be identified as triple and quartic interaction terms, as indicated in Figure 17.6. Finally, the term $\lambda v^4/4$ is just a constant, and has no physical consequences. Hence after spontaneous symmetry breaking, and having expanded the field about the vacuum state, the Lagrangian can be written as

$$\mathcal{L}(\eta) = \tfrac{1}{2}(\partial_\mu \eta)(\partial^\mu \eta) - \tfrac{1}{2}m_\eta^2 \eta^2 - V(\eta), \quad \text{with} \quad V(\eta) = \lambda v \eta^3 + \tfrac{1}{4}\lambda \eta^4. \tag{17.20}$$

It is important to realise that the Lagrangian of (17.20) is the same as the original Lagrangian of (17.18), but is now expressed as excitations about the minimum at $\phi = +v$. In principle, the same physical predictions can be obtained by using either form. However, in order to use perturbation theory, it is necessary to express the fields as small perturbations about the vacuum state.

## 17.5.2 Symmetry breaking for a complex scalar field

The idea of spontaneous symmetry breaking, introduced above in the context of a real scalar field, can be applied to the *complex* scalar field,

$$\phi = \frac{1}{\sqrt{2}}(\phi_1 + i\phi_2),$$

for which the corresponding Lagrangian is

$$\mathcal{L} = (\partial_\mu\phi)^*(\partial^\mu\phi) - V(\phi) \quad \text{with} \quad V(\phi) = \mu^2(\phi^*\phi) + \lambda(\phi^*\phi)^2. \tag{17.21}$$

When expressed in terms of the two (real) scalar fields $\phi_1$ and $\phi_2$ this is just

$$\mathcal{L} = \tfrac{1}{2}(\partial_\mu\phi_1)(\partial^\mu\phi_1) + \tfrac{1}{2}(\partial_\mu\phi_2)(\partial^\mu\phi_2) - \tfrac{1}{2}\mu^2(\phi_1^2 + \phi_2^2) - \tfrac{1}{4}\lambda(\phi_1^2 + \phi_2^2)^2. \tag{17.22}$$

As before, for the potential to have a finite minimum, $\lambda > 0$. The Lagrangian of (17.21) is invariant under the transformation $\phi \to \phi' = e^{i\alpha}\phi$, because $\phi'^*\phi' = \phi^*\phi$, and therefore possesses a *global* U(1) symmetry. The shape of the potential depends on the sign of $\mu^2$, as shown in Figure 17.7. When $\mu^2 > 0$, the minimum of the potential occurs when both fields are zero. If $\mu^2 < 0$, the potential has an infinite set of minima defined by

$$\phi_1^2 + \phi_2^2 = \frac{-\mu^2}{\lambda} = v^2,$$

as indicated by the dashed circle in Figure 17.7. The physical vacuum state will correspond to a particular point on this circle, breaking the *global* U(1) symmetry of the Lagrangian. Without loss of generality, the vacuum state can be chosen to

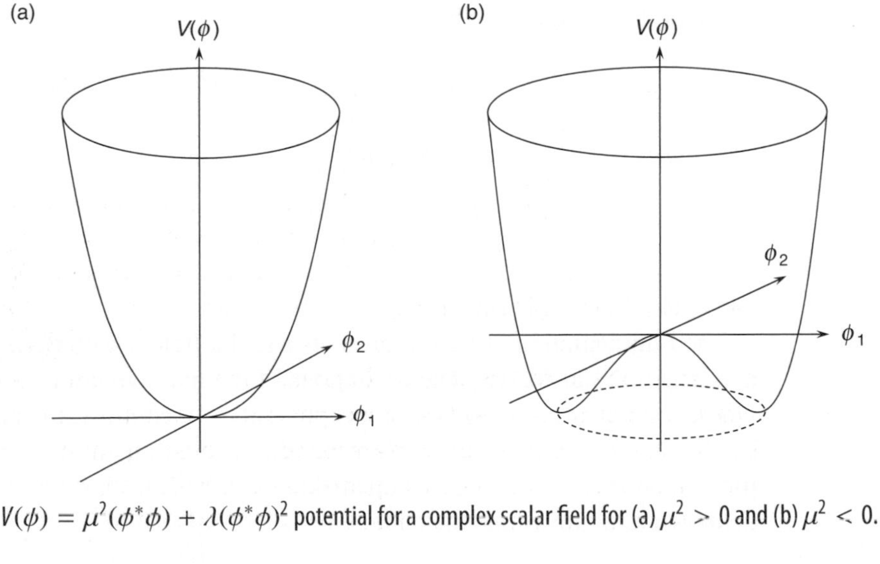


Fig. 17.7 The $V(\phi) = \mu^2(\phi^*\phi) + \lambda(\phi^*\phi)^2$ potential for a complex scalar field for (a) $\mu^2 > 0$ and (b) $\mu^2 < 0$.

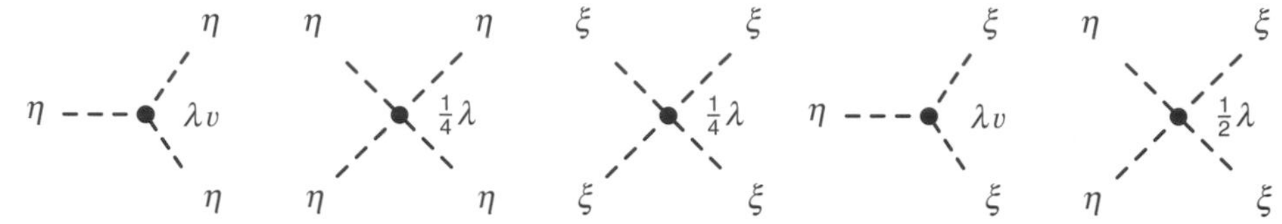


Fig. 17.8 The scalar interactions obtained by breaking the symmetry for a complex scalar field.

be in the real direction, $(\phi_1, \phi_2) = (v, 0)$, and the complex scaler field $\phi$ can be expanded about the vacuum state by writing $\phi_1(x) = \eta(x) + v$ and $\phi_2(x) = \xi(x)$,

$$\phi = \tfrac{1}{\sqrt{2}}(\eta + v + i\xi).$$

The Lagrangian of (17.22), written in terms of the fields $\eta$ and $\xi$, is

$$\mathcal{L} = \tfrac{1}{2}(\partial_\mu\eta)(\partial^\mu\eta) + \tfrac{1}{2}(\partial_\mu\xi)(\partial^\mu\xi) - V(\eta, \xi),$$

where the potential $V(\eta, \xi)$ is given by

$$V(\eta, \xi) = \mu^2\phi^2 + \lambda\phi^4 \quad \text{with} \quad \phi^2 = \phi\phi^* = \tfrac{1}{2}\left[(v + \eta)^2 + \xi^2\right].$$

The potential can be written in terms of the fields $\eta$ and $\xi$ using $\mu^2 = -\lambda v^2$,

$$\begin{aligned} V(\eta, \xi) &= \mu^2\phi^2 + \lambda\phi^4 \\ &= -\tfrac{1}{2}\lambda v^2 \left\{(v + \eta)^2 + \xi^2\right\} + \tfrac{1}{4}\lambda\left\{(v + \eta)^2 + \xi^2\right\}^2 \\ &= -\tfrac{1}{4}\lambda v^4 + \lambda v^2\eta^2 + \lambda v\eta^3 + \tfrac{1}{4}\lambda\eta^4 + \tfrac{1}{4}\lambda\xi^4 + \lambda v\eta\xi^2 + \tfrac{1}{2}\lambda\eta^2\xi^2. \end{aligned}$$

The term which is quadratic in the field $\eta$ can be identified as a mass, and the terms with either three or four powers of the fields can be identified as interaction terms. Thus the Lagrangian can be written as

$$\mathcal{L} = \tfrac{1}{2}(\partial_\mu\eta)(\partial^\mu\eta) - \tfrac{1}{2}m_\eta^2\eta^2 + \tfrac{1}{2}(\partial_\mu\xi)(\partial^\mu\xi) - V_{int}(\eta, \xi), \tag{17.23}$$

with $m_\eta = \sqrt{2\lambda v^2}$ and interactions given by

$$V_{int}(\eta, \xi) = \lambda v\eta^3 + \tfrac{1}{4}\lambda\eta^4 + \tfrac{1}{4}\lambda\xi^4 + \lambda v\eta\xi^2 + \tfrac{1}{2}\lambda\eta^2\xi^2. \tag{17.24}$$

These interaction terms correspond to triple and quartic couplings of the fields $\eta$ and $\xi$, as shown in Figure 17.8.

The Lagrangian of (17.23) represents a scalar field $\eta$ with mass $m_\eta = \sqrt{2\lambda v^2}$ and a *massless* scalar field $\xi$. The excitations of the massive field $\eta$ are in the direction where the potential is (to first order) quadratic. In contrast, the particles described by the massless scalar field $\xi$ correspond to excitations in the direction where the potential does not change, as indicated in Figure 17.9. This massless scalar particle is known as a Goldstone boson.

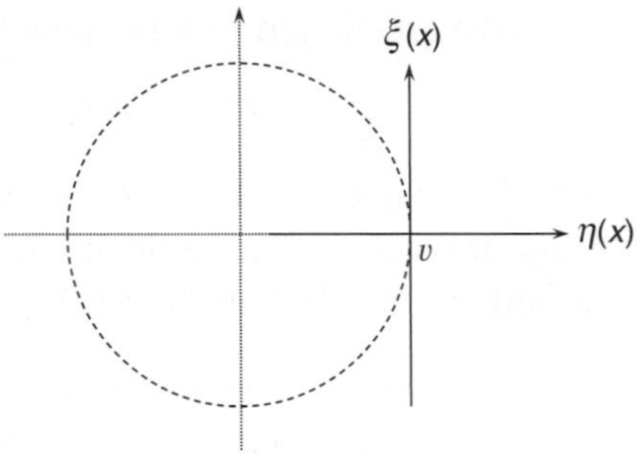


Fig. 17.9 The fields $\eta(x)$ and $\xi(x)$ in terms of the vacuum expectation value at $\phi = (v, 0)$.

### 17.5.3 The Higgs mechanism

In the Higgs mechanism, the spontaneous symmetry breaking of a complex scalar field with a potential,

$$V(\phi) = \mu^2\phi^2 + \lambda\phi^4, \tag{17.25}$$

is embedded in a theory with a *local* gauge symmetry. In this section, the example of a U(1) local gauge symmetry is used to introduce the main ideas.

Because of the presence of the derivatives in (17.21), the Lagrangian for a complex scalar field $\phi$ is *not* invariant under the U(1) *local* gauge transformation

$$\phi(x) \rightarrow \phi'(x) = e^{ig\chi(x)}\phi(x). \tag{17.26}$$

The required U(1) local gauge invariance can be achieved by replacing the derivatives in the Lagrangian with the corresponding covariant derivatives

$$\partial_\mu \rightarrow D_\mu = \partial_\mu + igB_\mu.$$

The resulting Lagrangian,

$$\mathcal{L} = (D_\mu\phi)^*(D^\mu\phi) - V(\phi^2),$$

is gauge invariant (see Problem 17.7) provided the new gauge field $B_\mu$, which appears in the covariant derivative, transforms as

$$B_\mu \rightarrow B'_\mu = B_\mu - \partial_\mu\chi(x). \tag{17.27}$$

Just as was the case for Dirac Lagrangian (see Section 17.3), the required local gauge invariance implies the existence a new gauge field with well-defined gauge transformation properties. The combined Lagrangian for the complex scalar field $\phi$ and the gauge field $B$ is

$$\mathcal{L} = -\tfrac{1}{4}F^{\mu\nu}F_{\mu\nu} + (D_\mu\phi)^*(D^\mu\phi) - \mu^2\phi^2 - \lambda\phi^4, \tag{17.28}$$

where $F^{\mu\nu}F_{\mu\nu}$ is the kinetic term for the new field with

$$F^{\mu\nu} = \partial^\mu B^\nu - \partial^\nu B^\mu.$$

The gauge field $B$ is required to be massless, since the mass term $\frac{1}{2}m_B B_\mu B^\mu$ would break the gauge invariance. The term involving the covariant derivatives, when written out in full, is

$$\begin{aligned}(D_\mu\phi)^*(D^\mu\phi) &= (\partial_\mu - igB_\mu)\phi^*(\partial^\mu + igB^\mu)\phi \\ &= (\partial_\mu\phi)^*(\partial^\mu\phi) - igB_\mu\phi^*(\partial^\mu\phi) + ig(\partial_\mu\phi^*)B^\mu\phi + g^2B_\mu B^\mu\phi^*\phi\end{aligned}$$

and the full expression for the Lagrangian is

$$\begin{aligned}\mathcal{L} = &-\tfrac{1}{4}F^{\mu\nu}F_{\mu\nu} + (\partial_\mu\phi)^*(\partial^\mu\phi) - \mu^2\phi^2 - \lambda\phi^4 \\ &- igB_\mu\phi^*(\partial^\mu\phi) + ig(\partial_\mu\phi^*)B^\mu\phi + g^2B_\mu B^\mu\phi^*\phi.\end{aligned} \tag{17.29}$$

For the case where the potential for the scalar field of (17.25) has $\mu^2 < 0$, the vacuum state is degenerate and the choice of the physical vacuum state spontaneously breaks the symmetry of the Lagrangian of (17.29). As before, the physical vacuum state is chosen to be $\phi_1 + i\phi_2 = v$, and the complex scalar field $\phi$ is expanded about the vacuum state by writing

$$\phi(x) = \tfrac{1}{\sqrt{2}}(v + \eta(x) + i\xi(x)). \tag{17.30}$$

Substituting (17.30) into (17.29) leads to (see Problem 17.6)

$$\mathcal{L} = \underbrace{\tfrac{1}{2}(\partial_\mu\eta)(\partial^\mu\eta) - \lambda v^2\eta^2}_{\text{massive }\eta} + \underbrace{\tfrac{1}{2}(\partial_\mu\xi)(\partial^\mu\xi)}_{\text{massless }\xi} - \underbrace{\tfrac{1}{4}F_{\mu\nu}F^{\mu\nu} + \tfrac{1}{2}g^2v^2B_\mu B^\mu}_{\text{massive gauge field}} - V_{int} + gvB_\mu(\partial^\mu\xi), \tag{17.31}$$

where $V_{int}(\eta, \xi, B)$ contains the three- and four-point interaction terms of the fields $\eta$, $\xi$ and $B$. As before, the breaking of the symmetry of the Lagrangian produces a massive scalar field $\eta$ and a massless Goldstone boson $\xi$. In addition, the previously massless gauge field $B$ has acquired a mass term $\frac{1}{2}g^2v^2B_\mu B^\mu$, achieving the aim of giving a mass to the gauge boson of the local gauge symmetry. Again it should be emphasised that this is exactly the same Lagrangian as (17.28), but with the complex scalar field expanded about the vacuum state at $\phi_1 + i\phi_2 = v$; by expanding the scalar fields about the vacuum where the fields have a non-zero vacuum expectation value, the underlying gauge symmetry of the Lagrangian has been hidden, but has not been removed.

However, there appear to be two problems with (17.31). The original Lagrangian contained four degrees of freedom, one for each of the scalar fields $\phi_1$ and $\phi_2$, and the two transverse polarisation states for the massless gauge field $B$. In the Lagrangian of (17.31), the gauge boson has become massive and therefore has the additional longitudinal polarisation state; somehow in the process of spontaneous symmetry breaking an additional degree of freedom appears to have been

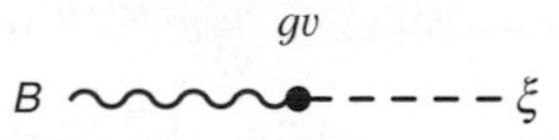


Fig. 17.10 The coupling between the gauge field $B$ and the Goldstone field $\xi$.

introduced. Furthermore, the $gvB_\mu(\partial^\mu\xi)$ term appears to represent a direct coupling between the Goldstone field $\xi$ and the gauge field $B$. It would appear that the spin-1 gauge field can transform into a spin-0 scalar field, as indicated in Figure 17.10. This term is somewhat reminiscent of the off-diagonal mass term encountered in the discussion of the neutral kaon system, which coupled the $K^0$ and $\overline{K}^0$ flavour states, suggesting that the fields appearing in (17.31) are not the physical fields. This coupling to the Goldstone field ultimately will be associated with the longitudinal polarisation state of the massive gauge boson.

The Goldstone field $\xi$ in (17.31) can be eliminated from the Lagrangian by making the appropriate gauge transformation. By writing

$$\tfrac{1}{2}(\partial_\mu\xi)(\partial^\mu\xi) + gvB_\mu(\partial^\mu\xi) + \tfrac{1}{2}g^2v^2B_\mu B^\mu = \tfrac{1}{2}g^2v^2\left[B_\mu + \frac{1}{gv}(\partial_\mu\xi)\right]^2,$$

and making the gauge transformation,

$$B_\mu(x) \rightarrow B'_\mu(x) = B_\mu(x) + \frac{1}{gv}\partial_\mu\xi(x), \tag{17.32}$$

the Lagrangian of (17.31) becomes

$$\mathcal{L} = \underbrace{\tfrac{1}{2}(\partial^\mu\eta)(\partial_\mu\eta) - \lambda v^2\eta^2}_{\text{massive }\eta} + \underbrace{-\tfrac{1}{4}F_{\mu\nu}F^{\mu\nu} + \tfrac{1}{2}g^2v^2B^{\mu\prime}B'_\mu}_{\text{massive gauge field}} - V_{int}.$$

Since the original Lagrangian was constructed to be invariant under local U(1) gauge transformations, the physical predications of the theory are unchanged by the gauge transformation of (17.32).

Thus, with the appropriate choice of gauge, the Goldstone field $\xi$ no longer appears in the Lagrangian. This choice of gauge corresponds to taking $\chi(x) = -\xi(x)/gv$ in (17.27). The corresponding gauge transformation of the original complex scalar field $\phi(x)$ is therefore

$$\phi(x) \rightarrow \phi'(x) = e^{-ig\frac{\xi(x)}{gv}}\phi(x) = e^{-i\xi(x)/v}\phi(x). \tag{17.33}$$

After symmetry breaking, the complex scalar field was expanded about the physical vacuum by writing $\phi(x) = \frac{1}{\sqrt{2}}(v + \eta(x) + i\xi(x))$, which to first order in the fields can be expressed as

$$\phi(x) \approx \frac{1}{\sqrt{2}}\left[v + \eta(x)\right]e^{i\xi(x)/v}.$$

The effect of the gauge transformation of (17.33) on the original complex scalar field is therefore

$$\phi(x) \to \phi'(x) = \frac{1}{\sqrt{2}} e^{-i\xi(x)/v} \left[v + \eta(x)\right] e^{i\xi(x)/v} = \frac{1}{\sqrt{2}}(v + \eta(x)).$$

Hence, the gauge in which the Goldstone field $\xi(x)$ is eliminated from the Lagrangian, which is known as the *Unitary gauge*, corresponds to choosing the complex scalar field $\phi(x)$ to be entirely real,

$$\phi(x) = \frac{1}{\sqrt{2}}(v + \eta(x)) \equiv \frac{1}{\sqrt{2}}(v + h(x)).$$

Here the field $\eta(x)$ has been written as the Higgs field $h(x)$ to emphasise that this is the physical field in the unitary gauge. It is important to remember that the physical predictions of the theory do not depend on the choice of gauge, but in the unitary gauge the fields appearing in the Lagrangian correspond to the physical particles; there are no "mixing" terms like $B_\mu(\partial^\mu \xi)$. The degree of freedom corresponding to the Goldstone field $\xi(x)$ no longer appears in the Lagrangian; it has been replaced by the degree of freedom corresponding to the longitudinal polarisation state of the now massive gauge field $B$. Sometimes it is said that the Goldstone boson has been "eaten" by the gauge field. Writing $\mu^2 = -\lambda v^2$, and working in the unitary gauge where $\phi(x) = \frac{1}{\sqrt{2}}(v + h(x))$, the Lagrangian of (17.28) can be written

$$\begin{aligned}
\mathcal{L} &= (D_\mu\phi)^*(D^\mu\phi) - \tfrac{1}{4}F_{\mu\nu}F^{\mu\nu} - \mu^2\phi^2 - \lambda\phi^4 \\
&= \tfrac{1}{2}(\partial_\mu - igB_\mu)(v+h)(\partial^\mu + igB^\mu)(v+h) - \tfrac{1}{4}F_{\mu\nu}F^{\mu\nu} - \tfrac{1}{2}\mu^2(v+h)^2 - \tfrac{1}{4}\lambda(v+h)^4 \\
&= \tfrac{1}{2}(\partial_\mu h)(\partial^\mu h) + \tfrac{1}{2}g^2 B_\mu B^\mu (v+h)^2 - \tfrac{1}{4}F_{\mu\nu}F^{\mu\nu} - \lambda v^2 h^2 - \lambda v h^3 - \tfrac{1}{4}\lambda h^4 + \tfrac{1}{4}\lambda v^4.
\end{aligned}$$

Gathering up the terms (and ignoring the $\lambda v^4/4$ constant) gives

$$\begin{aligned}
\mathcal{L} = \underbrace{\tfrac{1}{2}(\partial_\mu h)(\partial^\mu h) - \lambda v^2 h^2}_{\text{massive } h \text{ scalar}} &- \underbrace{\tfrac{1}{4}F_{\mu\nu}F^{\mu\nu} + \tfrac{1}{2}g^2v^2 B_\mu B^\mu}_{\text{massive gauge boson}} \\
+ \underbrace{g^2 v B_\mu B^\mu h + \tfrac{1}{2}g^2 B_\mu B^\mu h^2}_{h,B \text{ interactions}} &- \underbrace{\lambda v h^3 - \tfrac{1}{4}\lambda h^4}_{h \text{ self-interactions}}. 
\end{aligned} \tag{17.34}$$

This Lagrangian describes a massive scalar Higgs field $h$ and a massive gauge boson $B$ associated with the U(1) local gauge symmetry. It contains interaction terms between the Higgs boson and the gauge boson, and Higgs boson self-interaction terms, indicated in Figure 17.11. The mass of the gauge boson,

$$m_B = g\,v,$$

is related to the strength of the gauge coupling and the vacuum expectation value of the Higgs field. The mass of the Higgs boson is given by

$$m_H = \sqrt{2\lambda}\,v.$$

It should be noted that the vacuum expectation value $v$ sets the scale for the masses of both the gauge boson and the Higgs boson.

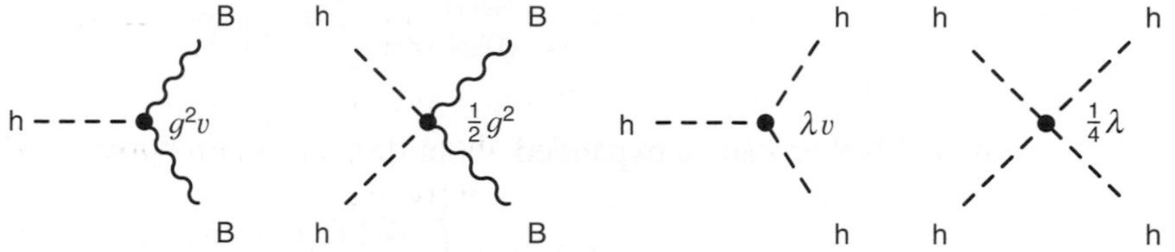


Fig. 17.11 The interaction terms arising from the Higgs mechanism for a U(1) local gauge theory.

### 17.5.4 The Standard Model Higgs

In the above example, the Higgs mechanism was used to generate a mass for the gauge boson corresponding to a U(1) local gauge symmetry. In the Salam–Weinberg model, the Higgs mechanism is embedded in the $U(1)_Y \times SU(2)_L$ local gauge symmetry of the electroweak sector of the Standard Model. Three Goldstone bosons will be required to provide the longitudinal degrees of freedom of the $W^+$, $W^-$ and Z bosons. In addition, after symmetry breaking, there will be (at least) one massive scalar particle corresponding to the field excitations in the direction picked out by the choice of the physical vacuum. The simplest Higgs model, which has the necessary four degrees of freedom, consists of two complex scalar fields.

Because the Higgs mechanism is required to generate the masses of the electroweak gauge bosons, one of the scalar fields must be neutral, written as $\phi^0$, and the other must be charged such that $\phi^+$ and $(\phi^+)^* = \phi^-$ give the longitudinal degrees of freedom of the $W^+$ and $W^-$. The minimal Higgs model consists of two complex scalar fields, placed in a weak isospin doublet

$$\phi = \begin{pmatrix} \phi^+ \\ \phi^0 \end{pmatrix} = \frac{1}{\sqrt{2}} \begin{pmatrix} \phi_1 + i\phi_2 \\ \phi_3 + i\phi_4 \end{pmatrix}. \tag{17.35}$$

As usual, the upper and lower components of the doublet differ by one unit of charge. The Lagrangian for this doublet of complex scalar fields is

$$\mathcal{L} = (\partial_\mu \phi)^\dagger (\partial^\mu \phi) - V(\phi), \tag{17.36}$$

with the Higgs potential,

$$V(\phi) = \mu^2 \phi^\dagger \phi + \lambda(\phi^\dagger \phi)^2.$$

For $\mu^2 < 0$, the potential has an infinite set of degenerate minima satisfying

$$\phi^\dagger \phi = \tfrac{1}{2}(\phi_1^2 + \phi_2^2 + \phi_3^2 + \phi_4^2) = \frac{v^2}{2} = -\frac{\mu^2}{2\lambda}.$$

After symmetry breaking, the neutral photon is required to remain massless, and therefore the minimum of the potential must correspond to a non-zero vacuum expectation value only of the neutral scalar field $\phi^0$,

$$\langle 0|\phi|0\rangle = \frac{1}{\sqrt{2}}\begin{pmatrix} 0 \\ v \end{pmatrix}.$$

The fields then can be expanded about this minimum by writing

$$\phi(x) = \frac{1}{\sqrt{2}}\begin{pmatrix} \phi_1(x) + i\phi_2(x) \\ v + \eta(x) + i\phi_4(x) \end{pmatrix}.$$

After the spontaneous breaking of the symmetry, there will be a massive scalar and three massless Goldstone bosons, which will ultimately give the longitudinal degrees of freedom of the $W^{\pm}$ and Z bosons. Rather than repeating the derivation given in Section 17.5.3 and "gauging-away" the Goldstone fields, here the Higgs doublet is immediately written in the unitary gauge,

$$\phi(x) = \frac{1}{\sqrt{2}}\begin{pmatrix} 0 \\ v + h(x) \end{pmatrix}.$$

The resulting Lagrangian is known as the Salam–Weinberg model. All that remains is to identify the masses of gauge bosons and the interaction terms.

The mass terms can be identified by writing the Lagrangian of (17.36) such that it respects the $SU(2)_L \times U(1)_Y$ local gauge symmetry of the electroweak model by replacing the derivatives with the appropriate covariant derivatives (discussed further in Appendix F),

$$\partial_\mu \to D_\mu = \partial_\mu + ig_W \mathbf{T} \cdot \mathbf{W}_\mu + ig' \frac{Y}{2} B_\mu, \tag{17.37}$$

where $\mathbf{T} = \frac{1}{2}\boldsymbol{\sigma}$ are the three generators of the SU(2) symmetry. In Chapter 15, the weak hypercharge of the Glashow–Salam–Weinberg (GSW) model was identified as $Y = 2\left(Q - I_W^{(3)}\right)$. Here, the lower component of the Higgs doublet is neutral and has $I_W^{(3)} = -\frac{1}{2}$, and thus the Higgs doublet has hypercharge $Y = 1$. Hence, the effect of the covariant derivative of (17.37) acting on the Higgs doublet $\phi$ is

$$D_\mu \phi = \frac{1}{2}\left[2\partial_\mu + \left(ig_W \boldsymbol{\sigma} \cdot \mathbf{W}_\mu + ig' B_\mu\right)\right]\phi,$$

where $D_\mu$ is a $2 \times 2$ matrix acting on the two component weak isospin doublet and the identity matrix multiplying the $\partial_\mu$ and $B_\mu$ terms is implicit in this expression.

The term in the Lagrangian that generates the masses of the gauge bosons is $(D_\mu \phi)^\dagger (D^\mu \phi)$. In the Unitary gauge $D_\mu \phi$ is given by

$$\begin{aligned} D_\mu \phi &= \frac{1}{2\sqrt{2}}\begin{pmatrix} 2\partial_\mu + ig_W W_\mu^{(3)} + ig' B_\mu & ig_W[W_\mu^{(1)} - iW_\mu^{(2)}] \\ ig_W[W_\mu^{(1)} + iW_\mu^{(2)}] & 2\partial_\mu - ig_W W_\mu^{(3)} + ig' B_\mu \end{pmatrix}\begin{pmatrix} 0 \\ v + h \end{pmatrix} \\ &= \frac{1}{2\sqrt{2}}\begin{pmatrix} ig_W(W_\mu^{(1)} - iW_\mu^{(2)})(v + h) \\ (2\partial_\mu - ig_W W_\mu^{(3)} + ig' B_\mu)(v + h) \end{pmatrix}. \end{aligned}$$

Taking the Hermitian conjugate gives $(D_\mu\phi)^\dagger$, from which

$$(D_\mu\phi)^\dagger(D^\mu\phi) = \tfrac{1}{2}(\partial_\mu h)(\partial^\mu h) + \tfrac{1}{8}g_W^2(W_\mu^{(1)} + iW_\mu^{(2)})(W^{(1)\mu} - iW^{(2)\mu})(v+h)^2 + \tfrac{1}{8}(g_W W_\mu^{(3)} - g'B_\mu)(g_W W^{(3)\mu} - g'B^\mu)(v+h)^2. \quad (17.38)$$

The gauge bosons masses are determined by the terms in $(D_\mu\phi)^\dagger(D^\mu\phi)$ that are quadratic in the gauge boson fields, i.e.

$$\tfrac{1}{8}v^2 g_W^2\left(W_\mu^{(1)}W^{(1)\mu} + W_\mu^{(2)}W^{(2)\mu}\right) + \tfrac{1}{8}v^2\left(g_W W_\mu^{(3)} - g'B_\mu\right)\left(g_W W^{(3)\mu} - g'B^\mu\right).$$

In the Lagrangian, the mass terms for the $W^{(1)}$ and $W^{(2)}$ spin-1 fields will appear as

$$\tfrac{1}{2}m_W^2 W_\mu^{(1)}W^{(1)\mu} \quad \text{and} \quad \tfrac{1}{2}m_W^2 W_\mu^{(2)}W^{(2)\mu},$$

and therefore the mass of the W boson is

$$m_W = \tfrac{1}{2}g_W v. \quad (17.39)$$

The mass of the W boson is therefore determined by the coupling constant of the $SU(2)_L$ gauge interaction $g_W$ and the vacuum expectation value of the Higgs field.

The terms in the Lagrangian of (17.38) which are quadratic in the neutral $W^{(3)}$ and $B$ fields can be written as

$$\frac{v^2}{8}\left(g_W W_\mu^{(3)} - g'B_\mu\right)\left(g_W W^{(3)\mu} - g'B^\mu\right) = \frac{v^2}{8}\begin{pmatrix} W_\mu^{(3)} & B_\mu \end{pmatrix}\begin{pmatrix} g_W^2 & -g_W g' \\ -g_W g' & g'^2 \end{pmatrix}\begin{pmatrix} W^{(3)\mu} \\ B^\mu \end{pmatrix}$$
$$= \frac{v^2}{8}\begin{pmatrix} W_\mu^{(3)} & B_\mu \end{pmatrix}\mathbf{M}\begin{pmatrix} W^{(3)\mu} \\ B^\mu \end{pmatrix}, \quad (17.40)$$

where $\mathbf{M}$ is the *non-diagonal* mass matrix. The off-diagonal elements of $\mathbf{M}$ couple together the $W^{(3)}$ and $B$ fields, allowing them to mix. Again this is reminiscent of the non-diagonal mass matrix encountered in the discussion of the neutral kaon system (see Section 14.4.3). The *physical* boson fields, which propagate as independent eigenstates of the free particle Hamiltonian, correspond to the basis in which the mass matrix is diagonal. The masses of the physical gauge bosons are given by the eigenvalues of $\mathbf{M}$, obtained from characteristic equation $\det(\mathbf{M} - \lambda I) = 0$,

$$(g_W^2 - \lambda)(g'^2 - \lambda) - g_W^2 g'^2 = 0,$$

giving

$$\lambda = 0 \quad \text{or} \quad \lambda = g_W^2 + g'^2. \quad (17.41)$$

Hence, in the diagonal basis the mass matrix of (17.40) is

$$\frac{1}{8}v^2\begin{pmatrix} A_\mu & Z_\mu \end{pmatrix}\begin{pmatrix} 0 & 0 \\ 0 & g_W^2 + g'^2 \end{pmatrix}\begin{pmatrix} A^\mu \\ Z^\mu \end{pmatrix},$$

where the $A_\mu$ and $Z_\mu$ are the physical fields corresponding to the eigenvectors of **M**. In the diagonal basis, the term in the Lagrangian representing the masses of the $A$ and $Z$ will be

$$\tfrac{1}{2}\begin{pmatrix} A_\mu & Z_\mu \end{pmatrix}\begin{pmatrix} m_A^2 & 0 \\ 0 & m_Z^2 \end{pmatrix}\begin{pmatrix} A^\mu \\ Z^\mu \end{pmatrix},$$

from which the masses of the physical gauge bosons can be identified as

$$m_A = 0 \quad \text{and} \quad m_Z = \tfrac{1}{2}v\sqrt{g_W^2 + g'^2}. \tag{17.42}$$

Hence, in the physical basis there is a massless neutral gauge boson $A$ which can be identified as the photon and a massive neutral gauge boson which can be identified as the Z. The physical fields, which correspond to the normalised eigenvectors of the mass matrix, are

$$A_\mu = \frac{g' W_\mu^{(3)} + g_W B_\mu}{\sqrt{g_W^2 + g'^2}} \quad \text{with} \quad m_A = 0, \tag{17.43}$$

$$Z_\mu = \frac{g_W W_\mu^{(3)} - g' B_\mu}{\sqrt{g_W^2 + g'^2}} \quad \text{with} \quad m_Z = \tfrac{1}{2}v\sqrt{g_W^2 + g'^2}. \tag{17.44}$$

Thus, the physical fields are mixtures of the massless bosons associated with the $U(1)_Y$ and $SU(2)_L$ local gauge symmetries. The combination corresponding to the Z boson, which is associated with the neutral Goldstone boson of the broken symmetry, has acquired mass through the Higgs mechanism and the field corresponding to the photon has remained massless. By writing the ratio of the couplings of the $U(1)_Y$ and $SU(2)_L$ gauge symmetries as

$$\frac{g'}{g_W} = \tan\theta_W, \tag{17.45}$$

the relationship between the physical fields and underlying fields of (17.43) and (17.44) can be written as

$$\begin{aligned} A_\mu &= \cos\theta_W B_\mu + \sin\theta_W W_\mu^{(3)}, \\ Z_\mu &= -\sin\theta_W B_\mu + \cos\theta_W W_\mu^{(3)}, \end{aligned}$$

which are exactly the relations that were asserted in Section 15.3. Furthermore, by using (17.45), the mass of Z boson in (17.42) can be expressed as

$$m_Z = \frac{1}{2}\frac{g_W}{\cos\theta_W}v.$$

Therefore, when combined with the corresponding expression for the W-boson mass given in (17.39), the Glashow–Salam–Weinberg model predicts

$$\frac{m_W}{m_Z} = \cos\theta_W.$$

The experimental verification of this relation, described in Chapter 16, provides a compelling argument for the reality of the Higgs mechanism.

The GSW model is described by just four parameters, the $SU(2)_L \times U(1)_Y$ gauge couplings $g_W$ and $g'$, and the two free parameters of the Higgs potential $\mu$ and $\lambda$, which are related to the vacuum expectation value of the Higgs field $v$ and the mass of the Higgs boson $m_H$ by

$$v^2 = \frac{-\mu^2}{\lambda} \quad \text{and} \quad m_H^2 = 2\lambda v^2.$$

By using the relation $m_W = \frac{1}{2}g_W v$ and the measured values for $m_W$ and $g_W$, the vacuum expectation value of the Higgs field is found to be

$$v = 246\,\text{GeV}.$$

The parameter $\lambda$ in the Higgs potential can be obtained from the mass of the Higgs boson as measured at the LHC (see Section 17.7).

## Couplings to the gauge bosons

In the $(D_\mu\phi)^\dagger(D^\mu\phi)$ term in the Lagrangian of (17.38), the gauge boson fields appear in the form of $VV(v+h)^2$, where $V = W^\pm, Z$. The $VVv^2$ terms determine the masses of the gauge bosons and the $VVh$ and $VVhh$ terms give rise to triple and quartic couplings between one or two Higgs bosons and the gauge bosons. From (15.12), the $W^+$ and $W^-$ fields are the linear combinations

$$W^\pm = \frac{1}{\sqrt{2}}\left(W^{(1)} \mp iW^{(2)}\right).$$

Hence the second term on the RHS (17.38) can be written in terms of the physical $W^+$ and $W^-$ fields,

$$\tfrac{1}{4}g_W^2 W_\mu^- W^{+\mu}(v+h)^2 = \tfrac{1}{4}g_W^2 v^2 W_\mu^- W^{+\mu} + \tfrac{1}{2}g_W^2 v W_\mu^- W^{+\mu} h + \tfrac{1}{4}g_W^2 W_\mu^- W^{+\mu} hh.$$

Here the first term gives the masses to the $W^+$ and $W^-$. The $hW^+W^-$ and $hhW^+W^-$ terms give rise to the triple and quartic couplings of the Higgs boson to the gauge bosons. The coupling strength at the $hW^+W^-$ vertex of Figure 17.12 is therefore

$$g_{HWW} = \tfrac{1}{2}g_W^2 v \equiv g_W m_W.$$

Hence the coupling of the Higgs boson to the W boson is proportional to the W-boson mass. Likewise, the coupling of the Higgs boson to the Z boson, $g_{HZZ} = g_Z m_Z$, is proportional to $m_Z$.

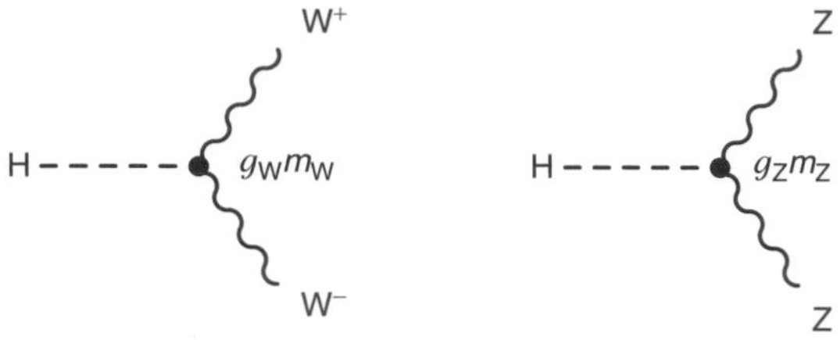


**Fig. 17.12** The trilinear couplings of the Higgs boson to the W and Z, where $g_Z = g_W/\cos\theta_W$.

### 17.5.5 Fermion masses

The Higgs mechanism for the spontaneous symmetry breaking of the $U(1)_Y \times SU(2)_L$ gauge group of the Standard Model generates the masses of the W and Z bosons. Remarkably, it also can be used to generate the masses of the fermions. Because of the different transformation properties of left- and right-handed chiral states, the fermion mass term in the Dirac Lagrangian,

$$-m\overline{\psi}\psi = -m\left(\overline{\psi}_R\psi_L + \overline{\psi}_L\psi_R\right),$$

does not respect the $SU(2)_L \times U(1)_Y$ gauge symmetry, and therefore cannot be present in the Lagrangian of the Standard Model.

In the Standard Model, left-handed chiral fermions are placed in SU(2) doublets, here written $L$, and right-handed fermions are placed in SU(2) singlets, here denoted $R$. Because the two complex scalar fields of the Higgs mechanism are placed in an SU(2) doublet $\phi(x)$, an infinitesimal SU(2) local gauge transformation has the effect,

$$\phi \to \phi' = (I + ig_W\boldsymbol{\epsilon}(x)\cdot\mathbf{T})\phi.$$

Exactly the same local gauge transformation applies to the left-handed doublet of fermion fields $L$. Therefore, the effect of the infinitesimal SU(2) gauge transformation on $\overline{L} \equiv L^\dagger\gamma^0$ is

$$\overline{L} \to \overline{L}' = \overline{L}(I - ig_W\boldsymbol{\epsilon}(x)\cdot\mathbf{T}).$$

Consequently, the combination $\overline{L}\phi$ is invariant under the $SU(2)_L$ gauge transformations. When combined with a right-handed singlet, $\overline{L}\phi R$, it is invariant under $SU(2)_L$ *and* $U(1)_Y$ gauge transformations; as is its Hermitian conjugate $(\overline{L}\phi R)^\dagger = \overline{R}\phi^\dagger L$. Hence, a term in the Lagrangian of the form $-g_f(\overline{L}\phi R + \overline{R}\phi^\dagger L)$ satisfies the $SU(2)_L \times U(1)_Y$ gauge symmetry of the Standard Model. For the $SU(2)_L$ doublet containing the electron, this corresponds to

$$\mathcal{L}_e = -g_e\left[\begin{pmatrix}\overline{\nu}_e & \overline{e}\end{pmatrix}_L\begin{pmatrix}\phi^+\\ \phi^0\end{pmatrix}e_R + \overline{e}_R\begin{pmatrix}\phi^{+*} & \phi^{0*}\end{pmatrix}\begin{pmatrix}\nu_e\\ e\end{pmatrix}_L\right], \tag{17.46}$$

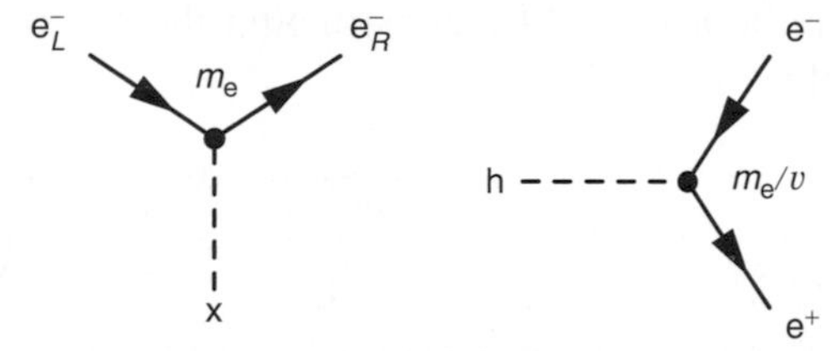


Fig. 17.13 Left: the interaction between a massless chiral electron and the non-zero expectation value of the Higgs field. Right: the interaction vertex for the coupling of the Higgs boson to an electron.

where $g_e$ is a constant known as the Yukawa coupling of the electron to the Higgs field. After spontaneously symmetry breaking, the Higgs doublet in the unitary gauge is

$$\phi(x) = \frac{1}{\sqrt{2}}\begin{pmatrix} 0 \\ v + h(x) \end{pmatrix},$$

and thus (17.46) becomes

$$\mathcal{L}_e = -\frac{g_e}{\sqrt{2}}v\left(\bar{e}_L e_R + \bar{e}_R e_L\right) - \frac{g_e}{\sqrt{2}}h\left(\bar{e}_L e_R + \bar{e}_R e_L\right). \tag{17.47}$$

The first term in (17.47) has exactly the form required for the fermion masses, but has now been introduced in gauge invariant manner. The Yukawa coupling $g_e$ is not predicted by the Higgs mechanism, but can be chosen to be consistent with the observed electron mass,

$$g_e = \sqrt{2}\frac{m_e}{v}.$$

In this case, (17.47) becomes

$$\mathcal{L}_e = -m_e\bar{e}e - \frac{m_e}{v}\bar{e}eh. \tag{17.48}$$

The first term in (17.48), which gives the mass of the electron, represents the coupling of electron to the Higgs *field* through its non-zero vacuum expectation value. The second term in (17.48) gives rise to a coupling between the electron and the Higgs *boson* itself. These two terms are illustrated in Figure 17.13, where the fermion masses arise from the coupling of left-handed and right-handed massless chiral fermions though the interaction with the non-zero expectation value of the Higgs field.

Because the non-zero vacuum expectation value occurs in the lower (neutral) component of the Higgs doublet, the combination of fields $\overline{L}\phi R + \overline{R}\phi^\dagger L$ only can generate the masses for the fermion in the lower component of an $SU(2)_L$ doublet. Thus it can be used to generate the masses of the charged leptons and the down-type quarks, but not the up-type quarks or the neutrinos. Putting aside the question of neutrino masses, a mechanism is required to give masses to the up-type quarks.

This can be achieved by constructing the conjugate doublet $\phi_c$ formed from the four fields in (17.35),

$$\phi_c = -i\sigma_2\phi^* = \begin{pmatrix} -\phi^{0*} \\ \phi^- \end{pmatrix} = \frac{1}{\sqrt{2}}\begin{pmatrix} -\phi_3 + i\phi_4 \\ \phi_1 - i\phi_2 \end{pmatrix}.$$

Because of the particular properties of SU(2), see Section 9.5, the conjugate doublet $\phi_c$ transforms in exactly the same way as the doublet $\phi$. This is analogous to the representation of up- and down-quarks and anti-up and anti-down in SU(2) isospin symmetry. A gauge invariant mass term for the up-type quarks can be constructed from $\overline{L}\phi_c R + \overline{R}\phi_c^\dagger L$, for example

$$\mathcal{L}_\mathrm{u} = g_\mathrm{u}\begin{pmatrix} \overline{\mathrm{u}} & \overline{\mathrm{d}} \end{pmatrix}_L \begin{pmatrix} -\phi^{0*} \\ \phi^- \end{pmatrix} \mathrm{u}_R + \text{Hermitian conjugate},$$

which after symmetry breaking becomes

$$\mathcal{L}_\mathrm{u} = -\frac{g_\mathrm{u}}{\sqrt{2}} v\left(\overline{\mathrm{u}}_L \mathrm{u}_R + \overline{\mathrm{u}}_R \mathrm{u}_L\right) - \frac{g_\mathrm{u}}{\sqrt{2}} h\left(\overline{\mathrm{u}}_L \mathrm{u}_R + \overline{\mathrm{u}}_R \mathrm{u}_L\right),$$

with the Yukawa coupling $g_\mathrm{u} = \sqrt{2}m_\mathrm{u}/v$, giving

$$\mathcal{L}_\mathrm{u} = -m_\mathrm{u}\overline{\mathrm{u}}\mathrm{u} - \frac{m_\mathrm{u}}{v}\overline{\mathrm{u}}\mathrm{u}h.$$

Hence for all *Dirac* fermions, gauge invariant mass terms can be constructed from either

$$\mathcal{L} = -g_\mathrm{f}\left[\overline{L}\phi R + (\overline{L}\phi R)^\dagger\right] \quad \text{or} \quad \mathcal{L} = g_\mathrm{f}\left[\overline{L}\phi_c R + (\overline{L}\phi_c R)^\dagger\right],$$

giving rise to both the masses of the fermions and the interactions between the Higgs boson and the fermion. The Yukawa couplings of the fermions to the Higgs field are given by

$$g_\mathrm{f} = \sqrt{2}\frac{m_\mathrm{f}}{v},$$

where the vacuum expectation value of the Higgs field is $v = 246\,\mathrm{GeV}$. Interestingly, for the top quark with $m_\mathrm{t} \sim 173.5 \pm 1.0\,\mathrm{GeV}$, the Yukawa coupling is almost exactly unity. Whilst this may be a coincidence, it is perhaps natural that the Yukawa couplings of the fermions are $\mathcal{O}(1)$. If the neutrino masses are also associated with the Higgs mechanism, it is perhaps surprising that they are so small, with Yukawa couplings of $\lesssim 10^{-12}$. This might suggest that the mechanism which generates the neutrino masses differs from that for other fermions. One interesting possibility is the seesaw mechanism described in the addendum to this chapter.

## 17.6 Properties of the Higgs boson

The Standard Model Higgs boson H is a neutral scalar particle. Its mass is a free parameter of the Standard Model that is given by $m_{\mathrm{H}} = 2\lambda v^2$. The Higgs boson couples to all fermions with a coupling strength proportional to the fermion mass. From (17.48), the Feynman rule for the interaction vertex with a fermion of mass $m_{\mathrm{f}}$ can be identified as

$$-i\frac{m_{\mathrm{f}}}{v} \equiv -i\frac{m_{\mathrm{f}}}{2m_{\mathrm{W}}}g_{\mathrm{W}}. \qquad (17.49)$$

The Higgs boson therefore can decay via $\mathrm{H} \to \mathrm{f\bar{f}}$ for all kinematically allowed decays modes with $m_{\mathrm{H}} > 2m_{\mathrm{f}}$. If it is sufficiently massive, the Higgs boson can also decay via $\mathrm{H} \to \mathrm{W^+W^-}$ or $\mathrm{H} \to \mathrm{ZZ}$. The Feynman diagrams and coupling strengths for these lowest-order decay modes are shown in Figure 17.14. In each case, the resulting matrix element is proportional to the mass of the particle coupling to the Higgs boson. The proportionality of the Higgs boson couplings to mass determines the dominant processes through which it is produced and decays; the Higgs boson couples preferentially to the most massive particles that are kinematically accessible.

### 17.6.1 Higgs decay

In principle, the Higgs boson can decay to all Standard Model particles. However, because of the proportionality of the coupling to the mass of the particles involved, the largest branching ratios are to the more massive particles. For a Higgs boson mass of 125 GeV, the largest branching ratio is to bottom quarks, $BR(\mathrm{H} \to \mathrm{b\bar{b}}) = 57.8\%$. The corresponding partial decay width $\Gamma(\mathrm{H} \to \mathrm{b\bar{b}})$ can be calculated from the Feynman rule for the $\mathrm{Hb\bar{b}}$ interaction vertex of (17.49) and the spinors for the quark and antiquark. Because the Higgs boson is a scalar particle, no polarisation four-vector is required; it is simply described by a plane wave. Consequently, the matrix element for the Feynman diagram shown in Figure 17.15 is

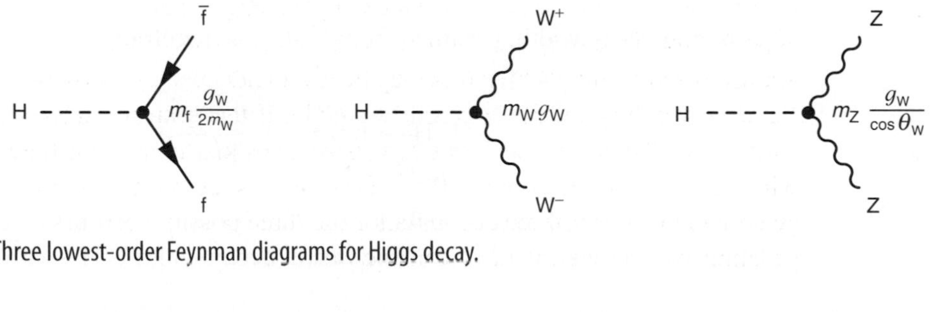


Fig. 17.14 Three lowest-order Feynman diagrams for Higgs decay.

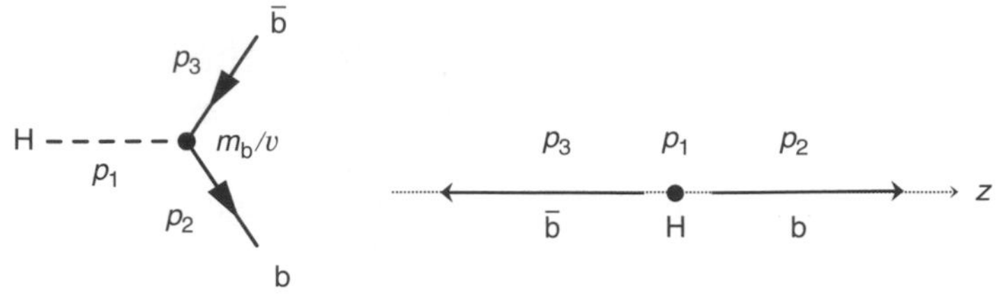


Fig. 17.15 The Feynman diagram for $h \to b\bar{b}$ and the four-momenta of the particles in the Higgs rest frame.

$$\mathcal{M} = \frac{m_b}{v}\bar{u}(p_2)v(p_3) = \frac{m_b}{v}u^\dagger\gamma^0 v. \tag{17.50}$$

Without loss of generality, the b-quark momentum can be taken to be in the $z$-axis. Because $m_H \gg m_b$, the final-state quarks are highly relativistic and therefore have four-momenta $p_2 \approx (E,0,0,E)$ and $p_3 \approx (E,0,0,-E)$, where $E = m_H/2$. In the ultra-relativistic limit, the spinors for the two possible helicity states for each of the b-quark ($\theta = 0, \phi = 0$) and the $\bar{b}$-antiquark ($\theta = \pi, \phi = \pi$) are

$$u_\uparrow(p_2) = \sqrt{E}\begin{pmatrix}1\\0\\1\\0\end{pmatrix},\ u_\downarrow(p_2) = \sqrt{E}\begin{pmatrix}0\\1\\0\\-1\end{pmatrix},\ v_\uparrow(p_3) = \sqrt{E}\begin{pmatrix}1\\0\\-1\\0\end{pmatrix}\ v_\downarrow(p_4) = \sqrt{E}\begin{pmatrix}0\\-1\\0\\-1\end{pmatrix}.$$

From the $u^\dagger\gamma^0 v$ form of the matrix element of (17.50), it can be seen immediately that only two of the four possible helicity combinations give non-zero matrix elements, these are

$$\mathcal{M}_{\uparrow\uparrow} = -\mathcal{M}_{\downarrow\downarrow} = \frac{m_b}{v}2E.$$

In both cases, the non-zero matrix elements correspond to spin configurations where the $b\bar{b}$ are produced in a spin-0 state. Because the Higgs is a spin-0 scalar, it decays isotropically and matrix element has no angular dependence. Furthermore, since the Higgs boson exists in a single spin state, the spin-averaged matrix element squared is simply,

$$\langle|\mathcal{M}|^2\rangle = |\mathcal{M}_{\uparrow\uparrow}|^2 + |\mathcal{M}_{\downarrow\downarrow}|^2 = \frac{m_b^2}{v^2}8E^2 = \frac{2m_b^2 m_H^2}{v^2}.$$

The partial decay width, obtained from (3.49), is therefore

$$\Gamma(H \to b\bar{b}) = 3 \times \frac{m_b^2 m_H}{8\pi v^2}, \tag{17.51}$$

where the factor of three accounts for the three possible colours of the $b\bar{b}$ pair. For a Higgs boson mass of 125 GeV, the partial decay width $\Gamma(H \to b\bar{b})$ is $\mathcal{O}(2\,\text{MeV})$.

**Table 17.1** The predicted branching ratios of the Higgs boson for $m_H = 125$ GeV.

| Decay mode | Branching ratio |
|---|---|
| $H \to b\bar{b}$ | 57.8% |
| $H \to WW^*$ | 21.6% |
| $H \to \tau^+\tau^-$ | 6.4% |
| $H \to gg$ | 8.6% |
| $H \to c\bar{c}$ | 2.9% |
| $H \to ZZ^*$ | 2.7% |
| $H \to \gamma\gamma$ | 0.2% |

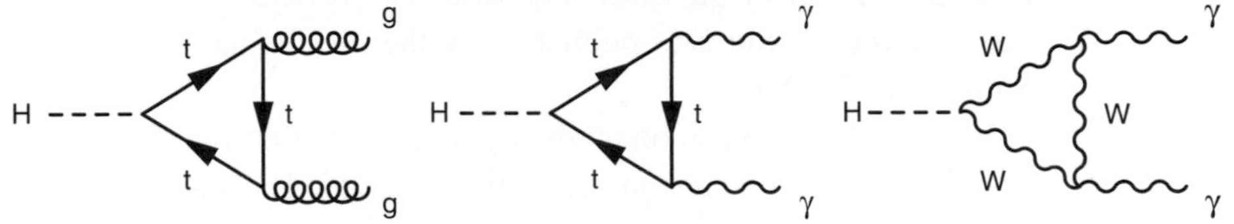


Fig. 17.16 The Feynman diagrams for the decays $H \to gg$ and $H \to \gamma\gamma$.

From (17.51), it can be seen that the partial decay rate to fermions is proportional to the square of the fermion mass, and therefore

$$\Gamma(H \to b\bar{b}) : \Gamma(H \to c\bar{c}) : \Gamma(H \to \tau^+\tau^-) \sim 3m_b^2 : 3m_c^2 : m_\tau^2. \tag{17.52}$$

It should be noted that quark masses run with $q^2$ in a similar manner to the running of $\alpha_S$. Hence the masses appearing in (17.51) are the appropriate values at $q^2 = m_H^2$, where the charm and bottom quark masses are approximately $m_c(m_H^2) \approx 0.6\,\text{GeV}$ and $m_b(m_H^2) \approx 3.0\,\text{GeV}$.

The branching ratios for a Standard Model Higgs boson with $m_H = 125\,\text{GeV}$ are listed in Table 17.1. Despite the fact that $m_H < 2m_W$, the second largest branching ratio is for the decay $H \to WW^*$, where the star indicates that one of the W bosons is produced off-mass-shell with $q^2 < m_W^2$. From the form of the W-boson propagator of (16.27), the presence of the off-shell W boson will tend to suppress the matrix element. Nevertheless, the large coupling of the Higgs boson to the massive W boson, $g_W m_W$, means that the branching ratio is relatively large. The Higgs boson also can decay to massless particles, $H \to gg$ and $H \to \gamma\gamma$, through loops of virtual top quarks and W bosons, as shown in Figure 17.16. Because the masses of the particles in these loops are large, these decays can compete with the decays to fermions and the off-mass-shell gauge bosons.

## 17.7 The discovery of the Higgs boson

Prior to the turn-on of the Large Hadron Collider at CERN, the window for a Standard Model Higgs was relatively narrow. The absence of a signal from the direct searches at LEP implied that $m_H > 114\,\mathrm{GeV}$. At the same time, the limits on the size of the quantum loop corrections from the precision electroweak measurements at LEP and the Tevatron suggested that $m_H \lesssim 150\,\mathrm{GeV}$ and that $m_H$ was unlikely to be greater than 200 GeV.

One of the main aims of the LHC was the discovery of the Higgs boson (assuming it existed). The LHC is not only the highest-energy particle collider ever built, it is also the highest-luminosity proton–proton collider to date. During 2010–2011 it operated at a centre-of-mass energy of 7 TeV and during 2012 at 8 TeV. Compelling evidence of the discovery of a new particle compatible with the Standard Model Higgs boson was published by the ATLAS and CMS experiments in the Summer of 2012.

The Higgs boson can be produced at the LHC through a number of different processes, two of which are shown in Figure 17.17. Because the Higgs boson couples preferentially to mass, the largest cross section at the LHC is through gluon–gluon fusion via a loop of virtual top quarks. The cross section for this process can be written in terms of the underlying cross section for gg → H and the gluon PDFs,

$$\sigma(\mathrm{pp} \to \mathrm{h}X) \sim \int_0^1 \int_0^1 g(x_1) g(x_2) \sigma(\mathrm{gg} \to \mathrm{H})\, \mathrm{d}x_1 \mathrm{d}x_2.$$

Consequently, the detailed knowledge of the PDFs for the proton is an essential component in the calculation of the expected Higgs boson production rate at the LHC. Fortunately, the proton PDFs are well known and the related uncertainties on the various Higgs production cross sections are less than 10%. Although the gluon–gluon fusion process has the largest cross section, from the experimental

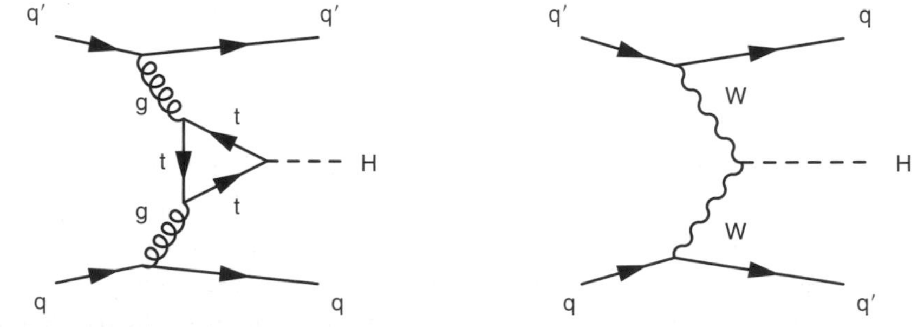


Fig. 17.17 Two of the most important Feynman diagram for Higgs boson production in pp collisions at the LHC. The gluon–gluon fusion process has a significantly higher cross section.

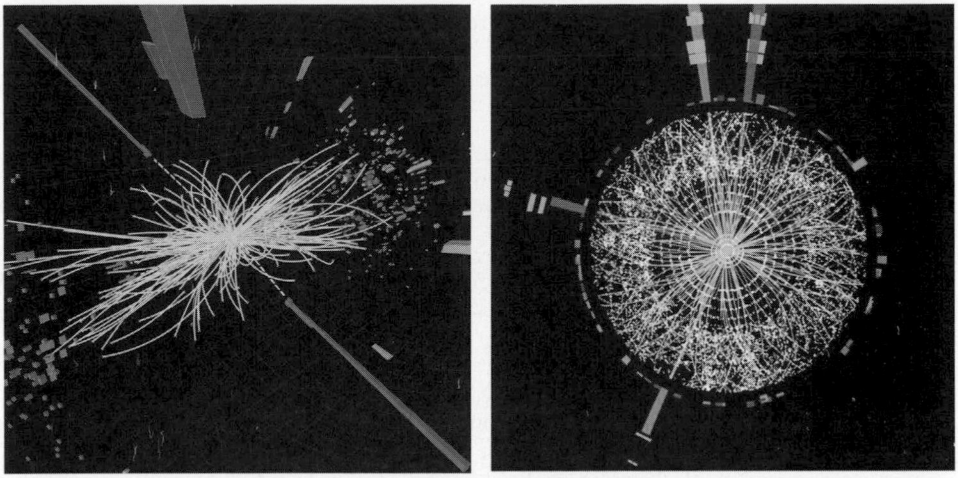

Fig. 17.18 Left: a candidate $H \to \gamma\gamma$ event in the CMS detector. Right: a candidate $H \to ZZ^* \to e^+e^-e^+e^-$ in the ATLAS detector. Reproduced with kind permission from the ATLAS and CMS collaborations, © 2012 CERN.

perspective the vector boson fusion process (shown in Figure 17.17) is also important. This is because it results in more easily identifiable final states consisting of just the decay products of the Higgs boson and two forward jets from the break-up of the colliding protons. In contrast, the gluon–gluon fusion process is accompanied by QCD radiation from the colour field, making the identification of the Higgs boson final states less easy.

In proton–proton collisions at a centre-of-mass energy of $\sim$ 8 TeV, the total production cross section for a Higgs boson with $m_H = 125\,\text{GeV}$ is approximately 20 pb. The first observations of the Higgs boson were based on approximately $20\,\text{fb}^{-1}$ of data (ATLAS and CMS combined). This data sample corresponded to a total of approximately $N = \sigma\mathcal{L} = 400\,000$ produced Higgs bosons. Whilst this number might seem large, it is a *tiny* fraction of the total number of interactions recorded at the LHC, most of which involve the QCD production of multi-jet final states. Consequently, it is difficult to distinguish the decays of the Higgs boson producing final states with jets from the large QCD background. For this reason, the most sensitive searches for the Higgs boson at the LHC are in decay channels with distinctive final-state topologies, such as $H \to \gamma\gamma$, $H \to ZZ^* \to \ell^+\ell^-\ell'^+\ell'^-$ (where $\ell = e$ or $\mu$) and $H \to WW^* \to e\nu_e\mu\nu_\mu$. Despite the relatively low branching ratios for these decay modes, the experimental signatures are sufficiently clear for them to be distinguished from the backgrounds from other processes. For example, Figure 17.18 shows a candidate $H \to \gamma\gamma$ event in the CMS detector (left-hand plot). The dashed lines point to the two large energy deposits in the electromagnetic calorimeter from two high-energy photons, which are easily identifiable. Similarly, the right-hand plot of Figure 17.18 shows a candidate $H \to ZZ^* \to e^+e^-e^+e^-$ event in the ATLAS detector. Here the four charged-particle tracks, pointing to four large energy deposits in the electromagnetic calorimeter, are clearly identifiable as high-energy electrons.

### 17.7.1 Results

The ATLAS and CMS experiments searched for the Higgs boson in several final states, $\gamma\gamma$, $ZZ^*$, $WW^*$, $\tau^+\tau^-$ and $b\overline{b}$. In both experiments, the most significant evidence for the Higgs boson was observed in the two most sensitive decay channels, $H \to \gamma\gamma$ and $H \to ZZ^* \to 4\ell$. In both these decay channels, the mass of the Higgs boson candidate can be reconstructed on an event-by-event basis from the invariant mass of its decay products. The left-hand plot of Figure 17.19 shows the distribution of the reconstructed invariant mass of the two photons in candidate $H \to \gamma\gamma$ events in the ATLAS detector. In this plot, each observed event is entered into the histogram with a weight of between zero and one, reflecting the estimated probability of it being compatible with the kinematics of Higgs production and decay. Compared to the expected background, an excess of events is observed at $m_{\gamma\gamma} \approx 126\,\text{GeV}$. The CMS experiment observed a similar excess. The right-hand plot of Figure 17.19 shows the distribution of the invariant masses of the four charged leptons in the CMS $H \to ZZ^* \to 4\ell$ search. The peak at 91 GeV is from Z-boson production. The peak at about 125 GeV can be attributed to the Higgs boson. Whilst the numbers of events are relatively small, the expected background in this region is also small. The ATLAS experiment observed a comparable excess of $H \to ZZ^* \to 4\ell$ candidates at the same mass.

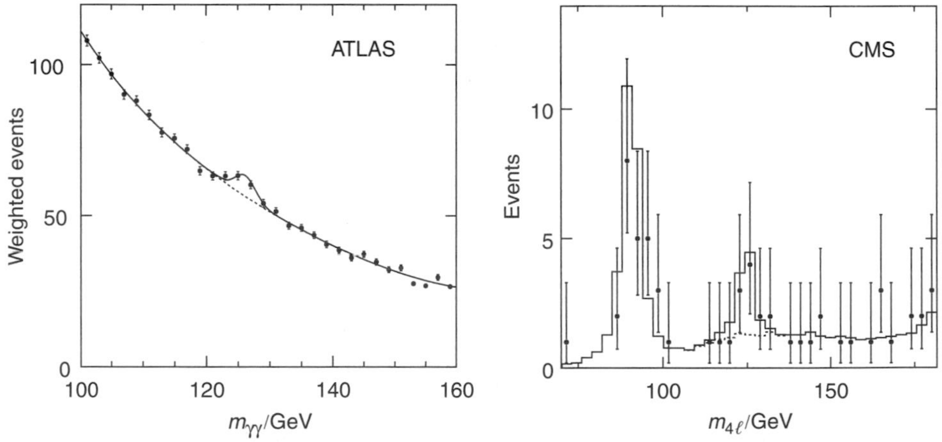


**Fig. 17.19** Left: the reconstructed invariant mass distribution of the two photons from candidate $H \to \gamma\gamma$ decays in the ATLAS experiment, adapted from Aad *et al.* (2012). Right: the distribution of the reconstructed invariant masses of the four leptons in candidate $H \to ZZ^* \to 4\ell$ events in the CMS experiment, adapted from Chatrchyan *et al.* (2012). In both plots the solid line shows the expected distribution from background and the observed Higgs signal and the dashed line indicates the expectation from background events alone.

The combined results of the ATLAS and CMS experiments provided statistically compelling evidence for the discovery of a new particle compatible with the expected properties of the Higgs boson. Until it has been demonstrated that the observed particle is a scalar, it is not possible to say conclusively that the Higgs boson has been discovered. However, even at the time of writing, it seems almost certain that the Higgs boson has been discovered; its production cross section is consistent with the Standard Model expectation and its mass is compatible with the indirect determinations from its presence in quantum loop corrections, as inferred from the precision electroweak measurements. Consistent measurements of the mass of the new particle were obtained by the ATLAS ($m = 126.0 \pm 0.6\,\text{GeV}$) and the CMS ($m = 125.3 \pm 0.6\,\text{GeV}$) experiments. On the reasonable assumption that the new particle is the Higgs boson, it can be concluded that

$$m_{\text{H}} \simeq 125.7 \pm 0.5\,\text{GeV}.$$

Since the discovery of the W and Z bosons in the mid 1980s, the search for the Higgs boson has been the highest priority in particle physics. Its discovery finally completed the particle spectrum of the Standard Model.

### 17.7.2 Outlook

The discovery of the Higgs boson is not the end of the story. The use of a single Higgs doublet in the Standard Model is the most economical choice, but it is not the only possibility. In supersymmetry (see Section 18.2.2), which is a popular extension to the Standard Model, there are (at least) two complex doublets of scalar fields, which give rise to five physical Higgs bosons. Furthermore, it is not clear whether the observed Higgs boson is a fundamental scalar particle or whether it might be composite. In the coming years, the measurements of the spin and branching ratios of the Higgs boson will further test the predictions of the Standard Model. Perhaps more importantly, a detailed understanding of all the properties of the Higgs boson may well open up completely new avenues in our understanding of the Universe and point to what lies beyond the Standard Model.

## Summary

The Higgs mechanism is an essential part of the Standard Model. It is based on a doublet of complex scalar fields with the Higgs potential $V(\phi) = \mu^2(\phi^\dagger\phi) + \lambda(\phi^\dagger\phi)^2$ where $\mu^2 < 0$. As a result, the vacuum state of the Universe is degenerate. The spontaneous breaking of this symmetry, when combined with the underlying $\text{SU}(2)_L \times \text{U}(1)_Y$ gauge symmetry of the electroweak model, provides masses to the W and Z gauge bosons with

$$m_W = m_Z \cos\theta_W = \tfrac{1}{2} g_W\, v,$$

where $v$ is the vacuum expectation value of the Higgs field. The value of

$$v = 246\,\text{GeV},$$

sets the mass scale for the electroweak and Higgs bosons. The interaction between the fermion fields and the non-zero expectation value of the Higgs field, provides a gauge-invariant mechanism for generating the masses of the Standard Model fermions.

In 2012, the discovery of the Higgs boson at the LHC with mass

$$m_H \simeq 126\,\text{GeV},$$

completed the spectrum of Standard Model particles. Following the discovery of the Higgs boson, it is hoped that the studies of its properties will provide clues to physics beyond the Standard Model, which is the main topic of the final chapter of this book.

## 17.8 *Addendum: Neutrino masses

The right-handed chiral neutrino states $\nu_R$ do not participate in any of the interactions of the Standard Model; they do not couple to the gluons or electroweak gauge bosons. Consequently, there is no direct evidence that they exist. However, from the studies of neutrino oscillations it is known that neutrinos do have mass, and therefore there must be a corresponding mass term in the Lagrangian. In the Standard Model, neutrino masses can be introduced in exactly the same way as for the up-type quarks using the conjugate Higgs doublet. In this case, after spontaneous symmetry breaking, the gauge invariant *Dirac mass* term for the neutrino is

$$\mathcal{L}_D = -m_D\,(\overline{\nu_R}\nu_L + \overline{\nu_L}\nu_R).$$

If this is the origin of neutrino masses, then right-handed chiral neutrinos exist. However, the neutrino masses are very much smaller than the masses of the other fermions, suggesting that another mechanism for generating neutrino mass might be present.

Because the *right-handed neutrinos* and *left-handed antineutrinos* transform as singlets under the Standard Model gauge transformations, any additional terms in the Lagrangian formed from these fields alone can be added to the Lagrangian without breaking the gauge invariance of the Standard Model. The left-handed antineutrinos appear in the Lagrangian as the CP conjugate fields defined by

$$\psi^c = \hat{C}\hat{P}\psi = i\gamma^2\gamma^0\psi^*,$$

where the CP conjugate field for the right-handed neutrino, written $\nu_R^c$, corresponds to a left-handed antineutrino. Therefore the Majorana mass term

$$\mathcal{L}_M = -\tfrac{1}{2}M\,(\overline{\nu_R^c}\nu_R + \overline{\nu_R}\nu_R^c),$$

which is formed from right-handed neutrino fields and left-handed antineutrino fields, respects the local gauge invariance of the Standard Model. Consequently, it can be added to the Standard Model Lagrangian. However, there is a price to pay; the Majorana mass term provides a direct coupling between between a particle and an antiparticle. For example, the corresponding Majorana mass term for the electron would allow $e^+ \leftrightarrow e^-$ transitions, violating charge conservation. This problem does not exist for the neutrinos. Furthermore, because neutrinos are neutral, it is possible that they are their own antiparticles, in which case they are referred to as *Majorana neutrinos* as opposed to Dirac neutrinos.

### 17.8.1 The seesaw mechanism

The most general renormalisable Lagrangian for the neutrino masses includes both the Dirac and Majorana mass terms, indicated in Figure 17.20. Because $\overline{\nu_L}\nu_R$ is equivalent to $\overline{\nu_R^c}\nu_L^c$, the Dirac mass term can be written

$$\mathcal{L}_D = -\tfrac{1}{2}m_D\,(\overline{\nu_L}\nu_R + \overline{\nu_R^c}\nu_L^c) + \ h.c.,$$

where $h.c.$ stands for the corresponding Hermitian conjugate. This term admits the possibility that neutrino masses arise from the spontaneous symmetry breaking of the Higgs mechanism. If in addition, the automatically gauge-invariant Majorana mass term is added by hand, the Lagrangian for the combined Dirac and Majorana masses is

$$\mathcal{L}_{DM} = -\tfrac{1}{2}\left[m_D\,\overline{\nu_L}\nu_R + m_D\,\overline{\nu_R^c}\nu_L^c + M\,\overline{\nu_R^c}\nu_R\right] + \ h.c.$$

or, equivalently,

$$\mathcal{L}_{DM} = -\tfrac{1}{2}\begin{pmatrix}\overline{\nu_L} & \overline{\nu_R^c}\end{pmatrix}\begin{pmatrix}0 & m_D\\ m_D & M\end{pmatrix}\begin{pmatrix}\nu_L^c\\ \nu_R\end{pmatrix} + \ h.c. \tag{17.53}$$

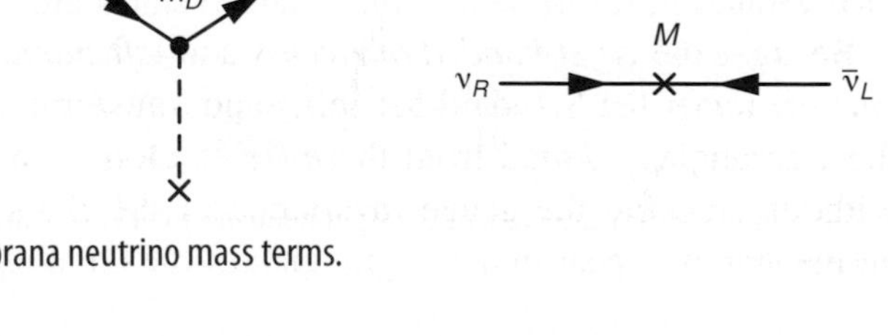


Fig. 17.20 The Dirac and Majorana neutrino mass terms.

The physical states of this system can be obtained from the basis in which the mass matrix is diagonal, analogous to the procedure for identifying the physical states of the neutral kaon system and the neutral gauge bosons of the $U(1)_Y \times SU(2)_L$ gauge symmetry. Hence, with the Lagrangian including Dirac and Majorana mass terms, the masses of the physical neutrino states are the eigenvalues of mass matrix $\mathbf{M}$ in (17.53). These can be found from the characteristic equation $\det(\mathbf{M} - \lambda I) = 0$, which implies $\lambda^2 - M\lambda - m_D^2 = 0$. Hence, in this model, the masses of the physical neutrinos would be

$$m_\pm = \lambda_\pm = \frac{M \pm \sqrt{M^2 + 4m_D^2}}{2} = \frac{M \pm M\sqrt{1 + 4m_D^2/M^2}}{2}.$$

If the Majorana mass $M$ is taken to be much greater than the Dirac mass $m_D$, then

$$m_\pm \approx \tfrac{1}{2}M \pm \tfrac{1}{2}\left(M + \frac{2m_D^2}{M}\right), \qquad (17.54)$$

giving a light neutrino state[2] ($\nu$) and heavy neutrino state ($N$) with masses

$$|m_\nu| \approx \frac{m_D^2}{M} \quad \text{and} \quad m_N \approx M.$$

In the seesaw mechanism, it is hypothesised that the Dirac mass terms for the neutrinos are of a similar size to the masses of the other fermions, i.e. $\mathcal{O}(1\,\text{GeV})$. The Majorana mass $M$ is then made sufficiently large that the lighter of the two physical neutrino states has a mass $m_\nu \sim 0.01\,\text{eV}$. In this way, the masses of the lighter neutrino states can be made to be very small, even when the Dirac mass term is of the same order of magnitude as the other fermions. For this to work, the Majorana mass must be very large, $M \sim 10^{11}\,\text{GeV}$.

If a Majorana mass term exists, the seesaw mechanism predicts that for each of the three neutrino generations, there is a very light neutrino with a mass much smaller than the other Standard Model fermions and a very massive neutrino state $m_N \approx M$. The physical neutrino states, which are obtained from the eigenvalues of the mass matrix, are

$$\nu = \cos\theta(\nu_L + \nu_L^c) - \sin\theta(\nu_R + \nu_R^c) \quad \text{and} \quad N = \cos\theta(\nu_R + \nu_R^c) + \sin\theta(\nu_L + \nu_L^c),$$

where $\tan\theta \approx m_D/M$. Since the left-handed chiral components of the light neutrino are multiplied by $\cos\theta$, the effect of introducing the Majorana mass term is to reduce the weak charged-current couplings of the light neutrino states by a factor $\cos\theta$. However, for $M \gg m_D$, the neutrino states are

$$\nu \approx (\nu_L + \nu_L^c) - \frac{m_D}{M}(\nu_R + \nu_R^c) \quad \text{and} \quad N \approx (\nu_R + \nu_R^c) + \frac{m_D}{M}(\nu_L + \nu_L^c),$$

[2] The minus sign for the mass of the light neutrino in (17.54) can be absorbed in to the definition of the fields.

and the couplings of the light neutrinos to the weak charged-current are essentially the same as those of the Standard Model. Since the massive neutrino state is almost entirely right-handed, it would not participate in the weak charged- or neutral-currents.

The seesaw mechanism provides an interesting hypothesis for the smallness of neutrino masses, but it is just a hypothesis. It would be placed on firmer ground if neutrinos were shown to be Majorana particles. One experimental consequence of neutrinos being Majorana particles would be the possibility of observing the phenomenon of neutrinoless double $\beta$-decay, which is discussed in the following chapter.

## Problems

**17.1** By considering the form of the polarisation four-vector for a longitudinally polarised massive gauge bosons, explain why the $t$-channel neutrino-exchange diagram for $e^+e^- \to W^+W^-$, when taken in isolation, is badly behaved at high centre-of-mass energies.

**17.2** The Lagrangian for the Dirac equation is

$$\mathcal{L} = i\overline{\psi}\gamma_\mu\partial^\mu\psi - m\overline{\psi}\psi,$$

Treating the eight fields $\psi_i$ and $\overline{\psi}_i$ as independent, show that the Euler-Lagrange equation for the component $\psi_i$ leads to

$$i\partial_\mu\overline{\psi}\gamma^\mu + m\overline{\psi} = 0.$$

**17.3** Verify that the Lagrangian for the free electromagnetic field,

$$\mathcal{L} = -\tfrac{1}{4}F^{\mu\nu}F_{\mu\nu},$$

is invariant under the gauge transformation $A_\mu \to A'_\mu = A_\mu - \partial_\mu\chi$.

**17.4** The Lagrangian for the electromagnetic field in the presence of a current $j^\mu$ is

$$\mathcal{L} = -\tfrac{1}{4}F^{\mu\nu}F_{\mu\nu} - j^\mu A_\mu.$$

By writing this as

$$\begin{aligned}\mathcal{L} &= -\tfrac{1}{4}(\partial^\mu A^\nu - \partial^\nu A^\mu)(\partial_\mu A_\nu - \partial_\nu A_\mu) - j^\mu A_\mu \\ &= -\tfrac{1}{2}(\partial^\mu A^\nu)(\partial_\mu A_\nu) + \tfrac{1}{2}(\partial^\nu A^\mu)(\partial_\mu A_\nu) - j^\mu A_\mu,\end{aligned}$$

show that the Euler–Lagrange equation gives the covariant form of Maxwell's equations,

$$\partial_\mu F^{\mu\nu} = j^\nu.$$

**17.5** Explain why the Higgs potential can contain terms with only even powers of the field $\phi$.

**17.6** Verify that substituting (17.30) into (17.29) leads to

$$\mathcal{L} = \tfrac{1}{2}(\partial^\mu\eta)(\partial_\mu\eta) - \lambda v^2\eta^2 + \tfrac{1}{2}(\partial^\mu\xi)(\partial_\mu\xi), -\tfrac{1}{4}F_{\mu\nu}F^{\mu\nu} + \tfrac{1}{2}g^2v^2B^\mu B_\mu - V_{int} + gvB_\mu(\partial^\mu\xi).$$

**17.7** Show that the Lagrangian for a complex scalar field $\phi$,

$$\mathcal{L} = (D_\mu\phi)^*(D^\mu\phi),$$

with the covariant derivative $D_\mu = \partial_\mu + igB_\mu$, is invariant under local U(1) gauge transformations,

$$\phi(x) \to \phi'(x) = e^{ig\chi(x)}\phi(x),$$

provided the gauge field transforms as

$$B_\mu \to B'_\mu = B_\mu - \partial_\mu\chi(x).$$

**17.8** From the mass matrix of (17.40) and its eigenvalues (17.41), show that the eigenstates in the diagonal basis are

$$A_\mu = \frac{g'W^{(3)}_\mu + gB_\mu}{\sqrt{g^2 + g'B_\mu}} \quad \text{and} \quad Z_\mu = \frac{gW^{(3)}_\mu - g'B_\mu}{\sqrt{g^2 + g'B_\mu}},$$

where $A_\mu$ and $Z_\mu$ correspond to the physical fields for the photon and Z.

**17.9** By considering the interaction terms in (17.38), show that the HZZ coupling is given by

$$g_{\mathrm{HZZ}} = \frac{g_{\mathrm{W}}}{\cos\theta_{\mathrm{W}}} m_{\mathrm{Z}}.$$

**17.10** For a Higgs boson with $m_{\mathrm{H}} > 2m_{\mathrm{W}}$, the dominant decay mode is into two on-shell W bosons, $\mathrm{H} \to \mathrm{W}^+\mathrm{W}^-$. The matrix element for this decay can be written

$$\mathcal{M} = -g_{\mathrm{W}} m_{\mathrm{W}} g_{\mu\nu}\epsilon^\mu(p_2)^*\epsilon^\nu(p_3)^*,$$

where $p_2$ and $p_3$ are respectively the four-momenta of the $\mathrm{W}^+$ and $\mathrm{W}^-$.

(a) Taking $\mathbf{p}_2$ to lie in the positive $z$-direction, consider the nine possible polarisation states of the $\mathrm{W}^+\mathrm{W}^-$ and show that the matrix element is non-zero only when both W bosons are left-handed ($\mathcal{M}_{\downarrow\downarrow}$), both W bosons are right-handed ($\mathcal{M}_{\uparrow\uparrow}$), or both are longitudinally polarised ($\mathcal{M}_{LL}$).

(b) Show that

$$\mathcal{M}_{\uparrow\uparrow} = \mathcal{M}_{\downarrow\downarrow} = -g_{\mathrm{W}} m_{\mathrm{W}} \quad \text{and} \quad \mathcal{M}_{LL} = \frac{g_{\mathrm{W}}}{m_{\mathrm{W}}}\left(\tfrac{1}{2}m_{\mathrm{H}}^2 - m_{\mathrm{W}}^2\right).$$

(c) Hence show that

$$\Gamma(\mathrm{H} \to \mathrm{W}^+\mathrm{W}^-) = \frac{G_{\mathrm{F}} m_{\mathrm{H}}^3}{8\pi\sqrt{2}}\sqrt{1 - 4\lambda^2}\left(1 - 4\lambda^2 + 12\lambda^4\right),$$

where $\lambda = m_{\mathrm{W}}/m_{\mathrm{H}}$.

**17.11** Assuming a total Higgs production cross section of 20 pb and an integrated luminosity of 10 $\mathrm{fb}^{-1}$, how many $\mathrm{H} \to \gamma\gamma$ and $\mathrm{H} \to \mu^+\mu^-\mu^+\mu^-$ events are expected in each of the ATLAS and CMS experiments.

**17.12** Draw the lowest-order Feynman diagrams for the processes $\mathrm{e}^+\mathrm{e}^- \to \mathrm{HZ}$ and $\mathrm{e}^+\mathrm{e}^- \to \mathrm{H}\nu_e\overline{\nu}_e$, which are the main Higgs production mechanism at a future high-energy linear collider.

**17.13** In the future, it might be possible to construct a muon collider where the Higgs boson can be produced directly through $\mu^+\mu^- \to \mathrm{H}$. Compare the cross sections for $\mathrm{e}^+\mathrm{e}^- \to \mathrm{H} \to \mathrm{b}\overline{\mathrm{b}}$, $\mu^+\mu^- \to \mathrm{H} \to \mathrm{b}\overline{\mathrm{b}}$ and $\mu^+\mu^- \to \gamma \to \mathrm{b}\overline{\mathrm{b}}$ at $\sqrt{s} = m_{\mathrm{H}}$.

# 18 The Standard Model and beyond

The success of the Standard Model of particle physics in describing the wide range of precise experimental measurements is a remarkable achievement. However, the Standard Model is just a model and there are many unanswered questions. This short concluding chapter provides a broad overview of the current state of our understanding of particle physics and describes some of the more important open issues.

## 18.1 The Standard Model

The ultimate theory of particle physics might consist of a (simple) equation with relatively few free parameters, from which everything else followed. Whilst the Standard Model (SM) is undoubtedly one of the great triumphs of modern physics, it is not this ultimate theory. It is a model constructed from a number of beautiful and profound theoretical ideas put together in a somewhat *ad hoc* fashion in order to reproduce the experimental data. The essential ingredients of the Standard Model, indicated in Figure 18.1, are: the Dirac equation of relativistic quantum mechanics that describes the dynamics of the fermions; Quantum Field Theory that provides a fundamental description of the particles and their interactions; the local gauge principle that determines the exact nature of these interactions; the Higgs mechanism of electroweak symmetry breaking that generates particle masses; and the wide-reaching body of experimental results that guide the way in which the Standard Model is constructed. The recent precision tests of the Standard Model and the discovery of the Higgs boson have firmly established the validity of the Standard Model at energies up to the electroweak scale. Despite this success, there are many unanswered questions.

### 18.1.1 The parameters of the Standard Model

If neutrinos are normal Dirac fermions, the Standard Model of particle physics has 25 (or 26) free parameters that have to be input by hand. These are: the masses of

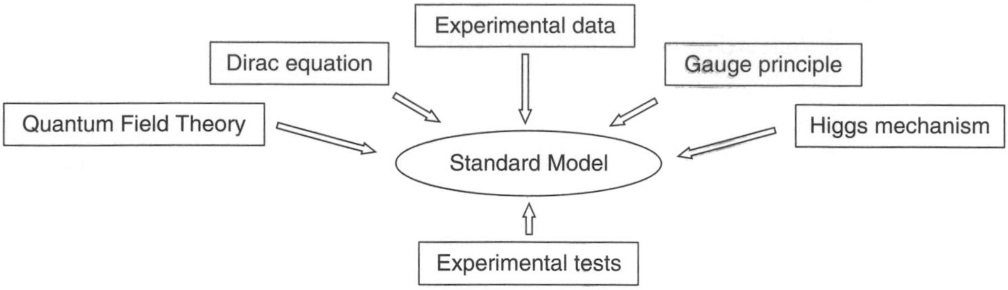


Fig. 18.1 The theoretical and experimental pillars of the Standard Model.

the *twelve* fermions (or perhaps more correctly the twelve Yukawa couplings to the Higgs field),

$$m_{\nu_1},\ m_{\nu_2},\ m_{\nu_3},\ m_{\mathrm{e}},\ m_{\mu},\ m_{\tau},\ m_{\mathrm{d}},\ m_{\mathrm{s}},\ m_{\mathrm{b}},\ m_{\mathrm{u}},\ m_{\mathrm{c}} \text{ and } m_{\mathrm{t}}\,;$$

the *three* coupling constants describing the strengths of the gauge interactions,

$$\alpha,\ G_{\mathrm{F}} \text{ and } \alpha_S\,,$$

or equivalently $g'$, $g_{\mathrm{W}}$ and $g_S$; the *two* parameters describing the Higgs potential, $\mu$ and $\lambda$, or equivalently its vacuum expectation value and the mass of the Higgs boson,

$$v \text{ and } m_{\mathrm{H}}\,;$$

and the *eight* mixing angles of the PMNS and CKM matrices, which can be parameterised by

$$\theta_{12},\ \theta_{13},\ \theta_{23},\ \delta, \text{ and } \lambda,\ A,\ \rho,\ \eta.$$

In principle, there is one further parameter in the Standard Model; the Lagrangian of QCD can contain a phase that would lead to CP violation in the strong interaction. Experimentally, this strong CP phase is known to be extremely small,

$$\theta_{CP} \simeq 0.$$

and is usually taken to be zero. If $\theta_{CP}$ is counted, then the Standard Model has 26 free parameters.

The relatively large number of free parameters is symptomatic of the Standard Model being just that; a model where the parameters are chosen to match the observations, rather than coming from a higher theoretical principle. Putting aside $\theta_{CP}$, of the 25 SM parameters, 14 are associated with the Higgs field, eight with the flavour sector and only three with the gauge interactions. Within each of these three broad areas, patterns emerge between the different parameters, suggesting the

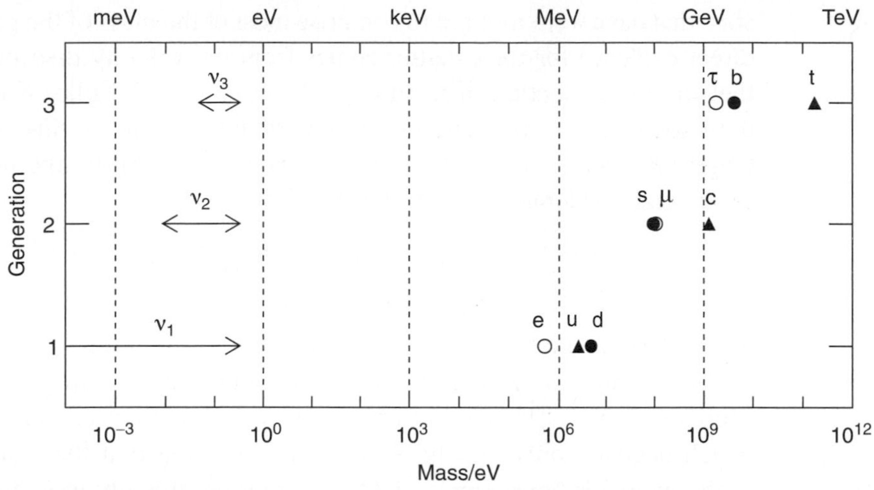


Fig. 18.2 The fermion masses shown by generation. The neutrino masses are displayed as approximate ranges of values assuming the normal hierarchy ($m_1 < m_2 < m_3$) and using the approximate upper limits on the sum of neutrino masses from cosmological constraints.

presence of some, as yet unknown, symmetry principle. For example, Figure 18.2 shows the observed masses of the fermions. With the exception of the neutrinos, the masses within a single generation are similar, and it is unlikely that this happens by chance. Likewise, the coupling constants of the three gauge interactions are of a similar order of magnitude, hinting that they might be different low-energy manifestations of a Grand Unified Theory (GUT) of the forces. These patterns provide hints for, as yet unknown, physics beyond the Standard Model.

## 18.2 Open questions in particle physics

The Standard Model is not the final theory of particle physics. However, there are many possibilities for the nature of physics beyond the Standard Model, for example, supersymmetry, large-scale extra dimensions, and ultimately perhaps even string theory. Here it is possible to give only a brief overview of a handful of the outstanding issues with the Standard Model and the possible solutions. The chosen topics focus on active areas of current experimental research.

### 18.2.1 What is dark matter?

The existence of *dark matter* in the Universe provides compelling evidence for physics beyond the Standard Model. Since the mid 1930s, it has been known that a significant fraction of the mass in the Universe is not bound up in the luminous stars that once were thought to comprise most of the mass of the galaxies. The most direct evidence for dark matter comes from the velocity distributions of stars as they orbit the galactic centre. In a spiral galaxy like the Milky Way, the majority of the luminous mass is located in the central bulge. Outside this central region, the tangental velocity of a star of mass $m$ should be given by the usual equation for centripetal acceleration in a gravitational field

$$\frac{mv^2}{r} \approx \frac{Gm}{r^2}M(r),$$

where $M(r)$ is the total mass within a radius $r$. Assuming that most of this mass is concentrated in the central bulge, the tangental velocities of the stars should decrease as $r^{-1/2}$. This is not consistent with the observed velocity distributions, which decrease only slowly with radius, implying that the distribution of mass in the galaxy is approximately $M(r) \propto r$. From this observation alone, it can be concluded that the mass of a galaxy has a significant non-luminous component, known as dark matter.

Further compelling evidence for the existence of dark matter is provided by a number of cosmological and astrophysical measurements related to the large-scale structure in the Universe and, in particular, the precision measurements of the small fluctuations in the cosmic microwave background (CMB) from the Cosmic Microwave Background Explorer (COBE) and Wilkinson Microwave Anisotropy Probe (WMAP) satellites. These and other observations have provided a firm experimental basis for the ΛCDM cosmological model, which is the standard model of cosmology. In the ΛCDM model, the total energy-matter density Ω of the Universe is consistent with the flat geometry of space-time predicted by inflationary models, with $\Omega = 1$. Within the ΛCDM model, only 5% of energy–matter density of the Universe is in the form of normal baryonic matter, $\Omega_B \simeq 0.05$. A further 23% is in the form of *cold dark matter* (CDM), $\Omega_C \simeq 0.23$, and the majority of the energy–matter density of the Universe is in the form of *dark energy*, $\Omega_\Lambda \simeq 0.72$. In the ΛCDM model, the dark energy is attributed to a non-zero cosmological constant of Einstein's equations of general relativity, $\Lambda \neq 0$, which tends to accelerate the expansion of the Universe.

It is a remarkable fact that our understanding of cosmology has reached the level of precision and sophistication where it now provides constraints on particle physics. Whilst the existence of dark energy does not (yet) impact our understanding of particle physics, the cosmological constraints on dark matter are highly relevant. The particle content of the Universe affects the way in which large-scale structure arises. Because lighter particles, such as neutrinos, remain relativistic throughout the expansion and cooling of the Universe, they affect the evolution of large-scale structure differently than massive particles, which become non-relativistic during the first few years after the Big Bang. On this basis, it is known that the majority of the energy–mass density associated with the non-baryonic dark matter is due to cold (non-relativistic) matter as opposed to hot (relativistic) particles. The cosmological measurements are sufficiently precise to constrain the sum of the neutrino masses to be approximately

$$\sum_{i=1}^{3} m_{\nu_i} \lesssim 1\,\mathrm{eV}.$$

The current experimental evidence indicates that only a small fraction of the cold dark matter is in the form of normal baryons, for example in low-mass brown dwarf stars. The success of the ΛCDM standard model of cosmology, therefore strongly suggests that a significant fraction of the cold dark matter in the Universe may be in the form of a new type of weakly interacting massive particle (WIMP), with a mass in the few GeV–TeV range. Such particles arise naturally in extensions to the Standard Model; for example, in many supersymmetric models the lightest supersymmetric particle is the stable weakly interacting neutralino $\tilde{\chi}_1^0$. Regardless of the precise nature of the dark matter, the direct detection of WIMPs is one of the main goals in particle physics at this time. WIMPs can either be observed through their production at the LHC or through the direct detection of the WIMPs that are believed to pervade our galaxy.

## Direct detection of dark matter

The direct detection of the galactic WIMP halo (assuming it exists) is extremely challenging. The WIMPs are predicted to have a Maxwell–Boltzmann velocity distribution with a root-mean-square (rms) velocity in the range 200–250 km s$^{-1}$, which corresponds to a mean kinetic energy of approximately $\langle T_\chi \rangle \approx 3\times 10^{-7} m_\chi$, where $m_\chi$ is the mass of the WIMP in GeV. WIMPs would interact with normal matter through the elastic scattering with nuclei, $\chi + A \to \chi + A$. Dark matter experiments attempt to detect the recoil of a nucleus after such a scattering process. However, the maximum kinetic energy transferred to a nucleus of mass number $A$ is only

$$T_{\max} \approx \frac{4Am_\chi m_\mathrm{p}}{(m_\chi + Am_\mathrm{p})^2} T_\chi \sim 1.2\times 10^{-6} \frac{Am_\chi^2 m_\mathrm{p}}{(m_\chi + Am_\mathrm{p})^2}.$$

Consequently, for WIMP masses greater than 10 GeV, the recoil energies are typically in the range of 1 – 10 keV. By the usual standards of particle physics, this is a very low energy and the possible detection techniques reflect this. There are two main ways of detecting the nuclear recoil. The ionisation produced by the recoiling nucleus can be detected from scintillation light in sodium iodide crystals or liquid noble gas detectors. Alternatively, in cryogenic detectors consisting of very pure silicon or germanium crystals cooled to low temperatures, WIMPs can be detected from the phonons produced by the particle interactions and also from the ionisation produced by the recoiling nucleus.

From the energy–matter density associated with the CDM, the local number density of WIMPs is expected to be about $n \sim 0.3\,/m_\chi[\text{GeV}]\,\text{cm}^{-3}$. This relatively low number density, combined with the low velocities of the WIMPs and the smallness of weak interaction cross sections, means that the expected event rates are very small; typically just a few events per year in the current 10 kg-scale detectors. Furthermore, because the nuclear recoil energies are so low, backgrounds from natural radioactivity have to be controlled carefully.

Despite the occasional tantalising hints for a signal, at the time of writing there has been no confirmed direct detection of dark matter. Nevertheless, for many favoured scenarios (including supersymmetry), the sensitivities of the current experiments are only just beginning to reach that required to observe a possible signal and the results from the experiments in the coming decade are eagerly awaited.

### 18.2.2 Does supersymmetry exist?

Supersymmetry (SUSY) is a popular extension to the Standard Model. In SUSY each Standard Model particle has a super-partner "sparticle" which differs by half a unit of spin. The super-partner of each chiral fermion is a spin-0 scalar (sfermion) and the super-partners of the spin-1 gauge fields are spin-half gauginos. The partners of the spin-0 Higgs field are a weak isospin doublet of spin-half Higgsinos, $\tilde{H}^0_{1,2}$ and $\tilde{H}^\pm$. The physical fields in the minimal supersymmetric model are listed in Table 18.1. The physical chargino and neutralino states are, in general, mixtures of the Higgsinos and gauginos. In many supersymmetric models, the lightest neutralino $\tilde{\chi}^0_1$ is a weakly interacting stable particle, and is a possible WIMP candidate for the dark matter in the Universe.

**Table 18.1** The Standard Model particles and their possible super-partners in the minimal supersymmetric model.

| Particle | | Spin | | Super-particle | | Spin |
|---|---|---|---|---|---|---|
| Quark | q | $\frac{1}{2}$ | | Squark | $\tilde{q}_L, \tilde{q}_R$ | 0 |
| Lepton | $\ell^\pm$ | $\frac{1}{2}$ | | Slepton | $\tilde{\ell}^\pm_L, \tilde{\ell}^\pm_R$ | 0 |
| Neutrino | $\nu$ | $\frac{1}{2}$ | | Sneutrino | $\tilde{\nu}_L, \tilde{\nu}_R(?)$ | 0 |
| Gluon | g | 1 | | Gluino | $\tilde{g}$ | $\frac{1}{2}$ |
| Photon | $\gamma$ | 1 | $\tilde{\gamma}$ | | | |
| Z boson | Z | 1 | $\tilde{Z}$ | Neutralino | $\tilde{\chi}^0_1, \tilde{\chi}^0_2, \tilde{\chi}^0_3, \tilde{\chi}^0_4$ | $\frac{1}{2}$ |
| Higgs | H | 0 | $\tilde{H}^0_1, \tilde{H}^0_2$; $\tilde{H}^\pm$ | | | |
| W boson | $W^\pm$ | 1 | $\tilde{W}^\pm$ | Chargino | $\tilde{\chi}^\pm_1, \tilde{\chi}^\pm_2$ | $\frac{1}{2}$ |

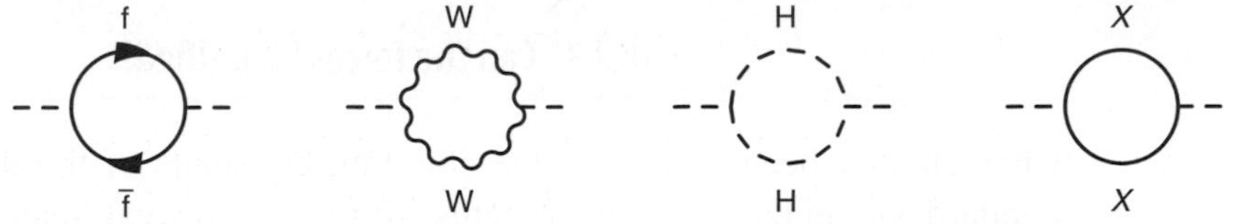


Fig. 18.3 Examples of loop corrections to the Higgs boson self-energy, where $X$ represents a new massive particle.

The possibility of explaining the dark matter in the Universe is not the prime motivation for supersymmetry. Just as quantum loop corrections contributed to the W-boson mass (see Section 16.4), quantum loops in the Higgs boson propagator, such as those indicated in Figure 18.3, contribute to the Higgs boson mass. This in itself is not a problem. However, if the Standard Model is part of theory that is valid up to very high mass scales, such as that of a Grand Unified Theory $\Lambda_{\mathrm{GUT}} \sim 10^{16}$ GeV or the Planck scale $\Lambda_{\mathrm{P}} \sim 10^{19}$ GeV, these corrections become very large. Because of these quantum corrections, which are quadratic in $\Lambda$, it is difficult to keep the Higgs mass at the electroweak scale of $10^2$ GeV. This is known as the Hierarchy problem. It can be solved by fine-tuning the new contributions to the Higgs mass such that they tend to cancel to a high degree of precision. However, supersymmetry provides a more natural solution to the Hierarchy problem; for every loop of particles there is a corresponding loop of sparticles, which provide a correction with the opposite sign. If the sparticle masses were the same as the particle masses, this cancellation would be exact. If supersymmetry were an exact symmetry of nature, the sparticles would have the same masses as the particles and already would have been discovered. Therefore, if supersymmetry exists, it is a broken symmetry and the mass scale of the SUSY particles is not known *a priori*. Nevertheless, there are theoretical arguments that favour a relatively low mass scale of $\mathcal{O}(1\,\mathrm{TeV})$.

The search for the production of SUSY particles is one of the main focuses of the search for new physics at the LHC. In most SUSY models, sparticles are predicted to decay into final states including the stable lightest supersymmetric particle (LSP), which being neutral escapes detection. For example, at the LHC the signature of squark pair production and subsequent decay, $\tilde{q}\tilde{q} \to qq\tilde{\chi}_1^0\tilde{\chi}_1^0$, is a pair of high-energy jets and a large component of missing transverse momentum from the unobserved neutralinos. At the time of writing, no evidence of the direct production of SUSY particles has been observed at the LHC; the ATLAS and CMS experiments have been able exclude squark and gluino masses below about 1 TeV. The limits on the slepton and gaugino masses are much weaker, since these particles are not produced directly in strong interactions. Whilst there is no current experimental evidence for SUSY, the first operation of the LHC at its full energy of $\sqrt{s} \sim 14$ TeV will provide discovery potential at significantly higher mass scales.

### 18.2.3 Can the forces be unified?

It has already been noted that the coupling constants of the three forces of the Standard Model have similar strengths. At the electroweak scale of $q^2 = m_Z^2$,

$$\alpha^{-1} : \alpha_W^{-1} : \alpha_S^{-1} \approx 128 : 30 : 9. \tag{18.1}$$

Furthermore, in Section 10.5 it was shown that the coupling constants of QED and QCD run with energy according to

$$\left[\alpha_i(q^2)\right]^{-1} = \left[\alpha_i(\mu^2)\right]^{-1} + \beta \ln\left(\frac{q^2}{\mu^2}\right),$$

where $\beta$ depends on the numbers of fermion and boson loops contributing to the gauge boson self-energy. In QED where the photon self-energy arises from fermion loops alone $\alpha$ increases with energy, whereas $\alpha_S$ decreases with energy due to the presence of gluon loops. Because of the weak boson self-interactions, which are a consequence of the SU(2) gauge symmetry, $\alpha_W$ also decreases with increasing energy scale, although not as rapidly as $\alpha_S$. The running of the different coupling constants therefore tends to bring their values together. It seems plausible that at some high-energy scale, the coupling constants associated with the U(1), SU(2) and SU(3) gauge symmetries converge to a single value. In the mid 1970s, it was suggested by Georgi and Glashow that the observed gauge symmetries of the Standard Model could be accommodated within a larger SU(5) symmetry group. In this Grand Unified Theory (GUT), the coupling constants of the Standard Model are found to converge (although not exactly) at an energy scale of about $10^{15}$ GeV, as shown in Figure 18.4a.

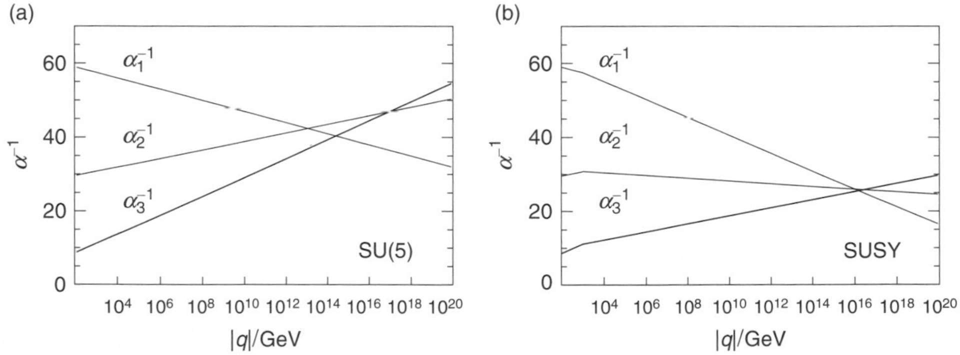


Fig. 18.4 An illustration of the running of the coupling constants in: (a) the SU(5) Grand Unified Theory and (b) a supersymmetric extension of SU(5) with new particles with masses of 1 TeV. It should be noted that, in the SU(5) model, the coupling constant of the U(1) symmetry is not $\alpha$ but $\alpha_1 = 5/3g'$.

The running of the coupling constants shown in Figure 18.4a assumes that only Standard Model particles enter the loops in the gauge boson propagators. If there is physics beyond the Standard Model with new particles at a mass scale $\Lambda$, these particles also would contribute to the gauge boson self-energy terms through extra fermionic and bosonic loops, modifying the running of the coupling constants for $q^2 > \Lambda^2$. For example, Figure 18.4b shows how supersymmetric particles at a scale of $\Lambda_{\text{SUSY}} = 1$ TeV would modify the evolution of the U(1), SU(2) and SU(3) couplings within the SU(5) GUT. Remarkably, the coupling constants converge to a single value of $\alpha_{\text{GUT}} \simeq 1/26$ at $|q| \sim 10^{16}$ GeV. In some sense, this convergence is inevitable since two non-parallel lines will always cross, and with the appropriate choice of the mass scale for new physics the three lines can always be made to meet at a single point. Nevertheless, it is interesting that the required mass scale turns out to be only 1 TeV.

It is now known that SU(5) is not the correct gauge group for a GUT; the predicted value for $\sin^2\theta_W$ is incompatible with the measured value. Despite this, the convergence of the coupling constants strongly suggests that the three forces of the Standard Model are the low-energy manifestations of some larger, as yet unknown, unified theory.

### 18.2.4 What is the nature of the Higgs boson?

The experimental study of the Higgs boson at the LHC is undoubtedly one of the most exciting areas in contemporary particle physics. Within the Standard Model, the Higgs boson is unique; it is the only fundamental scalar in the theory. Establishing the properties of the Higgs boson such as its spin, parity and branching ratios is essential to understand whether the observed particle is the Standard Model Higgs boson or something more exotic.

In the Standard Model, the Higgs mechanism assumes a doublet of complex scalar fields. Whilst this is the simplest choice, it is not unique. For example, supersymmetric extensions to the Standard Model require (at least) two doublets of complex scalar fields. In the two-Higgs doublet model (2HDM), three of the eight scalar fields are the Goldstone bosons that give mass to the W and Z bosons. The remaining five fields correspond to five physical Higgs bosons; two CP-even neutral scalars h and $H^0$, two charged scalar particles $H^\pm$, and a CP-odd neutral scalar $A^0$. In supersymmetry, the neutral Higgs boson (denoted h) must be light and can appear very much like the Standard Model Higgs boson, whereas the $H^\pm$, $A^0$ and $H^0$ can be very massive.

In supersymmetric models, the two Higgs doublets, which have different vacuum expectation values, respectively give the masses to the fermions in the upper and lower components of the weak isospin doublets. In this case, the couplings of the light Higgs boson to the fermions will differ from the Standard Model predictions, although the differences may be quite small. Consequently, the measurements of the branching ratios of the 125 GeV Higgs boson may reveal physics beyond the Standard Model. In the coming years, the study of the Higgs boson at the LHC will form one of the main thrusts of experimental particle physics. On a longer time scale, even more precise studies may be possible at a future $e^+e^-$ linear collider, such as the International Linear Collider (ILC) or the Compact Linear Collider (CLIC).

### 18.2.5 Flavour and the origin of CP violation

There are a number of fundamental questions related to the flavour sector of the Standard Model. Although it appears that there are exactly three generations of fermions, the Standard Model provides no explanation of *why* this is the case. Likewise, the Standard Model provides no clear explanation of why the CKM matrix is almost diagonal, and in contrast, the PMNS matrix is relatively "flat".

Furthermore, the complex phases in the CKM and PMNS matrix are the only places in the Standard Model where CP violation can be accommodated. Whilst CP violation in the quark sector has been studied in great depth, CP violation in the neutrino sector has yet to be observed. The measurement of the parameter $\delta$ in the PMNS matrix will be the focus of the next generation of long-baseline neutrino oscillations experiments.

However, even if CP violation is observed in neutrino oscillations, it seems quite possible that the CP violation in the Standard Model is insufficient to explain the observed matter–antimatter asymmetry of the Universe. One solution to this apparent problem is that (possibly large) CP-violating effects may occur in as yet undiscovered physics beyond the Standard Model. It is possible that such effects will be observed in the coming years, either directly or through loop corrections, in the decays of the vast numbers of b-quarks produced in the LHCb experiment at the LHC and the Belle-II experiment at KEK in Japan.

### 18.2.6 Are neutrinos Majorana particles?

The masses of the neutrinos are very different from the masses of the other fermions. If neutrinos are normal Dirac particles, this would imply an unnaturally small Yukawa coupling to the Higgs field. Whilst this is possible, the seesaw mechanism, described in Section 17.8.1, provides an attractive explanation for the smallness of the neutrino masses. Although the presence of a Majorana mass term in the Lagrangian would not automatically imply that neutrinos are Majorana particles, this would be a real possibility. In this case, the neutrinos would be their own antiparticles, $\nu \equiv \overline{\nu} \equiv \nu_M$.

Perhaps surprisingly, the observable effects of removing the distinction between neutrinos and antineutrinos are very small. In the Standard Model, the neutrino

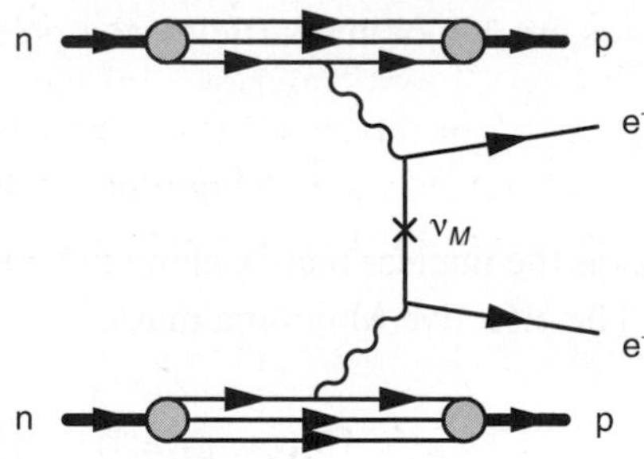


Fig. 18.5 The Feynman diagram for neutrinoless double β-decay.

produced in the decay $\pi^+ \to \mu^+\nu_\mu$ will always produce a $\mu^-$ in its subsequent $\nu_\mu n \to \mu^- p$ charged-current interactions (assuming it has not oscillated). Therefore in the Standard Model, the net number of leptons in the Universe, $L$ = N(leptons) − N(antileptons), is constant and lepton number $L$ is said to be conserved. If neutrinos were Majorana particles, the neutrino from the decay $\pi^+ \to \mu^+\nu_M$ could in principle interact as a RH chiral antiparticle $\nu_M p \to \mu^+ n$. The net effect would be a $\Delta L = -2$ change in lepton number and lepton number no longer would be a conserved quantity. However, because of the smallness of neutrino masses, the neutrino helicity states are almost identical to the chiral states and the fraction of lepton-number-violating processes would be suppressed by $O(m_\nu^2/m_\mu^2)$, which is too small to be observable. Consequently, experiments have focussed on the possibility of neutrinoless double β-decay, which can occur *only* if neutrinos are Majorana particles.

Certain even–even nuclei, where the usual $\beta^\pm$-decay or electron-capture processes are energetically forbidden, can decay to a more tightly bound even–even nucleus by the double β-decay process, $(Z, A) \to (Z + 2, A) + 2e^- + 2\overline{\nu}_e$, which can be thought of in terms of two simultaneous single β-decays. Whilst such $2\nu\beta\beta$-decays are rare, with half-lives in the range $\tau_{1/2} \sim 10^{19} - 10^{25}$ years, they have been observed for a number of isotopes. If neutrinos were Majorana particles, the lepton number violating *neutrinoless* double β-decays processes ($0\nu\beta\beta$) can occur through the Feynman diagram shown Figure 18.5. Experimentally $0\nu\beta\beta$ can be distinguished from the more common $2\nu\beta\beta$-decays from the energy spectrum of the electrons. Neutrinoless double β-decays would produce mono-energetic electrons with energy

$$E_e = \tfrac{1}{2}Q = \tfrac{1}{2}\left[M(Z, A) - M(Z + 2, A)\right],$$

where $M(Z, A)$ and $M(Z + 2, A)$ are the masses of the parent and daughter nuclei. In contrast, $2\nu\beta\beta$-decays produce a broad spectrum of electron energies with very few being produced close to the end point of $\tfrac{1}{2}Q$.

If neutrinos are Majorana particles, the predicted $0\nu\beta\beta$-decay rates are proportional to

$$\Gamma \propto G_F^4\, |m_{\beta\beta}|^2 \times |\mathcal{M}_{\rm nucl}|^2,$$

where $\mathcal{M}_{\rm nucl}$ is the nuclear matrix element and $m_{\beta\beta}$ is known as the effective Majorana mass. The effective Majorana mass,

$$m_{\beta\beta} = \sum_{i=1}^{3} U_{ei}^2\, m_{\nu_i}, \tag{18.2}$$

depends on the neutrino masses and the elements of the PMNS matrix. As a result, the predicted decay rates depend on the neutrino mass hierarchy, with the inverted hierarchy typically leading to larger predicted rates.

A number of experiments have searched for $0\nu\beta\beta$-decays in processes such as ${}^{76}_{32}\mathrm{Ge} \to {}^{76}_{34}\mathrm{Se} + e^- + e^-$ and ${}^{136}_{54}\mathrm{Xe} \to {}^{136}_{56}\mathrm{Ba} + e^- + e^-$. To date, there has been no confirmed observation of neutrinoless double β-decay, with the most stringent lifetime limits being set at $\tau_{1/2}^{0\nu\beta\beta} \gtrsim 10^{25}$ years. Nevertheless, the experiments are only just beginning to reach the required level of sensitivity where it might be possible observe $0\nu\beta\beta$-decay, even for the most optimistic values of $m_{\beta\beta}$. In the coming years, a number of larger experiments will start to search for neutrinoless double β-decay. A positive signal would represent a major discover, demonstrating that the neutrinos are fundamentally different from all other particles.

## 18.3 Closing words

Most of the theoretical concepts in the Standard Model were in place by the end of the 1960s. These ideas gained strong support with the discovery of the W and Z bosons at CERN in the mid 1980s. In the last decade of the twentieth century, the precision studies of the W and Z bosons provided tests of the predictions of the Standard Model at the quantum loop level. The start of the full operation of the LHC in 2010 represented a new stage in the experimental study of particle physics. With the discovery of the Higgs boson in 2012, the full spectrum of the Standard Model particles had been observed. This period of nearly 50 years from the late 1960s to 2012, represented a giant leap forward in our understanding of the Universe at the most fundamental level. I hope this book has helped you appreciate some of the profound theoretical ideas and the beautiful experimental measurements that have made the Standard Model of particle physics one of the central pillars of modern physics.

Despite it success, it should not be forgotten that the Standard Model is not the end of the story; there are just too many loose ends. The coming years will see the high-luminosity operation of the LHC at a centre-of-mass energy close to 14 TeV. In addition, a new generation of experiments will search for signatures for physics beyond the Standard Model. We may be standing at the threshold of new and potentially revolutionary discoveries. Only time will tell whether this will be the direct detection of dark matter, the demonstration that neutrinos are Majorana particles, the discovery of supersymmetry, or quite possibly something completely unexpected. The only certain thing is that interesting times lie ahead of us.

# Appendix A The Dirac delta-function

The Dirac delta-function is used in the development of the relativistic formulation of decay rates and interaction cross sections, described in Chapter 3. For this reason, a brief overview of the main properties of the Dirac delta-function is given here.

## A.1 Definition of the Dirac delta-function

The Dirac delta-function, written as $\delta(x)$, is defined to be an infinitesimally narrow peak of unit area, such that for all values of $x_1$ and $x_2$,

$$\int_{x_1}^{x_2} \delta(x-a)\,\mathrm{d}x = \begin{cases} 1 & \text{if } x_1 < a < x_2 \\ 0 & \text{otherwise} \end{cases}. \tag{A.1}$$

Figure A.1 shows a representation of $\delta(x-a)$ as a spike that is only non-zero at $x = a$. Because the integral of $\delta(x-a)$ is unity, it follows that

$$\int_{-\infty}^{+\infty} f(x)\delta(x-a)\,\mathrm{d}x = f(a). \tag{A.2}$$

The integral over $\delta(x-a)$ picks out the value of the integrand at $x = a$. This useful property can be used to express energy and momentum conservation in an integral form. For example, in the decay $a \to 1 + 2$, the only non-zero contribution to the integral

$$\int \cdots\ \delta(E_a - E_1 - E_2)\,\mathrm{d}E_1,$$

occurs for $E_a = E_1 + E_2$. Similarly, a three-dimensional delta-function $\delta^3(\mathbf{x})$ can be defined such that

$$\int \cdots\ \delta^3(\mathbf{p}_a - \mathbf{p}_1 - \mathbf{p}_2)\,\mathrm{d}^3\mathbf{p}_1,$$

is equivalent to imposing momentum conservation, $\mathbf{p}_a = \mathbf{p}_1 + \mathbf{p}_2$.

Whilst there is no unique functional form for the Dirac delta-function, it is sometimes helpful to think in terms of an explicit function that has the properties of

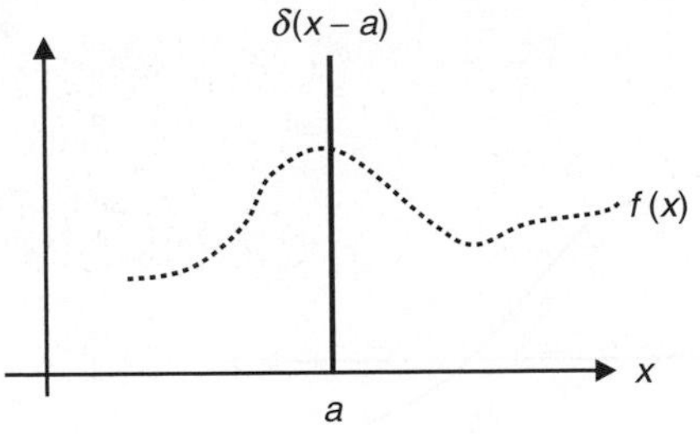


Fig. A.1 A graphical representation of $\delta(x-a)$.

(A.1). One possible representation of the Dirac delta-function is an infinitesimally narrow Gaussian distribution,

$$\delta(x-a) = \lim_{\sigma\to 0} \frac{1}{\sqrt{2\pi}\sigma} \exp\left(-\frac{(x-a)^2}{2\sigma^2}\right).$$

## A.2 Fourier transform of a delta-function

From (A.2) the Fourier transform of $\delta(x)$ can be written

$$g(k) = \mathcal{F}\{\delta(x)\} = \int_{-\infty}^{+\infty} \delta(x)e^{-ikx}\mathrm{d}x = e^0 = 1. \tag{A.3}$$

The Fourier transform of $\delta(x)$ is therefore a uniform distribution. This can be understood in terms of the Gaussian representation of the $\delta$-function; the Fourier transform of a Gaussian distribution of width $\sigma$ is a Gaussian of width $1/\sigma$. Hence, for the limiting case where $\sigma \to 0$, the Fourier transform has a width tending to infinity, corresponding to a uniform distribution. Using (A.3), the delta-function, $\delta(x-x_0)$, can be written as

$$\delta(x-x_0) = \frac{1}{2\pi}\int_{-\infty}^{+\infty} e^{+ik(x-x_0)}\,\mathrm{d}k,$$

and hence, from a simple relabelling of the variables, the integral

$$\int_{-\infty}^{+\infty} e^{i(k-k_0)x}\mathrm{d}x = 2\pi\delta(k-k_0). \tag{A.4}$$

## A.3 Delta-function of a function

In Chapter 3, the properties of the delta-function of a function are used to simplify the integrals that arise in the calculation of decay rates and cross sections

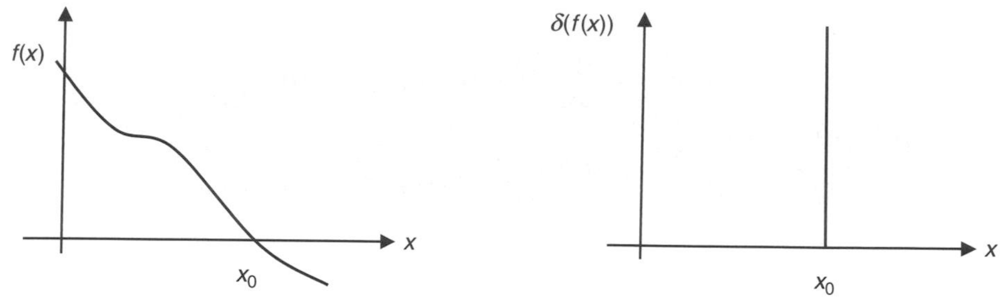


Fig. A.2 A function $f(x)$ and a graphical representation of $\delta(f(x))$.

in relativistic quantum mechanics. To derive the mathematical expression for the delta-function of a function $\delta(f(x))$ consider the function $f(x)$ that is zero at $x = x_0$, as shown in Figure A.2. Here $\delta(f(x))$ is only non-zero at $x_0$ but is no longer necessarily normalised to unit area. This can be seen by writing $y = f(x)$ and $\delta(f(x)) = \delta(y)$, where $f(x_0) = 0$. From the definition of the delta-function of (A.1)

$$\int_{y_1}^{y_2} \delta(y)\,\mathrm{d}y = \begin{cases} 1 & \text{if } y_1 < 0 < y_2 \\ 0 & \text{otherwise} \end{cases} .$$

Changing variable from $y$ to $x$ with $y = f(x)$ gives

$$\int_{x_1}^{x_2} \delta(f(x)) \frac{\mathrm{d}f}{\mathrm{d}x}\,\mathrm{d}x = \begin{cases} 1 & \text{if } x_1 < x_0 < x_2 \\ 0 & \text{otherwise} \end{cases} .$$

From (A.2) this equation can be written as

$$\left|\frac{\mathrm{d}f}{\mathrm{d}x}\right|_{x_0} \int_{x_1}^{x_2} \delta(f(x))\mathrm{d}x = \begin{cases} 1 & \text{if } x_1 < x_0 < x_2 \\ 0 & \text{otherwise} \end{cases} . \tag{A.5}$$

Writing the RHS of (A.5) as a delta function and using (A.1) gives

$$\left|\frac{\mathrm{d}f}{\mathrm{d}x}\right|_{x_0} \int_{x_1}^{x_2} \delta(f(x))\,\mathrm{d}x = \int_{x_1}^{x_2} \delta(x - x_0)\,\mathrm{d}x.$$

Hence the delta-function of a function is given by

$$\delta(f(x)) = \left|\frac{\mathrm{d}f}{\mathrm{d}x}\right|_{x_0}^{-1} \delta(x - x_0). \tag{A.6}$$

# Appendix B **Dirac equation**

The main properties of the Dirac equation and its solutions were developed in Chapter 4. Here, some of the more mathematically demanding aspects are considered. Firstly, the non-relativistic limit of the Dirac equation is used to identify the magnetic moment of a Dirac fermion and then the Lorentz transformation properties of the Dirac equation and its solutions are derived.

## B.1 Magnetic moment of a Dirac fermion

The magnetic moment $\mu$ of a Dirac particle can be identified by taking the non-relativistic limit of the Dirac equation in the presence of an electromagnetic field, where the energy associated with the magnetic moment is $U = -\boldsymbol{\mu}\cdot\mathbf{B}$. The Dirac equation for the four-component wavefunction $\psi$, written in operator form, is

$$\begin{aligned} \hat{H}_D\psi &= E\psi, \\ (\boldsymbol{\alpha}\cdot\hat{\mathbf{p}} + \beta m)\psi &= E\psi. \end{aligned} \tag{B.1}$$

In classical dynamics, the motion of a particle of charge $q$ in an electromagnetic field $A^\mu = (\phi, \mathbf{A})$ can be obtained from the minimal substitution

$$E \to E - q\phi \quad \text{and} \quad \mathbf{p} \to \mathbf{p} - q\mathbf{A}.$$

Applying this to the Dirac equation of (B.1) leads to

$$[\boldsymbol{\alpha}\cdot(\hat{\mathbf{p}} - q\mathbf{A}) + \beta m]\,\psi = (E - q\phi)\psi.$$

For the Pauli–Dirac representation of the $\alpha$ and $\beta$ matrices, this can be written

$$\begin{pmatrix} (E - m - q\phi)I & -\boldsymbol{\sigma}\cdot(\hat{\mathbf{p}} - q\mathbf{A}) \\ -\boldsymbol{\sigma}\cdot(\hat{\mathbf{p}} - q\mathbf{A}) & (E + m - q\phi)I \end{pmatrix}\begin{pmatrix} \psi_A \\ \psi_B \end{pmatrix} = 0, \tag{B.2}$$

where $I$ is the $2\times 2$ identity matrix and the four-component spinor $\psi$ has been split into the upper and lower two components $\psi_A$ and $\psi_B$. Equation (B.2) gives coupled equations for $\psi_A$ in terms of $\psi_B$,

$$[\sigma \cdot (\hat{\mathbf{p}} - q\mathbf{A})]\, \psi_B = (E - m - q\phi)\psi_A, \tag{B.3}$$

$$[\sigma \cdot (\hat{\mathbf{p}} - q\mathbf{A})]\, \psi_A = (E + m - q\phi)\psi_B. \tag{B.4}$$

In the non-relativistic limit, where $E \approx m \gg q\phi$, (B.4) can be written

$$\psi_B \approx \frac{1}{2m}\sigma \cdot (\hat{\mathbf{p}} - q\mathbf{A})\psi_A.$$

Substituting this expression for $\psi_B$ into (B.3) gives

$$(E - m - q\phi)\psi_A = \frac{1}{2m}\left[\sigma \cdot (\hat{\mathbf{p}} - q\mathbf{A})\right]^2 \psi_A. \tag{B.5}$$

For arbitrary three-vectors **a** and **b**,

$$\sigma \cdot \mathbf{a} = \begin{pmatrix} a_z & a_x - ia_y \\ a_x + ia_y & -a_z \end{pmatrix} \quad \text{and} \quad \sigma \cdot \mathbf{b} = \begin{pmatrix} b_z & b_x - ib_y \\ b_x + ib_y & -b_z \end{pmatrix},$$

from which it follows that

$$(\sigma \cdot \mathbf{a})(\sigma \cdot \mathbf{b}) \equiv (\mathbf{a} \cdot \mathbf{b})\, I + i\sigma \cdot (\mathbf{a} \times \mathbf{b}). \tag{B.6}$$

Applying this identity to the RHS of (B.5) gives

$$(E - m - q\phi)\psi_A = \frac{1}{2m}\left[(\hat{\mathbf{p}} - q\mathbf{A})^2 - iq\sigma \cdot (\hat{\mathbf{p}} \times \mathbf{A} + \mathbf{A} \times \hat{\mathbf{p}})\right]\psi_A.$$

Writing the momentum operator as $\hat{\mathbf{p}} = -i\nabla$, leads to

$$(E - m - q\phi)\psi_A = \frac{1}{2m}\left[(\hat{\mathbf{p}} - q\mathbf{A})^2 - q\sigma \cdot (\nabla \times \mathbf{A} + \mathbf{A} \times \nabla)\right]\psi_A. \tag{B.7}$$

Remembering that the differential operator acts on everything to the right, the term

$$\nabla \times (\mathbf{A}\psi_A) = (\nabla \times \mathbf{A})\psi_A + (\nabla\psi_A) \times \mathbf{A},$$

and therefore the last term in (B.7) can be written

$$\begin{aligned} \nabla \times (\mathbf{A}\psi_A) + \mathbf{A} \times (\nabla\psi_A) &= (\nabla \times \mathbf{A})\psi_A + (\nabla\psi_A) \times \mathbf{A} + \mathbf{A} \times (\nabla\psi_A) \\ &= (\nabla \times \mathbf{A})\psi_A \\ &\equiv \mathbf{B}\psi_A, \end{aligned}$$

where the last step follows from the relation between the magnetic flux density **B** and the vector potential, $\mathbf{B} = \nabla \times \mathbf{A}$. Hence (B.7) can be written

$$E\psi_A = \left[m + \frac{1}{2m}(\hat{\mathbf{p}} - q\mathbf{A})^2 + q\phi - \frac{q}{2m}(\sigma \cdot \mathbf{B})\right]\psi_A.$$

This is the non-relativistic limit of the energy of a Dirac particle in an electromagnetic field, from which the potential energy due to the magnetic interaction between the particle and the magnetic field can be seen to be

$$U = -\frac{q}{2m}(\sigma \cdot \mathbf{B}) \equiv -\boldsymbol{\mu} \cdot \mathbf{B},$$

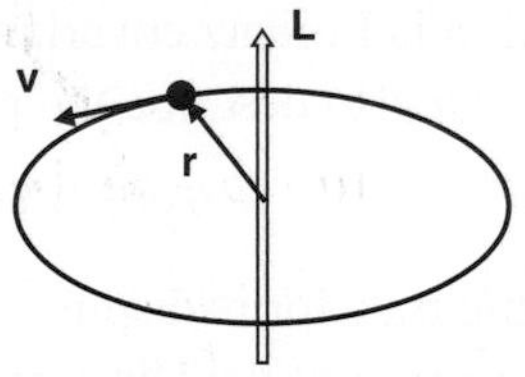


Fig. B.1 The classical magnetic moment of a current loop with angular momentum $L = mvr$.

from which the magnetic moment of the Dirac fermion can be identified as

$$\boldsymbol{\mu} = \frac{q}{2m}\boldsymbol{\sigma}.$$

Because the intrinsic spin of a fermion can be written as $\mathbf{S} = \frac{1}{2}\boldsymbol{\sigma}$, the intrinsic magnetic moment is

$$\boldsymbol{\mu} = \frac{q}{m}\mathbf{S}.$$

Classically, the magnetic moment associated with a current loop, as shown in Figure B.1, is given by the current multiplied by the area of the loop $\pi r^2$, hence

$$\boldsymbol{\mu} = \pi r^2 \frac{q v}{2\pi r}\hat{\mathbf{z}} = \frac{q}{2m}\mathbf{L},$$

where $|\mathbf{L}| = mvr$. Thus, the relationship between the magnetic moment $\boldsymbol{\mu}$ and the intrinsic angular momentum $\mathbf{S}$ of a Dirac fermion differs from the corresponding expression in classical physics by a factor of two. This is usually expressed in terms of the gyromagnetic ratio $g$ defined such that

$$\boldsymbol{\mu} = g\frac{q}{2m}\mathbf{S},$$

where the Dirac equation predicts $g = 2$.

## B.2 Covariance of the Dirac equation

The idea of Lorentz covariance is familiar from electromagnetism. Maxwell's equations for the electric and magnetic fields, $\mathbf{E}$ and $\mathbf{B}$, are Lorentz covariant, meaning that the equations are the same in all frames, although the fields are different.

Here consider a frame $\Sigma$ with coordinates $x^\mu$. In this frame, a particle with wavefunction $\psi(x)$ satisfies the Dirac equation,

$$i\gamma^\mu \partial_\mu \psi(x) = m\psi(x). \tag{B.8}$$

The Dirac equation is Lorentz covariant if an observer in a different frame $\Sigma'$, with coordinates $x'^{\nu}$, can also describe the particle by the Dirac equation

$$i\gamma^{\nu}\partial'_{\nu}\psi'(x') = m\psi'(x'), \tag{B.9}$$

where $\psi'(x')$ is the transformed spinor in $\Sigma'$ and the $\gamma$-matrices take the same form as in $\Sigma$. The covariance of the Dirac equation can be established if it can be shown that the spinors in $\Sigma$ and $\Sigma'$ can be related by the transformation,

$$\psi'(x') = S\psi(x),$$

where $S$ is a $4 \times 4$ matrix.

The (covariant) four-derivative $\partial'_{\nu}$ is related to $\partial_{\mu}$ by the Lorentz transformation

$$\partial'_{\nu} = \Lambda_{\nu}{}^{\mu}\partial_{\mu}.$$

Hence the required form of the Dirac equation in $\Sigma'$, given in (B.9), can be written

$$i\gamma^{\nu}\Lambda_{\nu}{}^{\mu}\partial_{\mu}S\psi(x) = mS\psi(x).$$

Because $S$ is a constant matrix, $\partial_{\mu}S\psi = S\partial_{\mu}\psi$, and thus

$$i\gamma^{\nu}\Lambda_{\nu}{}^{\mu}S\left[\partial_{\mu}\psi(x)\right] = mS\psi(x).$$

This can be compared to the Dirac equation as expressed in $\Sigma$, which when multiplied by $S$ is

$$iS\gamma^{\mu}\left[\partial_{\mu}\psi(x)\right] = mS\psi(x).$$

Hence the Lorentz covariance of the Dirac equation will be demonstrated if a matrix $S$ can be found such that

$$S\gamma^{\mu} = \gamma^{\nu}\Lambda_{\nu}{}^{\mu}S, \tag{B.10}$$

which would imply that the Dirac equation is valid in all inertial frames.

Rather than trying to find the general form for $S$, consider the case where the frame $\Sigma'$ is travelling with velocity $+\mathrm{v}\hat{\mathbf{z}}$ relative to the frame $\Sigma$. In this case

$$\Lambda_{\nu}{}^{\mu} = \begin{pmatrix} \gamma & 0 & 0 & +\beta\gamma \\ 0 & 1 & 0 & 0 \\ 0 & 0 & 1 & 0 \\ +\beta\gamma & 0 & 0 & \gamma \end{pmatrix},$$

where $\beta = \mathrm{v}/c$ and $\gamma^2 = (1-\beta^2)^{-1}$. In this case, the four relations of (B.10), one for each of $\mu = 0, 1, 2, 3$, can be written as

$$S\gamma^0 = \gamma\gamma^0 S + \beta\gamma\gamma^3 S, \tag{B.11}$$
$$S\gamma^1 = \gamma^1 S, \tag{B.12}$$
$$S\gamma^2 = \gamma^2 S, \tag{B.13}$$
$$S\gamma^3 = \beta\gamma\gamma^0 S + \gamma\gamma^3 S. \tag{B.14}$$

It is not difficult to show that the matrix

$$S = aI - b\gamma^0\gamma^3 \quad \text{with} \quad a = \sqrt{\tfrac{1}{2}(\gamma+1)} \quad \text{and} \quad b = \sqrt{\tfrac{1}{2}(\gamma-1)}, \tag{B.15}$$

satisfies these four equations and, as required, tends to the identity matrix as $\gamma \to 1$. Thus $S$ describes the transformation properties of Dirac spinors for a Lorentz transformation in the $+z$ direction. With this transformation, the mathematical form of the Dirac equation is preserved, demonstrating its Lorentz covariance.

Now consider the effect of this transformation on the spinor for a particle at rest in the frame $\Sigma$ as given by (4.42),

$$u_1(p) = u_1(m, \mathbf{0}) = \sqrt{2m}\begin{pmatrix} 1 \\ 0 \\ 0 \\ 0 \end{pmatrix}.$$

In the Dirac–Pauli representation, the matrix $S = aI - b\gamma^0\gamma^3$ is

$$S = \begin{pmatrix} a & 0 & -b & 0 \\ 0 & a & 0 & +b \\ -b & 0 & a & 0 \\ 0 & +b & 0 & a \end{pmatrix}.$$

Therefore, in the frame $\Sigma'$ the spinor of the particle is

$$u_1'(p') = S\,u_1(p) = \sqrt{2m}\begin{pmatrix} a \\ 0 \\ -b \\ 0 \end{pmatrix} = \sqrt{E'+m}\begin{pmatrix} 1 \\ 0 \\ -b/a \\ 0 \end{pmatrix},$$

where the last step follows from $2ma^2 = m(\gamma+1) = E'+m$. The ratio $b/a$, expressed in terms of $\beta$ and $\gamma$, is

$$\frac{b}{a} = \sqrt{\frac{\gamma-1}{\gamma+1}} = \sqrt{\frac{\gamma^2-1}{(\gamma+1)^2}} = \frac{\beta\gamma}{(\gamma+1)}.$$

In the frame $\Sigma'$, which is moving with a velocity $\mathbf{v} = \mathrm{v}\hat{\mathbf{z}}$ with respect to the frame $\Sigma$, the velocity of the particle is $\mathbf{v}' = -\mathrm{v}\hat{\mathbf{z}}'$ and therefore $p_z' = -m\gamma\beta$ and $E' = m\gamma$. Consequently

$$\frac{b}{a} = -\frac{p_z'}{E'+m},$$

and the spinor in the frame $\Sigma'$ can be written

$$u'_1(p') = \sqrt{E'+m}\begin{pmatrix} 1 \\ 0 \\ \frac{p'_z}{E'+m} \\ 0 \end{pmatrix},$$

which is, as expected, the corresponding general solution to the Dirac equation for a particle with momentum in the $z$-direction, as given by (4.51).

## B.3 Four-vector current

Section 11.3 introduced the concept of bilinear covariants, formed from two spinors, which have well-defined Lorentz transformation properties. For example, it was stated that $\overline{\psi}\phi$ transforms as a scalar and $\overline{\psi}\gamma^\mu\phi$ transforms as four-vector. Having identified the transformation properties of Dirac spinors, it is now possible to justify these statements.

For the two frames $\Sigma$ and $\Sigma'$, defined above, the spinor $\psi(x)$ in the frame $\Sigma$ transforms to $\psi'(x') = S\psi(x)$ in the frame $\Sigma'$, where $S$ is the matrix of (B.15). The adjoint spinor in the primed frame can be written

$$\overline{\psi}' = \psi'^\dagger\gamma^0 = (S\psi)^\dagger\gamma^0 = \psi^\dagger S^\dagger\gamma^0.$$

Hence the spinor product $\overline{\psi}'\phi'$ can be written

$$\overline{\psi}'\phi' = \psi^\dagger S^\dagger\gamma^0 S\phi, \tag{B.16}$$

For a Lorentz transformation along the $z$-axis, $S = aI - b\gamma^0\gamma^3$, and its Hermitian conjugate is given by

$$S^\dagger = aI - b\gamma^{3\dagger}\gamma^{0\dagger} = aI - b\gamma^0\gamma^3.$$

Therefore

$$\begin{aligned} S^\dagger\gamma^0 S &= (aI - b\gamma^0\gamma^3)\gamma^0(aI - b\gamma^0\gamma^3) \\ &= (aI - b\gamma^0\gamma^3)(aI + b\gamma^0\gamma^3)\gamma^0 \\ &= (a^2 I - b^2\gamma^0\gamma^3\gamma^0\gamma^3)\gamma^0 \\ &= (a^2 - b^2)\gamma^0 \\ &= \gamma^0. \end{aligned}$$

Hence (B.16) becomes

$$\overline{\psi}'\phi' = \psi^\dagger S^\dagger\gamma^0 S\phi = \psi^\dagger\gamma^0\phi = \overline{\psi}\phi,$$

from which it can be concluded that the spinor combination $\overline{\psi}\phi$ forms a Lorentz-invariant scalar.

The proof that $j^{\nu} = \overline{\psi}\gamma^{\nu}\phi$ transforms as a four-vector, requires some care and familiarity with the manipulation of index notation. In the derivation that follows it should be remembered that $S$ and the $\gamma$-matrices are $4 \times 4$ matrices and the order in which they appear in the various expressions needs to be retained. In contrast, the individual elements of the tensor for the Lorentz transformation $\Lambda^{\rho}{}_{\mu}$ are just numbers which can be written in any position in an expression. With this in mind, consider the Lorentz transformation properties of

$$\overline{\psi}'\gamma^{\mu}\phi' = (\psi^{\dagger}S^{\dagger}\gamma^{0})\gamma^{\mu}S\phi. \tag{B.17}$$

This is most easily simplified by finding an expression for $\gamma^{\mu}S$ in terms of $S\gamma^{\mu}$, and then using $S^{\dagger}\gamma^{0}S = \gamma^{0}$. Multiplying (B.10) by $\Lambda^{\rho}{}_{\mu}$, gives

$$\Lambda^{\rho}{}_{\mu}S\gamma^{\mu} = \Lambda^{\rho}{}_{\mu}\Lambda_{\nu}{}^{\mu}\gamma^{\nu}S.$$

It is straightforward to show that $\Lambda^{\rho}{}_{\mu}\Lambda_{\nu}{}^{\mu} = \delta^{\rho}_{\nu}$, and therefore

$$\Lambda^{\rho}{}_{\mu}S\gamma^{\mu} = \gamma^{\rho}S. \tag{B.18}$$

Relabelling the indices allows $\gamma^{\mu}S$ in (B.17) to be written as $\gamma^{\mu}S = \Lambda^{\mu}{}_{\nu}S\gamma^{\nu}$, such that

$$\begin{aligned}
\overline{\psi}'\gamma^{\mu}\phi' &= \psi^{\dagger}S^{\dagger}\gamma^{0}(\Lambda^{\mu}{}_{\nu}S\gamma^{\nu})\phi \\
&= \Lambda^{\mu}{}_{\nu}\psi^{\dagger}(S^{\dagger}\gamma^{0}S)\gamma^{\nu}\phi \\
&= \Lambda^{\mu}{}_{\nu}\psi^{\dagger}\gamma^{0}\gamma^{\nu}\phi \\
&= \Lambda^{\mu}{}_{\nu}\overline{\psi}\gamma^{\nu}\phi.
\end{aligned}$$

Therefore $j'^{\mu} = \overline{\psi}'\gamma^{\mu}\phi'$ is related to $j^{\nu} = \overline{\psi}\gamma^{\nu}\phi$ by

$$j'^{\mu} = \Lambda^{\mu}{}_{\nu}j^{\nu}, \tag{B.19}$$

proving that $j^{\nu} = \overline{\psi}\gamma^{\nu}\phi$ transforms as a (contravariant) four-vector.

## Problems

**B.1*** Show that the matrix

$$S = aI - b\gamma^{0}\gamma^{3} \quad \text{with} \quad a = \sqrt{\tfrac{1}{2}(\gamma+1)} \quad \text{and} \quad b = \sqrt{\tfrac{1}{2}(\gamma-1)},$$

satisfies the equations of (*B*.11)-(*B*.14).

**B.2*** By considering the Lorentz transformations along the $z$-axis

$$\boldsymbol{\Lambda} = \Lambda^{\mu}{}_{\nu} = \begin{pmatrix} \gamma & 0 & 0 & -\beta\gamma \\ 0 & 1 & 0 & 0 \\ 0 & 0 & 1 & 0 \\ -\beta\gamma & 0 & 0 & \gamma \end{pmatrix} \quad \text{and} \quad \Lambda_{\mu}{}^{\nu} = \begin{pmatrix} \gamma & 0 & 0 & +\beta\gamma \\ 0 & 1 & 0 & 0 \\ 0 & 0 & 1 & 0 \\ +\beta\gamma & 0 & 0 & \gamma \end{pmatrix},$$

show that

$$\Lambda^{\rho}{}_{\mu}\Lambda_{\nu}{}^{\mu} = \delta^{\rho}_{\nu},$$

where $\delta^{\rho}_{\nu}$ is the Kronecker delta-function.

# Appendix C The low-mass hadrons

A large number of hadronic states have been observed and a full listing can be found in Beringer *et al.* (2012). The more commonly encountered light $L = 0$ baryon states (with zero internal orbital angular momentum) are listed in Table C.1, which gives the quark content, mass, lifetime and main decay modes of the listed baryons. Where a baryon decays by the strong interaction, the lifetime is too short to be measured directly, typically $\sim 10^{-23}$ s. In this case the table gives the total decay width $\Gamma = 1/\tau$, which can be determined from the observed invariant mass distribution of the decay products. For example, the invariant mass distribution of the p and $\pi^+$ from the decay $\Delta^{++} \to p\pi^+$ will be a Lorentzian centred on $m(\Delta^{++})$ with FWHM equal to $\Gamma(\Delta^{++})$. Table C.2 gives the corresponding information for the most commonly encountered $L = 0$ meson states.

**Table C.1** The first half of the table shows a number of $L = 0, J^P = \frac{1}{2}^+$ baryons. The second half lists the baryons in the $L = 0$ baryon decuplet with $J^P = \frac{3}{2}^+$.

| Baryon | Quark content | Mass/MeV | Lifetime/Width | Main decay modes |
|---|---|---|---|---|
| p | uud | 938.3 | $> 2.9 \times 10^{29}$ yrs | |
| n | ddu | 939.6 | 880.1 s | $pe^-\bar{\nu}_e$ |
| $\Lambda$ | uds | 1115.7 | $2.6 \times 10^{-10}$ s | $p\pi^-$, $n\pi^0$ |
| $\Sigma^+$ | uus | 1189.4 | $8.0 \times 10^{-11}$ s | $p\pi^0$, $n\pi^+$ |
| $\Sigma^0$ | uds | 1192.6 | $7.4 \times 10^{-20}$ s | $\Lambda\gamma$ |
| $\Sigma^-$ | dds | 1197.4 | $1.5 \times 10^{-10}$ s | $n\pi^-$ |
| $\Xi^0$ | uss | 1314.9 | $2.9 \times 10^{-10}$ s | $\Lambda\pi^0$ |
| $\Xi^-$ | dss | 1321.7 | $1.6 \times 10^{-10}$ s | $\Lambda\pi^-$ |
| $\Lambda_c^+$ | udc | 2286.5 | $2.0 \times 10^{-13}$ s | many |
| $\Sigma_c$ | uuc, udc, ddc | 2453 | 2.2 MeV | $\Lambda_c^+\pi$ |
| $\Xi_c^+$ | usc | 2467.8 | $4.4 \times 10^{-13}$ s | $\Xi + \pi$s and others |
| $\Xi_c^0$ | dsc | 2470.9 | $1.1 \times 10^{-13}$ s | $\Xi^-$ and Ks |
| $\Omega_c^0$ | ssc | 2695.2 | $6.9 \times 10^{-14}$ s | $\Sigma^+$, $\Omega^-$ and Ks |
| $\Lambda_b^0$ | udb | 5619.4 | $1.4 \times 10^{-12}$ s | $\Lambda_c^+ + X$ |
| $\Delta$ | uuu, uud, udd, ddd | 1232 | 117 MeV | $N\pi$ |
| $\Sigma^*$ | uus, uds, dds | 1385 | 36 MeV | $\Lambda\pi$, $\Sigma\pi$ |
| $\Xi^*$ | uss, dss | 1533 | 9 MeV | $\Xi\pi$ |
| $\Omega^-$ | sss | 1672.5 | $8.2 \times 10^{-11}$ s | $\Lambda K^-$, $\Xi^0\pi^-$, $\Xi^-\pi^0$ |

**Table C.2** The first half of the table lists the main properties of a number of commonly encountered $L = 0$ pseudoscalar mesons with $J^P = 0^-$. The second half lists the properties of a number of $L = 0$ vector mesons with $J^P = 1^-$, including the $\Upsilon$ states.

| Meson | Quark content | Mass/MeV | Lifetime/Width | Main decay modes |
|---|---|---|---|---|
| $\pi^\pm$ | $u\bar{d}$, $d\bar{u}$ | 139.6 | $2.6 \times 10^{-8}$ s | $\mu^+\nu_\mu$ |
| $\pi^0$ | $(u\bar{u} - d\bar{d})$ | 135.0 | $8.4 \times 10^{-17}$ s | $\gamma\gamma$ |
| $\eta$ | $(u\bar{u} + d\bar{d} - 2s\bar{s})$ | 547.9 | 130(7) keV | $\gamma\gamma$, $\pi\pi\pi$ |
| $\eta'$ | $(u\bar{u} + d\bar{d} + s\bar{s})$ | 957.8 | 0.199(9) MeV | $\pi\pi\eta$, $\rho^0\gamma$ |
| $K^\pm$ | $u\bar{s}$, $s\bar{u}$ | 493.7 | $1.2 \times 10^{-8}$ s | $\mu^+\nu_\mu$, $\pi^0\pi^+$ |
| $K_S$ | $(d\bar{s} + s\bar{d})$ | 497.6 | $8.9 \times 10^{-11}$ s | $\pi\pi$ |
| $K_L$ | $(d\bar{s} - s\bar{d})$ | 497.6 | $5.1 \times 10^{-8}$ s | $\pi\pi\pi$, $\pi^\pm\ell^\mp\nu_\ell$, |
| $D^\pm$ | $c\bar{d}$, $d\bar{c}$ | 1869.6 | $1.0 \times 10^{-12}$ s | $\overline{K}^0\ell^+\nu_\ell$, $K^-\pi$s, $K_S\pi$s |
| $D^0$, $\overline{D}^0$ | $c\bar{u}$, $u\bar{c}$ | 1864.9 | $4.1 \times 10^{-13}$ s | $K^0X$, $\overline{K}^0X$, $K^-\ell^+\nu_\ell$ |
| $D_s^\pm$ | $c\bar{s}$, $s\bar{c}$ | 1968.5 | $5.0 \times 10^{-13}$ s | many |
| $\eta_c(1S)$ | $c\bar{c}$ | 2981 | 29.7(1) MeV | hadrons |
| $B^\pm$ | $u\bar{b}$, $b\bar{u}$ | 5279.3 | $1.6 \times 10^{-12}$ s | many |
| $B^0$, $\overline{B}^0$ | $d\bar{b}$, $b\bar{d}$ | 5279.6 | $1.5 \times 10^{-12}$ s | many |
| $B_s^0$, $\overline{B}_s^0$ | $s\bar{b}$, $b\bar{s}$ | 5366.8 | $1.5 \times 10^{-12}$ s | many |
| $B_c^\pm$ | $c\bar{b}$, $b\bar{c}$ | 6277 | $4.5 \times 10^{-13}$ s | many |
| $\rho^\pm, \rho^0$ | $u\bar{d}$, $d\bar{u}$, $(u\bar{u} - d\bar{d})$ | 774.5 | 149.1 MeV | $\pi\pi$ |
| $\omega$ | $(u\bar{u} + d\bar{d})$ | 782.7 | 8.5 MeV | $\pi^+\pi^-\pi^0$ |
| $\phi$ | $s\bar{s}$ | 1019.5 | 4.3 MeV | $K^+K^-$, $K_LK_S$, $\rho\pi$ |
| $K^{*\pm}$ | $u\bar{s}$, $s\bar{u}$ | 891.7 | 50.8 MeV | $K\pi$ |
| $K^{*0}$, $\overline{K}^{*0}$ | $d\bar{s}$, $s\bar{d}$) | 895.9 | 46 MeV | $K\pi$ |
| $D^{*0}$, $\overline{D}^{*0}$ | $c\bar{u}$, $u\bar{c}$ | 2007.0 | < 2.1 MeV | $D^0\pi^0$, $D^0\gamma$ |
| $D^{*\pm}$ | $c\bar{d}$, $d\bar{c}$ | 2010.3 | 96 keV | $D^0\pi^+$, $D^+\pi^0$ |
| $D_s^{*\pm}$ | $c\bar{s}$, $s\bar{c}$ | 2112.3 | < 1.9 MeV | $D_s^\pm\gamma$, $D_s^\pm\pi^0$ |
| $J/\psi(1S)$ | $c\bar{c}$ | 3096.9 | 92.9 keV | hadrons, $e^+e^-$, $\mu^+\mu^-$ |
| $B^*$ | $u\bar{b}$, $b\bar{u}$, $d\bar{b}$, $d\bar{u}$ | 5425.2 | | $B\gamma$ |
| $\Upsilon(1S)$ | $b\bar{b}$ | 9460.3 | 54 keV | hadrons, $\ell^+\ell^-$ |
| $\Upsilon(2S)$ | $b\bar{b}$ | 10023.3 | 31 keV | hadrons, $\Upsilon(1S) + X$ |
| $\Upsilon(3S)$ | $b\bar{b}$ | 10355.2 | 20 keV | hadrons, $\Upsilon(2S) + X$ |
| $\Upsilon(4S)$ | $b\bar{b}$ | 10579.4 | 20.5 MeV | $B^+B^-$, $B^0\overline{B}^0$ |

# Appendix D Gauge boson polarisation states

The polarisation states of the spin-1 bosons appear in the Feynman rules for the description of real external gauge bosons and in the spin sums implicit in the propagators for virtual gauge bosons. This appendix develops the description of gauge boson polarisation states, first for the massless photon and then for the massive W and Z bosons. Finally the gauge invariance of the electromagnetic interaction is used to determine the completeness relations for real and virtual photons.

## D.1 Classical electromagnetism

In Heaviside–Lorentz units, Maxwell's equations for the electric field strength $\mathbf{E}(\mathbf{x}, t)$ and magnetic flux density $\mathbf{B}(\mathbf{x}, t)$ are

$$\boldsymbol{\nabla}\cdot\mathbf{E} = \rho, \quad \boldsymbol{\nabla}\times\mathbf{E} = -\frac{\partial \mathbf{B}}{\partial t}, \quad \boldsymbol{\nabla}\cdot\mathbf{B} = 0 \quad \text{and} \quad \boldsymbol{\nabla}\times\mathbf{B} = \mathbf{J} + \frac{\partial \mathbf{E}}{\partial t}$$

where $\rho(\mathbf{x}, t)$ and $\mathbf{J}(\mathbf{x}, t)$ are respectively the electric charge and current densities. It is an experimentally established fact that charge is conserved. This implies that the charge and current density satisfy the continuity equation (see, for example, Section 2.3.2)

$$\frac{\partial \rho}{\partial t} + \nabla\cdot\mathbf{J} = 0.$$

If charge is conserved in all frames, this implies that $j^{\mu} = (\rho, \mathbf{J})$ must be a four-vector, and that the continuity equation can be written in the manifestly Lorentz invariant form

$$\partial_{\mu} j^{\mu} = 0.$$

The fields can be expressed in terms of the electric scalar potential $\phi$ and magnetic vector potential $\mathbf{A}$ such that

$$\mathbf{B} = \boldsymbol{\nabla}\times\mathbf{A} \quad \text{and} \quad \mathbf{E} = -\frac{\partial \mathbf{A}}{\partial t} - \boldsymbol{\nabla}\phi. \tag{D.1}$$

By defining the four-vector potential $A^\mu = (\phi, \mathbf{A})$, Maxwell's equations can be written compactly as

$$\partial_\mu F^{\mu\nu} = j^\nu, \tag{D.2}$$

where the field-strength tensor $F^{\mu\nu}$ is defined as

$$F^{\mu\nu} = \partial^\mu A^\nu - \partial^\nu A^\mu = \begin{pmatrix} 0 & -E_x & -E_y & -E_z \\ E_x & 0 & -B_z & B_y \\ E_y & B_z & 0 & -B_x \\ E_z & -B_y & B_x & 0 \end{pmatrix}. \tag{D.3}$$

The expressions of (D.2) and (D.3), which are equivalent to Maxwell's equations, can be written as

$$\partial_\mu \partial^\mu A^\nu - \partial^\nu \partial_\mu A^\mu = j^\nu. \tag{D.4}$$

Because $j^\nu$ is a four-vector, with well-defined Lorentz transformation properties, (D.4) implies that $A^\mu$ must also be a (contravariant) four-vector.

### D.1.1 Gauge invariance

The electric and magnetic fields defined by (D.1) are unchanged by the gauge transformation

$$\mathbf{A} \to \mathbf{A}' = \mathbf{A} + \boldsymbol{\nabla}\chi \quad \text{and} \quad \phi \to \phi' - \frac{\partial \chi}{\partial t},$$

where $\chi(\mathbf{x}, t)$ is any finite differentiable function. In four-vector notation, with $A_\mu = (\phi, -\mathbf{A})$ and $\partial_\mu = (\partial/\partial t, +\boldsymbol{\nabla})$, the gauge invariance of the electromagnetic fields can be expressed as

$$A_\mu \to A'_\mu = A_\mu - \partial_\mu \chi. \tag{D.5}$$

The freedom to chose the gauge can be exploited in order to simplify the covariant formulation of Maxwell's equations. From (D.5),

$$\partial^\mu A'_\mu = \partial^\mu (A_\mu - \partial_\mu \chi) = \partial^\mu A_\mu - \Box\chi.$$

The gauge where $\chi$ is chosen such that $\Box\chi = \partial^\mu A_\mu$, and therefore $\partial^\mu A'_\mu = 0$, is known as the Lorenz gauge (after Ludvig Lorenz, not Hendrik Lorentz). Dropping the primes on the fields, the Lorenz gauge condition is

$$\text{Lorenz gauge condition:} \quad \partial^\mu A_\mu = 0. \tag{D.6}$$

In the Lorenz gauge, the covariant form of Maxwell's equations of (D.4) becomes

$$\text{Lorenz gauge:} \quad \Box A^\mu \equiv \partial_\nu \partial^\nu A^\mu = j^\mu. \tag{D.7}$$

## D.2 Photon polarisation states

In the absence of charges $j^\mu = 0$ and in the Lorenz gauge, the free photon field satisfies

$$\Box A^\mu = 0. \tag{D.8}$$

This equation has plane wave solutions of the form

$$A^\mu = \epsilon^\mu(q)e^{-iq\cdot x}, \tag{D.9}$$

where $\epsilon^\mu(q)$ is a four vector describing the polarisation of the electromagnetic field. Substituting (D.9) into (D.8) gives

$$\Box A^\mu = -q^2\epsilon^\mu e^{-iq\cdot x} = 0.$$

Hence the plane wave solutions to the covariant form of Maxwell's equations for the free photon field must have $q^2 = 0$. Consequently (D.8) can be identified as the wave equation for a *massless* particle. Since a spin-1 boson has three degrees of freedom, it is not immediately obvious that it should be described by the four-vector $A^\mu$ that has four degrees of freedom. However, the four components of $\epsilon^\mu$ are not independent because the field $A^\mu$ satisfies the Lorenz gauge condition $\partial_\mu A^\mu = 0$, which implies

$$0 = \partial_\mu(\epsilon^\mu e^{-iq\cdot x}) = \epsilon^\mu\partial_\mu(e^{-iq\cdot x}) = -i\epsilon^\mu q_\mu e^{-iq\cdot x}.$$

Therefore

$$q_\mu\epsilon^\mu = 0, \tag{D.10}$$

from which it can be concluded that only three of the components of $\epsilon^\mu(q)$ are independent.

Having imposed the Lorenz condition, there is still the freedom to make the further gauge transformation

$$A_\mu \to A'_\mu = A_\mu - \partial_\mu\Lambda(x), \tag{D.11}$$

where $\Lambda(x)$ is any function which satisfies $\Box\Lambda = 0$. This gauge transformation leaves both the field equation of (D.7) and the Lorenz condition of (D.6) unchanged. Consider the gauge transformation

$$\Lambda(x) = -iae^{-iq\cdot x},$$

where $a$ is an arbitrary constant. This satisfies $\Box\Lambda = 0$ because $\Box\Lambda = -q^2\Lambda$ and $q^2 = 0$ for the massless spin-1 boson described by (D.8). With this choice of $\Lambda$, the plane wave of (D.9) becomes

$$
\begin{aligned}
A_\mu \rightarrow A'_\mu = A_\mu - \partial_\mu \Lambda &= \epsilon_\mu e^{-iq\cdot x} + ia\partial_\mu e^{-iq\cdot x} \\
&= \epsilon_\mu e^{-iq\cdot x} + ia(-iq_\mu)e^{-iq\cdot x} \\
&= (\epsilon_\mu + aq_\mu)e^{-iqx}.
\end{aligned}
$$

Therefore, the physical electromagnetic fields are unchanged by

$$\epsilon_\mu \rightarrow \epsilon'_\mu = \epsilon_\mu + aq_\mu, \tag{D.12}$$

and hence, any polarisation vectors that differ by a multiple of the four-momentum of the photon, correspond to the same physical photon.

In the *Coulomb gauge*, $a$ is chosen such that the time-like component of the polarisation is zero and the Lorenz condition of (D.10) becomes

$$\boldsymbol{\epsilon} \cdot \mathbf{q} = 0.$$

Hence in the Coulomb gauge (sometimes referred to as the transverse gauge), the polarisation of the photon is transverse to its direction of motion, and there are only two independent polarisation states. For a photon travelling in the $z$-direction, these can be written as the linearly polarised states

$$\epsilon_1^\mu = (0,1,0,0) \quad \text{and} \quad \epsilon_2^\mu = (0,0,1,0).$$

This choice is not unique, and the circularly polarised states

$$\epsilon_-^\mu = \frac{1}{\sqrt{2}}(0,1,-i,0) \quad \text{and} \quad \epsilon_+^\mu = -\frac{1}{\sqrt{2}}(0,1,i,0) \tag{D.13}$$

are often used; these are the states which correspond to the $z$-component of the spin of the photon being $S_z = \pm 1$.

## D.3 Polarisation states of massive spin-1 particles

The Lagrangian for a massless spin-1 field is

$$\mathcal{L}_0 = -\tfrac{1}{4}F^{\mu\nu}F_{\mu\nu},$$

where $F^{\mu\nu}$ is the field-strength tensor $F^{\mu\nu} = (\partial^\mu B^\nu - \partial^\nu B^\mu)$. The corresponding Lagrangian for a massive spin-1 boson is the Proca Lagrangian of (17.9). This includes an additional mass term which is quadratic in the fields

$$\mathcal{L}_m = -\tfrac{1}{4}F^{\mu\nu}F_{\mu\nu} + \tfrac{1}{2}m^2 B^\mu B_\mu.$$

From the Euler–Lagrange equation, the corresponding free field equation is

$$(\Box + m^2)B^\mu - \partial^\mu(\partial_\nu B^\nu) = 0. \tag{D.14}$$

The same equation can be obtained by noting that Klein–Gordon equation for massless and massive spin-0 scalars are respectively

$$\Box\phi = 0 \quad \text{and} \quad (\Box + m^2)\phi = 0.$$

This suggests that the equation of motion for a massive particle can be obtained from that for a massless particle by making the replacement $\Box \to \Box + m^2$. In the absence of sources, the equation of motion for the free photon is

$$\Box A^\mu - \partial^\mu(\partial_\nu A^\nu) = 0.$$

The corresponding equation for a massive spin-1 field $B^\mu$, obtained by making the substitution $\Box \to \Box + m^2$, is

$$(\Box + m^2)B^\mu - \partial^\mu(\partial_\nu B^\nu) = 0,$$

which is the Proca equation of (D.14).

Taking the derivative of the Proca equation, by acting on it with $\partial_\mu$, gives

$$\begin{aligned}(\Box + m^2)\partial_\mu B^\mu - \partial_\mu\partial^\mu(\partial_\nu B^\nu) &= 0\\ (\Box + m^2)\partial_\mu B^\mu - \Box(\partial_\nu B^\nu) &= 0\\ m^2\partial_\mu B^\mu &= 0,\end{aligned}$$

and therefore

$$\partial_\mu B^\mu = 0. \tag{D.15}$$

Thus, the Lorenz condition is *automatically* satisfied for a *massive* spin-1 boson. Consequently, there is no freedom to choose the gauge; from Chapter 17 it should come as no surprise that the mass term has broken the gauge invariance. Using the Lorenz condition, the Proca equation of(D.14) becomes

$$(\Box + m^2)B^\mu = 0. \tag{D.16}$$

For a massive particle with four momentum $q$, $q^2 = m^2$ and therefore

$$\Box e^{-iq\cdot x} = -q^2 e^{-iq\cdot x} = -m^2 e^{-iq\cdot x}.$$

Hence (D.16) has plane wave solutions

$$B^\mu = \epsilon^\mu e^{-iq\cdot x}. \tag{D.17}$$

The Lorenz condition of (D.15) implies that the four-vector polarisation of the plane wave solutions satisfies

$$q_\mu\epsilon^\mu = 0. \tag{D.18}$$

This constraint means that of the four degrees of freedom in $\epsilon^\mu$, only three are independent. Since there is no further freedom to choose the gauge, a spin-1 boson described by (D.17) has three independent polarisation states. For a spin-1 boson travelling in the $z$-direction, two of these can be chosen to be the two circular polarisation states of (D.13),

$$\epsilon_-^\mu = \tfrac{1}{\sqrt{2}}(0,1,-i,0) \quad \text{and} \quad \epsilon_+^\mu = -\tfrac{1}{\sqrt{2}}(0,1,i,0).$$

The third polarisation state, which is orthogonal to the circular polarisations states, can be written in the form

$$\epsilon_L^\mu \propto (\alpha, 0, 0, \beta).$$

The relationship between $\alpha$ and $\beta$ is fixed by (D.18) which implies $\alpha E - \beta p_z = 0$ and hence the longitudinal polarisation state is

$$\epsilon_L^\mu = \tfrac{1}{m}(p_z, 0, 0, E), \tag{D.19}$$

where the normalisation is such that in the rest frame of the gauge boson, the longitudinal polarisation state is just $(0,0,0,1)$.

## D.4 Polarisation sums

In Section 6.5.1 the completeness relation for Dirac spinors was shown to be

$$\sum_{s=1}^{2} u_s \bar{u}_s = (\gamma^\mu p_\mu + m) = \not{p} + m.$$

In the trace formalism this relation was used to perform spin sums in the calculation of matrix elements. The corresponding completeness relation for the sum over the polarisation states of a spin-1 boson,

$$\sum_\lambda \epsilon_\lambda^{*\mu} \epsilon_\lambda^\nu,$$

is used in similar calculations involving real gauge bosons in the initial- and final-state. Here, the main focus is on justifying the sums over the polarisation states of *virtual* gauge bosons that gives rise to the form of the propagator in the description of an interaction by the exchange of spin-1 particles.

### D.4.1 Polarisation sums for massive gauge bosons

The completeness relation for massive spin-1 bosons can be obtained by using the three polarisation states given in (D.13) and (D.19). For a boson propagating in the $z$-direction, the completeness relation can be expressed in tensor form as

$$\sum_\lambda \epsilon_\lambda^{*\mu}\epsilon_\lambda^\nu = \frac{1}{2}\begin{pmatrix} 0 & 0 & 0 & 0 \\ 0 & 1 & -i & 0 \\ 0 & +i & 1 & 0 \\ 0 & 0 & 0 & 0 \end{pmatrix} + \frac{1}{m^2}\begin{pmatrix} \mathrm{p}^2 & 0 & 0 & E\mathrm{p} \\ 0 & 0 & 0 & 0 \\ 0 & 0 & 0 & 0 \\ E\mathrm{p} & 0 & 0 & E^2 \end{pmatrix} + \frac{1}{2}\begin{pmatrix} 0 & 0 & 0 & 0 \\ 0 & 1 & +i & 0 \\ 0 & -i & 1 & 0 \\ 0 & 0 & 0 & 0 \end{pmatrix}$$

$$= \begin{pmatrix} 0 & 0 & 0 & 0 \\ 0 & 1 & 0 & 0 \\ 0 & 0 & 1 & 0 \\ 0 & 0 & 0 & 0 \end{pmatrix} + \frac{1}{m^2}\begin{pmatrix} E^2 - m^2 & 0 & 0 & E\mathrm{p} \\ 0 & 0 & 0 & 0 \\ 0 & 0 & 0 & 0 \\ E\mathrm{p} & 0 & 0 & \mathrm{p}^2 + m^2 \end{pmatrix}.$$

Writing $E$ and p as the components of the four-momentum $q$ gives

$$\sum_\lambda \epsilon_\lambda^{*\mu}\epsilon_\lambda^\nu = \begin{pmatrix} -1 & 0 & 0 & 0 \\ 0 & 1 & 0 & 0 \\ 0 & 0 & 1 & 0 \\ 0 & 0 & 0 & 1 \end{pmatrix} + \frac{1}{m^2}\begin{pmatrix} q^0q^0 & 0 & 0 & q^0q^3 \\ 0 & 0 & 0 & 0 \\ 0 & 0 & 0 & 0 \\ q^0q^3 & 0 & 0 & q^3q^3 \end{pmatrix},$$

which can be generalised to

$$\sum_\lambda \epsilon_\lambda^{*\mu}\epsilon_\lambda^\nu = -g^{\mu\nu} + \frac{q^\mu q^\nu}{m^2}. \tag{D.20}$$

### D.4.2 Polarisation sums for external photons

The situation for photons is complicated by the freedom of the choice of gauge. Consider a real photon travelling in the $z$-direction. In the Coulomb gauge, the transverse polarisation vectors can be written $\epsilon_1 = (0, 1, 0, 0)$ and $\epsilon_2 = (0, 0, 1, 0)$. Hence the sum over these transverse polarisation states is

$$\sum_\lambda \epsilon_\lambda^{*\mu}\epsilon_\lambda^\nu = \begin{pmatrix} 0 & 0 & 0 & 0 \\ 0 & 1 & 0 & 0 \\ 0 & 0 & 1 & 0 \\ 0 & 0 & 0 & 0 \end{pmatrix}.$$

This can be extended to a real photon with four-momentum $q$, travelling in an arbitrary direction. In this case, the sum over the two polarisation states *in the Coulomb gauge* can be seen to be

$$\sum_T \epsilon_T^{*i}\epsilon_T^j = \delta_{ij} - \frac{q^iq^j}{|\mathbf{q}|^2},$$

where $i, j = 1, 2, 3$ are the *space-like* indices of the polarisation vector and the sum is over the two transverse polarisation states of the photon. Since the polarisation four-vector $\epsilon^\mu$ and $\epsilon^\mu + aq^\mu$ describe the same photon, this result is gauge dependent.

The general completeness relation, expressed in terms of all four components of the polarisation four-vector, needs to take into account the gauge freedom associated

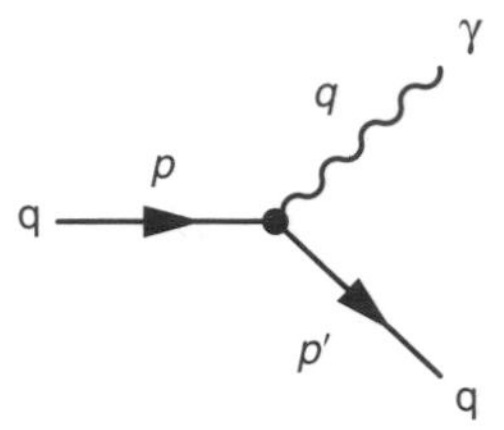


Fig. D.1 The Feynman diagram for $q \to q\gamma$ which might form part of the process electromagnetic decay $\rho^0 \to \pi^0\gamma$.

with a massless photon. For example, consider the quark-level Feynman diagram for the emission of a real photon in the decay $\rho^0 \to \pi^0\gamma$. At the quark level, the matrix element for the diagram of Figure D.1 is

$$\mathcal{M} = eQ_q\,[\overline{u}(p')\gamma_\mu u(p)]\,\epsilon_\lambda^\mu(q),$$

which can be written in the form

$$\mathcal{M} = j_\mu \epsilon_\lambda^\mu(q), \tag{D.21}$$

where $j_\mu$ is the four-vector associated with the fermion current (including the charge). The spin-summed matrix element squared is

$$\sum_\lambda |\mathcal{M}|^2 = \sum_\lambda j_\mu j_\nu^* \epsilon_\lambda^\mu \epsilon_\lambda^{*\nu}. \tag{D.22}$$

In the Coulomb gauge, where there are only two transverse polarisation states, the spin-summed matrix element squared takes the form

$$\sum_{T=1}^{2} |\mathcal{M}|^2 = j_\mu j_\nu^* \sum_{T=1}^{2} \epsilon_T^\mu \epsilon_T^{*\nu}. \tag{D.23}$$

In the frame where the photon is travelling in the $z$-direction, the four-vectors for the two transverse polarisation states can be written $\epsilon_1 = (0, 1, 0, 0)$ and $\epsilon_2 = (0, 0, 1, 0)$. In this case, the sum over the two transverse polarisation states in (D.23) gives

$$\sum_{T=1}^{2} |\mathcal{M}|^2 = j_1 j_1^* + j_2 j_2^*.$$

This expression can be written as a sum over all four components of $j_\mu$ using

$$\sum_{T=1}^{2} |\mathcal{M}|^2 = j_1 j_1^* + j_2 j_2^* = -g^{\mu\nu} j_\mu j_\nu^* + j_0 j_0^* - j_3 j_3^*. \tag{D.24}$$

From the gauge invariance of (D.12), it was seen that the polarisation vectors $\epsilon^\mu$ and $\epsilon^\mu + aq^\mu$ describe the same photon. Therefore, the matrix element of (D.21) must be invariant under the gauge transformation $\epsilon^\mu \to \epsilon^\mu + aq^\mu$, and hence

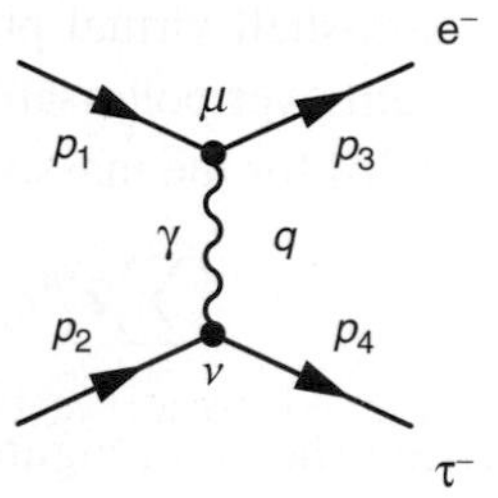


Fig. D.2 The Feynman diagram for the QED scattering process $e^-\tau^- \to e^-\tau^-$.

$$\mathcal{M} = j_\mu \epsilon_\lambda'^\mu(q) = j_\mu \epsilon_\lambda^\mu(q) + j_\mu q^\mu,$$

which implies

$$j_\mu q^\mu = 0. \quad \text{(D.25)}$$

For a photon with $q^\mu = (\mathrm{q}, 0, 0, +\mathrm{q})$, it can be concluded that $\mathrm{q} j_0 - \mathrm{q} j_3 = 0$. Consequently, the gauge invariance of electromagnetism here implies that $j_0 = j_3$ and therefore (D.24) is equivalent to

$$\sum_{T=1}^{2} |\mathcal{M}|^2 = -g^{\mu\nu} j_\mu j_\nu^*.$$

From the comparison with (D.23), it can be seen that the completeness relation for the spin sum over the polarisation states of a *real* initial- or final-state photon is

$$\sum_{\lambda=1}^{2} \epsilon_\lambda^\mu \epsilon_\lambda^{*\nu} = -g^{\mu\nu}.$$

### D.4.3 The photon propagator

The completeness relation derived above applies to external on-shell photons. The situation with off-mass-shell virtual photons is more complicated. Because virtual photons have $q^2 \neq 0$, it is not possible to simply neglect the longitudinal and scalar polarisation states. In Section 5.3, the matrix element for the QED scattering process $e^-\tau^- \to e^-\tau^-$, shown in Figure D.2, was expressed in Equation (5.19) as the sum over the polarisation states of the virtual photon and the four-vector currents for the electron and tau-lepton,

$$\mathcal{M} \propto \sum_{\lambda=1}^{4} j_\mu^{(e)} j_\nu^{(\tau)} \frac{\epsilon_\lambda^{*\mu}(q)\epsilon_\lambda^\nu(q)}{q^2}.$$

Because an off mass-shell virtual photon can be considered to have an invariant mass $q^2 \neq 0$, the sum over polarisation states can be obtained from the completeness relation of (D.20) for the massive spin-1 particles, replacing $m^2$ with $q^2$,

$$\sum_\lambda \epsilon_\lambda^{*\mu} \epsilon_\lambda^\nu = -g_{\mu\nu} + \frac{q^\mu q^\nu}{q^2}. \tag{D.26}$$

Because the diagram shown in Figure D.2 is the only lowest-order QED Feynman diagram for the process $e^-\tau^- \to e^-\tau^-$, the matrix element for this diagram *alone* must be gauge invariant and therefore must be unchanged by $\epsilon^\mu \to \epsilon^\mu + aq^\mu$ and $\epsilon^\nu \to \epsilon^\nu + bq^\nu$, hence

$$\sum_{\lambda=1}^{4} j_\mu^{(e)} j_\nu^{(\tau)} \epsilon_\lambda^{*\mu} \epsilon_\lambda^\nu = \sum_{\lambda=1}^{4} j_\mu^{(e)} j_\nu^{(\tau)} (\epsilon_\lambda^{*\mu} + aq^\mu)(\epsilon_\lambda^\nu + bq^\nu) \qquad \forall\, a \text{ and } b,$$

implying that

$$j_\mu^{(e)} j_\nu^{(\tau)} q^\mu q^\nu = 0.$$

Thus it can be concluded that the $q^\mu q^\nu / q^2$ term in (D.26) does not contribute to the matrix element. In this case, the spin sum associated with the photon propagator (for this single matrix element) can be written

$$\sum_\lambda \epsilon_\lambda^\mu \epsilon_\lambda^{*\nu} = -g^{\mu\nu},$$

and the Feynman rule for the photon propagator itself is simply

$$-i\frac{g^{\mu\nu}}{q^2}. \tag{D.27}$$

This form of the photon propagator is appropriate for all lowest-order diagrams where the fermions are on-shell.

In higher-order diagrams, where the virtual photon connects virtual fermions, only the sum of the amplitudes for all diagrams is gauge invariant. In this case, the matrix element for a *particular diagram* may not be gauge invariant and photon propagator must be written

$$-\frac{i}{q^2}\left[g^{\mu\nu} + (1-\xi)\frac{q^\mu q^\nu}{q^2}\right],$$

where $\xi$ is a gauge-dependent parameter. For most practical purposes, calculations are performed in the so-called Feynman gauge with $\xi = 1$, and the photon propagator is again given by (D.27). In general, for the calculation of higher-order diagrams in QED and QCD, the correct treatment of the choice of gauge requires care and represents a topic that goes well beyond the level of this book. Nevertheless, the use of the simple Feynman gauge form of the propagator for the photon and gluon propagators is appropriate for all the lowest-order diagrams encountered here.

# Appendix E Noether's theorem

Symmetries and conservation laws are central to the development of the Standard Model of particle physics. This short appendix gives an example of application of Noether's theorem to the global U(1) symmetry of the Dirac equation, where the conserved current is found to be the four-vector formed from the probability density and probability current.

Noether's theorem relates a symmetry of the Lagrangian to a conserved current. For example, the Lagrangian for the free Dirac field,

$$\mathcal{L} = i\overline{\psi}\gamma^\mu\partial_\mu\psi - m\overline{\psi}\psi, \tag{E.1}$$

is unchanged by the global U(1) phase transformation,

$$\psi \to \psi' = e^{i\theta}\psi.$$

The invariance of the Lagrangian can be expressed in terms of the infinitesimal global U(1) phase transformation,

$$\psi \to \psi' = (1 + i\varepsilon)\psi \quad \text{and} \quad \overline{\psi} \to \overline{\psi}' = (1 - i\varepsilon)\overline{\psi},$$

for which the changes in the fields and their derivatives are

$$\delta\psi = i\varepsilon\psi, \quad \delta(\partial_\mu\psi) = i\varepsilon(\partial_\mu\psi), \quad \delta\overline{\psi} = -i\varepsilon\overline{\psi} \quad \text{and} \quad \delta(\partial_\mu\overline{\psi}) = -i\varepsilon(\partial_\mu\overline{\psi}).$$

The U(1) global symmetry of the Lagrangian implies that

$$\delta\mathcal{L} = \frac{\partial\mathcal{L}}{\partial\psi}\delta\psi + \frac{\partial\mathcal{L}}{\partial(\partial_\mu\psi)}\delta(\partial_\mu\psi) + \frac{\partial\mathcal{L}}{\partial\overline{\psi}}\delta\overline{\psi} + \frac{\partial\mathcal{L}}{\partial(\partial_\mu\overline{\psi})}\delta(\partial_\mu\overline{\psi}) = 0, \tag{E.2}$$

from which

$$i\varepsilon\frac{\partial\mathcal{L}}{\partial\psi}\psi + i\varepsilon\frac{\partial\mathcal{L}}{\partial(\partial_\mu\psi)}(\partial_\mu\psi) - i\varepsilon\frac{\partial\mathcal{L}}{\partial\overline{\psi}}\overline{\psi} - i\varepsilon\frac{\partial\mathcal{L}}{\partial(\partial_\mu\overline{\psi})}(\partial_\mu\overline{\psi}) = 0. \tag{E.3}$$

The term involving the derivative with respect to $(\partial_\mu\psi)$ can be expressed as

$$\frac{\partial\mathcal{L}}{\partial(\partial_\mu\psi)}(\partial_\mu\psi) = \partial_\mu\left(\frac{\partial\mathcal{L}}{\partial(\partial_\mu\psi)}\psi\right) - \left[\partial_\mu\left(\frac{\partial\mathcal{L}}{\partial(\partial_\mu\psi)}\right)\right]\psi,$$

and thus (E.3) becomes

$$i\varepsilon\left[\frac{\partial\mathcal{L}}{\partial\psi} - \partial_\mu\left(\frac{\partial\mathcal{L}}{\partial(\partial_\mu\psi)}\right)\right]\psi + i\varepsilon\partial_\mu\left(\frac{\partial\mathcal{L}}{\partial(\partial_\mu\psi)}\psi\right) - \{\overline{\psi}\text{ terms}\} = 0. \tag{E.4}$$

From the Euler–Lagrange equation of (17.3), the term in square brackets is zero, as is the corresponding term for $\overline{\psi}$ and thus

$$i\varepsilon\partial_\mu\left[\frac{\partial\mathcal{L}}{\partial(\partial_\mu\psi)}\psi - \overline{\psi}\frac{\partial\mathcal{L}}{\partial(\partial_\mu\overline{\psi})}\right] = 0.$$

This can be recognised as a continuity equation of the form

$$\partial_\mu j^\mu = 0,$$

where the conserved current associated with the global U(1) symmetry of the Lagrangian is

$$j^\mu = (\rho, \mathbf{J}) = -i\left(\frac{\partial\mathcal{L}}{\partial(\partial_\mu\psi)}\psi - \overline{\psi}\frac{\partial\mathcal{L}}{\partial(\partial_\mu\overline{\psi})}\right). \tag{E.5}$$

The factor of $-i$ appearing in this expression ensures that the associated density is real and positive. The partial derivatives appearing in (E.5) can be obtained from the Lagrangian of (E.1),

$$\frac{\partial\mathcal{L}}{\partial(\partial_\mu\psi)} = i\overline{\psi}\gamma^\mu \quad\text{and}\quad \frac{\partial\mathcal{L}}{\partial(\partial_\mu\overline{\psi})} = 0,$$

from which it immediately can be seen that the conserved current associated with the U(1) global symmetry of the Dirac equation is

$$j^\mu = (\rho, \mathbf{J}) = \overline{\psi}\gamma^\mu\psi.$$

This is just the probability density and probability current associated with a Dirac spinor, originally identified in Section 4.3.

## Problem

**E.1*** The Lagrangian for a complex scalar field

$$\mathcal{L}_S = \frac{1}{2}(\partial_\mu\phi)^*(\partial^\mu\phi) - \frac{1}{2}m^2\phi^*\phi,$$

possesses a global U(1) symmetry. Use Noether's theorem to identify the conserved current.

# Appendix F Non-Abelian gauge theories

Local gauge theories are referred to as non-Abelian if the generators of the associated symmetry group do not commute. In a non-Abelian gauge theory, the transformation properties of the fields imply the existence of gauge boson self-interactions. In this Appendix, the example of the SU(2) local gauge theory of the weak interaction is used to introduce these concepts. Whilst the algebra is quite involved, the main ideas can be understood readily.

The generators of the SU(2) and SU(3) gauge groups of the Standard Model do not commute. For example, the three generators of the SU(2) symmetry group, $\mathbf{T} = \{T_1, T_2, T_3\}$, can be expressed in terms of the Pauli spin matrices as $\mathbf{T} = \boldsymbol{\sigma}/2$ and satisfy the commutation relations

$$[T_i, T_j] = \tfrac{1}{4}[\sigma_i, \sigma_j] = \tfrac{1}{4}2i\epsilon_{ijk}\sigma_k = i\epsilon_{ijk}T_k,$$

where $\epsilon_{ijk}$ is the totally antisymmetric Levi–Civita tensor. In general, the commutation relations for a particular group can be written in terms of the structure constants $f_{ijk}$ of the group defined by

$$[T_i, T_j] = if_{ijk}T_k. \tag{F.1}$$

In a non-Abelian gauge theory, the transformation properties of the associated gauge fields are not independent and additional self-interaction terms have to be added to the field-strength tensor for it to be gauge invariant. These self-interaction terms lead to the triple and quartic gauge bosons vertices encountered in the discussions of QCD and the electroweak interaction.

An infinitesimal SU(2) local gauge transformation can be written,

$$\varphi(x) = \begin{pmatrix} \nu_e(x) \\ e(x) \end{pmatrix} \rightarrow \varphi'(x) = (I + ig_W\boldsymbol{\alpha}(x)\cdot\mathbf{T})\,\varphi(x).$$

The corresponding infinitesimal transformation of the doublet of adjoint spinors $\overline{\varphi} = \varphi^\dagger\gamma^0$ is

$$\overline{\varphi}(x) \rightarrow \overline{\varphi}'(x) = \overline{\varphi}(x)\,(I - ig_W\boldsymbol{\alpha}(x)\cdot\mathbf{T})\,.$$

Taking the fermion masses to be zero (they can be introduced latter through the Higgs mechanism), the Lagrangian giving the Dirac equations for both the electron and the electron neutrino can be written as

$$\mathcal{L} = i\overline{\varphi}\gamma_\mu\partial^\mu\varphi \equiv i\overline{\nu}_e\gamma_\mu\partial^\mu\nu_e + i\overline{e}\gamma_\mu\partial^\mu e. \tag{F.2}$$

The gauge invariance of the Lagrangian can be ensured by replacing the derivatives $\partial^\mu$ with the corresponding covariant derivatives, defined by

$$\partial^\mu \to D^\mu = \partial^\mu + ig_W\mathbf{W}^\mu \cdot \mathbf{T},$$

where $\mathbf{W} = \{W_1, W_2, W_3\}$ are the three gauge fields of the SU(2) symmetry. Hence the Lagrangian of (F.2) becomes

$$\mathcal{L} = i\overline{\varphi}\gamma_\mu(\partial^\mu + ig_W\mathbf{W}^\mu \cdot \mathbf{T})\varphi. \tag{F.3}$$

This Lagrangian includes the interactions between the fermions and $W$ fields, but is only gauge invariant if the $W$ fields have the correct transformation properties.

The transformation properties of the gauge fields can be determined by noting that the gauge invariance of the Lagrangian is ensured if $D_\mu\varphi$ transforms in the same way as $\varphi$ itself, i.e.

$$D'^\mu\varphi' = (I + ig_W\boldsymbol{\alpha} \cdot \mathbf{T})\, D^\mu\varphi. \tag{F.4}$$

If this is the case,

$$\overline{\varphi}' D'^\mu\varphi' = \overline{\varphi}(x)\,(I - ig_W\boldsymbol{\alpha}(x) \cdot \mathbf{T})\,(I + ig_W\boldsymbol{\alpha}(x) \cdot \mathbf{T})\, D^\mu\varphi = \overline{\varphi}D^\mu\varphi + O(g_W^2\alpha^2).$$

The condition that $D_\mu\varphi$ must transform in the same way as $\varphi$, defines the gauge transformation properties of the fields. Because

$$D'^\mu = \partial^\mu + ig_W\mathbf{W}'^\mu \cdot \mathbf{T},$$

then $D'^\mu\varphi'$ is given by

$$D'^\mu\varphi' = (\partial^\mu + ig_W\mathbf{W}'^\mu \cdot \mathbf{T})\,(1 + ig_W\boldsymbol{\alpha}(x) \cdot \mathbf{T})\,\varphi.$$

Hence the requirement of (F.4) becomes

$$(\partial^\mu + ig_W\mathbf{W}'^\mu \cdot \mathbf{T})\,(1 + ig_W\boldsymbol{\alpha}(x) \cdot \mathbf{T})\,\varphi = (I + ig_W\boldsymbol{\alpha}(\mathbf{x}) \cdot \mathbf{T})\,(\partial^\mu + ig_W\mathbf{W}^\mu \cdot \mathbf{T})\,\varphi.$$

Expanding both sides and cancelling the common terms leads to

$$\begin{aligned} ig_W(\partial^\mu\boldsymbol{\alpha}) \cdot \mathbf{T} + ig_W\mathbf{W}'^\mu \cdot \mathbf{T} - g_W^2(\mathbf{W}'^\mu \cdot \mathbf{T})(\boldsymbol{\alpha} \cdot \mathbf{T}) = {} & ig_W(\mathbf{W}^\mu \cdot \mathbf{T}) \\ & - g_W^2(\boldsymbol{\alpha} \cdot \mathbf{T})(\mathbf{W}^\mu \cdot \mathbf{T}). \end{aligned} \tag{F.5}$$

Simply following the prescription used for QED and writing

$$W_k^\mu \to W_k'^\mu = W_k^\mu - \partial^\mu\alpha_k, \tag{F.6}$$

does not restore the gauge invariance. This is because the generators of SU(2) do not commute and therefore $(\boldsymbol{\alpha} \cdot \mathbf{T})(\mathbf{W} \cdot \mathbf{T}) \neq (\mathbf{W} \cdot \mathbf{T})(\boldsymbol{\alpha} \cdot \mathbf{T})$. The required gauge invariance can be achieved by adding a term to (F.6) such that the transformations of the three $W$ fields are no longer independent,

$$W_k^\mu \to W_k'^\mu = W_k^\mu - \partial^\mu \alpha_k - g_\mathrm{W} a_{ijk} \alpha_i W_j^\mu, \tag{F.7}$$

where the $a_{ijk}$ are appropriate constants that need to be identified. Inserting this expression in to (F.5) leads to

$$\begin{aligned} -ig_\mathrm{W}^2 a_{ijk} \alpha_i W_j^\mu T_k &= -g_\mathrm{W}^2 \left[ (\boldsymbol{\alpha} \cdot \mathbf{T})(\mathbf{W}^\mu \cdot \mathbf{T}) - (\mathbf{W}^\mu \cdot \mathbf{T})(\boldsymbol{\alpha} \cdot \mathbf{T}) \right] + O(g_\mathrm{W}^2 \alpha^2) \\ &= -g_\mathrm{W}^2 \alpha_i W_j^\mu [T_i, T_j] = -g_\mathrm{W}^2 \alpha_i W_j^\mu \left( i f_{ijk} T_k \right), \end{aligned}$$

where the $f_{ijk}$ are the structure constants of the SU(2) group, defined in (F.1). Hence, for invariance of the Lagrangian under SU(2) local gauge transformations, the constants $a_{ijk}$ in (F.7) must equal $f_{ijk} = \epsilon_{ijk}$. Thus the fields must transform as

$$W_k^\mu \to W_k'^\mu = W_k^\mu - \partial^\mu \alpha_k - g_\mathrm{W} \epsilon_{ijk} \alpha_i W_j^\mu, \tag{F.8}$$

which also can be written in vector form,

$$\mathbf{W}^\mu \to \mathbf{W}'^\mu = \mathbf{W}^\mu - \partial^\mu \boldsymbol{\alpha} - g_\mathrm{W} \boldsymbol{\alpha} \times \mathbf{W}^\mu.$$

In general, in a non-Abelian gauge theory, the transformation properties of the fields includes a term which depends on the structure constants of the group, defined by the commutation relations $[T_i, T_j] = i f_{ijk} T_k$.

The transformation of the fields of (F.8) ensures that the Lagrangian of (F.3) is invariant under local SU(2) phase transformations. However, the kinetic term for the $W$ fields has yet to be included. Simply taking $-\frac{1}{4}\mathbf{F}_{\mu\nu} \cdot \mathbf{F}^{\mu\nu}$ with $\mathbf{F}^{\mu\nu} = \partial^\mu \mathbf{W}^\nu - \partial^\nu \mathbf{W}^\mu$ is not gauge invariant. The gauge-invariant form can be found by noting that the field strength tensor for QED can be written in terms of its covariant derivative $D^\mu = \partial^\mu + iqA^\mu$ as

$$F^{\mu\nu} = \partial^\mu A^\nu - \partial^\nu A^\mu = \frac{1}{iq}[D^\mu, D^\nu].$$

Repeating this for the covariant derivative of the SU(2) gauge symmetry leads to

$$\begin{aligned} \frac{1}{ig_\mathrm{W}}[D^\mu, D^\nu] &= \frac{1}{ig_\mathrm{W}} \left[ \partial^\mu + ig_\mathrm{W} \mathbf{T} \cdot \mathbf{W}^\mu, \partial^\nu + ig_\mathrm{W} \mathbf{T} \cdot \mathbf{W}^\nu \right] \\ &= \mathbf{T} \cdot (\partial^\mu \mathbf{W}^\nu - \partial^\nu \mathbf{W}^\mu) + ig_\mathrm{W} \left[ (\mathbf{T} \cdot \mathbf{W}^\mu)(\mathbf{T} \cdot \mathbf{W}^\nu) - (\mathbf{T} \cdot \mathbf{W}^\nu)(\mathbf{T} \cdot \mathbf{W}^\mu) \right]. \end{aligned} \tag{F.9}$$

Because the generators of SU(2) do not commute, the second term on the RHS of (F.9) is not zero. The term involving the generators can be simplified by writing

$$\begin{aligned}(\mathbf{T}\cdot\mathbf{W}^\mu)(\mathbf{T}\cdot\mathbf{W}^\nu) - (\mathbf{T}\cdot\mathbf{W}^\nu)(\mathbf{T}\cdot\mathbf{W}^\mu) &= W_i^\mu W_j^\nu [T_i, T_j] \\ &= i\epsilon_{ijk} W_i^\mu W_j^\nu T_k = i(\mathbf{W}^\mu \times \mathbf{W}^\nu)_k T_k \\ &= i\,\mathbf{T}\cdot(\mathbf{W}^\mu \times \mathbf{W}^\nu).\end{aligned}$$

Hence (F.9) can be written as the tensor

$$\begin{aligned}W^{\mu\nu} \equiv \frac{1}{ig_W}[D^\mu, D^\nu] &= \mathbf{T}\cdot[\partial^\mu \mathbf{W}^\nu - \partial^\nu \mathbf{W}^\mu - g_W \mathbf{W}^\mu \times \mathbf{W}^\nu] \\ &= \tfrac{1}{2}\boldsymbol{\sigma}\cdot\mathbf{W}^{\mu\nu},\end{aligned}$$

where the generators have been written in terms of the Pauli spin-matrices and $\mathbf{W}^{\mu\nu}$ is given by

$$\mathbf{W}^{\mu\nu} = \partial^\mu \mathbf{W}^\nu - \partial^\nu \mathbf{W}^\mu - g_W \mathbf{W}^\mu \times \mathbf{W}^\nu. \tag{F.10}$$

Hence the field strength tensor for the three $W$ fields can be identified as $W^{\mu\nu} = \frac{1}{2}\boldsymbol{\sigma}\cdot\mathbf{W}^{\mu\nu}$. A gauge-invariant term in the Lagrangian for the W-boson fields can be formed by taking the trace of $W^{\mu\nu}W_{\mu\nu}$,

$$\mathcal{L}_W \propto \mathrm{Tr}\left([\boldsymbol{\sigma}\cdot\mathbf{W}^{\mu\nu}][\boldsymbol{\sigma}\cdot\mathbf{W}_{\mu\nu}]\right).$$

Using the identity of (B.6) this reduces to

$$\mathcal{L}_W \propto \mathrm{Tr}\left(\mathbf{W}^{\mu\nu}\cdot\mathbf{W}_{\mu\nu} I\right) = 2\mathbf{W}^{\mu\nu}\cdot\mathbf{W}_{\mu\nu}.$$

The invariance of this expression under the gauge transformation $U$ is ensured by the transformation properties of the covariant derivative where $D^\mu \to D'^\mu = UD^\mu U^{-1}$. This implies that $\mathbf{W}'^{\mu\nu} = U\mathbf{W}^{\mu\nu}U^{-1}$ and the cyclic property of traces ensures that

$$\mathrm{Tr}\left(\mathbf{W}'^{\mu\nu}\cdot\mathbf{W}'_{\mu\nu}\right) = \mathrm{Tr}\left(\mathbf{W}^{\mu\nu}\cdot\mathbf{W}_{\mu\nu}\right).$$

Hence the gauge-invariant term in the Lagrangian for the $W$ fields alone can be identified as

$$\mathcal{L}_W = -\tfrac{1}{4}\mathbf{W}^{\mu\nu}\cdot\mathbf{W}_{\mu\nu},$$

where the factor of one quarter is included to reproduce the normal kinetic term for spin-1 gauge bosons. The $g_W \mathbf{W}^\mu \times \mathbf{W}^\nu = g_W \epsilon_{ijk} W_j^\mu W_k^\nu$ term in the field strength tensor of (F.10), means that the Lagrangian contains W-boson self-interaction terms. This can be made explicit by dividing the Lagrangian into kinetic and interaction parts, $\mathcal{L}_W = \mathcal{L}_{\rm kin} + \mathcal{L}_{\rm int}$, with

$$\begin{aligned}\mathcal{L}_{\rm kin} &= -\tfrac{1}{4}\left(\partial^\mu W_i^\nu - \partial^\nu W_i^\mu\right)\left(\partial_\mu W_{i\nu} - \partial_\nu W_{i\mu}\right), \\ \mathcal{L}_{\rm int} &= +\tfrac{1}{2}g_W \epsilon_{ijk}(\partial^\mu W_i^\nu - \partial^\nu W_i^\mu)W_{j\mu}W_{k\nu} - \tfrac{1}{4}g_W^2 \epsilon_{ijk}\epsilon_{imn} W_j^\mu W_k^\nu W_{m\mu} W_{n\nu}.\end{aligned}$$

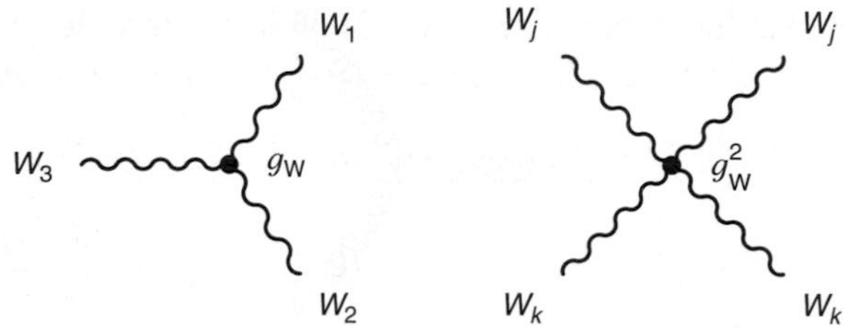


Fig. F.1 The Standard Model triple and quartic gauge boson vertices arising from the SU(2) local gauge invariance of the weak interaction.

The interaction term contains both triple gauge boson and quartic gauge boson vertices, as shown in Figure F.1. These self-interactions are a consequence of non-commuting generators of the SU(2) group, and the associated fields are referred to as Yang–Mills fields. In terms of the physical fields obtained from electroweak unification of the Standard Model, there are two triple gauge boson vertices, $\gamma W^+W^-$ and $ZW^+W^-$, and three quartic gauge boson vertices, $W^+W^-W^+W^-$, $W^+W^-ZZ$ and $W^+W^-\gamma\gamma$.

## Gluon self-interactions in QCD

The transformation properties of the gluon fields under the non-Abelian SU(3) local gauge transformation of QCD depends on the commutation relations between the generators

$$G_k^\mu \to G_k'^\mu = G_k^\mu - \partial^\mu \alpha_k - g_S f_{ijk}\alpha_i G_j^\mu, \tag{F.11}$$

where $f_{ijk}$ are the structure constants of the SU(3) group defined by

$$[T_i, T_j] = i f_{ijk} T_k.$$

The structure constants of QCD are completely antisymmetric with $f_{ijk} = -f_{ikj}$, and of the $8^3$ possible combinations of the indices $i$, $j$ and $k$, the only combinations which correspond to non-zero values are:

$$f_{123} = 1, \quad f_{147} = f_{246} = f_{257} = f_{345} = f_{516} = f_{637} = \tfrac{1}{2} \text{ and } f_{458} = f_{678} = \tfrac{\sqrt{3}}{2},$$

and their cyclic permutations. Without going into the details, the kinetic term in the Lagrangian for the eight gluon fields is given by

$$\mathcal{L} = -\tfrac{1}{4}\mathbf{G}^{\mu\nu} \cdot \mathbf{G}_{\mu\nu},$$

with

$$G_i^{\mu\nu} = \partial^\mu G_i^\nu - \partial^\nu G_i^\mu - g_S f_{ijk} G_j^\mu G_k^\nu.$$

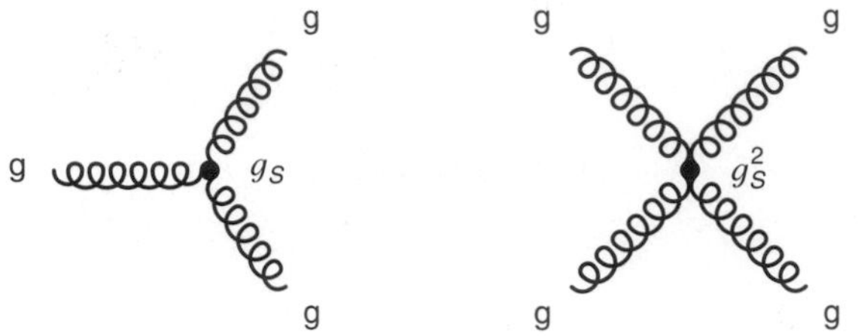


Fig. F.2 The triple and quartic gluon self-interaction vertices.

The $G_j^{\mu}G_k^{\nu}$ term in the field strength tensor gives rise to the triple and quartic gluon vertices, indicated in Figure F.2. It is the presence of these vertices, which are due to the non-Abelian nature of SU(3), that results in colour confinement.

# References


Aad, G. *et al*. 2012. *Phys. Lett.*, **B716**, 1–29.
Aaij, R. *et al*. 2012. *Phys. Lett.*, **B709**, 177–184.
Aaltonen, T. *et al*. 2011. *Phys. Rev.*, **D83**, 111101.
Aaltonen, T. *et al*. 2012. *Phys. Rev. Lett.*, **108**, 151803.
Aaron, F. D. *et al*. 2012. *JHEP*, **1209**, 061.
Abbiendi, G. *et al*. 2001. *Eur. Phys. J.*, **C19**, 587–651.
Abbiendi, G. *et al*. 2004. *Eur. Phys. J.*, **C33**, 173–212.
Abe, F. *et al*. 1999. *Phys. Rev.*, **D59**, 052002.
Abe, K. *et al*. 2005. *Phys. Rev.*, **D71**, 072003.
Abe, S. *et al*. 2008. *Phys. Rev. Lett.*, **100**, 221803.
Abramowicz, H. *et al*. 2010. *Eur. Phys. J.*, **C70**, 945–963.
Abulencia, A. *et al*. 2006. *Phys. Rev. Lett.*, **97**, 242003.
Achard, P. *et al*. 2006. *Eur. Phys. J.*, **C45**, 569–587.
Adachi, I. *et al*. 2012. *Phys. Rev. Lett.*, **108**, 171802.
Adamson, P. *et al*. 2011. *Phys. Rev. Lett.*, **106**, 181801.
Ahmad, Q. R. *et al*. 2002. *Phys. Rev. Lett.*, **89**, 011301.
Ahn, J. K. *et al*. 2012. *Phys. Rev. Lett.*, **108**, 191802.
An, F. P. *et al*. 2012. *Phys. Rev. Lett.*, **108**, 171803.
Anderson, C. D. 1933. *Phys. Rev.*, **43**, 491–494.
Angelopoulos, A. *et al*. 2000. *Eur. Phys. J.*, **C18**, 41–55.
Angelopoulos, A. *et al*. 2001. *Eur. Phys. J.*, **C22**, 55–79.
Aubert, B. *et al*. 2007. *Phys. Rev. Lett.*, **99**, 171803.
Bacino, W. *et al*. 1978. *Phys. Rev. Lett.*, **41**, 13–15.
Bahcall, J. N. and Pinsonneault, M. H. 2004. *Phys. Rev. Lett.*, **92**, 121301.
Bartel, W. *et al*. 1985. *Z. Phys.*, **C26**, 507–513.
Bartel, W. *et al*. 1968. *Phys. Lett.*, **B28**, 148–151.
Bayes, R. *et al*. 2011. *Phys. Rev. Lett.*, **106**, 041804.
Behrend, H. J. *et al*. 1987. *Phys. Lett.*, **B183**, 400–411.
Beringer, J. *et al*. (Particle Data Group). 2012. *Phys. Rev.*, **D86**, 010001.
Bethke, S. 2009. *Eur. Phys. J.*, **C64**, 689–703.
Bodek, A. *et al*. 1979. *Phys. Rev.*, **D20**, 1471–1552.
Breidenbach, M. *et al*. 1969. *Phys. Rev. Lett.*, **23**, 935–939.
Chatrchyan, S. *et al*. 2011. *Phys. Rev. Lett.*, **107**, 132001.
Chatrchyan, S. *et al*. 2012. *Phys. Lett.*, **B716**, 30–61.

Christenson, J. *et al.* 1964. *Phys. Rev. Lett.*, **13**, 138–140.
Cleveland, B. T. *et al.* 1998. *Astrophys. J.*, **496**, 505–526.
de Groot, J. G. H. *et al.* 1979. *Z. Phys.*, **C1**, 143–162.
Dirac, P. 1928. *Proc. R. Soc. Lond.*, **A117**, 610–634.
Fermi, E. 1934. *Z. Phys.*, **88**, 161–177.
Friedman, J. I. and Kendall, H. W. 1972. *Ann. Rev. Nucl. Sci.*, **22**, 203–254.
Fukada, S. *et al.* 2001. *Phys. Rev. Lett.*, **86**, 5651–5655.
Glashow, S. L., Iliopoulos, J. and Maiani, L. 1970. *Phys. Rev.*, **D2**, 1285–1292.
Halzen, F. L. and Martin, A. D. 1984. *Quarks and Leptons*. First edn. John Wiley and Sons.
Hughes, E. B. *et al.* 1965. *Phys. Rev.*, **B139**, 458–471.
LEP and SLD Collaborations. 2006. *Phys. Rept.*, **427**, 257–454.
Sakharov, A. D. 1967. *Pisma Zh. Eksp. Teor. Fiz.*, **5**, 32–35.
Sill, A. F. *et al.* 1993. *Phys. Rev.*, **D48**, 29–55.
Tzanov, M. *et al.* 2006. *Phys. Rev.*, **D74**, 012008.
Walker, R. C. *et al.* 1994. *Phys. Rev.*, **D49**, 5671–5689.
Whitlow, L. W. *et al.* 1992. *Phys. Lett.*, **B282**, 475–482.
Wu, C. S. *et al.* 1957. *Phys. Rev.*, **105**, 1413–1415.

# Further reading

This textbook provides a broad overview of the Standard Model of particle physics at an introductory level. It focuses on providing a contemporary perspective of both the underlying theory and the experimental results. The historical development of the subject is not covered here. Furthermore, at an introductory level, the material can only be taken so far. In particular, Quantum Field Theory is not covered; this is a subject in its own right. The small selection of books listed below either give a more historical view of particle physics or cover theoretical material at a more advanced graduate level.

## Introductory books with a more historical outlook

A. Bettini, *Introduction to Elementary Particle Physics*, Cambridge University Press (2008).

D. H. Perkins, *Introduction to High Energy Physics*, Cambridge University Press (2000).

## More advanced books covering a number aspects of theoretical particle physics

I. J. R. Aitchison and A. J. G. Hey, *Gauge Theories in Particle Physics*, Taylor and Francis (2004). A more advanced graduate-level textbook with the emphasis on gauge field theory.

J. D. Bjorken and S. D. Drell, *Relativistic Quantum Mechanics*, McGraw-Hill (1964). The classic textbook on relativistic quantum mechanics.

R. K. Ellis, W. J. Stirling and B. R. Webber, *QCD and Collider Physics*, Cambridge University Press (2003). A more advanced book on the phenomenology of QCD in hadron–hadron collisions.

H. Georgi, *Lie Algebras in Particle Physics*, Westview Press (1999). Covers the essentials of group theory as applied to particle physics.

F. Halzen and A. D. Martin, *Quarks and Leptons*, Wiley and Sons (1984). Gives a particularly good description of the parton model, QCD and the DGLAP evolution equations.

M. E. Peskin and D. V. Schroeder, *An Introduction to Quantum Field Theory*, Addison-Wesley (1995). A graduate level textbook providing an introduction to quantum field theory with many practical examples.

L. H. Ryder, *Quantum Field Theory*, Cambridge University Press (1996). Provides a relatively accessible introduction to quantum field theory.

## Other resources

Detailed listings of particle properties and reviews of topical areas in particle physics are provided by the Particle Data Group http://pdg.lbl.gov and published as *The Review of Particle Physics*, Beringer *et al.* (2012). Whilst the main audience is researchers in particle physics, there is a wealth of information here.

# Index